More information about this subseries at http://www.springer.com/series/7407

Lecture Notes in Computer Science 12808

Anna Lubiw · Mohammad Salavatipour ·
Meng He (Eds.)

Algorithms and Data Structures

17th International Symposium, WADS 2021
Virtual Event, August 9–11, 2021
Proceedings

Editors
Anna Lubiw
University of Waterloo
Waterloo, ON, Canada

Meng He
Dalhousie University
Halifax, Canada

Mohammad Salavatipour
University of Alberta
Edmonton, AB, Canada

ISSN 0302-9743 ISSN 1611-3349 (electronic)
Lecture Notes in Computer Science
ISBN 978-3-030-83507-1 ISBN 978-3-030-83508-8 (eBook)
https://doi.org/10.1007/978-3-030-83508-8

LNCS Sublibrary: SL1 – Theoretical Computer Science and General Issues

This Springer imprint is published by the registered company Springer Nature Switzerland AG
The registered company address is: Gewerbestrasse 11, 6330 Cham, Switzerland

Preface

This proceedings volume contains the papers presented at the 17th International Algorithms and Data Structures Symposium (WADS 2021), which was held on-line August 9–11, 2021, organized from Dalhousie University, Halifax, Nova Scotia, Canada. WADS, which alternates with the Scandinavian Symposium and Workshops on Algorithm Theory (SWAT), is a venue for researchers in the area of design and analysis of algorithms and data structures to present their work. In response to the call for papers, 123 papers were submitted to WADS this year. Each submission received at least three reviews. From these, the Program Committee selected 48 papers for presentation, of which 47 reached the final stage.

In addition, two invited talks were given by Vida Dujmović (University of Ottawa), and Ola Svensson (EPFL). Special issues of papers selected from WADS 2021 are planned for two journals, Algorithmica, and Computational Geometry: Theory and Applications.

The Alejandro López-Ortiz Best Paper Award for WADS 2021 was given to the paper "Better distance labeling for unweighted planar graphs," by Paweł Gawrychowski and Przemysław Uznański. An award for the best student presentation was decided at the conference and will be announced in a future proceedings. From the previous WADS conference, the 2019 Alejandro López-Ortiz Best Paper Award was given to the paper "Succinct Data Structures for Families of Interval Graphs," by Hüseyin Acan, Sankardeep Chakraborty, Seungbum Jo, and Srinivasa Rao Satti.

We thank the Program Committee for their hard work and good judgement and thank all the subreviewers who contributed to the reviewing process. We thank the organizing committee team at Dalhousie University for local arrangements. We gratefully acknowledge sponsorship from Elsevier, Springer, the Faculty of Computer Science at Dalhousie University, AARMS, Fields, and PIMS.

June 2021

Meng He
Anna Lubiw
Mohammad Salavatipour

The original version of the book was revised: a forgotten volume editor was added. The correction to the book is available at https://doi.org/10.1007/978-3-030-83508-8_48

Organization

Local Arrangements

Nathaniel Brown	Dalhousie University, Canada
Travis Gagie	Dalhousie University, Canada
Younan Gao	Dalhousie University, Canada
Meng He (Chair)	Dalhousie University, Canada
Zhen Liu	Dalhousie University, Canada
Michael St Denis	Dalhousie University, Canada

Steering Committee

Faith Ellen	University of Toronto, Canada
David Eppstein	University of California, Irvine, USA
Zachary Friggstad	University of Alberta, Canada
Ian Munro	University of Waterloo, Canada
Jörg Sack	Carleton University, Canada
Mohammad Salavatipour	University of Alberta, Canada

Program Committee

Mohammad Ali Abam	Sharif University of Technology, Iran
Ahmad Biniaz	University of Windsor, Canada
Anthony Bonato	Ryerson University, Canada
Parinya Chalermsook	Aalto University, Finland
Steven Chaplick	Maastricht University, The Netherlands
Giordano Da Lozzo	University of California, Irvine, USA
Khaled Elbassioni	Masdar Institute, UAE
Ruy Fabila-Monroy	Departamento de Matemáticas, Cinvestav, Mexico
Moran Feldman	University of Haifa, Israel
Travis Gagie	Dalhousie University, Canada
Meng He (Chair)	Dalhousie University, Canada
Pinar Heggernes	University of Bergen, Norway
Zhiyi Huang	The University of Hong Kong, Hong Kong
John Iacono	Université Libre de Bruxelles, Belgium
Shahin Kamali	University of Manitoba, Canada
Matthew Katz	Ben-Gurion University of the Negev, Israel
Guohui Lin	University of Alberta, Canada
Anna Lubiw (Chair)	University of Waterloo, Canada
Brendan Lucier	Microsoft Research, USA
Pat Morin	Carleton University, Canada
Yakov Nekrich	Michigan Technological University, USA

Yoshio Okamoto	University of Electro-Communications, Japan
Denis Pankratov	Concordia University, Canada
Venkatesh Raman	Institute of Mathematical Sciences, Chennai, India
Mohammad Salavatipour (Chair)	University of Alberta, Canada
Laura Sanita	Eindhoven University of Technology, The Netherlands
José A. Soto	Universidad de Chile, Chile
He Sun	University of Edinburgh, UK
Dimitrios Thilikos	Université de Montpellier, France
Aravindan Vijayaraghavan	Northwestern University, USA
Carola Wenk	Tulane University, USA

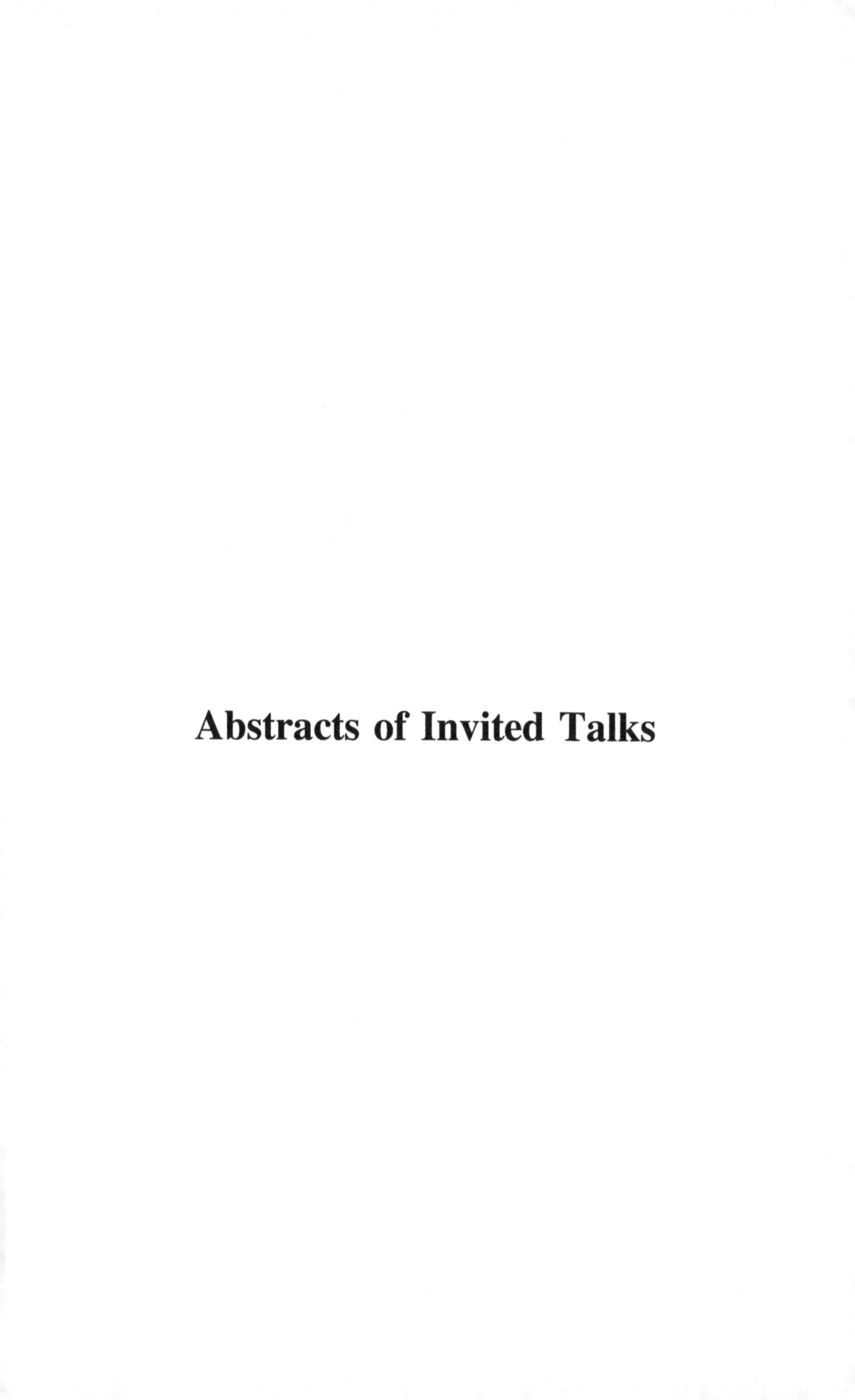

Abstracts of Invited Talks

Adjacency Labelling of Planar Graphs (and Beyond)[1]

Vida Dujmović

University of Ottawa, Canada
vdujmovi@uottawa.ca

Adjacency labelling schemes, which have been studied since the 1980's, ask for short labels for n-vertex graphs G such that the labels of two vertices u and v are sufficient to determine (quickly) if uv is an edge of G. One of the long-standing problems in the area was the optimal length of labels for planar graphs. The problem is closely related to the size of the smallest universal graph for all n-vertex planar graphs. In this talk I will show how we resolved this problem (up to lower order terms) with the help of a new graph theoretic tool: a product-structure theorem for planar graphs. This new tool and our result are applicable not only to planar graphs but also to bounded genus graphs, apex-minor-free graphs, bounded-degree graphs from minor closed families, and k-planar graphs.

[1] Supported by NSERC.

Algorithms for Explainable Clustering[1]

Ola Svensson

EPFL, Switzerland
ola.svensson@epfl.ch

An important topic in current machine learning research is to explain and/or interpret how models actually make their decisions. Motivated by this, Moshkovitz, Dasgupta, Rashtchian, and Frost recently formalized the problem of explainable clustering. A k-clustering is said to be explainable if it is given by a decision tree where each internal node splits data points with a threshold cut in a single dimension (feature), and each of the k leaves corresponds to a cluster.

In this talk, we see an algorithm that outputs an explainable clustering that loses at most a factor of $O(\log^2 k)$ compared to an optimal (not necessarily explainable) clustering for the k-medians objective, and a factor of $O(k \log^2 k)$ for the k-means objective. This improves over the previous best upper bounds of $O(k)$ and $O(k^2)$\$, respectively, and nearly matches the previous $\Omega(\log k)$ lower bound for k-medians and our new $\Omega(k)$ lower bound for k-means. Moreover, the algorithm is remarkably simple and, given an initial not necessarily explainable clustering, it is oblivious to the data points and runs in time $O(dk \log^2 k)$, independent of the number of data points n.

This is joint work with Buddhima Gamlath, Xinrui Jia, and Adam Polak.

[1] Supported by the Swiss National Science Foundation project 200021-184656 "Randomness in Problem Instances and Randomized Algorithms."

Contents

On the Spanning and Routing Ratios of the Directed Θ_6-Graph

Hugo A. Akitaya[1], Ahmad Biniaz[2], and Prosenjit Bose[3](✉)

[1] Department of Computer Science, University of Massachusetts Lowell, Lowell, USA
Hugo_Akitaya@uml.edu

[2] School of Computer Science, University of Windsor, Windsor, Canada
abiniaz@uwindsor.ca

[3] School of Computer Science, Carleton University, Ottawa, Canada
jit@scs.carleton.ca

Abstract. The family of Θ_k-graphs is an important class of sparse geometric spanners with a small spanning ratio. Although they are a well-studied class of geometric graphs, no bound is known on the spanning and routing ratios of the directed Θ_6-graph. We show that the directed Θ_6-graph of a point set P, denoted $\overrightarrow{\Theta}_6(P)$, is a 7-spanner and there exist point sets where the spanning ratio is at least $4 - \varepsilon$, for any $\varepsilon > 0$. It is known that the standard greedy Θ-routing algorithm may have an unbounded routing ratio on $\overrightarrow{\Theta}_6(P)$. We design a simple, online, local, memoryless routing algorithm on $\overrightarrow{\Theta}_6(P)$ whose routing ratio is at most 14 and show that no algorithm can have a routing ratio better than $6 - \varepsilon$.

Keywords: Spanners · Theta graphs · Routing algorithms

1 Introduction

A geometric graph $G = (V, E)$ is a graph whose vertex set V is a set of points in the plane and whose edge set E is a set of segments joining vertices. Typically, the edges are weighted with the Euclidean distance between their endpoints and we refer to such graphs as Euclidean geometric graphs. A spanning subgraph H of a weighted graph G is a t-spanner of G provided that the weight of the shortest path in H between any pair of vertices is at most t times the weight of the shortest path in G. The smallest constant t for which H is a t-spanner of G is known as the **spanning ratio** or the **stretch factor** of H.

There is a vast literature outlining different algorithms for constructing various geometric $(1 + \varepsilon)$-spanners of the complete Euclidean geometric graph (see [13,18] for a survey of the field). One can view a t-spanner H of a graph G as an approximation of G. From this perspective, there are many parameters that can be used to measure how good the approximation is. The obvious parameter is the spanning ratio, however, many other parameters have been studied in addition

Research supported in part by NSERC.

A. Lubiw et al. (Eds.): WADS 2021, LNCS 12808, pp. 1–14, 2021.
https://doi.org/10.1007/978-3-030-83508-8_1

to the spanning ratio such as the size, the weight, the maximum degree, connectivity, diameter to name a few. The study of spanners is a rich subfield and many of the challenges stem from the fact that these parameters are sometimes opposed to each other. For example, a spanner with high connectivity cannot have low maximum degree. As such, many different construction methods have been proposed which outline trade-offs between the various parameters.

A geometric graph H being a $(1+\varepsilon)$-spanner of the complete Euclidean geometric graph certifies the existence of a short path in H between every pair of vertices. Finding such a short path is as fundamental a problem as constructing a good spanner. Typically, most path-planning or routing algorithms are assumed to have access to the whole graph when computing a short path [12,15,17]. However, in many settings, the routing must be performed in an **online** manner. This presents different challenges since the whole graph is not available to the algorithm but the routing algorithm must explore the graph as it attempts to find a path. By providing the routing algorithm with a sufficient amount of memory or a large enough stream of random bits, one can successfully route online using a random walk [14,19] or Depth-First Search [15]. The situation is more challenging if the online routing algorithm is to be **memoryless** and **local**, i.e. the only information available to the algorithm, prior to deciding which edge to follow out of the current vertex, consists of the coordinates of the current vertex, the coordinates of the vertices adjacent to the current vertex and the coordinates of the destination vertex. The routing ratio of such an algorithm is analogous to the spanning ratio except that the ratio is with the weight of the path followed by the routing algorithm as opposed to the shortest path in the spanner. Thus, the routing ratio, by definition, is an upper bound on the spanning ratio. The main difficulty in designing these types of algorithms is that deterministic routing algorithms that are memoryless and local often fail by cycling [9].

Introduced independently by Clarkson [11] and Keil and Gutwin [16], Θ_k-graphs are an important class of $(1+\varepsilon)$-spanners of the complete Euclidean geometric graph for $\varepsilon > 0$. Θ_k-graphs have bounded spanning ratio [2,6,7,11,16,20] for all $k > 3$ and unbounded spanning ratio [1] for $k = 2, 3$. Informally, a Θ_k-graph is constructed in the following way: the plane around each vertex v is partitioned into k cones with apex v and cone angle $\theta = 2\pi/k$. In each cone, v is joined to the point whose projection on the bisector of the cone is closest to v. Although this naturally gives rise to a directed graph (where the previously described edges are directed away from v), much of the literature on Θ_k-graphs has focused on the underlying undirected graph. For example, the tightest upper and lower bounds on the spanning ratio for Θ_k-graphs are proven on the underlying undirected graphs (see [7] for a survey). Given a planar point set P, to avoid any confusion, we will denote the directed version of the Θ_k-graph as $\overrightarrow{\Theta}_k(P)$ and the underlying undirected graph as $\Theta_k(P)$. While it is harder to obtain routing algorithms for $\overrightarrow{\Theta}_k(P)$ because of the extra constraint imposed by the directed edges, $\overrightarrow{\Theta}_k(P)$ has the advantage of maximum out-degree bounded by k, which allows for local routing algorithms in ad-hoc networks where each node's storage is limited by a constant. In contrast, the $\Theta_k(P)$ graph can have maximum degree linear in $|P|$.

Note that the definition of $\overrightarrow{\Theta}_k$-graphs gives rise to a simple, online, local routing algorithm often referred to as **greedy** Θ-routing: when searching for a path from a vertex s to a vertex d, follow the edge from s in the cone that contains d. Repeat this procedure until the destination is reached. At each step, the only information used to make the routing decision is the location of the destination and the edge out of the current vertex that contains the destination. Thus, greedy Θ-routing is online, local and memoryless. Ruppert and Seidel [20] showed that greedy Θ-routing has a routing ratio of $1/(1-2\sin(\pi/k))$ for $k \geq 7$. For $3 < k < 7$, it was shown that the routing ratio is unbounded [5]. Intuitively, it seems that the routing ratio should be worse than the spanning ratio for all values of k, since an online routing algorithm must explore the graph while searching for a short path. Indeed, this is true for all values of $k \geq 7$, except when $k \equiv 0 \bmod 4$, in which case the upper bound on the routing ratio and the spanning ratio is $(\cos(\pi/k) + \sin(\pi/k))/(\cos(\pi/k) - \sin(\pi/k))$.

Recently, it was shown that $\overrightarrow{\Theta}_4$ has bounded routing ratio [6]. Although this is not claimed explicitly by the authors, a careful analysis of their proof shows that their result actually carries over to the directed setting. It was shown that the Half-Θ_6 graph – a subgraph of the Θ_6 graph whose edges only consist of those defined in even cones – has an optimal spanning ratio of 2 and an optimal routing ratio of $5/\sqrt{3}$ [4,8,10]. This is the first result we are aware of that shows a strict separation between the optimal spanning and routing ratios. However, the routing algorithm is defined on the undirected graph and the algorithm explicitly follows edges in the *wrong* direction. No tight bounds are known for the spanning ratios in $\overrightarrow{\Theta}_k$, except when $k \geq 7$ and $k \equiv 2 \bmod 4$, for which it is known that the spanning ratio is $1 + 2\sin(\pi/k)$, and this bound is tight in the worst case [7]. For a comprehensive overview of the current best known spanning and routing ratios for Θ_k, for $k \geq 7$, we refer the reader to [7] (Table 1).

1.1 Our Contributions

We focus on fundamental questions related to $\overrightarrow{\Theta}_6(P)$. All that is known is that it is strongly-connected [5]. We show that $\overrightarrow{\Theta}_6(P)$ is a 7-spanner (Sect. 3). Although

Table 1. Partial summary of the best known upper bounds for spanning and routing ratios. Bold numbers indicate results from this paper. Results followed by * have a known matching lower bound. See Bose et al. [7] for other results on general Θ_k when $k \bmod 4 \neq 0$, or the full version of the paper for a more complete table including lower bounds.

	$\Theta_4/\overrightarrow{\Theta}_4$	$\Theta_6/\overrightarrow{\Theta}_6$	$\Theta_{4k}/\overrightarrow{\Theta}_{4k}$, $k > 1$
Spanning	17/17 [6]	2* [4]/**7**	$\frac{cos(\theta/2)+\sin(\theta/2)}{cos(\theta/2)-\sin(\theta/2)}$ [7]/ $\frac{cos(\theta/2)+\sin(\theta/2)}{cos(\theta/2)-\sin(\theta/2)}$ [7]
Routing	17/17 [6]	$\frac{5}{\sqrt{3}}$* [8]/**14**	$\frac{cos(\theta/2)+\sin(\theta/2)}{cos(\theta/2)-\sin(\theta/2)}$ [7]/ $\frac{cos(\theta/2)+\sin(\theta/2)}{cos(\theta/2)-\sin(\theta/2)}$ [7]

our proof is constructive, it cannot be converted into a local routing algorithm since the construction of the routing path between given points requires knowledge of the whole graph. However, we are able to successfully design an online, local, memoryless routing algorithm on $\overrightarrow{\Theta}_6(P)$ whose routing ratio is at most 14 (Sect. 4). Our algorithm is simple but different from greedy Θ-routing, and, the analysis of the routing ratio is non-trivial since our algorithm makes some decisions that are counter-intuitive. For example, even if there exists a greedy edge whose endpoint is close to the destination, under certain circumstances, our algorithm chooses to go to a vertex that is farther away in a cone that does not contain the destination. In essence, greed is not always good. We complement these upper bounds with the following lower bounds in the full version of this paper. We note that our lower bounds are proven on the strongest model (for any online local algorithm even with arbitrary memory) of online routing and our upper bound is designed on the weakest model (online, local, and memoryless). We summarize our main results below.

Theorem 1. *The spanning ratio of $\overrightarrow{\Theta}_6$ is at most 7 and there exists a point set P such that the spanning ratio of $\overrightarrow{\Theta}_6(P)$ is at least $4 - \varepsilon$ for any $\varepsilon > 0$.*

Theorem 2. *There exists an online, local, memoryless routing algorithm whose routing ratio on $\overrightarrow{\Theta}_6$ is at most 14. For any $\varepsilon > 0$ and any local routing algorithm A in $\overrightarrow{\Theta}_6(P)$, the routing ratio of A is at least $6 - \varepsilon$.*

2 Preliminaries

In this section, we outline some notation and definitions. Given two points a, b in the plane, $\|ab\|$ refers to their Euclidean distance. A convex polygon C is **regular** if all its edges are of the same length. By $\|C\|$, we refer to the side length of C. The boundary of C is denoted as $bd(C)$ and the interior of C is denoted as $int(C)$. We call a triangle (resp. hexagon) **aligned** if each of its edges is parallel to a line of slope $\sqrt{3}$, slope 0 or slope $-\sqrt{3}$. Given two distinct points u, v in the plane, the canonical triangle of u with respect to v, denoted ∇_u^v is the regular aligned triangle where u is one of the vertices and v is on the edge of the triangle opposite u. Note that ∇_u^v is congruent to ∇_v^u. Let $\hexagon{u}^v$ be the regular aligned hexagon centered at u that has v on its boundary. The lines through u having slope $\sqrt{3}$, slope 0 and slope $-\sqrt{3}$, respectively, partition the hexagon into 6 regular aligned triangles. Label these triangles $\triangle_{uv}^0, \ldots, \triangle_{uv}^5$ in counter-clockwise order with the convention that $\triangle_{uv}^0$ is the triangle below u with a horizontal base. When referring to these triangles or sets related to these triangles, indices are manipulated modulo 6. When it is clear from the context, to make notation a little less cumbersome, we drop the subscript uv (see Fig. 1). Note that $\triangle^i$ for the $i \in \{0, \ldots, 5\}$ that has v on its base is identical to ∇_u^v. This implies that $\|\triangle^i\| = \|\nabla_u^v\| = \|\hexagon{u}^v\|$. Finally, we note that a regular aligned hexagon defines a distance metric. Given two points u, v in the plane, the hexagonal distance between u and v, $d_{\hexagon}(u, v) = \|\hexagon{u}^v\| = \|\hexagon{v}^u\|$.

A directed edge (u, v) in a graph is an ordered pair and represents an edge directed from vertex u to vertex v. We refer to u as the **tail** of the edge and v as the **head** of the edge. To simplify the discussion and avoid situations where points are bordering on two cones, we make the following general position assumption on a point set P: no two points lie on a line of slope $\sqrt{3}$, slope 0 or slope $-\sqrt{3}$. Note that a slight rotation of the point set removes this, as such, this assumption does not take away from the generality of our results. Given a set of points P in the plane, the directed Θ_6-graph whose vertex set is P is denoted $\overrightarrow{\Theta_6}(P)$. A directed edge (a, b) exists in $\overrightarrow{\Theta_6}(P)$ provided that $\bigtriangledown_a^b$ does not contain any point of $P \setminus \{a, b\}$. An equivalent way to construct $\overrightarrow{\Theta_6}(P)$ is the following. For each $u \in P$, the lines through u with slopes $-\sqrt{3}, 0, \sqrt{3}$ partition the plane into 6 cones. We label these cones $\wedge_u^i$, $i \in \{0, \ldots, 5\}$ counterclockwise with $\wedge_u^0$ being the cone directly below u. For each cone $\wedge_u^i$, add edge (u, v) if $v \in \wedge_u^i$ is the closest point to u in the $d_{\bigcirc}$ metric. This makes explicit the fact that the maximum out-degree of $\overrightarrow{\Theta_6}(P)$ is 6.

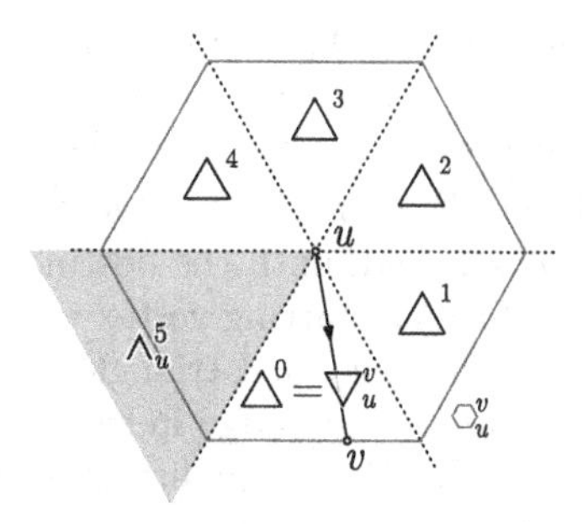

Fig. 1. Illustrations of our definitions.

2.1 The Routing Model

Given a graph $G = (V, E)$, with vertex set V and edge set E, an online, ℓ-local routing algorithm can be expressed as a function $f : V \times V \times H \times M \rightarrow V \times M$, where $M = \{0, 1\}^*$. The parameters of $f(u, d, G_\ell(u), m)$ are: u the current vertex, d the destination vertex, $G_\ell(u)$ the subgraph of G that consists of all paths rooted at u with length at most ℓ and m is a bit-string representing the memory. An invocation of the routing function $f(u, d, G_\ell(u), m)$ updates m and returns $v \in V$ such that the edge (u, v) should be followed out of u to reach destination d. This is the strongest model of online routing where the algorithm has infinite memory and is aware of the graph induced on the ℓ-neighborhood prior to making a routing decision. With this model, one can perform Depth-First Search on G. The algorithm is considered 1-local or **local** if $\ell = 1$. It is considered memoryless if $M = \emptyset$, that is, the algorithm has no memory or knowledge of where it started or where it has been. The weakest model is online, local and memoryless. For example, one cannot even perform Depth-First Search in this model. Although quite restrictive, our routing algorithm falls within the weakest model.

3 Upper Bound on the Spanning Ratio

In this section, we show that $\overrightarrow{\Theta}_6(P)$ is a 7-spanner. Given a destination vertex $d \in \overrightarrow{\Theta_6}(P)$, we define the **greedy edge** of vertex v with respect to d to be the outgoing edge of v in $\bigtriangledown_v^d$. Recall that the routing strategy of repeatedly following the greedy edge at every step until the destination is reached is called **greedy routing** or Θ-routing. The path found by the greedy routing algorithm is called the **greedy path**. Thus, the greedy path from s to d, denoted $\pi(s,d)$, is the path in $\overrightarrow{\Theta_6}(P)$ starting at s and where at every step, the greedy edge with respect to d is selected, until the destination d is reached.

Given a starting vertex s and a destination vertex d, by construction, we have that the canonical triangle, $\triangledown_s^d$, is contained in the hexagon $\hexagon_d^s$. Let (s,a) be the first greedy edge in $\pi(s,d)$. Then, since a is in $\triangledown_s^d$ we have that $d_{\hexagon}(a,d) < d_{\hexagon}(s,d)$. The inequality is strict since by our general position assumption a is contained in $int(\triangledown_s^d)$, or $a = d$. Therefore, at every step of the greedy routing algorithm, the hexagonal distance to the destination decreases. Since there are a finite number of points in P and the fact that the hexagonal distance to the destination is strictly decreasing at every step, the greedy algorithm terminates at d. We summarize this in the following lemma.

Lemma 1. *Given any pair of points $s, d \in P$, there always exists a greedy path from s to d in $\overrightarrow{\Theta_6}(P)$. Furthermore, let x be a vertex in $\pi(s,d)$ different from s and d. Then the following hold:*

- *$\hexagon_d^x$ is contained in $(\hexagon_d^s)$*
- *$d_{\hexagon}(x,d) < d_{\hexagon}(s,d)$*
- *$\pi(x,d)$ is contained in $\hexagon_d^x$*

Although the greedy routing algorithm always reaches its destination, its spanning ratio is not bounded by a constant [5]. The issue is that $\pi(s,d)$, although getting closer to d with respect to the hexagonal distance, can spiral around d many times (see Fig. 2).

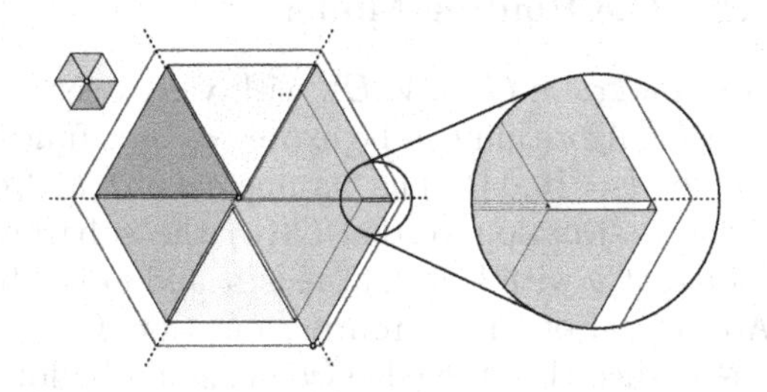

Fig. 2. From [5]. Colored triangles are interior-empty triangles $\triangledown_u^v$ that define an edge (u,v) of $\overrightarrow{\Theta_6}(P)$. Different colors encode different canonical triangles in $\{\triangle_{uv}^0, \ldots, \triangle_{uv}^5\}$. The spanning ratio of the greedy path from the perimeter to the center of the red hexagon is not bounded by a constant. (Color figure online)

However, if there happens to be an edge from d to s, i.e. $(d,s) \in \overrightarrow{\Theta_6}(P)$, then $\pi(s,d)$ can no longer spiral around d since $\triangledown_d^s$ is empty of points of P and acts as a barrier, as we shall prove in Lemma 2. This prevents the path from cutting across $\triangledown_d^s$. We then prove that if (d,s) is an edge of $\overrightarrow{\Theta_6}(P)$ then the spanning ratio of $\pi(s,d)$ is at most $6\,\|\triangledown_d^s\|$ (Corollary 1).

For $i \in \{0, \ldots, 5\}$, let $T_i = \{(a,b) \in \pi(s,d) \mid a \in \triangle_{ds}^i\}$. T_i is the set of all edges of $\pi(s,d)$ whose tail is in $\triangle^i$. Define the weight of T_i, denoted $\|T_i\|$, to be $\sum_{(a,b)\in T_i} \|\triangledown_a^b\|$. For ease of reference, label the sequence of vertices in $\pi(s,d)$ as $s = u_0, \ldots, u_k = d$ where k is the number of edges.

Lemma 2. *If (u_a, u_{a+1}) is an edge of $\pi(s,d)$ in T_i for $a \in \{0,\ldots,k-1\}$ and $i \in \{0,\ldots,5\}$, then u_{a+1} can only be in one of $\triangle_{ds}^{i-1}, \triangle_{ds}^{i}$ or $\triangle_{ds}^{i+1}$.*

Proof. Without loss of generality, assume that u_a is in $\triangle_{ds}^0$. Let $h^+(d)$ be the half-plane above the horizontal line through d. Since the edge of $\triangledown_{u_a}^d$ that contains d is horizontal and the interior of the triangle lies below the horizontal line through d, we have that $int(\triangledown_{u_a}^d) \cap h^+(d) = \emptyset$. Therefore, u_{a+1} cannot be in $\triangle^2, \triangle^3$ or $\triangle^4$, since the interiors of all those triangles are in $h^+(d)$. The lemma follows. □

Note that Lemma 2 immediately implies that the greedy path cannot spiral around d since that would require $\pi(s,d)$ to contain a point of P in $int(\triangledown_d^s)$, contradicting the existence of edge (d,s). This lets us bound the length of $\pi(s,d)$.

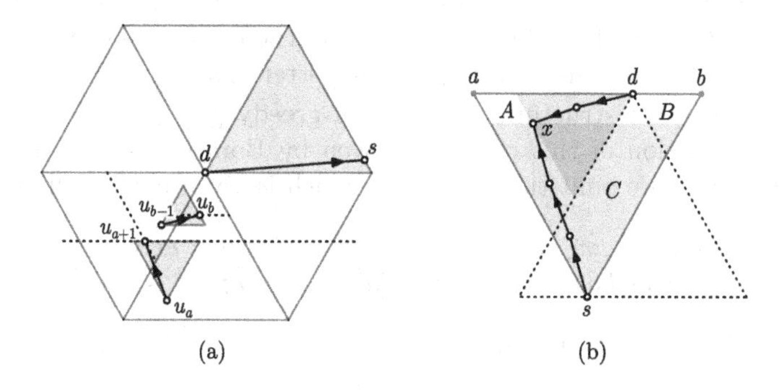

Fig. 3. (a) Illustration of Lemma 3. (b) Illustrations of Theorem 3.

Lemma 3. *Assume (d,s) is an edge of $\overrightarrow{\Theta_6}(P)$ and let u_a be a vertex of $\pi(s,d)$ in $\triangle_{ds}^i$. Let u_b be the next vertex in $\pi(s,d)$ after u_a that appears in $\triangle_{ds}^i$, i.e. $b > a$. Then, $int(\triangledown_{u_a}^{u_{a+1}}) \cap int(\triangledown_d^{u_b}) = \emptyset$.*

Proof. Without loss of generality, assume that u_a is in $\triangle_{ds}^0$. We have two cases: either $u_b = u_{a+1}$ or $u_b \neq u_{a+1}$. We begin with the former. If $u_b = u_{a+1}$ then the lemma holds trivially since $\triangledown_{u_a}^{u_{a+1}}$ and $\triangledown_d^{u_{a+1}}$ are separated by a horizontal line.

We now consider the case where $u_b \neq u_{a+1}$, i.e. $b > a+1$ (Fig. 3(a)). By Lemma 2 and u_b's definition, u_{a+1} must either be in $\triangle^1$ or $\triangle^5$. Without loss of generality, assume that u_{a+1} is in $\triangle^5$. Consider the edge (u_{b-1}, u_b) of $\pi(s,d)$. By Lemma 2, u_{b-1} must be in $\triangle^5$ since, by the existence of (d,s), the path cannot spiral around d and enter $\triangle^0$ from $\triangle^1$. By Lemma 1, u_{b-1} must be contained in $\triangledown_d^{u_{a+1}}$. Moreover, since (u_a, u_{a+1}) is an edge of the path, we have that $\triangledown_{u_a}^{u_{a+1}}$ is empty, which means that u_{b-1} lies above the horizontal line through u_{a+1}. This implies that u_b also lies above the horizontal line through u_{a+1} since the canonical triangle $\triangledown_{u_{b-1}}^{u_b}$ has a horizontal edge and lies above the horizontal line through u_{b-1}. Therefore, $int(\triangledown_{u_a}^{u_{a+1}}) \cap int(\triangledown_d^{u_b}) = \emptyset$. □

Lemma 4. *If $(d,s) \in \overrightarrow{\Theta_6}(P)$, then $\|T_i\| \leq \|\triangledown_d^s\|$, for $i \in \{0,\ldots,5\}$.*

Proof. We show the bound for T_0. Let $(a, b) \in T_0$. Let a' (resp. b') be the intersection of a horizontal line through a (resp. b) with the left side of $\triangle^0_{ds}$. Since $\triangledown^b_a$ is equilateral, $\|a'b'\| \geq \|ab\|$. If (a_1, b_1) and (a_2, b_2) are two edges in T_0, by Lemma 3, $a'_1b'_1$ and $a'_2b'_2$ do not overlap. Therefore, $\|T_0\|$ is at most $\|\triangledown^s_d\|$. □

We are now able to bound the length of $\pi(s, d)$ when $(d, s) \in \overrightarrow{\Theta_6}(P)$. As each edge of $\pi(s, d)$ appears in only one T_i, the bound follows from Lemma 4.

Corollary 1. *If $(d, s) \in \overrightarrow{\Theta_6}(P)$, then $\|\pi(s, d)\| \leq 6\,\|\triangledown^s_d\| = 6\,\left\|\hexagon_d^s\right\|$.*

Corollary 1 implies that $\overrightarrow{\Theta_6}(P)$'s spanning ratio is upper bounded by $12\sqrt{3}$. This follows from the fact that $\Theta_6(P)$'s spanning ratio is 2 and for each edge e in $\Theta_6(P)$ there is a directed path of length at most $6\sqrt{3}\,\|e\|$ from one endpoint of e to the other in $\overrightarrow{\Theta_6}(P)$ (the $\sqrt{3}$ term comes from the hexagonal distance metric).

A more careful analysis lets us prove a better spanning ratio. In order to do this, we uncover a structural property of greedy paths in $\overrightarrow{\Theta_6}(P)$. We note that a weaker version of this claim is proven by Bonichon et al. [3] (proof of Theorem 1). Thus, we omit the proof here which is given in the full version.

Theorem 3. *Between any pair of points $s, d \in P$, there exists an $x \in P$ in $\triangledown^d_s$ such that the following hold (note that if the interior of $\triangledown^d_s$ is empty then $x = d$):*

1. *$\pi(s, x)$ and $\pi(d, x)$ are both in $\triangledown^d_s$,*
2. *$\|\pi(s, x)\| \leq \|\triangledown^x_s\|$,*
3. *$\|\pi(d, x)\| \leq \|\triangledown^x_d\|$.*

Proof Sketch. See Fig. 3(b) for an example. We prove the claim by induction on the rank of pairs of points (s, d) as sorted order by $\|\triangledown^d_s\|$. The induction step builds the required paths using the greedy edge from s in d's direction and the path obtained by applying a stronger induction hypothesis. □

We now prove the main result of this section.

Theorem 4. *Between any pair of points $s, d \in P$, there exists a directed path $\delta(s, d)$ in $\overrightarrow{\Theta_6}(P)$ such that the length of $\delta(s, d)$ is at most $7\,\|sd\|$.*

Proof. Given a greedy path $\pi(u, v)$, the **reverse** path, denoted $\rho(v, u)$, is a directed path from v to u where every edge (x, y) in $\pi(u, v)$ is replaced with the greedy path $\pi(y, x)$. By Theorem 3, between any pair of points $s, d \in P$, there exists an $x \in \triangledown^d_s$ such that $\pi(s, x)$ and $\pi(d, x)$ are both in $\triangledown^d_s$, $\|\pi(s, x)\| \leq \|\triangledown^x_s\| \leq \left\|\triangledown^d_s\right\|$, and $\|\pi(d, x)\| \leq \|\triangledown^x_d\|$. Let $\delta(s, d)$ be the path resulting from the concatenation of $\pi(s, x)$ and $\rho(x, d)$. By construction, $\delta(s, d)$ is a directed path from s to d. Let A be one of the two triangles obtained from $\triangledown^d_s \backslash \triangledown^s_d$. Let a, b and s be the vertices of $\triangledown^d_s$ with a being incident to A. Without loss of generality, assume the orientation shown in Fig. 3(b) and that $x \in A$. Consider the triangle defined by s, a, d and let γ be the angle at s. By elementary trigonometry, we have that the spanning ratio is $\|\delta(s, d)\| \,/\, \|sd\| \leq (\sin(2\pi/3 - \gamma) + 6\sin\gamma)/\sin(\pi/3) \leq 7$, since the maximum is attained when $\gamma = \pi/3$. □

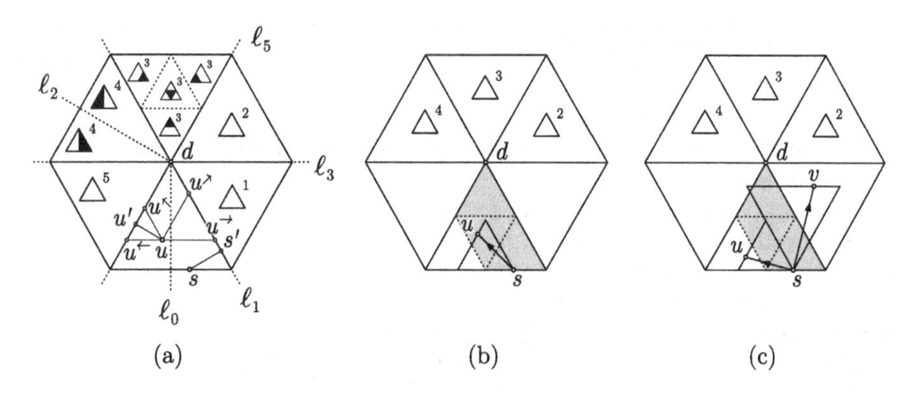

Fig. 4. (a) Examples of some of the notation used. (b) Example in which Algorithm 1 takes the non-greedy (s, u). (c) Example in which Algorithm 1 takes the greedy edge (s, v).

Although the proof of the spanning ratio of 7 for $\overrightarrow{\Theta_6}(P)$ is constructive, unfortunately, it does not provide an online routing algorithm. There are 3 main obstacles. First, in the proof, the path is constructed from both ends, where we build a greedy path from s to x and another from d to x. Second, the point x is not easily identifiable locally. And third, when finding the reverse path of an edge (a, b), one needs to know both a and b, which may not be information that is available if we are only aware of outgoing edges.

4 Routing Algorithm and Upper Bound on Routing Ratio

This section provides a routing algorithm in $\overrightarrow{\Theta}_6$. We first describe some notation used in this section. Similar to $\textcircled{u}^v$, we denote by $\hexagon_u^v$ the axis aligned hexagon rotated by $\pi/6$ that is centered at u and contains v on its boundary. For an example, see the shaded hexagon in Fig. 5. The following definitions refer to a hexagon $\hexagon_d^s$. Refer to Fig. 4 (a). Let ℓ_0 be the vertical line through d and ℓ_1, ℓ_2, ℓ_3, ℓ_4, and ℓ_5 be the lines through d with slopes $-\sqrt{3}$, $-\frac{1}{\sqrt{3}}$, 0, $\frac{1}{\sqrt{3}}$, and $\sqrt{3}$ respectively. For a point $u \in \triangle_{ds}^0$, we define point u' as the orthogonal projection of u on ℓ_1 or ℓ_5, whichever is closest to u. We also define points $u^{\rightarrow}$, $u^{\nearrow}$, $u^{\nwarrow}$, and $u^{\leftarrow}$ as the intersections between ℓ_1 or ℓ_5 and the rays from u with angles 0, $\pi/3$, $2\pi/3$, and π from the positive x-direction. We define $\blacktriangle^0$ and $\blacktriangle^0$ to be the left and right triangles obtained from partitioning $\triangle_{ds}^0$ with ℓ_0. We also partition $\triangle_{ds}^0$ into four congruent triangles with the line segments through two of the three midpoints of sides of $\triangle_{ds}^0$. Denote by $\blacktriangle^0$, $\blacktriangle^0$, $\blacktriangle^0$, $\blacktriangle^0$ the top, left, right, and middle triangles respectively. We apply the appropriate rotations to obtain the analogous definitions for $\triangle_{ds}^i$, $i \in \{1, \ldots, 5\}$. For example, $\blacktriangle^3$ in Fig. 4 (a) is the bottom triangle in $\triangle_{ds}^3$.

Algorithm 1: DirectedRoute(s, d, $N(s)$)

1. s is the current vertex, d is the destination and $N(s)$ is the 1-neighborhood of s.
2. Assume that s is in $\triangle^0_{ds}$ otherwise apply the appropriate rotations and/or reflection;
3. Let (s, v) be the edge from s in $\wedge^3_s$;
   // Greedy Edge
4. Let (s, u) be the edge from s in $\wedge^4_s$;
   // Non-greedy Edge if it exists
5. **if** u **exists and** $u \in \blacktriangle^0_{ds} \cup \blacktriangle^0_{ds} \cup \triangle^0_{ds}$ **then**
6. | Return u; // Take Non-greedy Edge
7. **else**
8. | Return v; // Take Greedy Edge
9. **end**

We define the **potential** of a point $p \in P$ as $\Phi(p) = \frac{\sqrt{3}}{2} \|\hexagon_d^{\,p}\|$. Let $(u_0 = s, u_1, \ldots, u_k = d)$ be the sequence of vertices visited by Algorithm 1. The following lemma shows that the potential decreases with each step of the algorithm. We provide its proof in the full version.

Lemma 5. *Let (u_a, u_{a+1}) be an edge taken by Algorithm 1. Then $\Phi(u_{a+1}) < \Phi(u_a)$, and $\|\hexagon_d^{u_{a+1}}\| < \|\hexagon_d^{u_a}\|$*

Since the potential of a point is a function of its position, Lemma 5 implies that no point is visited twice and the destination is always reached. We now bound the routing ratio. We apply a charging scheme for each edge taken by the algorithm based on its type. We classify the edges (u_a, u_{a+1}) taken by the algorithm as follows. If u_{a+1} is in $\blacktriangle^1_{ds}$ we call the edge a **long step**. Otherwise, we call the edge a **short step**.

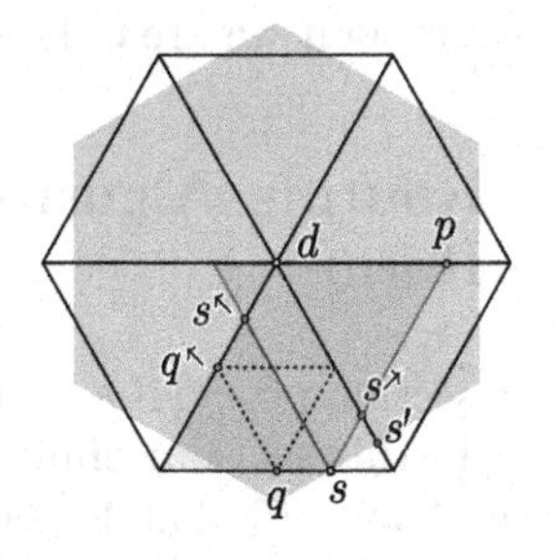

Fig. 5. The shaded hexagon $\hexagon_d^{s}$ contains points whose potentials are the same or lower than s. The red triangle is $\triangledown^d_s$ and the blue triangle is $\triangledown^{q^{\nwarrow}}_s$. (Color figure online)

Informal Overview of the Charging Scheme. Each step (u_a, u_{a+1}) will be associated with a decrease in potential $\Phi(u_a) - \Phi(u_{a+1})$ quantifying how much closer the current point u_{a+1} is to d than the previous point u_a. We show in Lemma 6 that for short steps, the decrease in potential is enough to pay for the size of the step $\|u_a u_{a+1}\|$. For long steps, the potential might decrease by an arbitrarily small amount. So in addition to the decrease in potential, we charge the step to a region of the hexagon $\hexagon_d^{s}$. The charged regions are axis-aligned trapezoids whose non-parallel edges are on ℓ_i, $i \in \{1, 2, 3\}$. Lemma 7 quantifies how the size of the charged trapezoids relates to the size of the step. We then show that the charged trapezoids have disjoint interiors, i.e., the same region cannot be double charged. This is what allows us to bound the cost of the path.

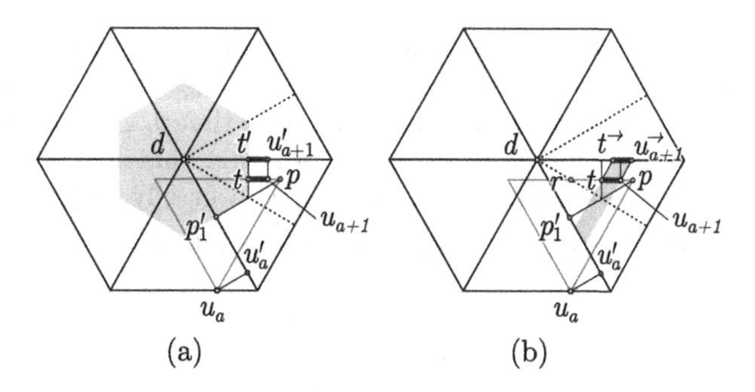

Fig. 6. Charging scheme for long steps.

For two points u and v in the interior of the same cone $\wedge^i_d$, define the **trapezoid** $\bigtriangleup^v_u$ as $\bigtriangledown^u_d \setminus \bigtriangledown^v_d$. Note that $\bigtriangleup^v_u = \emptyset$ if $\|\bigtriangledown^u_d\| \leq \|\bigtriangledown^v_d\|$, i.e. $\bigtriangledown^u_d \subset \bigtriangledown^v_d$.

Case 1 ((u_a, u_{a+1}) is short) Charge to the decrease in potential $\Phi(u_a) - \Phi(u_{a+1})$.
Case 2 ((u_a, u_{a+1}) is long) Charge to the decrease in potential $\Phi(u_a) - \Phi(u_{a+1})$ *and* to a region $\bigtriangleup^t_{u_{a+1}}$ where t is defined as follows. Refer to Fig. 6 (b). Let p be the upper right corner of $\bigtriangledown^{u_{a+1}}_{u_a}$ and r be the intersection between the upper edge of $\bigtriangledown^{u_{a+1}}_{u_a}$ and ℓ_2. We define t to be the midpoint of rp.

Lemmas 6 and 7 formalize the charging scheme. Due to space restrictions, their proofs are in the full version.

Lemma 6. *In Case 1 where (u_a, u_{a+1}) is a short step, the decrease in potential is at least half the size of the step, i.e.,* $\frac{\|u_a u_{a+1}\|}{2} \leq \Phi(u_a) - \Phi(u_{a+1})$.

We define the **length** $\left\|\bigtriangleup^t_{u_{a+1}}\right\|$ of a trapezoid $\bigtriangleup^t_{u_{a+1}}$ to be the length of one of its non parallel sides. Note that in the context of Case 2, $\left\|\bigtriangleup^t_{u_{a+1}}\right\| = \|tu_{a+1}\|$.

Lemma 7. *In Case 2 where (u_a, u_{a+1}) is a long step, the decrease in potential plus the length of the charged region is at least half the size of the step, i.e.,*

$$\frac{\|u_a u_{a+1}\|}{2} \leq \Phi(u_a) - \Phi(u_{a+1}) + \left\|\bigtriangleup^t_{u_{a+1}}\right\|.$$

Let $\mathcal{T}$ be the set of all charged trapezoids, and $\|\mathcal{T}\|$ be the sum of lengths of all trapezoids in $\mathcal{T}$. We show a property that allows us to upper bound $\|\mathcal{T}\|$.

Lemma 8. *Let $(u_a, \ldots, u_b)$ be a subsequence of steps taken by Algorithm 1 where all visited points are in the same cone of d. Without loss of generality, let this cone be $\wedge^0_d$, and let $u_a \in \blacktriangle^0_{du_a}$. If u_{b+1} is in $\wedge^5_d$, then $u_{b+1} \in \textcircled{d}^{u_a^{\nwarrow}}$. If u_{b+1} is in $\wedge^1_d$, then $u_{b+1} \in \textcircled{d}^{q^{\nearrow}}$ where q is the midpoint of the bottom edge of $\triangle^0_{du_a}$.*

Proof. Refer to Fig. 7 (a). If u_{b+1} is not in $\wedge_d^0$, then by Algorithm 1, (u_b, u_{b+1}) is greedy. By Lemma 2, u_{b+1} is either in $\triangle_{du_a}^1$ or $\triangle_{du_a}^5$. Since all edges of the subsequence are short edges, by definition, we have that u_{a+1} is in the pentagon $qq^{\nearrow}du_a^{\nwarrow}u_a$. A simple inductive argument shows that this implies that u_b is in the region $qq^{\nearrow}du_a^{\nwarrow}u_a$. Hence, since (u_b, u_{b+1}) is a greedy edge, we have that $\nabla_{u_b}^{u_{b+1}} \cap \triangle_{du_a}^1 \subset ⓓ^{u_a^{\nwarrow}}$ and $\nabla_{u_b}^{u_{b+1}} \cap \triangle_{du_a}^5 \subset ⓓ^{q^{\nearrow}}$ The lemma follows. □

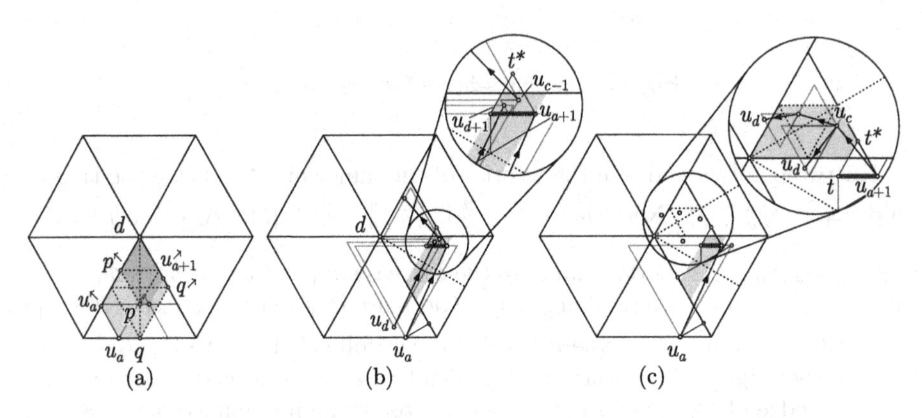

Fig. 7. (a) Illustration of Lemma 8. (b)–(c) Algorithm 1 cannot enter $\triangle_{u_{a+1}}^{t}$ once it leaves triangle $tu_{a+1}t^*$.

Lemma 9. *Let $\triangle_{u_{a+1}}^{t}$ be the trapezoid charged by a long step (u_a, u_{a+1}), and let $(u_{a+1}, \ldots, u_b)$, $a < b$ be the maximal subpath traversed by Algorithm 1 with $u_b \in \triangle_{u_{a+1}}^{t}$. Then, every step in the subpath is short, and every point in it ($u_i, i \in \{a+1, \ldots, b\}$) is in the equilateral triangle whose bottom edge is tu_{a+1}.*

Proof Sketch. The full proof is in the full version. Refer to Figs. 7 (b)–(c). Let $tu_{a+1}t^*$ be the equilateral triangle whose bottom edge is tu_{a+1}. From u_{a+1}, Algorithm 1 can only take short steps before leaving such triangle to a point u_c. We show that after u_c, the path output by the algorithm can never return to $tu_{a+1}t^*$. Then, no point visited after u_c can be in $\triangle_{u_{a+1}}^{t}$. □

Corollary 2. *The trapezoids in $\mathcal{T}$ are pairwise interior disjoint.*

Proof. For contradiction, assume that trapezoids $\triangle_{u_{a+1}}^{t_a}$ $\triangle_{u_{b+1}}^{t_b}$, charged by long steps (u_a, u_{a+1}) and (u_b, u_{b+1}) with $a < b$, intersect. Then, the larger base of $\triangle_{u_{b+1}}^{t_b}$ is between the two bases of $\triangle_{u_{a+1}}^{t_a}$. By construction of the trapezoids, u_b is in $\triangle_{u_{a+1}}^{t_a}$ contradicting Lemma 9. □

Theorem 5. *The routing ratio of Algorithm 1 is at most* 14.

Proof. By Lemmas 6 and 7, the length of the path returned by Algorithm 1 is at most

$$\sum_{i=1}^{k-1} \|u_i u_{i+1}\| \leq 2(\Phi(s) - \Phi(d) + \|\mathcal{T}\|).$$

By definition, $\Phi(s) < \|sd\|$ and $\Phi(d) = 0$. By Corollary 2, $\|\mathcal{T}\| \leq 6\|sd\|$ since the trapezoids can only fill the initial hexagon $⟨d⟩^s$. Then, $\sum_{i=1}^{k-1} \|u_i u_{i+1}\| \leq 14\|sd\|$ as required. □

5 Conclusions

We have provided upper and lower bounds for the spanning and routing ratios of $\overrightarrow{\Theta}_6(P)$. There are still gaps between the bounds as they are not matching. We believe that the actual bound is closer to the lower bounds, mainly because in the analysis of the upper bound of both the spanning and routing ratios, we account for the possibility that the path from source to destination goes around intermediate points and/or the destination. However, intuition seems to suggest that this does not actually happen and there is a shorter path that *cuts in* after going half-way around, which is the case in our lower bound constructions. We leave the closing of the gap between the upper and lower bounds as an open problem.

References

1. Aichholzer, O., et al.: Theta-3 is connected. Comput. Geom.: Theory Appl. **47**(9), 910–917 (2014)
2. Barba, L., Bose, P., De Carufel, J.L., van Renssen, A., Verdonschot, S.: On the stretch factor of the theta-4 graph. In: Dehne, F., Solis-Oba, R., Sack, J.R. (eds.) Algorithms and Data Structures. LNCS, vol. 8037, pp. 109–120. Springer, Heidelberg (2013). https://doi.org/10.1007/978-3-642-40104-6_10
3. Bonichon, N., Bose, P., Carmi, P., Kostitsyna, I., Lubiw, A., Verdonschot, S.: Gabriel triangulations and angle-monotone graphs: local routing and recognition. In: International Symposium on Graph Drawing and Network Visualization, pp. 519–531. Springer, Heidelberg (2016)
4. Bonichon, N., Gavoille, C., Hanusse, N., Ilcinkas, D.: Connections between theta-graphs, delaunay triangulations, and orthogonal surfaces. In: Thilikos, D.M. (ed.) Graph Theoretic Concepts in Computer Science. LNCS, vol. 6410, pp. 266–278. Springer, Heidelberg (2010). https://doi.org/10.1007/978-3-642-16926-7_25
5. Bose, P., De Carufel, J.-L., Devillers, O.: Expected complexity of routing in Θ_6 and half-Θ_6 graphs. arXiv preprint arXiv:1910.14289 (2019)
6. Bose, P., De Carufel, J.-L., Hill, D., Smid, M.H.M.: On the spanning and routing ratio of Theta-four. In: Symposium on Discrete Algorithms, pp. 2361–2370. SIAM (2019)
7. Bose, P., De Carufel, J.-L., Morin, P., van Renssen, A., Verdonschot, S.: Towards tight bounds on theta-graphs: more is not always better. Theoret. Comput. Sci. **616**, 70–93 (2016)

8. Bose, P., Fagerberg, R., van Renssen, A., Verdonschot, S.: Optimal local routing on delaunay triangulations defined by empty equilateral triangles. SIAM J. Comput. **44**(6), 1626–1649 (2015)
9. Bose, P., Morin, P.: Online routing in triangulations. SIAM J. Comput. **33**(4), 937–951 (2004)
10. Chew, P.: There are planar graphs almost as good as the complete graph. J. Comput. Syst. Sci. **39**(2), 205–219 (1989)
11. Clarkson, K.: Approximation algorithms for shortest path motion planning. In: Proceedings of the 19th Annual ACM Symposium on Theory of Computing (STOC 1987), pp. 56–65 (1987)
12. Dijkstra, E.: A note on two problems in connexion with graphs. Numer. Math. **1**, 269–271 (1959)
13. Eppstein, D.: Spanning trees and spanners. In: Handbook of Computational Geometry, pp. 425–461 (1999)
14. Grimmett, G.: Probability on Graphs: Random Processes on Graphs and Lattices, 2nd edn. Cambridge University Press, Cambridge (2018)
15. Hopcroft, J., Tarjan, R.: Algorithm 447: efficient algorithms for graph manipulation. Commun. ACM **16**(6), 372–378 (1973)
16. Keil, J.M., Gutwin, C.A.: Classes of graphs which approximate the complete Euclidean graph. Discret. Comput. Geom. **7**, 13–28 (1992). https://doi.org/10.1007/BF02187821
17. Lee, C.Y.: An algorithm for path connection and its applications. IRE Trans. Electron. Comput. **EC–10**(3), 346–365 (1961)
18. Narasimhan, G., Smid, M.: Geometric Spanner Networks. Cambridge University Press, Cambridge (2007)
19. Pearson, K.: The problem of the random walk. Nature **72**, 294 (1865)
20. Ruppert, J., Seidel, R.: Approximating the d-dimensional complete Euclidean graph. In: Proceedings of the 3rd Canadian Conference on Computational Geometry (CCCG 1991), pp. 207–210 (1991)

The Minimum Moving Spanning Tree Problem

Hugo A. Akitaya[1], Ahmad Biniaz[3(✉)], Prosenjit Bose[2], Jean-Lou De Carufel[4], Anil Maheshwari[2], Luís Fernando Schultz Xavier da Silveira[2], and Michiel Smid[2]

[1] Department of Computer Science, University of Massachusetts Lowell, Lowell, USA
[2] School of Computer Science, Carleton University, Ottawa, Canada
{jit,anil,michiel}@scs.carleton.ca, schultz@ime.usp.br
[3] School of Computer Science, University of Windsor, Windsor, Canada
abiniaz@uwindsor.ca
[4] School of Electrical Engineering and Computer Science, University of Ottawa, Ottawa, Canada
jdecaruf@uottawa.ca

Abstract. We investigate the problem of finding a spanning tree of a set of moving points in the plane that minimizes the maximum total weight (sum of Euclidean distances between edge endpoints) or the maximum bottleneck throughout the motion. The output is a single tree, i.e., it does not change combinatorially during the movement of the points. We call these trees the minimum moving spanning tree, and the minimum bottleneck moving spanning tree, respectively. We show that, although finding the minimum bottleneck moving spanning tree can be done in $O(n^2)$ time, it is NP-hard to compute the minimum moving spanning tree. We provide a simple $O(n^2)$-time 2-approximation and a $O(n \log n)$-time $(2 + \varepsilon)$-approximation for the latter problem.

Keywords: Minimum spanning tree · Moving points · NP-hardness · Convex distance function · Approximation algorithms

1 Introduction

The Euclidean minimum spanning tree (EMST) of a point set is the minimum weight graph that connects the given point set, where the weight of the graph is given by the sum of Euclidean distances between endpoints of edges. EMST is a classic data structure in computational geometry and it has found many uses in network design and in approximating NP-hard problems. In the visualization community, a series of methods generalize Euler diagrams to represent spatial data [2,8,9,16]. These approaches represent a set by a connected colored shape containing the points in the plane that are in the given set. In order to

Supported by NSERC.

A. Lubiw et al. (Eds.): WADS 2021, LNCS 12808, pp. 15–28, 2021.
https://doi.org/10.1007/978-3-030-83508-8_2

reduce visual clutter, approaches such as Kelp Diagrams [9] and colored spanning graphs [13] try to minimize the area (or "ink") of such colored shapes. Each shape can be considered as a generalization of the EMST of points in the set.

Motivated by visualizations of time-varying spatial data, we investigate a natural generalization of the minimum spanning tree (MST) and the minimum bottleneck spanning tree (MBST) for a set of moving points. In general it is desirable that visualizations are stable, i.e., small changes in the input should produce small changes in the output [17]. In this paper, we want to maintain all points connected throughout the motion by the same tree (the tree does not change topologically during the time frame) . Consider points in the plane moving on a straight line with constant speed over a time interval $[0, 1]$. The weight of an edge pq between points p and q is defined to be the Euclidean distance $\|pq\|$. Note that the weight of an edge changes over time. We define the *Minimum Moving Spanning Tree* (MMST) of a set of moving points to be a spanning tree that minimizes the maximum sum of weights of its edges during the time interval. Analogously, we define *Minimum Bottleneck Moving Spanning Tree* (MBMST) of a set of moving points to be a spanning tree that minimizes the maximum individual weight of edges in the tree during the time interval.

Apart from this motivation, the concepts of MMST and MBMST are relevant in the context of moving networks. Motivated by the increase in mobile data consumption, network architecture containing mobile nodes have been considered [14]. In this setting, the design of the topology of the networks is a challenge. Due to the mobility of the vertices, existing methods update the topology dynamically and the stability becomes important since there are costs associated with establishing new connections and handing over ongoing sessions. The MMST and MBMST offer stability in mobile networks.

Results and Organization. We study the problems of finding an MMST and an MBMST of a set of points moving linearly, each at constant speed. Section 2 provides formal definitions and proves that the distance function between points is convex in this setting. We use this property in an exact $O(n^2)$-time algorithm for the MBMST as shown in Sect. 3. Our algorithm computes the minimum bottleneck tree in a complete graph G on the moving points in which the weight of each edge is the maximum distance between the pairs of points during the time frame. In Sect. 4.1 we present an $O(n^2)$-time 2-approximation for MMST by computing the MST of G. In the full version of the paper we provide an example that shows our analysis for the approximation ratio is tight. In Sect. 4.2, we show that the MMST is equal to the minimum spanning tree of a point set in $\mathbb{R}^4$ with a non-Euclidean metric. Since this metric space has doubling dimension $O(1)$, we obtain an $O(n \log n)$-time $(2 + \varepsilon)$-approximation algorithm. Finally, we show that the problem of finding the MMST is NP-hard in Sect. 4.3 by reducing from the Partition problem.

Related work. Examples of other visualizations of time-varying spatial data are space-time cubes [15], that represent varying 2D data points with a third dimension, and motion rugs [6,21], that reduces the dimentionality of the movement of data points to 1D, presenting a 2D static overview visualizations. The representation of time-varying geometric sets were also the theme of a recent Dagstuhl Seminar 19192 “Visual Analytics for Sets over Time and Space” [10]. In the context of algorithms dealing with time-varying data Meulemans et al. [17] introduces a metric for stability, analysing the trade-off between quality and stability of results, and applying it to the EMST of moving points. Monma and Suri [18] study the number of topological changes that occur in the EMST when one point is allowed to move.

The problem of finding the MMST and MBMST of moving points can be seen as a bicriteria optimization problem if the points move linearly (as shown in Sect. 2.2). In this context, the addition of a new criterion could lead to an NP-hard problem, such as the bi-criteria shortest path problem in weighted graphs. Garey and Johnson show that given a source and target vertices, minimizing both length and weight of a path from source to target is NP-hard [11, p. 214]. Arkin et al. analyse other criteria combined with the shortest path problem [4], such as the total turn length and different norms for path length.

Maintaining the EMST and other geometric structures of a set of moving points have been investigated by several papers since 1985 [5]. Kinetic data structures have been proposed to maintain the EMST [1,20]. Research in this area have focused on bounds on the number of combinatorial changes in the EMST and efficient algorithms. To the best of our knowledge, the problem of finding the MMST and MBMST (a single tree that does not change during the movement of points) has not been investigated.

2 Preliminaries

In this section we formally define the minimum moving spanning tree and the minimum bottleneck moving spanning tree of a set of moving points. We then prove that, for points moving linearly, the distance function between a pair of points is convex.

2.1 Definitions

A *moving point* p in the plane is a continuous function $p : [0,1] \to \mathbb{R}^2$. We assume that p moves on a straight line segment in $\mathbb{R}^2$. We say that p is at $p(t)$ at time t. We are given a set $S = \{p_1, ..., p_n\}$ of moving points in the plane. A *moving spanning tree* T of S has S as its vertex set and weight function $w_T : [0,1] \to \mathbb{R}$ defined as $w_T(t) = \sum_{pq \in T} \|p(t)q(t)\|$. Let $\mathcal{T}(S)$ denote the set of all moving spanning trees of S. Let $w(T) = \sup_t w_T(t)$ be the weight of the moving spanning tree T. A minimum moving spanning tree (MMST) of S is a moving spanning tree of S with minimum weight. In other words an MMST is in

$$\arg\min_{T \in \mathcal{T}(S)} (w(T)) .$$

Let $b_T(t) = \sup_{pq \in T} \|p(t)q(t)\|$ denote the *bottleneck* of a tree T at time t. A minimum bottleneck moving spanning tree (MBMST) of S is a moving spanning tree of S that minimizes the bottleneck over all $t \in [0, 1]$. In other words an MBMST is in

$$\arg\min_{T \in \mathcal{T}(S)} \left(\max_t b_T(t) \right).$$

2.2 Convexity

Let p and q be two moving points in the plane. We assume that these points move along (possibly different) lines at (possibly different) constant velocities. Thus, for any real number t, we can write the positions of p and q at time t as

$$p(t) = (a_p + u_p t, b_p + v_p t)$$

and

$$q(t) = (a_q + u_q t, b_q + v_q t),$$

where a_p, u_p, b_p, v_p are constants associated with the point p. At time $t = 0$, p is at (a_p, b_p), and the velocity vector of p is (u_p, v_p). Let $d(t) = \|p(t)q(t)\|$ denote the Euclidean distance between p and q at time t. In the next lemma we prove that d is a convex function. The convexity of d is also implied by a result of Alt and Godau [3] that the free space diagram of any two line segments is convex.

Lemma 1. *The function d is convex.*

Proof. It suffices to prove that the second derivative of d is non-negative for all real numbers t. We can write

$$d(t) = \sqrt{At^2 + Bt + C},$$

where A, B, and C depend only on $a_p, u_p, b_p, v_p, a_q, u_q, b_q$, and v_q. Observe that $A \geq 0$. Since $d(t)$ represents a distance, $At^2 + Bt + C \geq 0$ for all t in $\mathbb{R}$. It follows that the discriminant of this quadratic function is non-positive, i.e.,

$$B^2 - 4AC \leq 0. \tag{1}$$

Let $\alpha = -B/2A$ and $\beta = C/A - B^2/(4A^2)$. Then

$$d(t) = \sqrt{A} \cdot \sqrt{(t - \alpha)^2 + \beta}.$$

The second derivative of the function $f(t) = \sqrt{t^2 + \beta}$ is given by

$$f''(t) = \frac{\beta}{(t^2 + \beta)^{3/2}}.$$

It follows from (1) that $\beta \geq 0$. Thus, $f''(t) \geq 0$ for all t in $\mathbb{R}$. Since $d(t) = \sqrt{A} \cdot f(t - \alpha)$, we have $d''(t) \geq 0$ for all t in $\mathbb{R}$, and in particular, for $t \in [0, 1]$. □

The convexity of the distance function between two moving points (Lemma 1) implies the following corollary.

Corollary 1. *The largest distance between two moving points is attained either at the start time or at the finish time.*

Let S be a set of n moving points in the plane. For two points p and q in S, we denote by $\|p(0)q(0)\|$ and $\|p(1)q(1)\|$ the distances between p and q at times $t = 0$ and $t = 1$, respectively. Moreover, we denote by $|pq|_{\max}$ the largest distance between p and q during time interval $[0, 1]$. By Corollary 1 we have

$$|pq|_{\max} = \max\{\|p(0)q(0)\|, \|p(1)q(1)\|\}. \tag{2}$$

3 Minimum Bottleneck Moving Spanning Tree

Since by Corollary 1 the largest length of an edge is attained either at time 0 or at time 1, it might be tempting to think that the MBMST of S is also attained at times 0 or 1. However the example in Fig. 1(a) shows that this may not be true. In this example we have four points a, b, c, and d that move from time 0 to time 1 as depicted in the figure. The MBST of these points at time 0 is the red tree R, and their MBST at time 1 is the blue tree B. Recall that $b_T(t)$ is the bottleneck of tree T at time t. Let $b(T) = \max_t b_T(t)$ be the *bottleneck* of T. In R the weight of ab at time 0 is 1 while its weight at time 1 is 3, and thus $b(R) = 3$. In B the weight of ad at time 1 is 1 while its weigh at time 0 is 3, and thus $b(B) = 3$. However, for this point set the tree $T = \{ac, cb, cd\}$ has bottleneck 2.

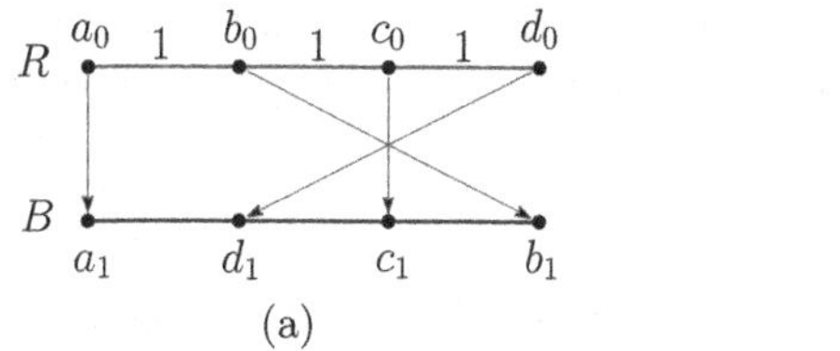

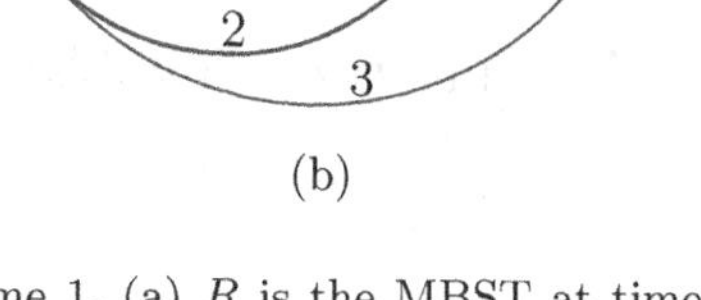

Fig. 1. Four points that move from time 0 to time 1. (a) R is the MBST at time 0, and B is the MBST at time 1. (b) The graph G; green edges form an MBST of this graph. (Color figure online)

Although the above example shows that the computation of the MBMST is not straightforward, we present a simple algorithm for finding the MBMST. Let G be the complete graph on points of S where the weight $w(pq)$ of every edge pq is the largest distance between p and q during time interval $[0, 1]$, that is, $w(pq) = |pq|_{\max}$; see Fig. 1(b).

Lemma 2. *The bottleneck of an MBMST of S is not smaller than the bottleneck of an MBST of G.*

Proof. Our proof is by contradiction. Let T^* be an MBMST of S and let T be an MBST of G. For the sake of contradiction assume that $b(T^*) < b(T)$, where we abuse the notation for simplicity making $b(T) = \max_{pq \in T} w(pq)$ the bottleneck of T. Let pq be a bottleneck edge of T, that is $b(T) = w(pq)$. Denote by T_p and T_q the two subtrees obtained by removing pq from T, and denote by V_p and V_q the vertex sets of these subtrees. Since the vertex set of T is the same as that of T^*, there is an edge, say rs, in T^* that connects a vertex of V_p to a vertex of V_q. Since the bottleneck of T^* is its largest edge-length in time interval $[0, 1]$, we have that $|rs|_{\max} \leqslant b(T^*)$. Since in G we have $w(rs) = |rs|_{\max}$, the following inequality is valid: $w(rs) = |rs|_{\max} \leqslant b(T^*) < b(T) = w(pq)$. Let T' be the spanning tree of G that is obtained by connecting T_p and T_q by rs. Then $b(T') \leqslant b(T^*)$. If we repeat this process for all bottleneck edges of T, then we obtain a tree T' whose bottleneck is strictly smaller than that of T. This contradicts the fact that T is an MBST of G. □

It is implied from Lemma 2 that any MBST of G is an MBMST of S. Since an MBST of a graph can be computed in time linear in the size of the graph [7], an MBST of G can be computed in $O(n^2)$ time. The following theorem summarizes our result in this section.

Theorem 1. *A minimum bottleneck moving spanning tree of n moving points in the plane can be computed in $O(n^2)$ time.*

4 Minimum Moving Spanning Tree

In this section we study the problem of computing an MMST of moving points. At the end of this section we prove that this problem is NP-hard. We start by proposing a 2-approximation algorithm for the MST problem. In the full version of the paper we show that our analysis of the approximation ratio is tight.

4.1 A 2-approximation Algorithm

Our algorithm is very simple and just computes a MST of the graph G that is constructed in Sect. 3.

Lemma 3. *The weight of any MST of G is at most two times the weight of any MMST of S.*

Proof. Let T be any MST of G and let T^* be any MMST of S. Let $w(T^*) = \sup_t w_T(t)$ be the weight of the moving spanning tree T^*. We abuse the notation for simplicity making $w(T) = \sum_{pq \in T} w(pq)$ the weight of the spanning tree T. We are going to show that $w(T) \leqslant 2 \cdot w(T^*)$. Let T' be a tree that is combinatorially equivalent to T^*, i.e., has the same topology as T^*. Assign to each edge pq of T' the weight $w(pq) = |pq|_{\max}$. After this weight assignment, T' is a spanning tree of G. Since T is a MST of G, we have $w(T) \leqslant w(T')$.

By Corollary 1 the largest distance between two points is achieved either at time 0 or at time 1. Let E_0^* be the set of edges of T^* whose endpoints largest distance is achieved at time 0. Define E_1^* analogously. Then $w(E_0^*) \leqslant w(T^*)$ and $w(E_1^*) \leqslant w(T^*)$. Moreover, $w(T') = w(E_0^*) + w(E_1^*)$. By combining these inequalities we get

$$w(T) \leqslant w(T') = w(E_0^*) + w(E_1^*) \leqslant w(T^*) + w(T^*) = 2 \cdot w(T^*).$$

□

A minimum spanning tree of G can be computed in $O(n^2)$ time using Prim's MST algorithm. The following theorem summarizes our result in this section.

Theorem 2. *There is an $O(n^2)$-time 2-approximation algorithm for computing the minimum moving spanning tree of n moving points in the plane.*

4.2 An $O(n \log n)$-time $(2+\varepsilon)$-approximation Algorithm

Section 4.1 showed that the weight of the minimum spanning tree of the graph G, defined in Sect. 3, gives a 2-approximation to the MMST. Since G has $\Theta(n^2)$ edges, it takes $\Theta(n^2)$ time to compute its MST. In this section, we prove that a $(1+\varepsilon)$-approximation to the minimum spanning tree of G can be computed in $O(n \log n)$ expected time. Thus, if we replace ε by $\varepsilon/2$, we obtain a $(2+\varepsilon)$-approximation to computing the MMST of a set of linearly moving points S.

For any point p in S, define the point

$$P = (p(0), p(1))$$

in $\mathbb{R}^4$. Doing this for all points in S, we obtain a set S' of n points in $\mathbb{R}^4$. For any two points P and Q in S', define their distance to be

$$\text{dist}(P, Q) = \max(\|p(0)q(0)\|, \|p(1)q(1)\|).$$

Since $\text{dist}(P, Q) = w(pq)$, the minimum spanning tree of our graph G has the same weight as the minimum spanning tree (under dist) of the point set S'.

Lemma 5 below states that dist satisfies the properties of a metric. Its proof uses the following lemma, which is probably well known.

Lemma 4. *Let V be an arbitrary set and let $d_1 : V \times V \to \mathbb{R}$ and $d_2 : V \times V \to \mathbb{R}$ be two functions, such that both (V, d_1) and (V, d_2) are metric spaces. Define the function $d : V \times V \to \mathbb{R}$ by*

$$d(a, b) = \max(d_1(a, b), d_2(a, b))$$

for all a and b in V. Then (V, d) is a metric space.

Proof. It is clear that, for all a and b in V, $d(a, a) = 0$, $d(a, b) > 0$ if $a \neq b$, and $d(a, b) = d(b, a)$. It remains to prove that the triangle inequality holds.

Let a, b, and c be elements of V. Then

$$\begin{aligned} d(a, b) &= \max(d_1(a, b), d_2(a, b)) \\ &\leq \max(d_1(a, c) + d_1(c, b), d_2(a, c) + d_2(c, b)). \end{aligned}$$

Using the inequality

$$\max(\alpha+\beta, \gamma+\delta) \leq \max(\alpha, \gamma) + \max(\beta, \delta),$$

it follows that

$$\begin{aligned} d(a,b) &\leq \max(d_1(a,c), d_2(a,c)) + \max(d_1(c,b), d_2(c,b)) \\ &= d(a,c) + d(c,b). \end{aligned}$$

□

Lemma 5. *The pair* (S', dist) *is a metric space.*

Proof. The proof follows from Lemma 4 and the definition of dist. □

The next lemma states that the metric space (S', dist) has bounded doubling dimension. We recall the definition. For any point P in S' and any real number $\rho > 0$, the ball with center P and radius ρ is the set

$$\text{ball}^{\text{dist}}(P, \rho) = \{Q \in S' : \text{dist}(P, Q) \leq \rho\}.$$

Let λ be the smallest integer such that for every real number $\rho > 0$, every ball of radius ρ can be covered by at most λ balls of radius $\rho/2$. The doubling dimension of (S', dist) is defined to be $\log \lambda$.

Lemma 6. *The doubling dimension of the metric space* (S', dist) *is* $O(1)$.

Proof. Recall that S' is a set of points in $\mathbb{R}^4$. We denote the Euclidean distance between two points P and Q of S' by $\|PQ\|$. The Euclidean ball with center P and radius ρ is denoted by $\text{ball}^e(P, \rho)$. Thus,

$$\text{ball}^e(P, \rho) = \{Q \in S' : |PQ| \leq \rho\}.$$

It is easy to verify that

$$\text{dist}(P, Q) \leq \|PQ\| \leq \sqrt{2} \cdot \text{dist}(P, Q). \tag{3}$$

Let P be a point in S', let $\rho > 0$ be a real number, and let $B^{\text{dist}} = \text{ball}^{\text{dist}}(P, \rho)$. We will prove that B^{dist} can be covered by $O(1)$ balls of radius $\rho/2$.

Let B^e be the Euclidean ball with center P and radius $\rho \cdot \sqrt{2}$. It follows from (3) that

$$B^{\text{dist}} \subseteq B^e.$$

It is well known that the doubling dimension of the Euclidean space $\mathbb{R}^4$ is bounded by a constant. Thus, by applying the definition of doubling dimension twice, we can cover B^e by $k = O(1)$ Euclidean balls $B_1^e, \ldots, B_k^e$ balls, each of radius $\rho \cdot \sqrt{2}/4 \leq \rho/2$. Let these balls have centers $C_1, \ldots, C_k$. For each $i = 1, \ldots, k$, define $B_i^{\text{dist}} = \text{ball}^{\text{dist}}(C_i, \rho/2)$. It follows from (3) that

$$B_i^e \subseteq B_i^{\text{dist}}.$$

Thus,

$$B^{\text{dist}} \subseteq B^e \subseteq \bigcup_{i=1}^{k} B_i^e \subseteq \bigcup_{i=1}^{k} B_i^{\text{dist}},$$

i.e., we have covered the ball B^{dist} by $k = O(1)$ balls of radius $\rho/2$. □

Lemma 7. *Let $\varepsilon > 0$ be a constant. In $O(n \log n)$ expected time, we can compute a $(1+\varepsilon)$-approximation to the minimum spanning tree of the metric space (S', dist).*

Proof. As (S', dist) has a constant doubling dimension (by Lemma 6), a result of Har-Peled and Mendel [12] implies that a $(1+\varepsilon)$-spanner of (S', dist) with $O(n)$ edges can be computed in $O(n \log n)$ expected time. Their algorithm assumes that any distance in the metric space can be computed in $O(1)$ time; this is the case for our distance function dist.

It is known that a minimum spanning tree of a $(1+\varepsilon)$-spanner is a $(1+\varepsilon)$-approximation to the minimum spanning tree. (See, e.g., [19, Theorem 1.3.1]).

Since the spanner has $O(n)$ edges, its minimum spanning tree can be computed in $O(n \log n)$ time using Prim's MST algorithm combined with a binary min-heap. □

As a consequence of Lemma 7 and the fact that $\text{dist}(P, Q) = w(pq)$, we have the following theorem.

Theorem 3. *In $O(n \log n)$ expected time, we can compute a $(2+\varepsilon)$-approximation for the minimum moving spanning tree of a set of linearly moving points in the plane.*

4.3 NP-hardness of MMST

Inspired by Arkin et al. [4], we reduce the Partition problem, which is known to be NP-hard [11], to the MMST problem. In one formulation of the Partition problem, we are given $n > 0$ positive integers $a_0, \ldots, a_{n-1}$ and must decide whether there is a subset $S \subseteq \{0, \ldots, n-1\}$ such that

$$\sum_{i \in S} a_i = \frac{1}{2} \sum_{i=0}^{n-1} a_i.$$

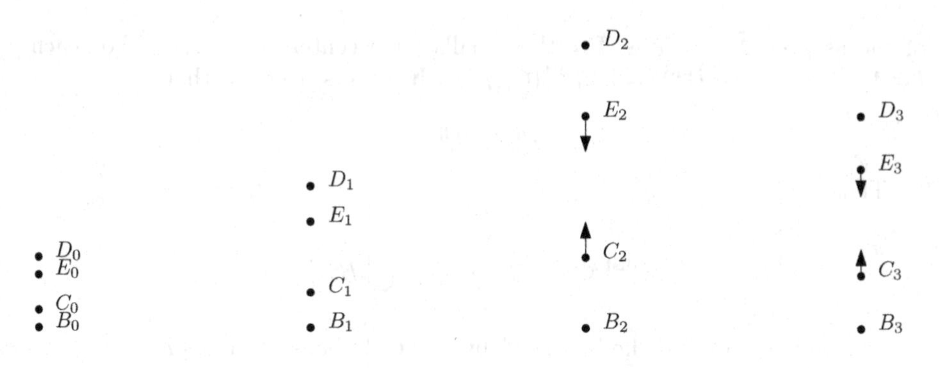

Fig. 2. The positions of the points in P at time $t = 1/4$ when $n = 4$ and $(a_0, a_1, a_2, a_3) = (1, 2, 4, 3)$. The velocities of C_2, E_2, C_3 and E_3 are depicted.

Construction. We construct an instance of a decision version of the MMST problem defined as follows. First we let $\ell = \max\{a_0, \ldots, a_{n-1}\}$ and then, for each $i \in \{0, \ldots, n-1\}$, we put the following points into our set P of moving points (Fig. 2):

- A_i, stationary at $(i\ell, 0)$;
- B_i, stationary at $(i\ell, \ell)$;
- C_i, moving from $(i\ell, \ell)$ to $(i\ell, \ell + a_i)$;
- D_i, stationary at $(i\ell, \ell + a_i)$; and
- E_i, moving from $(i\ell, \ell + a_i)$ to $(i\ell, \ell)$.

We then ask whether there is a moving spanning tree T with

$$w(T) \leq (2n-1)\ell + \frac{3}{2}\sum_{i=0}^{n-1} a_i.$$

Theorem 4. *The decision version of the MMST problem is weakly NP-hard.*

Proof. Let T be a moving spanning tree on vertex set P. Recall that $w_T(t)$ denotes the weight of T at time t. By Lemma 1, w_T is a convex function and the weight of T is indeed $w(T) = \max\{w_T(0), w_T(1)\}$.

Let K_0 be the set of edges A_iB_i for $i \in \{0, \ldots, n-1\}$ and A_iA_{i+1} for $i \in \{0, \ldots, n-2\}$ and let K_1 be the set of edges among B_i, C_i, D_i and E_i

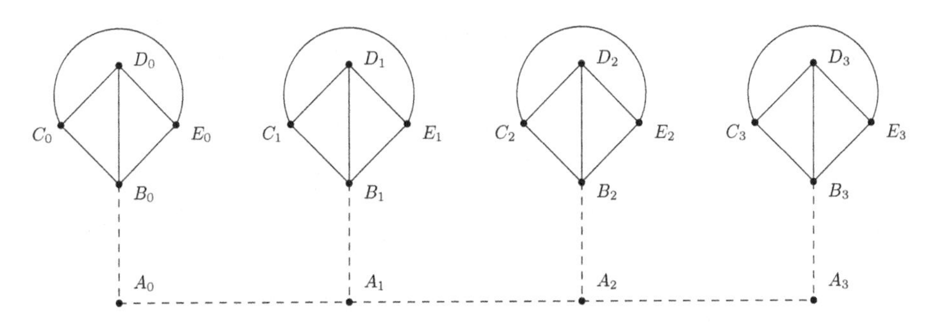

Fig. 3. The (topological) edges in K_0 (dashed) and in $K_1 \setminus K_0$ (solid).

for each $i \in \{0, \ldots, n-1\}$ together with K_0 (Fig. 3). We claim that there is a moving spanning tree T^* of minimum cost, i.e., an optimal solution to the MMST problem, whose edges are all in K_1. Assume the contrary for contradiction. Let T be an MMST whose intersection with K_1 is maximum. By assumption, T has at least an edge $e \notin K_1$. We now consider the two components obtained from deleting e from T. There must be at least one edge $e' \in K_1$ between the two components, since K_1 spans P. However, at any point in time, every edge in K_1 weights at most ℓ while every edge outside of K_1 weights at least ℓ, so if we bridge the two components with e', we will be left with a spanning tree T' with $w(T') \leq w(T)$ and with a larger intersection with K_1, contradicting the definition of T.

As every edge in K_0 is a bridge in the graph (P, K_1), the spanning tree T^* must contain K_0, so T^* consists of K_0 and, for each $i \in \{0, \ldots, n-1\}$, of a subtree T_i spanning $\{B_i, C_i, D_i, E_i\}$. The weights $w_{T_i}(0)$ and $w_{T_i}(1)$ must both be a multiple of a_i since so are the Euclidean distances between the vertices of T_i at these two times. There are two notable ways to build T_i: one is $T_i = \{B_iC_i, C_iD_i, D_iE_i\}$, which satisfies $w_{T_i}(0) = a_i$ and $w_{T_i}(1) = 2a_i$ and is thus called the $(1,2)$-tree; and the other is $T_i = \{B_iE_i, E_iD_i, D_iC_i\}$, which satisfies $w_{T_i}(0) = 2a_i$ and $w_{T_i}(1) = a_i$ and is thus called the $(2,1)$-tree.

We shall show that the $(1,2)$-tree or the $(2,1)$-tree have minimum weight among all moving spanning trees of $\{B_i, C_i, D_i, E_i\}$. Indeed, T_i is made of three edges and, since there are no three edges with weight zero at time 0, as can be seen in Fig. 4, we must have $w_{T_i}(0) \geq a_i$ and, similarly, $w_{T_i}(1) \geq a_i$. Furthermore, each edge between B_i, C_i, D_i and E_i adds up to at least a_i in terms of their weight at time 0 and at time 1. Therefore, $w_{T_i}(0) + w_{T_i}(1) \geq 3a_i$, so either $w_{T_i}(0) \geq 2a_i$, , or $w_{T_i}(1) \geq 2a_i$. As a result, we may assume, without loss of generality, that T_i is either the $(1,2)$-tree or the $(2,1)$-tree.

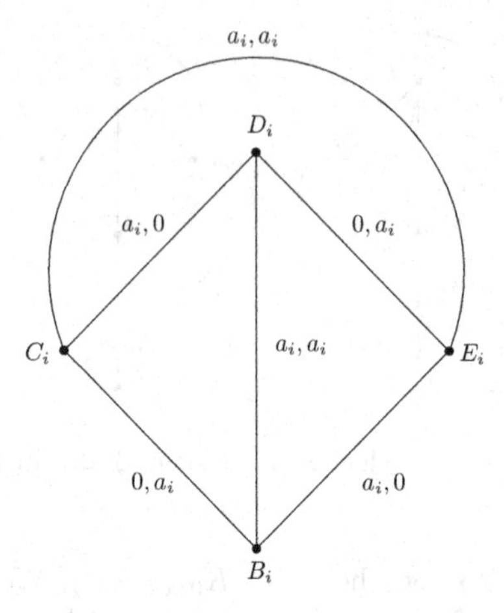

Fig. 4. Edges between B_i, C_i, D_i and E_i labeled with their weights at times 0 and 1.

Let now $S^* \subseteq \{0, \dots, n-1\}$ be the set of indices i such that T_i is the corresponding (2, 1)-tree. As $|K_0| = 2n - 1$, we have

$$w_{T^*}(0) = (2n-1)\ell + \sum_{i=0}^{n-1} a_i + \sum_{i \in S^*} a_i,$$

while

$$w_{T^*}(1) = (2n-1)\ell + \sum_{i=0}^{n-1} a_i + \sum_{i \in \{0,\dots,n-1\} \setminus S^*} a_i.$$

Therefore, the cost of T^* is

$$(2n-1)\ell + \sum_{i=0}^{n-1} a_i + \max\left\{\sum_{i \in S^*} a_i, \sum_{i \in \{0,\dots,n-1\} \setminus S^*} a_i\right\}.$$

Because

$$\sum_{i \in S^*} a_i \geq \frac{1}{2}\sum_{i=0}^{n-1} a_i \quad \text{or} \quad \sum_{i \in \{0,\dots,n-1\} \setminus S^*} a_i \geq \frac{1}{2}\sum_{i=0}^{n-1} a_i,$$

then the following holds

$$w(T^*) \geq (2n-1)\ell + \frac{3}{2}\sum_{i=0}^{n-1} a_i. \tag{4}$$

We claim that (4) holds with equality if and only if our instance of the Partition problem has a solution, i.e., there is a set $S \subseteq \{0, \dots, n-1\}$ such that the sum of a_i for $i \in S$ is half of $a_0 + \cdots + a_{n-1}$. Indeed, if the equality holds, we can simply let $S = S^*$. To show the converse, we build a tree T from the solution S of the Partition problem. This tree contains K_0, the corresponding $(2, 1)$-trees for i in S and the corresponding $(1, 2)$-trees for $i \in \{0, \ldots, n-1\} \setminus S$, resulting in a weight of

$$w(T) = (2n - 1)\ell + \frac{3}{2} \sum_{i=0}^{n-1} a_i.$$

Because T^* is an MMST, $w(T^*) \leq w(T)$, so the equality holds. □

Acknowledgement. This research was carried out at the *Eighth Annual Workshop on Geometry and Graphs*, held at the Bellairs Research Institute in Barbados, January 31–February 7, 2020. The authors are grateful to the organizers and to the participants of this workshop. We thank Günther Rote for pointing us to the work of Arkin et al. [4].

References

1. Abam, M.A., Rahmati, Z., Zarei, A.: Kinetic pie delaunay graph and its applications. In: Fomin, F.V., Kaski, P. (eds.) Algorithm Theory. LNCS, vol. 7357, pp. 48–58. Springer, Heidelberg (2012). https://doi.org/10.1007/978-3-642-31155-0_5
2. Alper, B., Riche, N., Ramos, G., Czerwinski, M.: Design study of linesets, a novel set visualization technique. IEEE Trans. Vis. Comput. Graph. **17**(12), 2259–2267 (2011)
3. Alt, H., Godau, M.: Computing the Fréchet distance between two polygonal curves. Int. J. Comput. Geom. Appl. **5**, 75–91 (1995)
4. Arkin, E.M., Mitchell, J.S., Piatko, C.D.: Bicriteria shortest path problems in the plane. In: Proceedings of 3rd Canadian Conference on Computational Geometry, pp. 153–156 (1991)
5. Atallah, M.J.: Some dynamic computational geometry problems. Comput. Math. Appl. **11**(12), 1171–1181 (1985)
6. Buchmüller, J., Jäckle, D., Cakmak, E., Brandes, U., Keim, D.A.: Motionrugs: visualizing collective trends in space and time. IEEE Trans. Vis. Comput. Graph. **25**(1), 76–86 (2018)
7. Camerini, P.M.: The min-max spanning tree problem and some extensions. Inf. Process. Lett. **7**(1), 10–14 (1978)
8. Collins, C., Penn, G., Carpendale, S.: Bubble sets: revealing set relations with isocontours over existing visualizations. IEEE Trans. Vis. Comput. Graph. **15**(6), 1009–1016 (2009). Proceedings of the IEEE Conference on Information Visualization
9. Dinkla, K., van Kreveld, M.J., Speckmann, B., Westenberg, M.A.: Kelp diagrams: point set membership visualization. Comput. Graph. Forum **31**(3pt1), 875–884 (2012)
10. Fabrikant, S.I., Miksch, S., Wolff, A.: Visual analytics for sets over time and space (Dagstuhl Seminar 19192). Dagstuhl Rep. **9**(5), 31–57 (2019)
11. Garey, M.R., Johnson, D.S.: Computers and Intractability, vol. 174. Freeman, San Francisco (1979)

12. Har-Peled, S., Mendel, M.: Fast construction of nets in low-dimensional metrics and their applications. SIAM J. Comput. **35**(5), 1148–1184 (2006)
13. Hurtado, F., et al.: Colored spanning graphs for set visualization. Comput. Geom. **68**, 262–276 (2018). Special Issue in Memory of Ferran Hurtado
14. Jaffry, S., Hussain, R., Gui, X., Hasan, S.F.: A comprehensive survey on moving networks. arXiv preprint arXiv:2003.09979 (2020)
15. Kraak, M.-J.: The space-time cube revisited from a geovisualization perspective. In: Proceedings of 21st International Cartographic Conference, pp. 1988–1996 (2003)
16. Meulemans, W., Riche, N.H., Speckmann, B., Alper, B., Dwyer, T.: KelpFusion: a hybrid set visualization technique. IEEE Trans. Vis. Comput. Graph. **19**(11), 1846–1858 (2013)
17. Meulemans, W., Speckmann, B., Verbeek, K., Wulms, J.: A framework for algorithm stability and its application to kinetic Euclidean MSTs. In: Bender, M., Farach-Colton, M., Mosteiro, M. (eds.) Theoretical Informatics. LNCS, vol. 10807, pp. 805–819. Springer, Cham (2018). https://doi.org/10.1007/978-3-319-77404-6_58
18. Monma, C.L., Suri, S.: Transitions in geometric minimum spanning trees. Discret. Comput. Geom. **8**, 265–293 (1992)
19. Narasimhan, G., Smid, M.: Geometric Spanner Networks. Cambridge University Press, Cambridge (2007)
20. Rahmati, Z., Zarei, A.: Kinetic Euclidean minimum spanning tree in the plane. J. Discret. Algorithms **16**, 2–11 (2012). Selected papers from the 22nd International Workshop on Combinatorial Algorithms (IWOCA 2011)
21. Wulms, J., Buchmüller, J., Meulemans, W., Verbeek, K., Speckmann, B.: Spatially and temporally coherent visual summaries. arXiv preprint arXiv:1912.00719 (2019)

Scheduling with Testing on Multiple Identical Parallel Machines

Susanne Albers[1] and Alexander Eckl[1,2(✉)]

[1] Department of Informatics, Technical University of Munich,
Boltzmannstr. 3, 85748 Garching, Germany
albers@in.tum.de, alexander.eckl@tum.de
[2] Advanced Optimization in a Networked Economy, Technical University of Munich,
Arcisstraße 21, 80333 Munich, Germany

Abstract. Scheduling with testing is a recent online problem within the framework of explorable uncertainty motivated by environments where some preliminary action can influence the duration of a task. Jobs have an unknown processing time that can be explored by running a test. Alternatively, jobs can be executed for the duration of a given upper limit. We consider this problem within the setting of multiple identical parallel machines and present competitive deterministic algorithms and lower bounds for the objective of minimizing the makespan of the schedule. In the non-preemptive setting, we present the SBS algorithm whose competitive ratio approaches 3.1016 if the number of machines becomes large. We compare this result with a simple greedy strategy and a lower bound which approaches 2. In the case of uniform testing times, we can improve the SBS algorithm to be 3-competitive. For the preemptive case we provide a 2-competitive algorithm and a tight lower bound which approaches the same value.

Keywords: Online scheduling · Identical parallel machines · Explorable uncertainty · Makespan minimization · Competitive analysis

1 Introduction

One of the most fundamental problems in online scheduling is makespan minimization on multiple parallel machines. An online sequence of n jobs with processing times p_j has to be assigned to m identical machines. The objective is to minimize the makespan of the schedule, i.e. the maximum load on any machine. In 1966, Graham [25] showed that the List Scheduling algorithm, which assigns every job to the currently least loaded machine, is $(2 - \frac{1}{m})$-competitive. Since then the upper bound has been improved multiple times, most recently to 1.9201

Funded by the Deutsche Forschungsgemeinschaft (DFG, German Research Foundation) – 277991500/GRK2201, and by the European Research Council, Grant Agreement No. 691672.

A. Lubiw et al. (Eds.): WADS 2021, LNCS 12808, pp. 29–42, 2021.
https://doi.org/10.1007/978-3-030-83508-8_3

by Fleischer and Wahl [22]. At the same time, the lower bound has also been the focus of a lot of research, the current best result is 1.88 by Rudin [37].

We consider this classical problem in the framework of *explorable uncertainty*, where part the input is initially unknown to the algorithm and can be explored by investing resources which are added as costs to the objective function. Let n jobs be given. Every job j has a processing time p_j and an upper bound u_j. It holds $0 \leq p_j \leq u_j$ for all j. Each job also has a testing time $t_j \geq 0$. A job can be executed on one of m identical machines in one of two modes: It can either be run untested, which takes time u_j, or be tested and then executed, which takes a total time of $t_j + p_j$. The number of jobs n, as well as all testing times t_j and upper bounds u_j are known to the algorithm in the beginning. In particular, an algorithm can sort/order the jobs in a convenient way based on these parameters. The processing time p_j for job j is revealed once the test t_j is completed. This *scheduling with testing* setting has been recently studied by Dürr et al. [13], and Albers and Eckl [3] on a single machine.

We differentiate between *preemptive* and *non-preemptive* settings: If preemption is allowed, a job may be interrupted at any time, and then continued later on a possibly different machine. No two machines may work on the same job at the same time. In case a job is tested, any section of the test must be scheduled earlier than any section of the actual job processing. In the non-preemptive setting, a job assigned to a machine has be fully scheduled without interruption on this machine, independent of whether it is tested or not. We also introduce the notion of *test-preemptive* scheduling, where a job can only be interrupted right after its test is completed.

Scheduling with testing is well-motivated by real world settings where a preliminary evaluation or operation can be executed to improve the duration or difficulty of a task. Examples for the case of multiple machines include e.g. a manufacturing plan where a number of jobs with uncertain length have to be assigned to multiple workers, or a distributed computing setting where tasks with unknown parameters have to be allocated to remote computing nodes by a central scheduler. Several examples for applicable settings for scheduling with testing can also be found in [3,13].

In summary, we study the classical problem of makespan minimization on identical parallel machines in the framework of explorable uncertainty. We use competitive analysis to compare the value of an algorithm with an optimal offline solution. The setting closely relates to online machine scheduling problems studied previously in the literature. We investigate deterministic algorithms and lower bounds for the preemptive and non-preemptive variations of this problem.

1.1 Related Work

Scheduling with testing describes the setting where jobs with uncertain processing times have to be scheduled tested or untested on a given number of machines. The problem has been first studied by Dürr et al. [13,14] for the special case of scheduling jobs on a single machine with uniform testing times $t_j \equiv 1$. They presented several algorithms and lower bounds for the objectives of the sum of completion times and the makespan. More recently, Albers and Eckl [3] considered the one machine case with testing times $t_j \geq 0$, presenting generalized algorithms for both objectives. In this paper, we consider scheduling with testing on identical parallel machines, a natural generalization of the previously studied one machine case.

Makespan minimization in online scheduling with identical machines has been studied extensively in the past decades, ever since Graham [25] established his $(2 - \frac{1}{m})$-competitive List Scheduling algorithm in 1966. In the deterministic setting, a series of publications improved Graham's result to competitive ratios of $2 - \frac{1}{m} - \varepsilon_m$ [23] where $\varepsilon_m \to 0$ for large m, 1.985 [8], 1.945 [31], and 1.923 [1], before Fleischer and Wahl [22] presented the current best result of 1.9201. In terms of the deterministic lower bound for general m, research has been just as fruitful. The bound was improved from 1.707 [19], to 1.837 [9], and 1.852 [1]. The best currently known bound of 1.88 is due to Rudin [37]. For the randomized variant, the lower bound has a current value of $\frac{e}{e-1} \approx 1.582$ [11,38], while the upper bound is 1.916 [2]. For the deterministic preemptive setting, Chen et al. [12] provide a tight bound of $\frac{e}{e-1}$ for large values of m.

More recently, various extension of this basic case have emerged. In *resource augmentation* settings the algorithm receives some extra resources like machines with higher speed [30], parallel schedules [6,33], or a reordering buffer [15,33]. A variation that is closely related to our setting is *semi-online scheduling*, where some additional piece of information is available to the online algorithm in advance. Possible pieces of information include for example the sum of all processing times [5,32,33], the value of the optimum [7], or information about the job order [26]. Refer also to the survey by Epstein [16] for an overview of makespan minimization in semi-online scheduling.

Scheduling with testing is directly related to *explorable uncertainty*, a research area concerned with obtaining additional information of unknown parameters through queries with a given cost. Kahan [29] pioneered this line of research in 1991 by studying approximation guarantees for the number of queries necessary to obtain the maximum and median value of a set of uncertain elements. Following this, a variety of problems have been studied in this setting, for example finding the median or k-smallest value [21,27,34], geometric tasks [10], caching [36], as well as combinatorial problems like minimum spanning tree [18,35], shortest path [20], and knapsack [24]. We refer to the survey by Erlebach and Hoffmann [17] for an overview. In the scheduling with testing model, the cost of the queries is added to the objective function. Similar settings are considered for example in Weitzman's pandora's box problem [40], or in the recent 'price of information' model by Singla [39].

1.2 Contribution

In this paper we provide the first results for makespan minimization on multiple machines with testing. We differentiate between general tests $t_j \geq 0$ and uniform tests $t_j = 1$, and consider non-preemptive as well as preemptive environments. In

Table 1, we illustrate our results for these cases. The parameter m corresponds to the number of machines in the instance.

Table 1. Overview of results

Setting	General tests	Uniform tests	Lower bound
Non-preemptive	$c(m) \xrightarrow[m\to\infty]{} 3.1016$	$c_1(m) \xrightarrow[m\to\infty]{} 3$	$\max(\varphi, 2-\frac{1}{m})$
Preemptive	2	2	$\max(\varphi, 2-\frac{2}{m}+\frac{1}{m^2})$

In the non-preemptive setting, we present our main algorithm with competitive ratio $c(m)$, which we refer to as the *SBS algorithm.* The function $c(m)$ is increasing in m and has a value of approximately 3.1016 for $m \to \infty$. For uniform tests, we can improve the algorithm to a competitive ratio of $c_1(m)$, which approaches 3 for large values of m. Additionally, we analyze a simple Greedy algorithm for general tests with a competitive ratio of $\varphi(2-\frac{1}{m})$, where $\varphi \approx 1.6180$ is the golden ratio. We also provide a lower bound with value $\max(\varphi, 2-\frac{1}{m})$. The values of $c(m)$, $c_1(m)$, the Greedy algorithm and the lower bound are summarized in Table 2. For all values of $m > 1$ the SBS algorithm has better ratios compared to Greedy. At the same time, the uniform version of the algorithm improves these results further. Though our algorithms work for any number of machines m, they all achieve the same ratio for $m = 1$ as was already proven in [13] and [3] for uniform and general tests, respectively.

If the scheduler is allowed to use preemption, we obtain a 2-approximation for both general and uniform tests. The result holds even in the more restrictive test-preemptive setting. The corresponding lower bound of $\max(\varphi, 2-\frac{2}{m}+\frac{1}{m^2})$ is tight when the number of machines becomes large.

We utilize various methods for our algorithms and lower bounds. The Greedy algorithm we present is a variation of the well-known List Scheduling algorithm introduced by Graham [25]. For the more involved SBS algorithm and its uniform version we employ testing rules for jobs based on the ratio between their upper bound and testing time similar to [3]. We additionally divide the schedule into phases based on these ratios, therefore sorting the jobs by the given parameters to guarantee competitiveness. In the preemptive setting, we divide the schedule into two independent phases, testing and execution, and use an offline algorithm

Table 2. Results in the non-preemptive setting for selected values of m

	1	2	3	4	5	10	100	∞
Greedy	1.6180	2.4271	2.6967	2.8316	2.9125	3.0743	3.2199	3.2361
SBS	1.6180	2.3806	2.6235	2.7439	2.8158	2.9591	3.0874	3.1016
Uniform-SBS	1.6180	2.3112	2.5412	2.6560	2.7248	2.8625	2.9862	3
Lower bound	1.6180	1.6180	1.6667	1.75	1.8	1.9	1.99	2

for makespan minimization to solve each instance separately. Lastly, the lower bounds we provide are loosely based on a common construction for the classical makespan minimization setting on multiple machines, where a large number of small jobs is followed by a single larger job.

The rest of the paper is structured in the following way: We start by giving some general definitions needed for later sections. In Sect. 2 we then first prove the competitive ratio of Greedy and the lower bound, before describing the main algorithm for the general case. At the end of the section, we then build a special version of the algorithm for the uniform case. In Sect. 3 we consider the preemptive setting and give an algorithm as well as a tight lower bound. We conclude the paper by describing some open problems.

1.3 Preliminary Definitions

We use the following notations throughout the document: For a job $j \in [n]$, the *optimal offline running time* of j, i.e. the time needed by the optimum to schedule j on a machine, is denoted as $\rho_j := \min(t_j + p_j, u_j)$, while the *algorithmic running time* of j, i.e. the time needed for an algorithm to run j on a machine, is given by

$$p_j^A := \begin{cases} t_j + p_j & \text{if } j \text{ is tested,} \\ u_j & \text{if } j \text{ is not tested.} \end{cases} \tag{1}$$

It is clear that $\rho_j \leq p_j^A$ for any job j. Additionally, it holds that $p_j \leq \rho_j$, since the processing times p_j are upper bounded by u_j.

At times, we may use the definition of the *minimal running time* of job j, which is given by $\tau_j := \min(t_j, u_j)$.

It is clear that any job must fulfill $\tau_j \leq \rho_j$. In total, we get the following estimation for the different running times:

$$\tau_j \leq \rho_j \leq p_j^A, \qquad \forall j \in [n] \tag{2}$$

Since an algorithm does not know the values p_j, the testing decisions for the jobs are non-trivial. A partial goal for any competitive algorithm is to define a testing scheme such that the algorithmic running times are not too large compared to the optimal offline running times. We provide the following result which was used previously in [3] and is based on Theorem 14 of [13]. The given testing scheme based on the ratio $r_j = u_j/t_j$ between upper bound and testing time is used multiple times within this paper.

Proposition 1. *Let job j be tested iff $r_j \geq \alpha$ for some $\alpha \geq 1$. Then:*

(a) $\forall j \in [n]$ *tested:* $p_j^A \leq \left(1 + \frac{1}{\alpha}\right) \rho_j$
(b) $\forall j \in [n]$ *not tested:* $p_j^A \leq \alpha \rho_j$

As a direct consequence of Proposition 1, an optimal testing scheme for a single job is given by setting the threshold α to the golden ratio $\varphi \approx 1.6180$ [13].

2 Non-preemptive Setting

In this section we assume that preemption is not allowed. Any job has to be assigned to one of m available machines. Since we only consider makespan minimization, we may assume that there is no idle time on the machines and the actual ordering of the executions on a machine does not influence the outcome of the objective. It is therefore sufficient to only consider the assignment of the jobs to the machines.

2.1 Lower Bound and Greedy Algorithm

We first prove a straightforward lower bound and extend the simple List Scheduling algorithm from the classical setting to our problem.

For the lower bound we choose negligibly small testing times coupled with very large upper bounds. This forces the algorithm to test all jobs and thus having to decide on a machine for a given job while having no information about its real execution time.

Theorem 1. *No online algorithm is better than* $(2 - \frac{1}{m})$*-competitive for the problem of makespan minimization on* m *identical machines with testing, even if all testing times are equal to* 1.

We note that $\varphi \approx 1.6180$ is always a lower bound for our problem (see [13]), which is relevant only for small values of $m \leq 2$. The proof of Theorem 1 is provided in the full version of this paper [4].

To prove a simple upper bound, we can generalize the List Scheduling algorithm to our problem variant as follows:

Consider the given jobs in any order. For a job j to be scheduled next, test j if and only if $u_j/t_j \geq \varphi$ and then execute it completely on the current least-loaded machine.

Theorem 2. *The extension of List Scheduling described above is* $\varphi(2 - \frac{1}{m})$*-competitive for minimizing the makespan on* m *identical machines with non-uniform testing, where* $\varphi \approx 1.6180$ *is the golden ratio. This analysis is tight.*

The proof structure is similar to the proof of List Scheduling and uses common lower bounds for makespan minimization. We again refer to the full version for all details.

2.2 SBS Algorithm

In this section we provide a 3.1016-competitive algorithm for the non-preemptive setting. It assigns jobs into three classes S_1, B, and S_2 based on their ratios between upper bounds and testing times.

Let $[n]$ be the set of all jobs. We define a threshold function $T(m)$ for all m and divide the jobs into disjoint sets $[n] = B \dot{\cup} S$, where S will be further subdivided into S_1 and S_2. The set B corresponds to jobs where the ratio $r_j = u_j/t_j$ between upper bound and testing time is large, while jobs in S have a small ratio. We define

$$B := \{j \in [n] : r_j \geq T(m)\},$$
$$S := [n] \setminus B.$$

For the set S, we would like the algorithm to be able to distinguish jobs based on their optimal offline running time ρ_j. Of course, without testing the algorithm does not know these values, so we instead use the minimal running time τ_j, which can be computed directly using offline input only, to divide the set S further.

We define $S_1 \subset S$, such that $|S_1| = \min(m, |S|)$ and $\forall j_1 \in S_1, j_2 \in S \setminus S_1$: $\tau_{j_1} \geq \tau_{j_2}$. In other words, S_1 is the set of at most m jobs in S with the largest minimal running times. If this definition of S_1 is not unique, we may choose any such set. We set $S_2 := S \setminus S_1$. It follows that if $|S| \leq m$, then $S_2 = \emptyset$.

The idea behind dividing S into two sets is to identify the m largest jobs according to minimal running time and schedule them first, each on a separate machine. This allows us to lower bound the runtime of the remaining jobs later in the schedule.

In Algorithm 1 we describe the SBS algorithm which solves the non-uniform case and works in three phases corresponding to the sets S_1, B and S_2:

Algorithm 1: SBS algorithm

```
1  B ← {j ∈ [n] : r_j ≥ T(m)};
2  S ← [n] \ B;
3  S_1 ← S' ⊂ S s.t. |S'| = min(m, |S|), τ_j1 ≥ τ_j2 ∀j1∈S', j2∈S\S';
4  S_2 ← S \ S_1;
5  foreach j ∈ S_1 do
6  |  if r_j ≥ φ then
7  |  |  test and run j on an empty machine;
8  |  else
9  |  |  run j untested on an empty machine;
10 |  end
11 end
12 foreach j ∈ B do
13 |  test and run j on the current least-loaded machine;
14 end
15 foreach j ∈ S_2 do
16 |  run j untested on the current least-loaded machine;
17 end
```

In order to have a non-trivial testing decision for jobs in S_1, it makes sense to require that $T(m) \geq \varphi$ for all m. More specifically, we will define the threshold function $T(m)$ in the non-uniform setting as follows:

$$T(m) = \frac{(3+\sqrt{5})m - 2 + \sqrt{(38+6\sqrt{5})m^2 - 4(11+\sqrt{5})m + 12}}{6m-2}$$

Theorem 3. *Let $T(m)$ be a parameter function of m defined as above. The SBS algorithm is $T(m)\left(\frac{3}{2}-\frac{1}{2m}\right)$-competitive for minimizing the makespan on m identical machines with non-uniform testing.*

The function $T(m)$ is increasing for all $m \geq 1$ and fulfills $T(1) = \varphi$ as well as approximately $T(m) \to 2.0678$ for $m \to \infty$. The competitive ratio of the algorithm is explicitly given by

$$c(m) = \frac{(3+\sqrt{5})m - 2 + \sqrt{(38+6\sqrt{5})m^2 - 4(11+\sqrt{5})m + 12}}{4m}.$$

For this function we have $c(1) = \varphi$ as well as approximately $c(m) \to 3.1016$ if m approaches infinity. Additionally, it holds that $c(m) < \varphi\left(2 - \frac{1}{m}\right)$ for all $m > 1$.

Proof. We assume w.l.o.g. that the job indices are sorted by non-increasing optimal offline running times $\rho_1 \geq \cdots \geq \rho_n$. We denote the last job to finish in the schedule of the algorithm as l and the minimum machine load before job l as t. It follows that the value of the algorithm is $t + p_l^A$.

The value of the optimum is at least as large as the average sum of the optimal offline running times, or

$$L := \frac{1}{m} \sum_{j \in [n]} \rho_j \leq \text{OPT}, \tag{3}$$

since in any schedule at least one machine must have a load of at least this average. At the same time, we know that the optimum has to schedule every job on some machine:

$$\rho_j \leq \text{OPT} \quad \forall j \in [n] \tag{4}$$

We also utilize another common lower bound in makespan minimization, which is the sum of the processing times of the m-th and $(m+1)$-th largest job. If there are at least $m+1$ jobs, then some machine has to schedule at least 2 of these jobs:

$$\rho_m + \rho_{m+1} \leq \text{OPT}. \tag{5}$$

Here, ρ_j is defined as 0 if the instance has less than j jobs.

We differentiate between jobs handled by the algorithm in different phases and bound the algorithmic running times against the optimal offline running times. We write $p_j^A \leq \alpha_j \rho_j$ and define different values for α_j depending on the set j belongs to. It holds that

$$\alpha_j = \begin{cases} \varphi & \text{if } j \in S_1, \\ 1 + \frac{1}{T(m)} & \text{if } j \in B, \\ T(m) & \text{if } j \in S_2, \end{cases} \tag{6}$$

by Proposition 1 and the testing strategy of the algorithm.

The objective value of the algorithm depends on the set job l belongs to, so we differentiate between three cases. The following proposition upper bounds the algorithmic value $\text{ALG} = t + p_l^A$ for each of these cases:

Proposition 2. *The value of the algorithm can be estimated as follows:*

$$ALG \leq \begin{cases} \varphi OPT & \text{if } l \in S_1, \\ \left(\varphi + \left(1 + \frac{1}{T(m)}\right)\left(1 - \frac{1}{m}\right)\right) OPT & \text{if } l \in B, \\ T(m)\left(\frac{3}{2} - \frac{1}{2m}\right) OPT & \text{if } l \in S_2. \end{cases}$$

To prove this proposition, we utilize the lower bounds (3)–(5) and the estimates (6) for the value of α_j. A critical step lies in the estimation of p_l^A for $l \in S_2$, where we are able to lower bound τ_l using the size of the m-th and $(m+1)$-th largest job because the algorithm already ran m jobs from S_1 in the beginning of the schedule. We refer to the full version [4] for a detailed proof.

It remains to take the maximum over all three cases and minimize the value in dependence of $T(m)$. The value in the case $l \in S_1$ is always less than the values given by the other cases, therefore we only want to minimize

$$\max\left(\varphi + \left(1 + \frac{1}{T(m)}\right)\left(1 - \frac{1}{m}\right), T(m)\left(\frac{3}{2} - \frac{1}{2m}\right)\right).$$

The left side of the maximum is decreasing in $T(m)$, while the right side is increasing. The minimal maximum is therefore attained when both sides are equal. It can be easily verified that for the given definition of the threshold function $T(m)$ both sides of the maximum are equal for all values of $m \geq 1$.

It follows that the final ratio can be estimated by $\frac{\text{ALG}}{\text{OPT}} \leq T(m)\left(\frac{3}{2} - \frac{1}{2m}\right)$. □

2.3 An Improved Algorithm for the Uniform Case

The previous section established an algorithm with a competitive ratio of approximately 3.1016. We now present an algorithm with a better ratio in the case when $t_j = 1$ for all jobs. We define the threshold function $T_1(m)$ as follows:

$$T_1(m) = \frac{2m - 1 + \sqrt{16m^2 - 14m + 3}}{3m - 1}$$

The *Uniform-SBS* algorithm works as follows: Sort the jobs by non-increasing u_j. Go through the sorted list of jobs and put the next job on the machine with the lowest current load. A job j is tested if $u_j \geq T_1(m)$, otherwise it is run untested.

Theorem 4. *Uniform-SBS is a $T_1(m)(\frac{3}{2} - \frac{1}{2m})$-competitive algorithm for uniform instances.*

For uniform jobs with $t_j = 1$, sorting by non-increasing upper bound u_j is consistent with sorting by non-increasing ratio r_j. Hence, Uniform-SBS is similar to the SBS algorithm reduced to the phases corresponding to the sets B and S, where S contains *all* small jobs. The reason behind running the m largest jobs of S first in the SBS algorithm was to upper bound the remaining jobs in S. For uniform testing times, this bound can be achieved *without* this special structure.

The function $T_1(m)$ is increasing for all $m \geq 1$ and fulfills $T_1(1) = \varphi$ as well as $T_1(m) \to 2$ for $m \to \infty$. Computing the competitive ratio explicitly yields

$$c_1(m) = \frac{2m - 1 + \sqrt{16m^2 - 14m + 3}}{2m}.$$

These values start from $c_1(1) = \varphi$ and approach $c_1(m) \to 3$ if $m \to \infty$. Additionally, it holds that $c_1(m) < c(m)$ for all $m > 1$. In other words, this special version of the algorithm is strictly better than the general SBS algorithm described in Sect. 2.2. We defer the proof of Theorem 4 to the full version of the paper [4].

3 Results with Preemption

In this section we assume that jobs can be preempted at any time during their execution. An interrupted job may be continued on a possibly different machine, but no two machines may work on the same job at the same time. Testing a job must be completely finished before any part of its execution can take place.

It makes sense to additionally consider the following stricter definition of preemption within scheduling with testing: Untested jobs must be run without interruption on a single machine. If a job is tested, its test must also be run without interruption on one machine. The execution after the test may then be run without interruption on a possibly different machine. We call this setting *test-preemptive*, referring to the fact that the only place where we might preempt a job is exactly when its test is completed. From an application point of view, the test-preemptive setting is a natural extension of the non-preemptive setting, allowing the scheduler to reconsider the assignment of a job after receiving more information through the test.

Clearly, the difficulty of settings within scheduling with testing increases in the following order: preemptive, test-preemptive and non-preemptive. We now present the 2-competitive *Two Phases* algorithm for the test-preemptive setting, which can be applied directly to the ordinary preemptive case. Additionally, we construct a lower bound of $2 - 2/m + 1/m^2$ for the ordinary preemptive case. This lower bound then also holds for test-preemption, and is therefore tight for both settings when the number of machines m approaches infinity.

The Two Phases algorithm for the test-preemptive setting works as follows: Let OFF denote an optimal offline algorithm for makespan minimization on m machines. In the first phase, the algorithm schedules all jobs for their minimal running time τ_j using the algorithm OFF. Herein, the algorithm tests all jobs except trivial jobs with $t_j > u_j$, where running the upper bound is optimal. In the second phase, all remaining jobs are already tested, hence the algorithm now knows all remaining processing times p_j. We then use the offline algorithm OFF again to schedule the remaining jobs optimally. Finally, the algorithm obliviously puts the second schedule on top of the first.

Theorem 5. *The Two Phases algorithm is 2-competitive for minimizing the makespan on m machines with testing in the test-preemptive setting.*

The proof makes use of the assumption that the algorithm has access to unlimited computational power, which is a common assumption in online optimization. If we do not give the online algorithm this power, the result is slightly worse, since offline makespan minimization is strongly NP-hard. We may then make use of the PTAS for offline makespan minimization by Hochbaum and Shmoys [28] to achieve a ratio of $2 + \varepsilon$ for any $\varepsilon > 0$, where the runtime of the algorithm increases exponentially with $1/\varepsilon$. All details of the proof can be found in the full version [4].

For the lower bound result we now consider the standard preemptive setting where a job can be interrupted at any time.

Theorem 6. *In the preemptive setting, no online algorithm for makespan minimization on m identical machines with testing can have a better competitive ratio than $2 - 2/m + 1/m^2$, even if all testing times are equal to* 1.

We note that $\varphi \approx 1.6180$ also remains a lower bound even for the preemptive case, since two machines cannot run the same job concurrently. It holds $2 - 2/m + 1/m^2 < \varphi$ only for values of $m \leq 4$.

Proof. Let us consider the following example: Let M be a sufficiently large number and let $m(m-1)$ small jobs be given with $t_j = 1, p_j = 0, u_j = M$ as well as one large job f with $t_f = 1, p_f = m-1, u_f = M$. As argued in the proof of Theorem 1, OPT has a value of m and we may assume that the algorithm tests every job.

In the preemptive setting we required that any execution of the actual processing time of a job can only happen after its test is completed, therefore any job j that finished testing at some time t is completed not earlier than $t+p_j$. The adversary decides the processing time of j by the following rule: If $t \geq m-1+1/m$ and job f has not yet been assigned, set $p_j = m-1$ (i.e. set $j = f$). Else, set $p_j = 0$.

If the adversary assigns job f at any point, then job f finished testing at time $t \geq m - 1 + 1/m$. It follows that

$$\mathrm{ALG} \geq t + p_f \geq m - 1 + \frac{1}{m} + m - 1 = 2m - 2 + \frac{1}{m}.$$

Hence the competitive ratio is at least $\frac{\mathrm{ALG}}{\mathrm{OPT}} \geq 2 - \frac{2}{m} + \frac{1}{m^2}$.

All that remains is to show that this assignment of f happens at some point during the runtime of the algorithm. Assume that this is not the case, i.e. all jobs finish testing earlier than $m-1+1/m$. The adversary sets all $p_j = 0$, hence it follows directly that all jobs are completely finished before $m - 1 + 1/m$. But this means that the algorithmic solution has a value of $\mathrm{ALG} < m - 1 + 1/m$.

Since $t_j = 1$ for all jobs, we know that the average load L fulfills

$$L \geq \frac{1}{m}(m(m-1)+1) = m - 1 + 1/m.$$

But L is a lower bound on the optimal value of the instance, even in the preemptive setting, contradicting $\mathrm{ALG} < m - 1 + 1/m$. □

4 Conclusion

We presented algorithms and lower bounds for the problem of scheduling with testing on multiple identical parallel machines with the objective of minimizing the makespan. Such settings arise whenever a preliminary action influences cost, duration or difficulty of a task. Our main results were a 3.1016-competitive algorithm for the non-preemptive case and a tight 2-competitive algorithm for the preemptive case if the number of machines becomes large.

Apart from closing the gaps between our ratios and the lower bounds, we propose the following consideration for future work: A natural generalization of our setting is to consider *fully-online* arrivals, where jobs arrive one by one and have to be scheduled immediately. It is clear that this setting is at least as hard as the problem considered in this paper. In the full version [4], we provide a simple lower bound with value 2 for this generalization that holds for all values of $m \geq 2$. An upper bound is clearly given by the Greedy algorithm we provided in Sect. 2. Finding further algorithms or lower bounds for this new setting is a compelling direction for future research.

References

1. Albers, S.: Better bounds for online scheduling. SIAM J. Comput. **29**(2), 459–473 (1999). https://doi.org/10.1137/S0097539797324874
2. Albers, S.: On randomized online scheduling. In: Proceedings of the Thirty-Fourth Annual ACM Symposium on Theory of Computing, pp. 134–143. STOC 2002, Association for Computing Machinery, New York, NY, USA (2002). https://doi.org/10.1145/509907.509930
3. Albers, S., Eckl, A.: Explorable uncertainty in scheduling with non-uniform testing times. In: Kaklamanis, C., Levin, A. (eds.) WAOA 2020. LNCS, vol. 12806, pp. 127–142. Springer, Cham (2021). https://doi.org/10.1007/978-3-030-80879-2_9
4. Albers, S., Eckl, A.: Scheduling with testing on multiple identical parallel machines (2021). arXiv: https://arxiv.org/abs/2105.02052
5. Albers, S., Hellwig, M.: Semi-online scheduling revisited. Theoret. Comput. Sci. **443**, 1–9 (2012). https://doi.org/10.1016/j.tcs.2012.03.031
6. Albers, S., Hellwig, M.: Online makespan minimization with parallel schedules. Algorithmica **78**(2), 492–520 (2017). https://doi.org/10.1007/s00453-016-0172-5
7. Azar, Y., Regev, O.: On-line bin-stretching. Theoret. Comput. Sci. **268**(1), 17–41 (2001). https://doi.org/10.1016/S0304-3975(00)00258-9
8. Bartal, Y., Fiat, A., Karloff, H., Vohra, R.: New algorithms for an ancient scheduling problem. In: Proceedings of the Twenty-Fourth Annual ACM Symposium on Theory of Computing, pp. 51–58. STOC 1992, Association for Computing Machinery, New York, NY, USA (1992). https://doi.org/10.1145/129712.129718
9. Bartal, Y., Karloff, H., Rabani, Y.: A better lower bound for on-line scheduling. Inf. Process. Lett. **50**(3), 113–116 (1994). https://doi.org/10.1016/0020-0190(94)00026-3
10. Bruce, R., Hoffmann, M., Krizanc, D., Raman, R.: Efficient update strategies for geometric computing with uncertainty. Theoret. Comput. Sci. **38**(4), 411–423 (2005). https://doi.org/10.1007/s00224-004-1180-4

11. Chen, B., van Vliet, A., Woeginger, G.J.: A lower bound for randomized on-line scheduling algorithms. Inf. Process. Lett. **51**(5), 219–222 (1994). https://doi.org/10.1016/0020-0190(94)00110-3
12. Chen, B., van Vliet, A., Woeginger, G.J.: An optimal algorithm for preemptive on line scheduling. In: van Leeuwen, J. (ed.) Algorithms ESA 1994. LNCS, vol. 855, pp. 300–306. Springer, Heidelberg (1994). https://doi.org/10.1007/BFb0049417
13. Durr, C., Erlebach, T., Megow, N., Meißner, J.: Scheduling with explorable uncertainty. In: Karlin, A.R. (ed.) 9th Innovations in Theoretical Computer Science Conference (ITCS 2018). Leibniz International Proceedings in Informatics (LIPIcs), vol. 94, pp. 30:1–30:14. Schloss Dagstuhl-Leibniz-Zentrum fuer Informatik, Dagstuhl, Germany (2018). https://doi.org/10.4230/LIPIcs.ITCS.2018.30
14. Dürr, C., Erlebach, T., Megow, N., Meißner, J.: An adversarial model for scheduling with testing. Algorithmica **82**(12), 3630–3675 (2020). https://doi.org/10.1007/s00453-020-00742-2
15. Englert, M., Özmen, D., Westermann, M.: The power of reordering for online minimum makespan scheduling. In: 2008 49th Annual IEEE Symposium on Foundations of Computer Science, pp. 603–612 (2008). https://doi.org/10.1109/FOCS.2008.46
16. Epstein, L.: A survey on makespan minimization in semi-online environments. J. Sched. **21**(3), 269–284 (2018). https://doi.org/10.1007/s10951-018-0567-z
17. Erlebach, T., Hoffmann, M.: Query-competitive algorithms for computing with uncertainty. Bull. EATCS **2**(116), 1–19 (2015)
18. Erlebach, T., Hoffmann, M., Krizanc, D., Mihal'ák, M., Raman, R.: Computing minimum spanning trees with uncertainty. In: Albers, S., Weil, P. (eds.) 25th International Symposium on Theoretical Aspects of Computer Science. LIPIcs, vol. 1, pp. 277–288. Schloss Dagstuhl-Leibniz-Zentrum fuer Informatik, Dagstuhl, Germany (2008). https://doi.org/10.4230/LIPIcs.STACS.2008.1358
19. Faigle, U., Kern, W., Turan, G.: On the performance of on-line algorithms for partition problems. Acta Cybern. **9**(2), 107–119 (1989)
20. Feder, T., Motwani, R., O'Callaghan, L., Olston, C., Panigrahy, R.: Computing shortest paths with uncertainty. J. Algorithms **62**(1), 1–18 (2007). https://doi.org/10.1016/j.jalgor.2004.07.005
21. Feder, T., Motwani, R., Panigrahy, R., Olston, C., Widom, J.: Computing the median with uncertainty. SIAM J. Comput. **32**(2), 538–547 (2003). https://doi.org/10.1137/S0097539701395668
22. Fleischer, R., Wahl, M.: On-line scheduling revisited. J. Sched. **3**(6), 343–353 (2000)
23. Galambos, G., Woeginger, G.J.: An on-line scheduling heuristic with better worst-case ratio than Graham's list scheduling. SIAM J. Comput. **22**(2), 349–355 (1993). https://doi.org/10.1137/0222026
24. Goerigk, M., Gupta, M., Ide, J., Schöbel, A., Sen, S.: The robust knapsack problem with queries. Comput. Oper. Res. **55**, 12–22 (2015). https://doi.org/10.1016/j.cor.2014.09.010
25. Graham, R.L.: Bounds for certain multiprocessing anomalies. Bell Syst. Techn. J. **45**(9), 1563–1581 (1966). https://doi.org/10.1002/j.1538-7305.1966.tb01709.x
26. Graham, R.L.: Bounds on multiprocessing timing anomalies. SIAM J. Appl. Math. **17**(2), 416–429 (1969). https://doi.org/10.1137/0117039
27. Gupta, M., Sabharwal, Y., Sen, S.: The update complexity of selection and related problems. In: Chakraborty, S., Kumar, A. (eds.) IARCS Annual Conference on Foundations of Software Technology and Theoretical Computer Science (FSTTCS 2011). Leibniz International Proceedings in Informatics (LIPIcs), vol. 13, pp. 325–338. Schloss Dagstuhl-Leibniz-Zentrum fuer Informatik, Dagstuhl, Germany (2011). https://doi.org/10.4230/LIPIcs.FSTTCS.2011.325

28. Hochbaum, D.S., Shmoys, D.B.: Using dual approximation algorithms for scheduling problems theoretical and practical results. J. ACM **34**(1), 144–162 (1987). https://doi.org/10.1145/7531.7535
29. Kahan, S.: A model for data in motion. In: Proceedings of the Twenty-third Annual ACM Symposium on Theory of Computing, pp. 265–277. STOC 1991, ACM, New York, NY, USA (1991). https://doi.org/10.1145/103418.103449
30. Kalyanasundaram, B., Pruhs, K.: Speed is as powerful as clairvoyance. J. ACM **47**(4), 617–643 (2000). https://doi.org/10.1145/347476.347479
31. Karger, D.R., Phillips, S.J., Torng, E.: A better algorithm for an ancient scheduling problem. J. Algorithms **20**(2), 400–430 (1996). https://doi.org/10.1006/jagm.1996.0019
32. Kellerer, H., Kotov, V., Gabay, M.: An efficient algorithm for semi-online multiprocessor scheduling with given total processing time. J. Sched. **18**(6), 623–630 (2015). https://doi.org/10.1007/s10951-015-0430-4
33. Kellerer, H., Kotov, V., Speranza, M.G., Tuza, Z.: Semi on-line algorithms for the partition problem. Oper. Res. Lett. **21**(5), 235–242 (1997). https://doi.org/10.1016/S0167-6377(98)00005-4
34. Khanna, S., Tan, W.C.: On computing functions with uncertainty. In: Proceedings of the Twentieth ACM SIGMOD-SIGACT-SIGART Symposium on Principles of Database Systems, pp. 171–182. PODS 2001, Association for Computing Machinery, New York, NY, USA (2001). https://doi.org/10.1145/375551.375577
35. Megow, N., Meißner, J., Skutella, M.: Randomization helps computing a minimum spanning tree under uncertainty. SIAM J. Comput. **46**(4), 1217–1240 (2017). https://doi.org/10.1137/16M1088375
36. Olston, C., Widom, J.: Offering a precision-performance tradeoff for aggregation queries over replicated data. In: 26th International Conference on Very Large Data Bases (VLDB 2000), pp. 144–155. VLDB 2000, Morgan Kaufmann Publishers Inc., San Francisco, CA, USA (2000)
37. Rudin III, J.F.: Improved bounds for the on-line scheduling problem. Ph.D. Thesis (2001)
38. Sgall, J.: A lower bound for randomized on-line multiprocessor scheduling. Inf. Process. Lett. **63**(1), 51–55 (1997). https://doi.org/10.1016/S0020-0190(97)00093-8
39. Singla, S.: The price of information in combinatorial optimization. In: Proceedings of the Twenty-Ninth Annual ACM-SIAM Symposium on Discrete Algorithms. p. 2523–2532. SODA '18, Society for Industrial and Applied Mathematics, USA (2018). https://doi.org/10.1137/1.9781611975031.161
40. Weitzman, M.L.: Optimal search for the best alternative. Econometrica **47**(3), 641–654 (1979). https://doi.org/10.2307/1910412

Online Makespan Minimization with Budgeted Uncertainty

Susanne Albers and Maximilian Janke(✉)

Department of Computer Science, Technical University of Munich, Munich, Germany
maximilian@janke.tech

Abstract. We study Online Makespan Minimization with uncertain job processing times. Jobs are assigned to m parallel and identical machines. Preemption is not allowed. Each job has a regular processing time while up to Γ jobs fail and require additional processing time. The goal is to minimize the makespan, the time it takes to process all jobs if these Γ failing jobs are chosen worst possible. This models real-world applications where acts of nature beyond control have to be accounted for.

So far Makespan Minimization With Budgeted Uncertainty has only been studied as an offline problem. We are first to provide a comprehensive analysis of the corresponding online problem.

We provide a lower bound of 2 for general deterministic algorithms showing that the problem is more difficult than its special case, classical Online Makespan Minimization. We further analyze Graham's Greedy strategy and show that it is precisely $\left(3 - \frac{2}{m}\right)$-competitive. This bound is tight. We finally provide a more sophisticated deterministic algorithm whose competitive ratio approaches 2.9052.

Keywords: Scheduling · Makespan Minimization · Online algorithm · Competitive analysis · Lower bound · Uncertainty · Budgeted Uncertainty

1 Introduction

Scheduling is universal in countless areas of computing, decision making and management: Machines simultaneously produce diverse specialized equipment; server hubs execute numerous types of programs at the same time; employees need to perform various tasks in parallel. Innumerous scheduling problems in the literature are derived from such applications and model fuzzy real-world problems using precise mathematical language and problem specifications. This unnatural precision leads to failure when classical approaches are applied to real-world environments. Real-world settings are less clear-cut with multiple sources of uncertainty that vastly affect results. Consider acts of nature beyond control and predictability occurring rarely but regularly: For example, machines malfunction and need to be repaired; programs exhibit bugs, which require restarts

Work supported by the European Research Council, Grant Agreement No. 691672, project APEG.

A. Lubiw et al. (Eds.): WADS 2021, LNCS 12808, pp. 43–56, 2021.
https://doi.org/10.1007/978-3-030-83508-8_4

and debugging; changes in working environment—such as a global pandemic—require new solutions such as working from home. Most tasks are eventually adapted to the situation and performed as efficiently as before. Still, a few remaining ones suddenly require a lot more time and effort in our daily schedules. Such errors in the expected difficulty of tasks easily render theoretical predictions void and motivated various models incorporating uncertainty in the literature.

A prolific line of research on online algorithms, [14,15,17,19,23,26,30] and references therein, considers explorable uncertainty. This type of uncertainty can be queried and explored by the online algorithm. Such approaches are sensible for many practical applications but fail if uncertainty is inherently unpredictable—or unexplorable—even to the offline algorithm.

Offline approaches, dating back to the1950 s [13], propose stochastic models; more recent approaches, particularly Budgeted Uncertainty, consider worst-case scenarios, which accustom risk-averse decision makers. To date, many offline problems, Scheduling [8,9,38], Bin Packing [34] and Linear Optimization [6,7] among them, have been analyzed under Budgeted Uncertainty assumptions.

Surprisingly, these studies never extended to online settings. Current online analyses measure the price of not having information regarding the future—an online algorithm is prepared for whatever may come—but are not the least skeptical about information once it is "obtained".

This paper studies the most basic scheduling problem of *Online Makespan Minimization* under *Budgeted Uncertainty* assumptions: Jobs have to be assigned to m identical and parallel machines. Preemption is not allowed. The goal is to minimize the time it takes to process them all, the *makespan*. Each job J_i is defined by two processing times. Its regular processing time $\tilde{p}_i$ is its time required under normal circumstances. Its additional time Δp_i has to be added in rare cases of failure for reparation, potential reduced performance or other application-specific slowdown. Since failures are the exception, it would be extremely pessimistic to assume that jobs in general take time $\tilde{p}_i + \Delta p_i$. Instead, one only accounts for at most Γ such failures. Given an assignment of jobs, we consider the maximum time required for processing these jobs in parallel if up to Γ failures occurred in total. This time is called the *uncertain makespan* or simply the *makespan*. The objective is to minimize this quantity.

For $\Gamma = 0$ the problem reduces to classical makespan minimization. Only regular processing times $\tilde{p}_i$ matter. Similarly, the "paranoid case" $\Gamma = \infty$, where every program has a bug and every machine constantly malfunctions, yields classical makespan minimization using *worst-case processing times* $p_i = \tilde{p}_i + \Delta p_i$.

To an *online algorithm* A jobs are revealed one by one and each has to be scheduled immediately and irrevocably before the next one is revealed. In particular, A never "learns" which jobs fail and cannot perform optimally on arbitrary input sequences. The (uncertain) makespan of A, denoted by $A(\mathcal{J})$, is then compared to the optimum makespan $\mathrm{OPT}(\mathcal{J})$. Online algorithm A is c-competitive, $c \geq 1$, if $A(\mathcal{J})$ exceeds $\mathrm{OPT}(\mathcal{J})$ by at most a factor c on all input sequences $\mathcal{J}$. The smallest such factor is the *competitive ratio* $c = \sup_{\mathcal{J}} \frac{A(\mathcal{J})}{\mathrm{OPT}(\mathcal{J})}$. The goal is to design online algorithms achieving small competitive ratios.

Related Work: Scheduling is a fundamental problem in theoretical computer science and has been studied extensively in both offline and online variants, see e.g. [4,16,20,22,24,31] and references therein. We thus only focus on results most relevant to our work beginning with robust scheduling. Early work studied arbitrary, mostly finite sets of scenarios, see e.g. [3,28,29,32]. More modern work [8,9,38] has adapted to the model of Budgeted Uncertainty from [6]. In particular, Bougeret et al. [9] habe provided a first 3-approximation for the offline version of the problem studied in this paper and a PTAS for constant Γ. This result has been recently improved by the same authors and Jansen [8]. They provide an EPTAS for general Γ and show that, assuming $P \neq NP$, the best possible approximation ratio for unrelated machines lies in the interval [2,3].

Online Makespan minimization is very thoroughly researched. We again only review results most relevant for our work. Already in the1960 s Graham [22] established that his famed Greedy strategy obtains a strong competitive ratio of precisely $2 - \frac{1}{m}$. It took nearly thirty years till a breakthrough of Galambos and Woeginger [21] sparked a fruitful line of research [1,4,27] leading to the currently best competitive ratio of 1.9201 from [20]. Chen et al. [10] have given an online algorithm whose competitiveness is at most $1 + \varepsilon$ times the best possible ratio but no explicit bounds on this ratio are obtained. For general m, lower bounds are given in [1,5,18,35]. The currently best lower bound is 1.885 due to [35].

More recent results on Online Makespan Minimization consider semi-online settings, which equip the online algorithm with additional capabilities or information ahead of time [11,16,36]; different approaches weaken the adversary's ability to determine the job-order [2,16]; yet other settings analyze more involved objective functions: particularly the model of vector scheduling [12,25] also considers jobs that have two or even more "processing times". Unlike in our model, these "times" represent multidimensional resource requirements.

A related line of research on online algorithms considers "explorable uncertainty". In 1991, Kahan [26] has investigated the number of queries necessary to determine median and maximum of a set of "uncertain" numbers. This sparked a long line of research covering many problems, such as finding the median or general rank-k-elements [19,23,26,30], caching [33] or most recently scheduling [14,15]. The latter work, in fact, adds the cost of querying to the general cost paid by the algorithm for performing its task at hand. We refer to the survey by Erlebach et al. [17] and references therein for more results on this topic.

Modern work of Singla [37] introduces the 'price of information' into classical stochastic uncertainty settings, which results in a model highly related to the Budgeted Uncertainty model.

Our Contribution: We study online makespan minimization with Budgeted Uncertainty in depth. First, we give tight bounds on the competitive ratio of the Greedy strategy under Budgeted Uncertainty, which shows surprising parallels to the traditional result. It is precisely $(3 - \frac{2}{m})$-competitive. This already beats the oldest published offline approximation ratio [9], while being a much simpler approach at the same time. The lower bound showing that Greedy is not better than $(3 - \frac{2}{m})$-competitive can be chosen such that $\tilde{p}_i = 0$ for all i raising the question whether this important special case is as hard the general problem.

Next, we provide a better deterministic algorithm particularly suited for large values of m and Γ. Our algorithm adapts the proven strategy from [20] and earlier work of prioritizing schedules exhibiting a steep load profile. These profiles are highly desirable and generally pose no problem to the online algorithm. It is then shown that difficult input sequences leading to less desirable profiles cannot be efficiently scheduled by any algorithm, including the optimum offline algorithm. This ameliorates the possibly higher makespan of the online algorithm on these difficult sequences.

Precise competitive ratios for small m and Γ, Fig. 2, attest a strong performance unless Γ is extremely small and m is big. The latter setting is rather unnatural since one expects the number of errors to scale with the number of machines (and jobs). For large m and Γ the competitive ratio rapidly approaches 2.9052. In general, the algorithm outperforms the Greedy strategy.

We end with a lower bound of 2 for the competitive ratio achievable by any deterministic algorithm. The general model of Budgeted Uncertainty is therefore strictly more difficult than the classical model without uncertainty where even Graham's Greedy strategy is $(2 - \frac{1}{m})$-competitive.

2 Problem Definition

Consider any input sequence $\mathcal{J} = J_1, \dots J_n$. Job J_i is defined by a pair $(\tilde{p}_i, \Delta p_i)$ of non-negative real numbers where $\tilde{p}_i$ is the *regular processing time* of J_i, while Δp_i is its *additional time*. The time job J_i takes to be processed in the worst case is $p_i = \tilde{p}_i + \Delta p_i$, its *robust time*.

A schedule is simply a function $\sigma\colon \mathcal{J} \to \mathcal{M}$ mapping job $J \in \mathcal{J}$ to the machine $\sigma(J) = M \in \mathcal{M}$ processing it. The *regular load* of a machine $M \in \mathcal{M}$ is $\tilde{l}_M = \sum_{\sigma(J_i)=M} \tilde{p}_i$, the time it takes said machine to process all jobs in the best-case. The additional times we have to account for in the worst-case is the *additional load* Δl_M, the sum of the Γ largest additional times of jobs (or all additional times, if less than Γ jobs are scheduled on M). Formally $\Delta l_M = \max\left(\sum_{J_i \in \mathcal{J}'} \Delta p_i \mid \mathcal{J}' \subseteq \sigma^{-1}(M), |\mathcal{J}'| \leq \Gamma\right)$. Let us fix any set $\mathcal{J}'(M)$ where the maximum in the previous term is obtained. We break ties by preferring jobs J_i that came later, i.e. with larger indices i, and by choosing $\mathcal{J}'(M)$ of minimal size. We say for each job $J_i \in \mathcal{J}'(M)$ that J_i *fails in* σ. This allows us to write $\Delta l_M = \sum_{J_i \text{ fails } M} \Delta p_i$. Finally, the *robust load* l_M *of a machine* M is the maximum time machine M may require if up to Γ jobs fail: Formally $l_M = \tilde{l}_M + \Delta l_M = \sum_{J_i \in \sigma^{-1}(M)} \tilde{p}_i + \sum_{J_i \text{ fails } M} \Delta p_i$.

Given any algorithm A, which outputs the schedule σ, its *(robust) makespan* is then $A(\mathcal{J}) = \max_M l_M$. The goal is to design algorithms exhibiting low makespans. Technically, in the classical problem only Γ jobs fail in total. For the analysis a more general, equivalent version leads to a better intuition: Since the definition of each Δl_M only makes use of the fact that at most Γ jobs fail on M, the problem stays equivalent if we allowed up to Γ jobs to fail per machine.

Let OPT be any (fixed) offline algorithm that on any input sequence $\mathcal{J}$ outputs an optimum schedule. By some abuse of notation we also denote the optimum makespan $\text{OPT}(\mathcal{J})$ by OPT, if the corresponding sequence $\mathcal{J}$ is clear.

An Algorithm A is called an *online algorithm*, if it assigns each job J_i in $\mathcal{J} = J_1, \ldots, J_n$ independent of future jobs, i.e. $J_{i+1}, \ldots, J_n$. It's *competitive ratio* is then $c = \sup_{\mathcal{J}} \frac{A(\mathcal{J})}{\text{OPT}(\mathcal{J})}$, the quantity we wish to minimize.

3 Graham's Greedy Strategy

In his seminal work [22] Graham analyzed the greedy strategy and showed that it is $\left(2 - \frac{1}{m}\right)$-competitive. In general, the scheduling literature differentiates between pre-greedy and post-greedy strategies. The former simply choose a least loaded machine; the latter choose a machine such that the resulting load is minimal. Ties can be broken arbitrarily. For classical Makespan Minimization these notions coincide and all greedy strategies are identical up to permutations of machines. For our problem, the pre-greedy strategies perform significantly worse, which is why we focus on post-greedy strategies.

We then are going to establish the following theorem.

Theorem 1. *The competitive ratio of the post-greedy strategy is precisely* $3 - \frac{2}{m}$.

In particular, we will also present a lower bound on which any neither the post-greedy nor the pre-greedy strategy does perform better. This lower bound, interestingly, can be chosen such that $\tilde{p} = 0$ for all jobs, which might be evidence that this case is not easier than the general case. A bound on which the pre-greedy strategy performs worse is omitted.

3.1 Upper Bound

Let us consider the case of classical makespan minimization, i.e. we assume that $\Delta p_i = 0$ for all jobs J_i. The core idea of Graham [22] was to consider the *average load* of any schedule, that is $\tilde{L} = \frac{1}{m} \sum_M \tilde{l}_M = \frac{1}{m} \sum_{J_i \in \mathcal{J}} \tilde{p}_i$. The second term shows that this average load is independent of the schedule considered. This has two important consequences. First, $\text{OPT} \geq \tilde{L}$ since even the optimal schedule cannot have all machine loads below average. On the other hand, no scheduler, not even the worst one, can bring all machine loads above the average load $\tilde{L}$. Graham thus argues that the least loaded machine considered by his greedy strategy has load at most $\tilde{L} \leq \text{OPT}$. Since the job placed on it cannot have processing time greater than OPT either, it thus cannot cause a load exceeding 2OPT. In other words, for classical makespan minimization the greedy strategy is 2-competitive.

In our setting, the core argument of Graham does not work anymore. For $\Delta p_i \neq 0$, the average robust load L is far from being independent of the schedule in question. In fact, it may differ by a factor of m. Before giving an example let us introduce the required notation. Consider the schedule computed by any algorithm A on input sequence $\mathcal{J}$ and let $L[A] = L[A, \mathcal{J}] = \frac{1}{m} \sum_M l_M$ denote its *average (robust) load*. Similarly, let $\Delta L[A] = \Delta L[A, \mathcal{J}] = \frac{1}{m} \sum_M \Delta l_M = L[A] - \tilde{L}$. Consider $m \cdot \Gamma$ (or more) jobs with processing vector $(\tilde{p}_i, \Delta p_i) = (0, 1/\Gamma)$. Let A be the strategy that always uses a machine of least load and let B be the algorithm which only uses one single machine. Then $L[A] = 1$ while $L[B] = \frac{1}{m}$. The average robust load thus highly depends upon the algorithm considered. Interestingly, we can bound said average load using the optimum makespan OPT.

Lemma 1. *The average (robust) load $L[A, \mathcal{J}]$ of any algorithm A on input $\mathcal{J} = J_1, \dots, J_n$ is at most $L[\mathrm{OPT}, \mathcal{J}] + (1 - \frac{1}{m}) \, \mathrm{OPT}(\mathcal{J})$.*

Proof. Let us fix the sequence $\mathcal{J}$ and omit it from the notation. Let T be the set of jobs that fail A but not OPT. If T is empty, all jobs that fail A also fail OPT and thus $L[A, \mathcal{J}] \leq L[\mathrm{OPT}, \mathcal{J}]$. The lemma follows. Else, consider $J_{\max} \in T$ of maximum additional processing time $\Delta p_{\max}$. Consider the machine M that contains $J_{\max}$ in the optimum schedule. Since $J_{\max}$ does not fail OPT there are Γ different jobs of additional processing time at least $\Delta p_{\max}$ assigned to M which fail OPT. Let G be the set of these jobs. We obtain

$$\begin{aligned}\Delta L[A, \mathcal{J}] - \Delta L[\mathrm{OPT}, \mathcal{J}] &\leq \frac{1}{m} \sum_{J_i \in T} \Delta p_i - \frac{1}{m} \sum_{J_i \in G} \Delta p_i \\ &\leq \frac{1}{m} \cdot (|T| - |G|) \Delta p_{\max}.\end{aligned}$$

At most Γ jobs can fail any machine, therefore $|T| \leq \Gamma m$. By definition $|G| = \Gamma$. Finally, $\Delta p_{\max} \leq \frac{\mathrm{OPT}}{\Gamma}$ since the Γ jobs in G all have additional processing time at least $\Delta p_{\max}$ while their total additional processing is at most OPT. Now the previous inequality yields

$$\Delta L[A, \mathcal{J}] - \Delta L[\mathrm{OPT}, \mathcal{J}] \leq \frac{\Gamma m - \Gamma}{m} \cdot \frac{\mathrm{OPT}}{\Gamma} = \left(1 - \frac{1}{m}\right) \mathrm{OPT}(\mathcal{J}).$$

Recall that the average regular load $\tilde{L} = \frac{1}{m} \sum_{J_i} \tilde{p}_i$ is the same for every algorithm. Therefore $L[A, \mathcal{J}] - L[\mathrm{OPT}, \mathcal{J}] = \Delta L[A, \mathcal{J}] - \Delta L[\mathrm{OPT}, \mathcal{J}]$. Together with the previous inequality this implies $L[A, \mathcal{J}] \leq L[\mathrm{OPT}, \mathcal{J}] + (1 - \frac{1}{m}) \, \mathrm{OPT}(\mathcal{J})$. □

Lemma 2. *The post-greedy strategy incurs makespan at most $(3 - \frac{2}{m}) \, \mathrm{OPT}(\mathcal{J})$.*

Proof. Let $\mathcal{J} = J_1, \dots, J_n$. Using induction over n we may assume that the statement of the lemma holds right before job J_n is scheduled. By definition of a post-greedy assignment it then suffices to see that there exists a machine M whose load will not exceed $(3 - 2/m) \, \mathrm{OPT}(\mathcal{J})$ if we assign J_n to it.

Let $(p_n, \Delta p_n)$ be the processing vector of J_n. Consider the greedy schedule of the first $n - 1$ jobs and preliminarily assign job J_n to any machine that causes it to fail, i.e. contains less than Γ jobs of processing time strictly exceeding Δp_n. If no such machine exists, schedule job J_n on an arbitrary machine. Let L be the average robust load of this schedule. We replace each job J_i by a job $\hat{J}_i$ whose regular processing time is $\hat{p}_i = \tilde{p}_i$ if the job does not fail this preliminary schedule and $\hat{p}_i = p_i = \tilde{p}_i + \Delta p_i$ else. Per definition that does not change the load of any machine. Also, $L = \frac{1}{m}\sum_i \hat{p}_i$. Now, we remove the last job $\hat{J}_n$ from the machine it was scheduled on and assign it to a least loaded machine M (with regards to the processing times $\hat{p}_i$). After removing the job $\hat{J}_n$ the average load of the schedule is $L - \frac{1}{m} \cdot \hat{p}_n$. Hence, after assigning $\hat{J}_n$ to the least loaded machine the makespan is at most $L + \left(1 - \frac{1}{m}\right)\hat{p}_n$. Now, replace the jobs $\hat{J}_i$ by their original variants J_i. This can only cause the load of M to decrease. Indeed, by our choice of preliminary assignment we had $\hat{p}_n = p_n$ unless there was no machine on which J_n could fail, so job J_n contributes at most $\hat{p}_n$ to the load of M. For other jobs that fail M it is clear that they continue to do so if we remove J_n. Thus, these jobs contribute processing time $p_i = \hat{p}_i$. Jobs that do not fail only contribute processing time $\tilde{p}_i \leq \hat{p}_i$.

We have shown that assigning J_n to machine M causes its load to be at most $L + \left(1 - \frac{1}{m}\right)\tilde{p}_n \leq \tilde{L}[\mathcal{J}, \mathrm{OPT}] + \left(1 - \frac{1}{m}\right)\mathrm{OPT} + \left(1 - \frac{1}{m}\right)\tilde{p}_n \leq \left(3 - \frac{2}{m}\right)\mathrm{OPT}$. The first inequality is due to Lemma 1 (there is some algorithm computing the preliminary schedule), the second due to the fact that the average load $\tilde{L}[\mathcal{J}, \mathrm{OPT}]$ of the optimum schedule as well as $\tilde{p}_n$, the robust processing time of any job, are both lower bounds for OPT. By definition a post-greedy strategy will not cause a makespan exceeding the one it could obtain by choosing M. □

3.2 Lower Bound

Lemma 3. *Neither pre-greedy nor post-greedy strategy can be better than* $\left(3 - \frac{2}{m}\right)$*-competitive, even when all jobs have regular processing time* $\tilde{p} = 0$.

Figure 1 illustrates the lower bound; the formal proof is left to the full version.

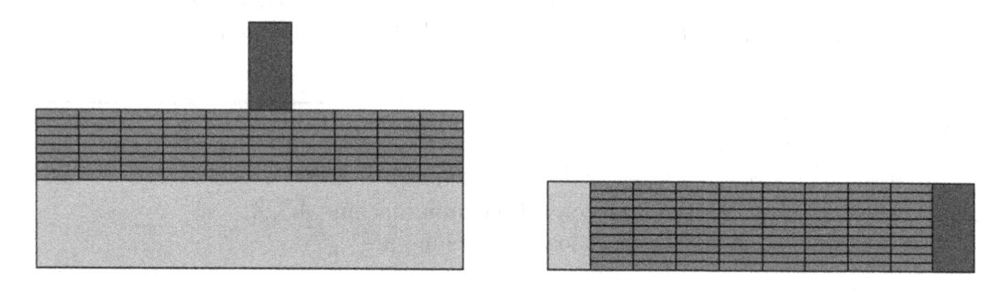

Fig. 1. The lower bound for greedy strategies, where all jobs have regular processing time $\tilde{p} = 0$. First, tiny sand-like jobs fill the greedy schedule (on the left) to a height of almost 1. Small jobs increase the height to $2 - 2/m$. Finally, a large job of size 1 causes a makespan of size $3 - 2/m$. The optimum schedule (on the right) places the sand-like jobs on a single machine. Since only Γ jobs fail the load is 1. The next $m - 1$ machines are filled with small jobs to a height of 1. The final machine captures the large job.

4 An Improved Deterministic Algorithm

The shortcoming of the Greedy strategy is that it creates 'critically flat' schedules. Jobs with high additional time tend to be spread thin onto separate machines when it would be better to cluster them. Moreover, a single large job J assigned to a flat schedule easily causes a high makespan. OPT commonly can 'sink J down' profiting a lot. Our algorithm thus tries to avoid flat schedules. When presented with one, it prefers to use a medium machine to make it steeper. Of course, recklessly using a medium machine on a dangerous schedule is folly. Said machine is only used if the algorithm can guarantee c-competitiveness. We specify the competitive ratio $c = c_{\Gamma,m}$, depending on both Γ and m, later.

For any input sequence $\mathcal{J} = J_1, \ldots, J_n$ and any time t consider the schedule of the algorithm right after after job J_t is scheduled. For $t = 0$ consider the empty schedule. We order machines by their robust loads, breaking ties arbitrarily but consistently. We call the $d = \left\lfloor \frac{c-2}{c} m \right\rfloor \approx 0.3116m$ least loaded machines *small*; the following d machines are *medium* and the remaining $m - 2d$ most-loaded ones are *large*. Let $\mathcal{M}^t_{\text{med}}$ be the set of medium machines and let $L^t_{\text{med}} = \frac{1}{|\mathcal{M}^t_{\text{med}}|} \sum_{M \in \mathcal{M}^t_{\text{med}}} l^t_M$ be their average robust load. Let M^t_{med} be the machine in $\mathcal{M}^t_{\text{med}}$ of least robust load $l^t_{\text{med}} = \min_{M \in \mathcal{M}^t_{\text{med}}} l^t_M$. Note that $l^t_{\text{med}} \leq L^t_{\text{med}}$. We use similar notation for the small and large machines: $\mathcal{M}^t_{\text{med}}$, L^t_{med}, l^t_{med}, $\mathcal{M}^t_{\text{large}}$, etc. with the index chosen accordingly. In particular, M^t_{small} denotes the least-loaded machine. Finally, let L^t denote the average (robust) load at time t.

We call the schedule at time $t \geq 0$ *steep* if $L^{t-1}_{\text{small}} \leq \left(1 - \frac{1}{2(c-1)}\right) L^{t-1}_{\text{large}}$ and *flat* else. Steep schedules are highly desirable. Our algorithm can and will always pick the least loaded machine. If the schedule is flat, our goal should be to make it steep again. Thus, first the least loaded medium machine M^{t-1}_{med} is sampled. If scheduling a job on machine M^{t-1}_{med} will not cause its load to exceed $\frac{c}{2} L^{t-1}$, i.e. if $l^{t-1}_{\text{med}} + p_t \leq \frac{c}{2} L^{t-1}$, we use M^{t-1}_{med}. This guarantees c-competitiveness, see Lemma 4. Else, the least loaded machine M_{small} has to be used. Seeing that this does not break c-competitiveness is the main challenge for the analysis.

Algorithm 1. *How to schedule job J_t with processing time p_t.*

1: **if** $L^{t-1}_{\text{small}} > \left(1 - \frac{1}{2(c-1)}\right) L^{t-1}_{\text{large}}$ and $l^{t-1}_{\text{med}} + p_t \leq \frac{c}{2} L^{t-1}$ **then**
2: Schedule job J_t on the least loaded medium machine M^{t-1}_{med};
3: **else** schedule job J_t on the least loaded machine M^{t-1}_{small}.

The values of c. Recall that $d = \left\lfloor \frac{c-2}{c} m \right\rfloor$. Let $\Gamma \geq 2$. The competitive ratio c is chosen minimally such that $c \geq \frac{7+\sqrt{17}}{4} \approx 2.7808$ and the following holds:

$$\left(1 - \frac{d}{2(c-1)m} - 2\frac{\Gamma+1}{c\Gamma}\right)\left(1 + \frac{c}{2m}\right)^d + 2\frac{\Gamma+1}{c\Gamma} \geq \frac{2}{c-1} \cdot \frac{m-1}{m}. \quad (1)$$

Unless m is chosen extremely small c is determined by Inequality (1) and fulfills it with equality. We show in the full version that c is below 2.9052 for $m, \Gamma \to \infty$.

The following is the main result of this paper.

Theorem 2. *The algorithm is c-competitive with $c < 2.9052$ for Γ large.*

Using a suitable data-structure that maintains the values L^{t-1}_{small}, L^{t-1}_{large} and l^{t-1}_{med} the algorithm can schedule each job efficiently in time $O(\log(m + \Gamma))$.

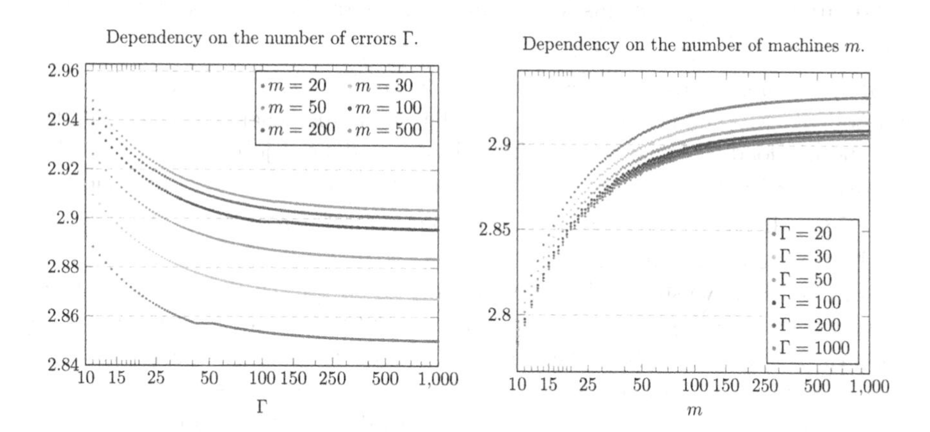

Fig. 2. The competitive ratios for different m and Γ. The ratio c is monotonously decreasing in Γ and tends to increase in m albeit not monotoneously due to the rounding involved in d. The x-axes are log-scaled. The graphs are colored.

Analysis of the Algorithm. Consider any input sequence $\mathcal{J} = J_1, \ldots, J_n$ and let $\text{OPT} = \text{OPT}(J_1, \ldots, J_n)$. By induction on n the makespan of the online algorithm did not exceed $c\text{OPT}$ before job J_n was scheduled. We need to show that J_n did not cause the makespan to exceed this value either. Let $L = L^n$ be the average robust load of the online schedule after all jobs in $\mathcal{J}$ have been scheduled.

Lemma 4. *We have $L \leq 2\text{OPT}$. In particular, if job J_n was scheduled on the least loaded medium machine M^{n-1}_{med} the makespan was at most $c\text{OPT}$.*

Proof. The first part follows from Lemma 1. Now, observe that if job J_n was assigned to machine M^{n-1}_{med} its load could not have exceeded $\frac{c}{2}L^{n-1} \leq c\text{OPT}$ afterwards per definition of the algorithm. □

We focus for simplicity on the case $m \to \infty$. The improvements required for small values of m are detailed at the end of the paragraph in Remark 1. Consider the case that J_n is scheduled on a least loaded machine. Let λ be its robust load. We need to bound $\lambda + p_n$, its load after receiving job J_n. Since $p_n \leq \text{OPT}$, we thus need to show that $\lambda \leq (c-1)\text{OPT}$. If $\lambda \leq \frac{c-1}{2}L$ this is a consequence of Lemma 4. Hence, we are left to consider the case where all machines have load at least $\lambda > \frac{c-1}{2}L$. We call such a schedule *critically flat.*[1] Our algorithm cannot

[1] The term 'critically flat' is not a misnomer, by Remark 2 such a schedule is, in particular, flat.

always prevent such schedules. The main part of our analysis shows that such schedules exhibit a highly specific structure.

Lemma 5. *If a schedule is critically flat, i.e. all machines have load* $\lambda > \frac{c-1}{2}L$, *then every machine contains a job of processing time at least* $\frac{\lambda}{2(c-1)}$.

We are going to prove this lemma in the next section. Let us first use it to conclude the analysis.

Proof of Theorem 2. By the previous analysis we only need to consider the case that job J_n is scheduled on a critically flat schedule where the least loaded machine has load λ. Since $\lambda \leq L \leq 2\mathrm{OPT}$ we are done if job J_n has processing time $p_n \leq \frac{\lambda}{2(c-1)} \leq \frac{\mathrm{OPT}}{c-1} \leq (c-2)\mathrm{OPT}$. The last inequality uses the condition $c \geq 2.7808$. If $p_n \leq \frac{\lambda}{2(c-1)}$ job J_n could only cause a makespan of $\lambda + p_n \leq 2\mathrm{OPT} + (c-2)\mathrm{OPT} = c\mathrm{OPT}$.

But else, we have shown that the sequence $J_1, \ldots, J_n$ contains $m+1$ jobs of processing time $\frac{\lambda}{2(c-1)}$. One of them is J_n while the remaining m exist due to Lemma 5. By the pigeonhole principle OPT needs to place two such jobs on a single machine, attaining a makespan of at least $2\frac{\lambda}{2(c-1)} = \frac{\lambda}{c-1}$. But this shows that $\lambda \leq (c-1)\mathrm{OPT}$ and thus $\lambda + p_n \leq (c-1)\mathrm{OPT} + \mathrm{OPT} = c\mathrm{OPT}$. □

Remark 1. Our previous definition of *critically flat* schedules given for $m \to \infty$ can be improved if m, the number of machines, is small. We then call the schedule *critically flat* if $\lambda > \frac{c-1}{2}\frac{m}{m-1}L^{n-1}$. Note that both definitions agree for $m \to \infty$. The previous arguments can be generalized to show that we are c-competitive if job J_n is not assigned to a schedule which is critically flat using this definition. The proof is left to the full version.

Understanding how Critically Flat Schedules are Formed. In our analysis we have to differentiate between *early* and *late* jobs. The latter will be scheduled on a full and flat schedule. For them to be assigned to a least loaded machine they will need to be fairly sizable. This requires the adversary to present quite large jobs constituting a lot of processing volume to achieve a critically flat schedule. Formally, we call a job J_t *late* if two conditions are met. We require job J_t to be scheduled on a flat schedule and we require machine M_{med}^{t-1} to have load at least λ before job J_t is scheduled. If a job J_t is not late or followed by a job which is not late, we call it *early*. In particular, a job can be both early and late. We next are going to consider late jobs.

Lemma 6. *If job* J_t *is late and caused a machine to first reach load* λ, *its (robust) processing time is at least* $p_t \geq \frac{\lambda}{2(c-1)}$.

Proof. Let $l = l_{\mathrm{med}}^{t-1}$ be the load of M_{med}^{t-1}. By assumption of J_t being late there holds $l \geq \lambda$. Since J_t caused a machine to first reach load λ it was not scheduled on M_{med}^{t-1} but on the least loaded machine M_{small}^{t-1} instead. Since the schedule was flat, job J_t must have had processing time exceeding $\frac{c}{2}L^{t-1} - l$, i.e. $p_t > \frac{c}{2}L^{t-1} - l$.

If the least loaded machine M_{small}^{t-1}, which J_t caused to reach load λ, had load less than $\lambda - \frac{\lambda}{2(c-1)}$, the statement of the lemma follows immediately. Else, all small machines had load at least $\lambda - \frac{\lambda}{2(c-1)}$ and all medium and large machines had load at least l. Thus $L^{t-1} \geq \frac{m-d}{m} l + \frac{d}{m}\left(\lambda - \frac{\lambda}{2(c-1)}\right)$. In particular

$$p_t > \frac{c}{2} L^{t-1} - l \geq \frac{c}{2}\left(\frac{m-d}{m} l + \frac{d}{m}\left(\lambda - \frac{\lambda}{2(c-1)}\right)\right) - l.$$

Note that d was chosen precisely maximal such that $d \leq \frac{c-2}{c} m$, or equivalently $\frac{c}{2}\frac{m-d}{m} - 1 \geq 0$. In fact, this inequality motivates the choice of d. The inequality implies that the previous term is non-decreasing in l and minimal for $l = \lambda$. Setting $l = \lambda$, we get that

$$p_t \geq \frac{c}{2}\left(1 - \frac{d}{2(c-1)m}\right)\lambda - \lambda > \left(\frac{c}{2} - \frac{c-2}{4(c-1)} - 1\right)\lambda \geq \frac{1}{2(c-1)}\lambda.$$

The first inequality uses that $d < \frac{c-2}{c} m$. The second inequality is simply an algebraic computation which uses that $c \geq \frac{7+\sqrt{17}}{4}$. □

The Critical Machines. We call a time t *early* or *late* if job J_t had this property. So far, we have treated late times. Now, we need to establish a corresponding result for early times. This requires us to understand a certain set of critical machines. Given any time t, let $\mathcal{M}_{\text{crit}}^{t-1}$ be the set of d most-loaded machines whose load has not yet reached λ before job J_t is scheduled. If less than d machines fulfill the latter condition, then all of them belong to $\mathcal{M}_{\text{crit}}^{t-1}$. Let $L_{\text{crit}}^{t-1} = \frac{1}{|\mathcal{M}_{\text{crit}}^{t-1}|} \sum_{M \in \mathcal{M}_{\text{crit}}^{t-1}} l_M^{t-1}$ be their average load. We use the convention $1/0 = \infty$, or in other words set $L_{\text{crit}}^{t-1} = \infty$ if $\mathcal{M}_{\text{crit}}^{t-1} = \emptyset$. The following lemma is fairly technical and will be proven in the full version.

Lemma 7. *Let s be the last early time, i.e. if $t > s$, then t is late. Then* $L_{\text{crit}}^{s} \leq \left(1 - \frac{1}{2(c-1)}\right)\lambda$.

Remark 2. The lemma implies that $\mathcal{M}_{\text{crit}}^{s} \neq \emptyset$. Using that $\mathcal{M}_{\text{crit}}^{n-1} = \emptyset$ we get that $s < n - 1$ or, equivalently, that critically flat schedules are always flat.

Recall that our algorithm considers certain machines to be small, large and medium. It turns out that if it was not for the online setting, i.e. if the algorithm had advance knowledge of the sequence, the critical machines are the true contestants for the label "medium". To be precise we will establish that the critical machines separate "large" machines of load at least λ from 'small' ones of load at most $\left(1 - \frac{1}{2(c-1)}\right)\lambda$. The following claim is the first step towards establishing this result in Lemma 8.

Claim. Assume that $L_{\text{crit}}^{t} \leq \left(1 - \frac{1}{2(c-1)}\right)\lambda$. If $M_{\text{med}}^{t-1} \in \mathcal{M}_{\text{crit}}^{t-1}$ then the schedule is steep. In particular, our algorithm never uses machine in $\mathcal{M}_{\text{crit}}^{t-1} \setminus \{M_{\text{small}}^{t-1}\}$.

Proof. Assume $M_{\text{med}}^{t-1} \in \mathcal{M}_{\text{crit}}^{t-1}$. Since $d-1$ machines lie strictly in between M_{med}^{t-1} and the machines in $\mathcal{M}_{\text{large}}^{t-1}$ the latter must be disjoint from $\mathcal{M}_{\text{crit}}^{t-1}$. Thus, they all had load λ. In particular, $L_{\text{large}}^{t-1} \geq \lambda$. Using this, we conclude that the schedule was steep and, consequently, that the algorithm used M_{small}^{t-1}:

$$L_{\text{small}}^{t-1} \leq L_{\text{small}}^{t} \leq L_{\text{crit}}^{t} \leq \left(1 - \frac{1}{2(c-1)}\right)\lambda \leq \left(1 - \frac{1}{2(c-1)}\right) L_{\text{large}}^{t-1}.$$

□

Lemma 8. *If t is early, $L_{\text{crit}}^{t-1} \leq \left(1 - \frac{1}{2(c-1)}\right)\lambda$. Moreover job J_t was either scheduled on a machine of load at least λ or on a machine of load at most $\left(1 - \frac{1}{2(c-1)}\right)\lambda$.*

Proof. Assume for contradiction sake that an early time t existed with $L_{\text{crit}}^{t-1} > \left(1 - \frac{1}{2(c-1)}\right)\lambda$. We may wlog. choose t maximal with that property. Then there holds $L_{\text{crit}}^{t} \leq \left(1 - \frac{1}{2(c-1)}\right)\lambda$ either by the maximality of t or by Lemma 7. Thus job J_t caused L_{crit} to decrease. This can only happen if job J_t was assigned to a machine in $\mathcal{M}_{\text{crit}}^{t-1}$ that was not M_{small}^{t-1}. But by the previous claim this does not happen, a contradiction.

For the second part observe that job J_t is either scheduled on a machine having load λ or on a machine of load at most $L_{\text{crit}}^{t-1} \leq \left(1 - \frac{1}{2(c-1)}\right)\lambda$, which is either the least loaded machine in $\mathcal{M}_{\text{crit}}$ or a machine of lesser load. □

We now conclude our analysis by proving the structural Lemma 5.

Proof of Lemma 5. Given any machine M, we show that job J_t that caused M to first reach (robust) load λ had processing time at least $\frac{\lambda}{2(c-1)}$. If job J_t was late, this already follows from Lemma 6. Else, by Lemma 8, job J_t was scheduled on a machine which had (robust) load $\left(1 - \frac{1}{2(c-1)}\right)\lambda$ before and λ after receiving job J_t. Thus job J_t had (robust) processing time at least $\frac{\lambda}{2(c-1)}$. □

4.1 Deterministic Lower Bounds

We can show that no general competitive ratio below 2 is possible unless very small numbers of machines are considered. We leave the proof to the full version.

Theorem 3. *No deterministic algorithm is better than 2-competitive for general m and $\Gamma = 2$.*

It may also be interesting to consider the debugging model, where all jobs have real processing time 0. In this case the classical lower bounds still apply if Γ is not chosen too small. For example, the lower bound of 1.852 for [1] holds for $\Gamma \geq 4$. On the other hand it is not clear that any algorithm performs better in tise case. Lemma 3 shows that Greedy does not.

References

1. Albers, S.: Better bounds for online scheduling. SIAM J. Comput. **29**(2), 459–473 (1999)
2. Albers, S., Janke, M.: Scheduling in the random-order model. In: 47th International Colloquium on Automata, Languages, and Programming (ICALP 2020). Schloss Dagstuhl-Leibniz-Zentrum für Informatik (2020)
3. Aloulou, M.A., Croce, F.D.: Complexity of single machine scheduling problems under scenario-based uncertainty. Oper. Res. Lett. **36**(3), 338–342 (2008)
4. Bartal, Y., Fiat, A., Karloff, H., Vohra, R.: New algorithms for an ancient scheduling problem. In: Proceedings of the Twenty-Fourth Annual ACM Symposium on Theory of Computing, pp. 51–58 (1992)
5. Bartal, Y., Karloff, H.J., Rabani, Y.: A better lower bound for on-line scheduling. Inf. Process. Lett. **50**(3), 113–116 (1994)
6. Bertsimas, D., Sim, M.: Robust discrete optimization and network flows. Math. Program. Ser. B **98**(1–3), 49–71 (2003). https://doi.org/10.1007/s10107-003-0396-4
7. Bertsimas, D., Sim, M.: The price of robustness. Oper. Res. **52**(1), 35–53 (2004)
8. Bougeret, M., Jansen, K., Poss, M., Rohwedder, L.: Approximation results for makespan minimization with budgeted uncertainty. In: Bampis, E., Megow, N. (eds.) WAOA 2019. LNCS, vol. 11926, pp. 60–71. Springer, Cham (2020). https://doi.org/10.1007/978-3-030-39479-0_5
9. Bougeret, M., Pessoa, A.A., Poss, M.: Robust scheduling with budgeted uncertainty. Discrete Appl. Math. **261**, 93–107 (2019)
10. Chen, L., Ye, D., Zhang, G.: Approximating the optimal algorithm for online scheduling problems via dynamic programming. Asia-Pac. J. Oper. Res. **32**(01), 1540011 (2015)
11. Cheng, T.E., Kellerer, H., Kotov, V.: Semi-on-line multiprocessor scheduling with given total processing time. Theor. Comput. Scie. **337**(1–3), 134–146 (2005)
12. Cohen, I., Im, S., Panigrahi, D.: Online two-dimensional load balancing. In: 47th International Colloquium on Automata, Languages, and Programming (ICALP 2020). Schloss Dagstuhl-Leibniz-Zentrum für Informatik (2020)
13. Dantzig, G.B.: Linear programming under uncertainty. Manage. Sci. **1**(3–4), 197–206 (1955)
14. Dürr, C., Erlebach, T., Megow, N., Meißner, J.: Scheduling with explorable uncertainty. In: 9th Innovations in Theoretical Computer Science Conference (ITCS 2018). Schloss Dagstuhl-Leibniz-Zentrum für Informatik (2018)
15. Dürr, C., Erlebach, T., Megow, N., Meißner, J.: An adversarial model for scheduling with testing. Algorithmica **82**(12), 3630–3675 (2020). https://doi.org/10.1007/s00453-020-00742-2
16. Englert, M., Özmen, D., Westermann, M.: The power of reordering for online minimum makespan scheduling. In: 2008 49th Annual IEEE Symposium on Foundations of Computer Science, pp. 603–612. IEEE (2008)
17. Erlebach, T., Hoffmann, M., Kammer, F.: Query-competitive algorithms for cheapest set problems under uncertainty. Theor. Comput. Sci. **613**, 51–64 (2016)
18. Faigle, U., Kern, W., Turán, G.: On the performance of on-line algorithms for partition problems. Acta Cybernetica **9**(2), 107–119 (1989)
19. Feder, T., Motwani, R., Panigrahy, R., Olston, C., Widom, J.: Computing the median with uncertainty. In: Proceedings of the Thirty-Second Annual ACM Symposium on Theory of Computing, pp. 602–607 (2000)
20. Fleischer, R., Wahl, M.: On-line scheduling revisited. J. Sched. **3**(6), 343–353 (2000)

21. Galambos, G., Woeginger, G.J.: An on-line scheduling heuristic with better worst-case ratio than Graham's list scheduling. SIAM J. Comput. **22**(2), 349–355 (1993)
22. Graham, R.L.: Bounds for certain multiprocessing anomalies. Bell Syst. Tech. J. **45**(9), 1563–1581 (1966)
23. Gupta, M., Sabharwal, Y., Sen, S.: The update complexity of selection and related problems. arXiv preprint arXiv:1108.5525 (2011)
24. Hochbaum, D.S., Shmoys, D.B.: Using dual approximation algorithms for scheduling problems theoretical and practical results. J. ACM (JACM) **34**(1), 144–162 (1987)
25. Im, S., Kell, N., Kulkarni, J., Panigrahi, D.: Tight bounds for online vector scheduling. In: 2015 IEEE 56th Annual Symposium on Foundations of Computer Science, pp. 525–544. IEEE (2015)
26. Kahan, S.: A model for data in motion. In: Proceedings of the Twenty-Third Annual ACM Symposium on Theory of Computing, pp. 265–277 (1991)
27. Karger, D.R., Phillips, S.J., Torng, E.: A better algorithm for an ancient scheduling problem. J. Algor. **20**(2), 400–430 (1996)
28. Kasperski, A., Kurpisz, A., Zieliński, P.: Approximating a two-machine flow shop scheduling under discrete scenario uncertainty. Eur. J. Oper. Res. **217**(1), 36–43 (2012)
29. Kasperski, A., Kurpisz, A., Zieliński, P.: Parallel machine scheduling under uncertainty. In: Greco, S., Bouchon-Meunier, B., Coletti, G., Fedrizzi, M., Matarazzo, B., Yager, R.R. (eds.) IPMU 2012. CCIS, vol. 300, pp. 74–83. Springer, Heidelberg (2012). https://doi.org/10.1007/978-3-642-31724-8_9
30. Khanna, S., Tan, W.C.: On computing functions with uncertainty. In: Proceedings of the Twentieth ACM SIGMOD-SIGACT-SIGART Symposium on Principles of Database Systems, pp. 171–182 (2001)
31. Lenstra, J.K., Shmoys, D.B., Tardos, E.: Approximation algorithms for scheduling unrelated parallel machines. Math. Program. **46**(1–3), 259–271 (1990). https://doi.org/10.1007/BF01585745
32. Mastrolilli, M., Mutsanas, N., Svensson, O.: Approximating single machine scheduling with scenarios. In: Goel, A., Jansen, K., Rolim, J.D.P., Rubinfeld, R. (eds.) APPROX/RANDOM-2008. LNCS, vol. 5171, pp. 153–164. Springer, Heidelberg (2008). https://doi.org/10.1007/978-3-540-85363-3_13
33. Olston, C., Widom, J.: Offering a Precision-Performance Tradeoff for Aggregation Queries Over Replicated Data. Tech. rep, Stanford (2000)
34. Basu Roy, A., Bougeret, M., Goldberg, N., Poss, M.: Approximating robust bin packing with budgeted uncertainty. In: Friggstad, Z., Sack, J.-R., Salavatipour, M.R. (eds.) WADS 2019. LNCS, vol. 11646, pp. 71–84. Springer, Cham (2019). https://doi.org/10.1007/978-3-030-24766-9_6
35. Rudin, J.F.: Improved bounds for the on-line scheduling problem (2001)
36. Sanders, P., Sivadasan, N., Skutella, M.: Online scheduling with bounded migration. Math. Oper. Res. **34**(2), 481–498 (2009)
37. Singla, S.: The price of information in combinatorial optimization. In: Proceedings of the Twenty-Ninth Annual ACM-SIAM Symposium on Discrete Algorithms, pp. 2523–2532. SIAM (2018)
38. Tadayon, B., Smith, J.C.: Algorithms and complexity analysis for robust single-machine scheduling problems. J. Sched. **18**(6), 575–592 (2015). https://doi.org/10.1007/s10951-015-0418-0

Pattern Matching in Doubling Spaces

Corentin Allair[1] and Antoine Vigneron[2(✉)]

[1] École Polytechnique, Paris, France
corentin.allair@polytechnique.edu
[2] School of Electrical and Computer Engineering, UNIST, Ulsan, Republic of Korea
antoine@unist.ac.kr

Abstract. We consider the problem of matching a metric space (X, d_X) of size k with a subspace of a metric space (Y, d_Y) of size $n \geqslant k$, assuming that these two spaces have constant doubling dimension δ. More precisely, given an input parameter $\rho \geqslant 1$, the ρ-distortion problem is to find a one-to-one mapping from X to Y that distorts distances by a factor at most ρ. We first show by a reduction from k-clique that, in doubling dimension $\log_2 3$, this problem is NP-hard and W[1]-hard. Then we provide a near-linear time approximation algorithm for fixed k: Given an approximation ratio $0 < \varepsilon \leqslant 1$, and a positive instance of the ρ-distortion problem, our algorithm returns a solution to the $(1+\varepsilon)\rho$-distortion problem in time $(\rho/\varepsilon)^{O(1)} n \log n$. We also show how to extend these results to the minimum distortion problem, which is an optimization version of the ρ-distortion problem where we allow scaling. For doubling spaces, we prove the same hardness results, and for fixed k, we give a $(1+\varepsilon)$-approximation algorithm running in time $(\operatorname{dist}(X, Y)/\varepsilon)^{O(1)} n^2 \log n$, where $\operatorname{dist}(X, Y)$ denotes the minimum distortion between X and Y.

Keywords: Pattern matching · Approximation algorithms · Doubling spaces

1 Introduction

A metric space has *doubling dimension* δ if any ball can be covered by at most 2^δ balls of half its radius. When $\delta = O(1)$, we say that this space is *doubling.* (See Sect. 2.) For instance, the Euclidean space $\mathbb{R}^d$ has doubling dimension $O(d)$, hence doubling spaces are generalizations of fixed-dimensional Euclidean spaces.

In this paper, we study pattern matching problems in doubling spaces. Given two doubling spaces (X, d_X) and (Y, d_Y) of doubling dimension δ, and sizes $|X| = k$ and $|Y| = n$, where $k \leqslant n$, our goal is to find a subspace of Y that

This work was supported by Basic Science Research Program through the National Research Foundation of Korea (NRF) funded by the Ministry of Education (2017R1D1A1B04036529). Work by C. Allair was conducted during an internship in UNIST.

A. Lubiw et al. (Eds.): WADS 2021, LNCS 12808, pp. 57–70, 2021.
https://doi.org/10.1007/978-3-030-83508-8_5

resembles the *pattern* X. More precisely, we consider the *ρ-distortion problem* and the *minimum distortion problem*, which we describe below.

Given $\rho \geqslant 1$, the ρ-distortion problem is to find, if it exists, a mapping $\sigma : X \to Y$ such that

$$(1/\rho)d_X(x,x') \leqslant d_Y(\sigma(x),\sigma(x')) \leqslant \rho d_X(x,x') \tag{1}$$

for all $x, x' \in X$. It follows from this definition that σ is injective.

The ρ-distortion problem is analogous to the problem of matching two point-sets in Euclidean space under rigid transformations, which are compositions of translations and rotations. If, in addition, we allow scaling, then an analogous problem in general metric spaces is the minimum distortion problem. The goal is to minimize the distortion $\text{dist}(\sigma) = \text{expansion}(\sigma) \times \text{expansion}(\sigma^{-1})$ over all injections $\sigma : X \to Y$, where $\text{expansion}(\sigma) = \max_{\substack{x,x' \in X \\ x \neq x'}} \frac{d_Y(\sigma(x),\sigma(x'))}{d_X(x,x')}$ and $\text{expansion}(\sigma^{-1}) = \max_{\substack{x,x' \in X \\ x \neq x'}} \frac{d_X(x,x')}{d_Y(\sigma(x),\sigma(x'))}$. The minimum of $\text{dist}(\sigma)$ over all injections $\sigma : X \to Y$ is denoted $\text{dist}(X,Y)$, and it is easy to see that $\text{dist}(X,Y) \geqslant 1$. The minimum distortion problem was introduced by Kenyon et al. [20] in the case where $k = n$, and thus σ is a bijection.

Motivated by applications to natural language processing, bioinformatics and computer vision, Ding and Ye [13] recently proposed a practical algorithm for a pattern matching problem in doubling spaces. However, this algorithm may only return an approximation of a local minimum, and its time bound is not given as a function of the input size. (The time bound depends on a factor γ that counts the number of rounds of the algorithm, which is not bounded in terms of the input parameters.) One of our goals is thus to provide a provably efficient algorithm for pattern matching in doubling spaces. Another motivation for our work is that, even though the complexity of the minimum distortion problem has been studied for several types of metrics, it appears that no result was previously known for two doubling metrics. (See the comparison with previous work below.)

Our Results. We first give a hardness result: We show that for any $\rho \geqslant 1$, the k-clique problem reduces to ρ-distortion in doubling dimension $\log_2 3$. It implies that the ρ-distortion problem is NP-hard, and is W[1]-hard when parameterized by k (Corollary 2). It also shows that this problem cannot be solved in time $f(k) \cdot n^{o(k)}$ for any computable function f, unless the exponential time hypothesis (ETH) is false (Corollary 3).

On the positive side, we present a near-linear time approximation algorithm for small values of k. More precisely, if $\rho \geqslant 1$ and $0 < \varepsilon \leqslant 1$, our algorithm returns in $2^{O(k^2 \log k)}(\rho^2/\varepsilon)^{2k\delta} n \log n$ time a solution to the $(1+\varepsilon)\rho$-distortion problem whenever a solution to the ρ-distortion problem exists. In this time bound, it is reasonable to assume that ρ is a small constant, say $\rho \leqslant 10$, as a larger value would mean that we allow a relative error of more than 900% in the quality of the matching, which is probably too much for most applications.

We also show how to extend these results to the minimum distortion problem. In particular, we show that the minimum distortion problem cannot be solved in time $f(k) \cdot n^{o(k)}$ for any computable function f, unless ETH is false, and we give a $2^{O(k^2 \log k)}(\mathrm{dist}(X,Y)^{2k\delta}/\varepsilon^{2k\delta+O(1)})n^2 \log n$ time, $(1+\varepsilon)$-approximation algorithm. Here again, it is reasonable to assume that $\mathrm{dist}(X,Y) = O(1)$, and then for any fixed k, this algorithm is an FPTAS with running time $(1/\varepsilon)^{O(1)} n^2 \log n$.

Comparison with Previous Work. One of the main differences between our results and previous work on point pattern matching under rigid transformations, or on the minimum distortion problem, is that we parameterize the problem by k, and hence k is regarded as a small number. The advantage is that the dependency of our time bounds in n are low (near-linear or near-quadratic). However, we obtain an exponential dependency in k.

Geometric point pattern matching problems have been studied extensively. (See for instance the survey by Alt and Guibas [2].) In the fixed-dimensional Euclidean space $\mathbb{R}^d$, these problems are usually tractable, as the space of transformations has a constant number of degrees of freedom. For instance, when $k = n$, we may want to decide whether X and Y are congruent, which means that there is a rigid transformation μ such that $\mu(X) = Y$. Alt et al. [3] showed how to find such a transformation in time $O(n^{d-2} \log n)$, when it exists. In practice, however, we cannot expect that point coordinates are known exactly, so it is unlikely that an exact match exists. We may thus want to find the smallest $\varepsilon > 0$ such that each point of X is brought to distance at most ε from a point in Y. (In other words, we allow an additive error ε.) Chew et al. gave an $O(k^3 n^2 \log^2(nk))$-time algorithm to solve this problem in the plane under rigid transformations [8].

As mentioned above, to the best of our knowledge, the only work published so far on pattern matching in doubling spaces presents a practical algorithm for matching two doubling spaces [13]. However it has not been proven to return a good approximation of the optimal solution in the worst case. Other problems studied in doubling spaces include approximate near-neighbor searching [4,9,19], spanners [5], routing [7,16], TSP [25], clustering [15], Steiner forest [6] ...

The minimum distortion problem has been studied under various metrics, when $k = n$. Kenyon et al. [20] gave a polynomial-time algorithm for line metrics (1-dimensional point sets) when $\mathrm{dist}(X,Y) < 5 + 2\sqrt{6}$. They also gave an algorithm that computes $\mathrm{dist}(X,Y)$ when d_X is the metric associated with an unweighted graph over X and d_Y is the metric associated with a bounded degree tree over Y, with a running time exponential in the maximum degree and doubly exponential in $\mathrm{dist}(X,Y)$. For general metrics, the minimum distortion is hard to approximate within a factor less than $\log^{1/4-\gamma} n$, for any $\gamma > 0$ [22]. Hall and Papadimitriou [18] showed that even for line metrics, the distortion is hard to approximate when it is large.

When $k \leqslant n$, Fellows et al. [14] showed that the problem of deciding whether $\mathrm{dist}(X,Y) \leqslant D$ is fixed-parameter tractable when parameterized by Δ and D, where d_X is the metric associated with an unweighted graph over X, and d_Y is the metric associated with a tree of degree at most Δ over Y. For two unweighted graph metrics, Cygan et al. [11] showed that the problem cannot be solved in time $2^{o(n \log n)}$ unless ETH is false. When X is an arbitrary finite metric space and Y is a subset of the real line, Nayyeri and Raichel [24] showed that a constant-factor approximation of $\text{dist}(X,Y)$ can be computed in time $\Phi(X)^{O((\text{dist}(x,y))^2)}(kn)^{O(1)}$, where $\Phi(X)$ is the spread of X. (See Sect. 2.)

In summary, the previously known theoretical results on the minimum distortion problem are either hardness results, or algorithms for cases where d_Y is a subset of a line metric or a tree metric. The algorithms presented in this paper, on the other hand, apply when d_X and d_Y are doubling metrics, which are generalizations of fixed-dimensional Euclidean metrics.

Our Approach. In Sect. 3, we present hardness results on the ρ-distortion problem. We reduce an instance $G(V,E)$ of k-clique to an instance of ρ-distortion consisting of two metric spaces (X,d_X) and (Y,d_Y) of sizes k and km, respectively, where $m=|V|$. The pattern (X,d_X) is an ultrametric, with exponentially increasing distances. The space (Y,d_Y) consists of k rings, each ring consisting of m points regularly spaced on a circle of perimeter 1. Each of these m points is associated with a vertex of the k-clique instance. The distances between the rings increase exponentially, and the input graph is encoded by having slightly longer edges for pairs of vertices lying in different rings that correspond to edges in the input graph. We prove that these two spaces have doubling dimension $\log_2 3$, and that this instance of ρ-distortion is equivalent to the k-clique instance we started from.

In Sect. 4, we give a self-contained description of a first approximation algorithm for the ρ-distortion problem that runs in time $2^{O(k^2\log k)}(\rho^2/\varepsilon)^{3k\delta}n+O(kn\log\Phi(Y))$ where $\Phi(Y)$ is the *spread* of Y. (See Sect. 2.) We first construct a *navigating net* over Y [23]. A navigating net is essentially a coordinate-free quadtree that records a metric space. It represents Y at all resolutions r where r is a powers of 2. At each scale r, our navigating net records an r-*net* Y_r of Y, which is a subset of Y whose points are a distance at least r apart, and such that the radius-r balls centered at Y_r cover Y.

Let r_X be the smallest scale that is at least ρ times the diameter of X. Our algorithm constructs, for each point $y\in Y_{r_X}$, a sparse set of matchings whose images are in the radius-$3r_X$ ball centered at y. (By sparse, we mean that any two such matchings send at least one point of X to two points of Y that are a distance at least $\varepsilon r_X/(2\rho^2)$ apart.) The union of these sets of matchings over all $y\in Y_{r_X}$ is denoted $L(X,\varepsilon,r_X)$, and we show that any solution to the ρ-distortion problem is close to at least one matching in $L(X,\varepsilon,r_X)$.

We compute $L(X,\varepsilon,r_X)$ recursively from $L(P,\beta,r_P)$ and $L(Q,\beta,r_Q)$ for $\beta=\varepsilon/(8k-8)$, where P and Q form a partition of X. More precisely, we obtain P and Q by running Kruskal's algorithm on X, and stopping at the second last step. It ensure that P and Q are well separated, and it follows that any ρ-matching $\bar{\sigma}:X\to Y$ can be approximated by a combination of two $(1+\beta)\rho$-matchings $\bar{\sigma}^P_\beta:P\to Y$ and $\bar{\sigma}^Q_\beta:Q\to Y$ recorded in $L(P,\beta,r_P)$ and $L(Q,\beta,r_Q)$, respectively. (See Fig. 1.) After computing $L(X,\varepsilon,r_X)$, we simply return one of the matchings that it records, if any.

In the full version of this paper [1, Section 5], we show how to improve this time bound to $2^{O(k^2\log k)}(\rho^2/\varepsilon)^{2k\delta}n\log n$. We achieve it using the approximate near-neighbor (ANN) data structure by Cole and Gottlieb [9]. First, this data

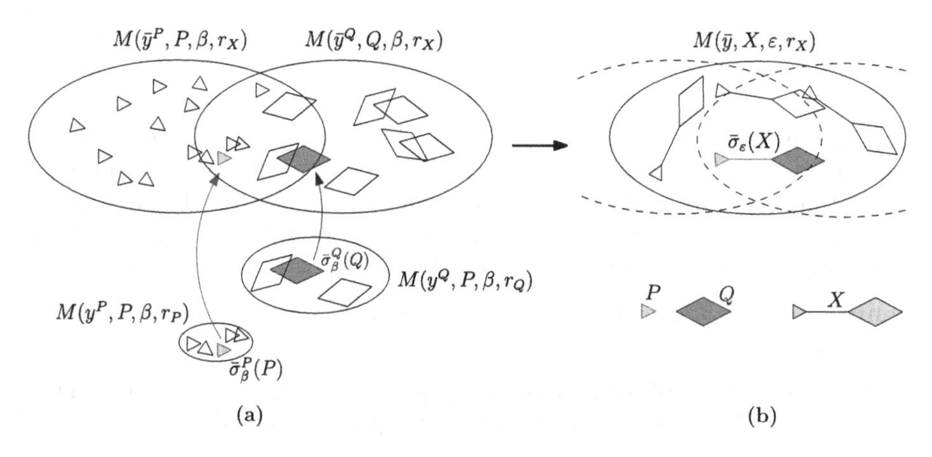

Fig. 1. Recursive construction of $L(X, \varepsilon, r_X)$. The pattern X is split into P and Q. (a) Bottom-up phase: We compute $L(P, \beta, r_X)$ and $L(Q, \beta, r_X)$ from $L(P, \beta, r_P)$ and $L(P, \beta, r_Q)$. (b) An approximate matching $\bar{\sigma}_\varepsilon : X \to Y$ in $L(X, \varepsilon, r_X)$ is obtained by combining two matchings $\bar{\sigma}^P_\beta : P \to Y$ and $\bar{\sigma}^Q_\beta : Q \to Y$.

structure allows us to efficiently prune the sets of matchings recorded at layer r_X, by only inserting a new matching if it is far enough from all previously inserted matching. It can be checked by performing a constant number of ANN queries in the set of matchings. As the space of matchings is doubling (Corollary 1), it takes logarithmic time. This saves a factor $k(\rho^2/\varepsilon)^{k\delta}/\log n$ in our time bound. Second, instead of computing the whole navigating net, which takes time $O(n \log \Phi(Y))$, we show how to compute any layer in $O(n \log n)$ time using ANN queries. As our algorithm only requires k layers of the navigating net, it removes the dependency on $\Phi(Y)$ from the time bound.

Finally, we show how to extend our results on the ρ-distortion problem to the minimum distortion problem [1, Section 6]. For the hardness result, it suffices to add an extra point to X and Y that is far enough from the other points, in order to make the reduction work. For the algorithms, we use the reduction by Kenyon et al. [20] of the minimum distortion problem to the ρ-distortion problem, which we speed-up using exponential search, and using a well-separated pairs decomposition, which allows us to reduce the number of candidate values for $\operatorname{dist}(X, Y)$.

2 Notation and Preliminary

Let (S, d_S) be a finite metric space. The ball $\mathrm{b}(x, r)$ centered at x with radius r is the set of points $x' \in S$ such that $d_S(x, x') \leqslant r$. The minimum and maximum interpoint distances in S are denoted $\operatorname{dmin}(S)$ and $\operatorname{diam}(S)$, respectively. In other words, $\operatorname{diam}(S)$ is the diameter of S. The *spread* of S is the ratio $\Phi(S) = \operatorname{diam}(S)/\operatorname{dmin}(S)$. The distance from a point x to a subset T

of S is $d_S(x,T) = \min_{t \in T} d_S(x,t)$. The distance between two sets T and U is $d_S(T,U) = \min_{t \in T, u \in U} d_S(t,u)$.

A metric space (S, d_S) has *doubling dimension* δ if any ball of radius r is contained in the union of at most 2^δ balls of radius $r/2$. When $\delta = O(1)$, we say that this space is *doubling.* This notion of dimension generalizes the dimension of a Euclidean space: In particular, the Euclidean space $\mathbb{R}^d$ has doubling dimension $O(d)$ [17]. In this paper, we will consider spaces of constant doubling dimension, so we assume that $\delta = O(1)$. We will need the following packing lemma:

Lemma 1 ([21]). *If a metric space (S, d_S) has doubling dimension δ, then $|S| \leqslant (4\Phi(S))^\delta$.*

We will also make use of the fact that a product of doubling metrics is doubling. A proof can be found in the full version of this paper [1, Lemma 2]

Lemma 2. *Let $(S_1, d_1), \ldots, (S_k, d_k)$ be metric spaces with doubling dimensions $\delta_1, \ldots, \delta_k$, respectively. Then the product metric (S, d_S) where $S = S_1 \times \cdots \times S_k$ and $d_S((u_1, \ldots, u_k), (v_1, \ldots, v_k)) = \max(d_1(u_1, v_1), \ldots, d_k(u_k, v_k))$ for all $(u_1, \ldots, u_k)$, $(v_1, \ldots, v_k) \in S$ has doubling dimension $\delta_1 + \cdots + \delta_k$.*

We call a mapping σ satisfying Eq. (1) a *ρ-matching* from X to Y. The distance between two matchings σ and σ' from X to Y is $d_M(\sigma, \sigma') = \max_{x \in X} d_Y(\sigma(x), \sigma'(x))$. We denote by $x_1, x_2, \ldots, x_k$ the k elements of X. So a matching $\sigma : X \to Y$ can be identified with a sequence of n points $(y_1, \ldots, y_k)$ where $y_1 = \sigma(x_1), \ldots, y_k = \sigma(x_k)$. In other words, the space of matchings from X to Y can be identified with (Y^k, d_M). Then it follows from Lemma 2 that:

Corollary 1. *The space of matchings from X to Y has doubling dimension $k\delta$.*

3 Reduction from k-Clique

Given an integer $k \geqslant 1$ and a graph $G(V, E)$, the *k-clique* problem is to decide whether there exists a subset $C \subseteq V$ of k vertices such that any two of these vertices are connected by an edge in E. This subset C is called a k-clique. In this section, we present a reduction from the k-clique problem to the ρ-distortion problem. So let $G(V, E)$ be an instance of k-clique with m vertices. We denote $V = \{v_1, \ldots, v_m\}$, and we assume that $m \geqslant 24$. For any $\rho \geqslant 1$, we will show how to construct an equivalent instance of the ρ-distortion problem consisting of two metric spaces (X, d_X) and (Y, d_Y) of respective sizes k and km, and of doubling dimension $\log_2 3$.

Let us first build Y, and its associated metric d_Y. We define an m-point *ring gadget* as a set $R_i = \{p_{i1}, .., p_{im}\}$ of m points spaced regularly on a circle of perimeter 1, so that the distance between any two points is the usual distance along this circle:

$$d_Y(p_{ij}, p_{ij'}) = (1/m) \cdot \min(j' - j, m + j - j') \text{ whenever } 1 \leqslant j \leqslant j' \leqslant m. \quad (2)$$

We define $Y = R_1 \cup \cdots \cup R_k$ as the disjoint union of k ring gadgets. The index j of the point $p_{ij} \in R_i$ corresponds to the vertex v_j of G. The distances between points in different rings are defined as follows. For any $i, i' \in \{1, \ldots, k\}$ such that $i \neq i'$, and for any $j, j' \in \{1, \ldots, m\}$,

$$d_Y(p_{ij}, p_{i'j'}) = \begin{cases} 2^{\max(i,i')} & \text{if } (v_j, v_{j'}) \in E, \text{ and} \\ 2^{\max(i,i')} - (1/m) & \text{otherwise.} \end{cases} \tag{3}$$

Thus, the distance between two vertices in different rings is a power of 2 if and only if their associated vertices in G are connected by an edge. The distance between two points in the same ring is given by Eq. (2).

The pattern set X consists of k distinct points $x_1, \ldots, x_k$, associated with the metric $d_X(x_i, x_{i'}) = 2^{\max(i,i')}\rho$ for all $i \neq i'$. The two spaces (X, d_X) and (Y, d_Y) are metric spaces with doubling dimension $\log_2 3$. A proof is given in the full version of this paper [1, Lemma 4–6].

So given an instance $G = (V, E)$ of the k-clique problem, we construct the metric spaces (X, d_X) and (Y, d_Y) as described above. These two metric spaces form an instance of the ρ-distortion problem. We need to show that these two instances of k-clique and ρ-distortion are equivalent:

Theorem 1. *The graph G admits a k-clique if and only if (X, d_X), (Y, d_Y) is a positive instance of the ρ-distortion problem.*

A proof of the theorem above is given in the full version of this paper [1, Theorem 9]. The main idea is that, in order for a mapping with distortion ρ to exist, we need to map each point x_i to a point y_i in the ring R_i, such that the lengths of the edges connecting the y_i's are powers of 2. Then the vertices in G corresponding to these points form a clique.

The construction of (X, d_X) and (Y, d_Y) from G is performed in polynomial time. It is also an FPT-reduction with parameter k. As k-clique is NP-complete and $W[1]$-hard, it follows that:

Corollary 2. *The ρ-distortion problem for doubling spaces of dimension* $\log_2 3$ *is NP-hard, and is $W[1]$-hard when parameterized by k.*

Unless the exponential time hypothesis (ETH) is false, our reduction also shows that that the ρ-distortion problem cannot be solved in time $n^{o(k)}$. More precisely, it follows from a hardness result on k-clique [12, Theorem 14.21] that:

Corollary 3. *The ρ-distortion problem for doubling spaces of dimension* $\log_2 3$ *cannot be solved in time $f(k) \cdot n^{o(k)}$ for any computable function f, unless ETH is false.*

4 Approximation Algorithm for the ρ-Distortion Problem

In this section, we present an approximation algorithm for the ρ-distortion problem. We assume that the doubling dimension is constant, that is, $\delta = O(1)$. As we saw in Sect. 3, the ρ-distortion problem is hard, so we relax the problem slightly: Given a parameter $0 < \varepsilon \leqslant 1$, the (ρ, ε)-*distortion problem* is to find a $(1+\varepsilon)\rho$-matching whenever a ρ-matching exists. If there is no ρ-matching, then our algorithm either returns a $(1+\varepsilon)\rho$-matching, or it does not return any result.

Navigating Nets. Our algorithm records Y in a *navigating net*, which is a data structure representing Y at different resolutions. This structure was introduced by Krauthgamer and Lee [23]. We will use a slightly modified version of it. Our version of the navigating net has the advantages that each layer is an r-net of Y (see definition below), and that it can easily be computed in $O(n \log \Phi(Y))$ time. On the other hand, it does not allow efficient deletions. Several other variations exist [9,19].

For any $r \geqslant 0$, an r-*net* of Y is a subset $Y_r \subseteq Y$ such that $\text{dmin}(Y_r) \geqslant r$ and $Y \subseteq \bigcup_{y \in Y_r} \text{b}(y, r)$. An r-net can be constructed incrementally by repeatedly adding new points that lie outside of the current union of balls, until Y is completely covered. Intuitively, an r-net represents Y at resolution r.

A *scale* is a rational number $r = 2^i$ such that $i \in \mathbb{Z}$. Our navigating net records a sequence of r-nets $Y_{r_{\min}}, Y_{2r_{\min}}, \ldots, Y_{r_{\max}}$ such that $r_{\min}$ and $r_{\max}$ are scales satisfying the inequalities $\text{dmin}(Y)/2 < r_{\min} \leqslant \text{dmin}(Y)$ and $\text{diam}(Y)/2 \leqslant r_{\max} < 2\,\text{diam}(Y)$. At the largest scale $r_{\max}$, the $r_{\max}$-net $Y_{r_{\max}}$ consists of a single point y_{root}, and thus $Y = \text{b}(y_{\text{root}}, r_{\max})$. At the lowest scale, we set $Y_{r_{\min}} = Y$. So the navigating net represents Y at all scales r such that $r_{\min} \leqslant r \leqslant r_{\max}$ by an r-net Y_r. All scales $r > r_{\max}$ are represented by $Y_r = Y_{r_{\max}} = \{y_{\text{root}}\}$, but we do not construct these copies of $Y_{r_{\max}}$ explicitly.

We may assume that $\log(\Phi(Y)) \geqslant 1$ as otherwise, $|Y| = O(1)$ by Lemma 1, and the ρ-distortion problem can be solved in $O(1)$ time by brute force. So we only construct Y_r at $O(\log(\Phi(Y))$ different scales r such that $\text{dmin}(Y)/2 < r_{\min} \leqslant r \leqslant r_{\max} < 2\,\text{diam}(Y)$.

We construct a graph over these r-nets. First, at each scale r such that $r_{\min} \leqslant r \leqslant r_{\max}$, we connect any two nodes y, $y' \in Y_r$ such that $d_Y(y, y') \leqslant 6r$ by an edge, that we call a *horizontal* edge. At each scale r such that $2r_{\min} \leqslant r \leqslant r_{\max}$, we also connect with a *vertical* edge each $y \in Y_{r/2}$ to a node $p(y) \in Y_r$, called the *parent* of y, such that $d_Y(y, p(y)) \leqslant r$. At least one such node $p(y)$ exists since Y_r is an r-net. We call y a *child* of $p(y)$. More generally, we say that y is a *descendant* of y' if y is a child of y', or y is a child of a descendant of y'. Conversely, we say that y' is an *ancestor* of y if y is a descendant of y'. When $r' > r_{\max}$, the node $y_{\text{root}} \in Y_{r'}$ is an ancestor of any node at a lower level.

These r-nets, together with the horizontal and vertical edges, form our navigating net. We will need the 4 lemma below, whose proofs are given in the full version of this paper [1, Lemma 12–15].

Lemma 3. *Each node of the navigating net has degree* $O(1)$.

Lemma 4. *The navigating net of* Y *can be computed in* $O(n \log \Phi(Y))$ *time.*

We associate the ball $\text{b}(y, 2r)$ with each node y in an r-net Y_r. These balls have the following properties.

Lemma 5. *At any scale r, and for any subset $S \subseteq Y$ such that* $\text{diam}(S) \leqslant r$, *there exists a node $y \in Y_r$ such that $S \subseteq \text{b}(y, 2r)$.*

Lemma 6. *Let r and r' be two scales such that $r < r'$. For any $y \in Y_r$, and for any ancestor $y' \in Y_{r'}$ of y, we have $b(y, 2r) \subseteq \text{b}(y', 2r')$.*

In summary, the balls $\text{b}(y, 2r)$, connected by the vertical edges, form a tree such that each ball is contained in each of its ancestors. The horizontal edges will help us traverse this tree within a given level.

Splitting the Pattern. Our algorithm proceeds recursively, by partitioning X into two well-separated subsets P and Q at each stage. More precisely, we will split X as follows. (Remember that $k = |X|$.)

Lemma 7. *If $k \geqslant 2$, we can partition X into two non-empty subsets P and Q such that* $\text{diam}(X) \leqslant (k-1) \cdot d_X(P, Q)$.

Proof. We obtain P and Q by running Kruskal's algorithm [10] for computing a minimum spanning tree of X, and stopping at the second-last step. So starting from the forest $(X, \emptyset)$, we repeatedly insert the shortest edge that connects any two trees of the current forest, until we are left with exactly two trees P and Q. At the last step, P and Q are then connected with an edge of length $\ell = d_X(P, Q)$, which is not shorter than any edge in the spanning trees we constructed for P and Q. Thus, $\text{diam}(P) \leqslant (|P| - 1)\ell$ and $\text{diam}(Q) \leqslant (|Q| - 1)\ell$. It follows that $\text{diam}(X) \leqslant (|P| - 1)\ell + \ell + (|Q| - 1)\ell = (k - 1)\ell$.

Recording Approximate Matchings. Our algorithm for the (ρ, ε)-distortion problem records a collection of approximate matchings for some layers of the navigating net. More precisely, for some non-empty subset W of X, for some $0 < \beta \leqslant 1$ and for some scales $r \geqslant \rho \,\text{diam}(W)$, we will construct a data structure $L(W, \beta, r)$ that records at least one $(1 + \beta)\rho$-matching from W to Y if a ρ-matching from W to Y exists. In particular, we will compute $L(X, \varepsilon, r_X)$ at an appropriate scale r_X, which records a solution to the (ρ, ε)-distortion problem if there is one. We first give three invariants of $L(W, \beta, r)$, and then we show how to compute $L(X, \varepsilon, r_X)$ recursively.

The data structure $L(W, \beta, r)$ records a set $M(y, W, \beta, r)$ of matchings at each node $y \in Y_r$. These sets satisfy the following properties.

Property 1. Let r be a scale such that $r \geqslant \rho \,\text{diam}(W)$.

(a) For any $y \in Y_r$, each matching $\sigma_\beta \in M(y, W, \beta, r)$ is a $(1 + \beta)\rho$-matching from W to Y such that $\sigma_\beta(W) \subseteq \text{b}(y, 3r)$.
(b) For any $y \in Y_r$ and any two distinct σ_β, $\sigma'_\beta \in M(y, W, \beta, r)$, we have $d_M(\sigma_\beta, \sigma'_\beta) \geqslant \beta r/(2\rho^2)$.
(c) For any ρ-matching $\sigma : W \to Y$, there exist $y \in Y_r$ and $\sigma_\beta \in M(y, W, \beta, r)$ such that $d_M(\sigma, \sigma_\beta) \leqslant \beta r/\rho^2$.

These properties imply the following bound on the sizes of these sets.

Lemma 8. *For any $y \in Y_r$, we have $|M(y, W, \beta, r)| = O((\rho^2/\beta)^{k\delta})$.*

Proof. Property 1a implies that any two matchings in $M(y, W, \beta, r)$ are at distance at most $6r$ from each other, that is, $\mathrm{diam}(M(y, W, \beta, r)) \leqslant 6r$. Property 1b means that $\mathrm{dmin}(M(y, W, \beta, r)) \geqslant \beta r/(2\rho^2)$, hence $\Phi(M(y, W, \beta, r)) \leqslant 12\rho^2/\beta$. Then by Lemma 1 and Corollary 1, we have $|M(y, W, \beta, r)| \leqslant (48\rho^2/\beta)^{k\delta}$.

Recursive Construction. Our algorithm constructs $L(W, \beta, r)$ recursively, for some subsets W of X and some values β and r. We start with the base case, then we show how to compute $L(W, \beta, r')$ from $L(W, \beta, r)$ when $\rho\,\mathrm{diam}(W) \leqslant r < r'$. Finally, we show how to recursively compute $L(X, \varepsilon, r_X)$ at an appropriate scale r_X by splitting X into P and Q according to Lemma 7. (See Fig. 1.)

Base case where $|W| = 1$ and $r = r_{\min}$. We have $W = \{x_i\}$, so a $(1 + \beta)\rho$-matching simply maps x_i to any element of Y. Therefore, at scale $r = r_{\min}$, we have $Y_r = Y$, so for each $y \in Y$, we record the matching that sends x_i to y in $M(y, W, \beta, r_{\min})$.

Bottom-up Construction. Suppose that $L(W, \beta, r)$ has been computed for some scale $r \geqslant \rho\,\mathrm{diam}(W)$. Given a larger scale $r' > r$, we now show how to construct $L(W, \beta, r')$.

For each $y' \in Y_{r'}$, we proceed as follows. Initially, we set $M(y', W, \beta, r') = \emptyset$. Then for each descendant $y \in Y_r$ of y', and for each matching $\sigma_\beta \in M(y, W, \beta, r)$, we check by brute force whether $d_M(\sigma_\beta, \sigma'_\beta) \geqslant \beta r'/(2\rho^2)$ for each matching σ'_β that was previously inserted into $M(y', W, \beta, r')$. If it is the case, we insert σ_β into $M(y', W, \beta, r')$, and otherwise we discard σ_β. It ensures that Property 1b holds for $M(y', W, \beta, r')$.

We now prove that Property 1a holds. Let $\sigma_\beta \in M(y', W, \beta, r')$. By construction, $\sigma_\beta \in M(y, W, \beta, r)$ for some descendant y of y'. As Property 1a holds at scale r, it follows that σ_β is a $(1 + \beta)\rho$-matching, and that $\sigma_\beta(W) \subseteq \mathrm{b}(y, 3r)$. Since y is a descendant of y', by Lemma 6, we have $\mathrm{b}(y, 2r) \subseteq \mathrm{b}(y', 2r')$, and thus $\mathrm{b}(y, 3r) \subseteq \mathrm{b}(y', 2r' + r) \subseteq \mathrm{b}(y', 3r')$. It follows that $\sigma_\beta(W) \subseteq \mathrm{b}(y', 3r')$, and thus Property 1a holds for $M(y', W, \beta, r')$.

Finally, we prove that Property 1c holds as well. Let $\sigma : W \to Y$ be a ρ-matching. As Property 1c holds for $L(W, \beta, r)$, there must be a node $y \in Y_r$ and $\sigma_\beta \in M(y, W, \beta, r)$ such that $d_M(\sigma, \sigma_\beta) \leqslant \beta r/\rho^2$. Let $y' \in Y_{r'}$ be the ancestor of y at scale r'. If σ_β was inserted into $M(y', W, \beta, r')$, then we are done because $d_M(\sigma, \sigma_\beta) \leqslant \beta r/\rho^2 \leqslant \beta r'/\rho^2$. Otherwise, we must have inserted a matching σ'_β such that $d_M(\sigma_\beta, \sigma'_\beta) < \beta r'/(2\rho^2)$. It follows that $d_M(\sigma, \sigma'_\beta) \leqslant d_M(\sigma, \sigma_\beta) + d_M(\sigma_\beta, \sigma'_\beta) \leqslant \beta(r + r'/2)/\rho^2 \leqslant \beta r'/\rho^2$, which completes the proof that Property 1 holds for $L(W, \beta, r')$.

We now analyze this algorithm. First we need to find all the descendants of each node $y' \in Y_{r'}$. We can do this by traversing the navigating net, which takes time $O(n \log \Phi(Y))$ as there are $O(\log \Phi(Y))$ levels in the navigating net. By Lemma 8, we have $|M(y', W, \beta, r')| = O((\rho^2/\beta)^{k\delta})$. Therefore, each time we attempt to insert a matching σ_β into $M(y', W, \beta, r')$, we compare it with $O((\rho^2/\beta)^{k\delta})$ previously inserted matchings, so it takes $O(k(\rho^2/\beta)^{k\delta})$ time as the distance between two matchings can be computed in $O(k)$ time. Since $|Y_r| \leqslant n$, and each set $M(y, W, \beta, r)$ has cardinality $O((\rho^2/\beta)^{k\delta})$, we spend $O(nk(\rho^2/\beta)^{2k\delta})$ time for computing $L(W, \beta, r')$. So we just proved the following.

Lemma 9. *Let r and r' be two scales such that $\rho\,\mathrm{diam}(W) \leqslant r < r'$. Given $L(W, \beta, r)$, we can compute $L(W, \beta, r')$ in $O(nk(\rho^2/\beta)^{2k\delta} + n \log \Phi(Y))$ time.*

Computing $L(X, \varepsilon, r_X)$ by splitting X. Suppose that $k \geqslant 2$. Let r_X be the smallest scale that is at least as large as $\rho\,\mathrm{diam}(X)$, hence $r_X = 2^{\lceil \log_2(\rho\,\mathrm{diam}(X)) \rceil}$. In particular, we have $r_X/2 < \rho\,\mathrm{diam}(X) \leqslant r_X$. If $r_X < r_{\min}$, then we have $\rho\,\mathrm{diam}(X) < r_{\min} \leqslant \mathrm{dmin}(Y)$, and there cannot be any ρ-matching, so our algorithm does not return any matching. Therefore, from now on, we may assume that $r_X \geqslant r_{\min}$.

Let P and Q be the sets obtained by splitting X as described in Lemma 7, so $\mathrm{diam}(X) \leqslant (k-1) \cdot d_X(P, Q)$. Let $\beta = \varepsilon/(8k-8)$, and suppose that $L(P, \beta, r_X)$ and $L(Q, \beta, r_X)$ have been computed earlier. We now show how to compute $L(X, \varepsilon, r_X)$.

For any two matchings $\sigma^P : P \to Y$ and $\sigma^Q : Q \to Y$, we denote by $\sigma^P \cdot \sigma^Q$ the matching from X to Y whose restrictions to P and Q are σ^P and σ^Q, respectively. In other words, if $\sigma = \sigma^P \cdot \sigma^Q$, then we have $\sigma(x) = \sigma^P(x)$ for all $x \in P$ and $\sigma(x) = \sigma^Q(x)$ for all $x \in Q$.

We compute $L(X, \varepsilon, r_X)$ as follows. For each node $y \in Y_{r_X}$, we consider all the pairs of matchings consisting of a matching $\sigma_\beta^P \in M(y^P, P, \beta, r_X)$ and a matching $\sigma_\beta^Q \in M(y^Q, Q, \beta, r_X)$, where y^P and y^Q are in Y_{r_X} and are at distance at most $6r_X$ from y. Then we consider $\sigma_\varepsilon = \sigma_\beta^P \cdot \sigma_\beta^Q$ as a candidate for being inserted into $M(y, X, \varepsilon, r_X)$. We first check whether σ_ε is a $(1+\varepsilon)\rho$-matching and $\sigma_\varepsilon(X) \subseteq \mathrm{b}(y, 3r_X)$. If it is the case, and if σ_ε is at distance at least $\varepsilon r_X/(2\rho^2)$ from any matching previously inserted into $M(y, X, \varepsilon, r_X)$, we insert σ_ε into $M(y, X, \varepsilon, r_X)$.

We first prove that this algorithm is correct. So we must prove that Property 1 holds for each set $M(y, X, \varepsilon, r_X)$. Property 1a follows from the fact that we only insert σ_ε if it is a $(1+\varepsilon)\rho$-matching and if $\sigma_\varepsilon(X) \subseteq \mathrm{b}(y, 3r_X)$. Property 1b follows from fact that we only insert σ_ε if it is at distance at least $\varepsilon r_X/(2\rho^2)$ from all the previously inserted matchings.

We now prove that Property 1c holds. So given a ρ-matching $\bar\sigma : X \to Y$, we want to prove that there exist $\bar y \in Y_{r_X}$ and $\bar\sigma_\varepsilon \in M(\bar y, X, \varepsilon, r_X)$ such that $d_M(\bar\sigma, \bar\sigma_\varepsilon) \leqslant \varepsilon r_X/\rho^2$. Let $\bar\sigma^P$ be the restriction of $\bar\sigma$ to P. In other words, $\bar\sigma^P : P \to Y$ is defined by $\bar\sigma^P(x) = \bar\sigma(x)$ for all $x \in P$. Similarly, let $\bar\sigma^Q$ be the restriction of $\bar\sigma$ to Q. Then $\bar\sigma^P$ and $\bar\sigma^Q$ are ρ-matchings. As Property 1c holds for $L(P, \beta, r_X)$, there exist $\bar y^P \in Y_{r_X}$ and $\bar\sigma_\beta^P \in M(\bar y^P, P, \beta, r_X)$ such that $d_Y(\bar\sigma^P, \bar\sigma_\beta^P) \leqslant \beta r_X/\rho^2$. Similarly, there exist $\bar y^Q \in Y_{r_X}$ and $\bar\sigma_\beta^Q \in M(\bar y^Q, Q, \beta, r_X)$ such that $d_Y(\bar\sigma^Q, \bar\sigma_\beta^Q) \leqslant \beta r_X/\rho^2$.

The matching $\bar{\sigma}_\varepsilon = \bar{\sigma}^P_\beta \cdot \bar{\sigma}^Q_\beta$ satisfies $d_M(\bar{\sigma}, \bar{\sigma}_\varepsilon) \leqslant \beta r_X/\rho^2 < \varepsilon r_X/\rho^2$. The lemma below, which is proved in the full version of this paper [1, Lemma 20], shows that it is a $(1+\varepsilon)\rho$-matching.

Lemma 10. *The matching $\bar{\sigma}_\varepsilon$ is a $(1+\varepsilon)\rho$-matching from X to Y.*

As $\bar{\sigma}$ is a ρ-matching, we have $\mathrm{diam}(\bar{\sigma}(X)) \leqslant \rho\,\mathrm{diam}(X)$ and thus $\mathrm{diam}(\bar{\sigma}(X)) \leqslant r_X$. By Lemma 5, it implies that there exists $\bar{y} \in Y_{r_X}$ such that $\bar{\sigma}(X) \subseteq \mathrm{b}(\bar{y}, 2r_X)$, and thus $\bar{\sigma}^P(P) \subseteq \mathrm{b}(\bar{y}, 2r_X)$. As $d_M(\bar{\sigma}^P, \bar{\sigma}^P_\beta) \leqslant \beta r_X/\rho^2 \leqslant r_X/8$, we have $\bar{\sigma}^P_\beta(P) \subseteq \mathrm{b}(\bar{y}, 17r_X/8)$. By Property 1a, we also have $\bar{\sigma}^P_\beta(P) \subseteq \mathrm{b}(\bar{y}^P, 3r_X)$, so the balls $\mathrm{b}(\bar{y}, 17r_X/8)$ and $\mathrm{b}(\bar{y}^P, 3r_X)$ intersect, which implies that $d_Y(\bar{y}, \bar{y}^P) < 6r_X$. The same proof shows that $d_Y(\bar{y}, \bar{y}^Q) < 6r_X$.

Therefore, our algorithm considers $\sigma_\varepsilon = \sigma^P_\beta \cdot \sigma^Q_\beta$ as a candidate solution for each $\sigma^P_\beta \in M(\bar{y}^P, P, \beta, r_X)$ and each $\sigma^Q_\beta \in M(\bar{y}^Q, Q, \beta, r_X)$. In particular, we must have considered the matching $\bar{\sigma}_\varepsilon = \bar{\sigma}^P_\beta \cdot \bar{\sigma}^Q_\beta$. As $d_M(\bar{\sigma}, \bar{\sigma}_\varepsilon) < \varepsilon r_X/\rho^2$ and $\bar{\sigma}(X) \subseteq b(\bar{y}, 2r_X)$, we have $\bar{\sigma}_\varepsilon(X) \subseteq b(\bar{y}, 3r_X)$, and thus we must have attempted to insert $\bar{\sigma}_\varepsilon$ into $M(\bar{y}, X, \varepsilon, r_X)$. If $\bar{\sigma}_\varepsilon$ was inserted into $M(\bar{y}, X, \varepsilon, r_X)$, then we are done. Otherwise, it means that there exists $\sigma'_\varepsilon \in M(\bar{y}, X, \varepsilon, r_X)$ such that $d_M(\bar{\sigma}_\varepsilon, \sigma'_\varepsilon) < \varepsilon r_X/(2\rho^2)$. Since $d_M(\bar{\sigma}, \bar{\sigma}_\varepsilon) \leqslant \beta r_X/\rho^2 < \varepsilon r_X/(2\rho^2)$, it follows that $d_M(\bar{\sigma}, \sigma'_\varepsilon) < \varepsilon r_X/\rho^2$. In any case, it shows that Property 1c holds. So we obtain the following result.

Lemma 11. *Suppose that $k \geqslant 2$ and $\beta = \varepsilon/(8k-8)$. Then we can compute $L(X, \varepsilon, r_X)$ from $L(P, \beta, r_X)$ and $L(Q, \beta, r_X)$ in $O(nk(\rho^2/\beta)^{3k\delta})$ time.*

Proof. The discussion above shows that our algorithm is correct. We still need to analyze its running time. Let $y \in Y_{r_X}$. As $d_Y(y, y^P) \leqslant 6r_X$ and $d_Y(y, y^Q) \leqslant 6r_X$, the nodes y^P and y^Q are connected to y by horizontal edges. As the nodes of the navigating net have constant degree, it implies that there are $O(1)$ pairs (y^P, y^Q) to consider. By Lemma 8, there are $O((\rho^2/\beta)^{k\delta})$ matchings in $M(y^P, P, \beta, r_X)$ and $M(y^Q, Q, \beta, r_X)$. Therefore, we consider $O((\rho^2/\beta)^{2k\delta})$ matchings σ_ε when constructing $M(y, X, \varepsilon, r_X)$. Each of these matchings is then compared with previously inserted matchings. We have $|M(y, X, \varepsilon, r_X)| = O((\rho^2/\varepsilon)^{k\delta}) = O((\rho^2/\beta)^{k\delta})$ by Lemma 8. We can compute the distance between two matchings in $O(k)$ time, so it takes $O(k(\rho^2/\beta)^{3k\delta})$ time to compute $M(y, X, \varepsilon, r_X)$. As there are at most n nodes $y \in Y_{r_X}$, the overall time bound is $O(nk(\rho^2/\beta)^{3k\delta})$.

Putting Everything Together. We can now describe our algorithm for the (ρ, ε)-distortion problem. We first compute the navigating net in $O(n \log \Phi(Y))$ time. If $r_X < r_{\min}$, then $\rho\,\mathrm{diam}\,X < \mathrm{dmin}(Y)$, and thus there is no ρ-matching. Otherwise, we recursively compute $L(X, \varepsilon, r_X)$, as described below. If $L(X, \varepsilon, r_X)$ contains at least one matching, then we return one of them, which by Property 1a is a $(1+\varepsilon)\rho$-matching. By Property 1c, if there is a solution to the ρ-distortion problem, then at least one $(1+\varepsilon)\rho$-matching must be recorded in $L(X, \varepsilon, r_X)$, which shows that this algorithm indeed solves the (ρ, ε)-distortion problem.

We now explain how to compute $L(X, \varepsilon, r_X)$ recursively. The base case is $k = 1$ and $r_X = r_{\min}$. As explained above, it can be done in $O(n)$ time by recording a trivial matching at each leaf node.

When $k \geqslant 2$, we split X into P and Q as described in Lemma 7. Then we compute recursively $L(P, \beta, r_P)$ and $L(Q, \beta, r_Q)$ where $\beta = \varepsilon/(8k-8)$ and r_P and r_Q are the smallest scales at least as large as $\rho\,\mathrm{diam}(P)$ and $\rho\,\mathrm{diam}(Q)$, respectively. We compute $L(P, \beta, r_X)$ and $L(Q, \beta, r_X)$ using Lemma 9, which takes time $O(nk(\rho^2/\beta)^{2k\delta} + n \log \Phi(Y))$. Then we obtain $L(X, \varepsilon, r_X)$ by Lemma 11 in $O(nk(\rho^2/\beta)^{3k\delta})$ time.

So the running time $T(n, k, \rho, \varepsilon)$ of our algorithm satisfies the relation

$$T(n, k, \rho, \varepsilon) = T(n, k_1, \rho, \beta) + T(n, k_2, \rho, \beta) + O(nk(\rho^2/\beta)^{3k\delta} + n \log \Phi(Y)),$$

where $k_1 + k_2 = k$, $1 \leqslant k_1 \leqslant k - 1$ and $\beta = \varepsilon/(8k - 8)$. This expression expands into a sum of $k-1$ terms $O(nk(\rho^2(8k)^k/\varepsilon)^{3k\delta} + n \log \Phi(Y))$ and k terms $O(n)$ for the base cases. Hence we have $T(n, k, \rho, \varepsilon) = 2^{O(k^2 \log k)} n(\rho^2/\varepsilon)^{3k\delta} + O(kn \log \Phi(Y))$. We just proved the following:

Theorem 2. *The (ρ, ε)-distortion problem can be solved in time* $2^{O(k^2 \log k)}(\rho^2/\varepsilon)^{3k\delta} n + O(kn \log \Phi(Y))$.

In the full version of this paper [1, Section 5], we show how to improve this time bound to $2^{O(k^2 \log k)}(\rho^2/\varepsilon)^{2k\delta} n \log n$. A brief description is also given above, at the end of Sect. 1 in this extended abstract.

Acknowledgments. We thank the anonymous referees for their helpful comments.

References

1. Allair, C., Vigneron, A.: Pattern matching in doubling spaces. CoRR abs/2012.10919 (2020)
2. Alt, H., Guibas, L.J.: Discrete geometric shapes: Matching, interpolation, and approximation. In: Handbook of Computational Geometry, pp. 121–153. Elsevier (2000)
3. Alt, H., Mehlhorn, K., Wagener, H., Welzl, E.: Congruence, similarity, and symmetries of geometric objects. Discrete Comput. Geom. **3**(3), 237–256 (1988). https://doi.org/10.1007/BF02187910
4. Arya, S., Mount, D.M., Vigneron, A., Xia, J.: Space-time tradeoffs for proximity searching in doubling spaces. In: Halperin, D., Mehlhorn, K. (eds.) ESA 2008. LNCS, vol. 5193, pp. 112–123. Springer, Berlin, Heidelberg (2008). https://doi.org/10.1007/978-3-540-87744-8_10
5. Borradaile, G., Le, H., Wulff-Nilsen, C.: Greedy spanners are optimal in doubling metrics. In: Proceedings of Symposium on Discrete Algorithms, pp. 2371–2379 (2019)
6. Chan, T.H., Hu, S., Jiang, S.H.: A PTAS for the Steiner forest problem in doubling metrics. SIAM J. Comput. **47**(4), 1705–1734 (2018)
7. Chan, T.H., Li, M., Ning, L., Solomon, S.: New doubling spanners: better and simpler. SIAM J. Comput. **44**(1), 37–53 (2015)

8. Chew, L.P., Goodrich, M.T., Huttenlocher, D.P., Kedem, K., Kleinberg, J.M., Kravets, D.: Geometric pattern matching under Euclidean motion. Comput. Geom. **7**, 113–124 (1997)
9. Cole, R., Gottlieb, L.: Searching dynamic point sets in spaces with bounded doubling dimension. In: Proceedings of ACM Symposium on Theory of Computing, pp. 574–583 (2006)
10. Cormen, T.H., Leiserson, C.E., Rivest, R.L., Stein, C.: Introduction to Algorithms, 3rd edn. MIT Press, Cambridge (2009)
11. Cygan, M., et al.: Tight lower bounds on graph embedding problems. J. ACM **64**(3), 18:1-18:22 (2017)
12. Cygan, M., et al.: Parameterized Algorithms. Springer, Cham (2015). https://doi.org/10.1007/978-3-319-21275-3_15
13. Ding, H., Ye, M.: On geometric alignment in low doubling dimension. In: Proceedings of AAAI Conference on Artificial Intelligence, pp. 1460–1467 (2019)
14. Fellows, M.R., Fomin, F.V., Lokshtanov, D., Losievskaja, E., Rosamond, F.A., Saurabh, S.: Distortion is fixed parameter tractable. In: Albers, S., Marchetti-Spaccamela, A., Matias, Y., Nikoletseas, S., Thomas, W. (eds.) ICALP 2009. LNCS, vol. 5555, pp. 463–474. Springer, Berlin, Heidelberg (2009). https://doi.org/10.1007/978-3-642-02927-1_39
15. Friggstad, Z., Rezapour, M., Salavatipour, M.R.: Local search yields a PTAS for k-means in doubling metrics. SIAM J. Comput. **48**(2), 452–480 (2019)
16. Gottlieb, L., Roditty, L.: Improved algorithms for fully dynamic geometric spanners and geometric routing. In: Proceedings of ACM-SIAM Symposium on Discrete Algorithms, pp. 591–600 (2008)
17. Gupta, A., Krauthgamer, R., Lee, J.R.: Bounded geometries, fractals, and low-distortion embeddings. In: Proceedings of IEEE Symposium on Foundations of Computer Science, pp. 534–543 (2003)
18. Hall, A., Papadimitriou, C.: Approximating the distortion. In: Chekuri, C., Jansen, K., Rolim, J.D.P., Trevisan, L. (eds.) APPROX 2005, RANDOM 2005. LNCS, vol. 3624, pp. 111–122. Springer, Berlin, Heidelberg (2005). https://doi.org/10.1007/11538462_10
19. Har-Peled, S., Mendel, M.: Fast construction of nets in low-dimensional metrics and their applications. SIAM J. Comput. **35**(5), 1148–1184 (2006)
20. Kenyon, C., Rabani, Y., Sinclair, A.: Low distortion maps between point sets. SIAM J. Comput. **39**(4), 1617–1636 (2009)
21. Kerber, M., Nigmetov, A.: Metric spaces with expensive distances. CoRR abs/1901.08805 (2019)
22. Khot, S., Saket, R.: Hardness of embedding metric spaces of equal size. In: Charikar, M., Jansen, K., Reingold, O., Rolim, J.D.P. (eds.) APPROX 2007, RANDOM 2007. LNCS, vol. 4627, pp. 218–227. Springer, Berlin, Heidelberg (2007). https://doi.org/10.1007/978-3-540-74208-1_16
23. Krauthgamer, R., Lee, J.R.: Navigating nets: simple algorithms for proximity search. In: Proceedings of ACM-SIAM Symposium on Discrete Algorithms, pp. 798–807 (2004)
24. Nayyeri, A., Raichel, B.: Reality distortion: Exact and approximate algorithms for embedding into the line. In: Proceedings of IEEE Symposium on Foundations of Computer Science, pp. 729–747 (2015)
25. Talwar, K.: Bypassing the embedding: algorithms for low dimensional metrics. In: Proceedings of ACM Symposium on Theory of Computing, pp. 281–290 (2004)

Reachability Problems for Transmission Graphs

Shinwoo An and Eunjin Oh(✉)

Department of Computer Science and Engineering, POSTECH, Pohang-si, Korea
{shinwooan,eunjin.oh}@postech.ac.kr

Abstract. Let P be a set of n points in the plane where each point p of P is associated with a radius $r_p > 0$. The transmission graph $G = (P, E)$ of P is defined as the directed graph such that E contains an edge from p to q if and only if $|pq| \leq r_p$ for any two points p and q in P, where $|pq|$ denotes the Euclidean distance between p and q. In this paper, we present a data structure of size $O(n^{5/3})$ such that for any two points in P, we can check in $O(n^{2/3})$ time if there is a path in G between the two points. This is the first data structure for answering reachability queries whose performance depends only on n but not on the radius ratio.

Keywords: Reachability · Intersection graph · Directed graph

1 Introduction

Consider a set S of unit disks in the plane. The *intersection graph* for S is defined as the undirected graph whose vertices correspond to the disks in S such that two vertices are connected by an edge if and only if the two disks corresponding to them intersect. It can be used as a model for broadcast networks: The disks of S represent transmitter-receiver stations with the same transmission power. One can view the broadcast range of a transmitter as a unit disk.

One straightforward way to deal with the intersection graph for S is to construct the intersection graph explicitly, and then run algorithms designed for general graphs. However, the intersection graph for S has complexity $\Theta(n^2)$ in the worst case even though it can be (implicitly) represented as n disks. Therefore, it is natural to seek faster algorithms for an intersection graph implicitly represented as its underlying set of disks. For instance, the shortest path between two vertices in a unit-disk intersection graph can be computed in near linear time [21]. For more examples, refer to [3,5,12].

A *transmission graph* is a *directed* intersection graph, which is introduced to model broadcast networks in the case that transmitter-receiver stations have different transmission power [18,20]. Let P be a set of n points in the plane where each point p of P is associated with a radius $r_p > 0$. The *transmission graph*

This work was supported by the National Research Foundation of Korea (NRF) grant funded by the Korea government (MSIT) (No. 2020R1C1C1012742).

A. Lubiw et al. (Eds.): WADS 2021, LNCS 12808, pp. 71–84, 2021.
https://doi.org/10.1007/978-3-030-83508-8_6

$G = (V, E)$ of P is an weighted directed graph whose vertex set corresponds to P. There is an edge (p, q) in E for two points p and q in P if and only if the Euclidean distance $|pq|$ between p and q is at most r_p. The weight of an edge (p, q) is defined as $|pq|$. It is sometimes convenient to consider a point p of P as the disk of radius r_p centered at p. We call it the *associated disk* of p, and denote it by D_p. We say p is *reachable* to q if there is a p-q path in G.

In this paper, we consider the *reachability* problem for transmission graphs: Given a set P of points associated with radii, check if a point of P is reachable to another point of P in the transmission graph. In the context of broadcast networks, this problem asks if a transmission station can transmit information to a receiver. We consider three versions of the reachability problem: the single-source reachability problem, (discrete) reachability oracles, and continuous reachability oracles. The *single-source reachability problem* asks to compute all vertices reachable from a given source node $p \in P$ in the transmission graph of P. Indeed, we consider the more general problem that asks to compute a t-spanner of size $O(n)$. Once we have a t-spanner of size $O(n)$, we can compute all vertices reachable from a given source node in linear time. A *(discrete) reachability oracle* is a data structure for P so that, given any two query points p and q in P, we can check if p is reachable to q in G efficiently. A *continuous reachability oracle* is a data structure for P for answering reachability queries that takes two points in the plane, one in P and one not necessarily in P, as a query.

Previous Work. The reachability problems and shortest-path problems have been extensively studied not only for general graphs but also for special classes of graphs; directed planar graphs [9], Euclidean spanners [8,17], and disk-intersection graphs [3,5]. In the following, we introduce several results for transmission graphs of disks in the plane. Let Ψ be the ratio between the largest and the smallest radii associated with the points in P.

- ***t*-Spanners (Single-source reachability problem).** One can solve the single-source reachability problem for transmission graphs in $O(n \log^4 n)$ time by constructing a dynamic data structures for weighted nearest neighbor queries [4,14]. Kaplan et al. [13] presented two algorithms for the more general problem that asks to compute a t-spanner of size $O(n)$ for any constant $t > 1$, one with $O(n \log^4 n)$ time and one with $O(n \log n + n \log \Psi)$ time. Recently, Ashur and Carmi [2] also considered this problem, and presented an $O(n^2 \log n)$-time algorithm for computing a t-spanner of which every node has a constant in-degree, and the total weight is bounded by a function of n and Ψ. Also, spanners for transmission graphs in an arbitrary metric space also have been considered [18,19].
- **Discrete reachability oracles.** Kaplan et al. [11] presented three reachability oracles: one for $\Psi < \sqrt{3}$, two for an arbitrary $\Psi > 1$. For an arbitrary Ψ, their first reachability oracle has performance which polynomially depends on Ψ, and the second one has performance which polylogarmically depends on Ψ. More specifically, the first data structure uses space $O(\Psi^3 n^{1/2})$, and has query time $O(\Psi^5 n^{3/2})$. The second one uses space $\tilde{O}_{n,\Psi}(n^{5/3})$, and has

query time $\tilde{O}_{n,\Psi}(n^{2/3})$, where $\tilde{O}_{n,\Psi}$ hides polylogarithmic factors in Ψ and n. This data structure is randomized in the sense that it allows to answer all queries correctly with high probability.
- **Continuous reachability oracles.** Kaplan et al. [13] shows that a discrete reachability oracle for the transmission graph G of P can be extended to a continuous reachability oracle. More specifically, given a discrete reachability oracle for G with space $S(n)$ and query time $Q(n)$, one can obtain in $O(n \log n \log \Psi)$ time a continuous reachability oracle for G with space $S(n) + O(n \log \Psi)$ and query time $O(Q(n) + \log n \log \Psi)$.

Our Results. As mentioned above, we improve the previously best-known results of the three versions of the reachability problem for transmission graphs.

- **t-Spanners (Single-source reachability problem).** We first present an $O(n \log^3 n)$-time algorithm for computing a t-spanner for a constant $t > 0$ in Sect. 2, which improves the running time of the algorithm by [13] by a factor of $O(\log n)$. Our construction is based on the Θ-graph and grid-like range tree introduced by [16]. This algorithm is also used for computing reachability oracles in Sects. 3 and 4.
- **Discrete reachability oracles.** We present two discrete reachability oracles for the transmission graph of P. The first one described in Sect. 3 uses space $O(n^{5/3})$ and has query time $O(n^{2/3})$, and can be computed in $O(n^{5/3})$ time. This is the first reachability oracle for a transmission graph whose performance is independent of Ψ.
  We omit the second one in this paper, which will be described in the full version of the paper. Its performance parameters depend polylogarithmically on the radius ratio Ψ. More specifically, it uses space $\tilde{O}_\Psi(n^{5/2})$, and has query time $\tilde{O}_\Psi(n^{3/2})$. It can be constructed in $\tilde{O}_\Psi(n^{5/2})$, where $\tilde{O}_\Psi(\cdot)$ hides polylogarithmic factors in Ψ. To obtain this, we combine two reachability oracles given by [11] whose performance parameters using a balanced separator of smaller size introduced by [7].
- **Continuous reachability oracles.** We also present a *continuous reachability oracle* with space $O(n^{5/3})$, query time $O(n^{2/3})$, and preprocessing time $O(n^{5/3} \log^2 n)$ in Sect. 4, which is the first continuous reachability oracle whose performance is independent of Ψ. Instead of using the approach in [13], we use auxiliary data structures whose performance is independent of Ψ together with the reachability oracle described in Sect. 3.

Due to lack of space, some proofs and details are omitted. The missing proofs and details can be found in the full version of this paper.

2 Improved Algorithm for Computing a t-Spanner

Let P be a set of n points associated with radii, and $G = (P, E)$ be the transmission graph of P. A subgraph H of G is called a *t-spanner* of G if for every pair

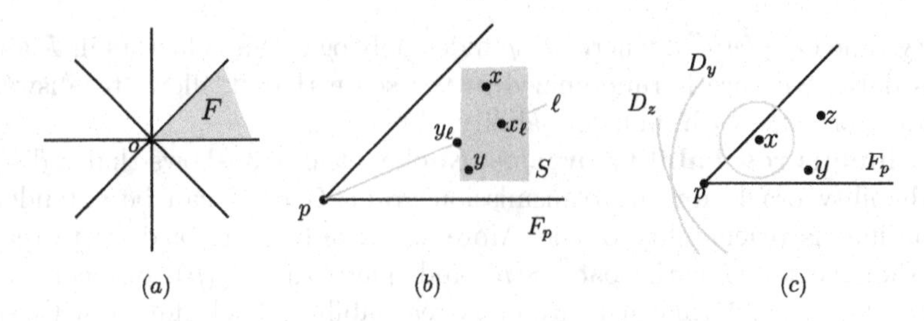

Fig. 1. Theta graph construction for $k = 8$. (a) The k cones of $\mathcal{F}$ subdivides the plane. (b) $\text{N}_F(p) = y$, and $\text{N}_S(p) = y$. (c) The edge (y, p) is picked.

of vertices of G, the distance in H between them is at most t times the distance in G between them. A *sparse* t-spanner is useful for constructing a reachability oracle efficiently; a t-spanner preserves the reachability information of G, and it allows us to investigate a small number of edges. Therefore, we first consider the problem of constructing a t-spanner of G in this section, and we use it for constructing a reachability oracle in Sect. 3.

In this section, we present an $O(n \log^3 n)$-time algorithm for computing a t-spanner of G of size $O(n)$ for any constant $t > 1$. This improves the running time of the algorithm proposed by Kaplan et al. [11], which runs in $O(n \log^4 n)$ time.[1] The spanner constructed by Kaplan et al. is a variant of the Yao graph. They first show that a variant of the Yao graph is a t-spanner for G, and then show how to construct it efficiently.

2.1 Theta Graphs and t-Spanners of Transmission Graphs

Our spanner construction is based on the *Θ-graph*, which is a geometric spanner similar to the Yao graph. Let $k > 0$ be a constant, which will be specified later, depending on t. Imagine that we subdivide the plane into k interior-disjoint cones with opening angle $2\pi/k$ which have the origin as their apexes. Let $\mathcal{F}$ be the set of such cones. See Fig. 1(a). For a cone $F \in \mathcal{F}$ and a point $p \in P$, let F_p denote the translated cone of F so that the apex of F_p lies on p. For each point $p \in P$, we pick k incoming edges for p, one for each cone of $\mathcal{F}$, as follows.

For a point q contained in F_p, let q_ℓ denote the orthogonal projection of q on the angle-bisector of F_p. Also, we let $d_F(p, q)$ be the Euclidean distance between p and q_ℓ, and let $\text{N}_F(p)$ denote the point q in F_p with $(q, p) \in E$ that minimizes $d_F(p, q)$. See Fig. 1(b). Note that $\text{N}_F(p)$ might not exist. For each cone $F \in \mathcal{F}$ and each point $p \in P$, we choose $(\text{N}_F(p), p)$. See Fig. 1(c). Let H_k be the graph consisting of the points in P and the chosen edges. If it is clear from the context, we simply use H to denote H_k.

Lemma 1. *For an integer $k > 8$, H_k is a $\tan(\frac{\pi}{4} + \frac{2\pi}{k})$-spanner of G.*

[1] Kaplan et al. mentioned that this algorithm takes an $O(n \log^5 n)$ time. However, this can be improved automatically into $O(n \log^4 n)$ using a data structure of [4].

Note that $t = \tan(\frac{\pi}{4} + \frac{2\pi}{k}) > 1$ converges to $\tan(\frac{\pi}{4}) = 1$ as $k \to \infty$. Therefore, for any constant $t > 1$, we can find a constant k such that H_k is a t-spanner of the transmission graph.

2.2 Efficient Algorithm for Computing the t-Spanner

In this section, we give an $O(n \log^3 n)$-time algorithm to construct H_k for a constant $k > 6$. To compute all edges of H_k, for each point $p \in P$ and each cone $F \in \mathcal{F}$, consider the translated cone F_p of F so that the apex lies on p, and compute $\mathrm{N}_F(p)$. We show how to do this for a cone $F \in \mathcal{F}$ only. The others can be handled analogously. Without loss of generality, we assume that the counterclockwise angle from the positive x-axis to two rays of F are 0 and $2\pi/k$, respectively. Let ℓ_1 and ℓ_2 be two lines orthogonal to the two rays, respectively.

Approach of Kaplan et al. The spanner constructed by Kaplan et al. [11] is a variation of the Yao graph. For each cone $F \in \mathcal{F}$ and a point $p \in P$, they pick the closest point in F_p to p among all points q with $p \in D_q$. Since they choose the closest point in a cone with respect to the Euclidean distance, they need to fit grid cells into a cone. To resolve this, they use various data structures including a compressed quadtree, a power diagram, a well-separated pair decomposition, and a dynamic nearest neighbor search data structure.

Our Approach. Instead, our construction is based on the Θ-graph. Recall that we pick the closest point in a cone with respect to $d_F(\cdot, \cdot)$ instead of the Euclidean distance. The order of the points of $F_p \cap P$ sorted with respect to $d_F(p, \cdot)$ is indeed the order of them sorted with respect to their projection points onto the angle-bisector of F.

In the following, we present an $O(n \log^3 n)$-time algorithms for computing all edges of H_k constructed for F. To do this, we use grid-like range trees proposed by Moidu et al. [16] together with a power diagram. With a slight abuse of notation, for a region S contained in F_p, let $\mathrm{N}_S(p)$ be the point q of $S \cap P$ with $(q, p) \in E$ that minimizes $d_F(p, q)$. See Fig. 1(b).

Data Structures. We construct the *two-level grid-like range tree* introduced by Moidu et al. [16] with respect to ℓ_1 and ℓ_2. It is a two-level balanced binary search tree. The first-level tree T_1 is a balanced binary search tree on the ℓ_1-projections of the points of P. Each node α in the first-level tree corresponds to a slab $I(\alpha)$ orthogonal to ℓ_1. It is also associated with the second-level tree T_α which is a binary search tree, not necessarily balanced, on the points of $P \cap I(\alpha)$. Unlike the standard range tree [6], T_α is obtained from a balanced binary search tree T_2 on the ℓ_2-projections of the points of P. More specifically, we remove the subtrees rooted at all nodes of T_2 whose corresponding parallelograms contain no point in $P \cap I(\alpha)$ in their union, and contract all nodes which have only one child. Then T_α is not necessarily balanced but a full binary tree of depth $O(\log n)$.

Given a point p of P, there are $O(\log^2 n)$ interior-disjoint parallelograms whose union contains all points of $P \cap F_p$. We denote the set of these parallelograms by $\mathcal{B}_p$. By construction, the cells of $\mathcal{B}_p$ are aligned for any point $p \in P$ so that we can consider them as a grid of size $O(\log n) \times O(\log n)$. See Fig. 2.

Lemma 2 [16]. *The two-level grid-like range tree on a set of n points in the plane can be computed in $O(n \log n)$ time. Moreover, its size is $O(n \log n)$.*

Then for each node v of the second-level trees, we construct a balanced binary search tree of the ℓ-projections of $P \cap B(v)$ as the third-level tree, where ℓ denotes the angle bisector of F. For a node β of the third-level trees, let $P(\beta)$ denote the set of the points stored in the subtree rooted at β. We construct the power diagram of $P(\beta)$. The *power diagram* is a weighted version of the Voronoi diagram. More specifically, the *power distance* between a point p and a disk D_q is defined as $|pq|^2 - r_q^2$. The power diagram partitions the plane into n regions such that all points in a same region have the same closest disk in power distance. The power diagram of n disks can be constructed in $O(n \log n)$ time with $O(n)$ space. Also, we can locate the disk D that minimizes the power distance from a query point p in $O(\log n)$ time. As a consequence, we can determine in $O(\log n)$ time if the query point p is in the union of disks by checking if $p \in D$ [10,14].

The construction time of the first, second, and third-level trees is $O(n \log^3 n)$ in total. Then we construct the power diagram for each node of a third-level tree in a *bottom-up* fashion. In particular, we start from constructing the power diagrams of the leaf nodes. For each internal node, we compute its power diagram by merging the power diagram of its two children. Therefore, we can construct the power diagrams for all nodes of a third-level tree in $O(m \log m)$ time, where m denotes the number of points corresponding to the root of the third-level tree. Since the sum of m's over all third-level trees is $O(n \log^2 n)$, the whole data structure can be constructed in $O(n \log^3 n)$ time.

Query Algorithm. For each cell $B \in \mathcal{B}_p$, we compute $\text{N}_B(p)$ in $O(\log^2 n)$ as follows. We start from the root of the third-level tree associated with B. We check if there is a point $q \in P(\beta)$ with $(q, p) \in E$ using the power diagram stored in the root node. If it does not exist, $\text{N}_B(p)$ does not exist. Otherwise, we traverse the third-level tree until we reach a leaf node. For each node β we encounter during the traversal, we consider the left child of β, say β_L. We check if there is a point $q \in P(\beta_L)$ with $(q, p) \in E$ using the power diagram stored in β_L. If it exists, we move to β_L. Otherwise, we move to the right child of β. We do this until we reach a leaf node, which stores $\text{N}_B(p)$.

In the following, we show how to choose $O(\log n)$ cells of $\mathcal{B}_p$, one of which contains $\text{N}_B(p)$. The cells of $\mathcal{B}_p$ are aligned along ℓ_1 and ℓ_2. They can be considered as a grid of $O(\log n) \times O(\log n)$ cells. We represent each row (parallel to ℓ_1) by integers $1, \ldots, O(\log n)$, and each column (parallel to ℓ_2) by integers $1, \ldots, O(\log n)$. We represent each cell of $\mathcal{B}_p$ by a pair $B(i, j)$ of indices such that i is the row-index and j is the column-index of the cell. For illustration, see Fig. 2(a). A cell $B = B(i, j)$ is said to be *useful* if $\text{N}_B(p)$ exists. Also, a useful cell $B = B(i,j)$ is called an *extreme cell* of $\mathcal{B}_p$ if no cell $B(i',j')$ is useful for indices i' and j' such that $i - j = i' - j'$ and $i' < i$.

Lemma 3. *The cell of $\mathcal{B}_p$ containing* $\mathrm{N}_F(p)$ *is an extreme cell. Moreover, the number of extreme cells of $\mathcal{B}_p$ is $O(\log n)$.*

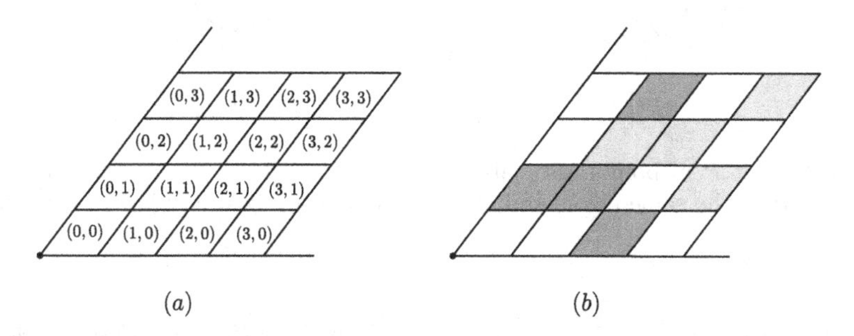

Fig. 2. (a) Index of the grid-like range tree (b) useful cells are colored with gray or red, and extreme cells are colored red. (Color figure online)

To compute $\mathrm{N}_F(p)$, we first compute $\mathcal{B}_p$ in $O(\log^2 n)$ time. For each cell $B \in \mathcal{B}_p$, we check if it is useful using the power diagram of $P \cap B$, which is stored in the root node of the third-level tree in $O(\log^3 n)$ time in total. Then we choose $O(\log n)$ extreme cells among the useful cells of $\mathcal{B}_p$. For each cell B of them, we compute $\mathrm{N}_B(p)$ in $O(\log^2 n)$ time, and thus the total query time is $O(\log^3 n)$.

Theorem 1. *Given a point set P and a constant $t > 1$, we can construct a t-spanner of the transmission graph of P within $O(n \log^3 n)$ time.*

Also, we can compute a BFS tree of G using H.

Theorem 2. *Let P be a set of n points, each associated with a radius. Given a t-spanner H of the transmission graph G of P as in Theorem 1, we can construct a BFS tree of G within $O(n \log n)$ time.*

3 Reachability Oracle for Unbounded Radius Ratio

In this section, we present a data structure of size $O(n^{5/3})$ so that given any two points p and q in P, we can check if p is reachable from q in $O(n^{2/3})$ time. Moreover, this data structure can be constructed in $O(n^{5/3})$ time. Note that this result is independent to the radius ratio Ψ.

We say a set of disks is *k-thick* if for any point p in the plane, there are at most k disks that contains p. Similarly, we say a transmission graph is *k-thick* if its underlying disk set is k-thick.

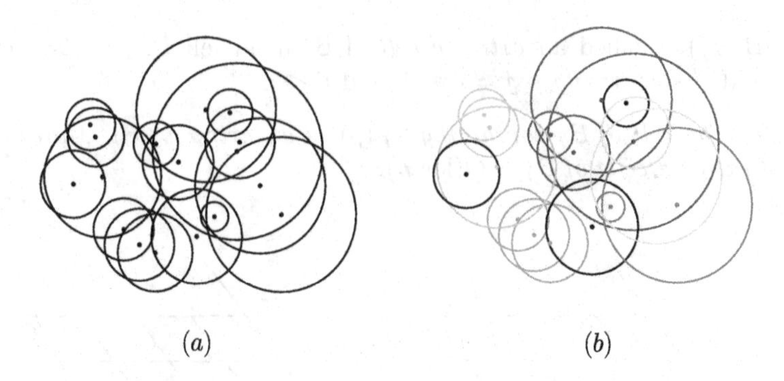

Fig. 3. (a) A set P of points associated with radii. (b) The disks in the same chain are colored with the same color, and the points in R are colored black. (Color figure online)

Lemma 4 *[15, Theorem 5.1]. For any set $\mathcal{D}$ of disks that is k-thick, there is a circle S intersecting $O(\sqrt{kn})$ disks of $\mathcal{D}$ such that the number of disks of $\mathcal{D}$ with $|S_{\text{in}}|, |S_{\text{out}}| \leq \frac{2n}{3}$, where S_{in} and S_{out} denote the set of disks of $\mathcal{D}$ contained in the interior of S and the exterior of S, respectively. In this case, We call S a* separating circle. *Moreover, we can compute S, S_{in} and S_{out} in linear time.*

Consider a separating circle S of the disk set induced by P. By Lemma 4, P is partitioned into three sets $S_{\text{in}}, S_{\text{out}}$, and $S_{\text{cross}} = \{p \in P \mid D_p \cap S \neq \emptyset\}$ such that every path in G connecting a point of S_{in} and a point of S_{out} visits a point in S_{cross}. We call S_{cross} a *separator* of G (or P). Using separators, we build a *separation tree* by repeatedly applying the algorithm in Lemma 4. As we will see in Sect. 3.2, the separation tree enables us to construct a reachability oracle efficiently. However, the transmission graph of a set of n points is n-thick in the worst case, and in this case, Lemma 4 does not give a non-trivial bound.

To resolve this, we partition P into $O(n^{2/3})$ *chains*, each consisting of $O(n^{1/3})$ points of P, and the *remaining set* R of points of P not belonging to any of the chains. Then we show that R is $O(n^{1/3})$-thick, and thus Lemma 4 gives an efficient reachability oracle for the subgraph of G induced by R. Additionally, we construct an auxiliary data structure for each chain.

3.1 Chain

We call a sequence $\langle p_1, \ldots, p_k \rangle$ of points of P sorted in the ascending order of their associated radii a *chain* if $(p_j, p_i) \in E$ for all indices i and j with $1 \leq i < j$. In other words, $|p_i p_j| \leq r_{p_j}$. In this section, we construct $O(n^{1/3})$-length chains as many as possible so that the remaining set R is k-thick for a small k.

To compute chains, we need a dynamic data structure for a set $\mathcal{D}$ of disks, dynamically changing by insertions and deletions, such that for a query point, we can check if there is a disk of $\mathcal{D}$ that contains the query point. This can be obtained using dynamic 3-D halfspace lower envelope data structure, which is given by [4], together with the standard lifting transformation. In particular, this data structure can be built in $O(n \log n)$ time and its insertion time, deletion time and query time are $O(\log^2 n)$, $O(\log^4 n)$ and $O(\log^2 n)$, respectively. For the convenience, we denote this data structure by $\mathsf{DNN}(\mathcal{D})$.

Lemma 5. *Let $\mathcal{D}$ be a set of disks, and p be a point in the plane. Given $\mathsf{DNN}(\mathcal{D})$, we can check if there are $n^{1/3}$ disks of $\mathcal{D}$ containing p in $O(n^{1/3} \log^4 n)$ time. Moreover, if they exist, we can return them, and delete them from $\mathcal{D}$ and $\mathsf{DNN}(\mathcal{D})$ within the same time bound.*

Let $\mathcal{D}$ be the set of disks induced by P, and we construct $\mathsf{DNN}(\mathcal{D})$. We choose the smallest disk D_p of $\mathcal{D}$ and remove D_p from $\mathcal{D}$. Then we update $\mathsf{DNN}(\mathcal{D})$ accordingly. We check if there are $n^{1/3}$ disks of $\mathcal{D}$ containing the center p of D_p by applying the algorithm in Lemma 5. If it returns $n^{1/3}$ disks, let L_p be the set consisting of p and the centers of those disks. Since $\mathcal{D}$ is updated, we can apply this procedure again. We do this until $\mathcal{D}$ is empty. As a result of this repetition, we obtain sets L_p's of points of P. Note that the disks induced by L_p contain p, and the number of L_p's is $O(n^{2/3})$.

Next, for each set L_p, we consider six interior-disjoint cones with opening angle $\pi/3$ with apex p. For each cone F, we sort the points of $L_p \cap F$ in the ascending order of their associated radii. Then we claim that the sorted list is a chain, and thus we obtain six chains for each set L_p. Therefore, we have $O(n^{2/3})$ chains in total.

Lemma 6. *The sequence of the points of $L_p \cap F$ sorted in the ascending order of their associated radii is a chain.*

Therefore, we have a set $\mathcal{C}$ of $O(n^{2/3})$ chains of length $O(n^{1/3})$. We call the set of points of P not contained in any of the chains of $\mathcal{C}$ the *remaining set.* Also, we use $\mathcal{R}$ to denote the subgraph of G induced by R, and call it the *remaining graph.*

Lemma 7. *The graph $\mathcal{R}$ is $6n^{1/3}$-thick.*

Proof. We first claim that the remaining set R does not have a $n^{1/3}$-length chain. Assume to the contrary that there is a $n^{1/3}$-length chain C, and let p be the first point in C. At some moment in the course of the algorithm, D_p becomes the smallest disk of $\mathcal{D}$. At this moment, all disks associated with the points in C are contained in $\mathcal{D}$. That is, at least $n^{1/3}$ disks of $\mathcal{D}$ contain p, and thus p must be contained in a chain of $\mathcal{C}$, which contradicts that p is a point of R.

Then we show that R is $6n^{1/3}$-thick. For any point x in the plane, we consider six interior-disjoint cones with opening angle $\pi/3$ with apex x. For a cone F, consider the list L of the points p of $R \cap F$ with $r_p \geq |px|$ sorted in the ascending order of their associated radii. The proof of Lemma 6 implies that L is a chain. By the claim mentioned above, the size of L is less than $n^{1/3}$. Now consider the union of the lists for all of the six cones, which has size less than $6n^{1/3}$. Notice that it is the set of all points $p \in P$ with $r_p \geq |px|$, and thus the lemma holds. □

By Lemma 5, we can compute all L_p's in $O(n^{4/3} \log^4 n)$ time, and for each L_p, we can compute six chains in $O(n^{1/3} \log n)$ time. Since the number of L_p's is $O(n^{2/3})$, the total time for computing all chains of $\mathcal{C}$ is $O(n^{4/3} \log^4 n)$ time.

3.2 Separation Tree of $\mathcal{R}$

In this section, we build a reachability oracle for $\mathcal{R}$, which is similar to the reachability oracle proposed by Kaplan et al. [11, Section 4.2]. In this case, since $\mathcal{R}$ is $O(n^{1/3})$-thick, we can derive a better result. Then Lemma 4 shows that there is a separator of size $O(n^{2/3})$. Recall that R is the vertex set of $\mathcal{R}$.

Data Structure. We construct the separation tree T of R recursively as follows. We compute a separator S_{cross} of $\mathcal{R}$ and two subsets S_{in} and S_{out} separated by S_{cross}. We recursively construct the separation trees of S_{in} and S_{out}. Then we make a new node v, and connect v with the roots of the separation trees of S_{in} and S_{out}. We let G_v denote the subgraph of G induced by R.

For each node v, we store the reachability information as follows: For every point $p \in G_v$, we store two lists of points of S_{cross} which is reachable to p and which is reachable from p within G_v. In particular, we construct a 2-spanner of G_v. Then, for each point $s \in S_{\text{cross}}$, we apply the BFS algorithm in Sect. 2 from s. Also, we reverse the spanner and again apply the BFS algorithm from s.

Query Algorithm. Given two query points $p, q \in R$, we want to check if q is reachable from p in $\mathcal{R}$. To do this, we observe the following. Let v and u be the two nodes of the separation tree T such that the separators of G_v and G_u contain p and q, respectively. They are uniquely defined because each point of R is contained in exactly one separator stored in T. Let L be the path of T from the lowest common ancestor of v and u to the root. Consider a path π from p to q in $\mathcal{R}$, if it exists. By construction, there is a node w in L such that the separator of G_w intersects π. Among them, consider the node closest to the root node. Then G_w contains π. Therefore, it suffices to check if q is reachable from p in G_x for every node x in L.

To use this observation, we first compute v, u and L in $O(\log n)$ time. Then for each node w of L, we check if there is a point x in separator such that p is reachable to x and q is reachable from x in $O(m)$ time, where m denotes the size of the separator of G_w. We return YES if and only if there is such a point x. Since the size of the separators stored in each node is geometrically increasing along L, the total size is dominated by the size of the separator of R, which is $O(n^{2/3})$. Therefore, our query algorithm takes $O(n^{2/3})$ time.

Lemma 8. *We can construct a separation tree T of $\mathcal{R}$ with associated reachability information in $O(n^{5/3})$ time and $O(n^{5/3})$ space. Then, we can query whether there is path from p to q in $\mathcal{R}$ within $O(n^{2/3})$ time.*

3.3 Chain Indices

In this section, we construct a reachability oracle for each chain $C \in \mathcal{C}$: Given any two points p and q in P, we can check if there is a path from p to q intersects C. For each chain $C = \langle p_1, ..., p_t \rangle$, we can construct the oracle in $O(n)$ time once we have a 2-spanner of G. To do this, we need the following lemma. See Fig. 4.

Lemma 9. *For two points p and q in P, let i be the largest index such that p is reachable to p_i, and j be the smallest index such that p_j is reachable to q. Then, there is a p-q path that intersects C if and only if $j \leq i$.*

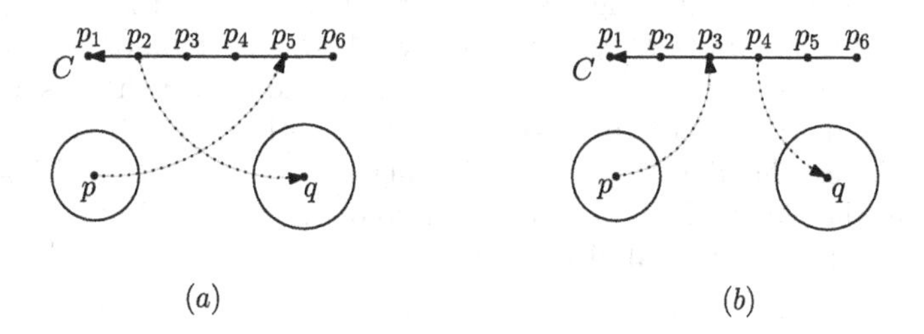

Fig. 4. (a) There is a p-q path via C if $j(q) \leq i(p)$. (b) There is no p-q path that intersects C if $j(q) > i(p)$

For every point $q \in P$, we store the largest index $i(q)$ such that q is reachable to $p_{i(q)}$, and store the smallest index $j(q)$ such that $p_{j(q)}$ is reachable to q.

Lemma 10. *We can compute the indices $i(\cdot)$ and $j(\cdot)$ for every $q \in P$ and every $C \in \mathcal{C}$ in $O(n^{5/3})$ time. Also, the total number of indices we store is $O(n^{5/3})$.*

3.4 Reachability Oracles

Given two points $p, q \in P$, we can check if p is reachable from q as follows. Suppose there is a p-q path π. If there is a chain C that intersects π at a point of C, say p_k. Then $j(q) \leq k \leq i(p)$ for the indices $j(q)$ and $i(p)$ stored in C by Lemma 9. In this case, we can find such a chain C in $O(n^{2/3})$ time by computing indices $j(q)$ and $i(p)$ for all chains of $\mathcal{C}$. Otherwise, no chain of $\mathcal{C}$ intersects π. Then π is contained in $\mathcal{R}$, and thus we can use the reachability oracle for $\mathcal{R}$ described in Sect. 3.2. This takes $O(n^{2/3})$ by Lemma 8.

Theorem 3. *Given a set P of points associated with radii, we can compute a reachability oracle for the transmission graph of P in $O(n^{5/3})$ time. The reachability oracle has size $O(n^{5/3})$ and supports the query time $O(n^{2/3})$.*

4 Continuous Reachability Oracle

In this section, we present a continuous reachability oracle which its complexity is independent of the radius ratio Ψ. In particular, our data structure has size $O(n^{5/3})$ so that for any two point $s \in P$ and $t \in \mathbb{R}^2$, we can check if s is reachable to t in $O(n^{2/3} \log^2 n)$ time. Also, this data structure can be constructed in $O(n^{5/3})$ time. If t is reachable from s, there is a point $p \in P$ reachable from s with $t \in D_p$. In this case, we define a *s-t path* in G as the concatenation of a s-p path in G and the segment connecting p and t.

Consider two query points $s \in P$ and $t \in \mathbb{R}^2$. If there is a s-t path π, we denote the vertex incident to t in π by $p(\pi)$. We construct auxiliary data structures for R and $\mathcal{C}$ to handle the following two cases. We first consider the case that there is a s-t path π with $p(\pi) \in R$. In this case, we choose a set R_0 of $O(1)$ points in R so that there is a s-t path π' with $p(\pi') \in R_0$ if and only if there is a s-t path π with $p(\pi) \in R$. If it is not the case, for any s-t path π, $p(\pi)$ is contained in a chain of $\mathcal{C}$. We can handle this by investigating every chain of $\mathcal{C}$, and finding the first point in the chain whose associated disk contains t. In addition to this, we construct the discrete reachability oracle for G described in Sect. 3.

The Remaining Set *R*, Revisited. We construct the data structure so that we can check if there is a s-t path π with $p(\pi) \in R$. To do this, wee construct the $O(n)$-sized data structure proposed by Afshani and Chan [1] such that for P and a query point t, we can find all points in P whose associated disks contain $t \in \mathbb{R}^2$ in $O(\log n + k)$ time, where k is the number of disks that contain t. Moreover, it can be constructed in $O(n \log n)$ time. Since R is $O(n^{1/3})$-thick, this query time is bounded by $O(n^{1/3})$.

Given two points $p \in P$ and $t \in \mathbb{R}^2$, we compute a set R_t of $O(n^{1/3})$ points of P whose associated disks contain t within $O(n^{1/3})$ time. Then we choose a subset R_0 of R_t of size $O(1)$ such that there is a s-t path π' with $p(\pi') \in R_0$ if and only if there is a s-t path π with $p(\pi) \in R$.

Lemma 11. *Assume that we are given a point $t \in \mathbb{R}^2$ and a set R_t of points of R whose associated disks contain t. We can compute a $O(1)$-sized set $R_0 \subset R_t$ such that $R_0 \cap D_p \neq \emptyset$ for every point $p \in R_t$ in $O(|R_t|)$ time.*

Then we can answer the continuous reachability query using the discrete reachability oracle for all points $q \in R_0$ in $O(n^{2/3})$ time by Theorem 3.

The Set $\mathcal{C}$ of Chains, Revisited. We construct a data structure for each chain $C \in \mathcal{C}$ so that we can compute the first point in C which contains t. To do this, we construct a balanced binary search tree of the indices in $[1, t]$ for $C = \langle p_1, \ldots, p_t \rangle$. For each node u of the binary search tree, we construct the power diagram of the points stored in the subtree rooted at u. Note that this data structure is a variation of the third-level tree proposed in Sect. 2.2. We sort the points along their indices here, while we sort the points along ℓ-projections in Sect. 2.2. Therefore, as we showed in Sect. 2.2, the construction takes $O(m \log m)$

time, and we can compute the first point in C which contains t within $O(\log^2 m)$ time for each chain C, where $m = |C|$. In this way, we can construct the auxiliary data structures for all chains of $\mathcal{C}$ in $O(n \log n)$ time.

Given two points $s \in P$ and $t \in \mathbb{R}^2$, we can check if s is reachable to t as follows. Suppose there is a s-t path π. If $p(\pi)$ is contained in a chain $C \in \mathcal{C}$, let k be the index of $p(\pi)$ in $C = \langle p_1, \ldots, p_t \rangle$, that is, $p_k = p(\pi)$. We let $j(t)$ denote the index of the first point in C which contains t, and let $i(s)$ denote the index of the last point in C which is reachable from s. Recall that $i(s)$ is stored in the discrete reachability oracle, and $j(t)$ can be computed using the auxiliary data structure for C as mentioned above. Then there is a s-t path π with $p(\pi) \in C$ if and only if $j(t) \leq k \leq i(s)$. We do this for all chains in $\mathcal{C}$. Since we can compute the first point that contains t for every chain of $\mathcal{C}$ in $O(n^{2/3} \log^2 n)$ time, the total query time is $O(n^{2/3} \log^2 n)$ time. Therefore, we have the following theorem.

Theorem 4. *Given a set P of points associated with radii, we can compute a continuous reachability oracle for the transmission graph of P in $O(n^{5/3})$ time. The reachability oracle has size $O(n^{5/3})$ and supports the query time $O(n^{2/3} \log^2 n)$.*

References

1. Afshani, P., Chan, T.: Optimal halfspace range reporting in three dimensions. In: Proceedings of the Twentieth Annual ACM-SIAM Symposium on Discrete Algorithms (SODA 2009), pp. 180–186 (2009)
2. Ashur, S., Carmi, P.: t-Spanners for transmission graphs using the path-greedy algorithm. In: 36th European Workshop on Computational Geometry (EuroCG 2020), Book of Abstracts, pp. 60:1–60:6 (2020)
3. Cabello, S., Jejčič, M.: Shortest paths in intersection graphs of unit disks. Comput. Geom. **48**(4), 360–367 (2015)
4. Chan, T.M.: Dynamic geometric data structures via shallow cuttings. In: Proceedings of the 35th International Symposium on Computational Geometry (SoCG 2019), pp. 24:1–24:13 (2019)
5. Chan, T.M., Skrepetos, D.: Approximate shortest paths and distance oracles in weighted unit-disk graphs. J. Comput. Geom. **10**(2), 3–20 (2019)
6. de Berg, M., Cheong, O., van Kreveld, M., Overmars, M.: Computational Geometry: Algorithms and Applications. Springer, Heidelberg (2008). https://doi.org/10.1007/978-3-540-77974-2
7. Fox, J., Pach, J.: Separator theorems and turán-type results for planar intersection graphs. Adv. Math. **219**(3), 1070–1080 (2008)
8. Gudmundsson, J., Levcopoulos, C., Narasimhan, G., Smid, M.: Approximate distance oracles for geometric spanners. ACM Trans. Algorithms **4**(1), 1–34 (2008)
9. Holm, J., Rotenberg, E., Thorup, M.: Planar reachability in linear space and constant time. In: Proceedings of the 56th Annual Symposium on Foundations of Computer Science (FOCS 2015), pp. 370–389 (2015)
10. Imai, H., Iri, M., Murota, K.: Voronoi diagram in the Laguerre geometry and its applications. SIAM J. Comput. **14**(1), 93–105 (1985)
11. Kaplan, H., Mulzer, W., Roditty, L., Seiferth, P.: Spanners and reachability oracles for directed transmission graphs. In; Proceedings of the 31st International Symposium on Computational Geometry (SoCG 2015), pp. 156–170 (2015)

12. Kaplan, H., Mulzer, W., Roditty, L., Seiferth, P.: Routing in unit disk graphs. Algorithmica **80**(3), 830–848 (2018)
13. Kaplan, H., Mulzer, W., Roditty, L., Seiferth, P.: Spanners for directed transmission graphs. SIAM J. Comput. **47**(4), 1585–1609 (2018)
14. Kaplan, H., Mulzer, W., Roditty, L., Seiferth, P., Sharir, M.: Dynamic planar Voronoi diagrams for general distance functions and their algorithmic applications. In: Proceedings of the Twenty-Eighth Annual ACM-SIAM Symposium on Discrete Algorithms (SODA 2017), pp. 2495–2504 (2017)
15. Miller, G., Teng, S.-H., Varasis, S.: A unified geometric approach to graph separators. In: Proceedings of the 32nd Annual Symposium on Foundations of Computer Science (FOCS 1991), pp. 538–547 (1991)
16. Moidu, N., Agarwal, J., Kothapalli, K.: Planar convex hull range query and related problems. In: Proceedings of the 25th Canadian Conference on Computational Geometry (CCCG 2013) (2013)
17. Oh, E.: Shortest-path queries in geometric networks. In: Proceedings of the 31st International Symposium on Algorithms and Computation (ISAAC 2020), pp. 52:1–52:15 (2020)
18. Peleg, D., Roditty, L.: Localized spanner construction for ad hoc networks with variable transmission range. ACM Trans. Sens. Netw. **7**(3), 1–14 (2010)
19. Peleg, D., Roditty, L.: Relaxed spanners for directed disk graphs. Algorithmica **65**(1), 146–158 (2013)
20. Von Rickenbach, P., Wattenhofer, R., Zollinger, A.: Algorithmic models of interference in wireless ad hoc and sensor networks. IEEE/ACM Trans. Network. **17**(1), 172–185 (2009)
21. Wang, H., Xue, J.: Near-optimal algorithms for shortest paths in weighted unit-disk graphs. Discrete Comput. Geom. **64**(4), 1141–1166 (2020)

On Minimum Generalized Manhattan Connections

Antonios Antoniadis[1], Margarita Capretto[2], Parinya Chalermsook[3], Christoph Damerius[4(✉)], Peter Kling[4], Lukas Nölke[5], Nidia Obscura Acosta[3], and Joachim Spoerhase[3]

[1] University of Twente, Enschede, The Netherlands
a.antoniadis@utwente.nl
[2] Universidad Nacional de Rosario, Rosario, Argentina
[3] Aalto University, Espoo, Finland
{parinya.chalermsook,nidia.obscuraacosta,joachim.spoerhase}@aalto.fi
[4] University of Hamburg, Hamburg, Germany
{christoph.damerius,peter.kling}@uni-hamburg.de
[5] University of Bremen, Bremen, Germany
noelke@uni-bremen.de

Abstract. We consider minimum-cardinality Manhattan connected sets with arbitrary demands: Given a collection of points P in the plane, together with a subset of pairs of points in P (which we call *demands*), find a minimum-cardinality superset of P such that every demand pair is connected by a path whose length is the ℓ_1-distance of the pair. This problem is a variant of three well-studied problems that have arisen in computational geometry, data structures, and network design: (i) It is a node-cost variant of the classical Manhattan network problem, (ii) it is an extension of the binary search tree problem to arbitrary demands, and (iii) it is a special case of the directed Steiner forest problem. Since the problem inherits basic structural properties from the context of binary search trees, an $O(\log n)$-approximation is trivial. We show that the problem is NP-hard and present an $O(\sqrt{\log n})$-approximation algorithm. Moreover, we provide an $O(\log \log n)$-approximation algorithm for complete k-partite demands as well as improved results for unit-disk demands and several generalizations. Our results crucially rely on a new lower bound on the optimal cost that could potentially be useful in the context of BSTs.

Keywords: Manhattan networks · Binary search tree · NP-hardness

1 Introduction

Given a collection of points $P \subset \mathbb{R}^2$ on the plane, the *Manhattan Graph* G_P of P is an undirected graph with vertex set $V(G_P) = P$ and arcs $E(G_P)$ that connect any vertically- or horizontally-aligned points. Point p is said to be *Manhattan-connected (M-connected)* to point q if G_P contains a shortest rectilinear path

The full version of this paper [1] can be found at https://arxiv.org/abs/2010.14338.

A. Lubiw et al. (Eds.): WADS 2021, LNCS 12808, pp. 85–100, 2021.
https://doi.org/10.1007/978-3-030-83508-8_7

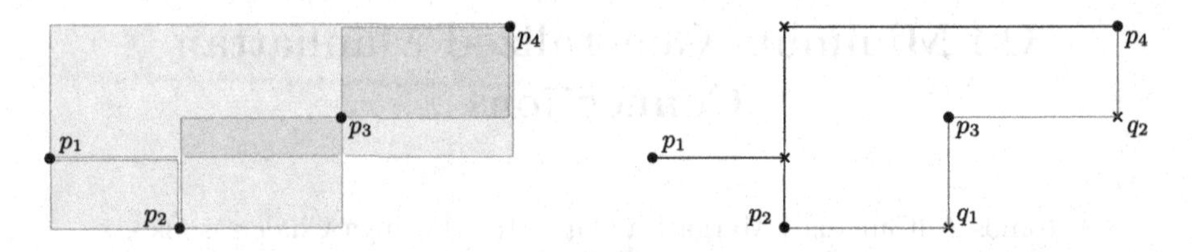

Fig. 1. Left: A Manhattan instance with input points in P drawn as black disks and demands in D drawn as orange rectangles. Right: The Manhattan Graph $G_{P\cup Q}$ of a feasible solution Q, with points in Q drawn as crosses. Points p_2 and p_4 are Manhattan-connected via the red path p_2 - q_1 - p_3 - q_2 - p_4. Points p_1 and p_3 are not Manhattan-connected but also not a demand pair in D. (Color figure online)

from p to q (i.e. a path of length $||p - q||_1$). In this paper, we initiate the study of the following problem: Given points $P \subset \mathbb{R}^2$ and demands $D \subseteq P \times P$, we want to find a smallest set $Q \subset \mathbb{R}^2$ such that every pair of vertices in D is M-connected in $G_{P\cup Q}$. We call this problem MINIMUM GENERALIZED MANHATTAN CONNECTIONS (MINGMCONN), see Fig. 1 for an illustration. Variants of this problem have appeared and received a lot of attention in many areas of theoretical computer science, including data structures, approximation algorithms, and computational geometry. Below, we briefly discuss them, as well as the implications of our results in those contexts.

Binary Search Trees (BSTs). The *Dynamic Optimality Conjecture* [19] is one of the most fundamental open problems in dynamic data structures, postulating the existence of an $O(1)$-competitive binary search tree. Despite continuing efforts and important progress for several decades (see, e.g., [3,7,10,11,18] and references therein), the conjecture has so far remained elusive, with the best known competitive ratio of $O(\log\log n)$ obtained by Tango trees [11]. Even in the offline setting, the best known algorithm is also a $O(\log\log n)$-approximation; the problem is not even known to be NP-hard. Demaine, Harmon, Iacono, Kane, and Pătraşcu [10] showed that approximating BST is equivalent (up to a constant in the approximation factor) to approximating the node-cost Manhattan problem with "evolving demand" (that is, points added to the solution create demands to all existing points).[1]

The long-standing nature of the $O(\log\log n)$ upper bound could suggest the lower bound answer. However, the understanding of lower bound techniques for BSTs has been completely lacking: It is not even known whether the problem is NP-hard! Our work is inspired by the following question. *Is it NP-hard to (exactly) compute a minimum-cost binary search tree?* We are, unfortunately, unable to answer this question. In this paper, we instead present a proof that a natural generalization of the problem from the geometric point of view (which

[1] In fact, the problem stated in [10] is called MINASS which appears different from Manhattan problem, but they can be shown to be equivalent, see the Appendix of the full version [1].

is exactly our MinGMConn) is NP-hard[2]. We believe that our construction and its analysis could be useful in further study of the BST problem from the perspective of lower bounds.

Edge-Cost Manhattan Problem. Closely related to MinGMConn is the edge-cost variant of Manhattan Network [15]: Given $P \subset \mathbb{R}^2$, our goal is to compute $Q \subset \mathbb{R}^2$ such that every pair in P is M-connected in $G_{P \cup Q}$, while minimizing the total lengths of the edges used for the connections. The problem is motivated by various applications in city planning, network layouts, distributed algorithms and VLSI circuit design, and has received attention in the computational geometry community. Since the edge-cost variant is NP-hard [6], the focus has been on approximation algorithms. Several groups of researchers presented 2-approximation algorithms [5,16], and this has remained the best known approximation ratio. Generalizations of the edge-cost variant have been proposed and studied in two directions: In [9], the authors generalize the Manhattan problem to higher dimension $\mathbb{R}^d$ for $d \geqslant 2$. The arbitrary-demand case was suggested in [5]. An $O(\log\ n)$-approximation algorithm was presented in [8], which remains the best known ratio. Our MinGMConn problem can be seen as an analogue of [8] in the node-cost setting. We present an improved approximation ratio of $O(\sqrt{\log\ n})$, therefore, raising the possibility of similar improvements in the edge-cost variants.

Directed Steiner Forests (DSF). MinGMConn is a special case of *node-cost directed Steiner forest (DSF)*: Given a directed graph $G = (V, E)$ and pairs of terminals $\mathcal{D} \subseteq V \times V$, find a minimum cardinality subset $S \subseteq V$ such that $G[S]$ contains a path from s to t for all $(s,t) \in \mathcal{D}$. DSF is known to be highly intractable, with hardness $2^{\log^{1-\epsilon}|V|}$ unless $\mathsf{NP} \subseteq \mathsf{DTIME}(n^{polylog\ n})$ [12]. The best known approximation ratios are slightly sub-linear [4,13]. Manhattan problems can be thought of as natural, tractable special cases of DSF, with approximability between constant and logarithmic regimes. For more details, see [9].

1.1 Our Contributions

In this paper, we present both hardness and algorithmic results for MinGMConn.

Theorem 1. *The* MinGMConn *problem is NP-hard, even if no two points in the input are horizontally or vertically aligned.*

This result can be thought of as a first step towards developing structural understanding of Manhattan connectivity w.r.t. lower bounds. We believe such understanding would come in handy in future study of binary search trees in the geometric view.

Next, we present algorithmic results. Due to the BST structures, an $O(\log\ n)$-approximation is trivial. The main ingredient in obtaining a sub-logarithmic

[2] Demaine et al. [10] prove NP-hardness for MinGMConn with uniform demands but allow the input to contain multiple points on the same row. Their result is incomparable to ours.

approximation is an approximation algorithm for the case of "few" x-coordinates. Formally, we say an input instance is *s-thin* if points in P lie on at most s different x-coordinates.

Theorem 2. *There exists an efficient $O(\log s)$-approximation algorithm for s-thin instances of* MinGMConn.

In fact, our algorithm produces solutions with $O(\log s \cdot \mathsf{IS}(P, D))$ points, where $\mathsf{IS}(P, D) \leqslant \mathsf{OPT}(P, D)$ is the cardinality of a *boundary independent set* (a notion introduced below). This is tight up to a constant factor, as there is an input (P, D) on s different columns such that $\mathsf{OPT}(P, D) = \Omega(\mathsf{IS}(P, D) \log s)$; see the Appendix of [1].

This theorem, along with the boundary independent set analysis, turns out to be an important building block for our approximation result, which achieves an approximation ratio that is sublogarithmic in n.

Theorem 3. *There is an $O(\sqrt{\log n})$-approximation algorithm for* MinGMConn.

This improves over the trivial $O(\log n)$-approximation and may grant some new hope with regards to an improvement over this factor for the edge-cost variants.

We provide improved approximation ratios for several settings when the graph formed by the demands has a special structure. For example, we obtain an $O(\log \log n)$-approximation algorithm for MinGMConn when the demands form a complete k-partite graph, and an $O(1)$-approximation for unit-disk demands; see the Appendix of [1].

1.2 Overview of Techniques

The NP-hardness proof is based on a reduction to 3-SAT. In contrast to the uniform case of MinGMConn (where there is a demand for each two input points), the non-uniform case allows us to encode the structure of a 3-SAT formula in a geometrical manner: we can use demand rectangles to form certain "paths" (see Fig. 2). We exploit this observation in the reduction design by translating clauses and variables into *gadgets*, rectangular areas with specific placement of input points and demands (see Fig. 3). Variable gadgets are placed between clause gadgets and a dedicated *starting point.* The crux is to design the instance such that a natural solution to the intra- and inter-gadget demands connects the starting point to either the positive or the negative part of each variable gadget. And, the M-paths leaving a variable gadget from that part can all reach *only* clauses with a *positive* appearance or *only* clauses with a *negative* appearance of that variable respectively. We refer to such solutions as *boolean solutions*, as they naturally correspond to a variable assignment. Additional demands between the starting point and the clause gadgets are satisfied by a boolean solution if and only if it corresponds to a satisfying variable assignment. The main part of the proof is to show that any small-enough solution is a boolean solution.

In the study of any optimization (in particular, minimization) problem, one of the main difficulties is to come up with a strong lower bound on the cost of an optimal solution that can be leveraged by algorithms. For binary search trees, many such bounds were known, and the strongest known lower bound is called an *independent rectangle bound (IR)*. However, IR is provably too weak for the purpose of MinGMConn, that is, the gap between the optimum and IR can be as large as $\Omega(n)$. We propose to use a new bound, which we call *vertically separable demands (VS)*. This bound turns out to be relatively tight and plays an important role in both our hardness and algorithmic results. In the hardness result, we use our VS bound to argue about the cost of the optimum in the soundness case.

Our $O(\sqrt{\log n})$-approximation follows the high-level idea of [2], which presents a geometric $O(\log \log n)$-approximation for BST. Roughly speaking, it argues (implicitly) that two combinatorial properties, which we refer to as (A) and (B), are sufficient for the existence of an $O(\log \log n)$-approximation: (A) the lower bound function is "subadditive" with respect to a certain instance partitioning, and (B) the instance is "sparse" in the sense that for any input (P, D), there exists an equivalent input (P', D') such that $|P'| = O(\mathsf{OPT}(P, D))$. In the context of BST, (A) holds for the Wilber bound and (B) is almost trivial to show.

In the MinGMConn problem, we prove that Property (A) holds for the new VS bound. However, proving Property (B) seems to be very challenging. We instead show a corollary of Property (B): There is an $O(\log s)$-approximation algorithm for MinGMConn, where s is the number of columns containing at least one input point. The proof of this relaxed property is the main new ingredient of our algorithmic result and is stated in Theorem 2. Finally, we argue that this weaker property still suffices for an $O(\sqrt{\log n})$-approximation algorithm. For completeness, we discuss special cases where we prove that Property (B) holds and thus an $O(\log \log n)$-approximation exists. See the full version [1].

1.3 Outlook and Open Problems

Inspired by the study of structural properties of Manhattan connected sets and potential applications in BSTs, we initiate the study of MinGMConn by proving NP-hardness and giving several algorithmic results.

There are multiple interesting open problems. First, can we show that the BST problem is NP-hard? We hope that our construction and analysis using the new VS bound would be useful for this purpose. Another interesting open problem is to obtain a $o(\log n)$-approximation for the edge-cost variant of the generalized Manhattan network problem.

Finally, it can be shown that our VS bound is sandwiched between OPT and IR. It is an interesting question to study the tightness of the VS bound when estimating the value of an optimal solution. Is VS within a constant factor from the optimal cost of BST? Can we approximate the value of VS efficiently within a constant factor?

2 Model and Preliminaries

Let $P \subset \mathbb{R}^2$ be a set of points on the plane. We say that points $p, q \in P$ are *Manhattan-connected (M-connected)* in P if there is a sequence of points $p = x_0, x_1, \ldots, x_k = q$ such that (i) the points x_i and x_{i+1} are horizontally or vertically aligned for $i = 0, \ldots, k-1$, and (ii) the total length satisfies $\sum_{i=0}^{k-1} ||x_i - x_{i+1}||_1 = ||p - q||_1$.

In the *minimum generalized Manhattan connections* (MINGMCONN) *problem*, we are given a set of *input points* P and their placement in a rectangular grid with integer coordinates such that there are no two points in the same row or in the same column. Additionally, we are given a set $D \subseteq \{(p,q)|p,q \in P\}$ of *demands.* The goal is to find a set of points Q of minimum cardinality such that p and q are M-connected with respect to $P \cup Q$ for all $(p,q) \in D$. Denote by $\mathsf{OPT}(P,D)$ the size of such a point set. We differentiate between the points of P and Q by calling them *input points* and *auxiliary points*, respectively. Since being M-connected is a symmetrical relation, we typically assume $x(p) < x(q)$ for all $(p,q) \in D$. Here, $x(p)$ and $y(p)$ denote the x- and y-coordinate of a point p, respectively. In our analysis, we sometimes use the notations $[n] := \{1, 2, \ldots, n\}$ and $[n]_0 := [n] \cup \{0\}$, where $n \in \mathbb{N}$.

Connection to Binary Search Trees. In the *uniform* case, i.e. $D = \{(p,q)|p,q \in P\}$, this problem is intimately connected to the BINARY SEARCH TREE (BST) problem in the geometric model [10]. Here, we are given a point set P and the goal is to compute a minimum set Q such that *every pair* in $P \cup Q$ is M-connected in $P \cup Q$. Denote by $\mathsf{BST}(P)$ the optimal value of the BST problem.

Independent Rectangles and Vertically Separable Demands. Following Demaine et al. [10], we define the independent rectangle number which is a lower bound on $\mathsf{OPT}(P,D)$. For a demand $(p,q) \in D$, denote by $R(p,q)$ the (unique) axis-aligned closed rectangle that has p and q as two of its corners. We call it the *demand rectangle* corresponding to (p,q). Two rectangles $R(p,q), R(p',q')$ are called *non-conflicting* if none contains a corner of the other in its interior. We say a subset of demands $D' \subseteq D$ is *independent*, if all pairs of rectangles in D' are non-conflicting. Denote by $\mathsf{IR}(P,D)$ the maximum integer k such that there is an independent subset D' of size k. We refer to k as the *independent rectangle number.*

For uniform demands, the problem admits a 2-approximation. Here, the independent rectangle number plays a crucial role. Specifically, it was argued in Harmon's PhD thesis [17] that a natural greedy algorithm costs at most the independent rectangle number and thus yields a 2-approximation. In our generalized demand case, however, the independent rectangle number turns out to be a bad estimate on the value of an optimal solution. Instead, we consider the notion of vertically separable demands, used implicitly in [10].

We say that a subset of demands $D' \subseteq D$ is *vertically separable* if there exists an ordering $R_1, R_2, \ldots, R_k$ of its demand rectangles and vertical line segments $\ell_1, \ell_2 \ldots, \ell_k$ such that ℓ_i connects the respective interiors of top and bottom boundaries of R_i and does not intersect any R_j, for $j > i$. For an input (P, D), denote by $\mathsf{VS}(P, D)$ the maximum cardinality of such a subset. We call a set of demands D *monotone*, if either $y(p) < y(q)$ for all $(p, q) \in D$ or $y(p) > y(q)$ for all $(p, q) \in D$. We assume the former case holds as both are symmetrical. In the following, we argue that VS is indeed a lower bound on OPT (the proof can be found in the full version [1]).

Lemma 4 [10]. *Let (P, D) be an input for* MinGMConn. *If D is monotone, then $\mathsf{IR}(P, D) \leqslant \mathsf{VS}(P, D) \leqslant \mathsf{OPT}(P, D)$. In general $\frac{1}{2}\mathsf{IR}(P, D) \leqslant \mathsf{VS}(P, D) \leqslant 2 \cdot \mathsf{OPT}(P, D)$.*

The charging scheme described Lemma 4's proof injectively maps a demand rectangle R to a point of the optimal solution in R. This implies the following corollary.

Corollary 5. *Let D be a vertically separable, monotone set of demands and Q a feasible solution. If $|Q| = |D|$, there is a bijection $c\colon Q \to D$ such that $q \in R(c(q))$ for all $q \in Q$. In particular, for $Q' \subseteq Q$ there are at least $|Q'|$ demands from D that each covers some $q \in Q'$.*

In general, the independent rectangle number and the maximum size of a vertically separable set are incomparable. By Lemma 4, $\mathsf{IR}(P, D) \leqslant 2 \cdot \mathsf{VS}(P, D)$. However, $\mathsf{IR}(P, D)$ may be smaller than $\mathsf{VS}(P, D)$ up to a factor of n. To see this, consider n diagonally shifted copies of a demand, e.g. $R_i = R\big((i, i), (i+n, i+n)\big)$, for $i = 1, \ldots, n$. Here, $\mathsf{IR}(P, D) = 1$ and $\mathsf{VS}(P, D) = n$. Thus, the concept of vertical separability is more useful as a lower bound.

3 NP-Hardness

In this section, we show Theorem 1 by reducing the 3-SAT problem to MinGMConn. In 3-SAT, we are given a formula ϕ consisting of m clauses $C_1, C_2, \ldots, C_m$ over n variables $X_1, X_2, \ldots, X_n$, each clause consisting of three literals. The goal is to decide whether ϕ is satisfiable. For our reduction, we construct a MinGMConn instance (P_ϕ, D_ϕ) and a positive integer $\alpha = \alpha(\phi)$ such that (P_ϕ, D_ϕ) has an optimal solution of size α if and only if ϕ is satisfiable (see the full version [1]). This immediately implies Theorem 1. In the following, we identify a demand $d \in D_\phi$ with its demand rectangle $R(d)$. This allows us to speak, for example, of intersections of demands, corners of demands, or points covered by demands.

Our construction of the MinGMConn instance (P_ϕ, D_ϕ) is based on different *gadgets* and their connections among each other. A gadget can be thought of as a rectangle in the Euclidean plane that contains a specific set of input points and demands between these. In the following, we give a coarse overview of our construction, describing how gadgets are placed and how they interact (Fig. 2). Moreover, we try to convey the majority of the intuition behind our reduction. Because of space constraints the actual proof of the NP-hardness is given in the full version [1].

Overview of the Construction. For each clause C_j, we create a *clause gadget* GC_j and for each variable X_i, a *variable gadget* GX_i. Clause gadgets are arranged along a descending diagonal line, so all of GC_j is to the bottom-right of GC_{j-1}. Variable gadgets are arranged in the same manner. This avoids unwanted interference among different clause and variable gadgets, respectively. The variable gadgets are placed to the bottom-left of all clause gadgets.

For each positive occurrence of a variable X_i in a clause C_j, we place a dedicated *connection point* $p_{ij}^+ \in P_\phi$ as well as suitable *connection demands* from p_{ij}^+ to a dedicated inner point of GX_i and to a dedicated inner point of GC_j. Their purpose is to force optimal MinGMConn solutions to create specific M-paths (going first up and then right in a narrow corridor) connecting a variable to the clauses in which it appears positively. We call the area covered by these two demands a *(positive) variable-clause path.* Similarly, there are connection points $p_{ij}^- \in P_\phi$ with suitable demands for negative appearances of X_i in C_j, creating a *(negative) variable-clause path* (going first right and then up in a narrow corridor).

Finally, there is a *starting point* $S \in P_\phi$ to the bottom-left of all other points. It has a demand to a *clause point* c_j in the top-right of each clause gadget GC_j (an *SC demand*) and to a *variable point* x_i in the bottom-left of each variable gadget GX_i (an *SX demand*). The inside of clause gadgets simply provides different entrance points for the variable-clause paths, while the inside of variable gadgets forces an optimal solution to choose between using either only positive or only negative variable-clause paths. We will use these choices inside variable gadgets to identify an optimal solution for (P_ϕ, D_ϕ) with a variable assignment for ϕ.

The *clause gadget* GC_j for clause C_j contains the *clause point* c_j and three *(clause) literal points* $\ell_{j1}, \ell_{j2}, \ell_{j3}$. The clause point is in the top-right. The literal points represent the literals of C_j and form a descending diagonal within the gadget such that positive are above negative literals. For each literal point ℓ_{jk}, there is a demand (ℓ_{jk}, c_j). Moreover, if ℓ_{jk} is positive and corresponds to the variable X_i, then there is a *(positive) connection demand* (p_{ij}^+, ℓ_{jk}). Similarly, if ℓ_{jk} is negative, there is a *(negative) connection demand* (p_{ij}^-, ℓ_{jk}). Finally, there is the *SC demand* (S, c_j).

The *variable gadget* GX_i for variable X_i contains the *variable point* x_i, two *(variable) literal points* x_i^+, x_i^-, one *demand point* d_i, as well as n_i^+ positive and n_i^- negative *literal connectors* x_{ik}^+ and x_{ik}^-, respectively. Here, n_i^+ and n_i^- are from $[m]_0$ and denote the number of positive and negative occurrences of X_i in ϕ, respectively. The variable point is in the bottom-left. The literal connectors and the demand point form a descending diagonal in the top-right, with the positive literal connectors above and the negative literal connectors below the demand point. The literal points x_i^+, x_i^- lie in the interior of the rectangle spanned by x_i and d_i, close to the top-left and bottom-right corner respectively. They are moved slightly inward to avoid identical x- or y-coordinates. Inside the gadgets, we have demands of the form (x_i^+, x_{ik}^+) and (x_i^-, x_{ik}^-) between literal points and literal connectors, (x_i^+, d_i) and (x_i^-, d_i) between literal points and the demand point, as well as (x_i, d_i) between the variable point and the demand point (an *XD demand*). Towards the outside, we have the *positive/negative connection demands*between literal points and literal connectors (x_{ik}^+, p_{ij}^+) if the k-th positive literal of X_i occurs in C_j and (x_{ik}^-, p_{ij}^-) if the k-th negative literal of X_i occurs in C_j as well as the *SX demand* (S, x_i).

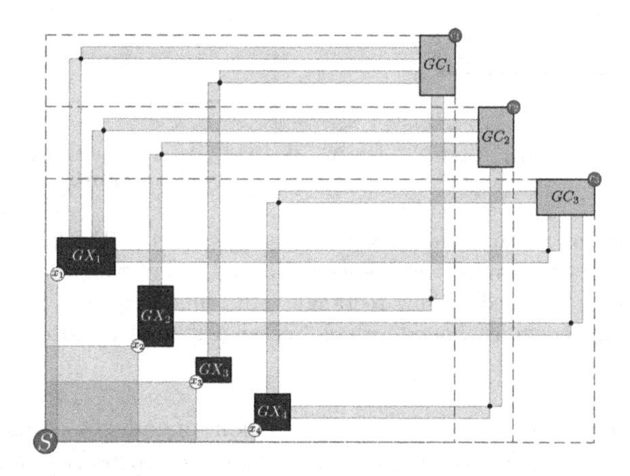

Fig. 2. MINGMCONN instance (P_ϕ, D_ϕ) for $\phi = (X_1 \vee \neg X_2 \vee X_3) \wedge (X_1 \vee X_2 \vee \neg X_4) \wedge (\neg X_1 \vee \neg X_2 \vee X_4)$. Input points are shown as (red, yellow, or black) disks. For clause and variable gadgets, we show only the clause points c_j and the variable points x_i; their remaining inner points and demands are illustrated in Fig. 3. The small black disks represent the connection points p_{ij}^+, p_{ij}^-. Non-SC demands are shown as shaded, orange rectangles, while SC demands are shown as dashed, red rectangles. (Color figure online)

Intuition of the Reduction. Our construction is such that non-SC demands (including those within gadgets) form a monotone, vertically separable demand set. Thus, for

$$D_{\overline{SC}} := \{d \in D_\phi | d \text{ is not an } SC \text{ demand}\} \quad \text{and} \quad \alpha = \alpha(\phi) := |D_{\overline{SC}}|, \qquad (1)$$

Lemma 4 implies that any solution Q_ϕ for (P_ϕ, D_ϕ) has size at least α.

The first part of the reduction shows that if ϕ is satisfiable, then there is an (optimal) solution Q_ϕ of size α. This is proven by constructing a family of *boolean solutions*. These are (partial) solutions Q_ϕ that can be identified with a variable assignment for ϕ and that have the following properties: Q_ϕ has size α and satisfies all non-SC demands. Additionally, it can satisfy an SC demand (S, c_j) only by going through some variable x_i, where such a path exists if and only if C_j is satisfied by the value assigned to X_i by (the variable assignment) Q_ϕ. In particular, if ϕ is satisfiable, there is a boolean solution Q_ϕ satisfying all SC demands. This implies that Q_ϕ is a solution to (P_ϕ, D_ϕ) of (optimal) size α.

We then provide the other direction of the reduction, stating that if there is a solution Q_ϕ for (P_ϕ, D_ϕ) of size α, then ϕ is satisfiable. Its proof is more

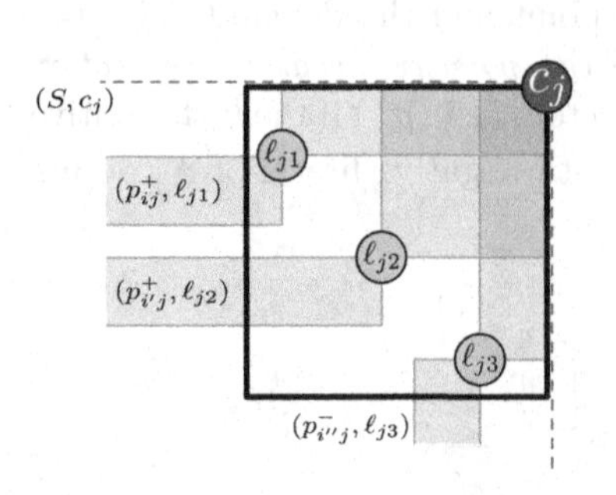

(a) Clause gadget GC_j for the clause $C_j = X_i \vee X_{i'} \vee \neg X_{i''}$.

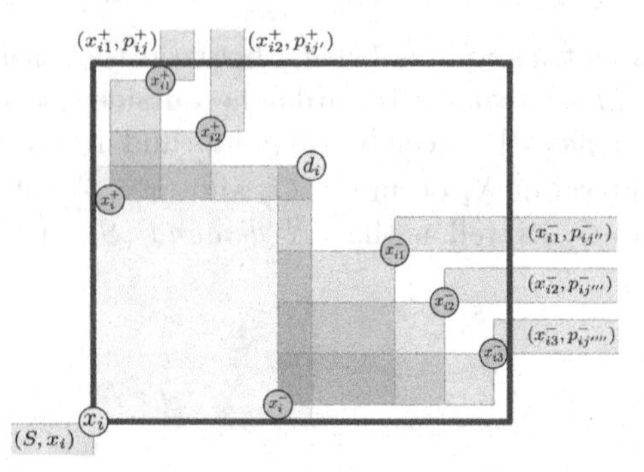

(b) Variable gadget GX_i for X_i occurring twice positively and thrice negatively.

Fig. 3. Examples for a clause and a variable gadgets. As in Fig. 2, input points are shown as circles and SC demands are shown as dashed, red rectangles. The XD demand (x_i, d_i) is shown as a shaded, yellow rectangle. All remaining (non-SC and non-XD) demands are again shown as shaded, orange rectangles. (Color figure online)

involved and is made possible by careful placement of gadgets, connection points, and demands. (See the full version [1] for the complete proof.) In a first step, we show that the small size of Q_ϕ implies that different parts of our construction each must be satisfied by only a few, dedicated points from Q_ϕ. For example, Q_ϕ has to use exactly n points to satisfy the n SX demands (S, x_i). Another result about "triangular" instances (e.g., the triangular grid formed by the n SX demands, see Fig. 2) states that, here, optimal solutions must lie on grid lines inside the "triangle". See the full version [1]. We conclude that any M-path from S to a clause point c_j must go through exactly one variable point x_i. Similarly, we show that the $6m$ connection demands (forming the $3m$ variable-clause paths) are satisfied by $6m$ points from Q_ϕ and, since they are so few, each of these points lies in the corner of a connection demand. This ensures that M-paths cannot cheat by, e.g., "jumping" between different variable-clause paths. More precisely, such a path can be entered only at the variable gadget where it starts and be left only at the clause gadget where it ends.

All that remains to show is that there cannot be two M-paths entering a variable gadget GX_i (which they must do via x_i) such that one leaves through a positive and the other through a negative variable-clause path. We can then interpret Q_ϕ as a boolean solution (the variable assignment for X_i being determined by whether M-paths leave GX_i through positive or through negative variable-clause paths). Since Q_ϕ satisfies all demands, in particular all SC demands, the corresponding variable assignment satisfies all clauses.

4 An Approximation Algorithm for s-Thin Instances

In this section, we present an approximation algorithm for s-thin instances (points in P lie on at most s distinct x-coordinates). In particular, we allow more than one point to share the same x-coordinate. However, we still require any two points to have distinct y-coordinates. We show an approximation ratio of $O(\log s)$, proving Theorem 2.

An *x-group* is a maximal subset of P having the same x-coordinate. Note that an $O(\log s)$-approximation for s-thin instances can be obtained via a natural "vertical" divide-and-conquer algorithm that recursively divides the s many x-groups in two subinstances with roughly $s/2$ many x-groups each. (Section 5 considered a more general version of this, subdividing into an arbitrary number of subinstances.) The analysis of this algorithm uses the number of input points as a lower bound on OPT. However, such a bound is not sufficient for our purpose of deriving an $O(\sqrt{\log n})$-approximation.

In this section, we present a different algorithm, based on "horizontal" divide-and-conquer (after a pre-processing step to sparsify the set of y-coordinates in the input via minimum hitting sets). Using horizontal rather than vertical divide-and-conquer may seem counter-intuitive at first glance as the number of y-coordinates in the input is generally unbounded in s. Interestingly enough, we can give a *stronger* guarantee for this algorithm by bounding the cost of the approximate solution against what we call a *boundary independent set.* Additionally, we show that the size of a such set is always upper bounded by the maximum number of vertically separable demands. This directly implies Theorem 2, since $2\,\mathsf{OPT}$ is an upper bound on the number of vertically separable demands (c.f. Lemma 4). Even more importantly, our stronger bound allows us to prove Theorem 3 in the next section since vertically separable demands fulfill the subadditivity property mentioned in the introduction. In the proof of Theorem 3, an arbitrary $O(\log s)$-approximation algorithm would not suffice.

By losing a factor 2 in the approximation ratio, we may assume that the demands are monotone (we can handle pairs with $x(p) < x(q)$ and $y(p) > y(q)$ symmetrically).

Definition 6 (Left & right demand segments). *Let (P, D) be an input instance. For each $R(p, q) \in Q$, denote by $\lambda(p, q)$ the vertical segment that connects $(x(p), y(p))$ and $(x(p), y(q))$. Similarly, denote by $\rho(p, q)$ the vertical segment that connects $(x(q), y(p))$ and $(x(q), y(q))$. That is, $\lambda(p, q)$ and $\rho(p, q)$ are simply the left and right boundaries of rectangle $R(p, q)$.*

Boundary Independent Sets. A *left boundary independent set* consists of pairwise non-overlapping segments $\lambda(p, q)$, a *right boundary independent set* of pairwise non-overlapping segments $\rho(p, q)$. A *boundary independent set* refers to either a left or a right boundary independent set. Denote by $\mathsf{IS}(P, D)$ the size of a maximum boundary independent set.

The following lemma implies that it suffices to work with boundary independent sets instead of vertical separability. The main advantage of doing so, is that (i) for IS, we do not have to identify any ordering of the demand subset, (ii) one can compute $\mathsf{IS}(P, D)$ efficiently, and (iii) we can exploit geometric properties of interval graphs, as we will do below.

Algorithm 1: HORIZONTALDC($P, D, \mathcal{R}$)

input : Instance (P, D) with rows $\mathcal{R}$
output : Feasible solution to (P, D) computed via horizontal divide-and-conquer

1 $Q \leftarrow \emptyset$; $m \leftarrow$ median of $\mathcal{R}$;
2 $D_m \leftarrow \{ (p,q) \in D \mid y(p) \leqslant m \leqslant y(q) \}$;
3 **foreach** $(p,q) \in D_m$ **do**
4 $\quad Q \leftarrow Q \cup \{(x(p), m), (x(q), m)\}$;
5 $D_t \leftarrow \{ (p,q) \in D \mid y(p) > m \}$; $\quad D_b \leftarrow \{ (p,q) \in D \mid y(q) < m \}$;
6 $P_t \leftarrow \{ p,q \mid (p,q) \in D_t \}$; $\quad P_b \leftarrow \{ p,q \mid (p,q) \in D_b \}$;
7 $\mathcal{R}_t \leftarrow \{ r \in \mathcal{R} \mid r > m \}$; $\quad \mathcal{R}_b \leftarrow \{ r \in \mathcal{R} \mid r < m \}$;
8 $Q \leftarrow Q \cup \text{HorizontalDC}(P_t, D_t, \mathcal{R}_t) \cup \text{HorizontalDC}(P_b, D_b, \mathcal{R}_b)$;
9 **return** Q;

Lemma 7. *For any instance (P, D) we have that $\mathsf{IS}(P, D) \leqslant \mathsf{VS}(P, D)$. Moreover, one can compute a maximum boundary independent set in polynomial time.*

Algorithm Description. Our algorithm, which we call HORIZONTALMANHATTAN, produces a Manhattan solution of cost $O(\log s) \cdot \mathsf{IS}(P, D)$, where s is the number of x-groups in P. The algorithm initially computes a set of "crucial rows" $\mathcal{R} \subseteq \mathbb{R}$ by computing a minimum hitting set in the interval set $\mathcal{I} = \{[y(p), y(q)] \mid (p,q) \in D\}$. In particular, the set $\mathcal{R}$ has the following property. For each $j \in \mathcal{R}$, let ℓ_j be a horizontal line drawn at y-coordinate j. Then the lines $\{\ell_j\}_{j \in \mathcal{R}}$ stab every rectangle in $\{R(p,q)\}_{(p,q) \in D}$. The following observation follows from the fact that the interval hitting set is equal to the maximum interval independent set.

Observation. $|\mathcal{R}| \leqslant \mathsf{IS}(P, D)$.

After computing $\mathcal{R}$, the algorithm calls a subroutine HORIZONTALDC (see Algorithm 1), which recursively adds points to each such row in a way that guarantees a feasible solution. We now proceed to the analysis of the algorithm.

Lemma 8 (Feasibility). *The algorithm* HORIZONTALMANHATTAN *produces a feasible solution in polynomial time.*

Lemma 9 (Cost). *For any s-thin instance (P, D), algorithm* HORIZONTALMANHATTAN *outputs a solution of cost $O(\log s) \cdot \mathsf{IS}(P, D)$.*

Proof. Let $r = |\mathcal{R}|$ be the number of rows computed in HORIZONTALMANHATTAN. Define $L = \{\lambda(p,q) \mid [y(p), y(q)] \in I\}$ to be the set of corresponding left sides of the demands in I. In particular, the segments in L are disjoint and $|L| = r$. We upper bound the cost of our solution as follows. For each added point, define a *witness interval*, witnessing its cost. The total number of points is then roughly bounded by the number of witness intervals, which we show to be $O(\log s)\mathsf{IS}(P, D)$.

We enumerate the recursion levels of Algorithm 1 from 1 to $\lceil \log r \rceil$ in a top-down fashion in the recursion tree. In each recursive call, at most s many points are added to Q in line 4—one for each distinct x-coordinate. Hence, during the first $\lceil \log r \rceil - \lceil \log s \rceil$ recursion levels at most $s \cdot 2^{\lceil \log r \rceil - \lceil \log s \rceil} = O(s \cdot \frac{r}{s}) = O(r)$ many points are added to Q in total. We associate each of these points with one unique left side in L in an arbitrary manner. For each of these points, we call its associated left side the witness of this point.

For any point added to Q in line 4 in one of the last $\lceil \log s \rceil$ recursion levels, pick the first (left or right) side of a demand rectangle that led to including this point. More precisely, if, in line 4, we add point $(x(p), m)$ to Q for the first time (which means that this point has not yet been added to Q via a different demand) then associate $\lambda(p, q)$ as a witness. Analogously, if we add $(x(q), m)$ for the first time then associate $\rho(p, q)$ as a witness.

Overall, we have associated to each point in the final solution a uniquely determined witness, which is a left or a right side of some demand rectangle. Note that any (left or right) side of a rectangle may be assigned as a witness to two solution points (once in the top recursion levels and once in the bottom levels). In such a case we create a duplicate of the respective side and consider them to be distinct witnesses.

Two witnesses added in the last $\lceil \log s \rceil$ recursion levels can intersect only if the recursive calls lie on the same root-to-leaf path in the recursion tree. Otherwise, they are separated by the median row of the lowest common ancestor in the recursion tree and cannot intersect. With this observation and the fact that the witnesses in L form an independent set, we can bound the maximum clique size in the intersection graph of all witnesses by $1 + \lceil \log s \rceil$.

This graph is an interval graph. Since interval graphs are perfect [14], there exists a $(1 + \lceil \log s \rceil)$-coloring in this graph. Hence, there exists an independent set of witnesses of size $1/(\lceil \log s \rceil + 1)$ times the size of the Manhattan solution. Taking all left or all right sides of demands in this independent set (whichever is larger) gives a *boundary* independent set of size at least $1/(2(1 + \lceil \log s \rceil))$ times the cost of the Manhattan solution. □

We conclude the section by noting that the proof of Theorem 2 directly follows by combining Lemmata 4, 7 and 9. As mentioned in the introduction, the factor $O(\log s)$ in Lemma 9 is tight in the strong sense that there is a MinGMConn instance (P, D) with s distinct x-coordinates such that $\mathsf{OPT}(P, D) = \Omega(\mathsf{IS}(P, D) \log s)$. See the appendix of the full version [1].

5 A Sublogarithmic Approximation Algorithm

In this section, we give an overview of how to leverage the $O(\log s)$-approximation for s-thin instances to design an $O(\sqrt{\log n})$-approximation algorithm for general instances.

Sub-instances. Let (P, D) be an instance of MinGMConn and let B be a bounding box for P, that is, $P \subseteq B$. Let $\mathcal{S} = \{S_1, \dots, S_s\}$ be a collection of

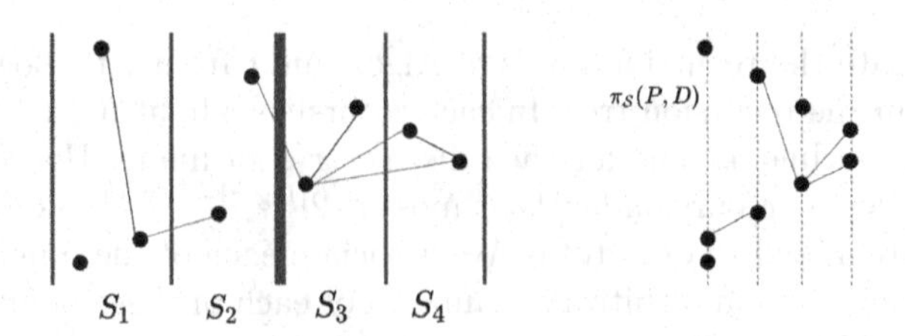

Fig. 4. An illustration of the inter-strip instance. Each strip S_i is collapsed into one column. The red demands are demands between pairs of points lying inside different strips. The black demands are demands that are handled by intra-strip instances. (Color figure online)

s vertical strips, ordered from left to right, that are obtained by drawing $s-1$ vertical lines that partition B. We naturally create $s+1$ sub-instances as follows. (See also Fig. 4.) First, we have s *intra-strip instances* $\{(P_i, D_i)\}_{i\in[s]}$ such that $P_i = P \cap S_i$ and $D_i = D \cap (S_i \times S_i)$. Next, we have the *inter-strip instance* $\pi_{\mathcal{S}}(D) = (P', D')$ where P' is obtained by collapsing each strip in $\mathcal{S}$ into a single column and D' is obtained from collapsing demands accordingly. For each point $p \in P$, denote by $\pi_{\mathcal{S}}(p)$ a copy of p in P' after collapsing. Note that this is a simplified description of the instances that avoids some technicalities. For a precise definition, see the appendix in the full version [1].

Sub-additivity of VS. The following is our sub-additivity property that we use crucially in our divide-and-conquer algorithm.

Lemma 10. *If* (P, D) *is an instance of* MinGMConn *with strip subdivision* $\mathcal{S}$*, then*

$$\mathsf{VS}(P, D) \geqslant \mathsf{VS}(\pi_{\mathcal{S}}(P, D)) + \sum_{S\in\mathcal{S}} \mathsf{VS}(P \cap S, D \cap (S \times S)).$$

Divide-and-Conquer. Choose the strips $\mathcal{S}$ so that $s = |\mathcal{S}| = 2^{\sqrt{\log n}}$. Thus the inter-strip instance admits an approximation of ratio $O(\log s) = O(\sqrt{\log n})$; in fact, we obtain a solution of cost $O(\sqrt{\log n})\mathsf{VS}(\pi_{\mathcal{S}}(P, D))$. We recursively solve each intra-strip instance (P_i, D_i), and combine the solutions from these $s+1$ sub-instances. (Details on how the solution can be combined are deferred to [1].)

We show by induction on the number of points that for any instance (P, D) the cost of the computed solution is $O(\sqrt{\log n})\mathsf{VS}(P, D)$. (Here, we do not take into account the cost incurred by combining the solutions to the sub-instances.) By induction hypothesis, we have for each (P_i, D_i) a solution of cost $O(\sqrt{\log n})\mathsf{VS}(P_i, D_i)$ since $|P_i| < |P|$. Note that we cannot use the induction hypothesis for the inter-strip instance since $|P'| = |P|$, which is why we need the $O(\log s)$-approximation algorithm. Using sub-additivity we obtain:

$$O(\sqrt{\log n})\Big(\mathsf{VS}(\pi_{\mathcal{S}}(P, D)) + \sum_{S\in\mathcal{S}} \mathsf{VS}(P \cap S, D \cap (S \times S))\Big) = O(\sqrt{\log n})\mathsf{VS}(P, D).$$

There is an additional cost incurred by combining the solutions of the sub-instances to a feasible solution of the current instance. In the full version [1], we argue that this can be done at a cost of $O(\mathsf{OPT})$ for each of the $\log n/\log s = \sqrt{\log n}$ many levels of the recursion. (This prevents us from further improving the approximation factor by picking $s = 2^{o(\sqrt{\log n})}$.)

References

1. Antoniadis, A., et al.: On minimum generalized Manhattan connections. CoRR abs/2010.14338 (2020). https://arxiv.org/abs/2010.14338
2. Chalermsook, P., Chuzhoy, J., Saranurak, T.: Pinning down the strong Wilber 1 bound for binary search trees. arXiv preprint arXiv:1912.02900 (2019)
3. Chalermsook, P., Goswami, M., Kozma, L., Mehlhorn, K., Saranurak, T.: Pattern-avoiding access in binary search trees. In: 2015 IEEE 56th Annual Symposium on Foundations of Computer Science, pp. 410–423. IEEE (2015)
4. Chekuri, C., Even, G., Gupta, A., Segev, D.: Set connectivity problems in undirected graphs and the directed Steiner network problem. ACM Trans. Algorithms (TALG) **7**(2), 1–17 (2011)
5. Chepoi, V., Nouioua, K., Vaxes, Y.: A rounding algorithm for approximating minimum Manhattan networks. Theoret. Comput. Sci. **390**(1), 56–69 (2008)
6. Chin, F.Y., Guo, Z., Sun, H.: Minimum Manhattan network is NP-complete. Discrete Comput. Geom. **45**(4), 701–722 (2011)
7. Cole, R.: On the dynamic finger conjecture for splay trees. part II: the proof. SIAM J. Comput. **30**(1), 44–85 (2000)
8. Das, A., Fleszar, K., Kobourov, S., Spoerhase, J., Veeramoni, S., Wolff, A.: Approximating the generalized minimum Manhattan network problem. Algorithmica **80**(4), 1170–1190 (2018)
9. Das, A., Gansner, E.R., Kaufmann, M., Kobourov, S., Spoerhase, J., Wolff, A.: Approximating minimum Manhattan networks in higher dimensions. Algorithmica **71**(1), 36–52 (2015)
10. Demaine, E.D., Harmon, D., Iacono, J., Kane, D., Pătraşcu, M.: The geometry of binary search trees. In: Proceedings of the Twentieth Annual ACM-SIAM Symposium on Discrete Algorithms, pp. 496–505. SIAM (2009)
11. Demaine, E.D., Harmon, D., Iacono, J., Pătraşcu, M.: Dynamic optimality-almost. SIAM J. Comput. **37**(1), 240–251 (2007)
12. Dodis, Y., Khanna, S.: Design networks with bounded pairwise distance. In: Proceedings of the Thirty-first Annual ACM Symposium on Theory of Computing, pp. 750–759 (1999)
13. Feldman, M., Kortsarz, G., Nutov, Z.: Improved approximation algorithms for directed Steiner forest. J. Comput. Syst. Sci. **78**(1), 279–292 (2012)
14. Golumbic, M.C.: Algorithmic Graph Theory and Perfect Graphs. Annals of Discrete Mathematics, vol. 57. North-Holland Publishing Co., NLD (2004)
15. Gudmundsson, J., Levcopoulos, C., Narasimhan, G.: Approximating a minimum Manhattan network. Nordic J. Comput. **8**(2), 219–232 (2001). http://dl.acm.org/citation.cfm?id=766533.766536
16. Guo, Z., Sun, H., Zhu, H.: A fast 2-approximation algorithm for the minimum Manhattan network problem. In: Fleischer, R., Xu, J. (eds.) AAIM 2008. LNCS, vol. 5034, pp. 212–223. Springer, Heidelberg (2008). https://doi.org/10.1007/978-3-540-68880-8_21

17. Harmon, D.D.K.: New bounds on optimal binary search trees. Ph.D. thesis, Massachusetts Institute of Technology (2006)
18. Iacono, J., Langerman, S.: Weighted dynamic finger in binary search trees. In: Proceedings of the Twenty-Seventh Annual ACM-SIAM Symposium on Discrete Algorithms, pp. 672–691. SIAM (2016)
19. Sleator, D.D., Tarjan, R.E.: Self-adjusting binary search trees. J. ACM (JACM) **32**(3), 652–686 (1985)

HalftimeHash: Modern Hashing Without 64-Bit Multipliers or Finite Fields

Jim Apple(✉)

Los Gatos, USA

Abstract. HalftimeHash is a new algorithm for hashing long strings. The goals are few collisions (different inputs that produce identical output hash values) and high performance.

Compared to the fastest universal hash functions on long strings (clhash and UMASH), HalftimeHash decreases collision probability while also increasing performance by over 50%, exceeding 16 bytes per cycle.

In addition, HalftimeHash does not use any widening 64-bit multiplications or any finite field arithmetic that could limit its portability.

Keywords: Universal hashing · Randomized algorithms

1 Introduction

A hash family is a map from a set of seeds S and a domain D to a codomain C. A hash family H is called is ε-almost universal ("ε-AU" or just "AU") when

$$\forall x, y \in D, x \neq y \implies \Pr_{s \in S}[H(s,x) = H(s,y)] \leq \varepsilon \in o(1)$$

The intuition behind this definition is that collisions can be made unlikely by picking randomly from a hash *family* independent of the input strings, rather than anchoring on a specific hash *function* such as MD5 that does not take a seed as an input. AU hash families are useful in hash tables, where collisions slow down operations and, in extreme cases, can turn linear algorithms into quadratic ones [2,10,18,25].

HalftimeHash is a new "universe collapsing" hash family, designed to hash long strings into short ones [1,11,21]. This differs from short-input families like SipHash or tabulation hashing, which are suitable for hashing short strings to a codomain of 64 bits [3,25]. Universe collapsing families are especially useful for composition with short-input families: when n long strings are to be handled by a hash-based algorithm, a universe-collapsing family that reduces them to hash values of length $c \lg n$ bits for some suitable $c > 2$ produces zero collisions with probability $1 - O(n^{2-c})$. A short-input hash family can then treat the hashed values as if they were the original input values [3,9,24,25]. This technique applies not only to hash tables, but also to message-authentication codes, load balancing in distributed systems, privacy amplification, randomized geometric algorithms, Bloom filters, and randomness extractors [4,9,12,13,22,23].

A. Lubiw et al. (Eds.): WADS 2021, LNCS 12808, pp. 101–114, 2021.
https://doi.org/10.1007/978-3-030-83508-8_8

On strings longer than 1 KB, HalftimeHash is typically 55% faster than clhash, the AU hash family that comes closest in performance.

HalftimeHash also has tunable output length and low probabilities of collision for applications that require them, such as one-time authentication [5]. The codomain has size 16, 24, 32, or 40 bytes, and ε varies depending on the codomain (see Fig. 2 and Sect. 6).

1.1 Portability

In addition to high speed on long strings, HalftimeHash is designed for a simple implementation that is easily portable between programming languages and machine ISA's. HalftimeHash uses less than 1200 lines of code in C++ and can take advantage of vector ISA extensions, including AVX-512, AVX2, SSE, and NEON.[1]

Additionally, no multiplications from $\mathbb{Z}_{2^{64}} \times \mathbb{Z}_{2^{64}}$ to $\mathbb{Z}_{2^{128}}$ are needed. This is in support of two portability goals – the first is portability to platforms or programming languages without native widening unsigned 64-bit multiplications. Languages like Java, Python, and Swift can do these long multiplications, but not without calling out to C or slipping into arbitrary-precision-integer code. The other reason HalftimeHash avoids 64-bit multiplications is portability to SIMD ISA extensions, which generally do not contain widening 64-bit multiplication.

1.2 Prior Almost-Universal Families

There are a number of fast hash algorithms that run at rates exceeding 8 bytes per cycle on modern x86-64 processors, including Fast Positive Hash, falkhash, xxh, MeowHash, and UMASH, and clhash [26]. Of these, only clhash and UMASH include claims of being AU; each of these uses finite fields and the x86-64 instruction for carryless (polynomial) multiplication.

Rather than tree hashing, hash families like clhash and UMASH use polynomial hashing (based on Horner's method) to hash variable-length strings down to fixed-size output. That approach requires 64-bit multiplication and also reduction modulo a prime (in $\mathbb{Z}$ or in $\mathbb{Z}_2[x]$), limiting its usability in SIMD ISA extensions.

1.3 Outline

The rest of this paper is organized as follows: Sect. 3 covers prior work that HalftimeHash builds upon. Section 4 introduces a new generalization of Nandi's "Encode, Hash, Combine" algorithm [20]. Section 5 discusses specific implementation choices in HalftimeHash to increase performance. Section 6 analyzes and tests HalftimeHash's performance.

[1] https://github.com/jbapple/HalftimeHash.

2 Notations and Conventions

Input string length n is measured in 32-bit *words*. "32-bit multiplication" means multiplying two unsigned 32-bit words and producing a single 64-bit word. "64-bit multiplication" similarly refers to the operation producing a 128-bit product. All machine integers are unsigned.

Sequences are denoted by angled brackets: "$\langle$", "$\rangle$", and $\lhd$ prepends a value onto a sequence. Subscripts indicate a numbered component of a sequence, starting at 0. Contiguous half-open subsequences are denoted "$x[y, z)$", meaning $\langle x_y, x_{y+1}, \ldots, x_{z-1}\rangle$.

$\bot$ is a new symbol not otherwise in the alphabet of words.

In the definition of ε-almost universal, ε is called the *collision probability* of H; it is inversely related to H's *output entropy*, $-\lg \varepsilon$. The seed is sometimes referred as *input entropy*, which is distinguished from the output entropy both because it is an explicit part of the input and because it is measured in words or bytes, not bits.

Each step of HalftimeHash applies various transforms to groups of input values. These groups are called *instances*. The processing of a transform on a single instance is called an *execution*.

Instances are logically contiguous but physically strided, for the purpose of simplifying SIMD processing. A physically contiguous set between two items in a single instance is called a *block*; the number of words in a block is called the *block size*. Because instances are logically contiguous, when possible, the analysis will elide references to the block size.

Tree hashing examples use a hash family parameter H that takes two words as input, but this can be easily extended to hash functions taking more than two words of input, in much the same way that binary trees are a special case of B-trees.

HalftimeHash produces output that is collision resistant among strings of the same length. Adding collision resistance between strings of *different* lengths to such a hash family requires only appending the length at the end of the output.

HalftimeHash variants will be specified by their number of output bytes: HalftimeHash16, HalftimeHash24, HalftimeHash32, or HalftimeHash40.

Except where otherwise mentioned, all benchmarks were run on an Intel i7-7800x (a Skylake X chip that supports AVX512), running Ubuntu 18.04, with Clang++ 11.0.1.

3 Prior Work

This section reviews hashing constructions that form components of HalftimeHash. In order to put these in context, a broad outline of HalftimeHash is in order.

HalftimeHash can be thought of as a tree-based, recursively-defined hash function. The leaves of the tree are the words of the unhashed input; the root is the output value. Every internal node has multiple inputs and a single output, corresponding with the child and parent nodes in the tree.

To a first approximation, a string is hashed by breaking it up into some number of contiguous parts, hashing each part, then combining those hash values. When the size of the input is low enough, rather than recurse, a construction called "Encode, Hash, Combine" (or "EHC") is used to hash the input.

3.1 Tree Hash

HalftimeHash's structure is based on a tree-like hash as described by Carter and Wegman [8, Section 3]. To hash a string, we use $\lceil \lg n \rceil$ randomly-selected keys k_i and a hash family H that hashes two words down to one. Then the tree hash T of a string $s[0, n)$ is defined recursively as:

$$\begin{aligned} T(k, \langle x \rangle) &\stackrel{\text{def}}{=} x \\ T(k, s[0, n)) &\stackrel{\text{def}}{=} H(k_{\lceil \lg n - 1 \rceil}, T(k, s[0, 2^{\lceil \lg n - 1 \rceil})), T(k, s[2^{\lceil \lg n - 1 \rceil}, n))) \end{aligned} \tag{1}$$

Carter and Wegman show that if H is ε-AU, T is $m\varepsilon$-AU for input that has length exactly 2^m. Later, Boesgaard et al. extended this proof to strings with lengths that are not a power of two [7].

3.2 NH

In HalftimeHash, NH, an almost-universal hash family, is used at the nodes of tree hash to hash small, fixed-length sequences [6]:

$$\sum_{i=0}^{m} (d_{2i} + s_{2i})(d_{2i+1} + s_{2i+1})$$

where $d, s \in \mathbb{Z}_{2^{32}}^{2m+2}$ are the input string and the input entropy, respectively. The $d_j + s_j$ additions are in the ring $\mathbb{Z}_{2^{32}}$, while all other operations are in the ring $\mathbb{Z}_{2^{64}}$. NH is 2^{-32}-AU. In fact, it satisfies a stronger property, 2^{-32}-AΔU [6]:

Definition 1. *A hash family H is said to be* ε-almost Δ-universal *(or just AΔU) when*

$$\forall x, y, \delta, \Pr_s[H(s, x) - H(s, y) = \delta] \leq \varepsilon \in o(1)$$

In tree nodes (though not in EHC, covered below), a variant of NH is used in which the last input pair is not hashed, thereby increasing performance:

$$\left(\sum_{i=0}^{m-1} (d_{2i} + s_{2i})(d_{2i+1} + s_{2i+1}) \right) + d_{2m} + 2^{32} d_{2m+1}$$

This hash family is still 2^{-32}-AU [7].

3.3 Encode, Hash, Combine

At the leaves of the tree hash, HalftimeHash uses the "Encode, Hash, Combine" algorithm [20]. EHC is parameterized by an erasure code with "minimum distance" k, which is a map on sequences of words such that any two input values that differ in *any* location produce encoded outputs that differ in *at least* $k > 1$ locations after encoding.

The EHC algorithm is:

1. A sequence of words is processed by an erasure code with minimum distance k, producing a longer encoded sequence.
2. Each word in the encoded sequence is hashed using an AΔU family with independently and randomly chosen input entropy.
3. A linear transformation T is applied to the resulting sequence of hash values. The codomain of T has dimension k, and T must have the property that any k columns of it are linearly independent.

Nandi proved that if the EHC matrix product is over a finite field, EHC is ε^k-AU. This AU collision probability could be achieved on the same input by instead running k copies of NH, but that would perform mk multiplications to hash m words, while EHC requires $m + k$ multiplications, excluding the multiplications implicit in applying T. That exclusion is the topic of Sect. 4.

4 Generalized EHC

At first glance, EHC might not look like it will reduce the number of multiplications needed, as the application of linear transformations usually requires multiplication. However, since T is not part of the randomness of the hash family, it can be designed to contain only values that are trivial to multiply by, such as powers of 2.

The constraint in [20] requires that any k columns of T form an invertible matrix. This is not feasible in linear transformations on $\mathbb{Z}_{2^{64}}$ in most useful dimensions. For instance, in HalftimeHash24, a 3×9 matrix T is used. Any such matrix will have at least one set of three columns with an even determinant, and which therefore has a non-trivial kernel.

Proof. Let U be a matrix over $\mathbb{Z}_2$ formed by reducing each entry of T modulo 2. Then $(\det T) \bmod 2 \equiv \det U$. Since there are only 7 unique non-zero columns of size 3 over $\mathbb{Z}_2$, by the pigeonhole principle, some two columns x, y of U must be equal. Any set of columns that includes both x and y has a determinant of $0 \bmod 2$. □

Let k be the minimum distance of the erasure code. While Nandi proved that EHC is ε^k-AU over a finite field, $\mathbb{Z}_{2^{64}}$ is not a finite field. However, there are similarities to a finite field, in that there are some elements in $\mathbb{Z}_{2^{64}}$ with inverses. Some other elements in $\mathbb{Z}_{2^{64}}$ are zero divisors, but only have one value that they can be multiplied by to produce 0. A variant of Nandi's proof is presented here as a warm-up to explain the similarities [20].

Lemma 1. *When the matrix product is taken over a field, if the hash function H used in step 2 is ε-$A\Delta U$, EHC is ε^k-$A\Delta U$.*

Proof. Let $\bar{H}$ be defined as $\bar{H}(s,x)_i \stackrel{\text{def}}{=} H(s_i, x_i)$. Let J be the encoding function that acts on x and y, producing an encoding of length e. Given that x and y differ, let F be k locations where $J(x)_i \neq J(y)_i$. Let $T|_F$ be the matrix formed by the columns of T where the column index is in F and let $\bar{H}|_F$ similarly be $\bar{H}$ restricted to the indices in F. Conditioning over the $e-k$ indices not in F, we want to bound

$$\Pr_s[T|_F\bar{H}|_F(s, J(x)) - T|_F\bar{H}|_F(s, J(y)) = \delta] \tag{2}$$

Since any k columns of T are independent, $T|_F$ is non-singular, and the equation is equivalent to $\bar{H}|_F(s, J(x)) - \bar{H}|_F(s, J(y)) = T|_F{}^{-1}\delta$, which implies

$$\bigwedge_{i \in F} H(s_i, J(x)_i) - H(s_i, J(y)_i) = \beta_i$$

where $\beta \stackrel{\text{def}}{=} T|_F{}^{-1}\delta$.

Since the s_i are all chosen independently, the probability of the conjunction is the product of the probabilities, showing

$$\begin{aligned} &\Pr_s[T|_F\bar{H}|_F(s, J(x)) - T|_F\bar{H}|_F(s, J(y)) = \delta] \\ &\leq \textstyle\prod_{i \in F} Pr_s[H(s_i, J(x)_i) - H(s_i, J(y)_i) = \beta_i] \end{aligned}$$

and since H is AΔU, this probability is ε^k. □

Note that this lemma depends on k being the minimum distance of the code. If the distance were less than k, then the matrix would be smaller, increasing the probability of collisions.

In the non-field ring $\mathbb{Z}_{2^{64}}$, the situation is altered. "Good" matrices are those in which the determinant of any k columns is divisible only by a small power of two. The intuition is that, since matrices in $\mathbb{Z}_{2^{64}}$ with odd determinants are invertible, the "closer" a determinant is to odd (meaning it is not divisible by large powers of two), the "closer" it is to invertible.

Theorem 1. *Let p be the largest power of 2 that divides the determinant of any k columns in T. The EHC step of HalftimeHash is $2^{k(p-32)}$-$A\Delta U$ when using NH as the hash family.*

Proof. In HalftimeHash, the proof of the lemma above unravels at the reliance upon the trivial kernel of $T|_F$. The columns of T in HalftimeHash are linearly independent, so the matrix $T|_F$ is injective in rings without zero dividers, but not necessarily injective in $\mathbb{Z}_{2^{64}}$.

However, even in $\mathbb{Z}_{2^{64}}$, the adjugate matrix $\text{adj}(A)$ has the property that $A \cdot \text{adj}(A) = \text{adj}(A) \cdot A = \det(A)I$. Let $\det(T|_F) = q2^{p'}$, where q is odd and $p' \leq p$. Now (2) reduces to

$$\begin{aligned}
&\Pr_s[T|_F\bar{H}|_F(s,x) - T|_F\bar{H}|_F(s,y) = \delta] \\
&\leq \Pr_s[\operatorname{adj}(T|_F)T|_F\bar{H}|_F(s,x) - \operatorname{adj}(T|_F)T|_F\bar{H}|_F(s,y) = \operatorname{adj}(T|_F)\delta] \\
&= \Pr_s[q2^{p'}\bar{H}|_F(s,x) - q2^{p'}\bar{H}|_F(s,y) = \operatorname{adj}(T|_F)\delta] \\
&= \Pr_s[2^{p'}\bar{H}|_F(s,x) - 2^{p'}\bar{H}|_F(s,y) = q^{-1}\operatorname{adj}(T|_F)\delta]
\end{aligned}$$

Now letting $\beta = q^{-1}\operatorname{adj}(T|_F)\delta$ and letting the modulo operator extend pointwise to vectors, we have

$$\begin{aligned}
&= \Pr_s[\bar{H}|_F(s,x) - \bar{H}|_F(s,y) \equiv \beta \bmod 2^{64-p'}] \\
&= \Pr_s\left[\bigwedge_{i\in F} H(s_i,x_i) - H(s_i,y_i) \equiv \beta_i \bmod 2^{64-p'}\right] \\
&= \prod_{i\in F}\Pr_s\left[H(s_i,x_i) - H(s_i,y_i) \equiv \beta_i \bmod 2^{64-p'}\right] \\
&= \left(2^{p'}2^{-32}\right)^{|F|} = 2^{k(p'-32)}
\end{aligned}$$

This quantity is highest when p' is at its maximum over all potential sets of columns F, and p' is at most p, by the definition of p. □

This generalized version of EHC is used in the implementation of HalftimeHash described in Sect. 5, with $p \leq 2^3$.

5 Implementation

This section describes the specific implementation choices made in HalftimeHash to ensure high output entropy and high performance. The algorithm performs the following steps:

- Generalized EHC on instances of the unhashed input, producing 2, 3, 4, or 5 output words (of 64 bits each) per input instance
- 2, 3, 4, or 5 executions of tree hash (with independently and randomly chosen input entropy) on the output of EHC, with NH at each internal node, producing a sequence of words logarithmic in the length of the input string, as described below in Eq. 3
- NH on the output of each tree hash, producing 16, 24, 32, or 40 bytes.

5.1 EHC

In addition to the trivial distance-2 erasure code of XOR'ing the words together and appending that as an additional word, HalftimeHash uses non-linear erasure codes discovered by Gabrielyan with minimum distance 3, 4, or 5 [14–16]. These codes can be computed without ady multiplications.

For the linear transformations, HalftimeHash uses matrices T selected so that the largest power of 2 that divides any determinant is 2^2 or 2^3. For instance, for the HalftimeHash24 variant, T has a p of 2^2:

$$\begin{pmatrix} 0&0&1&4&1&1&2&2&1 \\ 1&1&0&0&1&4&1&2&2 \\ 1&4&1&1&0&0&2&1&2 \end{pmatrix}$$

For other output widths, HalftimeHash uses

	HalftimeHash16	HalftimeHash32	HalftimeHash40
T	$\begin{pmatrix} 1&0&1&1&2&1&4 \\ 0&1&1&2&1&4&1 \end{pmatrix}$	$\begin{pmatrix} 0&0&0&1&1&4&2&4&1&1 \\ 0&1&2&0&0&1&1&2&4&1 \\ 2&0&1&0&4&0&1&1&1&1 \\ 1&1&0&1&0&0&4&1&2&8 \end{pmatrix}$	$\begin{pmatrix} 1&0&0&0&0&1&1&2&4 \\ 0&1&0&0&0&1&2&1&7 \\ 0&0&1&0&0&1&3&8&5 \\ 0&0&0&1&0&1&4&9&8 \\ 0&0&0&0&1&1&5&3&9 \end{pmatrix}$
p	2^2	2^3	2^3

The input group lengths for the EHC input are 6, 7, 7, and 5, as can be seen from the dimensions of the matrices: columns + 1 − rows. Note that each of these matrices contains coefficients that can be multiplied by with no more than two shifts and one addition.

5.2 Tree Hash

For the tree hashing at internal nodes (above the leaf nodes, which use EHC), $k \in \{2, 3, 4, 5\}$ tree hashes are executed with independently-chosen input entropy, producing output entropy of $-k \lg \varepsilon$. From the result from Carter and Wegman on the entropy of tree hash of a tree of height m, the resulting hash function is $m\varepsilon^k$-AU.

The key lemma they need is that almost universality is compossable:

Lemma 2 (Carter and Wegman). *If F is ε_F-AU, G is ε_G-AU, then*

- *$F \circ G$ where $F \circ G(\langle k_F, k_G \rangle, x) \stackrel{\text{def}}{=} F(k_F, G(k_G, x))$ is $(\varepsilon_F + \varepsilon_G)$-AU.*
- *$\langle F, G \rangle$ where $\langle F, G \rangle(\langle k_F, k_G \rangle, \langle x, y \rangle) \stackrel{\text{def}}{=} \langle F(k_F, x), G(k_G, y) \rangle$, is $\max(\varepsilon_F, \varepsilon_G)$-AU, even if $F = G$ and $k_F = k_G$.*

The approach in Badger of using Eq. 1 to handle words that are not in perfect trees can be increased in speed with the following method: For HalftimeHash, define $\widehat{T}$ as a family taking as input sequences of any length n and producing sequences of length $\lceil \lg n \rceil$ as follows, using Carter and Wegman's T defined in Sect. 3:

$$\begin{aligned} \widehat{T}_0(k, \langle\rangle) &\stackrel{\text{def}}{=} \langle \bot \rangle \\ \widehat{T}_0(k, \langle x \rangle) &\stackrel{\text{def}}{=} \langle x \rangle \\ \widehat{T}_{i+1}(k, s[0, n)) &\stackrel{\text{def}}{=} \begin{cases} \bot \triangleleft \widehat{T}_i(k, s[0, n)) & \text{if } 2^i > n \\ T(k, s[0, 2^i))) \triangleleft \widehat{T}_i(k, s[2^i, n)) & \text{if } 2^i \leq n \end{cases} \end{aligned} \tag{3}$$

There is one execution of T for every 1 in the binary representation of n. By an induction on $\lceil \lg n \rceil$ using the composition lemma, $\widehat{T}$ is $\varepsilon \lceil \lg n \rceil$-AU.

The output of $\widehat{T}$ is then hashed using an NH instance of size $\lceil \lg n \rceil$. This differs from Badger, where T is used to fully consume the input without the use of additional input entropy; T produces a single word per execution, while $\widehat{T}$ needs to be paired with NH post-processing in order to achieve that [7]. Empirically, $\widehat{T}$ has better performance than the Badger approach.

6 Performance

This section tests and analyzes HalftimeHash performance, including an analysis of the output entropy.

6.1 Analysis

The parameters used in this analysis are:

b the number of 64-bit words in a block. Blocks are used to take advantage of SIMD units.
d is the number of elements in each EHC instance before applying the encoding.
e is the number of blocks in EHC after applying the encoding.
f is fanout, the width of the NH instance at tree hash nodes.
k is the number of blocks produced by the Combine step of EHC. This is also the minimum distance of the erasure code, as described above.
p is the maximum power of 2 that divides a determinant of any $k \times k$ matrix made from columns of the matrix T; doubling p increases ε by a factor of 2^k.
w is the number of blocks in each item used in the Encode step of EHC.

In HalftimeHash24,

$$(b, d, e, f, k, p, w) = (8, 7, 9, 8, 3, 2^2, 3)$$

Each EHC execution reads in dw blocks, produces e blocks, uses ew words of input entropy, and performs ew multiplications.

For the tree hash portion of HalftimeHash, the height of the k trees drives multiple metrics. Each tree has $\lfloor n/bdw \rfloor$ blocks as input and every level execution forms a complete f-ary execution tree. The height of the tree is thus $h \stackrel{\text{def}}{=} \lfloor \log_f \lfloor n/bdw \rfloor \rfloor$.

Lemma 3. *The tree hash is* $2^{k \lg h - 32k}$*-AU.*

Proof. Carter and Wegman showed that tree hash has collision probability of $h\varepsilon$, where ε is the collision probability of a single node. Each tree node uses NH, so a single tree has collision probability $h2^{-32}$. A collision occurs for HalftimeHash at the tree hash stage if and only if all k trees collide, which has probability $\left(h2^{-32}\right)^k$, assuming that the EHC step didn't already induce a collision. □

The amount of input entropy needed is proportional to the height of the tree, with $f-1$ words needed for every level. HalftimeHash uses different input entropy for the k different trees, so the total number of 64-bit words of input entropy used in the tree hash step is $(f-1)hk$.

The number of multiplications performed is identical to the number of 64-bit words input, $kb\lfloor n/bdw \rfloor$.

The result of the tree hash is processed through NH, which uses $bfhk$ words of entropy and just as many multiplications.

There can also be as much as bdw words of data in the raw input that are not read by HalftimeHash, as they are less than the input size of one instance of EHC. Again, NH is used on this data, but now hashing k times, since this data has not gone through EHC. That requires $bdwk$ words of entropy and just as many multiplications.

For this previously-unread data, the number of words of entropy needed can be reduced by nearly a factor of k using the Toeplitz construction. Let r be the sequence of random words used to hash it. Instead of using $r[ibdw, (i+1)bdw)$ as the keys to hash component i with, HalftimeHash uses $r[i, bdw+i)$. This construction for multi-part hash output is AΔU [20,27].

6.2 Cumulative Analysis

The combined collision probability is $2^{-32k}\left(2^{kp}+h^k+1\right)$. For HalftimeHash24, and for strings less than an exabyte in length, this is more than 83 bits of entropy.

The combined input entropy needed (in words) is $ew+(f-1)hk+bfhk+bdw+k-1$ HalftimeHash24 requires 8.4 KB input entropy for strings of length up to one megabyte and 34KB entropy for strings of length up to one exabyte.

The number of multiplications is dominated by the EHC step, since the total is $(ew+k)b\lfloor n/bdw \rfloor + O(\log n)$ and ew is significantly larger than k. For a string of length 1MB, 84% of the multiplications happen in the EHC step. Intel's VTune tool show the same thing: 86% of the clock cycles are spent in the EHC step.[2]

6.3 Benchmarks

HalftimeHash passes all correctness and randomness tests in the SMHasher test suite; for a performance comparison, see Fig. 1 and [26].

Figure 2 displays the relationship between output entropy and throughput for HalftimeHash, UMASH, and clhash.[3] Adding more output entropy increases the number of non-linear arithmetic operations that any hash function has to perform [20]. The avoidance of doubling the number of multiplications for twice the output size is one of the primary reasons that HalftimeHash24, -32, and -40 are faster than running clhash or UMASH with 128-bit output. (The other is that carryless multiplication is not supported as a SIMD instruction.)

[2] Similarly, clhash and UMASH, which are based on 64-bit carryless NH, have their execution times dominated by the multiplications in their base step [17,19].

[3] UMASH and clhash are the fastest AU families for string hashing.

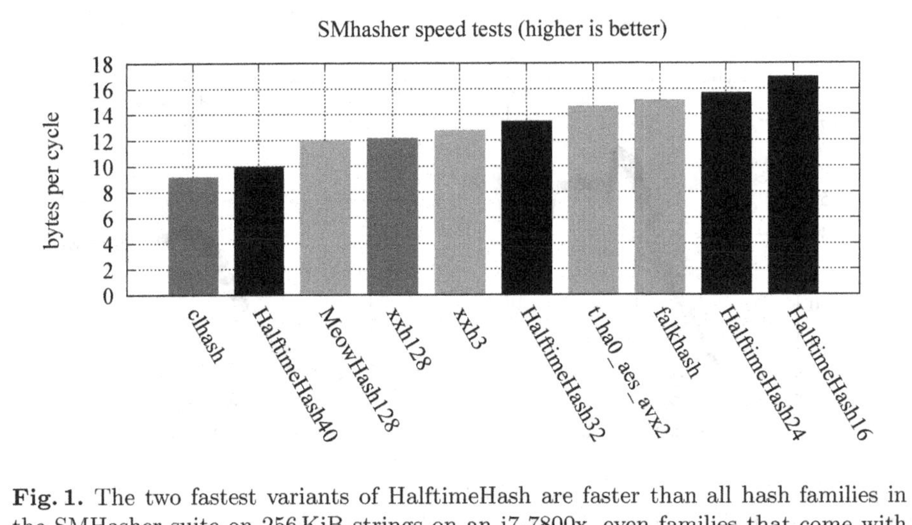

Fig. 1. The two fastest variants of HalftimeHash are faster than all hash families in the SMHasher suite on 256 KiB strings on an i7-7800x, even families that come with no AU guarantees [26]. Of the families here, only HalftimeHash and xxh128 pass all SMHasher tests, and only HalftimeHash and clhash are AU.

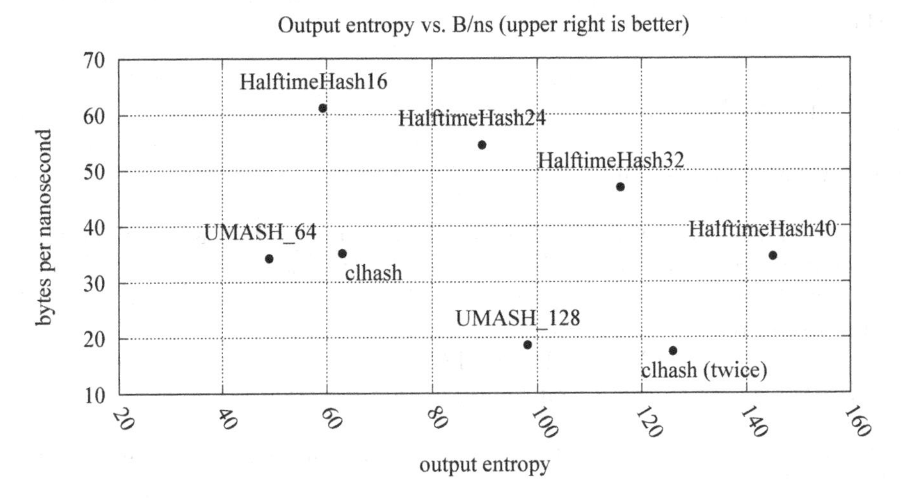

Fig. 2. Trade-offs for almost-universal string hashing functions on strings of size 250 KB on an i7-7800x. UMASH comes in two variants based on the output width in bits; clhash doesn't, but running clhash twice is included in the chart. For each clhash/UMASH version, at least one version of HalftimeHash is faster and has lower collision probability.

Figure 3 adds comparisons between clhash, UMASH, and HalftimeHash across input sizes and processor manufacturers. Although these two machines support different ISA vector extensions, the pattern is similar: for large enough input, HalftimeHash's throughput exceeds that of the carryless multiplication families.

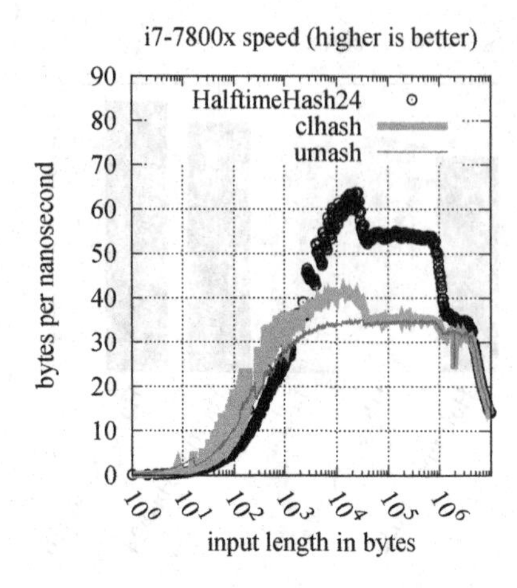

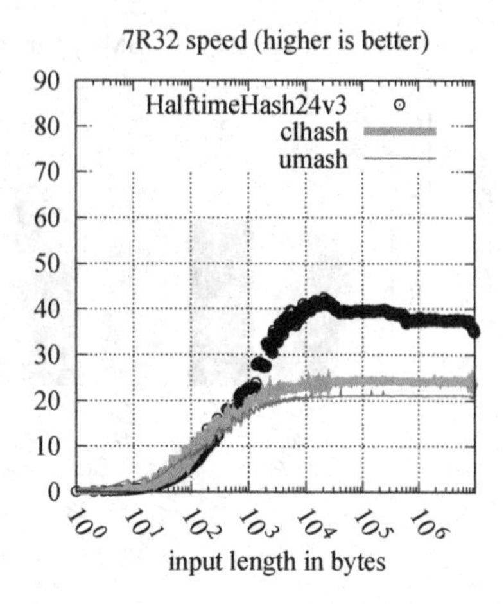

Fig. 3. Comparison of Intel (i7-7800x) and AMD (EC2 c5a.large, 7R32) performance. On both chips HalftimeHash24 is faster than clhash and UMASH for long strings. The "v3" after the name of the AMD HalftimeHash indicates block size: v3 means a 256-bit block size, while v4 (the default) means 512-bit block size. AMD chips do not support AVX-512, but still HalftimeHash with 256-bit blocks exceeds the speed of clmul-based hashing methods by up to a factor of 2.

7 Future Work

Areas of future research include:

- Tuning for JavaScript, which has no native 32-bit multiplication support, as it natively supports only double-precision floating point numbers
- Comparisons against hash algorithms in the Linux kernel, including Poly1305 and `crc32_pclmul_le_16`
- Benchmarks on POWER and ARM ISA's
- EHC benchmarks using 64-bit multiplication – carryless or integral.

Acknowledgments. Thanks to Daniel Lemire, Paul Khuong, and Guy Even for helpful discussions and feedback.

References

1. Alon, N., Dietzfelbinger, M., Miltersen, P.B., Petrank, E., Tardos, G.: Linear hash functions. J. ACM **46**(5), 667–683 (1999). https://doi.org/10.1145/324133.324179
2. Apple, J.: Ensure monotonic count (distinct x) performance (2015). https://issues.apache.org/jira/browse/IMPALA-2653. Accessed 22 Dec 2020
3. Aumasson, J.P., Bernstein, D.J.: SipHash: a fast short-input PRF. In: Galbraith, S., Nandi, M. (eds.) INDOCRYPT 2012. LNCS, vol. 7668, pp. 489–508. Springer, Heidelberg (2012). https://doi.org/10.1007/978-3-642-34931-7_28

4. Bernstein, D.J.: The Poly1305-AES message-authentication code. In: Gilbert, H., Handschuh, H. (eds.) FSE 2005. LNCS, vol. 3557, pp. 32–49. Springer, Heidelberg (2005). https://doi.org/10.1007/11502760_3
5. Bernstein, D.J.: Cryptography in NaCl. Network. Cryptogr. Libr. **3**, 385 (2009)
6. Black, J., Halevi, S., Krawczyk, H., Krovetz, T., Rogaway, P.: UMAC: fast and secure message authentication. In: Wiener, M. (ed.) CRYPTO 1999. LNCS, vol. 1666, pp. 216–233. Springer, Heidelberg (1999). https://doi.org/10.1007/3-540-48405-1_14
7. Boesgaard, M., Christensen, T., Zenner, E.: Badger - a fast and provably secure MAC. In: Ioannidis, J., Keromytis, A., Yung, M. (eds.) ACNS 2005. LNCS, vol. 3531, pp. 176–191. Springer, Heidelberg (2005). https://doi.org/10.1007/11496137_13
8. Carter, J.L., Wegman, M.N.: Universal classes of hash functions. J. Comput. Syst. Sci. **18**(2), 143–154 (1979). http://www.sciencedirect.com/science/article/pii/0022000079900448
9. Chung, K.M., Mitzenmacher, M., Vadhan, S.: Why simple hash functions work: exploiting the entropy in a data stream. Theory Comput. **9**(30), 897–945 (2013). http://www.theoryofcomputing.org/articles/v009a030
10. Crosby, S.A., Wallach, D.S.: Denial of service via algorithmic complexity attacks. In: Proceedings of the 12th USENIX Security Symposium, pp. 29–44 (2003)
11. Dietzfelbinger, M.: Universal hashing via integer arithmetic without primes, revisited. In: Böckenhauer, H.J., Komm, D., Unger, W. (eds.) Adventures Between Lower Bounds and Higher Altitudes. LNCS, vol. 11011, pp. 257–279. Springer, Cham (2018). https://doi.org/10.1007/978-3-319-98355-4_15
12. Dietzfelbinger, M., Hagerup, T., Katajainen, J., Penttonen, M.: A reliable randomized algorithm for the closest-pair problem. J. Algorithms **25**(1), 19–51 (1997)
13. Dodis, Y., Ostrovsky, R., Reyzin, L., Smith, A.: Fuzzy extractors: how to generate strong keys from biometrics and other noisy data. SIAM J. Comput. **38**(1), 97–139 (2008). https://doi.org/10.1137/060651380
14. Gabrielyan, E.: Erasure resilient (10,7) code (2005). https://docs.switzernet.com/people/emin-gabrielyan/051102-erasure-10-7-resilient/. Accessed 26 Nov 2020
15. Gabrielyan, E.: Erasure resilient MDS code with four redundant packets (2005). https://docs.switzernet.com/people/emin-gabrielyan/051103-erasure-9-5-resilient/. Accessed 26 Nov 2020
16. Gabrielyan, E.: Erausre resulient (9,7)-code (2005). https://docs.switzernet.com/people/emin-gabrielyan/051101-erasure-9-7-resilient/. Accessed 26 Nov 2020
17. Khuong, P.: UMASH: a fast and universal enough hash (2020). https://engineering.backtrace.io/2020-08-24-umash-fast-enough-almost-universal-fingerprinting/
18. Landau, J.: Exposure of HashMap iteration order allows for $O(n^2)$ blowup (2016). https://github.com/rust-lang/rust/issues/36481. Accessed 22 Dec 2020
19. Lemire, D., Kaser, O.: Faster 64-bit universal hashing using carry-less multiplications. J. Cryptogr. Eng. **6**(3), 171–185 (2015). https://doi.org/10.1007/s13389-015-0110-5
20. Nandi, M.: On the minimum number of multiplications necessary for universal hash functions. In: Cid, C., Rechberger, C. (eds.) FSE 2014. LNCS, vol. 8540, pp. 489–508. Springer, Heidelberg (2015). https://doi.org/10.1007/978-3-662-46706-0_25
21. Pagh, R., Rodler, F.F.: Cuckoo hashing. J. Algorithms **51**(2), 122–144 (2004). http://www.sciencedirect.com/science/article/pii/S0196677403001925

22. Renner, R., König, R.: Universally composable privacy amplification against quantum adversaries. In: Kilian, J. (ed.) TCC 2005. LNCS, vol. 3378, pp. 407–425. Springer, Heidelberg (2005). https://doi.org/10.1007/978-3-540-30576-7_22
23. Stoica, I., Morris, R., Karger, D., Kaashoek, M.F., Balakrishnan, H.: Chord: a scalable peer-to-peer lookup service for internet applications. ACM SIGCOMM Comput. Commun. Rev. **31**(4), 149–160 (2001)
24. Thorup, M.: String hashing for linear probing, pp. 655–664. Society for Industrial and Applied Mathematics (2009). https://epubs.siam.org/doi/abs/10.1137/1.9781611973068.72
25. Thorup, M.: Fast and powerful hashing using tabulation. CoRR abs/1505.01523 (2017). http://arxiv.org/abs/1505.01523
26. Urban, R., et al.: Smhasher (2020). https://github.com/rurban/smhasher
27. Woelfel, P.: Efficient strongly universal and optimally universal hashing. In: Kutyowski, M., Pacholski, L., Wierzbicki, T. (eds.) MFCS 1999. LNCS, vol. 1672, pp. 262–272. Springer, Heidelberg (1999). https://doi.org/10.1007/3-540-48340-3_24

Generalized Disk Graphs

Ívar Marrow Arnþórsson[1], Steven Chaplick[2], Jökull Snær Gylfason[1], Magnús M. Halldórsson[1], Jökull Máni Reynisson[1], and Tigran Tonoyan[3(✉)]

[1] Reykjavík University, Reykjavík, Iceland
{ivara16,jokull17,mmh,jokull16}@ru.is
[2] Maastricht University, Maastricht, The Netherlands
s.chaplick@maastrichtuniversity.nl
[3] Technion, Haifa, Israel

Abstract. A graph G is a *Generalized Disk Graph* if for some dimension $\eta \geq 1$, a non-decreasing sub-linear function f and natural number t, each vertex v_i can be assigned a length l_i and set $P_i \subseteq \mathbb{R}^\eta$ of t points such that $v_i v_j$ is an edge of G if and only if $l_i \leq l_j$ and $d(P_i, P_j) \leq l_i f(l_j/l_i) + l_i f(1)$, where $d(\cdot,\cdot)$ is the least distance between points in either set. Generalized disk graphs were introduced as a model of wireless network interference and have been shown to be dramatically more accurate than disk graphs or other previously known graph classes. However, their properties have not been studied extensively before.

We give a geometric representation of these graphs as intersection graphs of convex shapes, relate them to other geometric intersection graph classes, and solve several important optimization problems on these graphs using the geometric representation; either exactly (in two-dimensions) or approximately (in higher dimensions).

Keywords: Conflict graph · SINR model · Intersection graph

1 Introduction

The efficient use of wireless networks requires effective scheduling of its communication links. A clean approach is to model the wireless interference as a graph relation and to solve the corresponding scheduling problems – such as the minimum coloring problem or the maximum weighted independent set problem – in the graphs. The most natural graph formulation is to model each link in the plane by a two-dimensional disk of radius proportional to the power used. However, it is well known that such a formulation is a poor representation of reality, at least in a worst-case sense. Such limitations remain for any formulation involving fixed-sized 2-dimensional objects. A more faithful representation of wireless interference is the physical (or SINR) model, that models interference as a (geometrically defined) hypergraph. This, however, brings with it added

Work supported by grant 174484-051 from Icelandic Research Fund and grant 208348-0091 from Icelandic Student Innovation Fund.

A. Lubiw et al. (Eds.): WADS 2021, LNCS 12808, pp. 115–128, 2021.
https://doi.org/10.1007/978-3-030-83508-8_9

complications, since few optimization tools are available in hypergraph settings. Having a graph-based formulation is preferable for many reasons: simplicity, ease of analysis, and availability of effective algorithms.

A new graph class was proposed in [13] to better represent interference between wireless links. Whereas the adjacency predicate in disk graphs is a linear combination of the lengths l_i, l_j of the two links, it is here generalized to involve a function of the *relative* link lengths $l_i f(l_j/l_i)$ (when $l_i < l_j$), where f is a function parameter. This formulation attains a much greater fidelity to the SINR model than previously known, resulting in greatly improved approximation algorithms for the underlying wireless optimization problems in the SINR model.

Specifically, we say that a graph formulation *properly represents* a set of wireless links if any *feasible* subset of links (links whose transmissions can be correctly decoded under the SINR rules, when they transmit simultaneously) correctly forms an independent set in the graph. It is then a conservative representation: a coloring of the graph yields a valid TDMA (time-division multiplexed access) schedule of the wireless links. To capture how well the graph formulation works, we define its *performance gap* as the size of the largest clique in the graph whose corresponding links form a feasible set (in worst case, over all instances). The performance gap then indicates the worst-case *slowdown* that the optimal coloring of the graph has over the optimal TDMA schedule. Applying an approximation algorithm on the graph then yields another factor of slowdown.

Disk graph formulations have a performance gap proportional to Λ [19], the length diversity of the link set, or the number of different link lengths (rounded to powers of 2). In contrast, the graph class of [13] has a performance gap of only $\Theta(\log^* \Lambda)$, for an appropriately chosen function f. This results in the *only* sublinear approximation factors (in terms of Λ) known for a wide range of wireless scheduling problems and settings (see [4] for generalizations). Given the dominant status of this new graph class, it is natural to ask about its fundamental graph-theoretic properties and its relationship with other known classes.

The challenge of these *generalized disk graphs* (GDG), as we call them, over the geometric graph classes, is the increased level of abstraction, making reasoning and design of optimization algorithms more challenging. It is therefore an intriguing question if these graphs can be reformulated as geometric intersection graphs. Doing so would likely allow the placement of these graphs into the broader zoo of classes of geometric intersection graphs and unlock the potential to reuse many algorithmic results concerning optimization problems therein.

Our Results. We show that generalized disk graphs can indeed be represented as intersection graphs of convex geometric objects, by moving to the one higher dimension. This holds under modest restrictions on the functions used, covering the ones treated in [13]. In fact, the objects are *fat*, which allows us to immediately apply some known algorithms, and are *grounded*, i.e., the additional dimension is only used to give the "height" of the objects.

In particular, Chan's [7] polynomial-time approximation scheme (PTAS) for *maximum weighted independent set problem* (MWIS) carries over to GDGs. Moreover, we show that the dependence on the dimension in the time complexity can be reduced by using the fact that the objects are grounded. We also show that the result holds as well for an extension where each object consists of multiple nearby fat pieces (Theorem 4). This implies an improvement in the constant factor of the approximation ratio of [13] for the key problem of Maximum Weighted Independent Set of Links in the SINR model, a problem that underlies the solution methods for the vast majority of wireless scheduling problems [4]. This algorithm necessarily requires knowledge of the geometric representation.

We initiate a study of the structural and algorithmic properties of GDGs, with a particular focus on one-dimensional GDGs (represented in 2-D), namely when the nodes are embedded in a line. We relate 1-D GDGs to other classes of geometric intersection graphs, showing that they properly contain interval graphs, are properly contained in outerstring graphs [18], but are incomparable to max-point tolerance graphs [5,21], another class of grounded convex objects. We show that two core optimization problems can be solved exactly in polynomial time in 1-D when given the representation: maximum clique and maximum weight independent sets.

We also derive structural properties of general GDGs with implications to optimization problems: k-simpliciality and k-inductive independence. Our bounds improve on the (more general) bounds of [13] and match those of ordinary disk graphs. This leads to graph coloring approximations that match the best results known for disk graphs. Overall, we find GDGs to be a rich class, with a number of combinatorially desirable properties.

Related Work. GDGs were introduced in [13] to model wireless network interference, with the modeling extended in [4]. Adding another dimension to capture SINR effects has been successfully applied in [16] and [3] in different scenarios without links or explicit receivers.

In [12] an approximation scheme is given for MWIS of disks, which can be extended to higher dimensions and applies to the graphs we consider in this paper. However, a similar method given in [6], which is better suited to our graphs, implies a PTAS for MWIS of fat objects in arbitrary dimensions. We are able to project the intersection of the geometric shapes of our graphs to get fat objects and get improved time complexity for our graphs.

In [17] a polynomial-time algorithm for the independent set on outerstring graphs is given with time complexity $O(N^3)$, where N is the number of line segments needed to represent all strings in total. Since GDGs on one-dimensional point sets are indeed outerstring graphs, this result holds for them also. However, for the functions we study, a naive representation of the n-vertex graphs would involve $N \in \Omega(n^2)$ segments, resulting in a higher complexity than we obtain.

A related graph class is the set of the max point-tolerance graphs (MPT), which are the intersection graphs of intervals with an assigned point, such that two nodes are adjacent when the intersection of the corresponding intervals includes the point of both intervals. These have been shown to be equivalent to L-shapes grounded on a line with negative slopes [5]. In the same paper, the class of interval graphs was shown to be a subclass and a polynomial time algorithm for MWIS was given.

Notation and Definitions. We use $V(G)$ and $E(G)$ to denote the vertex set and edge set of graph G, respectively. The open (closed) neighborhood of a vertex v in G is denoted by $N_G(v)$ ($N_G[v]$), and with the subscript omitted when clear from the context.

A function f is *sub-linear* if $f(x) = o(x)$. For a tuple $\boldsymbol{x} = (x_1, \ldots, x_d) \in \mathbb{R}^\eta$ and $z \in \mathbb{R}$, we use $\boldsymbol{x} \circ z = (x_1, \ldots, x_d, z) \in \mathbb{R}^{\eta+1}$ to denote their concatenation. Let $d(x, y)$ denote the Euclidean distance between (points) x and y.

For ease of discussion, we use the term *cone* to refer to a right spherical cone of any dimension. For example, a 2-dimensional cone is an equilateral triangle, a 3-dimensional cone is a finite right circular cone, a 4-dimensional cone is a finite right spherical cone, and so on.

2 Geometric Representation

Each vertex v_i in a Generalized Disk Graph (GDG) has an associated *length* l_i and a *set* $P_i \subset \mathbb{R}^\eta$ *of* t *points*. The adjacency of two vertices depends on a function f, and therefore the class is parameterized by the dimension η, the cardinality t of the point sets P_i and the function f. We begin by defining the class of Generalized Disk Graphs given these parameters and then show that such a graph can be represented as a geometric intersection graph where each vertex is associated with t cone-like objects we call f-cones.

Definition 1. *Let $\eta, t \geq 1$ be integers and $f : \mathbb{R}_+ \to \mathbb{R}_+$ be a non-negative sub-linear function. The family $GDG_f^{\eta,t}$ consists of graphs $G = (V, E)$ where it is possible to associate a set $P_i = \{p_{i_1}, \ldots, p_{i_t}\} \subset \mathbb{R}^\eta$ and a length l_i to each vertex $v_i \in V$, such that for every pair $v_i, v_j \in V$ of vertices, v_i and v_j, with $l_i \leq l_j$, are adjacent in G if and only if*

$$d_{ij} \leq l_i \cdot f\left(l_j/l_i\right) + l_i \cdot f(1) \ ,$$

where $d_{ij} = \min\{d(p_{i_k}, p_{j_{k'}}) \mid 1 \leq k, k' \leq t\}$. If $t = 1$, we let $p_i = p_{i_1}$.

Note that there is slight difference between the definition above and that in [13], but they are equivalent modulo transformation $g(x) = f(x) + 1$. Observe that when f is the identity function $f(x) = x$, $GDG_f^{1,1}$ corresponds to interval graphs and $GDG_f^{2,1}$ corresponds to disk graphs[1]. As we shall see, the constant function $f(x) = 1$ leads to a graph class that is a known subclass of the MPT graphs.

We now define the $(\eta + 1)$-dimensional body we associate with each vertex.

Definition 2. *Let $G \in GDG_f^{\eta,t}$. For each vertex v_i, let $g_i(x) = x \cdot f(l_i/x)$, and let $\mathcal{C}_i = \{C_{i_1}, \ldots, C_{i_t}\}$ be the set of f-cones associated with v_i and given by*

$$C_{i_k} = \{\boldsymbol{x} \circ h \mid \boldsymbol{x} \in \mathbb{R}^\eta,\ d(p_{i_k}, \boldsymbol{x}) \leq g_i(h),\ 0 \leq h \leq l_i\} \ .$$

[1] We sometimes consider $f(x) = x$ as an example; it is not sublinear but it satisfies all important properties we need (e.g., it is grounded w.r.t. hyperplane $h = 1$).

Observe that in the definition, $g_i(0) = 0 \cdot f_i(l_i/0) = \lim_{x \to 0} x \cdot f_i(l_i/x) = 0$, is well-defined, since we have $f(x) = o(x)$.

We sometimes refer to the h coordinate as the *height.* If $t = 1$, there is only one f-cone, and we denote it by C_i. The sets $\mathcal{C}_i$ and $\mathcal{C}_j$ *intersect* when $(\bigcup_{C \in \mathcal{C}_i} C) \cap (\bigcup_{C' \in \mathcal{C}_j} C') \neq \emptyset$. We require three properties for the set $\mathcal{C}$ of f-cones associated with a graph $G \in GDG_f^{\eta,t}$:

1. **Grounded.** The set $\mathcal{C}$ is *grounded* if every cone in $\mathcal{C}$ intersects the hyperplane $Z_0 = \{\boldsymbol{x} \circ 0 \mid \boldsymbol{x} \in \mathbb{R}^\eta\}$, and for every point $\boldsymbol{x} \circ h \in \mathcal{C}$, we have $h \geq 0$. We assume that f is *non-negative and sub-linear*, in which case $\mathcal{C}$ is grounded. Moreover, $p_{i_k} \in Z_0$, for all i, k.
2. **Convex.** We require that each cone C_{i_k} in $\mathcal{C}$ is a convex body. It is easy to see that this requirement is satisfied when g_i is a concave function on $[0, \infty)$, for every i. We assume that f is *concave and twice differentiable* on $[0, \infty)$, in which case g_i is concave for $x \geq 0$, since $g_i''(x) = (l_i^2/x^3)f''(l_i/x)$, and it is well-known that for every twice-differentiable concave function, its second derivative is non-positive.
3. **Expanding.** The cones in $\mathcal{C}$ are *expanding* if each g_i is non-decreasing. Geometrically, this means that for each cone C_{i_k}, the η-dimensional balls $C_{i_k} \cap Z_h$, where $Z_h = \{\boldsymbol{x} \circ h \mid \boldsymbol{x} \in \mathbb{R}^\eta\}$, have diameter that does not decrease with increasing height h. The previous conditions – non-negativity, concavity, and differentiability of f – give us non-decreasing g_i. Since $g_i'(x) = f(l_i/x) - (l_i/x)f'(l_i/x)$, we have non-decreasing g_i when $f(x) \geq x \cdot f'(x)$. It is well known that for any concave differentiable function f, $f(x) \geq f(y) + f'(x)(x - y)$ for all x, y in its domain. This gives us $f(x) \geq f(0) + f'(x) \cdot x \geq x \cdot f'(x)$.

For the geometric representation of a graph $G \in GDG_f^{\eta,t}$ to fit these conditions, it is then sufficient that f is a non-negative, concave, twice-differentiable function. In the rest of the paper we assume that every f satisfies these conditions, i.e., *is well-behaved.* Observe that the key functions used in [13], $f(x) = x^\delta$, for $\delta \in [0, 1]$, and $f(x) = \log(x + 1)$ of any base, satisfy these criteria.

As noted above, when the f-cones of G satisfy these conditions, the intersection of the height-x hyperplane Z_x and an f-cone induces a η-dimensional ball (while the cone is $(\eta + 1)$-dimensional). We will show that the intersection of a pair of sets $\mathcal{C}_i$ and $\mathcal{C}_j$ can be determined by considering this set of balls, and hence the adjacency of two vertices in a GDG can be determined by considering the corresponding vertices in the induced ball graph. We define these objects for future reference.

Definition 3. *Let $G \in GDG_f^{\eta,t}$, and $v_i \in V(G)$. We let $\mathcal{D}_i^x(G) = Z_x \cap \mathcal{C}_i$ denote the set of balls at height x associated with vertex v_i. We also let $\mathcal{D}^x(G) = \bigcup_{v_i \in V(G)} \mathcal{D}_i^x(G)$. When clear from the context, we drop G and write $\mathcal{D}^x$, $\mathcal{D}_i^x$.*

Lemma 1. *Let $G \in GDG_f^{\eta,t}$, and let $v_i, v_j \in V(G)$ be vertices in G with $l_i \leq l_j$. It holds that $\mathcal{C}_i$ and $\mathcal{C}_j$ intersect iff $\mathcal{D}_i^{l_i}$ and $\mathcal{D}_j^{l_i}$ intersect.*

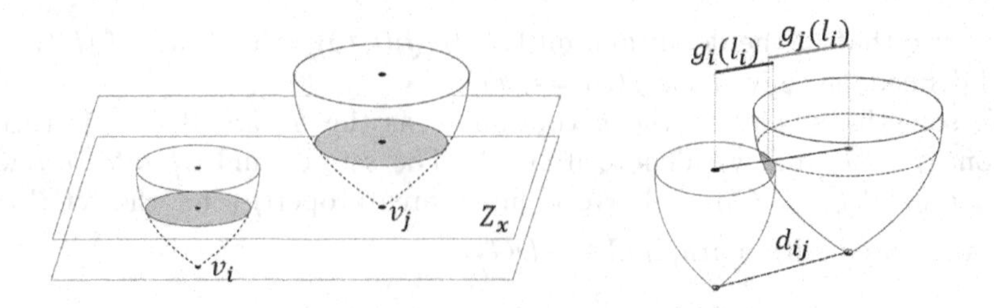

Fig. 1. Left: illustration of Definition 3; each plane Z_x induces a set of balls. Right: intersection of two f-cones is determined by the smaller height.

Proof. The sets $\mathcal{C}_i$ and $\mathcal{C}_j$ intersect iff $\mathcal{D}_i^x$ and $\mathcal{D}_j^x$ intersect for some x. It is clear that $x \leq l_i$, since otherwise $\mathcal{D}_i^x = \emptyset$. Since $x \leq l_i$ and the f-cones are expanding, it follows that $\mathcal{D}_i^x$ and $\mathcal{D}_j^x$ intersect iff $\mathcal{D}_i^{l_i}$ and $\mathcal{D}_j^{l_i}$ intersect (see Fig. 1).

With these preliminaries at hand, we are ready to prove the following representation theorem.

Theorem 1. *Each graph G in $GDG_f^{\eta,t}$ is the intersection graph of t-tuples of f-cones in $\mathbb{R}^{\eta+1}$.*

Proof. Let $G \in GDG_f^{\eta,t}$, and $v_i, v_j \in V(G)$ be such that $l_i \leq l_j$. By Lemma 1, the intersection of $\mathcal{C}_i$ and $\mathcal{C}_j$ is determined by the set of balls $\mathcal{D}_i^{l_i} \cup \mathcal{D}_j^{l_i}$. Let $D_i \in \mathcal{D}_i^{l_i}$ and $D_j \in \mathcal{D}_j^{l_i}$ be the balls with the minimum distance of all such pairs. The radii r_i of D_i and r_j of D_j are by definition $g_i(l_i)$ and $g_j(l_i)$, respectively. Hence D_i and D_j intersect iff $d_{i,j} \leq r_j + r_i = g_j(l_i) + g_i(l_i) = l_i f(l_j/l_i) + l_i f(1)$. This shows that $\mathcal{C}_i$ and $\mathcal{C}_j$ intersect iff v_i and v_j are adjacent in G.

If f is a homomorphism, i.e., $f(x_1 \cdot x_2) = f(x_1)f(x_2)$ for all x_1, x_2, then the conditions for geometric representation are greatly simplified. We then require only that f is non-negative, non-decreasing, and sub-linear. By defining the geometric sets differently (letting the height vary instead of the width) we can represent the vertices as sets of $(\eta + 1)$-dimensional (ordinary) cones satisfying all the required properties. We show this in the next theorem.

Theorem 2. *Let $G \in GDG_f^{\eta,t}$. If $f(a \cdot b) = f(a) \cdot f(b)$ then we can represent G geometrically by associating each vertex v_i with the set $\blacktriangledown_i = \{\triangledown_{i_1}, \ldots, \triangledown_{i_t}\}$ of cones where*

$$\triangledown_{i_k} = \{\boldsymbol{x} \circ h \mid \boldsymbol{x} \in \mathbb{R}^\eta, d(p_{i_k}, \boldsymbol{x}) \leq h \cdot f(l_i),\ 0 \leq h \leq l_i/f(l_i)\}.$$

Proof. The proof is similar to that of Theorem 1. Let $v_i, v_j \in V(G)$, with $l_i \leq l_j$. Since f is a homomorphism, the two closest cones of v_i and v_j intersect iff $d_{i,j} \leq l_i f(l_i)/f(l_i) + \frac{l_i}{f(l_i)} f(l_j) = l_i f(1) + l_i f(l_j/l_i)$, i.e., iff v_i, v_j is an edge in G.

We prefer to use this representation when dealing with $f(x) = x^\delta$ in particular. When $t = 1$ we write $\triangledown_i$ instead of $\triangledown_{i_1}$.

Several interesting properties of GDG representations can be obtained; e.g., they are invariant under scaling, and closed under stretching. Due to lack of space, we omit their routine and somewhat technical proofs. We conclude this section with an important property that relies on them.

A well studied geometric property is the concept of *fatness*. A non-trivial number of geometric algorithms require or work faster for fat objects. Definitions of fatness vary between applications and many different definitions exist, but they are often equivalent for convex objects; see [1,2,22] or the large number of citations in [10] for a variety of flavors of fatness. Commonly, an object is considered R-fat for $R \geq 1$ if its *slimness factor* is at most R. A natural definition of the slimness factor is the ratio between the side length of the smallest enclosing hypercube of the object and the side length of the largest enclosed hypercube. Using this definition, the closer the object is to having a slimness factor of 1 the "fatter" it is. We show that f-cones have a bounded slimness factor and thus benefit from many algorithms that exploit this property.

Lemma 2. *For every $G \in GDG_f^{\eta,t}$, there is an isomorphic graph $G' \in GDG_{\hat{f}}^{\eta,t}$ with a geometric representation by $\hat{f}$-cones with slimness factor at most $1 + \sqrt{\eta}$.*

Proof. Let $G \in GDG_f^d$. By the stretching/scaling invariance mentioned above, there is an isomorphic graph G' with $\hat{f}(x) = \frac{f(x)}{2f(1)}$ such that all vertices $v_i \in V(G')$ have $g_i(l_i) = \frac{l_i}{2}$. Each $\hat{f}$-cone in $\mathcal{C}_i$ is contained in a hypercube of side length l_i. Since each such $\hat{f}$-cone is convex it in turn contains a $(\eta + 1)$-dimensional cone of length l_i and diameter l_i. Hence, it is sufficient to consider the case where $\hat{f}$ is a constant function and thus each $\hat{f}$-cone is a $(\eta+1)$-dimensional cone. Consider a vertex v_i and assume (without loss of generality) that $l_i = 1$. Let $s \in [0, 1]$. Let C be a cone in $\mathcal{C}_i$. $B_s = C \cap Z_s$ is a η-dimensional ball of diameter s. Ball B_s contains a η-dimensional hypercube with diagonal s and side length $s/\sqrt{\eta}$ (by Euclidean distance in $\mathbb{R}^\eta$). This implies that C contains a $(\eta + 1)$-dimensional hypercube of side length $\min(1 - s, s/\sqrt{\eta})$. The latter is maximized by $s = \frac{\sqrt{\eta}}{1+\sqrt{\eta}}$, showing that C contains a hypercube of side length $\frac{1}{1+\sqrt{\eta}}$, and has slimness factor at most $1 + \sqrt{\eta}$ (Fig. 2).

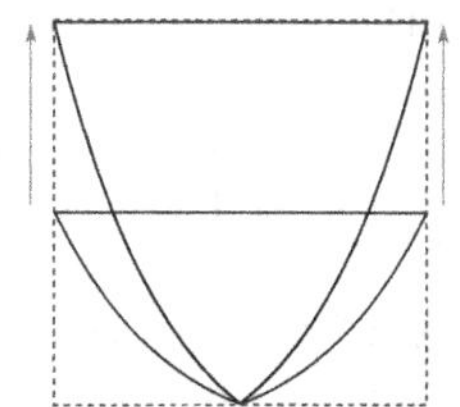
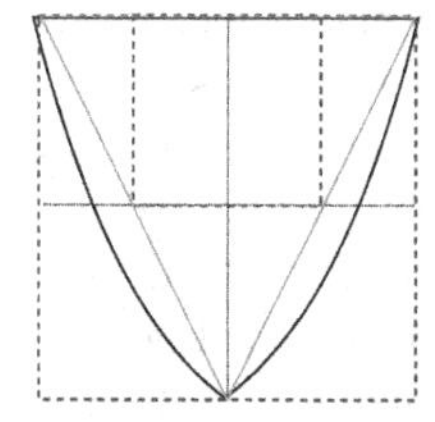

Fig. 2. On the left, we stretch a cone to fit the box completely. The figure on the right shows that the ratio between the enclosing and inscribed boxes is constant.

3 Approximation Scheme for Weighted Independent Sets

Recall that the in the *maximum weighted independent set* (MWIS) problem, we are given a graph G together with a positive weight assignment $w : V(G) \to \mathbb{R}_+$ to the vertices, and the goal is to compute an independent set I in G with maximum weight $\sum_{v \in I} w(v)$.

In the context of wireless network modeling, MWIS is perhaps the most prominent problem on GDGs. Even the Maximum Independent Set (MIS) problem (MWIS where $w(v) = 1$ for each vertex v) is NP-complete in GDGs, since it is NP-complete in unit disk graphs [8], which is a subclass of $GDG_f^{\eta,t}$ (give all f-cones the same length). However, the geometric representation allows us to use results from [6] to get a polynomial-time approximation scheme (PTAS). The definitions below are as in [6].

For a collection $\mathcal{C}$ of objects in $\mathbb{R}^m$, the *packing* problem is to find a largest subcollection of disjoint objects. This corresponds to solving the MIS problem in the geometric intersection graph of $\mathcal{C}$. The *size* of an object S is the side length of its smallest enclosing hypercube. Here we use the following definition of fatness. A collection $\mathcal{C}$ is *fat* if there is a constant c such that the following holds: For any r, and size-r axis-aligned box R, there is a set $T \subseteq \mathbb{R}^m$ of size c, such that for every object S of size at least r, $R \cap S \neq \emptyset$ implies $T \cap S \neq \emptyset$. Chan [6] gives a $(1 + \epsilon)$-approximation algorithm for the packing problem of fat objects in $\mathbb{R}^m$ that has runtime and space $n^{O(1/\epsilon^{m-1})}$. Our geometric representation of a graph $G \in GDG_f^{\eta}$ consists of a collection of fat objects, under the definition above. This can be seen by using the convexity of f-cones, as well as the fact that by Lemma 2 they have a constant aspect ratio, that is, the ratio between the size of the smallest enclosing and the largest enclosed hypercube. This directly implies a PTAS for MWIS for GDG_f^{η}, with runtime and space $n^{(1/\epsilon)^{\eta}}$, since the f-cones are in $\mathbb{R}^{\eta+1}$. Below, we improve on this by leveraging the groundedness of the f-cones and a closer examination of the algorithm.

An *r-grid interval* is an interval of the form $[ri, r(i + 1)]$ for an integer i, and an *r-grid cell* in $\mathbb{R}^m$ is the Cartesian product of m r-grid intervals. We will consider r-grid cells for $r = 2^{-l}$ over all integers l called *quadtree cells.* An object having size r is *k-aligned* if it is inside a quadtree cell of size at most kr. The algorithm is based on the following two ideas, where we let $k \approx m/\epsilon$: **1.** If the objects in $\mathcal{C}$ are k-aligned then an $n^{O(k^{m-1})}$ time and space dynamic programming algorithm gives an exact solution, **2.** There is a set $v_1, \ldots, v_k \in \mathbb{R}^m$ of vectors, s.t. for every object S, all but m of the shifts $S+v_i$ are $2k$-aligned. The complexity of the algorithm stems from the dynamic programming subroutine [6, Lemma 3.1], while the latter is dominated by $n^{O(K)}$, where K is the maximum size of a disjoint subcollection of objects of size at least r/k intersecting the boundary of an m-dimensional box R of size r.

The argument for general fat shapes is as follows: Since R can be covered with $2mk^{m-1}$ boxes of size r/k, fatness implies that any disjoint subcollection must have size at most $K \leq 2cmk^{m-1}$. This gives us the complexity $n^{O(K)} = n^{O(k^{m-1})}$ (details omitted due to lack of space). To improve this, consider a set S of disjoint

f-cones of size at least r/k in $\mathbb{R}^m$, as well as a box R of size r whose boundary the f-cones intersect. Since the f-cones are expanding and grounded, they also intersect the box that is obtained by translating R in the height coordinate until it becomes grounded. Now, observe that the collection of $(m-1)$-dimensional balls $\mathcal{D}_i^{r/k}$ obtained by intersecting the f-cones from S with the hyperplane $Z_{r/k}$ is a disjoint collection of $(m-1)$-dimensional balls of size at least r/k that intersects $R \cap Z_{r/k}$, which is just a $(m-1)$-dimensional box of size r. Since these balls are in $m-1$ dimensions and unit radius balls are fat, we can bound K by $2c'mk^{m-2}$, for a constant c'. This reduces the complexity of the algorithm to $n^{O(k^{m-2})}$. Note that $\eta = m-1$ and hence the complexity in terms of η is given by $n^{O(k^{\eta-1})}$. Thus, in our case, we can replace [6, Lemma 3.1] with the following lemma, which as mentioned above, leads to the improved overall complexity.

Lemma 3. *If all cones in $\mathcal{C}$ corresponding to $G \in GDG_f^{\eta,1}$ are k-aligned, then the weighted packing problem can be solved in $n^{O(k^{\eta-1})}$ time and space.*

Theorem 3. *Given a graph $G \in GDG_f^{\eta,1}$, we can find a $(1+\epsilon)$-factor approximation for the MWIS in $n^{O(1/\epsilon^{\eta-1})}$ time and space.*

For $t > 1$, we get a PTAS for MWIS if the representation of the graph is $O(1)$-*clustered*, as defined below. This definition is motivated by link scheduling in wireless networks, where the corresponding graph is 1-clustered [13].

Definition 4. *Graph $G \in GDG_f^{\eta,t}$, with $t > 1$, is* β-clustered *if for every vertex $v_i \in V(G)$ and each pair of points $p_j, p_k \in P_i$, we have $d(p_j, p_k) \leq \beta \cdot l_i$.*

Consider a β-clustered graph $G \in GDG_f^{\eta,t}$. Let us first show that the objects C_i (each a cardinality t set of f-cones) are fat. Since G is β-clustered, the size of C_i is at most βl_i. Consider a box R of size r, and a disjoint collection $\mathcal{C}$ of objects corresponding to a subset of vertices where each object has size at least r, consists of t f-conses, and intersects R. It follows from the bound on the size of objects that each f-cone in an object C_i in $\mathcal{C}$ has size at least r/β. Partition R into $O(\beta^{\eta+1})$ boxes of size r/β, which we call *small boxes*. We know that each object $C_i \in \mathcal{C}$ has at least one *representative* f-cone intersecting R, and hence intersecting a small box. Note that the set of representative cones is also disjoint. Since f-cones are fat, for each small box R', there is a set $T_{R'} \subseteq \mathbb{R}^{\eta+1}$ of c points, such that if a f-cone (of size r/β) intersects R' then it intersects $T_{R'}$. Taking the union of $T_{R'}$ over all R', we get a set T of $O(\beta^{\eta+1}) \cdot c$ points, such that if a collection $\mathcal{C}$ of disjoint objects of size at least r intersect R then they intersect T as well. If $\beta = O(1)$, this shows that the set of objects corresponding to a β-clustered GDG is fat. This allows us to apply the result of [6] to this case directly. It is also straightforward to extend the optimizations that led to Theorem 3 to $O(1)$-clustered GDGs.

Theorem 4. *Given a $O(1)$-clustered graph $G \in GDG_f^{\eta,t}$, we can find a $(1+\epsilon)$-approximation for the MWIS in $n^{O(1/\epsilon^{\eta-1})}$ time and space.*

4 Structural Properties

The following concept generalizes the perfect elimination ordering of chordal graphs. It was introduced in [14] and [23], but we shall use the definition of [15].

Definition 5. *A graph G is k-simplicial if there is an order $v_1 \dots v_n$ of the vertices such that for each vertex v_i, the set $\{v_j : v_j \in N(v_i),\ j > i\}$ can be partitioned into k cliques in G.*

GDGs were shown to be constant-simplicial in doubling metrics [13], with rather large constants. We give a simpler proof in Euclidean spaces, with smaller constants.

We additionally assume in this section (only) that f is *strictly sub-linear*, in the sense that for every $x \geq y \geq 1$, $f(x) \leq yf(x/y)$. This holds, e.g., for $f(x) \sim x^{\delta}$ ($\delta \in [0, 1]$).

Theorem 5. *Let $G \in GDG_f^{\eta,t}$. If $\eta = 1$ then G is $2t$-simplicial, and if $\eta = 2$ then G is $6t$-simplicial.*

Proof. Consider the case $\eta = 2$. It suffices to show that there is a vertex v whose neighborhood can be covered with $6t$ cliques, since we can remove that vertex and repeat the argument, to obtain the simplicial ordering. Let $v_k \in V(G)$ be such that $l_k = \min\{l_j : v_j \in V(G)\}$. By Lemma 1, we can focus on the intersection of disks in $\mathcal{D}^{l_k}$. Let $p \in P_k$. Partition Z_{l_k} into six $\pi/3$-angle sectors with rays centered at p, and let X be one sector (see Fig. 3). Consider the set V_X of vertices v_j, such that there is a $p_j \in P_j \cap X$, with the f-cones of v_k and v_j intersecting. It suffices to show that V_X induces a clique, since there are 6 sectors for each $p \in P_k$, and $|P_k| = t$. Let $v_i, v_j \in V_X$, with $l_i \leq l_j$, and $p_i \in P_i$ and $p_j \in P_j$ be the corresponding points in X. The segments $[p, p_i]$ and $[p, p_j]$ form an angle of size at most $\pi/3$. Since the segment $[p_i, p_j]$ is opposite this angle, $|p_i - p_j| \leq \max\{|p - p_i|, |p - p_j|\}$, and Using the intersection of f-cones,

$$|p_i - p_j| \leq \max\{l_k f(l_i/l_k) + l_k f(1), l_k f(l_j/l_k) + l_k f(1)\} \leq l_i f(l_j/l_i) + l_i f(1) ,$$

since f is strictly sub-linear and non-decreasing. Hence, v_i and v_j are adjacent, which also implies that V_X induces a clique, as required. The proof for $\eta = 1$ is similar and is omitted.

This gives us algorithmic bounds on the chromatic number $\chi(G)$ in terms of the clique number $\omega(G)$.

Corollary 1. *Graphs in $GDG_f^{1,t}$ ($GDG_f^{2,t}$) can be colored with at most $2t\omega(G)$ ($6t\omega(G)$) colors without knowledge of the geometric representation, resulting in a $2t$ ($5t$) approximation algorithm, respectively.*

Proof. An ordering $v_1, v_2, \dots v_n$ is *d-inductive* if for each vertex v_i, the set $N' = \{v_j : v_j \in N(v_i), j > i\}$ satisfies $|N'| \leq d$. A k-simplicial graph is necessarily d-inductive for $d \leq k(\omega(G) - 1) + 1 \leq k\omega(G)$-inductive, and a greedy coloring algorithm uses at most $d + 1$ colors.

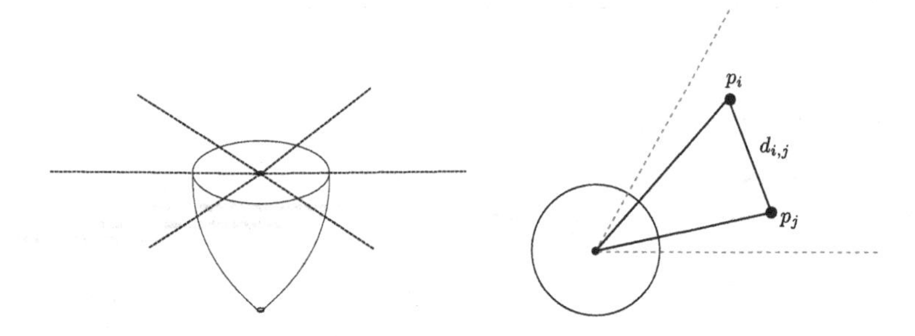

Fig. 3. The partitioning of the plane (left), and one part of the partition (right).

Given the representation, We can obtain a slightly better approximation by analyzing a related property. A graph is said to be *k-inductive independent* [23] if there is an ordering of the vertices so that each vertex has at most k mutually non-adjacent neighbors that follow it in the ordering. Such an ordering can be found in polynomial time when k is constant. This leads to k-approximation algorithms for coloring and weighted independent sets [23].

Theorem 6. *Graphs in $GDG_f^{2,t}$ are $5t$-inductive independent. There is a $5t$-approximation algorithm for MWIS and coloring graphs in $GDG_f^{2,t}$ without knowledge of the representation, when t is constant.*

The same approximation ratio has been shown to hold for disk graphs (see [11]) and, to our knowledge, no better results have been achieved.

5 One-Dimensional Case

We explore here relationship of $GDG_f^{1,1}$ to other classes, and show some exact algorithms for this restricted case.

An n-vertex graph H is an *outerstring graph* if it has an intersection representation of a set of n curves inside a disk such that one endpoint of every curve is on the boundary of the disk. Since there exists a homeomorphism from the disk to the halfspace, the geometric set corresponding to $G \in GDG_f^{1,1}$ is topologically equivalent to an outerstring graph.

Theorem 7. *$GDG_f^{1,1}$ is a subclass of outerstring graphs.*

A graph G is an *interval graph* if each vertex v_i can be associated with an interval $I_i \subseteq \mathbb{R}$, such that two vertices v_i and v_j are adjacent iff $I_i \cap I_j \neq \emptyset$.

Theorem 8. *Interval graphs are a subclass of $GDG_f^{1,1}$, for every increasing well-behaved function f.*

Proof. Let I be an interval graph with a given representation, and $(v_1, \ldots, v_k)$ be an ordering of vertices in increasing order of the right endpoints of the intervals. We construct a representation in $GDG_f^{1,1}$, by adding f-cones in this order. It

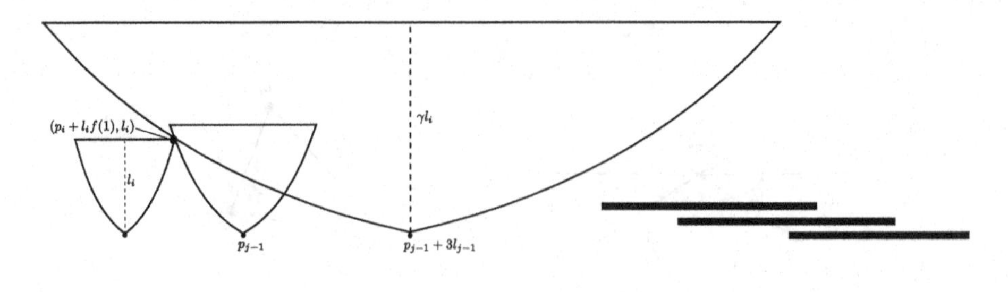

Fig. 4. How a GDG is constructed from an interval graph.

has the property that $l_i \geq l_j$, for $i > j$. The first vertex v_1 has position $p_1 = 0$ and length $l_1 = 1$. Consider a $j > 1$, and assume $l_{j-1} \geq l_i$, for $i < j - 1$. For vertex v_j, we let $p_j = p_{j-1} + cl_{j-1}$, where the $c > 1$ is such that $f(x) \leq cx/3$, for all $x \geq 0$. If v_j is not adjacent to any of $v_1, \ldots, v_{j-1}$, we set $l_j = l_{j-1}$, which ensures the non-adjacency: $g_{j-1}(l_{j-1}) + g_j(l_{j-1}) = 2f(1)l_{j-1} \leq (2c/3)l_{j-1} < d_{j,j-1}$. Otherwise, let v_i be the first vertex adjacent to v_j. We let $l_j = \gamma l_i$, for $\gamma = f^{-1}(d_{i,j}/l_i - f(1)) = f^{-1}(cl_{j-1}/l_i - f(1))$; since f is increasing and continuously differentiable, it has an increasing inverse (cf. The Inverse Function Theorem from standard Calculus). Thus, we have that $l_j \geq l_{j-1}$, as otherwise we would have $f^{-1}(cl_{j-1}/l_i - f(1))l_i < l_{j-1}$, and since $f(1) \leq c/3$, $l_i \leq l_{j-1}$, and f^{-1} is increasing, $f^{-1}((2c/3)l_{j-1}/l_i) \leq l_{j-1}/l_i$. We apply f on both sides, and since f is increasing, we get $(2c/3)l_{j-1}/l_i \leq f(l_{j-1}/l_i)$, which contradicts the choice of c. Thus, we get $l_j \geq l_i$, for $i < j$, and also $g_j(l_i) + g_i(l_i) = d_{i,j}$, which means that the right corner of the f-cone $\mathcal{C}_i$ is on the left edge of $\mathcal{C}_j$, and in particular, v_i and v_j are adjacent in $GDG_f^{1,1}$ (see Fig. 4). Let us show the correctness of the construction. Let v_k, v_m be a pair of vertices with $k < m$. Suppose that $v_k v_m \in E(I)$. If v_k is the least-index vertex adjacent to v_m then $\mathcal{C}_k, \mathcal{C}_m$ intersect, by construction. Otherwise, there is $s < k$, s.t. $v_s v_m \in E(I)$. We have $p_s < p_k < p_m$ and $l_s \leq l_k \leq l_m$, and $\mathcal{C}_s$ and $\mathcal{C}_m$ intersect. Clearly, then $\mathcal{C}_k$ and $\mathcal{C}_m$ intersect too, as required. Next, suppose that $v_k v_m \notin E(I)$. If v_m is not adjacent to any vertex v_s with $s < m$ then their cones do not intersect, by construction. Otherwise there is a vertex v_s with $s < m$ and $v_s v_m \in E(I)$. Since I is an interval graph, we must have $k < s$. Assume v_s is the vertex of least index adjacent to u_m. Then $\mathcal{C}_s$ and $\mathcal{C}_m$ intersect in exactly one point, i.e., the right corner point of $\mathcal{C}_s$. Since $p_k < p_s$ and $l_k \leq l_s$, the right corner of $\mathcal{C}_k$ is to the left of the right corner of $\mathcal{C}_s$, and at most on the same height. Thus, $\mathcal{C}_k$ and $\mathcal{C}_m$ cannot intersect.

It is known that interval graphs are a subclass of Max Point-Tolerance (MPT) graphs [5]. In the full version, we show that for $f(x) = x^\delta$, $\delta \in (0, 1)$, $K_{2,2,2} \in GDG_f^{1,1}$ and $K_{3,3} \notin GDG_f^{1,1}$. This implies, in particular that the classes $GDG_f^{1,1}$, for such f, are incompatible with MPT, since for MPT, the opposite inclusions have been demonstrated [5,20].

The Maximum Clique problem in unit disk graphs is solved in $O(n^{4.5})$ time by reducing it to MIS in a bipartite graph [8]. A similar approach gives us a $O(n^{3.5})$ time algorithm for Maximum Clique in $GDG_f^{1,1}$. In the full version, we also give a dynamic programming algorithm for exact MWIS in $GDG_f^{1,1}$.

Theorem 9. *Maximum Clique in* $GDG_f^{1,1}$ *is solvable in* $O(n^{3.5})$ *time.*

Proof. Let $v_i \in V(G)$ be such that $l_i = \min\{l_j : v_j \in V(G)\}$, and $H = G[N[v_i]]$. By Theorem 5, H is covered by at most two cliques, and thus $\bar{H}$ is bipartite, and we can use the Edmonds-Karp algorithm [9] to solve MIS in $\bar{H}$ in $O(n^{2.5})$ time. This solves Maximum Clique in H in $O(n^{2.5})$ time. Repeating this process for each vertex in G solves Maximum Clique in $O(n^{3.5})$ time.

Theorem 10. *MWIS in* $GDG_f^{1,1}$ *is solvable in* $O(n^3)$ *time.*

6 Conclusion and Open Questions

We have given a geometric formulation of GDGs that allows for improved optimization algorithms and the clarification of relations to other graph classes.

There are several questions that remain unanswered. We conjecture that for $\eta = 2$, $t = 1$, and $f(x) = x^\delta$ with $\delta \in (0, 1)$, the class $GDG_f^{2,1}$ properly contains disk graphs. We have seen the corresponding containment holds for the 1-dimensional analogs (i.e., interval graphs). We also believe that it does not matter which value of δ is chosen: that $GDG_f^{\eta,t}$ are all equal when $f(x) = x^\delta$ and $0 < \delta < 1$.

References

1. Agarwal, P.K., Katz, M.J., Sharir, M.: Computing depth orders for fat objects and related problems. Comput. Geom. **5**(4), 187–206 (1995)
2. Alt, H., et al.: Approximate motion planning and the complexity of the boundary of the union of simple geometric figures. Algorithmica **8**(1), 391–406 (1992)
3. Aronov, B., Bar-On, G., Katz, M.J.: Resolving SINR queries in a dynamic setting. In: Chatzigiannakis, I., Kaklamanis, C., Marx, D., Sannella, D. (eds.), 45th International Colloquium on Automata, Languages, and Programming, ICALP 2018, 9–13 July 2018, Prague, Czech Republic, volume 107 of LIPIcs, pp. 145:1–145:13. Schloss Dagstuhl - Leibniz-Zentrum für Informatik (2018)
4. Ásgeirsson, E.I., Halldórsson, M.M., Tonoyan, T.: Universal framework for wireless scheduling problems. In: 44th International Colloquium on Automata, Languages, and Programming, ICALP 2017, 10–14 July 2017, Warsaw, Poland, pp. 129:1–129:15 (2017)
5. Catanzaro, D., et al.: Max point-tolerance graphs. Discret. Appl. Math. **216**, 84–97 (2017)
6. Chan, T.M.: Polynomial-time approximation schemes for packing and piercing fat objects. J. Algorithms **46**(2), 178–189 (2003)
7. Chan, T.M., Har-Peled, S.: Approximation algorithms for maximum independent set of pseudo-disks. Discret. Comput. Geom. **48**(2), 373–392 (2012)

8. Clark, B.N., Colbourn, C.J., Johnson, D.S.: Unit disk graphs. Discret. Math. **86**(1–3), 165–177 (1990)
9. Edmonds, J., Karp, R.M.: Theoretical improvements in algorithmic efficiency for network flow problems. J. ACM **19**(2), 248–264 (1972)
10. Efrat, A., Katz, M.J., Nielsen, F., Sharir, M.: Dynamic data structures for fat objects and their applications. Comput. Geom. **15**(4), 215–227 (2000)
11. Erlebach, T., Fiala, J.: Independence and coloring problems on intersection graphs of disks. In: Bampis, E., Jansen, K., Kenyon, C. (eds.) Efficient Approximation and Online Algorithms. LNCS, vol. 3484, pp. 135–155. Springer, Berlin (2006). https://doi.org/10.1007/11671541_5
12. Erlebach, T., Jansen, K., Seidel, E.: Polynomial-time approximation schemes for geometric intersection graphs. SIAM J. Comput. **34**, 1302–1323 (2005)
13. Halldorsson, M.M., Tonoyan, T.: How well can graphs represent wireless interference? In: Proceedings of the Forty-Seventh Annual ACM Symposium on Theory of Computing. STOC 2015, pp. 635–644. Association for Computing Machinery, New York, NY, USA (2015)
14. Jamison, R.E., Mulder, H.M.: Tolerance intersection graphs on binary trees with constant tolerance 3. Discret. Math. **215**(1), 115–131 (2000)
15. Kammer, F., Tholey, T.: Approximation algorithms for intersection graphs. Algorithmica **68**(2), 312–336 (2012)
16. Kantor, E., Lotker, Z., Parter, M., Peleg, D.: The topology of wireless communication. J. ACM **62**(5), 37:1-37:32 (2015)
17. Keil, J.M., Mitchell, J.S.B., Pradhan, D., Vatshelle, M.: An algorithm for the maximum weight independent set problem on outerstring graphs. Comput. Geom. **60**, 19–25 (2017). The Twenty-Seventh Canadian Conference on Computational Geometry August 2015
18. Kratochvíl, J.: String graphs. I. The number of critical nonstring graphs is infinite. J. Comb. Theory, Ser. B **52**(1), 53–66 (1991)
19. Moscibroda, T., Wattenhofer, R.: The complexity of connectivity in wireless networks. In: INFOCOM, pp. 1–13. IEEE (2006)
20. Paul, S.: On characterizing proper-max-point tolerance graphs (2020)
21. Soto, M., Caro, C.T.: p-BOX: a new graph model. Discret. Math. Theor. Comput. Sci. **17**(1), 169–186 (2015)
22. van Kreveld, M.: On fat partitioning, fat covering and the union size of polygons. Comput. Geom. **9**(4), 197–210 (1998)
23. Ye, Y., Borodin, A.: Elimination graphs. ACM Trans. Algorithms **8**, 2 (2012)

A 4-Approximation of the $\frac{2\pi}{3}$-MST

Stav Ashur and Matthew J. Katz(✉)

Ben-Gurion University of the Negev, Beersheba, Israel
stavshe@post.bgu.ac.il, matya@cs.bgu.ac.il

Abstract. Bounded-angle (minimum) spanning trees were first introduced in the context of wireless networks with directional antennas. They are reminiscent of bounded-degree (minimum) spanning trees, which have received significant attention. Let P be a set of n points in the plane, and let $0 < \alpha < 2\pi$ be an angle. An α-spanning tree (α-ST) of P is a spanning tree of the complete Euclidean graph over P, with the following property: For each vertex $p_i \in P$, the (smallest) angle that is spanned by all the edges incident to p_i is at most α. An α-minimum spanning tree (α-MST) is an α-ST of P of minimum weight, where the weight of an α-ST is the sum of the lengths of its edges. In this paper, we consider the problem of computing an α-MST for the important case where $\alpha = \frac{2\pi}{3}$. We present a 4-approximation algorithm, thus improving upon the previous results of Aschner and Katz and Biniaz et al., who presented algorithms with approximation ratios 6 and $\frac{16}{3}$, respectively.

To obtain this result, we devise an $O(n)$-time algorithm that, given any Hamiltonian path Π of P, constructs a $\frac{2\pi}{3}$-ST $\mathcal{T}$ of P, such that $\mathcal{T}$'s weight is at most twice that of Π and, moreover, $\mathcal{T}$ is a 3-hop spanner of Π. This latter result is optimal in the sense that for any $\varepsilon > 0$ there exists a polygonal path for which every $\frac{2\pi}{3}$-ST (of the corresponding set of points) has weight greater than $2 - \varepsilon$ times the weight of the path.

Keywords: Bounded-angle spanning tree · Bounded-degree spanning tree · Hop-spanner

1 Introduction

Let $P = \{p_1, \ldots, p_n\}$ be a set of n points in the plane. An α-spanning tree (α-ST) of P, for an angle $0 < \alpha < 2\pi$, is a spanning tree of the complete Euclidean graph over P, with the following property: For each vertex $p_i \in P$, the (smallest) angle that is spanned by all the edges incident to p_i is at most α (see Fig. 1). An α-minimum spanning tree (α-MST) is then an α-ST of P of minimum weight, where the weight of an α-ST is the sum of the lengths of its edges.

Since there always exists a MST of P in which the degree of each vertex is at most 5 [12], the interesting range for α is $(0, \frac{8\pi}{5})$. The concept of bounded-angle (minimum) spanning tree (i.e., of an α-(M)ST) was introduced by Aschner and

M. Katz was supported by grant 1884/16 from the Israel Science Foundation.

A. Lubiw et al. (Eds.): WADS 2021, LNCS 12808, pp. 129–143, 2021.
https://doi.org/10.1007/978-3-030-83508-8_10

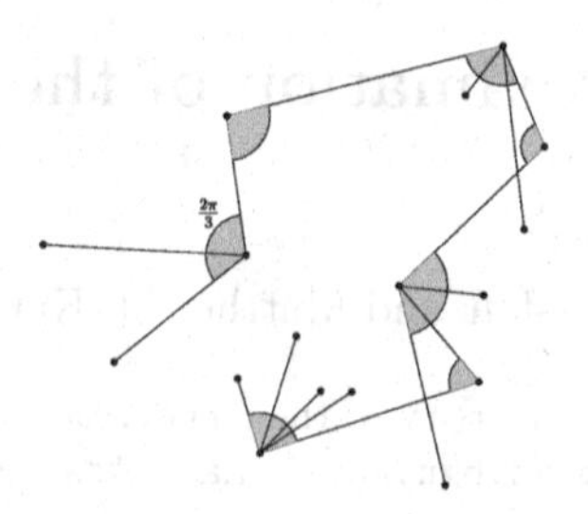

Fig. 1. A $\frac{2\pi}{3}$-ST.

Katz [3], who arrived at it through the study of wireless networks with directional antennas. However, it is interesting in its own right. The study of bounded-angle (minimum) spanning trees is also related to the study of bounded-degree (minimum) spanning trees, which received considerable attention (see, e.g., [8–11,13]). (A degree-k ST, is a spanning tree in which the degree of each vertex is at most k, and a degree-k MST is a degree-k ST of minimum weight.).

It is easy to see that an α-ST of P, for $\alpha < \frac{\pi}{3}$, does not always exist; think, for example, of the corners of an equilateral triangle. On the other hand, it is known (see [1,2,7]) that for any $\alpha \geq \frac{\pi}{3}$, there always exists an α-ST of P.

The next natural question is what is the status of the problem of computing an α-MST, for a given 'typical' angle α. Aschner and Katz [3] proved that (at least) for $\alpha = \pi$ and for $\alpha = \frac{2\pi}{3}$ the problem is NP-hard, and, therefore, it calls for efficient approximation algorithms.

Obviously, the weight of an α-MST of P, for any angle α, is at least the weight of a MST of P, so if we develop an algorithm for constructing an α-ST, for some angle α, and prove that the weight of the trees constructed by the algorithm never exceeds some constant c times the weight of the corresponding MSTs, then we have a c-approximation algorithm for computing an α-MST. Aschner et al. [4] showed that this approach is relevant only if $\alpha \geq \frac{\pi}{2}$, since for any $\alpha < \frac{\pi}{2}$, there exists a set of points for which the ratio between the weights of the α-MST and the MST is $\Omega(n)$.

In this paper, we focus on the important case where $\alpha = \frac{2\pi}{3}$. That is, we are interested in an algorithm for computing a 'good' approximation of $\frac{2\pi}{3}$-MST, where by good we mean that the weight of the output $\frac{2\pi}{3}$-ST is not much larger than that of a MST (and thus of a $\frac{2\pi}{3}$-MST). Aschner and Katz [3] presented a 6-approximation algorithm for the problem. Subsequently, Biniaz et al. [6] described an improved $\frac{16}{3}$-approximation algorithm. In this paper, we manage to reduce the approximation ratio to 4, by taking a different approach than the two previous algorithms.

Most of the paper is devoted to proving Theorem 7, which is of independent interest. Our main result, i.e., the 4-approximation algorithm, is obtained as an easy corollary of this theorem. Let Π denote the polygonal path $(p_1, ..., p_n)$. Then, Theorem 7 states that one can construct a $\frac{2\pi}{3}$-ST $\mathcal{T}$ of P, such that (i) the weight of $\mathcal{T}$, $\omega(\mathcal{T})$, is at most $2\omega(\Pi)$, and (ii) $\mathcal{T}$ is a 3-hop spanner of Π

(i.e., if there is an edge between p and q in Π, then there is a path consisting of at most 3 edges between p and q in $\mathcal{T}$). Notice that 2 is the best approximation ratio that one can hope for, since Biniaz et al. [6] showed that for any $\alpha < \pi$, the weight of an α-MST of a set of n points on the line, such that the distance between consecutive points is 1, is at least $2n-3$, whereas the weight of the MST (i.e., the polygonal path) is clearly $n-1$. (This lower bound is also mentioned without a proof in [3].).

We prove Theorem 7 by presenting an $O(n)$-time algorithm for constructing $\mathcal{T}$ and proving its correctness. The algorithm is very simple and easy to implement, but arriving at it and proving its correctness is far from trivial. One approach for constructing a $\frac{2\pi}{3}$-ST of P is to assign to each vertex of P an orientation, where an orientation of a vertex p is a cone of angle $\frac{2\pi}{3}$ with apex at p. The assignment of orientations induces a *transmission* graph G (over P), where $\{p_i, p_j\}$ is an edge of G if and only if p_j is in p_i's cone and p_i is in p_j's cone. Now, if G is connected, then by computing a minimum spanning tree of G one obtains a $\frac{2\pi}{3}$-ST of P. The challenge is of course to determine the orientations of the vertices, so that G is connected and the weight of a minimum spanning tree of G is bounded by a small constant times $\omega(\Pi)$.

Next, we describe some of the ideas underlying our algorithm for constructing $\mathcal{T}$. Assume for simplicity that n is even and consider the sequence of edges X obtained from Π by removing all the edges at even position (i.e., by removing the edges $\{p_2, p_3\}, \{p_4, p_5\}, \ldots$). For each edge $e = \{p, q\} \in X$, we consider the partition of the plane into four regions induced by e, see Fig. 2. This partition determines for each of e's vertices three 'allowable' orientations, see Fig. 3. Our algorithm assigns to each vertex of Π one of its three allowable orientations, such that the resulting transmission graph G contains the edges in X and at least one edge between any two adjacent edges in X. Finally, by keeping only the edges in X and a single edge between any two adjacent edges, we obtain $\mathcal{T}$. The novelty of the algorithm is in the way it assigns the orientations to the vertices to ensure that the resulting graph satisfies these conditions.

We now discuss the two previous results on computing an approximation of a $\frac{2\pi}{3}$-MST of P, and some of the related results. The first stage in the previous algorithms, as well as in the new one, is to compute a MST of P, MST(P), and from it a spanning path Π of P of weight at most $2\omega(\text{MST}(P))$. (Π is obtained by listing the vertices of P through a pre-order traversal of MST(P), where a vertex is added to the list when it is visited for the first time.) The algorithm of Aschner and Katz [3] operates on the path Π. It constructs a $\frac{2\pi}{3}$-ST of P from Π of weight at most $3\omega(\Pi)$, and thus of weight at most $6\omega(\text{MST}(P))$. The algorithm of Biniaz et al. [6] can operate only on non-crossing paths, so it first transforms Π to a non-crossing path Π' (through a sequence of $O(n^3)$ basic untangle operations), such that $\omega(\Pi') \leq \omega(\Pi)$. Then, it constructs a $\frac{2\pi}{3}$-ST of P from Π' of weight at most $\frac{8}{3}\omega(\Pi')$, and thus of weight at most $\frac{16}{3}\omega(\text{MST}(P))$. The new algorithm operates directly on Π, but in a completely different manner than its predecessors. It constructs a $\frac{2\pi}{3}$-ST of P from Π of weight at most $2\omega(\Pi)$, and thus of weight at most $4\omega(\text{MST}(P))$.

Notice that 4 is the best approximation ratio possible, for any such two-stage algorithm, provided the stages are independent. This is true since (i) Fekete et al. [9] showed that for any $\varepsilon > 0$ there exists a point set for which any spanning path has weight at least $2 - \varepsilon$ times the weight of a MST, and (ii) as mentioned above, for any $\varepsilon > 0$, there exists a point set and a corresponding spanning path for which any $\frac{2\pi}{3}$-ST has weight at least $2 - \varepsilon$ times the weight of the path.

As for other values of α, Aschner and Katz [3] presented a 16-approximation algorithm for computing a $\frac{\pi}{2}$-MST of P. The best known approximations of the degree-k MST, for $k = 2, 3, 4$, imply a 2-approximation of the π-MST, a 1.402-approximation of the $\frac{4\pi}{3}$-MST [8], and a 1.1381-approximation of the $\frac{3\pi}{2}$-MST [8,10].

2 Preliminaries

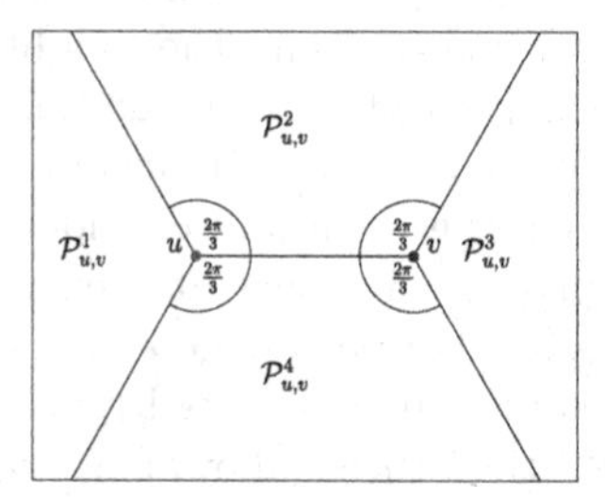

Fig. 2. The partition of the plane $\mathcal{P}_{u,v}$ induced by the ordered pair of points (u, v).

Definition 1. *Any ordered pair (u, v) of points in the plane, induces a partition of the plane into four regions, which we denote by $\mathcal{P}_{u,v}$; see Fig. 2. We denote the four regions by $\mathcal{P}^1_{u,v}$, $\mathcal{P}^2_{u,v}$, $\mathcal{P}^3_{u,v}$, and $\mathcal{P}^4_{u,v}$, as depicted in Fig. 2. Notice that the partitions $\mathcal{P}_{u,v}$ and $\mathcal{P}_{v,u}$ are identical, where $\mathcal{P}^1_{u,v} = \mathcal{P}^3_{v,u}$, $\mathcal{P}^2_{u,v} = \mathcal{P}^4_{v,u}$, etc. Sometimes, we prefer to consider the points u and v as an unordered pair of points, in which case we denote the partition induced by them as $\mathcal{P}_{\{u,v\}}$. In $\mathcal{P}_{\{u,v\}}$, we distinguish between the two* side *regions, which are $\mathcal{P}^1_{u,v}$ and $\mathcal{P}^3_{u,v}$ (alternatively, $\mathcal{P}^3_{v,u}$ and $\mathcal{P}^1_{v,u}$), and the two* center *regions, which are $\mathcal{P}^2_{u,v}$ and $\mathcal{P}^4_{u,v}$ (alternatively, $\mathcal{P}^4_{v,u}$ and $\mathcal{P}^2_{v,u}$).*

The *orientation* of a point u, is the orientation of a $\frac{2\pi}{3}$-cone with apex at u; we refer to this cone as the *transmission* cone of u. In the following definition, we define the three *basic* orientations of u with respect to another point v, based on $\mathcal{P}_{u,v}$; see Fig. 3.

Definition 2. *For a pair of points u and v, the three basic orientations of u with respect to v are:*

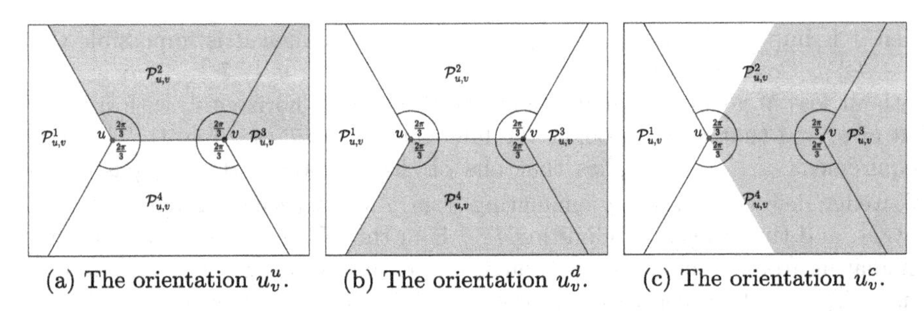

(a) The orientation u_v^u. (b) The orientation u_v^d. (c) The orientation u_v^c.

Fig. 3. The three basic orientations of u with respect to v; the superscripts u, d, and c stand for up, down, and center, respectively.

$\mathbf{u_v^u}$: *The only orientation of* u*, such that* $\mathcal{P}^2_{u,v}$ *is fully contained in the transmission cone of* u*,*

$\mathbf{u_v^d}$: *The only orientation of* u*, such that* $\mathcal{P}^4_{u,v}$ *is fully contained in the transmission cone of* u*, and*

$\mathbf{u_v^c}$: *The only orientation of* u*, such that* $\mathcal{P}^3_{u,v}$ *is fully contained in the transmission cone of* u*.*

Notice that in each of the basic orientations of u with respect to v, we have that v lies in u's cone. Therefore, for any assignment of basic orientation to u (with respect to v) and any assignment of basic orientation to v (with respect to u), the edge $\{u, v\}$ will be present in the resulting transmission graph.

Next, we prove three claims concerning the relationship between $\mathcal{P}_{\{u,v\}}$ and $\mathcal{P}_{\{x,y\}}$, where $\{u, v\}$ and $\{x, y\}$ are unordered pairs of points.

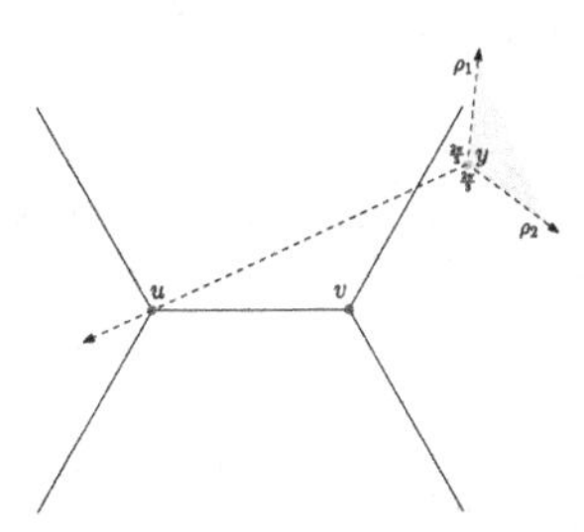

Fig. 4. Proof of Claim 1: If $y \in \mathcal{P}^3_{u,v}$ and $u \in \mathcal{P}^3_{x,y}$, then x must lie in the green region. (Color figure online)

Claim 1. *Let* $\{u, v\}$ *and* $\{x, y\}$ *be two unordered pairs of points. If* x *lies in one of the side regions of* $\mathcal{P}_{\{u,v\}}$ *and* y *lies in the other, then both* u *and* v *lie in the union of the center regions of* $\mathcal{P}_{\{x,y\}}$*.*

Proof. Assume, e.g., that $x \in \mathcal{P}^1_{u,v}$ and $y \in \mathcal{P}^3_{u,v}$. If u is not in one of the center regions of $\mathcal{P}_{\{x,y\}}$, then it is in one of the side regions of $\mathcal{P}_{\{x,y\}}$. But, if $u \in \mathcal{P}^1_{x,y}$, then it is impossible that $y \in \mathcal{P}^3_{u,v}$, and if $u \in \mathcal{P}^3_{x,y}$, then it is impossible that $x \in \mathcal{P}^1_{u,v}$. Consider for example the latter case, i.e., $u \in \mathcal{P}^3_{x,y}$, and assume, without loss of generality, that the line segment $\overline{uv}$ is horizontal, with u to the left of v, and that y is not below the line containing $\overline{uv}$ (see Fig. 4). Then, the requirement $u \in \mathcal{P}^3_{x,y}$ implies that one of the rays delimiting $\mathcal{P}^3_{x,y}$ is in the $\frac{2\pi}{3}$-wedge defined by the ray emanating from y and passing through u and the ray ρ_1, and the other ray delimiting $\mathcal{P}^3_{x,y}$ is in the $\frac{2\pi}{3}$-wedge defined by the ray emanating from y and passing through u and the ray ρ_2. So, $u \in \mathcal{P}^3_{x,y}$ implies that $\mathcal{P}^3_{x,y}$ and the green region in the figure are disjoint (when viewed as open regions), which, in turn, implies that x must lie in the green region. But this is impossible since the green region and $\mathcal{P}^1_{u,v}$ are disjoint.

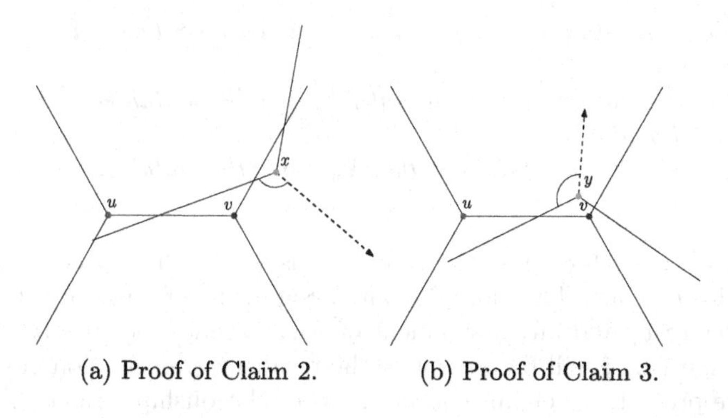

(a) Proof of Claim 2. (b) Proof of Claim 3.

Fig. 5. Left: If $u \in \mathcal{P}^1_{x,y}$ but $v \notin \mathcal{P}^1_{x,y}$, then y, which is on the dashed ray emanating from x, is necessarily in $\mathcal{P}^3_{u,v}$. Right: If $v \in \mathcal{P}^3_{x,y}$ but $u \notin \mathcal{P}^3_{x,y}$, then x, which is on the dashed ray emanating from y, is necessarily in $\mathcal{P}^2_{u,v}$. (Color figure online)

Claim 2. *Let $\{u, v\}$ and $\{x, y\}$ be two unordered pairs of points, such that x lies in one of the side regions of $\mathcal{P}_{\{u,v\}}$, say in the one adjacent to v, and y lies in one of the center regions of $\mathcal{P}_{\{u,v\}}$. Then, if u lies in the side region adjacent to x, then so does v.*

Proof. Assume that $u \in \mathcal{P}^1_{x,y}$ but $v \notin \mathcal{P}^1_{x,y}$. We show that this implies that $y \in \mathcal{P}^3_{u,v}$—a contradiction. Indeed, assume, without loss of generality, that the segment $\overline{uv}$ is horizontal, with u to the left of v, and that x is not below the line containing $\overline{uv}$ (see Fig. 5a). Since u and v are in different regions of $\mathcal{P}_{x,y}$, we know that the border between $\mathcal{P}^1_{x,y}$ (in which u resides) and $\mathcal{P}^4_{x,y}$ (in which v resides) crosses $\overline{uv}$. But, this implies that the dashed ray emanating from x is contained in $\mathcal{P}^3_{u,v}$, so y, which is somewhere on this ray, is in $\mathcal{P}^3_{u,v}$.

Claim 3. *Let $\{u, v\}$ and $\{x, y\}$ be two unordered pairs of points, such that x lies in one of the side regions of $\mathcal{P}_{\{u,v\}}$, say in the one adjacent to v, and y lies in one of the center regions of $\mathcal{P}_{\{u,v\}}$. Then, if v lies in the side region adjacent to y, then so does u.*

Proof. Assume that $v \in \mathcal{P}^3_{x,y}$ but $u \notin \mathcal{P}^3_{x,y}$. We show that this implies that x is in one of the center regions of $\mathcal{P}_{u,v}$—a contradiction. Indeed, assume, without loss of generality, that the segment $\overline{uv}$ is horizontal, with u to the left of v, and that $y \in \mathcal{P}^2_{u,v}$ (see Fig. 5b). Since u and v are in different regions of $\mathcal{P}_{x,y}$, we know that the border between $\mathcal{P}^3_{x,y}$ (in which v resides) and $\mathcal{P}^4_{x,y}$ (in which u resides) crosses $\overline{uv}$. But, this implies that the dashed ray emanating from y is contained in $\mathcal{P}^2_{u,v}$, so x, which is somewhere on this ray, is in $\mathcal{P}^2_{u,v}$.

3 Replacing an Arbitrary Path by a $\frac{2\pi}{3}$-Tree

Let $\{p_1, \ldots, p_n\}$ be a set of $n \geq 2$ points in the plane, and let Π denote the polygonal path $(p_1, \ldots, p_n)$. The *weight* of Π, $\omega(\Pi)$, is the sum of the lengths of the edges of Π, i.e., $\omega(\Pi) = \sum_{i=1}^{n-1} |p_i p_{i+1}|$. Let X and Y be the two natural matchings induced by Π, that is, $X = \{\{p_1, p_2\}, \{p_3, p_4\}, \ldots\}$ and $Y = \{\{p_2, p_3\}, \{p_4, p_5\}, \ldots\}$. Then, since $X \cap Y = \emptyset$, either $\omega(X)$ or $\omega(Y)$ is at most $\omega(\Pi)/2$. Assume, without loss of generality, that $\omega(X) \leq \omega(\Pi)/2$. Moreover, assume for now that n is even so X is a perfect matching.

In this section, we present an algorithm for replacing Π by a $\frac{2\pi}{3}$-tree, $\mathcal{T}$, such that $\omega(\mathcal{T}) \leq 2\omega(\Pi)$ and, moreover, $\mathcal{T}$ is a 3-hop spanner of Π (i.e., if there is an edge between p and q in Π, then there is a path consisting of at most three edges between p and q in $\mathcal{T}$). Our algorithm assigns to each of the vertices p of Π an orientation, which is one of the three basic orientations of p with respect to the vertex q matched to p in X.

In the subsequent description, we think of X as a sequence (rather than a set) of edges. Our algorithm consists of three phases.

3.1 Phase I

In the first phase of the algorithm, we iterate over the edges of X. When reaching the edge $\{p_i, p_{i+1}\}$, we examine it with respect to both its previous edge $\{p_{i-2}, p_{i-1}\}$ and its next edge $\{p_{i+2}, p_{i+3}\}$ in X. (The first edge is only examined w.r.t. its next edge, and the last edge is only examined w.r.t. its previous edge.) During the process, we either assign an orientation to one of p_i, p_{i+1}, to both of them, or to neither of them. In this phase, we only assign center orientations, i.e., u^c_v or v^c_u, where $\{u, v\}$ is an edge in X.

Let $e = \{u, v\}$ be the edge that is being considered and let $f = \{x, y\}$ be one of its (at most) two neighboring edges. We assign u the orientation u^c_v due to f if one of the following conditions holds:

1. One of f's vertices is in v's region (i.e., in the side region adjacent to v) and u is in the region of the other vertex of f; see Fig. 6a.
2. Both x and y are in v's region; see Fig. 6b.

Notice that it is possible that both conditions hold; see Fig. 6c. We say that u's orientation was *determined* by the second condition, only if the first condition

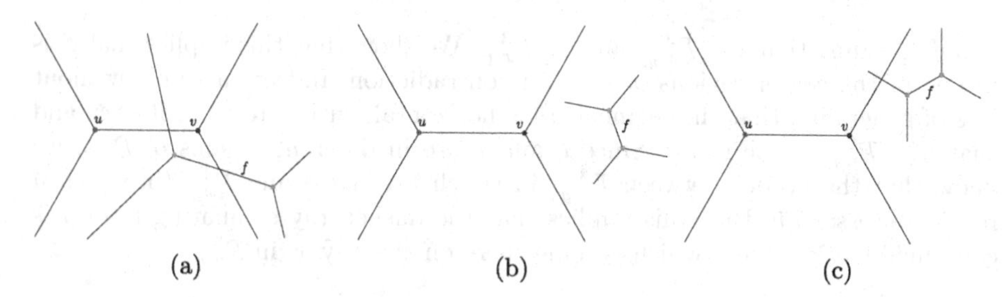

Fig. 6. The conditions by which we assign u the orientation u_v^c due to $f = \{x, y\}$. Left: u's orientation is determined by the first condition. Middle: u's orientation is determined by the second condition. Right: Both conditions hold, but we say that u's orientation is determined by the first condition.

does not hold; otherwise, we say that u's orientation was *determined* by the first condition.

Similarly, we assign v the orientation v_u^c due to f if one of the conditions above holds, when u is replaced by v.

The following series of claims deals with the outcome of examining an edge e with respect to a neighboring edge f.

Claim 4. *The orientation of at most one of the vertices of edge $e = \{u, v\}$ is determined, when e is examined with respect to a neighboring edge $f = \{x, y\}$.*

Proof. Assume that both u and v were oriented due to f and consider the conditions responsible for it. If the orientation of one of the vertices, say u, was determined by the second condition, then neither of the conditions can apply to v, since both conditions require that at least one of f's vertices is in u's region. If, however, the orientation of both u and v was determined by the first condition, then, without loss of generality, x is in u's region and y is in v's region, and by Claim 1 we conclude that u and v are in the center regions of $\mathcal{P}_{x,y}$, implying that neither of the vertices of e was oriented due to f.

Claim 5. *If the orientation of a vertex of edge $e = \{u, v\}$ is determined by the first condition, when e is examined with respect to a neighboring edge $f = \{x, y\}$, then the orientation of a vertex of f is determined by the first condition, when f is examined with respect to e, and these two vertices induce an edge of the transmission graph.*

Proof. Assume that, e.g., u's orientation is determined by the first condition (i.e., u is assigned the orientation u_v^c), when e is examined with respect to f. This means that there is a vertex of f, say x, that is in v's region, and that u is in y's region. Now, when we proceed to examine the edge f with respect to e, we find that u is in y's region and x is in v's region, so by the first condition we assign x the orientation x_y^c.

It remains to show that u and x induce and edge of the transmission graph. Indeed, x is in the transmission cone of u, since x is in v's region and u's cone contains v's region. Similarly, u is in the transmission cone of x, since u is in y's region and x's cone contains y's region.

Claim 6. *If the orientation of a vertex of edge $e = \{u, v\}$ is determined by the second condition, when e is examined with respect to a neighboring edge $f = \{x, y\}$, then neither of f's vertices is assigned an orientation due to e.*

Proof. If the orientation of, e.g., u is determined by the second condition, when e is examined with respect to f, then u is in one of the center regions of $\mathcal{P}_{\{x,y\}}$. Therefore, when f is examined with respect to e, the only condition that may hold is the first one. But if it does, then by Claim 5, the orientation of u is determined by the first condition, contrary to our assumption. We conclude that if the orientation of a vertex of e is determined by the second condition, then neither of f's vertices is assigned an orientation due to e.

3.2 Phase II

After completing the first phase, in which we iterated over the edges of X only once (i.e., a single round), we proceed to the second phase, in which we iterate over the edges of X again and again (i.e., multiple rounds). The second phase ends only after a full round is completed, in which no vertex is assigned an orientation. In a single round, we iterate over the edges of X, and for each pair of consecutive edges $e = \{u, v\}$ and $f = \{x, y\}$, where e precedes f, we assign orientations to the vertices of e and f, subject to the four rules listed below.

No reorienting: The orientation of a vertex is unmodifiable; that is, once the orientation of a vertex has been fixed (possibly already in the first phase), it cannot be changed.

Center orientation: A non-center orientation to a vertex u of an edge e is allowed, only if u is the second vertex of e to be assigned an orientation. Thus, if u is the first vertex of e to be assigned an orientation, then u must be assigned a center orientation.

Edge creation: Every operation that is performed must result in the creation of an edge of the transmission graph. This is achieved either by assigning orientations to two vertices simultaneously, or by orienting a vertex towards an already oriented vertex.

No double tapping: If one of e's vertices was already oriented due to f, where f is one of e's neighboring edges, then the other vertex of e will not be oriented due to f.

Notice that in this phase, unlike the previous one, the orientation decisions that we make when examining an edge e with respect to the next edge f, also depend on the orientations that some of the vertices of these edges may already have, and not only on the relative positions of these vertices.

In the full version of this paper [5], we illustrate several types of operations that are performed in the second phase, before proceeding to the third phase.

3.3 Phase III

In this phase we perform one final round, in which we orient all the vertices that were not yet oriented. More precisely, we iterate over the edges of X, considering each edge e with respect to the next edge f. When considering e, we orient its vertices that were not yet oriented, so that once we are done with e, both e itself and an edge connecting e and f are present in the transmission graph that is being constructed.

When considering the edge $e = \{u, v\}$ with respect to the next edge $f = \{x, y\}$, we know (by induction) that there already exists a transmission edge connecting e and the previous edge, so at most one of e's vertices was not yet oriented. If both vertices of e were already oriented, then either there already exists a transmission edge connecting e and f, or not. In the former case, proceed to the next edge of X (i.e., to f), and in the latter case, orient a vertex of f that was not yet oriented (there must be such a vertex), to obtain a transmission edge between e and f. We prove below that this is always possible.

If only one of e's vertices was already oriented, then let, e.g., u be the one that is not yet oriented. Now, if there already exists a transmission edge connecting e and f (i.e., v is connected to both the previous and the next edge of e), then assign u the orientation u_v^c (ensuring that e is a transmission edge). Otherwise, if one can assign an orientation to u, so that a transmission edge is created between u and an already oriented vertex of f, then do so. If this is impossible, then orient u and a vertex of f that was not yet oriented (there must be such a vertex), to obtain a transmission edge between e and f. We prove below that this is always possible.

3.4 Correctness

We first consider the more interesting case, where (i) one of the vertices of e, say u, is not yet oriented, (ii) there is no transmission edge between e and f, and (iii) it is impossible to orient u so that a transmission edge is created between u and an already oriented vertex of f. In this case, we need to prove that at least one of f's vertices is not yet oriented and that it is possible to orient both u and such a vertex of f to obtain a transmission edge between e and f.

We begin by showing that if both of f's vertices were already oriented, then either assumption (ii) or assumption (iii) does not hold. Indeed, by Claim 4 and the *No double tapping* rule of the second phase, one of f's vertices, say x, was oriented due to e. Now, if x was oriented during the first phase, then we distinguish between two cases according to the condition by which the orientation of x was determined.

x's orientation was determined by the first condition. In this case, by Claim 5, the edge $\{v, x\}$ is already in the transmission graph. In more detail, since u is not yet oriented, we must have that $x \in \mathcal{P}^1_{u,v}$ and $v \in \mathcal{P}^3_{x,y}$.

x's orientation was determined by the second condition. In this case, both u and v are in y's region and x is in one of the center regions of $\mathcal{P}_{\{u,v\}}$. So, by orienting u appropriately, one can obtain the transmission edge $\{u, x\}$.

If, however, x was oriented during the second phase, then by the *Edge creation* rule, an edge connecting e and f was already created.

We thus conclude that at least one of f's vertices is not yet oriented. We now consider, separately, the case where only one of f's vertices is not yet oriented and the case where both vertices of f are not yet oriented.

Only One of f's Vertices is Not Yet Oriented. Assume, without loss of generality, that y is the vertex of f that is already oriented. If y was oriented due to e, then by replacing x with y in the proof above, we get that either assumption (ii) or assumption (iii) does not hold. Therefore, we assume that y was oriented due to the edge following f, which implies that y was oriented in the first or second phase. Now, if u and x can be oriented to obtain the transmission edge $\{u, x\}$, then we are done. Otherwise, $u \in \mathcal{P}^1_{x,y}$ or $x \in \mathcal{P}^1_{u,v}$. We consider these cases below and show, for both of them, that a transmission edge between e and f can still be created.

$\mathbf{u} \in \mathcal{P}^1_{\mathbf{x},\mathbf{y}}$: Notice that since x is not yet oriented and y was oriented in the first or second phase, y's orientation is necessarily y^c_x. We consider each of the possible locations of v in $\mathcal{P}_{x,y}$, and show that regardless of v's location a transmission edge can be created.

1. If $v \in \mathcal{P}^1_{x,y}$, then y was oriented due to e during the first phase—contradiction.
2. If $v \in \mathcal{P}^3_{x,y}$, then, by Claim 1, x and y are in the center regions of $\mathcal{P}_{u,v}$, which allows us to orient u towards y to create the transmission edge $\{u, y\}$.
3. If v is in one of the center regions of $\mathcal{P}_{x,y}$, then we apply Claim 2 to show that we can orient x towards v to create the transmission edge $\{v, x\}$. Indeed, since (by assumption (iii)) we cannot orient u to create the transmission edge $\{u, y\}$, we know that $y \in \mathcal{P}^1_{u,v}$. So by Claim 2, we get that $x \in \mathcal{P}^1_{u,v}$. Therefore, since both x and y are in u's region, v's orientation was determined by the second condition during the first phase, and x can be oriented towards v to create the transmission edge $\{v, x\}$.

$\mathbf{x} \in \mathcal{P}^1_{\mathbf{u},\mathbf{v}}$: We first observe that if it is possible to create a transmission edge between v and x (i.e., $v \notin \mathcal{P}^1_{x,y}$), then it is possible to do so by assigning v a center orientation (since $x \in \mathcal{P}^1_{u,v}$), and we would have created the edge $\{v, x\}$ (by assigning v a center orientation and x an appropriate orientation) in the second phase, as y was oriented in the first or second phase. We assume therefore that it is impossible to create a transmission edge between v and x, which implies that $v \in \mathcal{P}^1_{x,y}$.

We now show that regardless of the location of y in $\mathcal{P}_{u,v}$, we get that $v \notin \mathcal{P}^1_{x,y}$—contradiction.

1. If $y \in \mathcal{P}^3_{u,v}$, then, by Claim 1, v is in a center region of $\mathcal{P}_{x,y}$.
2. If $y \in \mathcal{P}^1_{u,v}$, then an edge between v and y was created in the first phase (i.e., the orientations of both v and y were determined by the first condition of the first phase).
3. If y is in one of the center regions of $\mathcal{P}_{u,v}$, say $y \in \mathcal{P}^2_{u,v}$, then, by Claim 2 and using the assertion that $v \in \mathcal{P}^1_{x,y}$, we get that $u \in \mathcal{P}^1_{x,y}$ as well. Therefore, y

was assigned a center orientation in the first phase due to e, in contradiction to our assumption.

Both Vertices of f are Not Yet Oriented. If $x, y \in \mathcal{P}^1_{u,v}$, then v's orientation was determined by the second condition in the first phase (since if it were determined by the first condition, then we would already have an edge between e and f). Therefore, v's orientation is v^c_u and v is in one of the center regions of $\mathcal{P}_{x,y}$, and we orient either x or y towards v to create a transmission edge between e and f.

Assume, therefore, that at least one of f's vertices, say x, is not in $\mathcal{P}^1_{u,v}$. Now, if $u \notin \mathcal{P}^1_{x,y}$, then we orient u and x towards each other to create the edge $\{u, x\}$. So assume, in addition, that $u \in \mathcal{P}^1_{x,y}$. Under these assumptions, we show that regardless of the location of x in $\mathcal{P}_{u,v}$, $y \notin \mathcal{P}^1_{u,v}$, so u and y can be oriented towards each other to create the transmission edge $\{u, y\}$.

1. If x is in one of the center regions of $\mathcal{P}_{u,v}$, say $x \in \mathcal{P}^2_{u,v}$, then $y \notin \mathcal{P}^1_{u,v}$. Since, $y \in \mathcal{P}^1_{u,v}$, $x \in \mathcal{P}^2_{u,v}$ and $u \in \mathcal{P}^1_{x,y}$ implies (by Claim 3) that y was already oriented in the first phase.
2. If $x \in \mathcal{P}^3_{u,v}$, then again $y \notin \mathcal{P}^1_{u,v}$. Since, $y \in \mathcal{P}^1_{u,v}$ and $x \in \mathcal{P}^3_{u,v}$ implies (see Claim 1) that u is in one of the center regions of $\mathcal{P}_{x,y}$, contradicting the assumption $u \in \mathcal{P}^1_{x,y}$.

We now tend to the case where both vertices of e are already oriented, but there is no transmission edge between e and f. We first notice that this means that one of the vertices of e, say u, was oriented due to f. Moreover, u's orientation was determined by the second condition in the first phase, since otherwise an edge connecting e and f would already exist in the transmission graph. Next, we notice that at least one of the vertices of f was not yet oriented, since if both were oriented, then, again, one of them was oriented due to e and its orientation was determined by the second condition in the first phase. But, this implies that the first condition applies to both u and this vertex of f and that a transmission edge between them already exists.

Now, since u's orientation was determined by the second condition in the first phase, we know that it is in one of the center regions of $\mathcal{P}_{x,y}$. We can therefore orient the vertex of f that is not yet oriented towards u to create the required transmission edge.

At this point, the edge set of our transmission graph G contains X and at least one edge, for each pair e, f of consecutive edges of X, connecting a vertex of e and a vertex of f. Let $\mathcal{T}$ be the graph obtained from G by leaving only one (arbitrary) edge, for each pair of consecutive edges of X. Then, $\mathcal{T}$ is a $\frac{2\pi}{3}$-spanning tree of P. Denote by Y' the set of edges of $\mathcal{T}$ between (vertices of) consecutive edges of X. Then, $\omega(\mathcal{T}) = \omega(X) + \omega(Y') \leq \omega(X) + (2\omega(X) + \omega(Y)) = \omega(\Pi) + 2\omega(X) \leq 2\omega(\Pi)$. Moreover, $\mathcal{T}$ is a 3-hop spanner of Π, in the sense that if $\{p, q\}$ is an edge of Π, then there is a path between p and q in $\mathcal{T}$ consisting of at most 3 edges.

It was convenient to assume that X is a perfect matching, but it is possible of course that it is not. In the full version of this paper [5] we show how to deal with this case.

Running Time. The first and third phases of the algorithm each consist of a single round, whereas the second phase consists of $O(n)$ rounds. In each round we traverse the edges of X from first to last and spend $O(1)$ time at each edge. Thus, the running time of the first and third phases is $O(n)$, whereas the running time of the second phase is $O(n^2)$. We show below that the quadratic bound on the running time is due to our desire to keep the description simple, and that by slightly modifying the second phase we can reduce its running time to $O(n)$. The modification is based on the observation that beginning from the second round, an operation is performed when considering the pair e_i, e_{i+1} of edges of X (i.e., a transmission edge between them is created) if (i) an operation was performed in the previous round when considering e_{i+1} and e_{i+2}, or (ii) an operation was performed in the current round when considering e_{i-1} and e_i (or both).

Using this observation, we prove that two rounds are sufficient. Specifically, in the first round, we traverse the edges of X from first to last, i.e., a *forward* round, and in the second round, we traverse the edges of X from last to first, i.e., a *backward* round. In both rounds, in each iteration we consider the current edge and the following one, and check whether an operation can be performed (i.e., a transmission edge can be created), under the four rules listed in Sect. 3.2. We refer to such an operation as a *legal* operation.

We now prove that once we are done, no legal operation can be performed when considering a pair of adjacent edges of X. Indeed, let $g = \{p_i, p_{i+1}\}$, $f = \{p_{i+2}, p_{i+3}\}$, e, and d be four consecutive edges of X, and assume that after the backward round, one can still perform a legal operation when considering the pair e and f. Then, the operation became legal after an operation was performed when considering the pair f and g. Since, if it became legal after an operation was performed when considering the pair d and e, then we would have performed it during the backward round. However, by our assumption, no operation was performed during the backward round when considering the pair e and f, and therefore no operation was performed in this round when considering the pair f and g—contradiction.

The following theorem summarizes the main result of this section.

Theorem 7. *Let $P = \{p_1, \ldots, p_n\}$ be a set of n points in the plane, and let Π denote the polygonal path $(p_1, ..., p_n)$. Then, one can construct, in $O(n)$-time, a $\frac{2\pi}{3}$-spanning tree $\mathcal{T}$ of P, such that (i) $\omega(\mathcal{T}) \leq 2\omega(\Pi)$, and (ii) $\mathcal{T}$ is a 3-hop spanner of Π.*

Corollary 8. *Let $P = \{p_1, \ldots, p_n\}$ be a set of n points in the plane. Then, one can construct in $O(n \log n)$-time a $\frac{2\pi}{3}$-ST $\mathcal{T}$ of P, such that $\omega(\mathcal{T}) \leq 4\omega(\mathrm{MST}(P))$.*

4 Conclusion

Given a polygonal path Π, we have shown that it is possible to construct a $\frac{2\pi}{3}$-ST $\mathcal{T}$ of its corresponding set of points, whose weight is at most twice the weight of Π. Moreover, $\mathcal{T}$ is a 3-hop spanner of Π. As mentioned, this result is optimal in the sense that there exists a polygonal path for which it is impossible to construct a $\frac{2\pi}{3}$-ST of weight less than $2-\varepsilon$ times the path's weight, for any $\varepsilon > 0$. Consequently, we obtained a 4-approximation algorithm for computing a $\frac{2\pi}{3}$-MST of a set of points P, significantly improving the best previous approximation ratio of $\frac{16}{3}$ due to Biniaz et al. [6].

In general, the problem of computing an α-MST is a fascinating geometric problem (at least in our opinion). Moreover, it arises in the context of wireless networks with directional antennas of angle α. Thus, we believe that it (and its variants) will receive further attention in the future, similar to the older problem of computing a bounded-k MST.

References

1. Ackerman, E., Gelander, T., Pinchasi, R.: Ice-creams and wedge graphs. Comput. Geom. **46**(3), 213–218 (2013). http://dx.doi.org/10.1016/j.comgeo.2012.07.003
2. Aichholzer, O., et al.: Maximizing maximal angles for plane straight-line graphs. Comput. Geom. **46**(1), 17–28 (2013). http://dx.doi.org/10.1016/j.comgeo.2012.03.002
3. Aschner, R., Katz, M.J.: Bounded-angle spanning tree: modeling networks with angular constraints. Algorithmica **77**(2), 349–373 (2017). http://dx.doi.org/10.1007/s00453-015-0076-9
4. Aschner, R., Katz, M.J., Morgenstern, G.: Symmetric connectivity with directional antennas. Comput. Geom. **46**(9), 1017–1026 (2013). http://dx.doi.org10.1016/j.comgeo.2013.06.003/
5. Ashur, S., Katz, M.J.: A 4-approximation of the $\frac{2\pi}{3}$-MST. CoRR, abs/2010.11571 (2020). https://arxiv.org/abs/2010.11571
6. Biniaz, A., Bose, P., Lubiw, A., Maheshwari, A.: Bounded-angle minimum spanning trees. In: 17th Scandinavian Symposium and Workshops on Algorithm Theory, SWAT 2020, 22–24 June 2020, Tórshavn, Faroe Islands, pp. 14:1–14:22 (2020). https://dx.doi.org/10.4230/LIPIcs.SWAT.2020.14
7. Carmi, P., Katz, M.J., Lotker, Z., Rosén, A.: Connectivity guarantees for wireless networks with directional antennas. Comput. Geom. **44**(9), 477–485 (2011). http://dx.doi.org/10.1016/j.comgeo.2011.05.003
8. Chan, T.M.: Euclidean bounded-degree spanning tree ratios. Discret. Comput. Geom. **32**(2), 177–194 (2004). http://www.springerlink.com/index/10.1007/s00454-004-1117-3
9. Fekete, S.P., Khuller, S., Klemmstein, M., Raghavachari, B., Young, N.E.: A network-flow technique for finding low-weight bounded-degree spanning trees. J. Algorithms **24**(2), 310–324 (1997). http://dx.doi.org/10.1006/jagm.1997.0862
10. Jothi, R., Raghavachari, B.: Degree-bounded minimum spanning trees. Discret. Appl. Math. **157**(5), 960–970 (2009). http://dx.doi.org/10.1016/j.dam.2008.03.037
11. Khuller, S., Raghavachari, B., Young, N.E.: Low-degree spanning trees of small weight. SIAM J. Comput. **25**(2), 355–368 (1996). http://dx.doi.org/10.1137/S0097539794264585

12. Monma, C.L., Suri, S.: Transitions in geometric minimum spanning trees. Discret. Comput. Geom. **8**, 265–293 (1992). http://dx.doi.org/10.1007/BF02293049
13. Papadimitriou, C.H., Vazirani, U.V.: On two geometric problems related to the traveling salesman problem. J. Algorithms **5**(2), 231–246 (1984). http://dx.doi.org/10.1016/0196-6774(84)90029-4

Dynamic Dictionaries for Multisets and Counting Filters with Constant Time Operations

Ioana O. Bercea(✉) and Guy Even

Tel Aviv University, Tel Aviv, Israel
ioana@cs.umd.edu, guy@eng.tau.ac.il

Abstract. We resolve the open problem posed by Arbitman, Naor, and Segev [FOCS 2010] of designing a dynamic dictionary for multisets in the following setting: (1) The dictionary supports multiplicity queries and allows insertions and deletions to the multiset. (2) The dictionary is designed to support multisets of cardinality at most n (i.e., including multiplicities). (3) The space required for the dictionary is $(1 + o(1)) \cdot n \log \frac{u}{n} + \Theta(n)$ bits, where u denotes the cardinality of the universe of the elements. This space is $1 + o(1)$ times the information-theoretic lower bound for static dictionaries over multisets of cardinality n if $u = \omega(n)$. (4) All operations are completed in constant time in the worst case with high probability in the word RAM model.

A direct consequence of our construction is the first dynamic counting filter (i.e., a dynamic data structure that supports approximate multiplicity queries with a one-sided error) that, with high probability, supports operations in constant time and requires space that is $1 + o(1)$ times the information-theoretic lower bound for filters plus $O(n)$ bits.

The main technical component of our solution is based on efficiently storing variable-length bounded binary counters and its analysis via weighted balls-into-bins experiments in which the weight of a ball is logarithmic in its multiplicity.

Keywords: Ditionaries · Filters · Multisets

1 Introduction

We consider the dynamic dictionary problem for multisets. The special case of dictionaries for sets (i.e., multiplicities are ignored) is a fundamental problem in data structures and has been well studied [2,12,25,30]. In the case of multisets, elements can have arbitrary (adversarial) multiplicities and we are given an

This research was supported by a grant from the United States-Israel Binational Science Foundation (BSF), Jerusalem, Israel, and the United States National Science Foundation (NSF)

A full version of this paper can be found at [5].

A. Lubiw et al. (Eds.): WADS 2021, LNCS 12808, pp. 144–157, 2021.
https://doi.org/10.1007/978-3-030-83508-8_11

upper bound n on the total cardinality of the multiset (i.e., including multiplicities) at any point in time. The goal is to design a data structure that supports multiplicity queries (i.e., how many times does x appear in the multiset?) and allows insertions and deletions to the multiset (i.e., the dynamic setting).

A related problem is that of supporting *approximate* membership and multiplicity queries. Approximate set membership queries allow for one-sided errors in the form of false positives: given an error parameter $\varepsilon > 0$, the probability of returning a "yes" on an element not in the set is at most ε. Such data structures are known as *filters*. For multisets, the corresponding data structure is known as a *counting* filter (or a *spectral* filter). A counting filter returns a count that is at least the multiplicity of the element in the multiset and overcounts with probability bounded by ε. Counting filters have received significant attention over the years due to their applicability in practice [7,10,17]. One of the main applications of dictionaries for multisets is in designing dynamic filters and counting filters [2]. This application is based on Carter *et al.* [9] who showed that by hashing each element into a random fingerprint, one can reduce a counting filter to a dictionary for multisets by storing the fingerprints in the dictionary.

The lower bound on the space required for storing a dictionary follows from a simple counting argument (i.e., information theoretic lower bound). Namely, the space of a dictionary for multisets of cardinality n is at least $\log\binom{u+n}{n} = n\log(u/n) + \Theta(n)$ bits, where u is the size of the universe.[1, 2] In the case of filters, the lower bound is at least $n\log(1/\varepsilon) + \Theta(n)$ bits [23]. A data structure is *succinct* if the total number of bits it requires is $(1+o(1))\cdot\mathcal{B}$, where $\mathcal{B}$ denotes the lower bound on the space and the $o(1)$ term converges to zero as n tends to infinity. A data structure is *space-efficient* if it is succinct up to an additive $O(n)$ term in space.

For the design of both dictionaries and filters, the performance measures of interest are the space the data structure takes and the time it takes to perform the operations. The first goal is to design data structures for dictionaries over multisets that are space-efficient with high probability.[3] Our dictionary and counting filter are space-efficient for all ranges of parameters and succinct if the lower bound on the space satisfies $\mathcal{B} = \omega(n)$. Indeed, this is the case in a dictionary if $u = \omega(n)$ and in a filter if $\varepsilon = o(1)$.

The second goal is to support queries, insertions, and deletions in constant time in the word RAM model. The constant time guarantees should be in the worst case with high probability (see [1,2,8,21] for a discussion on the shortcomings of expected or amortized performance in practical scenarios). We assume that each memory access can read/write a word of $w = \log u$ contiguous bits.

---

[1] All logarithms are base 2 unless otherwise stated. $\ln x$ is used to denote the natural logarithm.

[2] This equality holds when u is significantly larger than n.

[3] By with high probability (whp), we mean with probability at least $1 - 1/n^{\Omega(1)}$. The constant in the exponent can be controlled by the designer and only affects the $o(1)$ term in the space of the dictionary or the filter.

The current best known dynamic dictionary for multisets was designed by Pagh, Pagh, and Rao [25] based on the dictionary for sets of Raman and Rao [30]. The dictionary is space-efficient and supports membership queries in constant time in the worst case. Insertions and deletions take amortized expected constant time and multiplicity queries take $O(\log n)$ in the worst case. In the case of sets, the state-of-the-art dynamic dictionary of Arbitman, Naor, and Segev [2] achieves the "best of both worlds": it is succinct and supports all operations in constant time whp. Arbitman *et al.* [2] pose the open problem of whether a similar result can be achieved for multisets.

Recently, progress on this problem was achieved by Bercea and Even [3] who designed a constant-time dynamic space-efficient dictionary for *random* multisets. In a random multiset, each element is sampled independently and uniformly at random from the universe (with repetitions). Multiplicities of elements in the dictionary in [3] are handled by storing duplicates. Namely, an element x with multiplicity $m(x)$ has $m(x)$ duplicate copies in the dictionary. The analysis employs ball-into-bins experiments in which the weight of a ball is linear in its multiplicity (more precisely, $\log(u/n) \cdot m(x)$). This analysis breaks if the multiset is arbitrary (i.e., not random).[4] Loosely speaking, in this paper, multiplicities are counted using variable-length counters. Thus, the analysis deals with ball-into-bins experiments in which the weight of a ball is *logarithmic* in its multiplicity (more precisely, $\log(u/n)+O(\log(m(x)))$). The design distinguishes between low and high multiplicities, where the threshold is $\log^3 n$, so that the length of the variable-length counters for the low multiplicities is $O(\log m(x)) = O(\log\log n)$ bits, and the length of counters for high multiplicities is $\log n$ bits.

1.1 Results

In the following theorem, *overflow* refers to the event that the space allocated in advance for the dictionary does not suffice. Such an event occurs if the random hash function fails to "balance loads".

Theorem 1 (dynamic multiset dictionary). *There exists a dynamic dictionary that maintains dynamic multisets of cardinality at most n from the universe* $\mathcal{U} = \{0,1\}^{\log_2 u}$ *with the following guarantees: (1) For every polynomial in n sequence of operations (multiplicity query, insertion, deletion), the dictionary does not overflow whp. (2) If the dictionary does not overflow, then every operation can be completed in constant time. (3) The required space is* $(1+o(1)) \cdot n\log(u/n) + O(n)$ *bits.*

Our dictionary construction considers a natural separation into the *sparse* and *dense* case based on the size of the universe relative to n. The sparse case, defined when $\log(u/n) = \omega(\log\log n)$, enables us to store additional $\Theta(\log\log n)$ bits per element without sacrificing space-efficiency. However, the encoding of the elements is longer, so fewer encodings can be packed in a word. In this

[4] For example, storing n copies of the same element would lead to almost all the elements being stored in the second level spare, causing the spare to overflow.

case, we propose a dictionary for multisets that is based on dynamic dictionaries that support both membership queries and satellite data (i.e., it stores (key, value) pairs where the key is the element and the value is its satellite data). We use two separate dictionaries: (1) One dictionary is used for the elements with multiplicity at most $\log^3 n$ (in which the satellite data is the multiplicity that is encoded using $O(\log\log n)$ bits). (2) The second dictionary is used for the elements with multiplicity at least $\log^3 n$ (in which the satellite data is the multiplicity that is encoded using $\log n$ bits). This construction is described in Sect. 3.

The dictionary for the *dense* case deals with the case in which $\log(u/n) = O(\log\log n)$.[5] Following [3], we hash distinct elements into a first level that consists of small space-efficient "bin dictionaries" of fixed capacity. The first level only stores elements of multiplicity strictly smaller than $\log^3 n$, just like in the dense case. However, we employ variable-length counters to encode multiplicities and store them in a separate structure called a "counter dictionary". We allocate one counter dictionary for each bin dictionary. The space (i.e., number of bits) of the counter dictionary is linear in the capacity of the associated bin dictionary (i.e., maximum number of elements that it can store). Namely, we spend a constant number of bits on average to encode the multiplicity of each element in the first level.

Elements that do not fit in the first level are stored in a secondary data structure called the *spare*. We prove that whp, the number of elements stored in the spare is $O(n/\log^3 n)$. Hence, even if a $\log n$-bit counter is attached to each element in the spare, then the spare still requires $o(n)$ bits. To bound the number of elements that are stored in the spare, we cast the process of hashing counters into counter dictionaries as a weighted balls-into-bins experiment in which balls have logarithmic weights (see Sect. 4.5).

As a corollary of Theorem 1, we obtain a counting filter with the following guarantees.[6]

Corollary 1 (dynamic counting filter). *There exists a dynamic counting filter for multisets of cardinality at most n from a universe* $\mathcal{U} = \{0,1\}^u$ *such that the following hold: (1) For every polynomial in n sequence of operations (multiplicity query, insertion, deletion), the filter does not overflow whp. (2) If the filter does not overflow, then every operation can be completed in constant time. (3) The required space is* $(1+o(1)) \cdot \log(1/\varepsilon) \cdot n + O(n)$ *bits. (4) For every multiplicity query, the probability of overcounting is bounded by* ε.

1.2 Related Work

The dictionary for multisets of Pagh *et al.* [25] is space-efficient and supports membership queries in constant time in the worst case. Insertions and deletions

[5] The dense case is especially relevant in practical approximate membership (filter) settings in which $u/n = 1/\varepsilon$ due to the reduction of Carter *et al.* [9].

[6] Note that we allow ε to be as small as n/u (below this threshold, we can simply use a dictionary instead).

take amortized expected constant time and multiplicity queries take $O(\log c)$ for a multiplicity of c. Multiplicities are represented "implicitly" by a binary counter whose operations (query, increment, decrement) are simulated as queries and updates to dictionaries on sets.[7] Increments and decrements to the counter take $O(1)$ bit probes (and hence $O(1)$ dictionary operations) but decoding the multiplicity takes $O(\log n)$ time in the worst case. We are not aware of any other dictionary constructions for multisets.[8]

Dynamic dictionaries for sets have been extensively studied [1,2,11–13,15, 19,28,30]. The dynamic dictionary for sets of Arbitman *et al.* [2] is succinct and supports operations in constant time whp. In [2], they pose the problem of designing a dynamic dictionary for multisets as an open question.

In terms of counting filters, several constructions do not come with worst case guarantees for storing arbitrary multisets [7,17]. The only previous counting filter with worst case guarantees we are aware of is the Spectral Bloom filter of Cohen and Matias [10] (with over 480 citations in Google Scholar). The construction is a generalization of the Bloom filter and hence requires $\Theta(\log(1/\varepsilon))$ memory accesses per operation. The space usage is similar to that of a Bloom filter and depends on the sum of logs of multiplicities. Consequently, when the multiset is a set, the required space is $1.44 \cdot \log(1/\varepsilon) \cdot n + \Theta(n)$.

1.3 Paper Organization

Preliminaries are in Sect. 2. The construction for the sparse case can be found in Sect. 3 and the one for the dense case is described and analyzed in Sect. 4. Corollary 1 is proved in the full version of the paper [5].

2 Preliminaries

For $k > 0$, let $[k]$ denote the set $\{0, \ldots, \lceil k \rceil - 1\}$. Let $\mathcal{U} \triangleq [u]$ denote the universe of all possible elements. We often abuse notation, and regard elements in $[u]$ as binary strings of length $\log u$. For a string $a \in \{0,1\}^*$, let $|a|$ denote the length of a in bits.

Definition 1 (multiset). *A* multiset $\mathcal{M}$ *over* $\mathcal{U}$ *is a function* $\mathcal{M} : \mathcal{U} \to \mathbb{N}$. *We refer to* $\mathcal{M}(x)$ *as the* multiplicity *of* x. *The* cardinality *of a multiset* $\mathcal{M}$ *is denoted by* $|\mathcal{M}|$ *and defined by* $|\mathcal{M}| \triangleq \sum_{x \in \mathcal{U}} \mathcal{M}(x)$. *The* support *of the multiset is denoted by* $\sigma(\mathcal{M})$ *and is defined by* $\sigma(\mathcal{M}) \triangleq \{x \mid \mathcal{M}(x) > 0\}$.

[7] To be more exact, for each bit of the counter, the construction in Pagh *et al.* [25] allocates a dictionary on sets such that the value of the bit can be retrieved by performing a lookup in the dictionary. Updating a bit of the counter is done by inserting or deleting elements in the associated dictionary.

[8] Data structures for predecessor and successor queries such as [29] can support multisets but they do not meet the required performance guarantees for multiplicity queries.

Operations over Dynamic Multisets. We consider the following operations: $\mathsf{insert}(x)$, $\mathsf{delete}(x)$, and $\mathsf{count}(x)$. Let $\mathcal{M}_t$ denote the multiset after t operations. A *dynamic multiset* $\{\mathcal{M}_t\}_t$ is specified by a sequence $\{\mathsf{op}_t\}_{t\geq 1}$ of as follows.[9]

$$\mathcal{M}_t(x) \triangleq \begin{cases} 0 & \text{if } t = 0 \\ \mathcal{M}_{t-1}(x) + 1 & \text{if } \mathsf{op}_t = \mathsf{insert}(x) \\ \mathcal{M}_{t-1}(x) - 1 & \text{if } \mathsf{op}_t = \mathsf{delete}(x) \\ \mathcal{M}_{t-1}(x) & \text{otherwise.} \end{cases}$$

We say that a dynamic multiset $\{\mathcal{M}_t\}_t$ has *cardinality* at most n if $|\mathcal{M}_t| \leq n$, for every t.

Dynamic Dictionary for Multisets. A *dynamic dictionary for multisets* maintains a dynamic multiset $\{\mathcal{M}_t\}_t$. The response to $\mathsf{count}(x)$ is simply $\mathcal{M}_t(x)$.

Dynamic Counting Filter. A *dynamic counting filter* maintains a dynamic multiset $\{\mathcal{M}_t\}_t$ and is parameterized by an error parameter $\varepsilon \in (0, 1)$. Let out_t denote the response to a $\mathsf{count}(x_t)$ at time t. We require that the output out_t satisfy the following conditions:

$$\mathsf{out}_t \geq \mathcal{M}_t(x_t) \tag{1}$$

$$\Pr\left[\mathsf{out}_t > \mathcal{M}_t(x_t)\right] \leq \varepsilon\ . \tag{2}$$

Namely, out_t is an approximation of $\mathcal{M}_t(x_t)$ with a one-sided error.

Definition 2 (overcounting). *Let* Err_t *denote the event that* $\mathsf{op}_t = \mathsf{count}(x_t)$, *and* $\mathsf{out}_t > \mathcal{M}_t(x_t)$.

Note that overcounting generalizes false positive events in filters over sets. Indeed, a false positive event occurs in a filter for sets if $\mathcal{M}_t(x_t) = 0$ and $\mathsf{out}_t > 0$.[10]

2.1 The Model

Memory Access Model. We assume that the data structures are implemented in the word RAM model in which every access to the memory accesses a word. Let w denote the memory word length in bits. We assume that $w = \log u$. See the full version of the paper [5] for a discussion on how the computations we perform over words are implemented in constant time.

Probability of Overflow. We prove that overflow occurs with probability at most $1/\mathsf{poly}(n)$ and that one can control the degree of the polynomial (the degree

[9] We require that $\mathsf{op}_t = \mathsf{delete}(x_t)$ only if $\mathcal{M}_{t-1}(x_t) > 0$.

[10] The probability space is induced only by the random choices (i.e., choice of hash functions) that the filter makes. Note also that if $\mathsf{op}_t = \mathsf{op}_{t'} = \mathsf{count}(x)$, then the events Err_t and $\mathrm{Err}_{t'}$ need not be independent.

of the polynomial only affects the $o(1)$ term in the size bound). The probability of an overflow depends only on the random choices that the dictionary makes.

Hash Functions. Our dictionary employs similar succinct hash functions as in Arbitman *et al.* [2] which have a small representation and can be evaluated in constant time. For simplicity, we first analyze the data structure assuming fully random hash functions (Sect. 4.5). In the full version of the paper [5], we prove that the same arguments hold when we use succinct hash functions and that the techniques in [2] used for sets can also be employed for multisets. The counting filter reduction additionally employs pairwise independent hash functions.

3 Dictionary for Multisets via Key-Value Dictionaries (Sparse Case)

In this section, we show how to design a multiset dictionary based on a dictionary on sets that supports attaching satellite data per element. Such a dictionary with satellite data supports the operations: query, insert, delete, retrieve, and update. A retrieve operation for x returns the satellite data of x. An update operation for x with new satellite data d stores d as the new satellite data of x. Loosely speaking, we use the satellite data to store a counter with $\Theta(\log\log n)$ bits. Hence, a succinct multiset dictionary is obtained from a succinct (set) dictionary only if $\log(u/n) = \omega(\log\log n)$.

Let $\mathsf{Dict}(\mathcal{U}, n, r)$ denote a dynamic dictionary for sets of cardinality at most n over a universe $\mathcal{U}$, where r bits of satellite data are attached to each element. One can design $\mathsf{Dict}(\mathcal{U}, n, r)$ from $\mathsf{Dict}(\mathcal{U}', n, 0)$, where $\mathcal{U}' \triangleq \mathcal{U} \times [2^s]$ if the first component of an element is a key. Namely, we require that the dataset $\mathcal{D}'(t) \subset \mathcal{U} \times [2^s]$ does not contain two elements (x_1, d_1) and (x_2, d_2) such that $x_1 = x_2$. An implementation of $\mathsf{Dict}(\mathcal{U}, n, r)$ (for $r = O(\log n)$) with constant time per operation can be obtained from the dictionary of Arbitman *et al.* [2] (see also [4]). The space of such an implementation is $(1 + o(1)) \cdot (\log(u/n) + r) \cdot n + O(n)$.

Let $\mathsf{MS\text{-}Dict}(\mathcal{U}, n)$ denote a dynamic dictionary for multisets over $\mathcal{U}$ of cardinality at most n. We propose a reduction that employs two dictionaries (with satellite data). The space for these dictionaries is allocated up front before the first element is inserted. (Hence, overflow of $\mathsf{MS\text{-}Dict}(n)$ occurs if one of these dictionaries overflows.)

Observation 1. *One can implement $\mathsf{MS\text{-}Dict}(\mathcal{U}, n)$ using two dynamic dictionaries: $D_1 = \mathsf{Dict}(\mathcal{U}, n, 3\log\log n)$ and $D_2 = \mathsf{Dict}(\mathcal{U}, n/(\log^3 n), \log n)$. Each operation over $\mathsf{MS\text{-}Dict}$ can be performed using a constant number of operations over D_1 and D_2.*

Proof (sketch). An element is *light* if its multiplicity is at most $\log^3 n$, otherwise it is *heavy*. Dictionary D_1 is used for storing the light elements, whereas dictionary D_2 is used for storing the heavy elements. The satellite data in both dictionaries is a binary counter of the multiplicity. Counters in D_1 are $3\log\log n$ bits long, whereas counters in D_2 are $\log n$ bits long.

Claim. If $\log(u/n) = \omega(\log\log n)$, then there exists a dynamic multiset dictionary that is succinct and supports operations in constant time in the worst case whp.

Proof. The implementation suggested in Observation 1 employs two dictionaries D_1 and D_2 (each with satellite data). The space of D_1 is $(1+o(1))\cdot((\log(u/n)+3\log\log n)\cdot n + O(n)$. The space of D_2 is $(1+o(1))\cdot((\log((u\log^3 n)/n)+\log n)\cdot \frac{n}{\log^3 n} + O(\frac{n}{\log^3 n}) = o(\log(u/n)\cdot n)$. Hence, the space of the multiset dictionary $\mathsf{MS\text{-}Dict}(n)$ is: $(1+o(1))\cdot((\log(u/n)+3\log\log n)\cdot n + O(n)$. In the sparse case $\log(u/n) = \omega(\log\log n)$. The lower bound on the space per element is $\log(u/n)$ bits, and hence the obtained $\mathsf{MS\text{-}Dict}(n)$ is succinct.

This completes the proof of Theorem 1 for the sparse case.

Remark. An alternative solution stores the multiplicities in an array separately from a dictionary that stores the support of the multiset. Let s denote the cardinality of the support of the multiset. Let $h : \mathcal{U} \to [s + o(s)]$ be a dynamic perfect hashing that requires $\Theta(s\log\log s)$ bits and supports operations in constant time (such as the one in [12]). Store the (variable-length) binary counter for x at index $h(x)$ in the array. The array can be implemented in space that is linear in the total length of the counters and supports query and update operations in constant time [6].

4 Dictionary for Multisets (Dense Case)

In this section, we prove Theorem 1 for the case in which $\log(u/n) = O(\log\log n)$, which we call the *dense* case. We refer to this dictionary construction as the *MS-Dictionary* (Multiset Dictionary) in the dense case.

The MS-Dictionary construction follows the same general structure as in [2,3,12]. Specifically, it consists of two levels of dictionaries. The first level is designed to store a $(1-o(1))$ fraction of the elements (Sect. 4.3). An element is stored in the first level provided that its multiplicity is at most $\log^3 n$ and there is enough capacity. Otherwise, the element is stored in the second level, which is called the *spare* (Sect. 4.4).

The first level of the MS-Dictionary consists of m *bin dictionaries* $\{\mathsf{BD}_i\}_{i\in[m]}$ together with m *counter dictionaries* $\{\mathsf{CD}_i\}_{i\in[m]}$. Each bin dictionary can store at most $n_B = (1+\delta)B$ distinct elements, where $\delta = o(1)$ and $B \triangleq n/m$ denotes the mean occupancy of bin dictionaries. We say that a bin dictionary is *full* if it stores n_B distinct elements.

Each counter dictionary stores variable-length binary counters. Each counter represents the multiplicity of an element in the associated bin dictionary. Each counter dictionary can store counters whose total length in bits is at most $12B$. We say that a counter dictionary is *full* if the total length of the counters stored in it is $12B$ bits. (The *length* of a counter that stores the value c is $\lceil\log_2(c+1)\rceil$ bits. The *encoding* of the counter is longer, namely $2(1+\lceil\log_2(c+1)\rceil)$ bits because we employ variable-length encoding.)

The following invariant specifies which elements are stored in the spare.

Invariant 2 *An element x such that $\mathcal{M}_t(x) > 0$ is stored in the spare at time t if: (1) $\mathcal{M}_t(x) \geq \log^3 n$, or (2) the bin dictionary corresponding to x is full, or (3) the counter dictionary corresponding to x is full.*

We emphasize that an element x cannot further stay in the spare if it does not satisfy Invariant 2. Namely, if the justification for storing x in the spare does not hold anymore, then it has to be transferred to first level. This transfer may be performed in a "lazy" fashion. Namely, instead of searching for elements in the spare that should be transferred to the first level, the transfer takes place when we stumble on them while trying to insert an element.

We denote the upper bound on the cardinality of the support of the multiset stored in the spare by n_S. We say that the spare *overflows* when more than n_S elements are stored in it.

4.1 Parametrization

The choice of parameters in the design of the MS-Dictionary for the dense case is summarized in Table 1.

Table 1. Setting of parameters in the MS-Dictionary in the dense case (i.e., $\log(u/n) = O(\log\log n)$).

Parameter value	Meaning
u	Cardinality of the universe $\mathcal{U}$
n	Maximum cardinality of the multiset $\mathcal{M}(t)$
$B \triangleq \frac{\log n}{\log(u/n)}$	Average number of elements per bin
$m \triangleq \frac{n}{B}$	Number of bins
$\delta \triangleq \Theta(\frac{\log\log n}{\sqrt{B}})$	Over-provisioning fraction per bin
$n_B \triangleq (1+\delta) \cdot B$	Maximum number of distinct elements stored in a bin
$n_s \triangleq \frac{3n}{\log^3 n}$	Maximum number of distinct elements stored in the spare

4.2 Hash Functions

We employ a permutation $\pi : \mathcal{U} \rightarrow \mathcal{U}$. We define $h^b : \mathcal{U} \rightarrow [m]$ to be the leftmost $\log m$ bits of the binary representation of $\pi(x)$ and by $h^r : \mathcal{U} \rightarrow [u/m]$ to be the remaining $\log(u/m)$ bits of $\pi(x)$. An element x is hashed to the bin dictionary of index $h^b(x)$. Hence storing x in the first level of the dictionary amounts to storing $h^r(x)$ in BD_i, where $i = h^b(x)$, and storing $\mathcal{M}_t(x)$ in CD_i. (This reduction in the universe size is often called "quotienting" [12,22,25,26]).

The overflow analysis in Sect. 4.5 assumes truly random permutations. In the full version of the paper [5], we discuss how one can replace this assumption with the succinct hash functions of Arbitman *et al.* [2].

4.3 The First Level of the Multiset Dictionary

The first level of the MS-Dictionary consists of bin dictionaries and counter dictionaries.

Bin Dictionaries. Each bin dictionary (BD) is a deterministic dictionary for sets of cardinality at most n_B that supports queries, insertions and deletions. Each bin dictionary can be implemented using global lookup tables [2] or Elias-Fano encoding [3]. Implementation via global lookup tables is succinct, whereas the Elias-Fano encoding requires $2 + \log(u/n)$ bits per element, and is succinct only if $\log(u/n) = \omega(1)$. Moreover, each BD fits in a constant number of words and performs queries, insertions and deletions in constant time.

Counter Dictionaries. Let $(x_1, \ldots x_\ell)$ denote the sequence of elements stored in BD_i. Let $\mathcal{M}(x_i)$ denote the multiplicity of x_i. The counter dictionary CD_i stores the sequence of multiplicities $(\mathcal{M}(x_1), \ldots, \mathcal{M}(x_\ell))$. Namely, the order of the element multiplicities stored in CD_i is the same order in which the corresponding elements are stored in BD_i. Multiplicities in CD_i are encoded using variable-length counters. We employ a trivial 2-bit alphabet to encode $0, 1$ and "end-of-counter" symbols for encoding the multiplicities. Hence, the *length* of a counter that stores the value c is $\lceil \log_2(c+1) \rceil$ bits while its *encoding* $2(1+\lceil \log_2 c \rceil)$ bits long. The contents of CD_i is simply a concatenation of the encoding of the counters. We allocate $2(12B + n_B) = O(B)$ bits per CD.[11]

The CD supports the operations of multiplicity query, increment and decrement. These operations are carried out naturally in constant time because each CD_i fits in $O(1)$ words. We note that an increment may cause the CD to be full, in which case x is deleted from the bin dictionary and is inserted into the spare together with its updated counter. Similarly, a decrement may zero the counter, in which case x is deleted from the bin dictionary (and hence its multiplicity is also deleted from the counter dictionary).

4.4 The Spare

Since the multiplicity of every element in the spare is at most n, the multiplicity can be represented by a $\log n$-bit counter. As in the dense case, the spare can be implemented using a dynamic dictionary $\mathsf{Dict}(\mathcal{U}, n_s, \log n)$. An additional requirement from that spare is that it supports moving elements back to the first level if their insertion no longer violates Invariant 2.

For this purpose, we propose to employ the dictionary of Arbitman *et al.* [1] that is a de-amortized construction of the cuckoo hash table of Pagh and Rodler [27]. Namely, each element is assigned two locations in an array. If upon insertion, both locations are occupied, then space for the new element is made by "relocating" an element occupying one of the two locations. Long chains of relocations are "postponed" by employing a queue of pending insertions. Thus,

[11] Note, however, that we define a CD to be full if the sum of counter lengths is $12B$ (even if we did not use all its space). The justification for this definition is to simplify the analysis.

each operation is guaranteed to perform in constant time in the worst case. The space that the dictionary occupies is $O(n_S(\log(u/n) + \log n) + O(n_s) = o(n)$.

The dynamic dictionary in [1] is used as a spare in the incremental filter in [2]. We use it a similar manner to maintain Invariant 2 in a "lazy" fashion. Namely, if an element x residing in the spare is no longer in violation of Invariant 2 (for instance, due to a deletion in the bin dictionary), we do not immediately move x from the spare back to its bin dictionary. Instead, we "delay" such an operation until x is examined during a chain of relocations. Specifically, during an insertion to the spare, for each relocated element, one checks if this element is still in violation of Invariant 2. If it is not, then it is deleted from the spare and inserted into the first level. This increases the time of operations only by a constant and does not affect the overflow probability of the spare.

4.5 Overflow Analysis

The event of an overflow occurs if more than n_S distinct elements are stored in the spare. In this section, we prove that overflow does not occur whp with respect to perfectly random hash functions.

Invariant 2 reduces the dynamic setting to the incremental setting in the sense that the number of elements in the spare at time t depends only on $\mathcal{D}(t)$ (and not on the complete history). The overflow analysis proceeds by proving that, for every t, the spare does not overflow whp. By applying a union bound, we conclude that overflow does not occur whp over a polynomial number of operations in the dynamic setting.

Recall that each component of the first level of the dictionary has capacity parameters: each bin dictionary has an upper bound of $n_B = (1 + \delta)B$ on the number of distinct elements it stores and each counter dictionary has an upper bound of $12B$ on the total length of the counters it stores. Additionally, the first level only stores elements whose multiplicity is strictly smaller than $\log^3 n$. According to Invariant 2, if the insertion of some element x exceeds these bounds, then x is moved to the spare.

We bound the number of elements that go to the spare due to failing one of the conditions of Invariant 2 separately. The number of elements whose multiplicity is at least $\log^3 n$ is at most $n/\log^3 n$. The number of distinct elements that are stored in the spare because their bin dictionary is full is at most $n/\log^3 n$ whp. The proof of this bound can be derived by modifying the proof of Claim 4.5 (see also [2]). We focus on the number of distinct elements whose counter dictionary is full.

Claim. The number of distinct elements whose corresponding CD is full is at most $n/\log^3 n$ whp.

Proof. Recall that there are $m = n/B$ counter dictionaries and that each CD stores the multiplicities of at most $n_B = (1 + \delta)B$ distinct elements of multiplicity strictly smaller than $\log^3 n$. In a full CD, the sum of the counter lengths reaches $12B$. We start by bounding the probability that the total length of the counters in a CD is at least $12B$.

Formally, consider a multiset $\mathcal{M}$ of cardinality n consisting of s distinct elements $\{x_i\}_{i\in[s]}$ with multiplicities $\{f_i\}_{i\in[s]}$ (note that $\sum_{i\in[s]} f_i = n$). The length of the counter for multiplicity f_i is $w_i \triangleq \lceil \log(f_i+1)\rceil$ (we refer to this quantity as *weight*). For $\beta \in [m]$, let $\mathcal{M}^\beta$ denote the sub-multiset of $\mathcal{M}$ consisting of the elements x_i such that $h^b(x_i) = \beta$. Let C_β denote the event that the weight of $\mathcal{M}^\beta$ is at least $12B$, namely $\sum_{x_i\in\mathcal{M}^\beta} w_i \geq 12B$. We begin by bounding the probability of event C_β occurring.

For $i \in [s]$, define the random variable $X_i \in \{0, w_i\}$, where $X_i = w_i$ if $h^b(x_i) = \beta$ and 0 otherwise. Since the values $\{(h^b(x_i), h^q(x_i))\}_i$ are sampled at random without replacement (i.e., obtained from a random permutation), the random variables $\{X_i\}_i$ are negatively associated. Let $\mu \triangleq \frac{1}{m}\cdot\sum_{i\in[s]} w_i$ denote the expected weight per CD. Since $w_i \leq \log(2(1+f_i))$, by the concavity of $\log(x)$, we have

$$\mu \leq \frac{s}{m}\log\frac{\sum_{i\in[s]}2(1+f_i)}{s} \leq \frac{s}{m}\log\left(2+\frac{2n}{s}\right) \leq 2B\ .$$

Since $w_i \leq \log\log^3 n$ (we omit the ceiling to improve readability), by Chernoff's bound:

$$\Pr\left[C_\beta\right] = \Pr\left[\sum_{i\in[s]} X_i \geq 6\cdot 2B\right] \leq 2^{-\frac{12B}{\log\log^3 n}} = 1/(\log n)^{\omega(1)}\ .$$

Let $I(C_\beta)$ denote the indicator variable for event C_β. Then $\mathbb{E}\left[\sum_\beta I(C_\beta)\right] \leq n/(\log n)^{\omega(1)}$. Moreover, the RVs $\{I(C_\beta)\}_\beta$ are negatively associated (more weight in bin b implies less weight in bin b'). By Chernoff's bound:

$$\Pr\left[\sum_b I(C_\beta) \geq \frac{n}{\log^5 n}\right] \leq O(2^{-n/(\log^5 n)})\ .$$

Whp, a bin is assigned at most $\log^2 n$ elements. We conclude that the number of elements that are stored in the spare due to events $\bigcup_b C_\beta$ is at most $n/(\log^3 n)$ whp.

4.6 Space Analysis

Each bin dictionary takes $n_B\log(u/n) + \Theta(n_B)$ bits, where $n_B = (1+\delta)B$, $B = n/m$ and $\delta = o(1)$. Each CD occupies $\Theta(B)$ bits. Therefore, the first level of the MS-Dictionary takes $(1+\delta)n\log(u/n) + \Theta(n)$ bits. The spare takes $O(n_S\log(u/n)) = o(n)$ bits, since $n_S = \Theta(n/\log^3 n)$. Therefore, the space the whole dictionary takes is $(1+o(1))\cdot\log(u/n) + \Theta(n)$ bits. This completes the proof of Theorem 1 for the dense case.

References

1. Arbitman, Y., Naor, M., Segev, G.: De-amortized cuckoo hashing: Provable worst-case performance and experimental results. In: Albers, S., Marchetti-Spaccamela, A., Matias, Y., Nikoletseas, S., Thomas, W. (eds.) Automata, Languages and Programming. ICALP 2009. Lecture Notes in Computer Science, vol. 5555. Springer, Berlin, Heidelberg (2009). https://doi.org/10.1007/978-3-642-02927-1_11
2. Arbitman, Y., Naor, M., Segev, G.: Backyard cuckoo hashing: constant worst-case operations with a succinct representation. In: 2010 IEEE 51st Annual Symposium on Foundations of Computer Science. pp. 787–796. IEEE (2010)
3. Bercea, I.O., Even, G.: A dynamic space-efficient filter with constant time operations. In: 17th Scandinavian Symposium and Workshops on Algorithm Theory, SWAT 2020, June 22–24, 2020, pp. 11:1–11:17. Tórshavn, Faroe Islands (2020). https://doi.org/10.4230/LIPIcs.SWAT.2020.11, https://doi.org/10.4230/LIPIcs.SWAT.2020.11
4. Bercea, I.O., Even, G.: Fully-dynamic space-efficient dictionaries and filters with constant number of memory accesses. CoRR abs/1911.05060 (2019). http://arxiv.org/abs/1911.05060
5. Bercea, I.O., Even, G.: A space-efficient dynamic dictionary for multisets with constant time operations. CoRR abs/2005.02143 (2020). https://arxiv.org/abs/2005.02143
6. Blandford, D.K., Blelloch, G.E.: Compact dictionaries for variable-length keys and data with applications. ACM Trans. Algorithms **4**(2) (2008). https://doi.org/10.1145/1361192.1361194
7. Bonomi, F., Mitzenmacher, M., Panigrahy, R., Singh, S., Varghese, G.: An improved construction for counting Bloom filters. In: Azar, Y., Erlebach, T. (eds.) Algorithms – ESA 2006. ESA 2006. Lecture Notes in Computer Science, vol. 4168. Springer, Berlin, Heidelberg (2006). https://doi.org/10.1007/11841036_61
8. Broder, A., Mitzenmacher, M.: Using multiple hash functions to improve ip lookups. In: Proceedings IEEE INFOCOM 2001. Conference on Computer Communications. Twentieth Annual Joint Conference of the IEEE Computer and Communications Society (Cat. No. 01CH37213). vol. 3, pp. 1454–1463. IEEE (2001)
9. Carter, L., Floyd, R., Gill, J., Markowsky, G., Wegman, M.: Exact and approximate membership testers. In: Proceedings of the Tenth Annual ACM Symposium on Theory of Computing, pp. 59–65. ACM (1978)
10. Cohen, S., Matias, Y.: Spectral Bloom filters. In: Proceedings of the 2003 ACM SIGMOD International Conference on Management of Data, pp. 241–252 (2003)
11. Dalal, K., Devroye, L., Malalla, E., McLeish, E.: Two-way chaining with reassignment. SIAM J. Comput. **35**(2), 327–340 (2005)
12. Demaine, E.D., auf der Heide, F.M., Pagh, R., Pǎtraşcu, M.: De dictionariis dynamicis pauco spatio utentibus. In: Correa, J.R., Hevia, A., Kiwi, M. (eds.) LATIN 2006: Theoretical Informatics. LATIN 2006. Lecture Notes in Computer Science, vol. 3887. Springer, Berlin, Heidelberg (2006). https://doi.org/10.1007/11682462_34
13. Dietzfelbinger, M., auf der Heide, F.M.: A new universal class of hash functions and dynamic hashing in real time. In: Paterson, M.S. (ed.) Automata, Languages and Programming. ICALP 1990. Lecture Notes in Computer Science, vol. 443. Springer, Berlin, Heidelberg (1990). https://doi.org/10.1007/BFb0032018

14. Dietzfelbinger, M., Rink, M.: Applications of a splitting trick. In: Albers, S., Marchetti-Spaccamela, A., Matias, Y., Nikoletseas, S., Thomas, W. (eds.) Automata, Languages and Programming. ICALP 2009. Lecture Notes in Computer Science, vol. 5555. Springer, Berlin, Heidelberg (2009). https://doi.org/10.1007/978-3-642-02927-1_30
15. Dietzfelbinger, M., Weidling, C.: Balanced allocation and dictionaries with tightly packed constant size bins. Theoret. Comput. Sci. **380**(1–2), 47–68 (2007)
16. Elias, P.: Efficient storage and retrieval by content and address of static files. J. ACM **21**(2), 246–260 (1974)
17. Fan, L., Cao, P., Almeida, J., Broder, A.Z.: Summary cache: a scalable wide-area web cache sharing protocol. IEEE/ACM Trans. Network. **8**(3), 281–293 (2000)
18. Fano, R.M.: On the Number of Bits Required to Implement an Associative Memory. Memorandum 61. Computer Structures Group, Project MAC, MIT, Cambridge, Mass (1971)
19. Fotakis, D., Pagh, R., Sanders, P., Spirakis, P.: Space efficient hash tables with worst case constant access time. Theory Comput. Syst. **38**(2), 229–248 (2005)
20. Kaplan, E., Naor, M., Reingold, O.: Derandomized constructions of k-wise (almost) independent permutations. Algorithmica **55**(1), 113–133 (2009)
21. Kirsch, A., Mitzenmacher, M.: Using a queue to de-amortize cuckoo hashing in hardware. In: Proceedings of the Forty-Fifth Annual Allerton Conference on Communication, Control, and Computing, vol. 75 (2007)
22. Knuth, D.E.: The Art of Computer Programming, vol. 3: Searching and sorting. Addison-Wisley, Reading MA (1973)
23. Lovett, S., Porat, E.: A lower bound for dynamic approximate membership data structures. In: 2010 IEEE 51st Annual Symposium on Foundations of Computer Science, pp. 797–804. IEEE (2010)
24. Naor, M., Reingold, O.: On the construction of pseudorandom permutations: Luby-Rackoff revisited. J. Cryptol. **12**(1), 29–66 (1999)
25. Pagh, A., Pagh, R., Rao, S.S.: An optimal Bloom filter replacement. In: SODA, pp. 823–829. SIAM (2005)
26. Pagh, R.: Low redundancy in static dictionaries with constant query time. SIAM J. Comput. **31**(2), 353–363 (2001)
27. Pagh, R., Rodler, F.F.: Cuckoo hashing. In: auf der Heide, F.M. (eds.) Algorithms – ESA 2001. ESA 2001. Lecture Notes in Computer Science, vol. 2161. Springer, Berlin, Heidelberg (2001). https://doi.org/10.1007/3-540-44676-1_10
28. Panigrahy, R.: Efficient hashing with lookups in two memory accesses. In: Proceedings of the Sixteenth Annual ACM-SIAM Symposium on Discrete Algorithms, pp. 830–839. Society for Industrial and Applied Mathematics (2005)
29. Pătraşcu, M., Thorup, M.: Dynamic integer sets with optimal rank, select, and predecessor search. In: 2014 IEEE 55th Annual Symposium on Foundations of Computer Science, pp. 166–175. IEEE (2014)
30. Raman, R., Rao, S.S.: Succinct dynamic dictionaries and trees. In: Baeten, J.C.M., Lenstra, J.K., Parrow, J., Woeginger, G.J. (eds.) Automata, Languages and Programming. ICALP 2003. Lecture Notes in Computer Science, vol. 2719. Springer, Berlin, Heidelberg (2003). https://doi.org/10.1007/3-540-45061-0_30
31. Schmidt, J.P., Siegel, A., Srinivasan, A.: Chernoff-Hoeffding bounds for applications with limited independence. SIAM J. Discret. Math. **8**(2), 223–250 (1995)
32. Siegel, A.: On universal classes of extremely random constant-time hash functions. SIAM J. Comput. **33**(3), 505–543 (2004)

The Neighborhood Polynomial of Chordal Graphs

Helena Bergold[1,2(✉)], Winfried Hochstättler[1], and Uwe Mayer[1]

[1] Fakultät für Mathematik und Informatik, FernUniversität in Hagen, Hagen, Germany
winfried.hochstaettler@fernuni-hagen.de

[2] Department of Computer Science, Freie Universität Berlin, Berlin, Germany
helena.bergold@fu-berlin.de

Abstract. In this paper, we study the neighborhood polynomial and the complexity of its computation for chordal graphs. The neighborhood polynomial of a graph is the generating function of subsets of its vertices that have a common neighbor. We introduce a parameter for chordal graphs called anchor width and an algorithm to compute the neighborhood polynomial which runs in polynomial time if the anchor width is polynomially bounded. The anchor width is the maximal number of different sub-cliques which appear as a common neighborhood. Furthermore we study the anchor width for chordal graphs and some subclasses such as chordal comparability graphs and chordal graphs with bounded leafage. The leafage of a chordal graphs is the minimum number of leaves in the host tree of a subtree representation. We show that the anchor width of a chordal graph is at most n^ℓ where ℓ denotes the leafage. This shows that for some subclasses computing the neighborhood polynomial is possible in polynomial time while it is NP-hard for general chordal graphs.

Keywords: Neighborhood polynomial · Domination polynomial · Chordal graph · Comparability graph · Leafage · Anchor width

1 Introduction

In this paper we study the neighborhood polynomial of graphs and give an algorithm to compute the polynomial for chordal graphs in polynomial time for some subclasses. Throughout the paper, all graphs are simple, finite and undirected. For a graph $G = (V, E)$, the *neighborhood* of a vertex $v \in V$ is the set of all adjacent vertices, denoted by $N_G(v) = \{u \in V \mid uv \in E\}$. The *neighborhood complex* of a graph G, first introduced by Lovász [13], consists of all subsets of vertices $W \subseteq V$ which have a common neighbor, that is $\mathcal{N}_G = \{U \subseteq V | \exists v \in V : U \subseteq N_G(v)\}$. This set-system is clearly hereditary, hence it is

The authors thank Kolja Knauer and Manfred Scheucher for helpful discussions and the anonymous reviewers for helpful comments. Helena Bergold was partially supported by DFG-GRK 2434.

A. Lubiw et al. (Eds.): WADS 2021, LNCS 12808, pp. 158–171, 2021.
https://doi.org/10.1007/978-3-030-83508-8_12

a simplicial complex. To count the number of sets with cardinality k in $\mathcal{N}_G$, we define the *neighborhood polynomial*

$$N_G(x) = \sum_{U \in \mathcal{N}_G} x^{|U|},$$

which is the generating function of the neighborhood complex $\mathcal{N}_G$. Since we only consider finite graphs, the sum is finite and $N_G(x)$ is a polynomial such as all other generating functions considered in this paper. We investigate the complexity of computing the neighborhood polynomial of some graph classes. In particular, we look at chordal graphs and subclasses like interval graphs, split graphs and chordal comparability graphs. In order to do this, we introduce the *anchor width* of a graph and develop an algorithm for computing the neighborhood polynomial in Sect. 3. We will see that the anchor width is the essential parameter for a polynomial runtime of our algorithm. If for any graph class the anchor width is polynomially bounded in the number of vertices, our algorithm is efficient. In particular our main result is the following theorem.

Theorem 1. *Let G be a chordal graph with n vertices and anchor width k. Computing the neighborhood polynomial takes at most $\mathcal{O}(n^3k + n^2k^2)$ time.*

In Sect. 4 we investigate the leafage $\ell(G)$ introduced in [12] of a chordal graph G on n vertices and show that the anchor width is at most $n^{\ell(G)}$ (cf. Theorem 2). For interval graphs, which are the graphs with leafage at most two, we give a family with quadratic anchor width.

2 Preliminaries

The neighborhood polynomial was introduced by Brown and Nowakowski [5] who investigated the effect of some elementary graph operations on the neighborhood polynomial. Given two graphs $G_1 = (V_1, E_1)$ and $G_2 = (V_2, E_2)$ on disjoint vertex sets, the *union* $G_1 \cup G_2$ of the graphs is the graph on the vertex set $V_1 \cup V_2$ with edge set $E_1 \cup E_2$. The *join* $G_1 + G_2$ of the two graphs is the graph on the vertex set $V_1 \cup V_2$ consisting of both graphs together with all possible edges between vertices in V_1 and vertices in V_2, that is $E = E_1 \cup E_2 \cup \{v_1v_2 \mid v_1 \in V_1, v_2 \in V_2\}$.

Proposition 1 ([5]). *Let G_1 and G_2 be two graphs on disjoint vertex sets. Then the neighborhood polynomial of the disjoint union $G_1 \cup G_2$ is*

$$N_{G_1 \cup G_2}(x) = N_{G_1}(x) + N_{G_2}(x) - 1.$$

Proposition 2 ([5]). *Let $G_1 = (V_1, E_1), G_2 = (V_2, E_2)$ be two graphs on disjoint vertex sets. Then the neighborhood polynomial of the join $G_1 + G_2$ is*

$$N_{G_1+G_2}(x) = (1+x)^{|V_2|} N_{G_1}(x) + (1+x)^{|V_1|} N_{G_2}(x) - N_{G_1}(x) N_{G_2}(x).$$

The two graph operations, disjoint union and join, are used to define cographs. *Cographs* are exactly the graphs which do not contain an induced P_4, a path on 4 vertices. They can be constructed recursively. Starting with a single vertex as a cograph, the disjoint union and the join of two cographs are cographs. For this and other well-known graph theoretic facts, we refer to [9]. The neighborhood polynomial of a single vertex graph is $N(K_1, x) = 1$ and the two operations disjoint union and join, given by the two formulas in Proposition 1 and Proposition 2 are computable in linear time. Note that $(1+x)^n$ can be computed in linear time using the binomial theorem. Corneil et al. [6] present a linear time algorithm to recognize cographs and give the corresponding recursive construction rules using disjoint union and join. Hence the neighborhood polynomial of a cograph is computable in quadratic time.

Another graph operation is attaching one vertex v to a subset of vertices of a graph G. This operation was studied by Alipour and Tittmann [1], who gave an explicit formula for a neighborhood polynomial after attaching a vertex to a subset of vertices. More formally for a graph $G = (V, E)$, a subset $U \subseteq V$ of vertices and an additional vertex $v \notin V$, we denote by $G_{U \triangleright v}$ the graph with vertex set $V \cup \{v\}$ and edge set $E \cup \{uv \mid u \in U\}$. We use the following notation for all $W \subseteq V$:

$$N_G^{\cap}(W) = \bigcap_{w \in W} N_G(w) \quad \text{and} \quad N_G^{\cup}(W) = \bigcup_{w \in W} N_G(w).$$

Proposition 3 ([1]). *Let $G = (V, E)$ be a graph, $U \subseteq V$ and $v \notin V$. Then the neighborhood polynomial of $G_{U \triangleright v}$ is*

$$N_{G_{U \triangleright v}}(x) = N_G(x) + \sum_{\substack{W \subseteq U, \\ W \neq \emptyset}} \phi_W + \sum_{\substack{W \subseteq U, \\ W \neq \emptyset}} (-1)^{|W|+1} x(1+x)^{|N_G^{\cap}(W)|},$$

where

$$\phi_W = \begin{cases} x^{|W|}, & \text{if } N_G^{\cap}(W) = \emptyset; \\ 0, & \text{otherwise.} \end{cases}$$

Using this formula, Alipour and Tittmann [1] showed that for a fixed integer k, computing the neighborhood polynomial of k-degenerate graphs is possible in polynomial time. A *k-degenerate graph* is a graph where every subgraph has a vertex v with $\deg(v) \leq k$. Using the degeneracy, we can pick one vertex of degree $\leq k$ after another and update the neighborhood polynomial by the formula of Proposition 3 in order to get a polynomial runtime. As a corollary it follows that there is a polynomial-time algorithm to compute the neighborhood polynomial for planar (or more general graphs of bounded genus) and k-regular graphs [1]. This update formula of Alipour and Tittman (see Proposition 3) was the starting point of our investigations for chordal graphs.

A graph G is said to be *chordal* if there is no induced cycle of length ≥ 4. Equivalently a graph is chordal if and only if it has a perfect elimination order. A *perfect elimination order* is an ordering of the vertices $v_1, \ldots, v_n$ such that for all i the neighborhood of v_i in $G[\{v_i, \ldots v_n\}]$ is a clique. Here for a subset $U \subseteq V$ the graph $G[U]$ denotes the subgraph of G induced by U. A vertex, whose neighborhood is a clique is called *simplicial.* It is well-known that every chordal graph has at least two simplicial vertices, which gives us the perfect elimination order (cf. [9]). In order to study the neighborhood polynomial of chordal graphs and subclasses, we make use of the perfect elimination order to build the chordal graph by attaching one vertex after another to a clique. We adapt the formula of Alipour and Tittmann (Proposition 3) to our use. To get some complexity results of computing the neighborhood polynomial, the connection to the domination polynomial is useful. For this we introduce dominating sets. A *dominating set* of a graph $G = (V, E)$ is a set of vertices $D \subseteq V$ such that

$$D \ \cup N_G^{\cup}(D) = V.$$

The family of all dominating sets of a graph G is denoted by $\mathcal{D}_G$ and the *domination polynomial* $D_G(x)$ is the generating function of $\mathcal{D}_G$ that is

$$D_G(x) = \sum_{U \in \mathcal{D}_G} x^{|U|}.$$

The following relation between domination polynomials and neighborhood polynomials holds. For a proof see for example [11] or [7].

Proposition 4. *For a graph $G = (V, E)$ and its complement graph $\overline{G}$ it holds:*

$$D_{\overline{G}}(x) + N_G(x) = (1 + x)^{|V|}.$$

The connection of these two polynomials can be used to determine the complexity of computing the neighborhood polynomial. In particular, the neighborhood polynomial is computable in polynomial time if and only if the domination polynomial of the complement graph is computable in polynomial time. Furthermore the contributions to the well-known graph problem DOMSET, imply some complexity results for the neighborhood polynomial. DOMSET is the problem of deciding whether a graph has a dominating set of size $\leq k$ for a given k.

Corollary 1. *Let $\mathcal{G}$ be a class of graphs and $\overline{\mathcal{G}}$ the class of the complement graphs of $\mathcal{G}$. If DOMSET is* NP*-complete on $\overline{\mathcal{G}}$, then computing the neighborhood polynomial on $\mathcal{G}$ is* NP*-hard.*

DOMSET is NP-hard on many graph classes such as chordal graphs [4]. Bertossi [3] showed that it is NP-hard on bipartite graphs and split graphs. *Split graphs* are the graphs where the vertex set can be partitioned into a clique and an independent set. Since split graphs are exactly the graphs which are chordal and co-chordal (i.e. the complement graph is chordal) [9], DOMSET is also NP-hard on co-chordal graphs. This together with Corollary 1 shows the NP-hardness of computing the neighborhood polynomial in split graphs (cf. [7]) and hence in chordal graphs.

3 Algorithm for Chordal Graphs

Our algorithm relies on the perfect elimination order of chordal graphs and comes from the vertex-attachment formula of Alipour and Tittmann, see Proposition 3. First, we adapt this formula to our special case where we attach a vertex to a clique. To study the new arising neighborhood sets after vertex attachment, we introduce anchor sets, which are subsets of a clique appearing as a common neighborhood of a set of vertices. The maximal number of anchor sets of a clique, which we denote as anchor width, is the essential parameter in this algorithms in order to get a polynomial runtime for our algorithm.

Let C be a clique in a graph $G = (V, E)$. We define the set of neighbors of the clique C, not including the clique itself as the *periphery* of C, denoted by

$$P_G(C) = N_G^{\cup}(C) \backslash C.$$

A subset $M \subseteq P_G(C)$ of the periphery is called *periphery set*. Note that the empty set is also a periphery set. We call a non-empty subset A of C *anchor set*, if it is the common neighborhood in C of some periphery set M. See Fig. 1 for an illustration. In general not all subsets of a clique are an anchor set. For every $M \subseteq P_G(C)$ we define the *corresponding anchor set* in C as

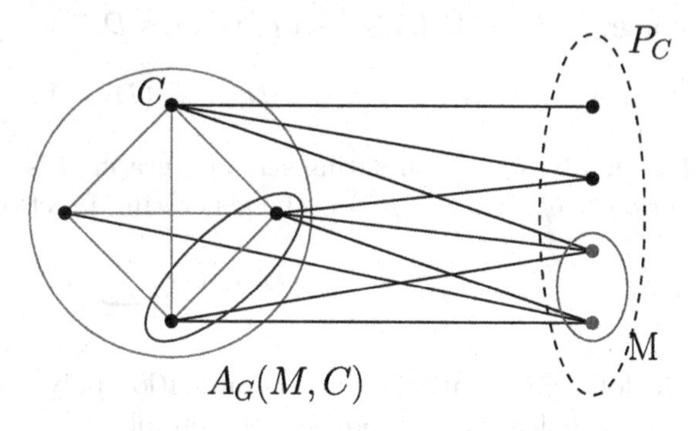

Fig. 1. An illustration of the introduced sets, the periphery P_C and a periphery set M with corresponding anchor set $A_G(M, C)$ of a clique C.

$$A_G(M, C) = N_G^{\cap}(M) \cap C$$

if the intersection is non-empty. Note that several periphery sets M and M' can correspond to the same anchor set $A_G(M, C) = A_G(M', C)$. For an anchor set A of C, the periphery sets $M \subseteq P_G(C)$ whose common neighborhood in C is A build the family

$$\mathcal{P}_G(A, C) = \{M \subseteq P_G(C) \mid A_G(M, C) = A\}.$$

The generating function of $\mathcal{P}_G(A, C)$ is called *periphery polynomial* and defined by

$$P_G(A, C, x) = \sum_{M \in \mathcal{P}_G(A,C)} x^{|M|}.$$

Note that $\mathcal{P}_G(A, C) = \emptyset$ and $P_G(A, C, x) = 0$ if A is not an anchor set of C. The family of all anchor sets of a clique C is

$$\mathcal{A}_G(C) = \{A \subseteq C \mid A \neq \emptyset \text{ and } \exists M \subseteq P_G(C) : A = A_G(M, C)\}.$$

Note that $C \in \mathcal{A}_G(C)$ for every clique C, since C is the anchor set of the empty periphery set. The *anchor width* of a graph G on n vertices is the smallest number k such that $|\mathcal{A}_G(C)| \leq k$ for all cliques C in G.

For a maximal clique C_{max} and a clique C contained in C_{max} the following relations hold. We omit the proof due to length restriction (see full version [2]).

Lemma 1. *Let C be a clique and C_{max} a maximal clique containing C in a graph $G = (V, E)$. Then the following conditions hold:*

(a) $C_{max} \backslash C \subseteq P_G(C) \subseteq P_G(C_{max}) \cup (C_{max} \backslash C)$
(b) $\mathcal{A}_G(C) = \{A \cap C \mid A \cap C \neq \emptyset \text{ and } A \in \mathcal{A}_G(C_{max})\}$
(c) For every $A \in \mathcal{A}_G(C)$ the periphery polynomial is

$$P_G(A, C, x) = (1 + x)^{|C_{max} \backslash C|} \sum_{\substack{A' \in \mathcal{A}_G(C_{max}) \\ A' \cap C = A}} P_G(A', C_{max}, x).$$

This shows that it is sufficient to provide the information about anchor sets and periphery polynomials for all maximal cliques of the graph. With this information we are able to compute the necessary information for all other cliques. Furthermore the anchor width only depends on the size of the anchor family of the maximal cliques.

In the following, we derive a formula for the neighborhood polynomial after vertex attachment using the periphery polynomial. For every set $U \subseteq V$ of vertices we define the *local neighborhood* $\mathcal{N}_G(U)$ of U as the family consisting of all vertex sets of G which have a common neighbor in U, that is

$$\mathcal{N}_G(U) = \{W \subseteq V \mid \exists v \in U : \ W \subseteq N_G(v)\}.$$

Note that $\mathcal{N}_G(V) = \mathcal{N}_G$. For every clique C, we can partition the local neighborhood $\mathcal{N}_G(C)$ by the following lemma into the disjoint sets

$$\mathcal{N}_G(A, C) = \{N \in \mathcal{N}_G(C) \mid N \cap P_G(C) \in \mathcal{P}_G(A, C)\}, \quad A \in \mathcal{A}_G(C).$$

Lemma 2. *For every clique C of the graph G, it holds*

$$\mathcal{N}_G(C) = \dot{\bigcup_{A \in \mathcal{A}_G(C)}} \mathcal{N}_G(A, C).$$

Proof. For every $N \in \mathcal{N}_G(C)$ there is by definition a $v \in C$ which is adjacent to every element in N. Hence the common neighborhood of $N \cap P_G(C)$ inside C is non-empty. This common neighborhood is an anchor set A. Since these anchor sets differ for different families $\mathcal{N}_G(A, C)$ the union is disjoint. □

Lemma 2 is useful since we only have to determine the generating functions of $\mathcal{N}_G(A, C)$ for every $A \in \mathcal{A}_G(C)$. Adding these generating functions, we maintain the generating function of the local neighborhood $\mathcal{N}_G(C)$. In the next lemma, we derive a formula to compute the generating function of $\mathcal{N}_G(A, C)$ for every $A \in \mathcal{A}_G(C)$.

Lemma 3. *For a given anchor set $A \in \mathcal{A}_G(C)$ of a clique C, the generating function of $\mathcal{N}_G(A, C)$ is*

$$N_G(A, C, x) = P_G(A, C, x)\left((1+x)^{|C|} - x^{|A|}(1+x)^{|C|-|A|}\right).$$

Proof. We count the number of sets with respect to the cardinality in $\mathcal{N}_G(A, C)$. Every $M \in \mathcal{P}_G(A, C)$ is in $\mathcal{N}_G(A, C)$. Hence $P_G(A, C, x)$ must be a summand of $N_G(A, C, x)$. Furthermore there are supersets N for all M which contribute to $N_G(A, C, x)$. Since we look at all $M \in \mathcal{P}_G(A, C)$, it is enough to look at supersets $N = M \cup X$, where X is a subset of C. In order to keep N in the local neighborhood $\mathcal{N}_G(C)$, we need a common neighbor in C. Since the common neighborhood of M inside C is the anchor set A, the common neighborhood of N must contain an element of A. Hence X cannot be the whole anchor set A. In particular, the possibilities to extend M are the elements of the family

$$\mathcal{X} = \{X \mid \exists a \in A : X \subseteq C \backslash \{a\}\}.$$

All sets in $\mathcal{X}$ consist of a disjoint union of a proper subset of A and a subset of $C \backslash A$. This leads to the generating function

$$\left((1+x)^{|A|} - x^{|A|}\right)(1+x)^{|C|-|A|} = (1+x)^{|C|} - x^{|A|}(1+x)^{|C|-|A|}$$

of $\mathcal{X}$. Note that the constant summand of this polynomial is 1. The generating function of $\mathcal{P}_G(A, C)$, which counts the different possibilities of M is counted by $P_G(A, C, x)$. □

This leads us to an update formula similar to Proposition 3 adapted to attaching a vertex to a clique.

Corollary 2. *Let $G = (V, E)$ be a graph and C a clique in the graph. The neighborhood polynomial of $G_{C \triangleright v}$ with vertex set $V \cup \{v\}$ is:*

$$\begin{aligned} N_{G_{C\triangleright v}}(x) &= N_G(x) + \phi_G(C) \\ &\quad + x \sum_{A \in \mathcal{A}_G(C)} P_G(A, C, x)\left((1+x)^{|C|} - x^{|A|}(1+x)^{|C|-|A|}\right), \end{aligned}$$

where

$$\phi_G(C) = \begin{cases} x^{|C|}, & \text{if } C \text{ is a maximal clique in } G; \\ 0, & \text{otherwise}. \end{cases}$$

Proof. Let $X \in \mathcal{N}_{G_{C\triangleright v}}$ be a neighborhood set in the graph $G_{C\triangleright v}$. We consider the following three cases:

- If $X \subseteq V$ and $X \nsubseteq N_G(v)$, then $X \in \mathcal{N}_G$ is in the neighborhood complex of G. Hence X is considered in the first summand $N_G(x)$ of the above formula.

- Now let $X \subseteq V$ and $X \subseteq N_G(v)$. If X is a proper subset of C, it already has a common neighbor in G, hence it is already counted in the first summand. Similarly this holds if $X = C$ and C is not a maximal clique in G, in other words C has a common neighbor in G. Thus the only case where a new neighborhood arises is if C is a maximal clique in G. In $G_{C \triangleright v}$ the common neighbor of C is v. This is counted in $\phi_G(C)$.
- Let us now consider the case $v \in X$, i.e. $X \not\subseteq V$. Since v is connected to all elements in C, we need to count all subsets $Y \subseteq V$ which have a common neighbor in C. This is equivalent to count the number of elements in $\mathcal{N}_G(C)$. Combining Lemma 2 and Lemma 3, we obtain

$$\sum_{A \in \mathcal{A}_G(C)} P_G(A, C, x) \left((1+x)^{|C|} - x^{|A|}(1+x)^{|C|-|A|} \right)$$

  as the generating function of $\mathcal{N}_G(C)$. In X there is one additional element v. Hence we have to multiply the polynomial with x.

Since the above cases are disjoint, this leads to the formula of the neighborhood polynomial as stated. □

With this formula, we are able to compute the neighborhood polynomial after attaching a vertex v to a clique C in a graph. In order to compute the neighborhood polynomial of a chordal graph G, we need the perfect elimination order $v_1, \ldots, v_n$. If the chordal graph is connected, we add the vertices in reverse order, starting with v_n and then adding v_i to the corresponding clique in $G[v_{i+1}, \ldots, v_n]$. For attaching one vertex, the neighborhood polynomial can be computed with Corollary 2. If the chordal graph is not connected we use the same procedure explained above for every connected component and compute the neighborhood polynomial by adding the polynomials of the connected components as in Proposition 1. In order to compute the formula of Corollary 2, we need the anchor family of C and the corresponding periphery polynomials $P_G(A, C, x)$ for every $A \in \mathcal{A}_G(C)$. As we have seen in Lemma 1 it is enough to store these informations for the maximal cliques and compute them in every step for the required clique C. The details of updating this information will be explained in the next paragraph.

We now study how to update the anchor families and periphery polynomials for the maximal cliques after attaching a vertex in order to have the correct ones in the next step. Fix a clique C of the graph G. The graph with attached vertex v to C is denoted by $G^+ = G_{C \triangleright v}$. We get exactly one new maximal clique $C^+ = C \cup \{v\}$ which we have to add to the list of maximal cliques in the graph. If C is a maximal clique in G we have to delete the C from the list of maximal cliques.

We determine the anchor sets and periphery polynomial of the new arising maximal clique C^+. The periphery of C^+ in G^+ is $P_{G^+}(C^+) = P_G(C)$ and the family of anchor sets is $\mathcal{A}_{G^+}(C^+) = \mathcal{A}_G(C) \cup \{C^+\}$ with the same periphery polynomials as in G that is $P_{G^+}(A, C^+, x) = P_G(A, C, x)$ for all $A \in \mathcal{A}_G(C)$ and $P_{G^+}(C^+, C^+, x) = 1$.

Now we go through the list of maximal cliques and update the necessary information if needed. The maximal cliques in G which have no intersection with C, do not change in G^+ and hence we do not have to update anything. Let C_{max} be a maximal clique in G with $C_{max} \cap C \neq \emptyset$. If $C = C_{max}$, we are done since this is not a maximal clique in G^+. Hence we assume $C \neq C_{max}$. The periphery of C_{max} in G^+ consists of the periphery of C_{max} in G together with the new element v. More formally it holds

$$P_{G^+}(C_{max}) = P_G(C) \cup \{v\}.$$

Now we identify the anchor sets of C_{max} in G^+. Every anchor set of C_{max} in G remains an anchor set in G^+. Since the new vertex v is attached to the subset $C_{max} \cap C$ of the considered clique C_{max}, this subset $C_{max} \cap C \neq \emptyset$ is a new anchor set in G^+, if it was not already an anchor set in G. Furthermore all subsets of C_{max} occurring as non-empty intersection of $C_{max} \cap C$ with an anchor set in $\mathcal{A}_G(C_{max})$ build an anchor set of C_{max} in G^+. The family of anchor sets of C_{max} in G^+ is:

$$\mathcal{A}_{G^+}(C_{max}) =$$
$$\mathcal{A}_G(C_{max}) \cup \{A \cap (C_{max} \cap C) \mid A \cap (C_{max} \cap C) \neq \emptyset \text{ and } A \in \mathcal{A}_G(C_{max})\}.$$

Since $C_{max} \in \mathcal{A}_G(C_{max})$, $C_{max} \cap C$ is an element of the second set in the above equation. Now we determine the periphery polynomial $P_{G^+}(A, C_{max}, x)$ for every anchor set $A \in \mathcal{A}_{G^+}(C_{max})$. Since $C \neq C_{max}$, the intersection $C_{max} \cap C$ is a proper subset of C_{max}. We consider the following three cases:

- If A is a proper subset of $C \cap C_{max}$, all corresponding periphery sets in G are a corresponding periphery set in G^+ and we can add v to every corresponding periphery set M in G, since the intersection with the neighborhood $N_{G^+}(v) = C$ does not change the anchor set. In this case we get:

$$P_{G^+}(A, C_{max}, x) = (1 + x) P_G(A, C_{max}, x).$$

- If $A = C \cap C_{max}$, the periphery sets in G with corresponding anchor set A which are counted in $P_G(A, C_{max}, x)$ still have the same anchor set in G^+. Furthermore v is a new periphery set with anchor set $A = C \cap C_{max}$ and all periphery sets which have a superset A' of A as corresponding anchor set, form together with v a periphery set with anchor set A. Hence the updated periphery polynomial is

$$P_{G^+}(A, C_{max}, x) =$$
$$P_G(A, C_{max}, x) + x \left(1 + \sum_{A' \supsetneq A, A' \in \mathcal{A}_G(C_{max})} P_G(A', C_{max}, x) \right). \quad (1)$$

- In the remaining case A is not a subset of $C \cap C_{max}$, hence v is not in a periphery set with anchor set A. So the periphery polynomial stays the same, which means

$$P_{G^+}(A, C_{max}, x) = P_G(A, C_{max}, x).$$

This concludes the algorithm to compute the neighborhood polynomial of a chordal graph. Recall that the anchor width of a graph G is the smallest k such that $|\mathcal{A}_G(C)| \leq k$ for all cliques of C.

The algorithm explained above leads to a polynomial time algorithm if the anchor width is polynomially bounded. For a detailed analysis of its complexity as claimed in Theorem 1 we refer to the full version [2].

4 Complexity of the Anchor Width

In this section, we will discuss some subclasses of chordal graphs and study their anchor width. We show that there are subclasses with polynomial bounded anchor width. We arrived at these graph classes starting from interval graphs, the first class for which we found a polynomial bound. For these subclasses the algorithm explained in Sect. 3 runs in polynomial time. In contrast to this result, we show that the anchor width of split graphs, a simple well-known subclass, is not polynomially bounded (see the following proposition). Hence the algorithm introduced in Sect. 3 might take super-polynomial time.

Proposition 5. *For all $n \in \mathbb{N}$ there is a split graph on n vertices with anchor width at least $2^{\frac{n}{2}} - 1$.*

Proof. We construct an infinite family of split graphs S_m on $n = 2m$ vertices such that the anchor width is $2^m - 1$. We start with a clique $C = \{c_1, \ldots, c_m\}$ of size m and attach vertices $p_1, \ldots, p_m$ such that every p_i $(1 \leq i \leq m)$ is adjacent to c_j for all $j \neq i$. This constructed graph is a split graph since C is a clique and $\{p_1, \ldots, p_m\}$ an independent set. All vertices p_i are therefore in the periphery P_C of C, hence $|P_C| = m = \frac{n}{2}$. All non-empty subsets of C are an anchor set, hence the anchor width of this graph is $2^m - 1 = 2^{\frac{n}{2}} - 1$. □

Another interesting subclass are the chordal comparability graphs. For these graphs we show a linear bound on the anchor width. A graph $G = (V, E)$ is a *comparability graph* if there is a poset $(V, \prec)$ such that two vertices $u, v \in V$ are adjacent in G if and only if $u \prec v$ or $v \prec u$.

Proposition 6. *The anchor width of a chordal comparability graph with n vertices is at most $2n$.*

Proof. Let G be a chordal comparability graph with corresponding poset $(V, \prec)$. Consider a maximal clique $C_{max} = \{c_1, \ldots, c_m\}$ in G. A clique in the graph corresponds to a chain in the poset. Hence the maximal clique C_{max} of size m corresponds to a maximal chain $c_1 \prec \ldots \prec c_m$ of length m in the poset. So whenever there is a vertex $v \notin C_{max}$, which is adjacent to $c_i \in C_{max}$ such that $v \prec c_i$ then v is adjacent to c_k for all $k \geq i$. Let i be the minimal element of the clique such that $v \prec c_i$. Since the clique is maximal, $i > 1$ and v is not comparable to c_{i-1}. Similar we get for every vertex $w \notin C_{max}$ which is connected to a vertex c_j of the clique with $w \succ c_j$ that w is connected to all elements c_k of the clique with $k \leq j$. Let c_j be the maximal element of the clique connected to

w, then $j < m$ and c_{j+1} is not comparable to w. Hence an anchor set in C_{max} is a chain of the form $c_i \prec c_{i+1} \prec \ldots \prec c_{j-1} \prec c_j$. Assume there is an anchor set with $1 < i < j < m$ and vertices v and w such that $v \prec c_i$ and $w \succ c_j$. Then v and w are connected by an edge since $w \succ c_j \succ c_i \succ v$ holds. And since w and c_{j+1} do not share an edge and analogously v and c_{i-1}, we get an induced cycle of length 4 which is not possible since the graph is chordal. In Fig. 2(a) the poset is illustrated by its Hasse diagram and gives an illustration of the contradiction. This shows that all anchor sets of C_{max} are of the form $c_1 \prec \ldots \prec c_{j-1} \prec c_j$ for $j \leq m$ or $c_i \prec c_{i+1} \prec \ldots \prec c_m$ for $i \geq 1$. We have at most $2m - 1 \leq 2n$ possibilities for those sets. □

Another interesting family of subclasses are the chordal graphs with bounded leafage. For those graphs we can show a polynomial upper bound of the anchor width. The leafage is a parameter which stems from the intersection graph representation of chordal graphs. An *intersection graph* is the graph consisting of one vertex for every set in the family. Two vertices are adjacent if and only if the corresponding sets have a non-empty intersection. An interval graph is an intersection graph of a family of subtrees of a path and chordal graphs are exactly the graphs which are the intersection graph of a family of subtrees of a *host tree* [8]. We call a representation of a chordal graph by a family of subtrees a *subtree representation*. Lin et al. [12] introduced the leafage of a chordal graph, which measures how close a chordal graph is to an interval graph. More precisely, the *leafage* $\ell(G)$ of a chordal graph G is defined as the minimum number of leaves of the host tree among all subtree configurations of G. We call a subtree representation *optimal* if it has the minimum number of leaves in the host tree. The interval graphs are exactly the chordal graphs with leafage at most 2. The split graphs S_m constructed earlier have leafage m and a host tree is the star $K_{1,m}$. Habib and Stacho present in [10] a polynomial-time algorithm in order to compute the leafage of a chordal graph. As mentioned in [12] we may restrict to host trees whose number of vertices is the number of maximal cliques of G.

Lemma 4. *There exists an optimal representation such that the vertices of the host tree are in one-to-one correspondence with the maximal cliques of the graph.*

Proof. Since every pairwise intersecting family of subtrees has the *Helly property*, i.e. the intersection of all subtrees of a clique is non-empty [9], there is at least one common vertex v_C in the host tree for every clique C of the chordal graph G. A vertex in the host tree cannot belong to different maximal cliques since their union has to form a clique as well and hence the cliques would not be maximal. Furthermore all subtrees intersecting in a vertex v of the host tree build a clique C. If C is not maximal, there is a maximal clique C_{max} containing C. Contracting the path from v to $v_{C_{max}}$ in the host tree does not increase the number of leaves. Thus, if we choose an optimal representation with few vertices as possible, the claim follows. □

We study the connection between the anchor width and the leafage of a chordal graph and show an upper bound of the anchor width. In the following, we describe a subtree of the host tree by its vertices.

Theorem 2. *For a chordal graph G with leafage $\ell = \ell(G)$ and n vertices, the anchor width is at most n^ℓ.*

Proof. Let $C = C_{max}$ be a maximal clique in the graph G. We consider an optimal subtree representation of G such that the vertices of the host tree T are in one-to-one correspondence with the maximal cliques of G (cf. Lemma 4). Let v_C be the vertex in the host tree which corresponds to the clique C of G. From v_C there is a unique path in the host tree to all ℓ leaves which we denote by $P_1, \ldots, P_\ell$.

For a periphery set $M \subseteq P_C$, the corresponding anchor set consists of those elements of the clique whose neighborhood contains M. For every $w \in M$, there is a tree T_w representing w in the subtree configuration. Since C is a maximal clique, these trees T_w do not contain v_C since otherwise w would belong to C. For every path P_i, we define a vertex v_i representing M on P_i as follows:

$$v_i \in \underset{v \in T_{w_i} \cap P_i}{\arg\min}\, dist(v, v_c),$$

where w_i is an element from the periphery such that

$$w_i \in \underset{w \in M}{\arg\max} \min_{v \in T_w \cap P_i} dist(v, v_c).$$

So for every $w \in M$ such that $T_w \cap P_i \neq \emptyset$, we choose the closest vertex v_w to v_C on the path P_i of the corresponding tree T_w. Among those vertices $\{v_w\}_w$, the vertex v_i is the vertex with maximal distance to v_C. If there is no subtree T_w of the periphery which has a non-empty intersection with the path P_i, we set $v_i = v_C$. Note that the v_i's are not necessarily distinct.

Now the anchor set $A = A_G(M, C)$ consists exactly of all subtrees of the clique C, which contain all $v_1, \ldots, v_\ell$ and v_C. If there is no such subtree corresponding to an element of the clique, there is no corresponding anchor set to M in C. The anchor set A is fully determined by the vertices v_i.

A chordal graph with n vertices has at most n maximal cliques. Hence the host tree has at most n vertices which gives at most n choices for every v_i. In total we have at most n^ℓ choices for the tuple $(v_1, \ldots, v_\ell)$ and hence at most n^ℓ different anchor sets. This shows the upper bound for the anchor width. □

Since interval graphs are the graphs with leafage at most 2, it follows:

Corollary 3. *The anchor width of interval graphs on n vertices is at most n^2.*

The magnitude of this bound is optimal since there is an infinite family of interval graphs on $n = 4m + 1$ vertices with a clique of size $2m + 1$ which has at least $m^2 = \left(\frac{(n-1)}{4}\right)^2$ different anchor sets. For the construction (see Fig. 2(b)), we take the path P on $2m + 1$ vertices $v_{-m}, \ldots, v_0, \ldots, v_m$ as a host tree. The subtrees corresponding to the clique C are the $2m + 1$ paths on the vertices

$$\{v_{-m}, \ldots, v_i\} \text{ for } i = 0, \ldots, m \quad \text{and} \quad \{v_i, \ldots, v_m\} \text{ for } i = -m + 1, \ldots, 0.$$

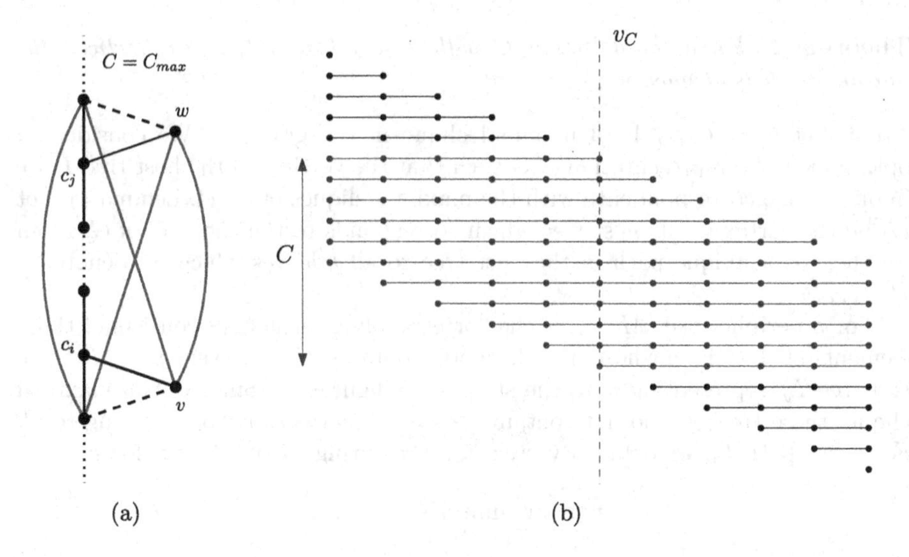

Fig. 2. (a) Hasse diagram of a poset corresponding to a comparability graph with induced C_4 which gives a contradiction in the proof of Proposition 6; (b) Construction of an interval graph with 21 vertices and a maximal clique of size 11 and 25 anchor sets.

The common intersection v_C of the clique is the vertex v_0. Furthermore we define the following subpaths, which are in the periphery of C:

$$\{v_{-m}, \ldots, v_i\} \text{ for } i = -m, \ldots, -1 \quad \text{and} \quad \{v_i, \ldots, v_m\} \text{ for } i = 1, \ldots, m.$$

For every choice of $i \in \{-m, \ldots, -1\}$ and $j \in \{1, \ldots, m\}$, we consider the two paths:

$$\{v_{-m}, \ldots, v_i\} \quad \text{and} \quad \{v_j, \ldots, v_m\}$$

of the periphery. The anchor set corresponding to this two-element periphery set consists of all paths in the host tree corresponding to a clique vertex which contain v_i and v_j. For every choice of i and j these anchor sets differ. Hence there are at least m^2 anchor sets.

5 Discussion

In this paper we studied an algorithm for computing the neighborhood polynomial of chordal graphs, which is in general an NP-hard problem. The runtime of the algorithm depends on the introduced parameter anchor width. If the anchor width of a subclass of chordal graphs is bounded, we have a polynomial-time algorithm to compute the neighborhood polynomial. In Sect. 4 we investigated some subclasses and showed that the anchor width is bounded for chordal graphs with bounded leafage and chordal comparability graphs. Furthermore we showed that the anchor width is not bounded for split graphs. It would be interesting to get further subclasses of chordal graphs with bounded anchor width. It might be possible to give an upper bound for the anchor width using the asteroidal number. In [12] it is shown that the leafage is an upper bound for the asteroidal number for all chordal graphs and they coincide for chordal graphs whose host tree is a subdivision of $K_{1,n}$ as shown in [14]. Furthermore an infinite family of graphs similar to the one for interval graphs, which shows that the magnitude of the upper bound is optimal, would be interesting.

On top of that there might be other problems on chordal graphs which are hard in general but polynomial solvable on those subclasses with bounded anchor width.

References

1. Alipour, M., Tittmann, P.: Graph operations and neighborhood polynomials. Discuss. Math. Graph. Theory. **41**, 697–711 (2021). https://doi.org/10.7151/dmgt.2347
2. Bergold, H., Hochstättler, W., Mayer, U.: The Neighborhood Polynomial of Chordal Graphs. arXiv:2008.08349 (2020)
3. Bertossi, A.A.: Dominating sets for split and bipartite graphs. Inf. Process. Lett. **19**(1), 37–40 (1984)
4. Booth, K.S., Johnson, J.H.: Dominating sets in chordal graphs. SIAM J. Comput. **11**(1), 191–199 (1982)
5. Brown, J.I., Nowakowski, R.J.: The neighbourhood polynomial of a graph. Australas. J. Comb. **42**, 55–68 (2008)
6. Corneil, D.G., Perl, Y., Stewart, L.K.: A linear recognition algorithm for cographs. SIAM J. Comput. **14**(4), 926–934 (1985)
7. Day, D.: On the neighbourhood polynomial. Master thesis, Dalhousie University (2017)
8. Gavril, F.: The intersection graphs of subtrees in trees are exactly the chordal graphs. J. Comb. Theory Ser. B **16**(1), 47–56 (1974)
9. Golumbic, M.C.: Algorithmic Graph Theory and Perfect Graphs (1980)
10. Habib, M., Stacho, J.: Polynomial-Time Algorithm for the Leafage of Chordal Graphs. In: Fiat, A., Sanders, P. (eds.) ESA 2009. LNCS, vol. 5757, pp. 290–300. Springer, Heidelberg (2009). https://doi.org/10.1007/978-3-642-04128-0_27
11. Heinrich, I., Tittmann, P.: Neighborhood and domination polynomials of graphs. Graphs Comb. **34**, 1203–1216 (2018)
12. Lin, I.-J., McKee, T.A., West, D.B.: The leafage of a chordal graph. Discuss. Math. Graph Theory **18**(1), 23–48 (1998)
13. Lovász, L.: Kneser's conjecture, chromatic number, and homotopy. J. Comb. Theory Ser. A **25**(3), 319–324 (1978)
14. Prisner, E.: Representing triangulated graphs in stars. Abh. Math. Semin. Univ. Hambg. **62**, 29–41 (1992)

Incomplete Directed Perfect Phylogeny in Linear Time

Giulia Bernardini[1,3(✉)], Paola Bonizzoni[1], and Paweł Gawrychowski[2]

[1] Università degli Studi di Milano - Bicocca, Milano, Italy
[2] Institute of Computer Science, University of Wrocław, Wrocław, Poland
[3] CWI, Amsterdam, The Netherlands
giulia.bernardini@cwi.nl

Abstract. Reconstructing the evolutionary history of a set of species is a central task in computational biology. In real data, it is often the case that some information is missing: the Incomplete Directed Perfect Phylogeny (IDPP) problem asks, given a collection of species described by a set of binary characters with some unknown states, to complete the missing states in such a way that the result can be explained with a directed perfect phylogeny. Pe'er et al. [SICOMP 2004] proposed a solution that takes $\tilde{\mathcal{O}}(nm)$ time (the $\tilde{\mathcal{O}}(\cdot)$ notation suppresses polylog factors) for n species and m characters. Their algorithm relies on pre-existing dynamic connectivity data structures: a computational study recently conducted by Fernández-Baca and Liu showed that, in this context, complex data structures perform worse than simpler ones with worse asymptotic bounds.

This gives us the motivation to look into the particular properties of the dynamic connectivity problem in this setting, so as to avoid the use of sophisticated data structures as a blackbox. Not only are we successful in doing so, and give a much simpler $\mathcal{O}(nm \log n)$-time algorithm for the IDPP problem; our insights into the specific structure of the problem lead to an asymptotically optimal $\mathcal{O}(nm)$-time algorithm.

1 Introduction

A rooted phylogenetic tree models the evolutionary history of a set of species: the leaves are in a one-to-one correspondence with the species, all of which have a common ancestor represented by the root. A standard way of describing the species is by a set of characters that can assume several possible states, so that each species is described by the states of its characters. Such a representation is naturally encoded by a matrix $\mathcal{A}$, $a_{i,j}$ being the state of character j in species i. When, for each possible character state, the set of all nodes that have the same state induces a connected subtree, a phylogeny is called *perfect*. The problem of reconstructing a perfect phylogeny from a set of species is known to be linearly-solvable in the case when the characters are binary [11], and it is NP-hard in the general case [2]. A popular variant of binary perfect phylogeny requires that the characters are directed, that is, on any path from the root to a leaf a character can change its state from 0 to 1, but the opposite cannot happen [5].

A. Lubiw et al. (Eds.): WADS 2021, LNCS 12808, pp. 172–185, 2021.
https://doi.org/10.1007/978-3-030-83508-8_13

In this paper, we study the Incomplete Directed Perfect Phylogeny problem (IDPP for short) introduced by Pe'er et al. [20]. The input of this problem is a matrix of binary character vectors in which some character states are unknown, and the question is whether it is possible to complete the missing states in such a way that the result can be explained with a directed perfect phylogeny.

Related work. Besides being relevant in its own right [1,18,19,22,24], the problem of handling phylogenies with missing data is crucial in various tasks of computational biology, like resolving genotypes with some missing information into haplotypes [13,17] and inferring tumor phylogenies from single-cell sequencing data with mutation losses [21]; a generalization of the perfect phylogeny model where a character can be gained only once and can be lost at most k times, called the k-Dollo model [3,4,6,12], has also been extensively studied. A deep understanding of the IDPP problem leading to new efficient solutions may thus highlight novel approaches for all such important tasks.

The algorithm proposed in [20] solves the IDPP problem for a matrix of n species and m characters in $\tilde{\mathcal{O}}(nm)$ time[1] with a graph-theoretic approach. A crucial step of such algorithm is to maintain the connected components of a graph under a sequence of edge deletions. The use of pre-existing dynamic connectivity data structures for this purpose is the bottleneck in the overall time complexity.

A connectivity data structure is *fully-dynamic* when both edge insertion and deletion are allowed, and *decremental* when only edge deletion is considered. A long line of results brought down the computational time required for updating the data structure after edge insertions and/or deletions, and for answering connectivity queries, to roughly logarithmic: the following table summarizes the existing results for a graph with N vertices and M edges. For fully-dynamic connectivity we report the update time required for a single edge insertion or deletion, while for decremental connectivity we report the overall time required to eventually delete all the edges. All the listed results, except for [14], assume that edge deletions can be interspersed with connectivity queries. The algorithm of Henzinger et al. [14], in contrast, deletes edges in batches (b_0 is the number of batches that do not result in a new component) and connectivity queries can be only asked between one batch of deletions and another.

Fully-dynamic	Update time	Query time
Holm et al. [15]	$\mathcal{O}(\log^2 N)$, amortized	$\mathcal{O}(\log N/\log\log N)$
Gibb et al. [10]	$\mathcal{O}(\log^4 N)$, worst case	$\mathcal{O}(\log N/\log\log N)$ w.h.p.
Huang et al. [16]	$\mathcal{O}(\log N(\log\log N)^2)$, expected amortized	$\mathcal{O}(\log N/\log\log\log N)$
Decremental	Total update time	Query time
Even et al. [8]	$\mathcal{O}(MN)$	$\mathcal{O}(1)$
Thorup [25]	$\mathcal{O}(M\log^2(N^2/M) + N\log^3 N\log\log N)$, expected	$\mathcal{O}(1)$
Henzinger et al. [14]	$\mathcal{O}(N^2\log N + b_0\min\{N^2, M\log N\})$	$\mathcal{O}(1)$

[1] The $\tilde{\mathcal{O}}(\cdot)$ notation suppresses polylog factors.

By plugging in a dynamic connectivity structure, the worst case running time of the algorithm of [20], given a matrix of n species and m characters, becomes deterministic $\mathcal{O}(nm \log^2(n+m))$ (using fully dynamic connectivity structure of Holm et al. [15]), expected $\mathcal{O}(nm \log((n+m)^2/nm) + (n+m)\log^3(n+m)\log\log(n+m))$ (using decremental connectivity structure of Thorup [25]), expected $\mathcal{O}(nm \log(n+m)(\log\log(n+m))^2)$ (using fully dynamic connectivity structure of Huang et al. [16]), or deterministic $\mathcal{O}((n+m)^2 \log(n+m))$ (using decremental structure of Henzinger et al. [14]). This should be compared with a lower bound of $\Omega(nm)$, following from the work of Gusfield on directed binary perfect phylogeny [11] (under the natural assumption that the input is given as a matrix). For $n = m$, the algorithm of [20] using [25] achieves this lower bound at the expense of randomisation (and being very complicated), while for the general case the asymptotically fastest solution is still at least one log factor away from the lower bound.

A closer look to the algorithm of [20], that we describe in more details in Sect. 2, reveals that it operates on bipartite graphs and only deletes vertices on one of the sides. It seems plausible that some of the known dynamic connectivity structures are actually asymptotically more efficient on such instances. However, all of them are very complex (with the result of Holm et al. [15] being the simplest, but definitely not simple), and this is not clear. Furthermore, recently Fernández-Baca and Liu [9] performed an experimental study of the algorithm of Pe'er et al. for IDPP [20] with the aim of assessing the impact of the underlying dynamic graph connectivity data structure on their solution. Specifically, they tested the use of the data structure of Holm et al. [15] against a simplified version of the same method, and showed that, in this context, simple data structures perform better than more sophisticated ones with better asymptotic bounds.

Our Results and Techniques. We are motivated to look for simpler, ad-hoc methods for the specific type of decremental connectivity that is used in IDPP: vertex deletion from just one side of a bipartite graph. We start by describing a simple structure that dynamically maintains the connected components of a bipartite graph with N vertices on each side, whilst vertices are removed from one of the sides. The starting point for our solution is an application of a particular version of the sparsification technique of Eppstein et al. [7]: we define a hierarchical decomposition of a bipartite graph, and maintain a forest representing the connected components of each subgraph in this decomposition. Recall that the original description of this technique focused on inserting and deleting edges, while we are interested in deleting vertices (and only from one side of the graph). We thus tweak the decomposition for our particular use case, obtaining an extremely simple data structure with $\mathcal{O}(N^2 \log N)$ total update time, which we show to imply an $\mathcal{O}(nm \log n)$ algorithm for IDPP.

The main technical part of our paper refines this solution to shave the logarithmic factor and thus obtain an asymptotically optimal algorithm. We stress that while Eppstein et al. [7] did manage to avoid any extra log factors by applying a more complex decomposition of the graph than a complete binary tree (used in the conference version of their paper), this does not seem to translate to our setting, as we operate on vertices instead of edges. The high-level idea of our solution is to amortize the time spent on updating the forest representing the components of every subgraph with the progress in disconnecting its vertices, and re-use the results from the subgraph on the previous level of the decomposition to update the subgraph on the next level. As a consequence, the IDPP problem can be solved in time linear in the input size. Under the natural assumption that the input is given as a matrix, this is asymptotically optimal [11].

Theorem 1. *Given an incomplete matrix* $\mathcal{A}_{n\times m}$*, the* IDPP *problem can be solved in time* $\mathcal{O}(nm)$.

Paper Organization. In Sect. 2 we provide a description of the algorithm of Pe'er et al. [20] and a series of preliminary observations. In Sect. 3 we show a simple and self-contained decremental connectivity data structure that considers the removal of vertices from one side of a bipartite graph. This structure implies an $\mathcal{O}(nm \log n)$ time solution for the IDPP problem. Finally, in Sect. 4 we present our main result and describe a decremental connectivity data structure for removing vertices from one side of a bipartite graph that implies a linear-time algorithm for IDPP.

2 Preliminaries

Let $G = (V, E)$ be a graph. The subgraph induced by a subset of vertices $V' \subseteq V$ is $G_{V'} = (V', E \cap (V' \times V'))$. We say that a forest $F = (V, E')$ represents the connected components of a graph G when the connected components of F and G are the same (note that we do not require $E' \subseteq E$). We denote by $S = \{s_1, \ldots, s_n\}$ the set of species and by $C = \{c_1, \ldots, c_m\}$ the set of characters. A matrix of character states $\mathcal{A}_{n\times m} = [a_{ij}]_{n\times m}$, where each entry is a state from $\{0, 1, ?\}$ and the rows correspond to the species, is said to be *incomplete*. The state a_{ij} is one, zero or ? depending on whether character j is present, absent or unknown for species i. A completion $\mathcal{B}_{n\times m}$ of such $\mathcal{A}_{n\times m}$ is obtained by replacing the ? entries of $\mathcal{A}_{n\times m}$ with either 0 or 1: formally, $\mathcal{B}_{n\times m}$ is a binary matrix with entries $b_{ij} = a_{ij}$ for each i, j such that $a_{ij} \neq ?$.

A *directed perfect phylogeny* for a binary matrix $\mathcal{B}_{n\times m}$ is a rooted tree $\mathcal{T}$ whose leaves are bijectively labelled by S and such that there is a surjection from the characters C to the internal nodes of $\mathcal{T}$ with the following property: if $c_j \in C$ is associated with a node x, then s_i is a leaf of the subtree rooted at x if and only if $b_{ij} = 1$. In particular, the term *directed* means that characters can be gained but not lost on any root-to-leaf path. We say that an incomplete matrix admits a directed perfect phylogenetic tree if there exists a completion of the matrix that has such a tree. The Incomplete Directed Perfect Phylogeny problem (IDPP for short), introduced in [20], asks, given an incomplete matrix $\mathcal{A}$, to find a directed perfect phylogenetic tree for $\mathcal{A}$, or determine that no such tree exists.

Algorithm 1: The high-level structure of Alg_A [20].

```
1 while there is at least one character in G(A) do
2     Find the connected components of G(A)
3     for each connected component K_i of G(A) with at least one character do
4         Compute the set U of all characters in K_i which are
          S(K_i)-semiuniversal in A
5         if U = ∅ then return FALSE
6         Deactivate every c ∈ U
7     return TRUE
```

The 1-set (resp. 0-set and ?-set) of a character c_j in an incomplete matrix $\mathcal{A}$ is the set of species $\{s_i | a_{ij} = 1\}$ (resp. $a_{ij} = 0$ and $a_{ij} = ?$). For a subset $S' \subseteq S$, a character c is S'-semiuniversal in $\mathcal{A}$ if its 0-set does not intersect S', that is, if $\mathcal{A}[s, c] \neq 0$ for all $s \in S'$. It is convenient to represent the character state matrix $\mathcal{A}$ as a graph: the vertices are $V = S \cup C$ and the edges are $S \times C$, partitioned into $E_1 \cup E_? \cup E_0$, with $E_x = \{(s_i, c_j) | a_{ij} = x\}$ for $x \in \{0, 1, ?\}$. The edges of $E_1, E_?, E_0$ are called *solid*, *optional*, and *forbidden*, respectively. We denote by $G(\mathcal{A}) = (S \cup C, E_1)$ the bipartite graph consisting only of the solid edges. A Σ is a subgraph induced by three vertices from S and two vertices from C, consisting of exactly four edges that form a path of length 4.

Previous solutions. Pe'er et al. [20] consider a graph representation of the input matrix $\mathcal{A}$, and show that finding a subset $D \subsetneq (E_1 \cup E_?)$ such that $E_1 \subseteq D$ and $(S \cup C, D)$ is Σ-free, or determining that no such D exists, is equivalent to solving IDPP. Their main algorithm exploits this characterization and the following properties: (i) if $\mathcal{A}$ admits a phylogenetic tree, then so does the matrix obtained by setting to 1 all the entries of column c, for each S-semiuniversal c; (ii) given a partition $(K_1, \ldots, K_r)$ of $S \cup C$, where each K_i is a connected component of $G(\mathcal{A})$, the matrix obtained by setting to 0 all entries corresponding to edges between K_i and K_j, for $i \neq j$, admits a phylogenetic tree if $\mathcal{A}$ does; and (iii) if there is a component K_i with no $S(K_i)$-semiuniversal characters, then for any $D \subsetneq (E_1 \cup E_?)$ such that $E_1 \subset D$, the graph $(S \cup C, D)$ is not Σ-free (and thus $\mathcal{A}$ has no phylogenetic tree). It follows that there is no interaction between the species and characters belonging to different connected components, and therefore the whole reasoning can be repeated on each such component separately.

We denote by $S(K)$ and $C(K)$ the set of species and characters, respectively, of a connected component K of $G(\mathcal{A})$; $\mathcal{A}|_K$ denotes the submatrix of $\mathcal{A}$ consisting of the species and characters in K. *Deactivating* a character c in $G(\mathcal{A})$ consists in deleting c and all its incident edges. At a high level Alg_A, the main algorithm of [20], works as follows. At each step, for each connected component K_i of $G(\mathcal{A})$, it computes the set U of $S(K_i)$-semiuniversal characters. If $U = \emptyset$, because of property (iii) $\mathcal{A}$ does not admit a phylogenetic tree, and the process halts. Otherwise, it sets to 1 the entries of $\mathcal{A}|_{K_i}$ corresponding to U, and sets to 0 the entries of $\mathcal{A}$ between vertices that lay in different connected components. It then deactivates all the characters in U and updates the connected components of $G(\mathcal{A})$ using some dynamic connectivity structure. Algorithm 1 summarizes this process: for the sake of clarity, we only included the steps that compute the information needed for determining whether $\mathcal{A}$ has a phylogenetic tree, and we left out the operations that actually construct the tree.

2.1 Preliminary Results

Our goal is to improve Alg_A by optimizing its bottleneck, that is maintaining the connected components of $G(\mathcal{A})$. We start by describing a data structure that conveniently represents the connected components of a bipartite graph G.

Lemma 1. *The connected components of a bipartite graph $G = (S \cup C, E)$ can be represented in $\mathcal{O}(|S| + |C|)$ space so that, given a vertex, we can access its component, including the size and a pointer to the list of species and characters inside, in constant time, and move a vertex to another component (or remove it from the graph) also in constant time.*

Proof. Each component of G is represented by a doubly-linked list of its vertices (more precisely, a list of species and a list of characters), and also stores the size of the list. An array of length $n + m$, indexed by the vertices of G, stores a pointer from each vertex to its component and a pointer from each vertex to its position in the list of its component. The components are, in turn, organised in a doubly-linked list. Such representation takes space linear in the number of vertices and allows us to access all the required information in constant time. Further, removing or moving a vertex to another component takes constant time. □

We denote by $\mathsf{cc}(G)$ the data structure of Lemma 1, which encodes the connected components of G. A graph $F = (V, E')$ consisting of a forest of *rooted stars* [23] can be straightforwardly obtained from $\mathsf{cc}(G)$ as follows. For each component K, we define the central vertex $v \in K$ to be the head of the doubly-linked list of characters of K in $\mathsf{cc}(G)$. Then, we add an edge (u, v) to E', for any $u \in K$ with $u \neq v$. This construction can be implemented in $\mathcal{O}(|V|)$ time. Although we do not require E' to be a subset of the edges of G, by construction the connected components of F and G are the same. The useful property is that we can use $\mathsf{cc}(G)$ to simulate access to the adjacency lists of F without constructing it explicitly, as stated by the following lemma.

Lemma 2. *Given a bipartite graph $G = (S \cup C, E)$ and $\mathsf{cc}(G)$, the access to the adjacency lists of a forest of rooted stars F with the same connected components as G can be simulated in constant time without constructing F explicitly.*

Proof. To simulate the access to the adjacency list of a vertex v, we first look up its component K in $\mathsf{cc}(G)$ and retrieve the head u of the doubly-linked list of characters of K. By Lemma 1, this operation requires constant time. If $u = v$, then the adjacency list of v is the list of vertices of K stored in $\mathsf{cc}(G)$. Otherwise, the adjacency list of v consists only of a single vertex u. □

Our intent is to solve the following special case of decremental connectivity.

Problem: (N_ℓ, N_r)-DC
Input: a bipartite graph $G = (S \cup C, E)$ with $N_\ell = |S|$ and $N_r = |C|$.
Update: deactivate a character $c \in C$.
Query: return the connected components of the subgraph induced by S and the remaining characters.

When analysing the complexity of (N_ℓ, N_r)-DC, we allow preprocessing the input graph G in $\mathcal{O}(N_\ell N_r)$ time, and assume that all characters will be eventually deactivated when analysing the total update time. We can of course deactivate multiple characters at once by deactivating them one-by-one. The overall time complexity of Algorithm 1 depends on the complexity of (N_ℓ, N_r)-DC as follows.

Lemma 3. *Consider an $n \times m$ incomplete matrix $\mathcal{A}$. If the (n, m)-*DC *problem can be solved in $f(n, m)$ total update time and $g(n, m)$ query time, then the* IDPP *problem can be solved for $\mathcal{A}$ in time $\mathcal{O}(nm + f(n, m) + \min\{n, m\} \cdot g(n, m))$.*

Proof. There are three nontrivial steps in every iteration of the while loop: finding the connected components in line 2, computing the semiuniversal characters of every connected component in line 4, and finally deactivating characters in line 6. Every character is deactivated at most once, so the overall complexity of all deactivations is $\mathcal{O}(f(n, m))$. We claim that in every iteration of the while loop, except possibly for the very last, (1) at least one character is deactivated, and (2) there exist two species that cease to belong to the same connected component. (1) is immediate, as otherwise we have a connected component K_i with no $S(K_i)$-semiuniversal characters and the algorithm terminates. To prove (2), assume otherwise, then we have a connected component K_i such that $S(K_i)$ does not change after deactivating all $S(K_i)$-semiuniversal characters. But then in the next iteration the set of $S(K_i)$-semiuniversal characters is empty and the algorithm terminates. (1) and (2) together imply that the number of iterations is bounded by $\min\{n, m\}$. The overall complexity of finding the connected components is thus $\mathcal{O}(\min\{n, m\} \cdot g(n, m))$.

It remains to bound the overall complexity of computing the semiuniversal characters by $\mathcal{O}(nm)$. This has been implicitly done in [20, proof of Theorem 12], but we provide a full explanation for completeness. For every character $c \in C$, we maintain the count of solid and optional edges connecting c (in the graph representation of $\mathcal{A}$) with the species that belong to its same connected component of $G(\mathcal{A})$ (recall that $G(\mathcal{A})$ consists only of the solid edges of the graph representation of $\mathcal{A}$). Assuming that we can indeed maintain these counts, in every iteration all the semiuniversal characters can be generated in $\mathcal{O}(m)$ time, so in $\mathcal{O}(\min\{n, m\} \cdot m) = \mathcal{O}(nm)$ overall time.

To update the counts, consider a connected component K that, after deactivating some characters, is split into possibly multiple smaller components $K_1, K_2, \ldots, K_k$. Note that we can indeed gather such information in $\mathcal{O}(n + m)$ time, assuming access to a representation of the connected components before and after the deactivation. We assume that the connected components are maintained with the representation described in Lemma 1, and therefore we can access a list of the vertices in every K_i. Then, we consider every pair $i, j \in \{1, 2, \ldots, k\}$ such that $i \neq j$, $C(K_i) \neq \emptyset$ and $S(K_j) \neq \emptyset$. We iterate over every $c \in K_i$ and $s \in K_j$, and if (s, c) is an edge in the graph of $\mathcal{A}$ (observe that it cannot be a solid edge, as K_i and K_j are distinct connected components) we decrease the count of c. By first preparing lists of components K_i such that $C(K_i) \neq \emptyset$ and $S(K_i) \neq \emptyset$, this can be implemented in time bounded by the number of considered possible edges (s, c), and every such possible edge is considered at most once during the whole execution. Therefore, the overall complexity of maintaining the counts is $\mathcal{O}(nm)$. Additionally, we need $\mathcal{O}(nm)$ time to initialise the (n, m)-DC structure. □

Before proceeding to design an efficient solution for the (N_ℓ, N_r)-DC problem, we show that it is in fact enough to consider the (N, N)-DC problem.

Lemma 4. *Assume that the* (N, N)-DC *problem can be solved in* $f(N)$ *total update time and* $g(N)$ *query time. Then, for any* $N' \geq N$, *both the* (N, N')-DC *problem and the* (N', N)-DC *problem can be solved in* $\mathcal{O}(N'/N \cdot f(N))$ *total update time and* $\mathcal{O}(N'/N \cdot g(N))$ *query time.*

Proof. We first consider the (N, N')-DC problem, in which $|S| < |C|$. We create $\lceil N'/N \rceil$ instances of (N, N)-DC by partitioning C into groups of N vertices (the last group might be smaller). In each instance we have the same set of species S. Deactivating a character $c \in C$ is implemented by deactivating it in the corresponding instance of (N, N)-DC. Overall, this takes $\mathcal{O}(N'/N \cdot f(N))$ time. To answer a query, we first query all the instances in $\mathcal{O}(N'/N \cdot g(N))$ time. The output of each instance can be converted to a forest of rooted stars with the same connected components in $\mathcal{O}(N)$ time. We take the union of all these forests to obtain an auxiliary graph with at most $\lceil N'/N \rceil \cdot (N-1) = \mathcal{O}(N')$ edges, and find its connected components in $\mathcal{O}(N')$ time. Assuming that $f(N) \geq N$, this takes $\mathcal{O}(N'/N \cdot f(N))$ overall time and gives us the connected components of the whole input graph.

Now we consider the (N', N)-DC problem. We create $\lceil N'/N \rceil$ instances of (N, N)-DC by partitioning S into groups of N vertices, and in each instance we have the same set of characters C. Thus, deactivating a character $c \in C$ is implemented by deactivating it in every instance. This takes $\mathcal{O}(N'/N \cdot f(N))$ total time. A query is implemented exactly as above by querying all the instances and combining the results in $\mathcal{O}(N'/N \cdot f(N))$ time. □

3 (N, N)-DC in $\mathcal{O}(N^2 \log N)$ Total Update Time and $\mathcal{O}(N)$ Time per Query

Our solution for the (N, N)-DC problem is based on a hierarchical decomposition of G into multiple smaller subgraphs as in the sparsification technique of Eppstein et al. [7] (as mentioned in the introduction, appropriately tweaked for our use case). The decomposition is represented by a complete binary tree

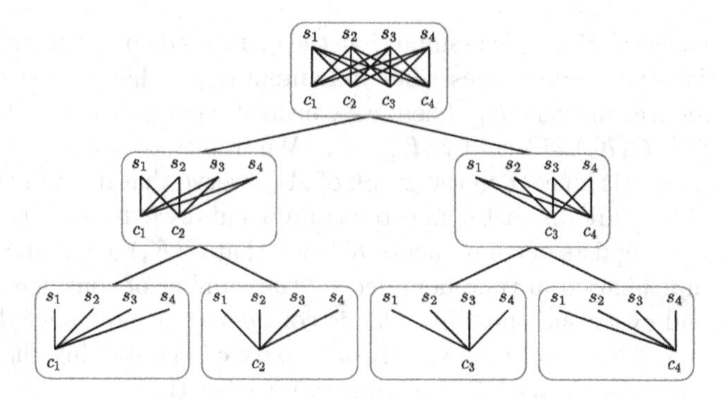

Fig. 1. The decomposition tree of $K_{4,4}$.

$\mathrm{DT}(G)$ of depth $\log N$. We identify the leaves of $\mathrm{DT}(G)$ with the characters C. Each node v corresponds to the set of characters C_v identified with the leaves in the subtree of v, and is responsible for the subgraph G_v of G induced by C_v and the whole set of species S. Thus, the root is responsible for the whole G, see Fig. 1. Each node v maintains $\mathsf{cc}(v)$, the connected components of G_v represented as per Lemma 1. We stress that, while $\mathsf{cc}(v)$ is explicitly maintained, we do not explicitly store G_v at every node v. Given G, the preprocessing required to construct $\mathrm{DT}(G)$ together with $\mathsf{cc}(v)$ for every node v takes $\mathcal{O}(N^2)$ time by the following argument. First, we construct $\mathsf{cc}(G_c)$ for every leaf c. This can be done in $\mathcal{O}(N)$ time per leaf by simply iterating the neighbours of c in G. We then proceed bottom-up and compute $\mathsf{cc}(v)$ for every inner node v in $\mathcal{O}(N)$ time using the following lemma.

Lemma 5. *Let v be an inner node of $\mathrm{DT}(G)$, and v_ℓ, v_r be its children. Given $\mathsf{cc}(v_\ell)$ and $\mathsf{cc}(v_r)$ we can compute $\mathsf{cc}(v)$ in $\mathcal{O}(N)$ time.*

Proof. We construct the forests of rooted stars representing the connected components of $\mathsf{cc}(v_\ell)$ and $\mathsf{cc}(v_r)$ in $\mathcal{O}(N)$ time and take their union. Then we find the connected components of this union in $\mathcal{O}(N)$ time and save them as $\mathsf{cc}(v)$. □

We proceed to explain how to solve the (N, N)-DC problem in $\mathcal{O}(N \log N)$ time per update and $\mathcal{O}(N)$ time per query. The query simply returns $\mathsf{cc}(r)$, where r is the root of $\mathrm{DT}(G)$. The update is implemented as follows. Deactivating a character c possibly affects $\mathsf{cc}(v)$ for all ancestors v of leaf c. In particular, $\mathsf{cc}(c)$ becomes a collection of isolated vertices and can be recomputed in $\mathcal{O}(1 + |S|) = \mathcal{O}(N)$ time. We iterate over all ancestors v, starting from the parent of c. For each such v, let v_ℓ and v_r be its left and right child, respectively. We can assume that $\mathsf{cc}(v_\ell)$ and $\mathsf{cc}(v_r)$ have been already correctly updated. We compute $\mathsf{cc}(v)$ from $\mathsf{cc}(v_\ell)$ and $\mathsf{cc}(v_r)$ by applying Lemma 5 in $\mathcal{O}(N)$ time. When summed over all the ancestors, the update time becomes $\mathcal{O}(N \log N)$, so $\mathcal{O}(N^2 \log N)$ over all deactivations. By Lemmas 3 and 4, this implies that, given an incomplete matrix $\mathcal{A}_{n \times m}$, the IDPP problem can be solved in time $\mathcal{O}(nm \log(\min\{n, m\}))$ without using any dynamic connectivity data structure as a blackbox.

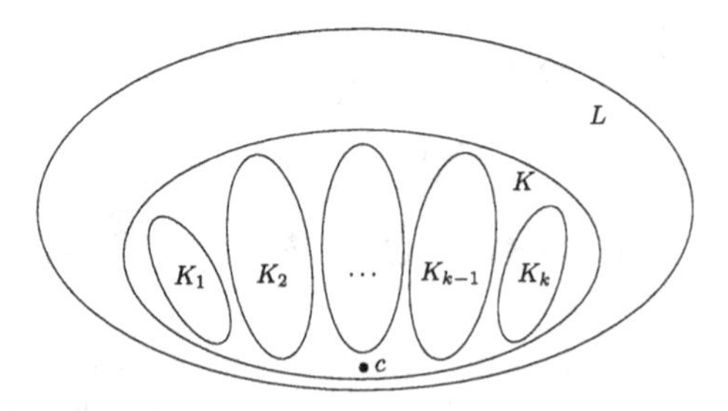

Fig. 2. After having removed c from K to obtain $K_1, K_2, \ldots, K_k$, we want to remove c from L.

4 (N, N)-DC in $\mathcal{O}(N^2)$ Total Update Time and $\mathcal{O}(N)$ Time per Query

Our faster solution is also based on a hierarchical decomposition $\mathrm{DT}(G)$ of G. As before, every node v stores $\mathsf{cc}(v)$, so a query simply returns $\mathsf{cc}(r)$. The difference is in implementing an update. We observe that, if for some ancestor v of a leaf c the only change to $\mathsf{cc}(v)$ is removing c from its connected component, then this also holds for all the subsequent ancestors, and therefore each of them can be updated in constant time. This suggests that we should try to amortise the cost of an update with the progress in splitting $\mathsf{cc}(v)$ into smaller components.

We will need to compare the situation before and after an update, and so we introduce the following notation. A node v of $\mathrm{DT}(G)$ is responsible for the subgraph G_v before the update and for the subgraph G'_v after the update; $\mathsf{cc}(v)$ and $\mathsf{cc}'(v)$ denote the connected components of G_v and G'_v, respectively. The crucial observation is that $\mathsf{cc}'(v)$ is obtained from $\mathsf{cc}(v)$ by removing c from its connected component and, possibly, splitting this connected component into multiple smaller ones, while leaving the others intact.

Deactivating a character c begins with updating naively $\mathsf{cc}(c)$ in $\mathcal{O}(N)$ time. Then we iterate over the ancestors of c in $\mathrm{DT}(G)$. Let v_{i+1} be the currently considered ancestor, v_i the ancestor considered in the previous iteration, and u_i be the other child of v_{i+1} (sibling of v_i). Let the component of G_{v_i} containing c be K. As observed above, the components of G'_{v_i} are the same as the components of G_{v_i}, except that K is replaced by possibly multiple components $K_1, K_2, \ldots, K_k$, where $\bigcup_{j=1}^{k} K_j = K \setminus \{c\}$. If $k = 1$ then we trivially remove c from its connected component in every G_{v_j}, for $j = i+1, i+2, \ldots$ and terminate the update, so we can assume that $k \geq 2$. We further assume that, after having updated the components of G_{v_i}, we obtained a list of pointers to $K_1, K_2, \ldots, K_k$. Let L be the connected component of c in $G_{v_{i+1}}$, with $K \subseteq L$ because the subgraphs are monotone with respect to inclusion on any leaf-to-root path. Now the goal is to transform $G_{v_{i+1}}$ into $G'_{v_{i+1}}$, to update its components (using $\mathsf{cc}'(v_i)$ and $\mathsf{cc}(u_i)$), and additionally to obtain a list of pointers to the components obtained by splitting L. See Fig. 2 for an illustration.

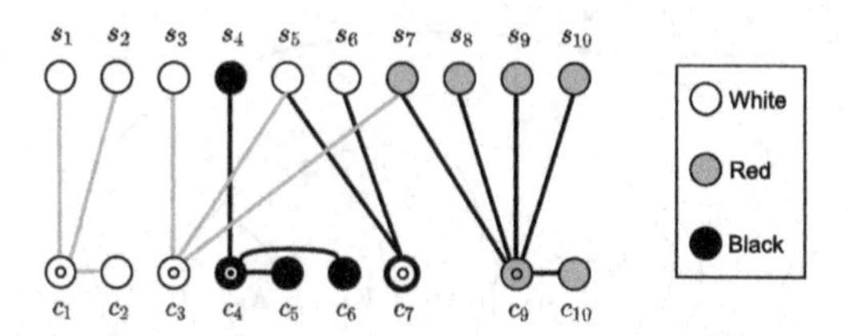

Fig. 3. The auxiliary graph implicitly constructed for a node v_{i+1} after deactivating c_8. Black edges are used for the star forest of v_i, grey edges for the star forest of u_i; an inner circle identifies the central vertices. K_1 is the component of c_9; c_7 is the next vertex to be considered in the visit, and it will eventually become red.

We start by initialising $G'_{v_{i+1}}$ to be $G_{v_{i+1}}$, and by removing c from L. As in the proof of Lemma 5, we will use an auxiliary graph consisting of the union of two star forests representing the connected components of G'_{v_i} and G_{u_i}, respectively. However, instead of explicitly constructing them, we simulate access to the adjacency lists of every vertex in both forests using $\mathsf{cc}'(v_i)$ and $\mathsf{cc}(u_i)$, as per Lemma 2. In turn, this allows us to simulate access to the adjacency list of every vertex in the auxiliary graph. See Fig. 3 for an example of the auxiliary graph.

By renaming the components we can assume that $|K_1| \geq |K_2|, |K_3|, \ldots, |K_k|$. In order to determine the new connected components after the removal of c, we will visit the vertices of L: when doing so, we will use different colours to represent vertices whose new connected component contains K_1 (red), vertices whose new component is different from the one of K_1 (black) and vertices whose new component is still unknown (white). Initially, the vertices of K_1 are red and all of the other vertices of the auxiliary graph are white. This initialisation is done implicitly, meaning that we will assume that all the vertices of K_1 are red and the rest are white without explicitly assigning the colours; whenever retrieving the colour of a vertex u, we first check if $u \in K_1$, and if so assume that it is red. This allows us to implement the initialisation in constant time instead of $\mathcal{O}(N)$ time. We will perform the visit of L by running the following search procedure from an arbitrarily chosen vertex of each K_j, for $j = 2, 3, \ldots, k$.

The search procedure run from a vertex x first checks if x is white, and immediately terminates otherwise. Then, it starts visiting the vertices of the connected component of x in the auxiliary graph: at any moment, each vertex in such component is either white or red. As soon as the search encounters a red vertex, it is terminated and all the vertices visited in the current invocation are explicitly coloured red. Otherwise, the procedure has identified a new connected component K' of $G'_{v_{i+1}}$. The vertices of K' are removed from L, all vertices of K' are coloured black in the auxiliary graph, and a new component K' of $G'_{v_{i+1}}$ is created in $\mathcal{O}(|K'|)$ time. See Fig. 3 for an example.

Lemma 6. *The total time spent on all calls to the search procedure in the current iteration is* $\mathcal{O}(|L| - |K_1|)$.

Proof. All vertices visited in the current iteration belong to L. The search is terminated as soon as we encounter a red vertex, and all vertices of K_1 are red from the beginning. Therefore, each run of the search procedure encounters at most one vertex of K_1, and we can account for traversing the edge leading to this vertex separately paying $\mathcal{O}(k-1) = \mathcal{O}(|L| - |K_1|)$ overall. It remains to bound the number of all other traversed edges. This is enough to bound the overall time of the traversal, because every edge is traversed at most twice, and the number of visited isolated vertices is at most $k - 1 = \mathcal{O}(|L| - |K_1|)$.

For any other edge $e = \{u, v\}$, we have $u, v \in L$ but $u, v \notin K_1$. These edges can be partitioned into two forests, depending on whether they originate from $\mathsf{cc}'(v_i)$ or $\mathsf{cc}(u_i)$. Consequently, we must analyse the total number of edges in a union of two forests spanning $L \setminus K_1$; but this is of course $\mathcal{O}(|L| - |K_1|)$. □

We now need to analyse the sum of $|L|-|K_1|$ over all the iterations. Because $\bigcup_{j=1}^{k} K_j \subseteq L$, we can split this expression into two parts:

1. $L \setminus \bigcup_{j=1}^{k} K_j$,
2. $\sum_{j=2}^{k} |K_j|$.

Because the sets $L \setminus \bigcup_{j=1}^{k} K_j$ considered in different iterations are disjoint, the first parts sum up to $\mathcal{O}(n)$. It remains to bound the sum of the second parts. This will be done by the following argument. Consider an arbitrary G_v corresponding to a subgraph induced by all the species and a subset of 2^d characters. Whenever its connected component K is split into smaller connected components $K_1, K_2, \ldots, K_k$ after deactivating a character c in the subtree of v, the second part $\sum_{j=2}^{k} |K_j|$ is distributed among the vertices of $\bigcup_{j=2}^{k} K_j$. That is, each vertex of $\bigcup_{j=2}^{k} K_j$ pays 1. Observe that the size of the connected component containing such a vertex decreases by a factor of at least 2, because $|K_2|, |K_3|, \ldots, |K_k| \leq |K|/2$. To bound the sum of second parts, we analyse the total cost paid by all the vertices of G_v due to deactivating the characters in the subtree of v (recall that in the end all such characters are deactivated).

Lemma 7. *The total cost paid by the vertices of G_v, over all 2^d deactivations affecting v, is $\mathcal{O}(N \cdot d)$.*

Proof. We claim that in the whole process there can be at most 2^{t+1} deactivations incurring a cost from $[N/2^{t+1}, N/2^t)$. Assume otherwise, then there exists a vertex x charged twice by such deactivations. As a result of the first deactivation, the size of the connected component containing x drops from less than $N/2^t$ to below $N/2^{t+1}$. Consequently, during the next deactivation that charges x the cost must be smaller than $N/2^{t+1}$, a contradiction. As we have 2^d deactivation overall, the total cost can be at most:

$$\sum_{t=0}^{d} 2^{t+1} \cdot N/2^t = \mathcal{O}(N \cdot d)$$

□

There are $N/2^d$ nodes of $\mathrm{DT}(G)$ affected by 2^d deactivations, making the sum of the second parts:

$$\sum_{d=0}^{\log n} N/2^d \cdot n \cdot d < N^2 \sum_{d=0}^{\infty} d/2^d = \mathcal{O}(N^2).$$

Overall, the total update time is $\mathcal{O}(N^2)$, so by Lemmas 3 and 4 we arrive at the main result of this paper.

Theorem 1. *Given an incomplete matrix* $\mathcal{A}_{n\times m}$*, the* IDPP *problem can be solved in time* $\mathcal{O}(nm)$*.*

Acknowledgements. This project has received funding from the European Union's Horizon 2020 research and innovation programme under the Marie Skłodowska-Curie grant agreement No 872539. GB was supported by the Netherlands Organisation for Scientific Research (NWO) under project OCENW.GROOT.2019.015 "Optimization for and with Machine Learning (OPTIMAL)".

References

1. Bashir, A., Ye, C., Price, A.L., Bafna, V.: Orthologous repeats and mammalian phylogenetic inference. Genome Res. **15**(7), 998–1006 (2005)
2. Bodlaender, H.L., Fellows, M.R., Hallett, M.T., Wareham, H.T., Warnow, T.J.: The hardness of perfect phylogeny, feasible register assignment and other problems on thin colored graphs. Theoret. Comput. Sci. **244**(1–2), 167–188 (2000)
3. Bonizzoni, P., Braghin, C., Dondi, R., Trucco, G.: The binary perfect phylogeny with persistent characters. Theoret. Comput. Sci. **454**, 51–63 (2012)
4. Bonizzoni, P., Ciccolella, S., Della Vedova, G., Soto, M.: Beyond perfect phylogeny: Multisample phylogeny reconstruction via ilp. In: 8th ACM-BCB, pp. 1–10 (2017)
5. Camin, J.H., Sokal, R.R.: A method for deducing branching sequences in phylogeny. Evolution, pp. 311–326 (1965)
6. El-Kebir, M.: Sphyr: tumor phylogeny estimation from single-cell sequencing data under loss and error. Bioinformatics **34**(17), i671–i679 (2018)
7. Eppstein, D., Galil, Z., Italiano, G.F., Nissenzweig, A.: Sparsification-a technique for speeding up dynamic graph algorithms. J. ACM **44**(5), 669–696 (1997)
8. Even, S., Shiloach, Y.: An on-line edge-deletion problem. J. ACM **28**(1), 1–4 (1981)
9. Fernández-Baca, D., Liu, L.: Tree compatibility, incomplete directed perfect phylogeny, and dynamic graph connectivity: An experimental study. Algorithms **12**(3), 53 (2019)
10. Gibb, D., Kapron, B., King, V., Thorn, N.: Dynamic graph connectivity with improved worst case update time and sublinear space. arXiv:1509.06464 (2015)
11. Gusfield, D.: Efficient algorithms for inferring evolutionary trees. Networks **21**(1), 19–28 (1991)
12. Gusfield, D.: Persistent phylogeny: a galled-tree and integer linear programming approach. In: 6th ACM-BCB, pp. 443–451 (2015)
13. Halperin, E., Karp, R.M.: Perfect phylogeny and haplotype assignment. In: Proceedings of the Eighth Annual International Conference on Resaerch in Computational Molecular Biology, pp. 10–19 (2004)

14. Henzinger, M.R., King, V., Warnow, T.: Constructing a tree from homeomorphic subtrees, with applications to computational evolutionary biology. Algorithmica **24**(1), 1–13 (1999)
15. Holm, J., De Lichtenberg, K., Thorup, M.: Poly-logarithmic deterministic fully-dynamic algorithms for connectivity, minimum spanning tree, 2-edge, and biconnectivity. J. ACM **48**(4), 723–760 (2001)
16. Huang, S.E., Huang, D., Kopelowitz, T., Pettie, S.: Fully dynamic connectivity in $O(\log n(\log\log n)^2)$ amortized expected time. In: 28th SODA, pp. 510–520. SIAM (2017)
17. Kimmel, G., Shamir, R.: The incomplete perfect phylogeny haplotype problem. J. Bioinform. Comput. Biol. **3**(02), 359–384 (2005)
18. Kirkpatrick, B., Stevens, K.: Perfect phylogeny problems with missing values. IEEE/ACM Trans. Comput. Biol. Bioinf. **11**(5), 928–941 (2014)
19. Nikaido, M., Rooney, A.P., Okada, N.: Phylogenetic relationships among cetartiodactyls based on insertions of short and long interpersed elements: hippopotamuses are the closest extant relatives of whales. Proc. Natl. Acad. Sci. **96**(18), 10261–10266 (1999)
20. Pe'er, I., Pupko, T., Shamir, R., Sharan, R.: Incomplete directed perfect phylogeny. SIAM J. Comput. **33**(3), 590–607 (2004)
21. Satas, G., Zaccaria, S., Mon, G., Raphael, B.J.: Scarlet: Single-cell tumor phylogeny inference with copy-number constrained mutation losses. Cell Syst. **10**(4), 323–332 (2020)
22. Satya, R.V., Mukherjee, A.: The undirected incomplete perfect phylogeny problem. IEEE/ACM Trans. Comput. Biol. Bioinf. **5**(4), 618–629 (2008)
23. Shiloach, Y., Vishkin, U.: An $o(\log n)$ parallel connectivity algorithm. J. Algorithms **3**(1), 57–67 (1982)
24. Stevens, K., Gusfield, D.: Reducing multi-state to binary perfect phylogeny with applications to missing, removable, inserted, and deleted data. In: Moulton, V., Singh, M. (eds.) Algorithms in Bioinformatics. WABI 2010. Lecture Notes in Computer Science, vol. 6293. Springer, Berlin, Heidelberg (2010). https://doi.org/10.1007/978-3-642-15294-8_23
25. Thorup, M.: Decremental dynamic connectivity. J. Algorithms **33**(2), 229–243 (1999)

Euclidean Maximum Matchings in the Plane—Local to Global

Ahmad Biniaz[1(✉)], Anil Maheshwari[2], and Michiel Smid[2]

[1] University of Windsor, Windsor, Canada
ahmad.biniaz@gmail.com
[2] Carleton University, Ottawa, Canada
{anil,michiel}@scs.carleton.ca

Abstract. Let M be a perfect matching on a set of points in the plane where every edge is a line segment between two points. We say that M is *globally maximum* if it is a maximum-length matching on all points. We say that M is *k-local maximum* if for any subset $M' = \{a_1b_1, \ldots, a_kb_k\}$ of k edges of M it holds that M' is a maximum-length matching on points $\{a_1, b_1, \ldots, a_k, b_k\}$. We show that local maximum matchings are good approximations of global ones.

Let μ_k be the infimum ratio of the length of any k-local maximum matching to the length of any global maximum matching, over all finite point sets in the Euclidean plane. It is known that $\mu_k \geqslant \frac{k-1}{k}$ for any $k \geqslant 2$. We show the following improved bounds for $k \in \{2, 3\}$: $\mu_2 \geqslant \sqrt{3/7}$ and $\mu_3 \geqslant 1/\sqrt{2}$. We also show that every pairwise crossing matching is unique and it is globally maximum.

Towards our proof of the lower bound for μ_2 we show the following result which is of independent interest: If we increase the radii of pairwise intersecting disks by factor $2/\sqrt{3}$, then the resulting disks have a common intersection.

Keywords: Planar points · Maximum matching · Global maximum · Local maximum · Pairwise crossing matching · Pairwise intersecting disks

1 Introduction

A maximum-weight matching in an edge-weighted graph is a matching in which the sum of edge weights is maximized. Maximum-weight matching is among well-studied structures in graph theory and combinatorial optimization. It has been studied from both combinatorial and computational points of view in both abstract and geometric settings, see for example [1,3,4,8,10–12,15,16,18,23,24, 30]. Over the years, it has found applications in several areas such as scheduling, facility location, and network switching. It has also been used as a key subroutine in other optimization algorithms, for example, network flow algorithms [13,25],

Supported by NSERC.

A. Lubiw et al. (Eds.): WADS 2021, LNCS 12808, pp. 186–199, 2021.
https://doi.org/10.1007/978-3-030-83508-8_14

maximum cut in planar graphs [19], and switch scheduling algorithms [27] to name a few. In the geometric setting, where vertices are represented by points in a Euclidean space and edges are line segments, the maximum-weight matching is usually referred to as the *maximum-length matching.*

Let P be a set of $2n$ distinct points in the plane, and let M be a perfect matching on P where every edge of M is a straight line segment. We say that M is *globally maximum* if it is a maximum-length matching on P. For an integer $k \leqslant n$ we say that M is *k-local maximum* if for any subset $M' = \{a_1b_1, \dots, a_kb_k\}$ of k edges of M it holds that M' is a maximum-length matching on points $\{a_1, b_1, \dots, a_k, b_k\}$; in other words M' is a maximum-length matching on the endpoints of its edges. Local maximum matchings appear in local search heuristics for approximating global maximum matchings, see e.g. [2].

It is obvious that any global maximum matching is locally maximum. On the other hand, local maximum matchings are known to be good approximations of global ones. Let μ_k be the infimum ratio of the length of any k-local maximum matching to the length of any global maximum matching, over all finite point sets in the Euclidean plane. For $k = 1$, the ratio μ_1 could be arbitrary small, because any matching is 1-local maximum. For $k \geqslant 2$, however, it is known that $\mu_k \geqslant \frac{k-1}{k}$ (see e.g. [2, Corollary 8]); this bound is independent of the Euclidean metric and it is valid for any edge-weighted complete graph. A similar bound is known for matroid intersection [26, Corollary 3.1]. We present improved bounds for μ_2 and μ_3; this is going to be the main topic of this paper.

1.1 Our Contributions

The general lower bound $\frac{k-1}{k}$ implies that $\mu_2 \geqslant 1/2$ and $\mu_3 \geqslant 2/3$. We use the geometry of the Euclidean plane and improve these bounds to $\mu_2 \geqslant \sqrt{3/7} \approx 0.654$ and $\mu_3 \geqslant 1/\sqrt{2} \approx 0.707$. In the discussion at the end of this paper we show that analogous ratios for local minimum matchings could be arbitrary large.

For an edge set E, we denote by $w(E)$ the total length of its edges. To obtain the lower bound $1/\sqrt{2}$ for μ_3 we prove that for any 3-local maximum matching M it holds that $w(M) \geqslant w(M^*)/\sqrt{2}$ where M^* is a global maximum matching for the endpoints of edges in M. To do so, we consider the set D of diametral disks of edges in M. A recent result of Bereg et al. [4] combined with Helly's theorem [21,29] implies that the disks in D have a common intersection. We take a point in this intersection and connect it to endpoints of all edges of M to obtain a star S. Then we show that $w(M^*) \leqslant w(S) \leqslant \sqrt{2} \cdot w(M)$, which proves the lower bound.

Our proof approach for showing the lower bound $\sqrt{3/7}$ for μ_2 is similar to that of μ_3. However, our proof consists of more technical ingredients. We show that for any 2-local maximum matching M it holds that $w(M) \geqslant \sqrt{3/7} \cdot w(M^*)$ where M^* is a global maximum matching for the endpoints of edges of M. Again we consider the set D of diametral disks of edges of M. A difficulty arises here because now the disks in D may not have a common intersection, although they pairwise intersect. To overcome this issue we enlarge the disks in D to obtain a new set of disks that have a common intersection. Then we take a point in this intersection and construct our star S as before, and we show that $w(M^*) \leqslant w(S) \leqslant \sqrt{7/3} \cdot w(M)$. To obtain this result we face two technical complications: (i) we need to show that the enlarged disks have a common intersection, and (ii) we need to bound the distance from the center of star S to endpoints of M. To overcome the first issue we prove that if we increase the radii of pairwise intersecting disks by factor $2/\sqrt{3}$ then the resulting disks have a common intersection; the factor $2/\sqrt{3}$ is the smallest that achieves this property. This result has the same flavor as the problem of stabbing pairwise intersecting disks with four points [6,7,20,31]. To overcome the second issue we prove a result in distance geometry.

In a related result, which is also of independent interest, we show that every pairwise crossing matching is unique and it is globally maximum. To show the maximality we transform our problem into an instance of the "multicommodity flows in planar graphs" that was studied by Okamura and Seymour [28] in 1981.

1.2 Some Related Works

From the computational point of view, Edmonds [11,12] gave a polynomial-time algorithm for computing weighted matchings in general graphs (the term *weighted matching* refers to both minimum-weight matching and maximum-weight matching). Edmonds' algorithm is a generalization of the Hungarian algorithm for weighted matching in bipartite graphs [23,24]. There are several implementations of Edmonds' algorithm (see e.g. [15,17,18,25]) with the best known running time $O(mn + n^2 \log n)$ [15,16] where n and m are the number of vertices and edges of the graph. One might expect faster algorithms for the "maximum-length matching" in the geometric setting where vertices are points in the plane and any two points are connected by a straight line segment; we are not aware of any such algorithm. For general graphs, there is a linear-time $(1-\varepsilon)$-approximation of maximum-weight matching [8].

The analysis of maximum-length matching ratios has received attention in the past. In a survey by Avis [3] it is shown that the matching obtained by a greedy algorithm (that picks the largest available edge) is a 1/2-approximation of the global maximum matching (even in arbitrary weighted graphs). Alon, Rajagopalan, Suri [1] studied non-crossing matchings, where edges are not allowed to cross each other. They showed that the ratio of the length of a maximum-length non-crossing matching to the length of a maximum-length matching is at least $2/\pi$; this ratio is the best possible. Similar ratios have been studied for non-crossing spanning trees, Hamiltonian paths and cycles [1,5,9]. Bereg et al. [4] showed the following combinatorial property of maximum-length matchings: the diametral disks, introduced by edges of a maximum-length matching, have a common intersection. A somewhat similar property was proved by Huemer et al. [22] for bi-colored points.

2 A Lower Bound for k-Local Maximum Matchings

For the sake of completeness, and to facilitate comparisons with our improved bounds, we repeat a proof of the general lower bound $\frac{k-1}{k}$, borrowed from [2].

Theorem 1. *Every k-local maximum matching is a $\frac{k-1}{k}$-approximation of a global maximum matching for any $k \geqslant 2$.*

Proof. Consider any k-local maximum matching M and a corresponding global maximum matching M^*. The union of M and M^* consists of even cycles and/or single edges which belong to both matchings. It suffices to show, for each cycle C, that the length of edges in $C \cap M$ is at least $\frac{k-1}{k}$ times that of edges in $C \cap M^*$.

Let $e_0, e_1, \ldots, e_{|C|-1}$ be the edges of C that appear in this order. Observe that $|C| \geqslant 4$, and that the edges of C alternate between M and M^*. Let C_M and C_{M^*} denote the sets of edges of C that belong to M and M^*, respectively. If $|C| \leqslant 2k$ then $w(C_M) = w(C_{M^*})$ because M is k-local maximum, and thus we are done. Assume that $|C| \geqslant 2k+2$. After a suitable shifting of indices we may assume that $C_M = \{e_i : i \text{ is even}\}$ and $C_{M^*} = \{e_i : i \text{ is odd}\}$. Since M is k-local maximum, for each even index i we have

$$w(e_i) + w(e_{i+2}) + \cdots + w(e_{i+2k-2}) \geqslant w(e_{i+1}) + w(e_{i+3}) + \cdots + w(e_{i+2k-3})$$

where all indices are taken modulo $|C|$. By summing this inequality over all even indices, every edge of C_M appears exactly k times and every edge of C_{M^*} appears exactly $k-1$ times, and thus we get $k \cdot w(C_M) \geqslant (k-1) \cdot w(C_{M^*})$. □

It is implied from Theorem 1 that $\mu_2 \geqslant 1/2$ and $\mu_3 \geqslant 2/3$. To establish stronger lower bounds, we need to incorporate more powerful ingredients. We use geometry of the Euclidean plane and improve both lower bounds.

3 Better Lower Bound for 3-Local Maximum Matchings

We describe our improved bound for 3-local maximum matchings first because it is easier to understand. Our Theorem 4 implies that $\mu_3 \geqslant 1/\sqrt{2}$. The proof of our theorem benefits from the following result of Bereg et al. [4] and Helly's theorem [21,29].

Theorem 2. (Bereg et al. [4]). *Consider any maximum matching of any set of six points in the plane. The diametral disks of the three edges in this matching have a nonempty intersection.*

Theorem 3. (Helly's theorem in $\mathcal{R}^2$). *If in a family of convex sets in the plane every triple of sets has a nonempty intersection, then the entire family has a nonempty intersection.*

Theorem 4. *Every 3-local Euclidean maximum matching is a $\frac{1}{\sqrt{2}}$-approximation of a global Euclidean maximum matching.*

Proof. Consider any 3-local maximum matching M. Let M^* be a global maximum matching for the endpoints of edges of M. Consider the set D of diametral disks introduced by edges of M. Since M is 3-local maximum, any three disks in D have a common intersection (by Theorem 2). With this property, it is implied

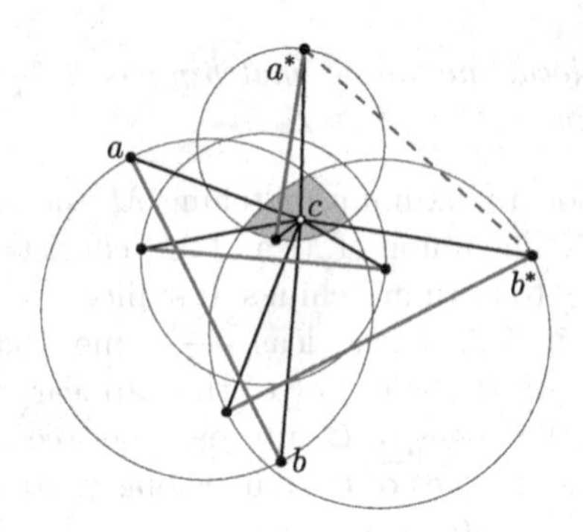

Fig. 1. Red edges belong to M, black edges belong to S, and blue edge belongs to M^*.

by Theorem 3 that the disks in D have a common intersection (the shaded region in Fig. 1). Let c be a point in this intersection. Let S be the star obtained by connecting c to all endpoints of edges of M as in Fig. 1. Since c is in the diametral disk of every edge $ab \in M$, it is at distance at most $|ab|/2$ from the midpoint of ab. By applying Lemma 1 (which will be proved in Sect. 4), with c playing the role of p and $r = 1$, we have

$$|ca| + |cb| \leqslant \sqrt{2} \cdot |ab|. \tag{1}$$

In Inequality (1), for every edge $ab \in M$, a unique pair of edges in S is charged to ab. Therefore, $w(S) \leqslant \sqrt{2} \cdot w(M)$. Now consider any edge $a^*b^* \in M^*$. By the triangle inequality we have that

$$|a^*b^*| \leqslant |ca^*| + |cb^*|. \tag{2}$$

In Inequality (2), every edge of M^* is charged to a unique pair of edges in S. Therefore, $w(M^*) \leqslant w(S)$. Combining the two resulting inequalities we have that $w(M) \geqslant w(M^*)/\sqrt{2}$. □

Remark 1. In 1995, Fingerhut [14] conjectured that for any maximum-length matching $\{(a_1, b_1), \ldots, (a_n, b_n)\}$ on any set of $2n$ points in the plane there exists a point c such that

$$|a_i c| + |b_i c| \leqslant \alpha \cdot |a_i b_i| \tag{3}$$

for all $i \in \{1, \ldots, n\}$, where $\alpha = 2/\sqrt{3}$. The smallest known value for α that satisfies Inequality (3) is $\alpha = \sqrt{2}$, which is implied by the result of [4]. A proof of this conjecture, combined with an argument similar to our proof of Theorem 4, would imply approximation ratio $\frac{\sqrt{3}}{2} \approx 0.866$ for 3-local maximum matchings.

4 Better Lower Bound for 2-Local Maximum Matchings

In this section we prove that $\mu_2 \geqslant \sqrt{3/7} \approx 0.65$, that is, 2-local maximum matchings are $\sqrt{3/7}$ approximations of global ones. Our proof approach employs an argument similar to that of 3-local maximum matchings. Here we are facing an obstacle because diametral disks that are introduced by edges of a 2-local maximum matching may not have a common intersection. To handle this issue, we require stronger tools. Our idea is to increase the radii of disks—while preserving their centers—to obtain a new set of disks that have a common intersection. Then we apply our argument on this new set of disks. This gives rise to somewhat lengthier analysis. Also, two technical complications arise because now we need to show that the new disks have a common intersection, and we need to bound the total distance from any point in new disks to the endpoints of the corresponding matching edges. The following lemmas play important roles in our proof.

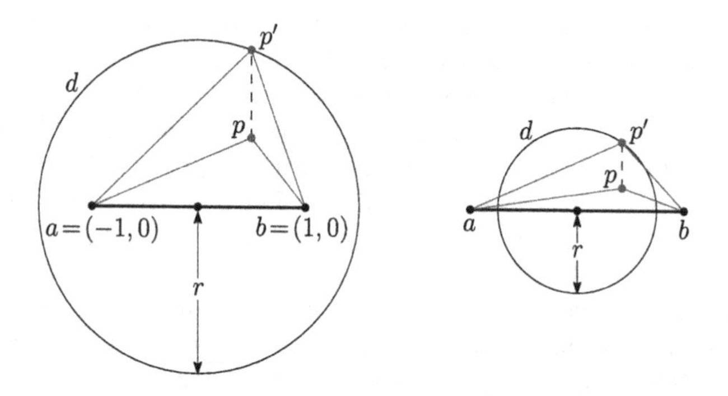

Fig. 2. Illustration of the proof of Lemma 1.

Lemma 1. *Let $r > 0$ be a real number. If ab is a line segment in the plane and p is a point at distance at most $\frac{r\cdot|ab|}{2}$ from the midpoint of ab then*

$$|pa| + |pb| \leqslant \sqrt{r^2+1}\cdot|ab|.$$

Proof. After scaling by factor $2/|ab|$ we will have $|ab| = 2$ and p at distance at most r from the midpoint of ab. After a suitable rotation and translation assume that $a = (-1, 0)$ and $b = (1, 0)$. Any point $p = (x, y)$ at distance at most r from the midpoint of ab lies in the disk d of radius r that is centered at $(0, 0)$ as in Fig. 2. Since $|ab| = 2$, it suffices to prove that $|pa| + |pb| \leqslant 2\sqrt{r^2+1}$. Without loss of generality we may assume that $x \geqslant 0$ and $y \geqslant 0$. Let p' be the vertical projection of p onto the boundary of d as in Fig. 2. Observe that $|pa| \leqslant |p'a|$ and $|pb| \leqslant |p'b|$. Thus the largest value of $|pa| + |pb|$ occurs when p is on the boundary of d. Therefore, for the purpose of this lemma we assume that p is on the boundary circle of d. The circle has equation $x^2 + y^2 = r^2$. Therefore, we can define $|pa| + |pb|$ as a function of x as follows where $0 \leqslant x \leqslant r$ (recall that x is the x-coordinate of p, and y is the y-coordinate of p).

$$\begin{aligned} f(x) = |pa| + |pb| &= \sqrt{(x+1)^2 + y^2} + \sqrt{(x-1)^2 + y^2} \\ &= \sqrt{x^2 + y^2 + 1 + 2x} + \sqrt{x^2 + y^2 + 1 - 2x} \\ &= \sqrt{r^2 + 1 + 2x} + \sqrt{r^2 + 1 - 2x}. \end{aligned}$$

We are interested in the largest value of $f(x)$ on interval $x \in [0, r]$. By computing its derivative it turns out that $f(x)$ is decreasing on this interval. Thus the largest value of $f(x)$ is achieved at $x = 0$, and it is $2\sqrt{r^2 + 1}$. □

Lemma 2. *Let a, p, b, q be the vertices of a convex quadrilateral that appear in this order along the boundary. If $|pa| = |pb|$ and $\angle aqb \geqslant 2\pi/3$ then $|pq| \leqslant \frac{2}{\sqrt{3}}|pa|$.*

Proof. After a suitable scaling, rotation, and reflection assume that $|pa| = 1$, ab is horizontal, and p lies below ab as in Fig. 3-left. Since $|pa| = 1$ in this new setting, it suffices to prove that $|pq| \leqslant 2/\sqrt{3}$. Consider the ray emanating from p and passing through q. Let q' be the point on this ray such that $\angle aq'b = 2\pi/3$, and observe that $|pq'| \geqslant |pq|$. Thus for the purpose of this lemma we can assume that $\angle aqb = 2\pi/3$. The locus of all points q, with $\angle aqb = 2\pi/3$, is a circular arc C with endpoints a and b. See Fig. 3-middle. Let c be the center of the circle that defines arc C. Since ab is horizontal and $|pa| = |pb|$, the center c lies on the vertical line through p. Let d be the disk of radius 1 centered at p. If c lies on or below p then C lies in d and consequently q is in d. In this case $|pq| \leqslant 1$, and we are done. Assume that c lies above p as in Fig. 3-middle. By the law of cosines we have $|pq| = \sqrt{|pc|^2 + |cq|^2 - 2|pc||cq|\cos\beta}$ where β is the angle between segments cp and cq. Since $|pc|$ and $|cq|$ are fixed for all points q on C, the largest value of $|pq|$ is attained at $\beta = \pi$. Again for the purpose of this lemma we can assume that $\beta = \pi$, in which case $|qa| = |qb|$. Let α denote the angle between segments pa and pb. Define $f(\alpha) = |pq|$ where $0 \leqslant \alpha \leqslant \pi$. Recall that $\angle aqb = 2\pi/3$. This setting is depicted in Fig. 3-right. By the law of sines we have

$$f(\alpha) = |pq| = \frac{\sin\left(\frac{\pi}{6} + \frac{\pi-\alpha}{2}\right)}{\sin\left(\frac{\pi}{3}\right)} = \frac{2\sin\left(\frac{4\pi-3\alpha}{6}\right)}{\sqrt{3}},$$

where $0 \leqslant \alpha \leqslant \pi$. By computing the derivative of $f(\alpha)$ it turns out that its largest value is attained at $\alpha = \pi/3$, and it is $2/\sqrt{3}$. □

Theorem 5. *Let D be a set of pairwise intersecting disks. Let D' be the set of disks obtained by increasing the radii of all disks in D by factor $2/\sqrt{3}$ while preserving their centers. Then all disks in D' have a common intersection. The factor $2/\sqrt{3}$ is tight.*

Proof. It suffices to show that any three disks in D' have a common intersection because afterwards Theorem 3 implies that all disks in D' have a common intersection. Consider any three disks d'_1, d'_2, d'_3 in D' that are centered at c_1, c_2, c_3, and let d_1, d_2, d_3 be their corresponding disks in D. If d_1, d_2, d_3 have a common intersection, so do d'_1, d'_2, and d'_3. Assume that d_1, d_2, d_3 do not have a common

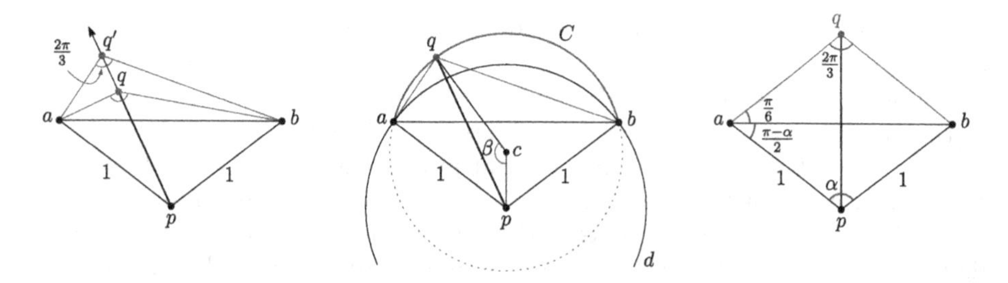

Fig. 3. Illustration of the proof of Lemma 2.

intersection, as depicted in Fig. 4. Let u be the innermost intersection point of boundaries of d_1 and d_2, v be the innermost intersection point of boundaries of d_2 and d_3, and w be the innermost intersection point of boundaries of d_3 and d_1, as in Fig. 4. We show that the Fermat point of triangle $\triangle uvw$ lies in all disks d'_1, d'_2, and d'_3. This would imply that these three disks have a common intersection. The Fermat point of a triangle is a point that minimizes the total distance to the three vertices of the triangle. If all angles of the triangle are less than $2\pi/3$ the Fermat point is inside the triangle and makes angle $2\pi/3$ with every two vertices of the triangle. If the triangle has a vertex of angle at least $2\pi/3$ the Fermat point is that vertex.

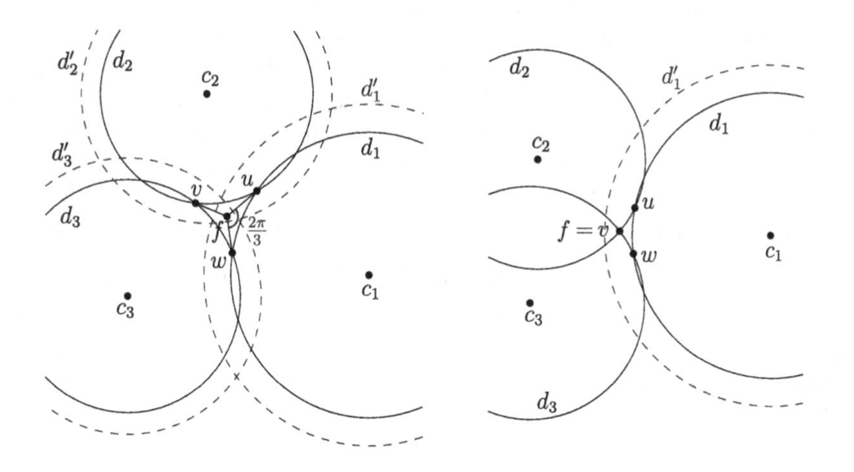

Fig. 4. Illustration of the proof of Theorem 5

Let f be the Fermat point of $\triangle uvw$. First assume that all angles of $\triangle uvw$ are less than $2\pi/3$, as in Fig. 4-left. In this case f is inside $\triangle uvw$ and $\angle ufw = \angle wfv = \angle vfu = 2\pi/3$. By Lemma 2 we have $|c_1 f| \leqslant \frac{2}{\sqrt{3}}|c_1 u|$ (w, c_1, u, f play the roles of a, p, b, q in the lemma, respectively). This and the fact that the radius of d'_1 is $\frac{2}{\sqrt{3}}|c_1 u|$ imply that f lies in d'_1. Analogously, we can show that f lies in d'_2 and d'_3. This finishes our proof for this case.

Now assume that one of the angles of $\triangle uvw$, say the angle $\angle uvw$ at v, is at least $2\pi/3$; see Fig. 4-right. In this case $f = v$. Since f is on the boundaries of d_2 and d_3, it lies in d'_2 and d'_3. By Lemma 2 we have $|c_1 f| \leqslant \frac{2}{\sqrt{3}}|c_1 u|$. Similarly to the previous case, this implies that f lies in d'_1. This finishes our proof.

The factor $2/\sqrt{3}$ in the theorem is tight in the sense that if we replace it by any smaller constant then the disks in D' may not have a common intersection. To verify this consider three disks of the same radius that pairwise touch (but do not properly intersect). For example assume that d_1, d_2, d_3 in Fig. 4-left have radius 1 and pairwise touch at u, v, and w. In this case d'_1, d'_2, d'_3 have radius $2/\sqrt{3}$. Moreover $\angle wc_1u = \angle uc_2v = \angle vc_3w = \pi/3$ and f is inside $\triangle uvw$. In this setting $|c_1f| = |c_2f| = |c_3f| = 2/\sqrt{3}$. This implies that f is the only point in the common intersection of d'_1, d'_2 and d'_3. Therefore, if the radii of these disks are less than $2/\sqrt{3}$ then they wouldn't have a common intersection. □

Theorem 6. *Every 2-local Euclidean maximum matching is a $\sqrt{3/7}$ approximation of a global Euclidean maximum matching.*

Proof. Our proof approach is somewhat similar to that of Theorem 4. Consider any 2-local maximum matching M. Let M^* be a global maximum matching for the endpoints of edges of M. It is well known that that the two diametral disks introduced by the two edges of any maximum matching, on any set of four points in the plane, intersect each other (see e.g. [4]). Consider the set D of diametral disks introduced by edges of M. Since M is 2-local maximum, any two disks in D intersect each other. However, all disks in D may not have a common intersection. We increase the radii of all disks in D by factor $2/\sqrt{3}$ while preserving their centers. Let D' be the resulting set of disks. By Theorem 5 the disks in D' have a common intersection. Let c be a point in this intersection. Let S be the star obtained by connecting c to all endpoints of edges of M. Consider any edge $ab \in M$, and let d be its diametral disk in D and d' be the corresponding disk in D'. The radius of d' is $\frac{2}{\sqrt{3}} \cdot \frac{|ab|}{2}$. Since c is in d', its distance from the center of d' (which is the midpoint of ab) is at most $\frac{2}{\sqrt{3}} \cdot \frac{|ab|}{2}$. By applying Lemma 1, with $p = c$ and $r = 2/\sqrt{3}$, we have $|ca| + |cb| \leqslant \sqrt{7/3} \cdot |ab|$. This implies that $w(S) \leqslant \sqrt{7/3} \cdot w(M)$. For any edge $a^*b^* \in M^*$, by the triangle inequality we have $|a^*b^*| \leqslant |ca^*| + |cb^*|$, and thus $w(M^*) \leqslant w(S)$. Therefore, $w(M) \geqslant \sqrt{3/7} \cdot w(M^*)$. □

5 Pairwise-Crossing Matchings are Globally Maximum

A pairwise crossing matching is a matching in which every pair of edges cross each other. It is easy to verify that any pairwise crossing matching is 2-local maximum. We claim that such matchings are in fact global maximum. We also claim that pairwise crossing matchings are unique. Both claims can be easily verified for points in convex position. In this section we prove these claims for points in general position, where no three points lie on a line.

Observation 1. *Let M be a pairwise crossing perfect matching on a point set P. Then for any edge $ab \in M$ it holds that the number of points of P on each side of the line through ab is $(|P|-2)/2$.*

Theorem 7. *A pairwise crossing perfect matching on a point set is unique if it exists.*

Proof. Consider any even-size point set P that has a pairwise crossing perfect matching. For the sake of contradiction assume that P admits two different perfect matchings M_1 and M_2 each of which is pairwise crossing. The union of M_1 and M_2 consists of connected components which are single edges (belong to both M_1 and M_2) and even cycles. Since $M_1 \neq M_2$, $M_1 \cup M_2$ contains some even cycles. Consider one such cycle, say C. Let C_1 and C_2 be the sets of edges of C that belong to M_1 and M_2 respectively. Observe that each of C_1 and C_2 is a pairwise crossing perfect matching for vertices of C.

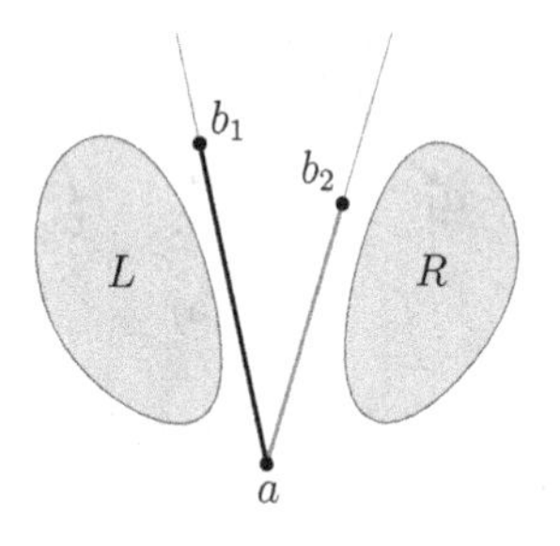

Let a denote the lowest vertex of C; a is a vertex of the convex hull of C. Let b_1 and b_2 be the vertices of C that are matched to a via C_1 and C_2 respectively. After a suitable reflection assume that b_2 is to the right side of the line through a and b_1 as in the figure to the right. Let L be the set of vertices of C that are to the left side of the line through ab_1, and let R be the set of vertices of C that are to the right side of the line through ab_2. Since C_1 is pairwise crossing, by Observation 1 we have $|L| = (|C|-2)/2$. Analogously we have $|R| = (|C|-2)/2$. Set $C' = L \cup R \cup \{a, b_1, b_2\}$, and observe that $C' \subseteq C$. Since the sets L, R, and $\{a, b_1, b_2\}$ are pairwise disjoint, $|C'| = |L| + |R| + 3 = |C| + 1$. This is a contradiction because C' is a subset of C. □

In Theorem 9 we prove that a pairwise crossing matching is globally maximum, i.e., it is a maximum-length matching for its endpoints. The following "edge-disjoint paths problem" that is studied by Okamura and Seymour [28] will come in handy for our proof of Theorem 9. To state this problem in a simple way, we borrow some terminology from [32].

Let $G = (V, E)$ be an embedded planar graph and let $N = \{(a_1, b_1), \ldots, (a_k, b_k)\}$ be a set of pairs of distinct vertices of V that lie on the outerface, as in Fig. 5(a). A problem instance is a pair (G, N) where the augmented graph $(V, E \cup \{a_1b_1, \ldots, a_kb_k\})$ is Eulerian (i.e. it has a closed trail containing all edges). We note that the augmented graph may not be planar. The problem is to decide whether there are edge-disjoint paths $P_1, \ldots, P_k$ in G such that each P_i connects a_i to b_i.[1] Okamura and Seymour [28] gave a necessary and sufficient condition for the existence of such paths; this condition is stated below in Theorem 8. A *cut* X is a nonempty proper subset of V. Let $c(X)$ be the number of edges in G with one endpoint in X and the other in $V \backslash X$, and let $d(X)$ be the number of pairs (a_i, b_i) with one element in X and the other in $V \backslash X$. A cut X is *essential*

[1] This problem has applications in multicommodity flows in planar graphs [28].

if the subgraphs of G induced by X and $V\setminus X$ are connected and neither set is disjoint with the outerface of G. If X is essential then each of X and $V\setminus X$ shares one single connected interval with the outerface; see Fig. 5(a).

Theorem 8 (Okamura and Seymour, 1981). *An instance (G, N) is solvable if and only if for any essential cut X it holds that $c(X) - d(X) \geqslant 0$.*

Wagner and Weihe [32] studied a computational version of the problem and presented a linear-time algorithm for finding edge-disjoint paths $P_1, \ldots, P_k$.

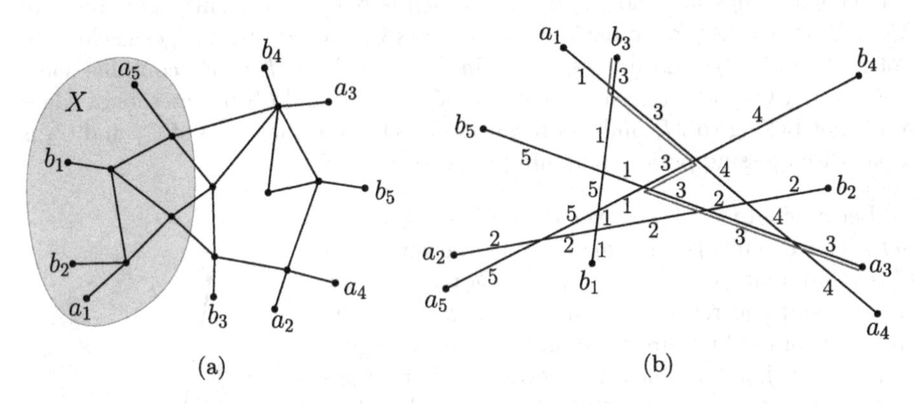

Fig. 5. (a) An essential cut X with $c(X) = 4$ and $d(X) = 2$. (b) Edge-disjoint paths between endpoints of edges of M^*.

Theorem 9. *Any pairwise crossing matching is globally maximum.*

Proof. Consider any matching M with pairwise crossing segments, and let P be the set of endpoints of edges of M. Let $\mathcal{A}$ be the arrangement defined by the segments of M. Notice that $w(\mathcal{A}) = w(M)$, where $w(\mathcal{A})$ is the total length of segments in $\mathcal{A}$. This arrangement is a planar graph where every vertex, that is a point of P, has degree 1 and every vertex, that is an intersection point of two segments of M, has degree 4 (assuming no three segments intersect at the same point). Now consider any perfect matching M^* on P; M^* could be a global maximum matching. Denote the edges of M^* by $a_1b_1, a_2b_2, \ldots$. To prove the theorem it suffices to show that $w(M^*) \leqslant w(\mathcal{A})$. To show this inequality, we prove existence of edge-disjoint paths between all pairs (a_i, b_i) in $\mathcal{A}$, as depicted in Fig. 5(b). We may assume that M and M^* are edge disjoint because shared edges have the same contribution to each side of the inequality.

Observe that the pair $(\mathcal{A}, M^*)$ is an instance of the problem of Okamura and Seymour [28] because the augmented graph is Eulerian (here we slightly abuse M^* to refer to a set of pairs). In the augmented graph, every point of P has degree 2, whereas the degree of every other vertex is the same as its degree in $\mathcal{A}$. Consider any essential cut X in $\mathcal{A}$. Set $X_P = X \cap P$. Consider the two sets X_P and $P\setminus X_P$. Denote the smaller set by Y_1 and the larger set by Y_2. Notice that $|Y_1 \cup Y_2| = |P|$, $|Y_1| \leqslant |P|/2$, and $|Y_2| \geqslant |P|/2$. We claim that no two points of Y_1 are matched to each other by an edge of M. To verify this claim we use contradiction. Assume that for two points a and b in Y_1 we have $ab \in M$. Since X is essential, each of Y_1 and Y_2 consists of some points of P that are consecutive on the outerface of $\mathcal{A}$. This and the fact that M is pairwise crossing imply that all points of Y_2 lie on one side of the line through ab. This contradicts Observation 1, and hence proves our claim.

The above claim implies that every point in Y_1 is matched to a point in Y_2 by an edge of M. Any such edge of M introduces at least one edge between X and $\mathcal{A} \setminus X$ in $\mathcal{A}$. Therefore $c(X) \geqslant |Y_1|$. Since every a_i and every b_i belong to P, the number of pairs (a_i, b_i) with one element in X and another one in $\mathcal{A} \setminus X$ is the same as the number of such pairs with one element in Y_1 and the other in Y_2. The number of such pairs cannot be more than $|Y_1|$, and thus $d(X) \leqslant |Y_1|$. To this end we have that $c(X) \geqslant d(X)$. Having this constraint, Theorem 8 implies that the instance $(\mathcal{A}, M^*)$ is solvable, and thus there are edge-disjoint paths between all pairs (a_i, b_i). By the triangle inequality, $w(M^*)$ is at most the total length of these edge-disjoint paths, which is at most $w(\mathcal{A})$. □

6 Discussion

We believe that 3-local Euclidean maximum matchings are "very good" approximations of global Euclidean maximum matchings. In particular we think that the lower bound on the length ratio should be closer to 1 than to $1/\sqrt{2}$. A natural open problem is to use the geometry of the Euclidean plane and improve the lower bounds on the length ratios for 2- and 3-local maximum matchings.

From the computational point of view, there are algorithms that compute a global maximum matching in polynomial time [15–18, 25] and there is a linear-time algorithm that gives a $(1 - \varepsilon)$-approximation [8]. It would be interesting to see how fast a k-local maximum matching can be computed. Theorem 1 suggests a local search strategy where repeatedly k-subsets of the current matching are tested for improvement. In its straightforward version this requires superlinear time. It would be interesting to see whether geometric insights could speed up the local search, maybe not (theoretically) matching the linear-time bound from [8], but leading to a practical and in particular simple algorithm.

We note that analogous ratios for minimum-length matchings could be arbitrary large. In the figure to the right $2n$ points are placed on a circle such that distances between consecutive points are alternating between 1 and arbitrary small constant ϵ. For a sufficiently large n, the red matching which has n edges of length 1, would be 2-local minimum (the two arcs in the figure are centered at a and b, and show that the length $|ab|$ is larger than the total length of two consecutive red edges). In this setting, the global minimum matching would have n edges of length ϵ. This shows that the ratio of the length of 2-local minimum matchings to that of global minimum matchings

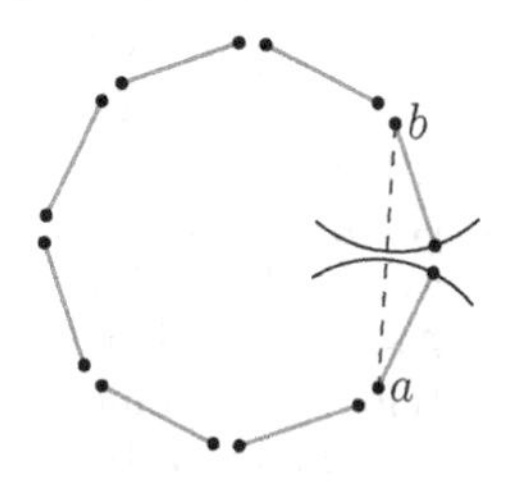

could be arbitrary large. By increasing the number of points (and hence flattening the perimeter of the circle) in this example, it can be shown that the length ratio of k-local minimum matchings could be arbitrary large, for any fixed $k \geqslant 2$.

References

1. Alon, N., Rajagopalan, S., Suri, S.: Long non-crossing configurations in the plane. Fundam. Inform. **22**(4), 385–394 (1995). Also in SoCG 1993
2. Arkin, E.M., Hassin, R.: On local search for weighted k-set packing. Math. Oper. Res. **23**(3), 640–648 (1998). Also in ESA 1997
3. Avis, D.: A survey of heuristics for the weighted matching problem. Networks **13**(4), 475–493 (1983)
4. Bereg, S., Chacón-Rivera, O., Flores-Peñaloza, D., Huemer, C., Pérez-Lantero, P., Seara, C.: On maximum-sum matchings of points (2019). arXiv:1911.10610
5. Biniaz, A., et al.: Maximum plane trees in multipartite geometric graphs. Algorithmica **81**(4), 1512–1534 (2019). Also in WADS 2017
6. Carmi, P., Katz, M.J., Morin, P.: Stabbing pairwise intersecting disks by four points (2018). arXiv:1812.06907
7. Danzer, L.: Zur Lösung des Gallaischen Problems über Kreisscheiben in der Euklidischen Ebene. Stud. Sci. Math. Hung. **21**(1–2), 111–134 (1986)
8. Duan, R., Pettie, S.: Linear-time approximation for maximum weight matching. J. ACM **61**(1), 1:1–1:23 (2014)
9. Dumitrescu, A., Tóth, C.D.: Long non-crossing configurations in the plane. Discrete Comput. Geom. **44**(4), 727–752 (2010). Also in STACS 2010
10. Dyer, M., Frieze, A., McDiarmid, C.: Partitioning heuristics for two geometric maximization problems. Oper. Res. Lett. **3**(5), 267–270 (1984)
11. Edmonds, J.: Maximum matching and a polyhedron with 0,1-vertices. J. Res. Natl. Bur. Stand. B **69**, 125–130 (1965)
12. Edmonds, J.: Paths, trees, and flowers. Canad. J. Math. **17**, 449–467 (1965)
13. Edmonds, J., Karp, R.M.: Theoretical improvements in algorithmic efficiency for network flow problems. J. ACM **19**(2), 248–264 (1972)
14. Eppstein, D.: Geometry junkyard. https://www.ics.uci.edu/~eppstein/junkyard/maxmatch.html
15. Gabow, H.N.: Data structures for weighted matching and nearest common ancestors with linking. In: Proceedings of the First Annual ACM-SIAM Symposium on Discrete Algorithms (SODA), pp. 434–443 (1990)
16. Gabow, H.N.: Data structures for weighted matching and extensions to b-matching and f-factors. ACM Trans. Algorithms **14**(3), 39:1–39:80 (2018)
17. Gabow, H.N., Galil, Z., Spencer, T.H.: Efficient implementation of graph algorithms using contraction. J. ACM **36**(3), 540–572 (1989)
18. Galil, Z., Micali, S., Gabow, H.N.: An O(EV log V) algorithm for finding a maximal weighted matching in general graphs. SIAM J. Comput. **15**(1), 120–130 (1986). Also in FOCS 1982
19. Hadlock, F.: Finding a maximum cut of a planar graph in polynomial time. SIAM J. Comput. **4**(3), 221–225 (1975)
20. Har-Peled, S., et al.: Stabbing pairwise intersecting disks by five points. In: 29th International Symposium on Algorithms and Computation, ISAAC, pp. 50:1–50:12 (2018)

21. Helly, E.: Über Mengen konvexer Körper mit gemeinschaftlichen Punkten. Jahresber. Dtsch. Math. Ver. **32**, 175–176 (1923)
22. Huemer, C., Pérez-Lantero, P., Seara, C., Silveira, R.I.: Matching points with disks with a common intersection. Discrete Math. **342**(7), 1885–1893 (2019)
23. Kuhn, H.W.: The Hungarian method for the assignment problem. Naval Res. Logist. Q. **2**, 83–97 (1955)
24. Kuhn, H.W.: Variants of the Hungarian method for assignment problems. Naval Res. Logist. Q. **3**, 253–258 (1956)
25. Lawler, E.: Combinatorial Optimization: Networks And Matroids. Holt, Rinehart and Winston, New York (1976)
26. Lee, J., Sviridenko, M., Vondrák, J.: Submodular maximization over multiple matroids via generalized exchange properties. Math. Oper. Res. **35**(4), 795–806 (2010)
27. McKeown, N., Anantharam, V., Walrand, J.C.: Achieving 100% throughput in an input-queued switch. In: Proceedings of the 15th IEEE INFOCOM, pp. 296–302 (1996)
28. Okamura, H., Seymour, P.D.: Multicommodity flows in planar graphs. J. Comb. Theory Ser. B **31**(1), 75–81 (1981)
29. Radon, J.: Mengen konvexer Körper, die einen gemeinsamen Punkt enthalten. Math. Ann. **83**(1), 113–115 (1921)
30. Rendl, F.: On the Euclidean assignment problem. J. Comput. Appl. Math. **23**(3), 257–265 (1988)
31. Stachó, L.: A solution of Gallai's problem on pinning down circles. Mat. Lapok **32**(1–3), 19–47 (1981). (1981/1984)
32. Wagner, D., Weihe, K.: A linear-time algorithm for edge-disjoint paths in planar graphs. Combinatorica **15**(1), 135–150 (1995). Also in ESA 1993

Solving Problems on Generalized Convex Graphs via Mim-Width

Flavia Bonomo-Braberman[1], Nick Brettell[2], Andrea Munaro[3], and Daniël Paulusma[4(✉)]

[1] ICC (CONICET-UBA) and Departamento de Computación, Universidad de Buenos Aires, Buenos Aires, Argentina
fbonomo@dc.uba.ar

[2] School of Mathematics and Statistics, Victoria University of Wellington, Wellington, New Zealand
nick.brettell@vuw.ac.nz

[3] School of Mathematics and Physics, Queen's University Belfast, Belfast, UK
a.munaro@qub.ac.uk

[4] Department of Computer Science, Durham University, Durham, UK
daniel.paulusma@durham.ac.uk

Abstract. A bipartite graph $G = (A, B, E)$ is $\mathcal{H}$-convex, for some family of graphs $\mathcal{H}$, if there exists a graph $H \in \mathcal{H}$ with $V(H) = A$ such that the set of neighbours in A of each $b \in B$ induces a connected subgraph of H. Many NP-complete problems become polynomial-time solvable for $\mathcal{H}$-convex graphs when $\mathcal{H}$ is the set of paths. In this case, the class of $\mathcal{H}$-convex graphs is known as the class of convex graphs. The underlying reason is that this class has bounded mim-width. We extend the latter result to families of $\mathcal{H}$-convex graphs where (i) $\mathcal{H}$ is the set of cycles, or (ii) $\mathcal{H}$ is the set of trees with bounded maximum degree and a bounded number of vertices of degree at least 3. As a consequence, we can reprove and strengthen a large number of results on generalized convex graphs known in the literature. To complement result (ii), we show that the mim-width of $\mathcal{H}$-convex graphs is unbounded if $\mathcal{H}$ is the set of trees with arbitrarily large maximum degree or an arbitrarily large number of vertices of degree at least 3. In this way we are able to determine complexity dichotomies for the aforementioned graph problems. Afterwards we perform a more refined width-parameter analysis, which shows even more clearly which width parameters are bounded for classes of $\mathcal{H}$-convex graphs.

1 Introduction

Many computationally hard graph problems can be solved efficiently if we place constraints on the input. Instead of solving individual problems in an ad hoc

Brettell and Paulusma received support from the Leverhulme Trust (RPG-2016-258). Bonomo received support from UBACyT (20020170100495BA and 20020160100095BA). Brettell also received support from a Rutherford Foundation Postdoctoral Fellowship, administered by the Royal Society Te Apārangi.

A. Lubiw et al. (Eds.): WADS 2021, LNCS 12808, pp. 200–214, 2021.
https://doi.org/10.1007/978-3-030-83508-8_15

way we may try to decompose the vertex set of the input graph into large sets of "similarly behaving" vertices and to exploit this decomposition for an algorithmic speed up that works for many problems simultaneously. This requires some notion of an "optimal" vertex decomposition, which depends on the type of vertex decomposition used and which may relate to the minimum number of sets or the maximum size of a set in a vertex decomposition. An optimal vertex decomposition gives us the "width" of the graph. A graph class has *bounded width* if every graph in the class has width at most some constant c. Boundedness of width is often the underlying reason why a graph-class-specific algorithm runs efficiently: in such a case, the proof that the algorithm is efficient for some special graph class reduces to a proof showing that the width of the class is bounded by some constant. We will give examples, but also refer to the surveys [16,19, 22,26,42] for further details and examples.

Width parameters differ in strength. A width parameter p *dominates* a width parameter q if there is a function f such that $p(G)$ is at most $f(q(G))$ for every graph G. If p dominates q but q does not dominate p, then we say that p is *more powerful* than q. If both p and q dominate each other, then p and q are *equivalent*. If neither p is more powerful than q nor q is more powerful than p, then p and q are *incomparable*. If p is more powerful than q, then the class of graphs for which p is bounded is larger than the class of graphs for which q is bounded and so efficient algorithms for bounded p have greater applicability with respect to the graphs under consideration. The *trade-off* is that fewer problems exhibit an efficient algorithm for the parameter p, compared to the parameter q.

The notion of powerfulness leads to a large hierarchy of width parameters, in which new width parameters continue to be defined. The well-known parameters boolean-width, clique-width, module-width and rank-width are equivalent to each other [10,34,38]. They are more powerful than the equivalent parameters branch-width and treewidth [14,39,42] but less powerful than mim-width [42], which is less powerful than sim-width [27]. To give another example, thinness is more powerful than path-width [33], but less powerful than mim-width and incomparable to clique-width or treewidth [4].

For each group of equivalent width parameters, a growing set of NP-complete problems is known to be tractable on graph classes of bounded width. However, there are still large families of graph classes for which boundedness of width is not known for many width parameters.

Our Focus. We consider the relatively new width parameter *mim-width*, which we define below. Recently, we showed in [7,8] that boundedness of mim-width is the underlying reason why some specific hereditary graph classes, characterized by two forbidden induced subgraphs, admit polynomial-time algorithms for a range of problems including k-Colouring and its generalization List k-Colouring (the algorithms are given in [13,15,20]). Here we prove that the same holds for certain *superclasses of convex graphs* known in the literature. Essentially all the known polynomial-time algorithms for such classes are obtained by reducing to the class of convex graphs. We show that our new approach via mim-width simplifies the analysis, unifies the sporadic approaches and explains the reductions to convex graphs.

Mim-width. A set of edges M in a graph G is a *matching* if no two edges of M share an endpoint. A matching M is *induced* if there is no edge in G between vertices of different edges of M. Let $(A, \overline{A})$ be a partition of the vertex set of a graph G. Then $G[A, \overline{A}]$ denotes the bipartite subgraph of G induced by the edges with one endpoint in A and the other in $\overline{A}$. Vatshelle [42] introduced the notion of *maximum induced matching width*, also called mim-width. Mim-width measures the extent to which it is possible to decompose a graph G along certain vertex partitions $(A, \overline{A})$ such that the size of a maximum induced matching in $G[A, \overline{A}]$ is small. The kind of vertex partitions permitted stem from classical branch decompositions. A *branch decomposition* for a graph G is a pair (T, δ), where T is a subcubic tree and δ is a bijection from $V(G)$ to the leaves of T. Every edge $e \in E(T)$ partitions the leaves of T into two classes, L_e and $\overline{L_e}$, depending on which component of $T - e$ they belong to. Hence, e induces a partition $(A_e, \overline{A_e})$ of $V(G)$, where $\delta(A_e) = L_e$ and $\delta(\overline{A_e}) = \overline{L_e}$. Let $\text{cutmim}_G(A_e, \overline{A_e})$ be the size of a maximum induced matching in $G[A_e, \overline{A_e}]$. Then the *mim-width* $\text{mimw}_G(T, \delta)$ of (T, δ) is the maximum value of $\text{cutmim}_G(A_e, \overline{A_e})$ over all edges $e \in E(T)$. The *mim-width* $\text{mimw}(G)$ of G is the minimum value of $\text{mimw}_G(T, \delta)$ over all branch decompositions (T, δ) for G. We refer to Fig. 1 for an example.

Computing the mim-width is NP-hard [40], and approximating the mim-width in polynomial time within a constant factor of the optimal is not possible unless NP = ZPP [40]. It is not known how to compute in polynomial time a branch decomposition for a graph G whose mim-width is bounded by some function in the mim-width of G. However, for graph classes of bounded mim-width this might be possible. In that case, the mim-width of $\mathcal{G}$ is said to be *quickly computable*. One can then try to develop a polynomial-time algorithm for the graph problem under consideration via dynamic programming over the computed branch decomposition. We give examples of such problems later.

Convex Graphs and Generalizations. A bipartite graph $G = (A, B, E)$ is *convex* if there exists a path P with $V(P) = A$ such that the neighbours in A of each $b \in B$ induce a connected subpath of P. Convex graphs generalize bipartite permutation graphs (see, e.g., [5]) and form a well-studied graph class.

Belmonte and Vatshelle [1] proved that the mim-width of convex graphs is bounded and quickly computable. We consider superclasses of convex graphs and research to what extent mim-width can play a role in obtaining polynomial-time algorithms for problems on these classes.

Let $\mathcal{H}$ be a family of graphs. A bipartite graph $G = (A, B, E)$ is $\mathcal{H}$-*convex* if there exists a graph $H \in \mathcal{H}$ with $V(H) = A$ such that the set of neighbours in A of each $b \in B$ induces a connected subgraph of H. If $\mathcal{H}$ consists of all paths, we obtain the class of convex graphs. A *caterpillar* is a tree T that contains a path P, the *backbone* of T, such that every vertex not on P has a neighbour on P. A caterpillar with a backbone consisting of one vertex is a *star*. A *comb* is a caterpillar such that every backbone vertex has exactly one neighbour outside the backbone. The *subdivision* of an edge uv replaces uv by a new vertex w and edges uw and wu. A *triad* is a tree that can be obtained from a 4-vertex star after a sequence of subdivisions. For $t, \Delta \geq 0$, a (t, Δ)-*tree* is a tree with maximum degree at most Δ and containing at most t vertices of degree at least 3; note that, for example, a triad is a $(1, 3)$-tree. If $\mathcal{H}$ consists of all cycles, all trees, all stars, all triads, all combs or all (t, Δ)-trees, then we obtain the class of *circular convex graphs*, *tree convex graphs*, *star convex graphs*, *triad convex graphs*, *comb convex graphs* or (t, Δ)-*tree convex graphs*, respectively. See Fig. 1 for an example.

To show the relationships between the above graph classes we need some extra terminology. Let $\mathcal{C}_{t,\Delta}$ be the class of (t, Δ)-tree convex graphs. For fixed t or Δ, we have increasing sequences $\mathcal{C}_{t,0} \subseteq \mathcal{C}_{t,1} \subseteq \cdots$ and $\mathcal{C}_{0,\Delta} \subseteq \mathcal{C}_{1,\Delta} \subseteq \cdots$. For $t \in \mathbb{N}$, the class of (t, ∞)-*tree convex graphs* is $\bigcup_{\Delta \in \mathbb{N}} \mathcal{C}_{t,\Delta}$, denoted by $\mathcal{C}_{t,\infty}$. Similarly, for $\Delta \in \mathbb{N}$, the class of (∞, Δ)-*tree convex graphs* is $\bigcup_{t \in \mathbb{N}} \mathcal{C}_{t,\Delta}$, denoted by $\mathcal{C}_{\infty,\Delta}$. Hence, $\mathcal{C}_{t,\infty}$ and $\mathcal{C}_{\infty,\Delta}$ are the set-theoretic limits of the increasing sequences $\{\mathcal{C}_{t,\Delta}\}_{\Delta \in \mathbb{N}}$ and $\{\mathcal{C}_{t,\Delta}\}_{t \in \mathbb{N}}$, respectively. The class of (∞, ∞)-*tree convex graphs* is $\bigcup_{t,\Delta \in \mathbb{N}} \mathcal{C}_{t,\Delta}$, which coincides with the class of tree convex graphs. Notice that the class of convex graphs coincides with $\mathcal{C}_{t,2}$, for any $t \in \mathbb{N} \cup \{\infty\}$, and with $\mathcal{C}_{0,\Delta}$, for any $\Delta \in \mathbb{N} \cup \{\infty\}$. The class of star convex graphs coincides with $\mathcal{C}_{1,\infty}$. Moreover, each triad convex graph belongs to $\mathcal{C}_{1,3}$ and each comb convex graph belongs to $\mathcal{C}_{\infty,3}$. A bipartite graph is *chordal bipartite* if every induced cycle in it has exactly four vertices. Every convex graph is chordal bipartite (see, e.g., [5]) and every chordal bipartite graph is tree convex (see [24,29]). In Fig. 2 we display these and other relationships, which directly follow from the definitions.

Brault-Baron et al. [6] proved that chordal bipartite graphs have unbounded mim-width. Hence, the result of [1] for convex graphs cannot be generalized to

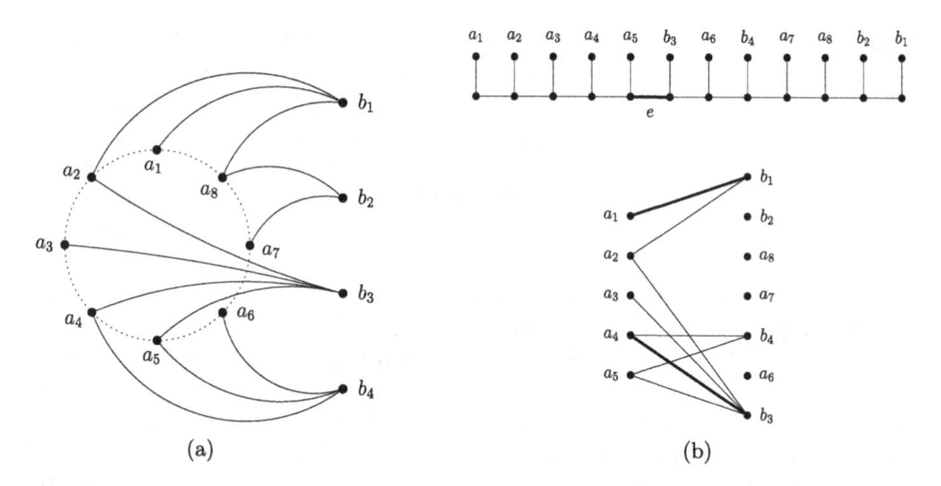

Fig. 1. (a) A circular convex graph $G = (A, B, E)$ with a circular ordering on A. (b) A branch decomposition (T, δ) for G, where T is a caterpillar with a specified edge e, together with the graph $G[A_e, \overline{A_e}]$. The bold edges in $G[A_e, \overline{A_e}]$ form an induced matching and it is easy to see that $\text{cutmim}_G(A_e, \overline{A_e}) = 2$.

chordal bipartite graphs. We determine the mim-width of the other classes in Fig. 2 but first discuss known algorithmic results for these classes.

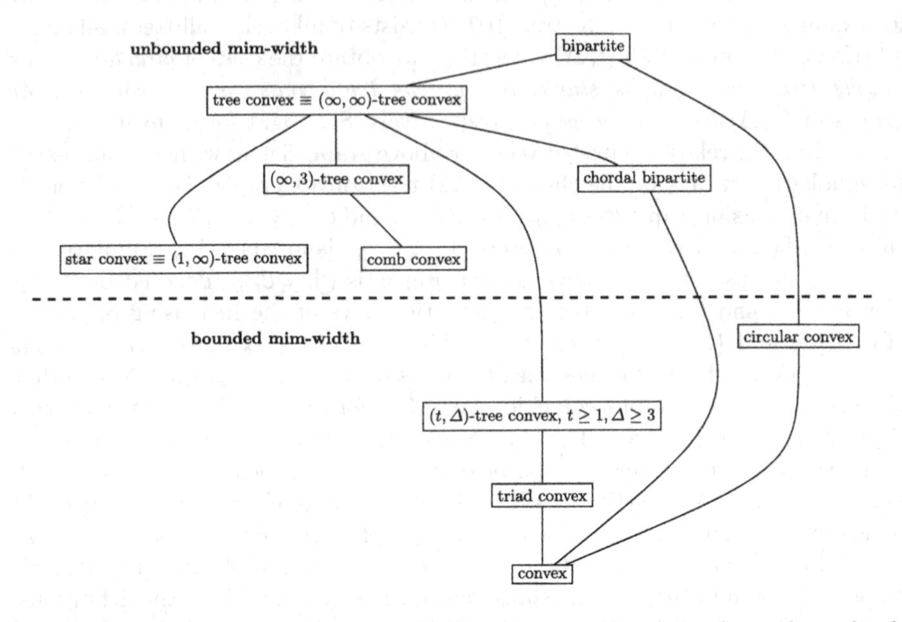

Fig. 2. The inclusion relations between the classes we consider. A line from a lower-level class to a higher one means the first class is contained in the second.

Known Results. Belmonte and Vatshelle [1] and Bui-Xuan et al. [11] proved that so-called Locally Checkable Vertex Subset and Vertex Partitioning (LC-VSVP) problems are polynomial-time solvable on graph classes whose mim-width is bounded and quickly computable. This result was extended by Bergougnoux and Kanté [2] to variants of such problems with additional constraints on connectivity or acyclicity. Each of the problems mentioned below is a special case of a Locally Checkable Vertex Subset (LCVS) problem possibly with one of the two extra constraints. Panda et al. [36] proved that INDUCED MATCHING is polynomial-time solvable for circular convex and triad convex graphs, but NP-complete for star convex and comb convex graphs. Pandey and Panda [37] proved that DOMINATING SET is polynomial-time solvable for circular convex, triad convex and $(1, \Delta)$-tree convex graphs for every $\Delta \geq 1$. Liu et al. [31] proved that CONNECTED DOMINATING SET is polynomial-time solvable for circular convex and triad convex graphs. Chen et al. [12] showed that (CONNECTED) DOMINATING SET and TOTAL DOMINATING SET are NP-complete for star convex and comb convex graphs. Lu et al. [32] proved that INDEPENDENT DOMINATING SET is polynomial-time solvable for circular convex and triad convex graphs. The latter result was shown already in [41] using a dynamic programming approach instead of a reduction to convex graphs [32]. Song et al. [41] showed in fact a stronger result, namely that INDEPENDENT

DOMINATING SET is polynomial-time solvable for (t, Δ)-tree convex graphs for every $t \geq 1$ and $\Delta \geq 3$. They also showed in [41] that INDEPENDENT DOMINATING SET is NP-complete for star convex and comb convex graphs. Hence, they obtained a dichotomy: INDEPENDENT DOMINATING SET is polynomial-time solvable for (t, Δ)-tree convex graphs for every $t \geq 1$ and $\Delta \geq 3$ but NP-complete for $(\infty, 3)$-tree convex graphs and $(1, \infty)$-tree convex graphs.

The same dichotomy (explicitly formulated in [44]) holds for FEEDBACK VERTEX SET and is obtained similarly. Namely, Jiang et al. [25] proved that this problem is polynomial-time solvable for triad convex graphs and mentioned that their algorithm can be generalized to (t, Δ)-tree convex graphs for every $t \geq 1$ and $\Delta \geq 3$. Jiang et al. [24] proved that FEEDBACK VERTEX SET is NP-complete for star convex and comb convex graphs. In addition, Liu et al. [30] proved that FEEDBACK VERTEX SET is polynomial-time solvable for circular convex graphs, whereas Jiang et al. [24] proved that the WEIGHTED FEEDBACK VERTEX SET problem is polynomial-time solvable for triad convex graphs.

It turns out that the above problems are polynomial-time solvable on circular convex graphs and subclasses of (t, Δ)-tree convex graphs, but NP-complete for star convex graphs and comb convex graphs. In contrast, Panda and Chaudhary [35] proved that DOMINATING INDUCED MATCHING is not only polynomial-time solvable on circular convex and triad convex graphs, but also on star convex graphs. Nevertheless, we notice *a common pattern*: many dominating set, induced matching and graph transversal type of problems are polynomial-time solvable for (t, Δ)-tree convex graphs, for every $t \geq 1$ and $\Delta \geq 3$, and NP-complete for comb convex graphs, and thus for $(\infty, 3)$-tree convex graphs, and star convex graphs, or equivalently, $(1, \infty)$-tree convex graphs. Moreover, essentially all the polynomial-time algorithms reduce the input to a convex graph.

Our Results. We simplify the analysis, unify the above approaches and explain the reductions to convex graphs, using mim-width. We prove three results that, together with the fact that chordal bipartite graphs have unbounded mim-width [6], explain the dotted line in Fig. 2. The first two results generalize the result of [1] for convex graphs. The third result gives two new reasons why tree convex graphs (that is, (∞, ∞)-tree convex graphs) have unbounded mim-width.

Theorem 1. *Let G be a circular convex graph. Then* $\mathrm{mimw}(G) \leq 2$. *Moreover, we can construct in polynomial time a branch decomposition (T, δ) for G with* $\mathrm{mimw}_G(T, \delta) \leq 2$.

Theorem 2. *Let G be a (t, Δ)-tree convex graph with $t, \Delta \in \mathbb{N}$ and $t \geq 1$ and $\Delta \geq 3$. Let*

$$f(t, \Delta) = \max\left\{2\left\lfloor\left(\frac{\Delta}{2}\right)^2\right\rfloor, 2\Delta - 1\right\} + t^2\Delta.$$

Then $\mathrm{mimw}(G) \leq f(t, \Delta)$. *Moreover, we can construct in polynomial time a branch decomposition (T, δ) for G with* $\mathrm{mimw}_G(T, \delta) \leq f(t, \Delta)$.

Theorem 3. *The class of star convex graphs and the class of comb convex graphs each has unbounded mim-width.*

Hence, we obtain a structural dichotomy (recall that star convex graphs are the $(1, \infty)$-tree convex graphs and that comb convex graphs are $(\infty, 3)$-tree convex):

Corollary 1. *Let $t, \Delta \in \mathbb{N} \cup \{\infty\}$ with $t \geq 1$ and $\Delta \geq 3$. The class of (t, Δ)-tree convex graphs has bounded mim-width if and only if $\{t, \Delta\} \cap \{\infty\} = \varnothing$.*

Algorithmic Consequences. As discussed, the following six problems were shown to be NP-complete for star convex and comb convex graphs, and thus for $(1, \infty)$-tree convex graphs and $(\infty, 3)$-tree convex graphs: Feedback Vertex Set [2,24]; Dominating Set, Connected Dominating Set, Total Dominating Set [12]; Independent Dominating Set [41]; Induced Matching [36]. These problems are examples of LCVS problems, possibly with connectivity or acyclicity constraints. Hence, they are polynomial-time solvable for every graph class whose mim-width is bounded and quickly computable [1,2,11]. Recall that the same holds for Weighted Feedback Vertex Set [23] and (Weighted) Subset Feedback Vertex Set [3]; these three problems generalize Feedback Vertex Set and are thus NP-complete for star convex graphs and comb convex graphs. Combining these results with Corollary 1 yields the following complexity dichotomy.

Corollary 2. *Let $t, \Delta \in \mathbb{N} \cup \{\infty\}$ with $t \geq 1$, $\Delta \geq 3$ and Π be one of the nine problems mentioned above, restricted to (t, Δ)-tree convex graphs. If $\{t, \Delta\} \cap \{\infty\} = \varnothing$, then Π is polynomial-time solvable; otherwise, Π is NP-complete.*

It is worth noting that this complexity dichotomy does not hold for all LCVS problems; recall that Dominating Induced Matching is polynomial-time solvable on star convex graphs [35]. Theorems 1 and 2, combined with the result of [11], imply that this problem is also polynomial-time solvable on circular convex graphs and (t, Δ)-tree convex graphs for every $t \geq 1$ and $\Delta \geq 3$.

Further Algorithmic Consequences. Theorems 1 and 2, combined with the result of [28], also generalize a result of Díaz et al. [17] for List k-Colouring on convex graphs to circular convex and (t, Δ)-tree convex graphs ($t \geq 1$, $\Delta \geq 3$).

Additional Structural Results. We prove Theorems 1–3 in Sects. 2–4, respectively. In Sect. 5 we perform a more refined analysis. We consider a hierarchy of width parameters and determine exactly which of the generalized convex classes considered in the previous sections have bounded width for each of these parameters. This does not yet yield any new algorithmic results. In the same section we also give some other research directions.

Preliminaries. Let $G = (V, E)$ be a graph. For $v \in V$, the *neighbourhood* $N_G(v)$ is the set of vertices adjacent to v. The *degree* $d(v)$ of a vertex $v \in V$ is the size $|N_G(v)|$. A vertex of degree k is a *k-vertex*. A graph is *subcubic* if every vertex has degree at most 3. We let $\Delta(G) = \max\{d(v) : v \in V\}$. For disjoint $S, T \subseteq V$, we say that S is *complete to* T if every vertex of S is adjacent to every vertex of T. For $S \subseteq V$, $G[S] = (S, \{uv : u, v \in S, uv \in E\})$ is the subgraph of G *induced* by S. The *disjoint union* $G + H$ of graphs G and H has vertex set $V(G) \cup V(H)$ and edge set $E(G) \cup E(H)$. A graph is *r-partite*, for $r \geq 2$, if its vertex set admits a partition into r classes such that every edge has its endpoints in different classes. A 2-partite graph is also called *bipartite*. A graph G is a *support* for a hypergraph $H = (V, \mathcal{S})$ if the vertices of G correspond to the vertices of H and, for each hyperedge $S \in \mathcal{S}$, the subgraph of G induced by S is connected. When a bipartite graph $G = (A, B, E)$ is viewed as a hypergraph $H = (A, \{N(b) : b \in B\})$, then a support T for H with $T \in \mathcal{H}$ is a witness that G is $\mathcal{H}$-convex.

2 The Proof of Theorem 1

We need the following known lemma on recognizing circular convex graphs.

Lemma 1. (see, e.g., Buchin et al. *[9]* **).** *Circular convex graphs can be recognized and a cycle support computed, if it exists, in polynomial time.*

For an integer $\ell \geq 1$, an *ℓ-caterpillar* is a subcubic tree T on 2ℓ vertices with $V(T) = \{s_1, \ldots, s_\ell, t_1, \ldots, t_\ell\}$, such that $E(T) = \{s_i t_i \ : \ 1 \leq i \leq \ell\} \cup \{s_i s_{i+1} \ : \ 1 \leq i \leq \ell - 1\}$. Note that we label the leaves of an ℓ-caterpillar $t_1, t_2, \ldots, t_\ell$, in this order. Given a total ordering $\prec$ of length ℓ, we say that (T, δ) is *obtained from* $\prec$ if T is an ℓ-caterpillar and δ is the natural bijection from the ℓ ordered elements to the leaves of T. We are now ready to prove Theorem 1.

Theorem 1 (restated). *Let G be a circular convex graph. Then* $\text{mimw}(G) \leq 2$. *Moreover, we can construct in polynomial time a branch decomposition (T, δ) for G with* $\text{mimw}_G(T, \delta) \leq 2$.

Proof. Let $G = (A, B, E)$ be a circular convex graph with a circular ordering on A. By Lemma 1, we construct in polynomial time such an ordering $a_1, \ldots, a_n$, where $n = |A|$ (see Fig. 1). Let $B_1 = N(a_n)$ and $B_2 = B \setminus B_1$. We obtain a total ordering $\prec$ on $V(G)$ by extending the ordering $a_1, \ldots, a_n$ as follows. Each $b \in B_1$ is inserted after a_n, breaking ties arbitrarily. Each $b \in B_2$ is inserted immediately after the largest element of A it is adjacent to (hence immediately after some a_i with $1 \leq i < n$), breaking ties arbitrarily.

Let T be the $|V(G)|$-caterpillar obtained from $\prec$. Below we will prove that $\text{mimw}_G(T, \delta) \leq 2$. Let $e \in E(T)$. We may assume without loss of generality that e is not incident to a leaf of T. Let M be a maximum induced matching of $G[A_e, \overline{A_e}]$. As e is not incident to a leaf, we may assume without loss of generality that each vertex in $\overline{A_e}$ is larger than any vertex in A_e in the ordering $\prec$.

We first observe that at most one edge of M has one endpoint in B_2. Indeed, suppose there exist two edges $xy, x'y' \in M$, each with one endpoint in B_2, say without loss of generality $\{y, y'\} \subseteq B_2$. Since each vertex in B_2 is adjacent only to smaller vertices, $\{y, y'\} \subseteq \overline{A_e}$ and $\{x, x'\} \subseteq A_e$. Without loss of generality, $y \prec y'$. However, $N(y)$ and $N(y')$ are intervals of the ordering and so either $x \in N(y')$ or $x' \in N(y)$, contradicting the fact that M is induced.

We now show that at most two edges in M have an endpoint in B_1 and, if exactly two such edges are in M, then no edge with an endpoint in B_2 is. First suppose that three edges of M have one endpoint in B_1 and let u_1, u_2, u_3

be these endpoints. Since $N(u_1)$, $N(u_2)$ and $N(u_3)$ are intervals of the circular ordering on A all containing a_n, one of these neighbourhoods is contained in the union of the other two, contradicting the fact that M is induced.

Finally suppose exactly two edges u_1v_1 and $u_2v_2 \in M$ have one endpoint in B_1 and thus their other endpoint in A. Let $\{u_1, u_2\} \subseteq \overline{A_e}$ and $\{v_1, v_2\} \subseteq A_e$. Then, as each vertex in $\overline{A_e}$ is larger than any vertex in A_e in $\prec$, we find that u_1 and u_2 belong to B_1 and thus $\{v_1, v_2\} \subseteq A$. Now if there is some edge $u_3v_3 \in M$ such that $u_3 \in B_2$, then $u_3 \in \overline{A_e}$. Recall that $N(u_1)$ and $N(u_2)$ are intervals of the circular ordering on A both containing a_n. Since M is induced, for each $i, j \in \{1, 2\}$, we have that $v_i \in N(u_j)$, if $i = j$, and $v_i \notin N(u_j)$, if $i \neq j$. This implies that one of v_1 and v_2 is larger than v_3 in $\prec$ and so it is contained in $N(u_3)$, contradicting the fact that M is induced. This concludes the proof. □

3 The Proof of Theorem 2

We need the following lemma on recognizing (t, Δ)-tree convex graphs[1].

Lemma 2. *For $t, \Delta \in \mathbb{N}$, (t, Δ)-tree convex graphs can be recognized and a (t, Δ)-tree support computed, if it exists, in $O(n^{t+3})$ time.*

Proof. Given a hypergraph $H = (V, \mathcal{S})$ together with degrees d_i for each $i \in V$, Buchin et al. [9] provided an $O(|V|^3 + |\mathcal{S}||V|^2)$ time algorithm that solves the following decision problem: Is there a tree support for H such that each vertex i of the tree has degree at most d_i? If it exists, the algorithm computes a tree support satisfying this property. Given as input a bipartite graph $G = (A, B, E)$, we consider the hypergraph $H = (A, \mathcal{S})$, where $\mathcal{S} = \{N(b) : b \in B\}$. For each of the $\binom{|A|}{t} = O(|A|^t)$ subsets $A' \subseteq A$ of size t we proceed as follows: we assign a degree Δ to each of its elements and a degree 2 to each element in $A \setminus A'$. We then apply the algorithm in [9] to the $O(|A|^t)$ instances thus constructed. If G is (t, Δ)-tree convex, then the algorithm returns a (t, Δ)-tree support for H. □

The proof of Theorem 2 heavily relies on the following result for mim-width.

Lemma 3. (Brettell et al. *Let G be a graph and $(X_1, \ldots, X_p)$ be a partition of $V(G)$ such that $\text{cutmim}_G(X_i, X_j) \leq c$ for all distinct $i, j \in \{1, \ldots, p\}$, and $p \geq 2$. Let $h = \max\left\{c\left\lfloor\left(\frac{p}{2}\right)^2\right\rfloor, \max_{i\in\{1,\ldots,p\}}\{\text{mimw}(G[X_i])\} + c(p-1)\right\}$. Then $\text{mimw}(G) \leq h$. Moreover, given a branch decomposition (T_i, δ_i) for $G[X_i]$ for each i, we can construct in $O(p)$ time a branch decomposition (T, δ) for G with $\text{mimw}_G(T, \delta) \leq h$.*

We also need the following lemma (proof omitted).

[1] Jiang et al. [24] proved that WEIGHTED FEEDBACK VERTEX SET is polynomial-time solvable for triad convex graphs if a triad support is given as input. They observed that an associated tree support can be constructed in linear time, but this does not imply that a triad support can be obtained. Lemma 2 shows that indeed a triad support can be obtained in polynomial time and need not be provided on input.

Lemma 4. *Let G be a $(1, \Delta)$-tree convex graph, for some $\Delta \geq 3$. Let $f(\Delta) = \max\left\{2\left\lfloor\left(\frac{\Delta}{2}\right)^2\right\rfloor, 2\Delta - 1\right\}$. Then* $\mathrm{mimw}(G) \leq f(\Delta)$, *and we can construct in polynomial time a branch decomposition (T, δ) for G with* $\mathrm{mimw}_G(T, \delta) \leq f(\Delta)$.

We are now ready to prove Theorem 2.

Theorem 2 (restated). *Let G be a (t, Δ)-tree convex graph with $t, \Delta \in \mathbb{N}$ and $t \geq 1$ and $\Delta \geq 3$. Let*

$$f(t, \Delta) = \max\left\{2\left\lfloor\left(\frac{\Delta}{2}\right)^2\right\rfloor, 2\Delta - 1\right\} + t^2\Delta.$$

Then $\mathrm{mimw}(G) \leq f(t, \Delta)$. *Moreover, we can construct in polynomial time a branch decomposition (T, δ) for G with* $\mathrm{mimw}_G(T, \delta) \leq f(t, \Delta)$.

Proof. We use induction on t. If $t = 1$, the result follows from Lemma 4. Let $t > 1$ and let $G = (A, B, E)$ be a (t, Δ)-tree convex graph. By Lemma 2, we can compute in polynomial time a (t, Δ)-tree T with $V(T) = A$ and such that, for each $v \in B$, $N_G(v)$ forms a subtree of T. Consider an edge $uv \in E(T)$ such that $T - uv$ is the disjoint union of a (t_1, Δ)-tree T_1 containing u and a (t_2, Δ)-tree T_2 containing v, where $\max\{t_1, t_2\} < t$ and $t_1, t_2 \geq 1$. Clearly such an edge can be found in linear time. For $i \in \{1, 2\}$, let $V(T_i) = A_i$. Clearly, $A = A_1 \cup A_2$. We now partition B into two classes as follows. The set B_1 contains all vertices in B with at least one neighbour in A_1, and $B_2 = B \setminus B_1$. In view of Lemma 3, we then consider the partition $(A_1 \cup B_1, A_2 \cup B_2)$ of $V(G)$. For $i \in \{1, 2\}$, $G[A_i \cup B_i]$ is a (t_i, Δ)-tree convex graph with $t_i < t$ and so, by the induction hypothesis, $\mathrm{mimw}(G[A_i \cup B_i]) \leq \max\left\{2\left\lfloor\left(\frac{\Delta}{2}\right)^2\right\rfloor, 2\Delta - 1\right\} + (t-1)^2\Delta$.

We now claim that $\mathrm{cutmim}_G(A_1 \cup B_1, A_2 \cup B_2) \leq \Delta(t - 1)$. Let M be a maximum induced matching in $G[A_1 \cup B_1, A_2 \cup B_2]$. Since no vertex in B_2 has a neighbour in A_1, all edges in M have one endpoint in B_1 and the other in A_2. We now consider the (t_2, Δ)-tree T_2 as a tree rooted at v, so that the nodes of T_2 inherit a corresponding ancestor/descendant relation. Since T_2 has maximum degree at most Δ and contains at most t_2 vertices of degree at least 3, it has at most $\Delta t_2 \leq \Delta(t - 1)$ leaves. Suppose, to the contrary, that $|M| > \Delta(t - 1)$. We first claim that there exist $xy, x'y' \in M$ with $\{y, y'\} \subseteq A_2$ and such that y' is a descendant of y. Indeed, for each leaf z of T_2, consider the unique z, v-path in T_2. There are at most $\Delta(t - 1)$ such paths and each vertex of T_2 is contained in one of them. By the pigeonhole principle, there exist two matching edges $xy, x'y' \in M$, with $\{y, y'\} \subseteq A_2$, such that y and y' belong to the same path; without loss of generality, y' is then a descendant of y, as claimed. Since $N_G(x')$ induces a subtree of T, the definition of $(A_1 \cup B_1, A_2 \cup B_2)$ implies that $N_G(x') \cap V(T_2)$ contains v and induces a subtree of T_2. But then this subtree contains y and so x' is adjacent to y as well, contradicting the fact that M is induced.

Combining the previous paragraphs and Lemma 3, we then obtain that

$$\begin{aligned} \text{mimw}(G) &\le \max\left\{\Delta(t-1), \max\left\{2\left\lfloor\left(\frac{\Delta}{2}\right)^2\right\rfloor, 2\Delta-1\right\} + (t-1)^2\Delta + \Delta(t-1)\right\} \\ &= \max\left\{2\left\lfloor\left(\frac{\Delta}{2}\right)^2\right\rfloor, 2\Delta-1\right\} + (t-1)^2\Delta + \Delta(t-1) \\ &\le \max\left\{2\left\lfloor\left(\frac{\Delta}{2}\right)^2\right\rfloor, 2\Delta-1\right\} + t^2\Delta. \end{aligned}$$

Finally, we compute a branch decomposition of G. We do this recursively by using Lemmas 3 and 4. □

4 The Proof of Theorem 3

For proving Theorem 3, we need the following lemma.

Lemma 5. (see Wang et al. [43]). Let $G = (A, B, E)$ be a bipartite graph and G' be the bipartite graph obtained from G by making k new vertices complete to B. If $k = 1$, then G' is star convex. If $k = |A|$, then G' is comb convex.

Theorem 3 (restated). *The class of star convex graphs and the class of comb convex graphs each has unbounded mim-width.*

Proof. We show that, for every integer ℓ, there exist star convex graphs and comb convex graphs with mim-width larger than ℓ. Therefore, let $\ell \in \mathbb{N}$. There exists a bipartite graph $G = (A, B, E)$ such that $\text{mimw}(G) > \ell$ (see, e.g., [7]). Let G' be the star convex graph obtained as in Lemma 5. Adding a vertex does not decrease the mim-width [42]. Then $\text{mimw}(G') \ge \text{mimw}(G) > \ell$. Let now G'' be the comb convex graph obtained as in Lemma 5. Then $\text{mimw}(G'') \ge \text{mimw}(G) > \ell$. □

5 A Refined Parameter Analysis and Final Remarks

We perform a more refined analysis on width parameters for the graph classes listed in Fig. 2. We will consider the graph width parameters listed in Fig. 3. Our results are summarized in Fig. 4. We omit the proofs but note that we provide a *complete* picture with respect to the width parameters and graph classes considered.

We are not aware of any new algorithmic implications. In particular, it would be interesting to research if there are natural problems that are NP-complete for graphs of bounded mim-width but polynomial-time solvable for graphs of bounded thinness or bounded linear mim-width. In addition, it would also be interesting to obtain dichotomies for more graph problems solvable in polynomial time for graph classes whose mim-width is bounded and quickly computable. For example, what is the complexity of List k-Colouring ($k \ge 3$) for star convex and comb convex graphs? We leave this for future research.

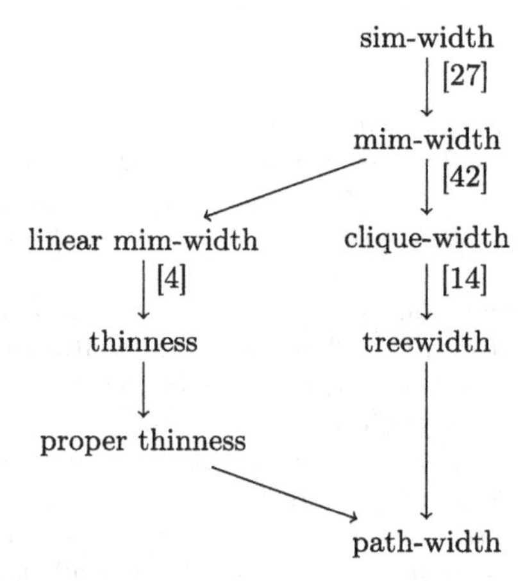

Fig. 3. The relationships between the different width parameters that we consider in Sect. 5. Parameter p is more powerful than parameter q if and only if there exists a directed path from p to q. To explain the incomparabilities, proper interval graphs have proper thinness 1 [33] and unbounded clique-width [18], whereas trees have treewidth 1 and unbounded linear mim-width [21]. Unreferenced arrows follow from the definitions of the width parameters involved except for the arrow from proper thinness to path-width whose proof we omitted.

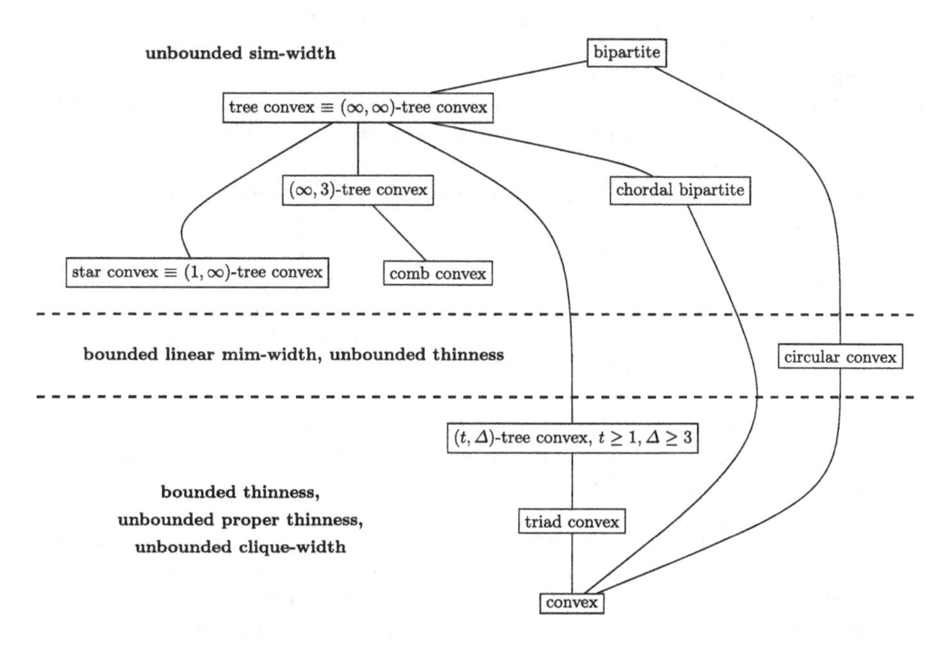

Fig. 4. The inclusion relations between the classes we consider. A line from a lower-level class to a higher one means the first class is contained in the second.

References

1. Belmonte, R., Vatshelle, M.: Graph classes with structured neighborhoods and algorithmic applications. Theoret. Comput. Sci. **511**, 54–65 (2013)
2. Bergougnoux, B., Kanté, M.M.: More applications of the d-neighbor equivalence connectivity and acyclicity constraints. Proc ESA 2009. LIPIcs **144**, 17:1-17:14 (2019)
3. Bergougnoux, B., Papadopoulos, C., Telle, J.A.: Node multiway cut and subset feedback vertex set on graphs of bounded MIM-width. In: Adler, I., Müller, H. (eds.) WG 2020. LNCS, vol. 12301, pp. 388–400. Springer, Cham (2020). https://doi.org/10.1007/978-3-030-60440-0_31
4. Bonomo, F., de Estrada, D.: On the thinness and proper thinness of a graph. Discrete Appl. Math. **261**, 78–92 (2019)
5. Brandstädt, A., Le, V.B., Spinrad, J.P.: Graph Classes: A Survey. SIAM Monographs on Discrete Mathematics and Applications, Philadelphia, PA (1999)
6. Brault-Baron, J., Capelli, F., Mengel, S.: Understanding model counting for beta-acyclic CNF-formulas. Proc. STACS 2015. LIPIcs **30**, 143–156 (2015)
7. Brettell, N., Horsfield, J., Munaro, A., Paesani, G., Paulusma, D.: Bounding the MIM-width of hereditary graph classes. Proc. IPEC 2020, LIPIcs **180**, 6:1-6:18 (2020)
8. Brettell, N., Horsfield, J., Munaro, A., Paulusma, D.: List k-colouring P_t-free graphs: a mim-width perspective. CoRR, abs/2008.01590 (2020)
9. Buchin, K., van Kreveld, M., Meijer, H., Speckmann, B., Verbeek, K.: On planar supports for hypergraphs. J. Graph Algorithms Appl. **15**, 533–549 (2011)
10. Bui-Xuan, B.M., Telle, J.A., Vatshelle, M.: Boolean-width of graphs. Theoret. Comput. Sci. **412**, 5187–5204 (2011)
11. Bui-Xuan, B.M., Telle, J.A., Vatshelle, M.: Fast dynamic programming for locally checkable vertex subset and vertex partitioning problems. Theoret. Comput. Sci. **511**, 66–76 (2013)
12. Chen, H., Lei, Z., Liu, T., Tang, Z., Wang, C., Xu, K.: Complexity of domination, hamiltonicity and treewidth for tree convex bipartite graphs. J. Comb. Optim. **32**, 1–16 (2015). https://doi.org/10.1007/s10878-015-9917-3
13. Chudnovsky, M., Spirkl, S., Zhong, M.: List 3-coloring P_t-free graphs with no induced 1-subdivision of $K_{1,s}$. Discrete Math. **343**, 1–5 (2020)
14. Courcelle, B., Olariu, S.: Upper bounds to the clique width of graphs. Discrete Appl. Math. **101**, 77–114 (2000)
15. Couturier, J.F., Golovach, P.A., Kratsch, D., Paulusma, D.: List coloring in the absence of a linear forest. Algorithmica **71**, 21–35 (2015)
16. Dabrowski, K.K., Johnson, M., Paulusma, D.: Clique-width for hereditary graph classes. Lond. Math. Soc. Lect. Note Ser. **456**, 1–56 (2019)
17. Díaz, J., Diner, Ö.Y., Serna, M.J., Serra, O.: On list k-coloring convex bipartite graphs. In: Gentile, C., Stecca, G., Ventura, P. (eds.) Graphs and Combinatorial Optimization: From Theory to Applications. AIRO Springer Series, vol. 5, pp. 15–26. Springer, Cham (2020). https://doi.org/10.1007/978-3-030-63072-0_2
18. Golumbic, M.C., Rotics, U.: On the clique-width of some perfect graph classes. Int. J. Found. Comput. Sci. **11**, 423–443 (2000)
19. Hliněný, P., Oum, S., Seese, D., Gottlob, G.: Width parameters beyond tree-width and their applications. Comput. J. **51**, 326–362 (2008)
20. Hoàng, C.T., Kamiński, M., Lozin, V.V., Sawada, J., Shu, X.: Deciding k-colorability of P_5-free graphs in polynomial time. Algorithmica **57**, 74–81 (2010)

21. Høgemo, S., Telle, J.A., Vågset, E.R.: Linear MIM-width of trees. In: Sau, I., Thilikos, D.M. (eds.) WG 2019. LNCS, vol. 11789, pp. 218–231. Springer, Cham (2019). https://doi.org/10.1007/978-3-030-30786-8_17
22. Jaffke, L.: Bounded Width Graph Classes in Parameterized Algorithms. Ph.D. thesis, University of Bergen (2020)
23. Jaffke, L., Kwon, O., Telle, J.A.: Mim-width II the feedback vertex set problem. Algorithmica **82**, 118–145 (2020)
24. Jiang, W., Liu, T., Wang, C., Xu, K.: Feedback vertex sets on restricted bipartite graphs. Theoret. Comput. Sci. **507**, 41–51 (2013)
25. Jiang, W., Liu, T., Xu, K.: Tractable feedback vertex sets in restricted bipartite graphs. In: Wang, W., Zhu, X., Du, D.-Z. (eds.) COCOA 2011. LNCS, vol. 6831, pp. 424–434. Springer, Heidelberg (2011). https://doi.org/10.1007/978-3-642-22616-8_33
26. Kamiński, M., Lozin, V.V., Milanič, M.: Recent developments on graphs of bounded clique-width. Discrete Appl. Math. **157**, 2747–2761 (2009)
27. Kang, D.Y., Kwon, O., Strømme, T.J.F., Telle, J.A.: A width parameter useful for chordal and co-comparability graphs. Theoret. Comput. Sci. **704**, 1–17 (2017)
28. Kwon, O.: Personal communication (2020)
29. Liu, T.: Restricted bipartite graphs: comparison and hardness results. In: Gu, Q., Hell, P., Yang, B. (eds.) AAIM 2014. LNCS, vol. 8546, pp. 241–252. Springer, Cham (2014). https://doi.org/10.1007/978-3-319-07956-1_22
30. Liu, T., Lu, M., Lu, Z., Xu, K.: Circular convex bipartite graphs: feedback vertex sets. Theoret. Comput. Sci. **556**, 55–62 (2014)
31. Liu, T., Lu, Z., Xu, K.: Tractable connected domination for restricted bipartite graphs. J. Comb. Optim. **29**(1), 247–256 (2014). https://doi.org/10.1007/s10878-014-9729-x
32. Lu, M., Liu, T., Xu, K.: Independent domination: reductions from circular- and triad-convex bipartite graphs to convex bipartite graphs. In: Fellows, M., Tan, X., Zhu, B. (eds.) AAIM/FAW -2013. LNCS, vol. 7924, pp. 142–152. Springer, Heidelberg (2013). https://doi.org/10.1007/978-3-642-38756-2_16
33. Mannino, C., Oriolo, G., Ricci, F., Chandran, S.: The stable set problem and the thinness of a graph. Oper. Res. Lett. **35**, 1–9 (2007)
34. Oum, S., Seymour, P.D.: Approximating clique-width and branch-width. J. Comb. Theory Ser. B **96**, 514–528 (2006)
35. Panda, B.S., Chaudhary, J.: Dominating induced matching in some subclasses of bipartite graphs. In: Pal, S., Vijayakumar, A. (eds.) CALDAM 2019. LNCS, vol. 11394, pp. 138–149. Springer, Cham (2019). https://doi.org/10.1007/978-3-030-11509-8_12
36. Panda, B.S., Pandey, A., Chaudhary, J., Dane, P., Kashyap, M.: Maximum weight induced matching in some subclasses of bipartite graphs. J. Comb. Optim. **40**(3), 713–732 (2020). https://doi.org/10.1007/s10878-020-00611-2
37. Pandey, A., Panda, B.: Domination in some subclasses of bipartite graphs. Discrete Appl. Math. **252**, 51–66 (2019)
38. Rao, M.: Clique-width of graphs defined by one-vertex extensions. Discrete Math. **308**, 6157–6165 (2008)
39. Robertson, N., Seymour, P.D.: Graph minors X. Obstructions to tree-decomposition. J. Comb. Theory Ser. B **52**, 153–190 (1991)
40. Sæther, S.H., Vatshelle, M.: Hardness of computing width parameters based on branch decompositions over the vertex set. Theoret. Comput. Sci. **615**, 120–125 (2016)

41. Song, Yu., Liu, T., Xu, K.: Independent domination on tree convex bipartite graphs. In: Snoeyink, J., Lu, P., Su, K., Wang, L. (eds.) AAIM/FAW -2012. LNCS, vol. 7285, pp. 129–138. Springer, Heidelberg (2012). https://doi.org/10.1007/978-3-642-29700-7_12
42. Vatshelle, M.: New Width Parameters of Graphs. Ph.D. thesis, University of Bergen (2012)
43. Wang, C., Chen, H., Lei, Z., Tang, Z., Liu, T., Xu, K.: Tree convex bipartite graphs: NP-complete domination, hamiltonicity and treewidth. In: Chen, J., Hopcroft, J.E., Wang, J. (eds.) FAW 2014. LNCS, vol. 8497, pp. 252–263. Springer, Cham (2014). https://doi.org/10.1007/978-3-319-08016-1_23
44. Wang, C., Liu, T., Jiang, W., Xu, K.: Feedback vertex sets on tree convex bipartite graphs. In: Lin, G. (ed.) COCOA 2012. LNCS, vol. 7402, pp. 95–102. Springer, Heidelberg (2012). https://doi.org/10.1007/978-3-642-31770-5_9

Improved Bounds on the Spanning Ratio of the Theta-5-Graph

Prosenjit Bose, Darryl Hill(✉), and Aurélien Ooms

School of Computer Science, Carleton University, Ottawa, Canada
jit@scs.carleton.ca

Abstract. We show an upper bound of $\frac{\sin\left(\frac{3\pi}{10}\right)}{\sin\left(\frac{2\pi}{5}\right)-\sin\left(\frac{3\pi}{10}\right)} < 5.70$ on the spanning ratio of Θ_5-graphs, improving on the previous best known upper bound of 9.96 [Bose, Morin, van Renssen, and Verdonschot. The Theta-5-graph is a spanner. *Computational Geometry*, 2015.]

Keywords: Theta graphs · Spanning ratio · Stretch factor · Geometric spanners.

1 Introduction

A geometric graph G is a graph whose vertex set is a set of points P in the plane, and where the weight of an edge uv is equal to the Euclidean distance $|uv|$ between u and v. Informally, a Θ_k-graph is a geometric graph built by dividing the area around each point of $v \in P$ into k equal angled cones, connecting v to the *closest* neighbor in each cone (we shall define closest later). Such graphs arise naturally in settings like wireless networks, where signals to anyone but your nearest neighbor are likely to be drowned out by interference. Moreover, the fact that signal strength fades quadratically with distance, and thus that power requirements are proportional to the square of the distance the signal has to travel, makes many small hops economically superior to one large hop, even if the sum of the distances is larger. The *spanning ratio* (sometimes called the *stretch factor*) of a geometric graph G is the maximum over all pairs $u, v \in P$ of the ratio between the length of the shortest path from u to v in G and the Euclidean distance from u to v. Using simple geometric observations and techniques, we give a new analysis of the spanning ratio of Θ_5-graphs, bringing down the best known upper bound from 9.96 [5] to 5.70.

Theorem 1. *Given a set P of points in the plane, the Θ_5-graph of P is a* 5.70*-spanner.*

Θ_k-graphs were introduced simultaneously by Keil and Gutwin [8,9], and Clarkson [7]. Both papers gave a spanning ratio of $1/(\cos\theta - \sin\theta)$, where

Research supported in part by NSERC, VILLUM Foundation grant 16582, and FRIA Grant 5203818F (FNRS).

A. Lubiw et al. (Eds.): WADS 2021, LNCS 12808, pp. 215–228, 2021.
https://doi.org/10.1007/978-3-030-83508-8_16

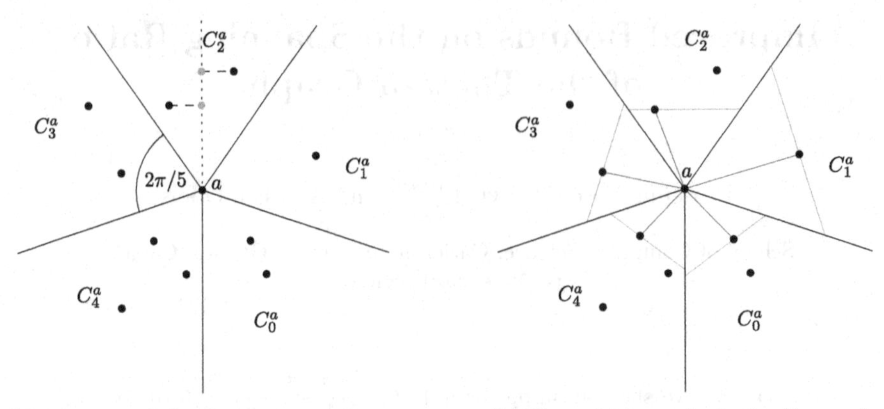

(a) Measure of the distance to point a. (b) The neighbors of a in the Θ_5-graph.

Fig. 1. The area around a point a is divided into cones with angle $2\pi/5$.

$\theta = 2\pi/k$ is the angle defined by the cones. Observe that this gives a constant spanning ratio for $k \geq 9$. When this ratio t is constant, we call the graph a t-spanner. Ruppert and Seidel [11] improved this to $1/(1 - 2\sin(\theta/2))$, which applies to Θ_k-graphs with $k \geq 7$. Chew [6] gave a tight bound of 2 for $k = 6$. Bose et al. [4] give the current best bounds on the spanning ratio of a large range of values of k. For $k = 5$, Bose et al. [5] showed an upper bound of 9.96, and a lower bound of 3.78. For $k = 4$, Bose, De Carufel, Hill, and Smid [3] showed a spanning ratio of 17, while Barba et al. [2] gave a lower bound of 7 on the spanning ratio. For $k = 3$, although Aichholzer et al. [1] showed Θ_3 to be connected, El Molla [10] showed that there is no constant t for which Θ_3 is a t-spanner.

In this paper we study the spanning ratio of Θ_5. We consider two arbitrary vertices, a and b, and show that there must exist a short path between them using induction on the rank of the Euclidean distance $|ab|$ among all distances between pairs of points in P. Our main result states that for all $a, b \in P$ the shortest path $\mathcal{P}(a, b)$ has length $|\mathcal{P}(a, b)| \leq K \cdot |ab|$, where $K = 5.70$.

We organize the rest of the paper as follows. In Sect. 2 we introduce concepts and notation, and give some assumptions about the positions of a and b that do not reduce the generality of our arguments. In Sect. 3 we solve all but a handful of cases using general arguments that simplify the analysis. The remaining cases are solved using ad-hoc methods, showing a spanning ratio of $K = 6.16$. In Sect. 4 we observe that only a single case requires $K \geq 6.16$. We analyze this case in detail to show that $|\mathcal{P}(a, b)| \leq K \cdot |ab|$ for all $K \geq 5.70$. Due to space constraints, some proofs have been omitted. All omitted proofs are available in the full version of the paper.

2 Preliminaries

Let $k \geq 3$ be an integer. Let P be set of points in the plane in general position, that is, all distances (as defined below) between pairs of points are unique and no two points have the same x-coordinate or y-coordinate. Construct the Θ_k-graph of P as follows. The vertex set is P. For each i with $0 \leq i < k$, let $\mathcal{R}_i$ be the ray emanating from the origin that makes an angle of $2\pi i/k$ with the negative y-axis.[1] All cone indices are manipulated mod k, i.e., $\mathcal{R}_k = \mathcal{R}_0$. For each vertex v we add at most k outgoing edges as follows: For each i with $0 \leq i < k$, let $\mathcal{R}_i^v$ be the ray emanating from v parallel to $\mathcal{R}_i$. Let C_i^v be the cone consisting of all points in the plane that are strictly between the rays $\mathcal{R}_i^v$ and $\mathcal{R}_{i+1}^v$ or on $\mathcal{R}_{i+1}^v$. If C_i^v contains at least one point of $P \setminus \{v\}$, then let w_i be the *closest* such point to v, where we define the closest point to be the point whose perpendicular projection onto the bisector of C_i^v minimizes the Euclidean distance to v. We add the directed edge vw_i to G. While the use of directed edges better illustrates this construction, in what follows we regard all edges of a Θ_5-graph as undirected. See Fig. 1 for an example of cones and construction.

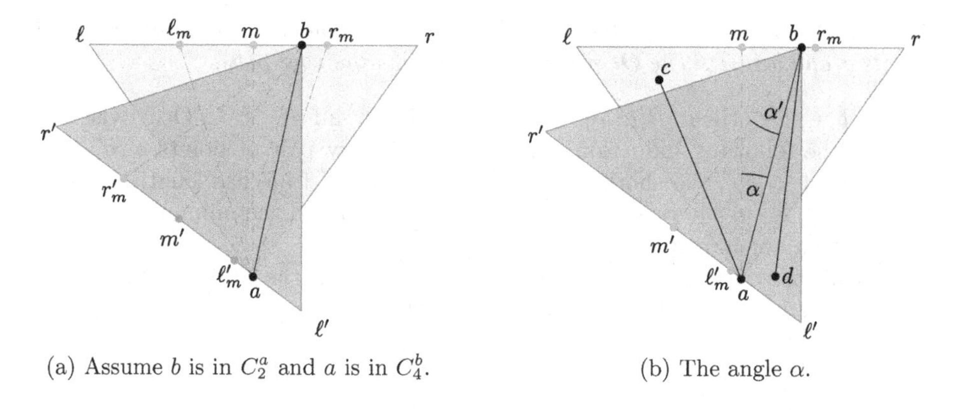

(a) Assume b is in C_2^a and a is in C_4^b. (b) The angle α.

Fig. 2. Vertices a and b and the canonical triangles T_{ab} and T_{ba}.

For the following description, refer to Fig. 2. Consider two vertices a and b of P. Given the Θ_5-graph of P, we define the *canonical triangle* T_{ab} to be the triangle bounded by the sides of the cone of a that contains b and the line through b perpendicular to the bisector of that cone. Note that to any pair of vertices a and b corresponds two canonical triangles, namely T_{ab} and T_{ba}. Without loss of generality assume that b is in C_2^a. Let ℓ be the leftmost vertex of the triangle T_{ab} and let r be the rightmost vertex of the triangle T_{ab}. Let m be the midpoint of ℓr. Note that a must be in C_4^b or C_0^b; since the cases are symmetric we consider the case where a is in C_4^b. Thus b is to the right of m. Let r_m be the intersection of ℓr and the bisector of $\angle ram$[2], and let ℓ_m be the intersection of ℓr and the bisector of $\angle ma\ell$. Let ℓ' and r' be the left and right endpoints of T_{ba} respectively

[1] Angle values are given counter-clockwise unless otherwise stated.

[2] In what follows we use $\triangle abc$ to denote the triangle defined by the points a, b, and c (given counter-clockwise). We use $\angle abc$ to denote the amplitude of the angle at b in that triangle.

(as seen from b facing a). Let m' be the midpoint of $\ell'r'$, and let ℓ'_m and r'_m be the intersections of $\ell'r'$ and the bisector of $\angle \ell'bm'$ and $\angle m'br'$ respectively. See Fig. 2a. Let $\alpha = \angle bam$ and let $\alpha' = \angle abm'$. Note that $\alpha + \alpha' = \pi/5$ since α and $\frac{2\pi}{5} - \alpha'$ are alternate interior angles. Thus either $\alpha \leq \pi/10$ or $\alpha' \leq \pi/10$. Without loss of generality, we assume $\alpha \leq \pi/10$. Let c be the closest neighbor to a in C_2^a, and let d be the closest neighbor to b in C_4^b. See Fig. 2b. For simplicity, we write "Θ_5" to mean "the Θ_5-graph of P".

We proceed by induction to bound the spanning ratio of Θ_5. We show that, for any pair of points $a, b \in P$, the length of a shortest path $|\mathcal{P}(a, b)|$ in Θ_5 is at most K times the Euclidean distance between its endpoints. The induction is on the rank of the Euclidean distance $|ab|$ among all distances between pairs of points in P. The exact bound on K is made explicit in the proof. Lemma 1 is sufficient for the base case to be reached by induction. The proof is left to the full version of the paper.

Lemma 1. *Let (a_0, b_0) be the pair of points in P that minimizes $|ab|$ over all points a and b in P. The Θ_5-graph of P contains the edge a_0b_0.*

If $ab \in \Theta_5$, then $|\mathcal{P}(a, b)| \leq K|ab|$ holds for all $K \geq 1$. Otherwise we assume the following induction hypothesis: for every pair of points $a', b' \in P$ where $|a'b'| < |ab|$, the shortest path $\mathcal{P}(a', b')$ from a' to b' has length at most $|\mathcal{P}(a', b')| \leq K \cdot |a'b'|$, for some $K \geq 1$. Our goal is to find the minimum value of K for which our inductive argument holds.

Recall that c is the closest point to a in C_2^a and d is the closest point to b in C_4^b. We restrict our analysis to the following three paths:

(1) $ac + \mathcal{P}(c, b)$,
(2) $bd + \mathcal{P}(d, a)$, and
(3) $ac + \mathcal{P}(c, d) + db$.

Depending on the particular arrangement of a, b, c, and d, we examine a subset of these and find a minimum value for K that satisfies at least one of the following inequalities:

(A) $|ac| + K \cdot |cb| \leq K \cdot |ab|$,
(B) $|bd| + K \cdot |da| \leq K \cdot |ab|$, and
(C) $|ac| + K \cdot |cd| + |db| \leq K \cdot |ab|$.

Observe that our inductive argument follows if any of these cases holds. For instance, if we prove (A) holds for some value K, it implies that $|cb| < |ab|$ (since all distances are positive), and thus $|\mathcal{P}(c, b)| \leq K \cdot |cb|$ by the induction hypothesis. Similar conclusions follow for statements (B) and (C). Thus we can combine (1)–(3) with (A)–(C) as follows.

(a) $|\mathcal{P}(a, b)| \leq |ac| + |\mathcal{P}(c, b)| \leq |ac| + K \cdot |cb| \leq K \cdot |ab|$.
(b) $|\mathcal{P}(a, b)| \leq |bd| + |\mathcal{P}(d, a)| \leq |bd| + K \cdot |da| \leq K \cdot |ab|$.
(c) $|\mathcal{P}(a, b)| \leq |ac| + |\mathcal{P}(c, d)| + |db| \leq |ac| + K \cdot |cd| + |db| \leq K \cdot |ab|$.

For any given arrangement of vertices we prove that at least one of (A), (B), or (C) holds true for some value K, and find the smallest value for which this is true. Our proof relies mainly on case analysis, but some of these cases have similar structure. We exploit this structure in Sect. 3 by designing two reusable lemmas. These lemmas, along with additional arguments, are then applied to different arrangements of a, b, c, and d. For all but one case we show that at least one of (a), (b), or (c) holds true for $K \geq 5.70$. The last case requires $K \geq 6.16$. We improve this further to $K \geq 5.70$, but due to the complexity of this last case, we dedicate Sect. 4 to its analysis.

3 Analysis

We first introduce two triangles T_2 and T_3 for which inequalities of the form of (A) and (B) hold for reasonable values of K (see Fig. 3). Note the triangles are numbered to correspond to the lemmas they appear in. We state these inequalities as lemmas whose repeated use simplifies the proof of our main result. The proofs are available in the full paper.

Lemma 2. *(Figure 3a) Let T_2 be a triangle with vertices (s, v, u) and corresponding interior angles $(\frac{\pi}{5}, \frac{\pi}{2}, \frac{3\pi}{10})$. Let t be a point on uv and let w be a point inside $\triangle stu$. Then $|sw| + K|wt| \leq K|st|$ for all $K \geq 4.53$.*

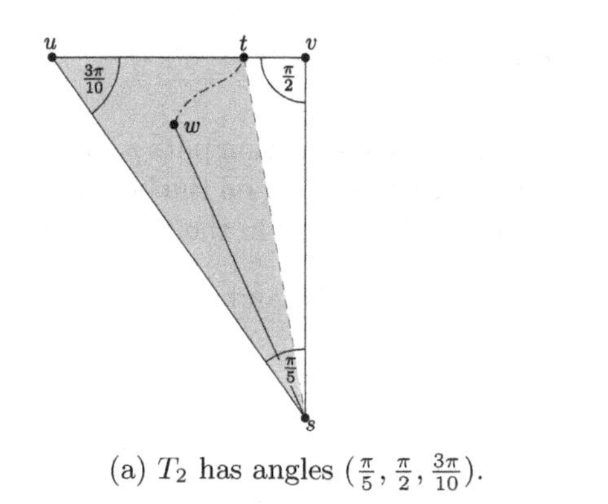

(a) T_2 has angles $(\frac{\pi}{5}, \frac{\pi}{2}, \frac{3\pi}{10})$.

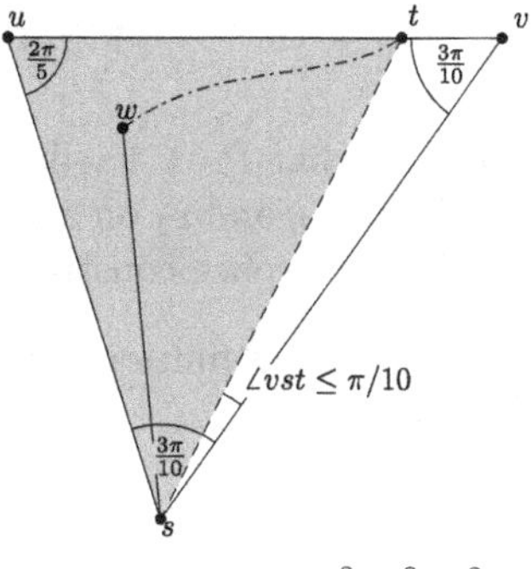

(b) T_3 has angles $(\frac{3\pi}{10}, \frac{3\pi}{10}, \frac{2\pi}{5})$.

Fig. 3. Triangles T_2 and T_3.

Lemma 3. *(Figure 3b) Let T_3 be a triangle with vertices (s, v, u) and corresponding interior angles $(\frac{3\pi}{10}, \frac{3\pi}{10}, \frac{2\pi}{5})$. Let t be a point on uv such that $\angle vst \leq \pi/10$ and let w be a point inside $\triangle stu$. Then $|sw| + K|wt| \leq K|st|$ for all $K \geq 5.70$.*

As in the definition of T_{ab} and T_{ba}, let c be the point closest to a in T_{ab} and let d be the point closest to b in T_{ba}. We proceed by case analysis depending on the location of the points c and d.

If c is to the right of ab or if d is to the right of ab, we can apply Lemma 2 to show the existence of a short path from a to b. When both c and d are left of ab, we use a more complicated argument requiring a new definition:

Definition 1. *(Figure 4) Given any pair of points (a, b) in P, let r' and r'_m be as in the definition of T_{ba}. We define P_{ab} to be the regular pentagon with vertices $(p_0, p_1, p_2 = r', p_3 = r'_m, p_4)$ where p_4 is above the line going through r' and r'_m (this uniquely defines the remaining points p_0 and p_1).*

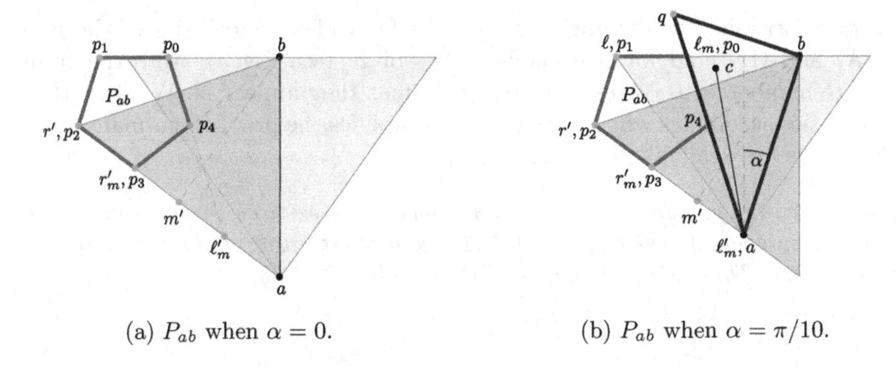

(a) P_{ab} when $\alpha = 0$. (b) P_{ab} when $\alpha = \pi/10$.

Fig. 4. The regular pentagon P_{ab}.

Observe that P_{ab} is fixed with respect to T_{ba}. This construction puts p_4 inside T_{ab} and puts p_0 and p_1 on a horizontal line with b, with p_0 lying on the boundary of T_{ab}. Due to space constraints, a formal proof can be found in the full paper.

Note 1. Given Definition 1 we have that $p_4 \in T_{ab}$, $p_0 \in \ell b$, and p_1 lies on the line through ℓ and b.

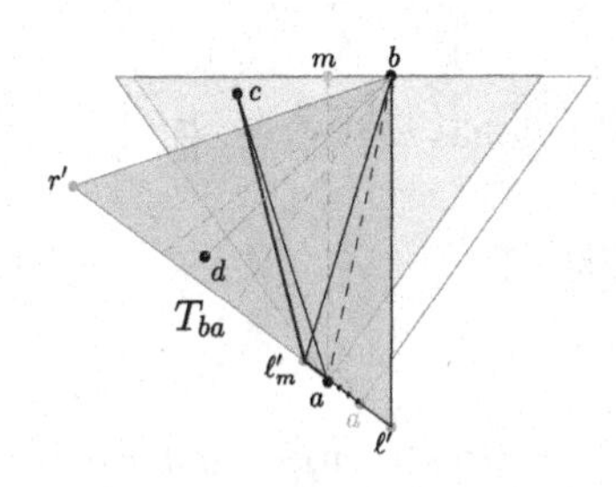

Fig. 5. Transformation 1.

Given this definition, we consider the following cases: When c is not in P_{ab} we prove $|ac| + |\mathcal{P}(c, b)| \leq 5.70|ab|$. When d is not in P_{ab} we prove $|bd| + |\mathcal{P}(d, a)| \leq 5.70|ab|$. When both c and d are in P_{ab} we analyze the length of the path $ac + \mathcal{P}(c, d) + db$. Lemma 12 gives us a bound of $6.16|ab|$ with a simple proof. Using a more technical analysis, we obtain a bound of $5.70|ab|$. This is proven in Lemma 16 in Sect. 4.

Some of the proofs use the simplifying assumption that $\alpha = \pi/10$. This is achieved through the following transformation: given a, b, c, $d \in P$ with T_{ab} and T_{ba} as defined earlier, we define:

Transformation 1. *Fix b, c, d, and T_{ba}, and translate a along $r'\ell'$.*

See Fig. 5. Observe that this transformation changes $|ac|$ and $|ab|$, but not $|bd|$, $|cd|$, or $|cb|$. The transformation also changes $|ad|$, but we do not use it in any case that depends on this value. In the full paper we prove the following lemma allowing the application of Transformation 1 without loss of generality in several cases.

Lemma 4. *Under Transformation 1, the values of $|bd|$, $|cd|$, and $|cb|$ are unchanged, and $\Psi = |ac| - K|ab|$ is maximized when $a = \ell'_m$ for all $K \geq 3.24$.*

Note that applying Transformation 1 with $a = \ell'_m$ is equivalent to assuming $\alpha = \pi/10$.

All these proofs can be combined in an analysis comprising *eight* cases depending on the location of c and d with respect to T_{ab}, T_{ba}, and P_{ab}, as illustrated in Algorithm 1:

Algorithm 1: Applying the Lemmas

1. If c is right of ab, Lemma 5.
2. If d is right of ab, Lemma 6.
3. Else both c and d are left of ab. We have the following cases:
   (a) If c is in T_{ba}, Lemma 7.
   (b) Else c is NOT in T_{ba} and:
      i. If c is NOT in P_{ab}, Lemma 8.
      ii. Else c is in P_{ab} and:
         - If d is right of am, Lemma 9.
         - If d is left of am and above c, Lemma 10
         - If d is below c (i.e. $d \notin T_{ab}$ such that bd and ac cross)
            - If d is NOT in P_{ab}, Lemma 11.
            - If d is in P_{ab}, Lemma 12 with $K \geq 6.16$ or Lemma 16 with $K \geq 5.70$.

One can check that all locations of c and d are covered. This proves our main theorem:

Theorem 1. *Given a set P of points in the plane, the Θ_5-graph of P is a 5.70-spanner.*

We use the remainder of the paper to prove each lemma.

Lemma 5. *If c is right of ab, then $|\mathcal{P}(a, b)| \leq K|ab|$ for $K \geq 4.53$.*

Proof. (Figure 6) Let $(s, t, w, u, v) = (a, b, c, r, m)$, thus these points correspond to triangle T_2 of Lemma 2. Thus $|ac| + K|cb| \leq K|ab|$ for all $K \geq 4.53$. The induction hypothesis and Lemma 2 imply that there is a path from a to b with length at most

$$|\mathcal{P}(a, b)| \leq |ac| + |\mathcal{P}(c, b)| \leq |ac| + K|cb| \leq K|ab|.$$

□

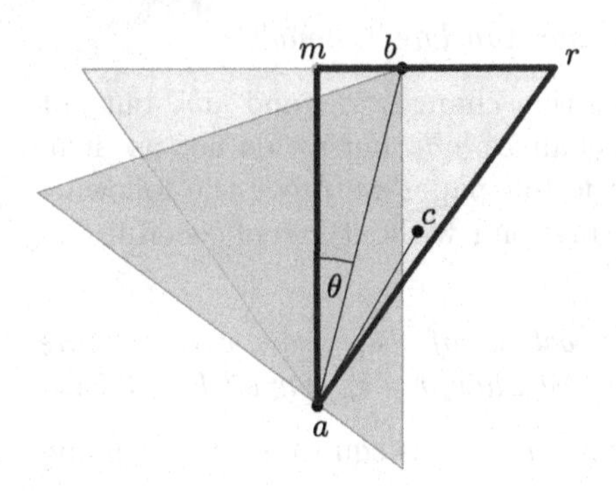

Fig. 6. Points (a, r, m) correspond to T_2 (in blue) with $t = b$ and $w = c$. (Color figure online)

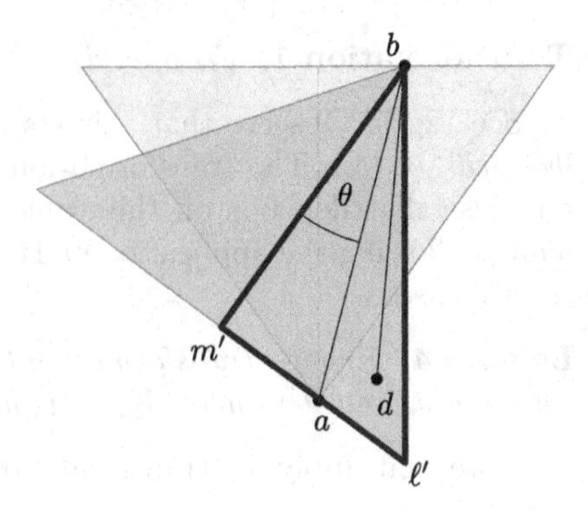

Fig. 7. Points (b, m', ℓ') correspond to T_2 (in blue) with $t = a$ and $w = d$. (Color figure online)

Lemma 6. *If d is right of ab, then $|\mathcal{P}(a, b)| \leq K|ab|$ for $K \geq 4.53$.*

Proof. (Figure 7) Let $(s, t, w, u, v) = (b, a, d, m', \ell')$, thus these points correspond to triangle T_2 from Lemma 2. Thus $|bd| + K|da| \leq K|ab|$ for $K \geq 4.53$ by Lemma 2. The induction hypothesis and Lemma 2 imply that there is a path from a to b with length at most

$$|\mathcal{P}(a, b)| \leq |bd| + |\mathcal{P}(d, a)| \leq |bd| + K|da| \leq K|ab|.$$

□

Lemma 7. *If c is left of ab and in $T_{ab} \cap T_{ba}$, then $|\mathcal{P}(a, b)| \leq K|ab|$ for $K \geq 5.70$.*

Proof. (Figure 8) Let p be the intersection of br' and $a\ell$, and let q be the intersection of the lines through $r'b$ and ar_m. Observe that $0 \leq \angle r_m ab \leq \pi/10$, thus $\angle r_m ab$ has the same range as $\angle vst$ from T_3 in Lemma 3. If we let points $(s, t, w, u, v) = (a, b, c, p, q)$, then these points correspond to the triangle T_3, and thus $|ac| + K|cb| \leq K|ab|$ for $K \geq 5.70$ by Lemma 3. Our induction hypothesis and Lemma 3 imply that there is a path from a to b with length

$$|\mathcal{P}(a, b)| \leq |ac| + |\mathcal{P}(c, b)| \leq |ac| + K|cb| \leq K|ab|.$$

□

Lemma 8. *If $c \in T_{ab} \setminus (T_{ba} \cup P_{ab})$, then $|\mathcal{P}(a, b)| \leq K|ab|$ for all $K \geq 4.53$.*

Proof. (Figure 4b) Let $\Phi = |ac| + K|cb| - K|ab|$. We apply Transformation 1. Since $c \notin T_{ba}$ it must be left of $b\ell'_m$, thus c remains left of ab. As a moves left, so does the left side of T_{ab}, which means that c remains inside T_{ab}. Thus Lemma 4 implies that Φ is maximized at $\alpha = \pi/10$, thus we assume this is the case.

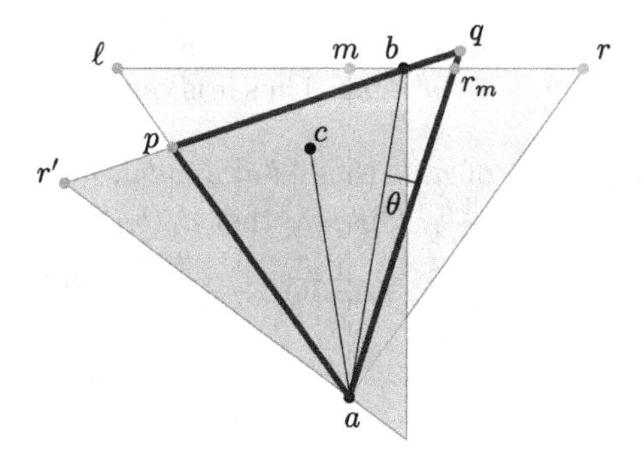

Fig. 8. Points (a, q, p) correspond to the triangle T_3 with angles $(\frac{3\pi}{10}, \frac{2\pi}{5}, \frac{3\pi}{10})$ as denoted by the blue triangle. Let $t = b$ and $w = c$, and $\theta = \frac{\pi}{10} - \alpha$, which falls in the range of $0 \leq \angle vsu \leq \pi/10$. (Color figure online)

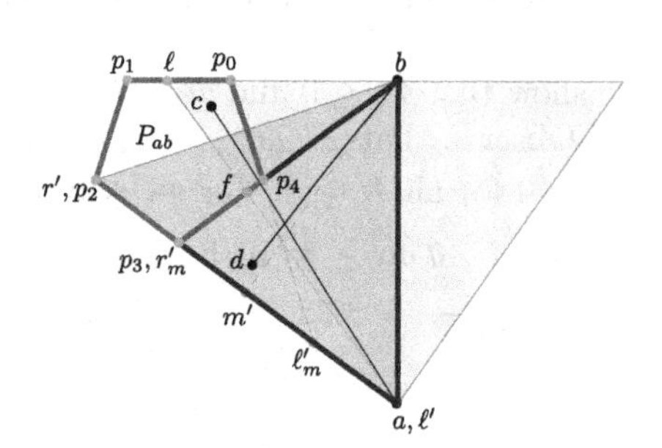

Fig. 9. We use the fact that p_4 lies in T_{ab} and apply T_3.

Observe that $\angle ba\ell_m = \pi/5$, and $\angle \ell_m ba = 2\pi/5 < \pi/2$. Let q be the intersection of the line through b orthogonal to ab and the line through a and ℓ_m. If we let $(s, t, w, u, v) = (a, b, c, q, b)$ then these points correspond to T_2. Then Lemma 2 tells us that $|ac| + K|cb| \leq K|ab|$ and thus $\Phi = |ac| + K|cb| - K|ab| \leq 0$ for all $K \geq 4.53$. □

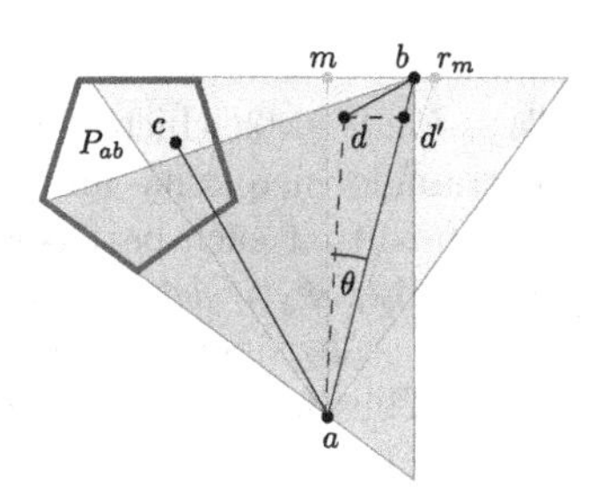

Fig. 10. The point c is in $P_{ab} \setminus T_{ba}$, and d is right of am.

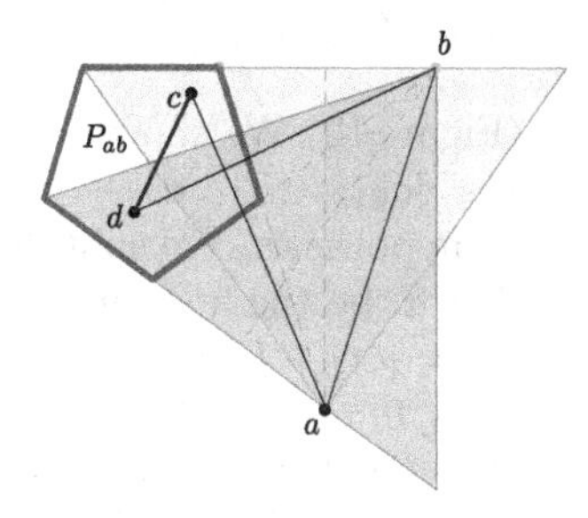

Fig. 11. The segments ac and bd cross and c and d are in P_{ab}.

Lemma 9. *If d is left of ab and right of am, then $|\mathcal{P}(a, b)| \leq K|ab|$ for $K \geq 3.24$.*

Proof. (Figure 10) We show $\Phi = |bd| + K|da| - K|ab| \leq 0$, which implies $|\mathcal{P}(a, b)| \leq |bd| + |\mathcal{P}(d, a)| \leq K|ab|$ by the triangle inequality and the induction hypothesis.

Let d' be the horizontal projection of d onto ab. Let $\Phi_1 = |bd| - K|bd'|$ and $\Phi_2 = K|da| - K|d'a|$, and note that $\Phi = \Phi_1 + \Phi_2$ since $d' \in ab$. Thus it is sufficient to show that $\Phi_1 \le 0$ and $\Phi_2 \le 0$.

Observe that $\angle d'da > \pi/2$, since d is right of am, thus $|d'a| > |da|$, and $\Phi_2 \le 0$ for all $K \ge 1$. For $\Phi_1 \le 0$ we need $K \ge \frac{|bd|}{|bd'|}$. Observe that $d_y(b, d') \le |bd'|$ and $\angle d'db \ge \pi/10$ because $d \in T_{ba}$. Thus $K \ge \frac{1}{\sin(\pi/10)} \ge \frac{|bd|}{d_y(b,d')}$, and $K \ge \frac{1}{\sin(\pi/10)} = 3.23\ldots$ is sufficient. □

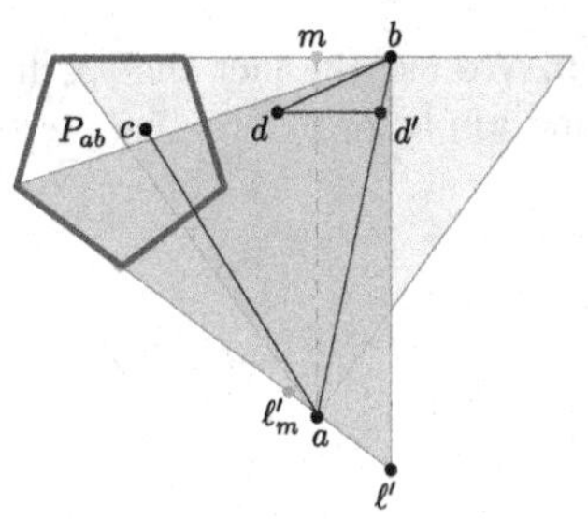

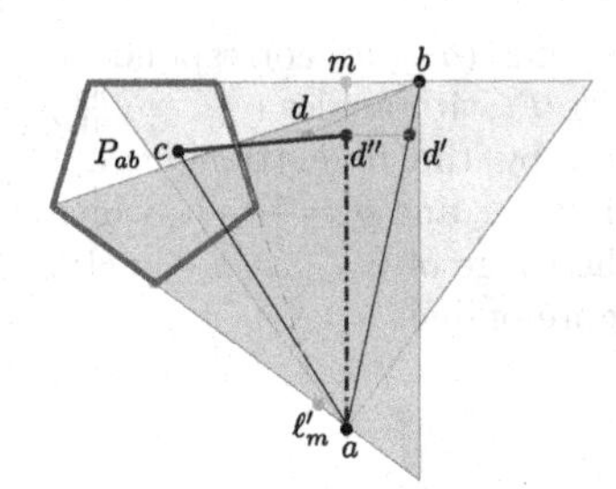

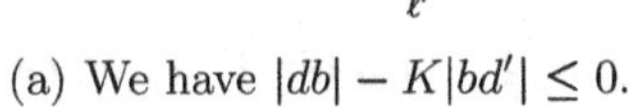
(a) We have $|db| - K|bd'| \le 0$.

(b) We have $|ac| + K|cd| - K|ad'| \le 0$.

Fig. 12. The point c is in $P_{ab} \setminus T_{ba}$, and d is left of am but above c.

Lemma 10. *If c is in $P_{ab} \setminus T_{ba}$, and d is left of am but above c, then $|\mathcal{P}(a, b)| \le K|ab|$ for all $K \ge 4.53$.*

Proof. (Figure 12) We show $\Phi = |ac| + K|cd| + |db| - K|ab| \le 0$, which implies $|\mathcal{P}(a, b)| \le |ac| + |\mathcal{P}(c, d)| + |db| \le K|ab|$ by the triangle inequality and the induction hypothesis. We split Φ into two parts, and show that each part is less than 0. Let d' be the horizontal projection of d onto ab. Let $\Phi_1 = |bd| - K|bd'|$, and let $\Phi_2 = |ac| + K|cd| - K|ad'|$. Observe that $\Phi = \Phi_1 + \Phi_2$ since $d' \in ab$.

For $\Phi_1 \le 0$, observe that $d_y(b, d) = d_y(b, d') \le |bd'|$. Thus let $\Phi'_1 = |bd| - K \cdot d_y(b, d) \ge \Phi_1$. Let $\theta = \angle d'db$, and observe that $\Phi'_1 = |bd|(1 - K \sin\theta)$. Note that $\theta \ge \pi/10$ since $d \in T_{ba}$, and thus $K \ge 3.24$ is sufficient.

For $\Phi_2 \le 0$, let d'' be the horizontal projection of d onto am. Since $\angle ad''d' = \pi/2$, $|ad''| \le |ad'|$. Since $c \notin T_{ba}$, $\angle cdd'' \ge 9\pi/10$, thus $|cd''| > |cd|$. Let $\Phi'_2 = |ac| + K|cd''| - K|ad''| \ge \Phi_2$. Let q be the horizontal projection of d'' onto $a\ell$. Let the points $(s, t, w, u, v) = (a, d'', c, q, d'')$ and thus these points correspond to T_2. Thus $|ac| + K|cd''| \le K|ad''|$ for all $K \ge 4.53$ by Lemma 2. □

Lemma 11. *If bd and ac cross with d left of ab and not in P_{ab}, then $|\mathcal{P}(a, b)| \le K|ab|$ for all $K \ge 5.70$.*

Proof. (Figure 9) Since ac and bd cross, d must be outside of T_{ab} (otherwise ad would be and edge of Θ_5, but not ac). We want to show that d is below br'_m.

Assuming this is the case, then $0 \leq \angle ab\ell' \leq \pi/10$, and thus $\angle ab\ell'$ is in the range of $0 \leq \angle vsu \leq \pi/10$. Let points $(s, t, w, u, v) = (b, a, d, r'_m, \ell')$, then these points correspond to triangle T_3 of Lemma 3. Thus $|bd| + K|da| \leq K|ab|$ for $K \geq 5.70$. Our induction hypothesis and Lemma 3 imply that there is a path from b to a with length at most

$$|\mathcal{P}(a,b)| \leq |bd| + |\mathcal{P}(d,a)| \leq |bd| + K|da| \leq K|ab|.$$

We are left with showing that d is below br'_m. Recall that P_{ab} is fixed with respect to T_{ba}. Since d is outside of T_{ab} and P_{ab}, if p_4p_0 is inside T_{ab}, d must be below br'_m. Since the slope of p_0p_4 is less than the slope of ℓa, it is sufficient to show that p_4 is inside T_{ab} which follows by Note 1. □

Lemma 12. *If ac and bd cross and both c and d are in P_{ab}, then $|\mathcal{P}(a,b)| \leq K|ab|$ for $K \geq 6.16$.*

Proof. (Figure 11) We show $\Phi = |ac| + K|cd| + |db| - K|ab| \leq 0$, which implies $|\mathcal{P}(a,b)| \leq |ac| + |\mathcal{P}(c,d)| + |db| \leq K|ab|$ by the triangle inequality and the induction hypothesis. Under Transformation 1, Lemma 4 implies that Φ is maximized when $\alpha = \pi/10$, so we assume this is the case. Since c, d, and P_{ab} are fixed, c and d are still inside P_{ab} after Transformation 1. Given that c and d are in P_{ab}, the furthest apart c and d can be is if they are both on a diagonal of P_{ab}. The length of one side of P_{ab} is at most $\frac{\sin(\pi/10)}{\sin(3\pi/10)}|ab|$. That means a diagonal of P_{ab}, and thus $|cd|$, has length at most $2\sin(3\pi/10)\frac{\sin(\pi/10)}{\sin(3\pi/10)}|ab| = 2\sin(\pi/10)|ab|$. At their longest, $|ac|$ and $|bd|$ each have length $\frac{\sin(2\pi/5)}{\sin(3\pi/10)}|ab|$ by the law of sines. We want

$$\Phi = |ac| + K|cd| + |db| - K|ab| \leq 0.$$

Solving for K gives

$$K \geq \frac{|ac| + |db|}{|ab| - |cd|} \geq \frac{2 \cdot \sin(2\pi/5)}{\sin(3\pi/10) \cdot (1 - 2 \cdot \sin(\pi/10))} = 6.15\ldots$$

□

4 Proving a Spanning Ratio of 5.70

In this section we present a lemma with a stronger bound for the case handled by Lemma 12. Proving this lemma requires a careful analysis of the locations of c and d and the tradeoffs between the values of $|ac| + |db|$ and $K|cd|$. Let $\Phi = |ac| + K|cd| + |db| - K|ab|$. For the rest of this section, assume we have applied Transformation 1, and thus $\alpha = \pi/10$ and Φ is maximized. Since P_{ab}, c and d are fixed, both c and d are still in P_{ab}. Let c' be the intersection of the line through a and c and the segment p_0p_1, and let d' be the intersection of the line through b and d and the segment p_3p_4. See Fig. 13. Let $\Phi' = |ac'| + K|c'd'| + |d'b| - K|ab|$, and let $\Phi'' = |ap_1| + K|p_1p_3| + |p_3b| - K|ab|$.

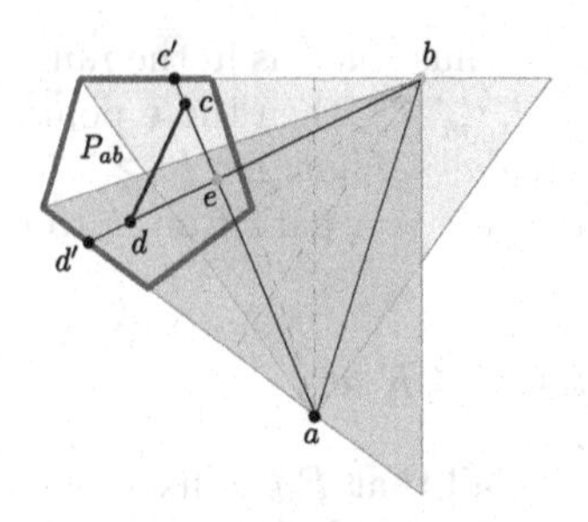

Fig. 13. Points c' and d' on P_{ab}.

We split the analysis into three steps that amount to proving the following lemmas:

Lemma 13. *For all* $K \geq 5.70$, $\Phi \leq \Phi'$.

Lemma 14. *For all* $K \geq 5.70$, $\Phi' \leq \Phi''$.

Lemma 15. *For all* $K \geq 5.70$, $\Phi'' \leq 0$.

The following lemma follows from these lemmas, the triangle inequality, and the induction hypothesis. It supersedes Lemma 12:

Lemma 16. *If ac and bd cross and both c and d are in P_{ab}, then $|\mathcal{P}(a,b)| \leq K|ab|$ for $K \geq 5.70$.*

Substituting Lemma 16 for Lemma 12 in the proof of Theorem 1 brings the spanning ratio of the Θ_5-graph down to 5.70. We are left with proving Lemmas 13, 14, and 15. The proof of Lemma 14 is left to the full paper. The proofs of Lemmas 13 and 15 are presented below.

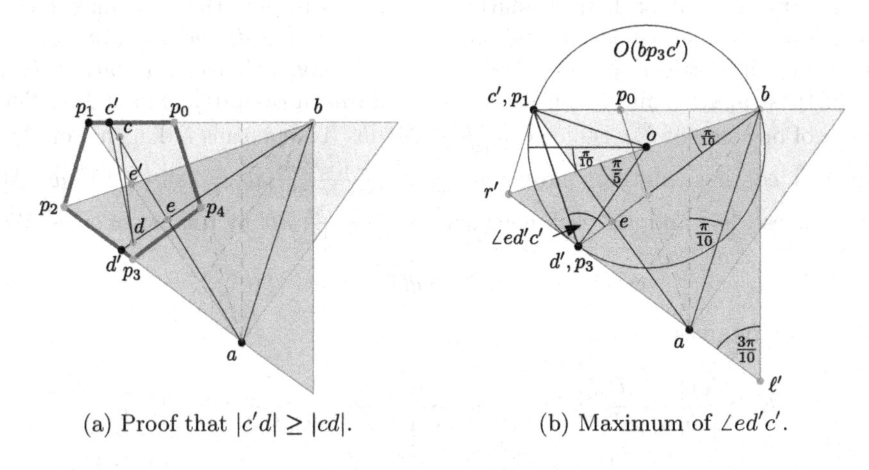

(a) Proof that $|c'd| \geq |cd|$.

(b) Maximum of $\angle ed'c'$.

Fig. 14. Finding the longest distance from a to b when c and d are in P_{ab},

4.1 Proof of Lemma 13

Lemma 13 states that $|ac| + |bd| + K|cd| - K|ab| \leq |ac'| + |bd'| + K|c'd'| - K|ab|$ for $K \geq 5.70$. See Fig. 14a. Let e be the intersection of ac and bd, and let e' be the intersection of br' and $a\ell$. Observe that $\angle \ell e'r' = 2\pi/5$, and thus we can see that $\angle dec \geq 2\pi/5$. This implies that $\angle dec$ cannot be the smallest angle in Δdec, since that would require $\angle dec \leq \pi/3$. Thus at least one of $\angle dce$ and $\angle edc$ is the smallest angle in Δdec. Since we have applied Transformation 1, and can thus assume that $\alpha = \pi/10$, the cases are symmetric. We can therefore, without loss of generality, assume that $\angle dce$ is the smallest angle in Δdec.

Lemma 13. *For all $K \geq 5.70$, $\Phi \leq \Phi'$.*

Proof (Proof of Lemma 13). Since c lies on ac' and d lies on bd', we have $|ac| \leq |ac'|$ and $|bd| \leq |bd'|$, and it is sufficient to show that $|cd| \leq |c'd'|$. We first show that $|cd| \leq |c'd|$. Since $\angle dce$ is the smallest angle in $\triangle dec$, $\angle dce < \pi/3$. That implies that $\angle c'cd > \pi/2$, which implies that $c'd$ is the longest side of triangle $\triangle cc'd$, and thus $|cd| \leq |c'd|$. See Fig. 14a.

We now show that $|c'd'| \geq |c'd|$. If $\angle c'dd' \geq \pi/2$, then $c'd'$ is the longest side of $\triangle c'dd'$, and $|c'd'| \geq |c'd|$ and we are done. Otherwise assume $\angle c'dd' < \pi/2$.

The law of sines tells us that $\frac{|c'd'|}{\sin \angle c'dd'} = \frac{|c'd|}{\sin \angle dd'c'}$. Since $\sin\theta$ is an increasing function for $0 \leq \theta < \pi/2$, showing that $\angle c'dd' \geq \angle dd'c'$ is sufficient to show $|c'd'| \geq |c'd|$, as it would imply both angles are $< \pi/2$. Observe that $\angle c'dd' \geq \angle c'ed'$ and $\angle ed'c' = \angle dd'c'$, thus it is sufficient to prove that $\angle c'ed' \geq \angle ed'c'$.

Observe that $\angle ced = \angle c'ed' \geq 2\pi/5$. We now find the maximum of $\angle dd'c' = \angle ed'c' \leq 2\pi/5$. Observe that if c' moves left, $\angle ed'c'$ increases, thus assume c' is at p_1. Let $O(bp_3c')$ be the circle through b, p_3, and c' with center o. Observe that o lies on br'. Observe that $\angle r'bd' = \pi/10$, thus $\angle r'op_3 = \pi/5$. Segment or' makes an angle of $\pi/10$ with the horizontal line through o. Thus od' makes an angle of $3\pi/10$ with the horizontal line through o, and thus the line tangent to $O(bp_3c')$ at p_3 is the line supporting $\ell'r'$, since $\ell'r'$ makes an angle of $3\pi/10$ with the vertical line through ℓ'. See Fig. 14b. That implies that $[p_2, p_3)$ lies outside of $O(bp_3c')$, which means for every point d', $\angle ed'c' \leq \angle ep_3c' = 2\pi/5$, and thus $\angle c'dd' \geq \angle dd'c'$ as required. □

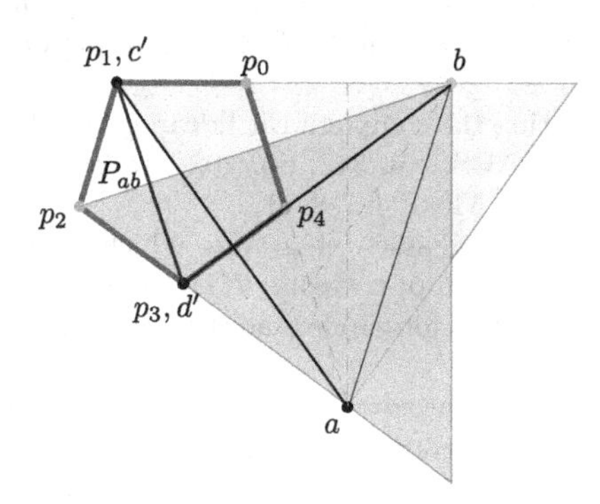

Fig. 15. An example of Φ''.

4.2 Proof of Lemma 15

Lemma 15. *For all $K \geq 5.70$, $\Phi'' \leq 0$.*

Proof (Proof of Lemma 15). (Figure 15) We apply Transformation 1 with $\alpha = \frac{\pi}{10}$ and assume that $|ab| = 1$. Then using the law of sines we get $|bp_3| = 1$, $|ap_1| = \frac{\sin(2\pi/5)}{\sin(3\pi/10)}$, and $|p_1p_3| = 2\sin(3\pi/10)\frac{\sin(\pi/10)}{\sin(3\pi/10)} = 2\sin(\pi/10)$. We want

$$\Phi'' = |ap_1| + K|p_1p_3| + |p_3b| - K|ab| \leq 0.$$

Solving for K gives

$$K \geq \frac{|ap_1| + |p_3b|}{|ab| - |p_1p_3|} = \frac{\frac{\sin(2\pi/5)}{\sin(3\pi/10)} + 1}{1 - 2\sin(\pi/10)} = 5.69\ldots$$

□

References

1. Aichholzer, O., et al.: Theta-3 is connected. Comput. Geom. **47**(9), 910–917 (2014)
2. Barba, L., Bose, P., De Carufel, J.L., van Renssen, A., Verdonschot, S.: On the stretch factor of the Theta-4 graph. In: Dehne, F., Solis-Oba, R., Sack, J.R. (eds.) WADS 2013. LNCS, vol. 8037, pp. 109–120. Springer, Heidelberg (2013). https://doi.org/10.1007/978-3-642-40104-6_10
3. Bose, P., De Carufel, J.L., Hill, D., Smid, M.: On the spanning and routing ratio of Theta-four. In: Proceedings of the Thirtieth Annual ACM-SIAM Symposium on Discrete Algorithms (SODA), pp. 2361–2370 (2019)
4. Bose, P., De Carufel, J.L., Morin, P., van Renssen, A., Verdonschot, S.: Towards tight bounds on Theta-graphs: more is not always better. Theoret. Comput. Sci. **616**, 70–93 (2016)
5. Bose, P., Morin, P., van Renssen, A., Verdonschot, S.: The Theta-5-graph is a spanner. Comput. Geom. **48**(2), 108–119 (2015)
6. Chew, L.P.: There are planar graphs almost as good as the complete graph. J. Comput. Syst. Sci. **39**(2), 205–219 (1989)
7. Clarkson, K.: Approximation algorithms for shortest path motion planning. In: Proceedings of the Nineteenth Annual ACM Symposium on Theory of Computing (STOC), pp. 56–65 (1987)
8. Keil, J.M.: Approximating the complete Euclidean graph. In: Karlsson, R., Lingas, A. (eds.) SWAT 1988. LNCS, vol. 318, pp. 208–213. Springer, Heidelberg (1988). https://doi.org/10.1007/3-540-19487-8_23
9. Keil, J.M., Gutwin, C.A.: Classes of graphs which approximate the complete Euclidean graph. Discret. Comput. Geom. **7**(1), 13–28 (1992)
10. El Molla, N.M.: Yao spanners for wireless ad-hoc networks. Ph.D. thesis, Villanova University (2009)
11. Ruppert, J., Seidel, R.: Approximating the d-dimensional complete Euclidean graph. In: Proceedings of the 3rd Canadian Conference on Computational Geometry (CCCG) (1991)

Computing Weighted Subset Transversals in H-Free Graphs

Nick Brettell[1], Matthew Johnson[2], and Daniël Paulusma[2](✉)

[1] School of Mathematics and Statistics, Victoria University of Wellington, Wellington, New Zealand
nick.brettell@vuw.ac.nz

[2] Department of Computer Science, Durham University, Durham, UK
{matthew.johnson2,daniel.paulusma}@durham.ac.uk

Abstract. For the Odd Cycle Transversal problem, the task is to find a small set S of vertices in a graph that intersects every cycle of odd length. The Subset Odd Cycle Transversal requires S to intersect only those odd cycles that include a vertex of a distinguished vertex subset T. If we are given weights for the vertices, we ask instead that S has small weight: this is the problem Weighted Subset Odd Cycle Transversal. We prove an almost-complete complexity dichotomy for Weighted Subset Odd Cycle Transversal for graphs that do not contain a graph H as an induced subgraph. Our general approach can also be used for Weighted Subset Feedback Vertex Set, which enables us to generalize a recent result of Papadopoulos and Tzimas.

1 Introduction

For a *transversal* problem, one seeks to find a small set of vertices within a given graph that intersects every subgraph of a specified kind. Two problems of this type are Feedback Vertex Set and Odd Cycle Transversal, where the objective is to find a small set S of vertices that intersects, respectively, every cycle and every cycle with an odd number of vertices. Equivalently, when S is deleted from the graph, what remains is a forest or a bipartite graph, respectively.

For a *subset transversal* problem, we are also given a vertex subset T and we must find a small set of vertices that intersects every subgraph of a specified kind *that also contains a vertex of* T. An *(odd)* T*-cycle* is a cycle of the graph (with an odd number of vertices) that intersects T. A set $S_T \subseteq V$ is a T*-feedback vertex set* or an *odd* T*-cycle transversal* of a graph $G = (V, E)$ if S_T has at least one vertex of, respectively, every T-cycle or every odd T-cycle; see also Fig. 1. A *(non-negative) weighting* of G is a function $w : V \to \mathbb{R}^+$. For $v \in V$, $w(v)$ is the *weight* of v, and for $S \subseteq V$, the weight $w(S)$ of S is the sum of the weights of the vertices in S. In a *weighted subset transversal* problem the task is to find a transversal whose weight is less than a prescribed bound. We study:

The research in this paper received support from the Leverhulme Trust (RPG-2016-258).

A. Lubiw et al. (Eds.): WADS 2021, LNCS 12808, pp. 229–242, 2021.
https://doi.org/10.1007/978-3-030-83508-8_17

WEIGHTED SUBSET FEEDBACK VERTEX SET
Instance: a graph G, a subset $T \subseteq V(G)$, a non-negative vertex weighting w of G and an integer $k \geq 1$.
Question: does G have a T-feedback vertex set S_T with $w(S_T) \leq k$?

WEIGHTED SUBSET ODD CYCLE TRANSVERSAL
Instance: a graph G, a subset $T \subseteq V(G)$, a non-negative vertex weighting w of G and an integer $k \geq 1$.
Question: does G have an odd T-cycle transversal S_T with $w(S_T) \leq k$?

Both problems are NP-complete even when the weighting function is 1 and $T = V$. We continue a systematic study of transversal problems on hereditary graph classes, focusing on the weighted subset variants. *Hereditary* graph classes can be characterized by a set of forbidden induced subgraphs. We begin with the case where this set has size 1: the class of graphs that, for some graph H, do not contain H as an induced subgraph; such a graph is said to be *H-free.*

Past Results. We first note some NP-completeness results for the special case where $w \equiv 1$ and $T = V$, which corresponds to the original problems FEEDBACK VERTEX SET and ODD CYCLE TRANSVERSAL. These results immediately imply NP-completeness for the weighted subset problems. By Poljak's construction [14], for every integer $g \geq 3$, FEEDBACK VERTEX SET is NP-complete for graphs of finite girth at least g (the girth of a graph is the length of its shortest cycle). There is an analogous result for ODD CYCLE TRANSVERSAL [4]. It has also been shown that FEEDBACK VERTEX SET [10] and ODD CYCLE TRANSVERSAL [4] are NP-complete for line graphs and, therefore, also for claw-free graphs. Thus the two problems are NP-complete for the class of H-free graphs whenever H contains a cycle or claw. Of course, a graph with no cycle is a forest, and a forest with no claw has no vertex of degree at least 3. Hence, we need now only focus on the case where H is a *linear forest*, that is, a collection of disjoint paths.

There is no linear forest H for which FEEDBACK VERTEX SET on H-free graphs is known to be NP-complete, but for ODD CYCLE TRANSVERSAL we can take $H = P_2 + P_5$ or $H = P_6$, as the latter problem is NP-complete even for (P_2+P_5, P_6)-free graphs [5]. It is known that SUBSET FEEDBACK VERTEX SET [6] and SUBSET ODD CYCLE TRANSVERSAL [3], which are the special cases with $w \equiv 1$,

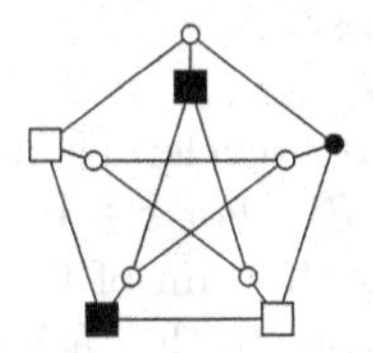

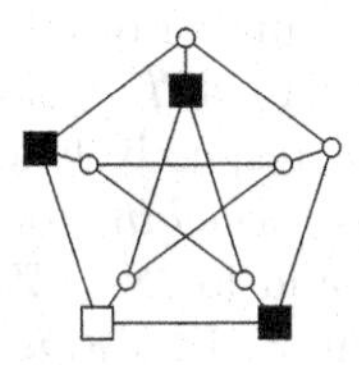

Fig. 1. Two examples (from [3]) of the Petersen graph with the set T indicated by square vertices. The set S_T of black vertices forms both an odd T-cycle transversal and a T-feedback vertex set. On the left, $S_T \cap T \neq \emptyset$. On the right, $S_T \subseteq T$.

are NP-complete for $2P_2$-free graphs; in fact, these results were proved for split graphs which form a proper subclass of $2P_2$-free graphs. For the weighted subset problems, there is just one additional case of NP-completeness currently known, from the interesting recent work of Papadopoulos and Tzimas [13] as part of the following dichotomy.

Theorem 1 ([13]). Weighted Subset Feedback Vertex Set *on sP_1-free graphs is polynomial-time solvable if $s \leq 4$ and* NP*-complete if $s \geq 5$.*

The unweighted version of Subset Feedback Vertex Set can be solved in polynomial time for sP_1-free graphs for every $s \geq 1$ [13]. In contrast, for many transversal problems, the complexities on the weighted and unweighted versions for H-free graphs align; see, for example Vertex Cover [7], Connected Vertex Cover [8] and (Independent) Dominating Set [9].

The other known polynomial-time algorithm for Weighted Subset Feedback Vertex Set on H-free graphs is for the case where $H = P_4$. This can be proven in two ways: Weighted Subset Feedback Vertex Set is polynomial-time solvable for permutation graphs [12] and also for graphs for which we can find a decomposition of constant mim-width [2]; both classes contain the class of P_4-free graphs. To the best of our knowledge, algorithms for Weighted Subset Odd Cycle Transversal on H-free graphs have not previously been studied.

We now mention the polynomial-time results on H-free graphs for the unweighted subset variants of the problems (which do not imply anything for the weighted subset versions). Both Subset Feedback Vertex Set and Subset Odd Cycle Transversal are polynomial-time solvable on H-free graphs if $H = P_4$ or $H = sP_1 + P_3$ [3,12]. Additionally, Feedback Vertex Set is polynomial-time solvable on P_5-free graphs [1] and sP_3-free graphs for every integer $s \geq 1$ [11], and both Feedback Vertex Set and Odd Cycle Transversal are polynomial-time solvable on sP_2-free graphs for every $s \geq 1$ [4].

Our Results. Our main result is the following almost-complete dichotomy. We write $H \subseteq_i G$, or $G \supseteq_i H$ to say that H is an induced subgraph of G.

Theorem 2. *Let H be a graph with $H \notin \{2P_1 + P_3, P_1 + P_4, 2P_1 + P_4\}$. Then* Weighted Subset Odd Cycle Transversal *on H-free graphs is polynomial-time solvable if $H \subseteq_i 3P_1 + P_2$, $P_1 + P_3$, or P_4, and is* NP*-complete otherwise.*

As a consequence, we obtain a dichotomy analogous to Theorem 1.

Corollary 1. *The* Weighted Subset Odd Cycle Transversal *problem on sP_1-free graphs is polynomial-time solvable if $s \leq 4$ and is* NP*-complete if $s \geq 5$.*

For the hardness part of Theorem 2 it suffices to show hardness for $H = 5P_1$; this follows from the same reduction used by Papadopoulos and Tzimas [13] to prove Theorem 1. The three tractable cases, where $H \in \{P_4, P_1 + P_3, 3P_1 + P_2\}$, are all new. Out of these cases, $H = 3P_1 + P_2$ is the most involved. For this case we use a different technique to that used in [13]. Although we also reduce

Table 1. The complexity of Feedback Vertex Set (FVS), Odd Cycle Transversal (OCT), and their subset (S) and weighted subset (WS) variants, when restricted to H-free graphs for linear forests H. All problems are NP-complete for H-free graphs when H is not a linear forest. The four blue cases (two for WSFVS, two for WSOCT) are the *algorithmic* contributions of this paper; see also Theorems 2 and 3.

	polynomial-time	unresolved	NP-complete
FVS	$H \subseteq_i P_5$ or sP_3 for $s \geq 1$	$H \supseteq_i P_1 + P_4$	none
OCT	$H = P_4$ or $H \subseteq_i sP_1 + P_3$ or sP_2 for $s \geq 1$	$H = sP_1 + P_5$ for $s \geq 0$ or $H = sP_1 + tP_2 + uP_3 + vP_4$ for $s, t, u \geq 0$, $v \geq 1$ with $\min\{s, t, u\} \geq 1$ if $v = 1$, or $H = sP_1 + tP_2 + uP_3$ for $s, t \geq 0$, $u \geq 1$ with $u \geq 2$ if $t = 0$	$H \supseteq_i P_6$ or $P_2 + P_5$
SFVS, SOCT	$H = P_4$ or $H \subseteq_i sP_1 + P_3$ for $s \geq 1$	$H = sP_1 + P_4$ for $s \geq 1$	$H \supseteq_i 2P_2$
WSFVS, WSOCT	$H \subseteq_i P_4, P_1 + P_3$, or $3P_1 + P_2$	$H \in \{2P_1 + P_3, P_1 + P_4, 2P_1 + P_4\}$	$H \supseteq_i 5P_1$ or $2P_2$

to the problem of finding a minimum weight vertex cut that separates two given terminals, our technique relies less on explicit distance-based arguments, and we devise a method for distinguishing cycles according to parity. Our technique also enables us to extend the result of [13] on Weighted Subset Feedback Vertex Set from $4P_1$-free graphs to $(3P_1+P_2)$-free graphs, leading to the same almost-complete dichotomy for Weighted Subset Feedback Vertex Set.

Theorem 3. *Let H be a graph with $H \notin \{2P_1 + P_3, P_1 + P_4, 2P_1 + P_4\}$. Then* Weighted Subset Feedback Vertex Set *on H-free graphs is polynomial-time solvable if $H \subseteq_i 3P_1 + P_2$, $P_1 + P_3$, or P_4, and is* NP*-complete otherwise.*

We refer to Table 1 for an overview of the current knowledge of the problems, including the results of this paper.

2 Preliminaries

Let $G = (V, E)$ be a graph. If $S \subseteq V$, then $G[S]$ denotes the subgraph of G induced by S, and $G - S$ is the graph $G[V \setminus S]$. The path on r vertices is denoted P_r. the *union* operation + creates the disjoint union $G_1 + G_2$ having vertex set $V(G_1) \cup V(G_2)$ and edge set $E(G_1) \cup E(G_2)$. By sG, we denote the disjoint union of s copies of G. Thus sP_1 denotes the graph whose vertices form an independent set of size s. A *(connected) component* of G is a maximal connected subgraph of G. The *neighbourhood* of a vertex $u \in V$ is the set $N_G(u) = \{v \mid uv \in E\}$. For $U \subseteq V$, we let $N_G(U) = \bigcup_{u \in U} N(u) \setminus U$. Let S and T be two disjoint vertex sets of a graph G. Then S is *complete* to T if every vertex of S is adjacent to every vertex of T, and S is *anti-complete* to T if there are no edges between S and T.

3 General Framework of the Polynomial Algorithms

We first explain our general approach with respect to odd cycle transversals. Afterwards we modify our terminology for feedback vertex sets, but we note that our approach can be easily extended to other kinds of transversals as well. So, consider an instance (G, T, w) of WEIGHTED SUBSET ODD CYCLE TRANSVERSAL. A subgraph of G with no odd T-cycles is *T-bipartite*. Note that a subset $S_T \subseteq V$ is an odd T-cycle transversal if and only if $G[V \setminus S_T]$ is T-bipartite. A *solution* for (G, T, w) is an odd T-cycle transversal S_T. From now on, whenever S_T is defined, we let $B_T = V(G) \setminus S_T$ denote the vertex set of the corresponding T-bipartite graph. If $u \in B_T$ belongs to at least one odd cycle of $G[B_T]$, then u is an *odd* vertex of B_T. Otherwise, when $u \in B_T$ is not in any odd cycle of $G[B_T]$, we say that u is an *even* vertex of B_T. Note that by definition every vertex in $T \cap B_T$ is even. We let $O(B_T)$ and $R(B_T)$ denote the sets of odd and even vertices of B_T (so $B_T = O(B_T) \cup R(B_T)$). A solution S_T is *neutral* if B_T consists of only even vertices; in this case S_T is an odd cycle transversal of G. We say that S_T is *T-full* if B_T contains no vertex of T. If S_T is neither neutral nor T-full, then S_T is a *mixed* solution. We can now outline our approach to finding minimum weight odd T-cycle transversals:

1. Compute a neutral solution of minimum weight.
2. Compute a T-full solution of minimum weight.
3. Compute a mixed solution of minimum weight.
4. From the three computed solutions, take one of overall minimum weight.

As mentioned, a neutral solution is a minimum-weight odd cycle transversal. Hence, in Step 1, we will use existing polynomial-time algorithms from the literature for computing such an odd cycle transversal (these algorithms must be for the weighted variant). Step 2 is trivial: we can just set $S_T := T$ (as w is non-negative). Hence, most of our attention will go to Step 3. For Step 3, we analyse the structure of the graphs $G[R(B_T)]$ and $G[O(B_T)]$ for a mixed solution S_T and how these graphs relate to each other.

For WEIGHTED SUBSET FEEDBACK VERTEX SET we follow exactly the same approach, but we use slightly different terminology. A subgraph of a graph $G = (V, E)$ is a *T-forest* if it has no T-cycles. Note that a subset $S_T \subseteq V$ is a T-feedback vertex set if and only if $G[V \setminus S_T]$ is a T-forest. We write $F_T = V \setminus S_T$ in this case. If $u \in F_T$ belongs to at least one cycle of $G[F_T]$, then u is a *cycle vertex* of F_T. Otherwise, if $u \in F_T$ is not in any cycle of $G[F_T]$, we say that u is a *forest vertex* of F_T. By definition every vertex in $T \cap F_T$ is a forest vertex.

We obtain our results for WEIGHTED SUBSET FEEDBACK VERTEX SET by a simplification of our algorithms for WEIGHTED ODD CYCLE TRANSVERSAL. Hence, to explain our approach fully, we will now give a polynomial-time algorithm for WEIGHTED ODD CYCLE TRANSVERSAL for $(3P_1 + P_2)$-free graphs.

4 Applying Our Framework on $(3P_1 + P_2)$-Free Graphs

We let $G = (V, E)$ be a $(3P_1 + P_2)$-free graph with a vertex weighting w, and let $T \subseteq V$. For Step 1, we need the polynomial-time algorithm of [4] for ODD

CYCLE TRANSVERSAL on sP_2-free graphs ($s \geq 1$), and thus on $(3P_1 + P_2)$-free graphs (take $s = 4$). The algorithm in [4] was for the unweighted case, but it can be easily adapted for the weighted case.[1]

Lemma 1. *For every integer $s \geq 1$,* WEIGHTED ODD CYCLE TRANSVERSAL *is polynomial-time solvable for sP_2-free graphs.*

As Step 2 is trivial, we focus on Step 3. We will reduce to a classical problem, well known to be polynomial-time solvable by standard network flow techniques.

WEIGHTED VERTEX CUT
Instance: a graph $G = (V, E)$, two distinct non-adjacent terminals t_1 and t_2, and a non-negative vertex weighting w.
Task: determine a set $S \subseteq V \setminus \{t_1, t_2\}$ of minimum weight such that t_1 and t_2 are in different connected components of $G - S$.

For a mixed solution S_T, we let $O = O(B_T)$ and $R = R(B_T)$; recall that $O \neq \emptyset$ and $R \cap T \neq \emptyset$. For our reduction to WEIGHTED VERTEX CUT, we need some **structural lemmas**. We first bound the number of components of $G[O]$.

Lemma 2. *Let $G = (V, E)$ be a $(3P_1 + P_2)$-free graph, and let $T \subseteq V$. For every mixed solution S_T, the graph $G[O]$ has at most two connected components.*

We now prove that $|R| \leq 8$. If $G[O]$ is disconnected, then even $|R| \leq 2$, as shown in Lemma 3. Otherwise we use Lemma 4 and the fact that $G[R]$ is bipartite.

Lemma 3. *Let $G = (V, E)$ be a $(3P_1 + P_2)$-free graph, and let $T \subseteq V$. For every mixed solution S_T, if $G[O]$ is disconnected, then R is a clique with $|R| \leq 2$.*

Lemma 4. *Let $G = (V, E)$ be a $(3P_1 + P_2)$-free graph and let $T \subseteq V$. For every mixed solution S_T, every independent set in $G[R]$ has size at most* 4.

We say that a vertex in O is a *connector* if it has a neighbour in R.

Lemma 5. *Let $G = (V, E)$ be a $(3P_1 + P_2)$-free graph, and let $T \subseteq V$. For every mixed solution S_T, if $G[O]$ has two connected components D_1 and D_2, then D_1 and D_2 each have at most one connector.*

Proof. By Lemma 3, R is a clique of size at most 2. For contradiction, suppose that, say, D_1 has two distinct connectors v_1 and v_2. Then v_1 and v_2 each have at most one neighbour in R, else the vertices of R would be in an odd cycle in $G[B_T]$, as R is a clique. Let u_1 be the neighbour of v_1 in R, and let u_2 be the neighbour of v_2 in R; note that $u_1 = u_2$ is possible.

An edge on a path P from v_1 to v_2 in D_1 does not belong to an odd cycle in $G[D_1]$; else there would be a path P' from v_1 to v_2 in $G[O]$ with a different parity

[1] Proofs of Lemmas 1–4 are omitted for space reasons. A full version of this paper can be found at https://arxiv.org/abs/2007.14514.

than P and one of the cycles $u_1v_1Pv_2u_2u_1$ or $u_1v_1P'v_2u_2u_1$ is odd, implying that u_1 and u_2 would not be even.

By definition, v_1 and v_2 belong to at least one odd cycle, which we denote by C_1 and C_2, respectively. Then $V(C_1) \cap V(C_2) = \emptyset$ and there is no edge between a vertex of C_1 and a vertex of C_2 except from possibly the edge v_1v_2; else there would be a path from v_1 to v_2 in $G[O]$ with an edge that belongs to an odd cycle (C_1 or C_2), a contradiction with what we found above. Note also that u_1 has no neighbours in $V(C_1)$ other than v_1; otherwise $G[B_T]$ would have an odd cycle containing u_1. Moreover, u_1 has no neighbours in $V(C_2)$ either, except v_2 if $u_1 = u_2$; otherwise $G[B_T]$ would contain an odd cycle containing u_1 and u_2.

We now let w_1 and x_1 be two adjacent vertices of C_1 that are not adjacent to u_1. Let w_2 be a vertex of C_2 not adjacent to u_1. Then, we found that $\{u_1, w_2, w_1, x_1\}$ induces a $2P_1 + P_2$ (see Fig. 2).

We continue by considering D_2, the other connected component of $G[O]$. By definition, D_2 has an odd cycle C'. As $|R| \leq 2$ and each vertex of R can have at most one neighbour on an odd cycle in $G[B_T]$, we find that C' contains a vertex v' not adjacent to any vertex of R, so v' is not adjacent to u_1. As v' and the vertices of $\{w_2, w_1, x_1\}$ belong to different connected components of $G[O]$, we find that v' is not adjacent to any vertex of $\{w_2, w_1, x_1\}$ either. However, now $\{u_1, v', w_2, w_1, x_1\}$ induces a $3P_1 + P_2$ (see also Fig. 2), a contradiction. □

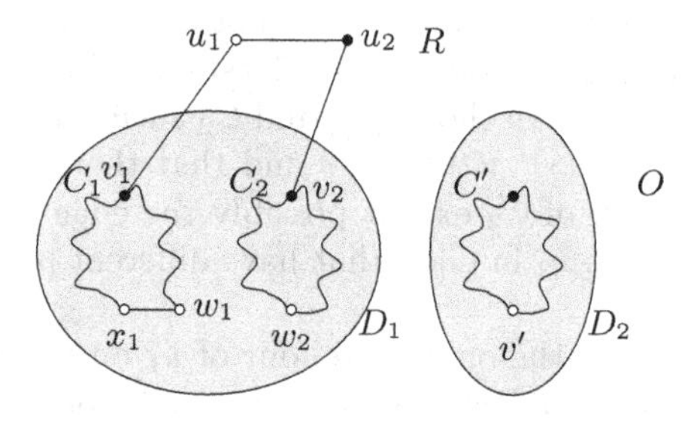

Fig. 2. An illustration for the proof of Lemma 5: the white vertices induce a $3P_1 + P_2$.

We need one more structural lemma about connectors, in the case where $G[O]$ is connected. Let R consist of two adjacent vertices u_1 and u_2. Let O (with $O \cap T = \emptyset$) be the disjoint union of two complete graphs K and L, each on an odd number of vertices that is at least 3, plus a single additional edge, such that:

1. u_1 is adjacent to exactly one vertex v_1 in K and to no vertex of L;
2. u_2 is adjacent to exactly one vertex v_2 in L and to no vertex of K; and
3. v_1 and v_2 are adjacent.

Note that $G[B_T]$ is indeed T-bipartite. We call the corresponding mixed solution S_T a *2-clique solution* (see Fig. 3).

Lemma 6. *Let $G = (V, E)$ be a $(3P_1 + P_2)$-free graph and let $T \subseteq V$. For every mixed solution S_T that is not a 2-clique solution, if $G[O]$ is connected, then O has no two connectors with a neighbour in the same connected component of $G[R]$.*

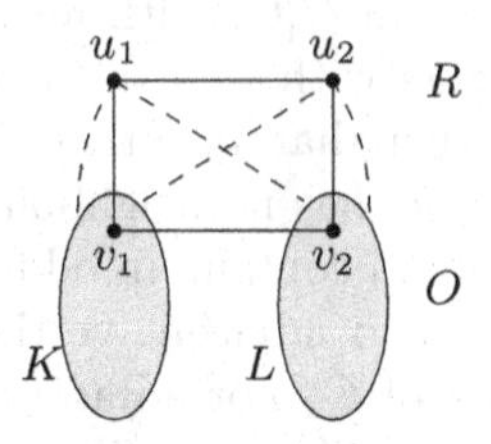

Fig. 3. The structure of B_T corresponding to a 2-clique solution S_T.

Proof. For some $p \geq 1$, let $F_1, \ldots, F_p$ be the set of components of $G[R]$. For contradiction, assume O has two distinct connectors v_1 and v_2, each with a neighbour in the same F_i, say, F_1. Let $u_1, u_2 \in V(F_1)$ be these two neighbours, where $u_1 = u_2$ is possible. Let Q be a path from u_1 to u_2 in F_1 (see Fig. 4). We make an important claim: *All paths from v_1 to v_2 in $G[O]$ have the same parity.* The reason is that if there exist paths P and P' from v_1 to v_2 in $G[O]$ that have different parity, then either the cycle $u_1v_1Pv_2u_2Qu_1$ or the cycle $u_1v_1P'v_2u_2Qu_1$ is odd. This would mean that u_1 and u_2 are not even.

By definition, v_1 and v_2 each belong to at least one odd cycle, which we denote by C_1 and C_2, respectively. We choose C_1 and C_2 such that they have minimum length. We note that $V(C_1) \cap V(C_2) = \emptyset$ and that there is no edge between a vertex of C_1 and a vertex of C_2 except possibly the edge v_1v_2; otherwise there would be paths from v_1 to v_2 in $G[O]$ that have different parity, a contradiction with the claim above.

We also note that v_1 is the only neighbour of u_1 on C_1; otherwise u_1 would belong to an odd cycle of $G[B_T]$. Similarly, v_2 is the only neighbour of u_2 on C_2. Moreover, u_1 has no neighbour on C_2 except v_2 if $u_1 = u_2$, and u_2 has no neighbour on C_1 except v_1 if $u_1 = u_2$. This can be seen as follows. For a contradiction, first suppose that, say, u_1 has a neighbour w on C_2 and $w \neq v_2$. As C_2 is an odd cycle, there exist two vertex-disjoint paths P and P' on C_2 from w to v_2 of different parity. Using the edges u_1w and u_2v_2 and the path Q from u_1 to u_2, this means that u_1 and u_2 are on odd cycle of $G[B_T]$. However, this is not possible as u_1 and u_2 are even. Hence, u_1 has no neighbour on $V(C_2) \setminus \{v_2\}$. By the same reasoning, u_2 has no neighbour on $V(C_1) \setminus \{v_1\}$. Now suppose that u_1 is adjacent to v_2 and that $u_1 \neq u_2$. Then u_1 is not adjacent to u_2, otherwise the vertices u_1, u_2 and v_2 would form a triangle, and consequently, u_1 and u_2 would not be even. Recall that $V(C_1) \cap V(C_2) = \emptyset$ and that there is no edge between a vertex of C_1 and a vertex of C_2. Hence, we can now take u_1, u_2, a vertex of $V(C_1) \setminus \{v_1\}$, and two adjacent vertices of $V(C_2) \setminus \{v_2\}$ (which exist as C_2 is a cycle) to find an induced $3P_1 + P_2$, a contradiction.

We now claim that C_1 and C_2 each have exactly three vertices. For contraction, assume that at least one of them, C_1 has length at least 5 and that in C_1, we have that x and y are the two neighbours of v_1. As C_1 has minimum length, x and y are not adjacent. Let t_1 and t_2 be adjacent vertices of C_2 distinct from v_2. Then $\{u_1, x, y, t_1, t_2\}$ induces a $3P_1 + P_2$ in G, a contradiction. Hence, C_1 and C_2 are triangles, say with vertices v_1, w_1, x_1 and v_2, w_2, x_2, respectively.

Now suppose $G[O]$ has a path from v_1 to v_2 on at least three vertices. Let s be the vertex adjacent to v_1 on this path. Then $s \notin \{w_1, x_1, w_2, x_2\}$ and s is not adjacent to any vertex of $\{w_1, x_1, w_2, x_2\}$ either; otherwise $G[O]$ contains two paths from v_1 to v_2 that are of different parity. As u_1 and s are not adjacent (else u_1 belongs to a triangle), we find that $\{s, u_1, w_2, w_1, x_1\}$ induces a $3P_1 + P_2$, a contradiction (see also Fig. 4). We conclude that as $G[O]$ is connected, v_1 and v_2 must be adjacent.

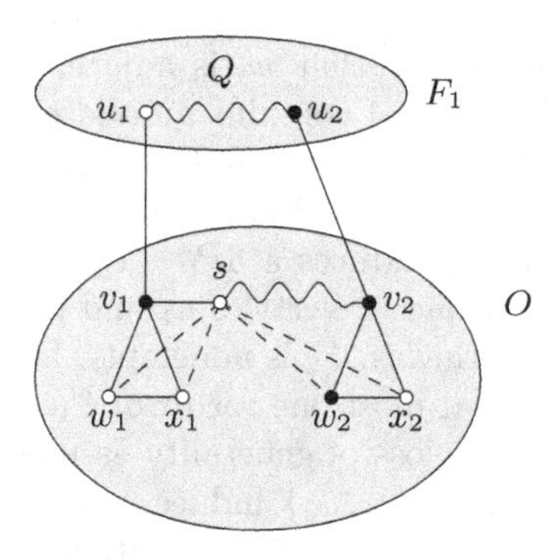

Fig. 4. The white vertices induce a $3P_1 + P_2$.

So far, we found that O contains two vertex-disjoint triangles on vertex sets $\{v_1, w_1, x_1\}$ and $\{v_2, w_2, x_2\}$, respectively, with v_1v_2 as the only edge between them. As v_1 is adjacent to v_2, we find that $u_1 \neq u_2$; otherwise $\{u_1, v_1, v_2\}$ would induce a triangle, which is not possible as $u_1 \in R$. Recall that u_1 is not adjacent to any vertex of $V(C_1) \cup V(C_2)$ except v_1, and similarly, u_2 is not adjacent to any vertex of $V(C_1) \cup V(C_2)$ except v_2. Then u_1 must be adjacent to u_2, as otherwise $\{u_1, u_2, w_1, w_2, x_2\}$ would induce a $3P_1 + P_2$.

Let $z \in O \setminus (V(C_1) \cup V(C_2))$. Suppose u_1 is adjacent to z. First assume z is adjacent to w_1 or x_1, say w_1. Then $u_1zw_1x_1v_1u_1$ is an odd cycle. Hence, this is not possible. Now assume z is adjacent to w_2 or x_2, say w_2. Then $u_1zw_2v_2u_2u_1$ is an odd cycle. This is not possible either. Hence, z is not adjacent to any vertex of $\{w_1, x_1, w_2, x_2\}$. Moreover, z is not adjacent to u_2, as otherwise $\{u_1, u_2, z\}$ induces a triangle in $G[B_T]$. However, $\{u_2, w_2, z, w_1, x_1\}$ now induces a $3P_1 + P_2$. Hence, u_1 is not adjacent to z. In other words, v_1 is the only neighbour of u_1 on O. By the same arguments, v_2 is the only neighbour of u_2 on O.

Let K be a maximal clique of O that contains C_1 and let L be a maximal clique of O that contains C_2. Note that K and L are vertex-disjoint, as for example, $w_1 \in K$ and $w_2 \in L$ are not adjacent. We claim that $O = K \cup L$. For contradiction, assume that r is a vertex of O that does not belong to K or L. As u_1 and u_2 are adjacent vertices that have no neighbours in $O \setminus \{v_1, v_2\}$, the $(3P_1 + P_2)$-freeness of G implies that $G[O \setminus \{v_1, v_2\}]$ is $3P_1$-free. As $K \setminus \{v_1\}$ and $L \setminus \{v_2\}$ induce the disjoint union of two complete graphs on at least two vertices, this means that r is adjacent to every vertex of $K \setminus \{v_1\}$ or to every vertex of $L \setminus \{v_2\}$, say r is adjacent to every vertex of $K \setminus \{v_1\}$. Then r has no neighbour r' in $L \setminus \{v_2\}$, as otherwise the cycle $v_1u_1u_2v_2r'rw_1v_1$ is an odd cycle in $G[B_T]$ that contains u_1 (and u_2). Moreover, as K is maximal and r is adjacent to every vertex of $K \setminus \{v_1\}$, we find that r and v_1 are not adjacent. Recall also that u_2 has v_2 as its only neighbour in O, hence u_2 is not adjacent to r. This means that $\{r, v_1, u_2, w_2, x_2\}$ induces a $3P_1 + P_2$, which is not possible. We conclude that $O = K \cup L$; consequently, both K and L have odd size.

We now consider the graph F_1 in more detail. Suppose F_1 contains another vertex $u_3 \notin \{u_1, u_2\}$. As F_1 is connected and bipartite (as $V(F_1) \subseteq R$), we may assume without loss of generality that u_3 is adjacent to u_1 but not to u_2. If u_3 has a neighbour K, then $G[B_T]$ contains an odd cycle that uses u_1, u_3 and one vertex of K (if the neighbour of u_3 in K is v_1) or three vertices of K (if the neighbour of u_3 in K is not v_1). Hence, u_3 has no neighbour in K. This means that $\{u_2, u_3, w_2, w_1, x_1\}$ induces a $3P_1 + P_2$, so u_3 cannot exist. Hence, F_1 consists only of the two adjacent vertices u_1 and u_2.

Now suppose that $p \geq 2$, that is, F_2 is nonempty. Let $u' \in V(F_2)$. As $u' \in R$, we find that u' is adjacent to at most one vertex of C_1 and to at most one vertex of C_2. Hence, we may without loss of generality assume that u' is not adjacent to w_1 and w_2. Then $\{u', w_1, w_2, u_1, u_2\}$ induce a $3P_1 + P_2$. We conclude that $R = \{u_1, u_2\}$. However, now S_T is a 2-clique solution of G, a contradiction. □

An algorithmic lemma, for finding a 2-clique solution of minimum weight:

Lemma 7. *Let $G = (V, E)$ be a $(3P_1 + P_2)$-free graph with a vertex weighting w, and let $T \subseteq V$. It is possible to find in polynomial time a 2-clique solution for (G, w, T) that has minimum weight.*

Proof. As the cliques K and L in B_T have size at least 3 for a 2-clique solution S_T, there are distinct vertices x_1, y_1 in $K \setminus \{v_1\}$ and distinct vertices x_2, y_2 in $L \setminus \{v_2\}$. The ordered 8-tuple $(u_1, u_2, v_1, v_2, x_1, y_1, x_2, y_2)$ is a *skeleton* of the 2-clique solution. We call the labelled subgraph of B_T that these vertices induce a *skeleton graph.*

In order to find a 2-clique solution of minimum weight in polynomial time, we consider all $\mathcal{O}(n^8)$ possible ordered 8-tuples $(u_1, u_2, v_1, v_2, x_1, y_1, x_2, y_2)$ of vertices of G and further investigate those that induce a skeleton graph. In this case, we note that if these vertices form the skeleton of a 2-clique solution S_T, then $R(B_T) = \{u_1, u_2\}$ and $O(B_T)$ is a subset of $V' = \{v_1, x_1, y_1\} \cup \{v_2, x_2, y_2\} \cup (N(v_1) \cap N(x_1) \cap N(y_1)) \cup (N(v_2) \cap N(x_2) \cap N(y_2))$. We further refine the definition of V' by deleting any vertex that cannot, by definition, belong to $O(B_T)$; that is, we remove every vertex that belongs to $T \cup (N(\{u_1, u_2\}) \setminus \{v_1, v_2\})$ or is a neighbour of both a vertex in $\{v_1, x_1, y_1\}$ and a vertex in $\{v_2, x_2, y_2\}$. We write

$G' = G[V']$. Note that u_1 and u_2 are not in G' (as they are not adjacent to any vertex in $\{x_1, x_2, y_1, y_2\}$), whereas $v_1, v_2, x_1, x_2, y_1, y_2$ all are in G'.

We now show that the sets $K' = \{v_1, x_1, y_1\} \cup (N(\{v_1, x_1, y_1\}) \cap V')$ and $L' = \{v_2, x_2, y_2\} \cup (N(\{v_2, x_2, y_2\}) \cap V')$ partition V', and moreover, that K' and L' are cliques. By definition, every vertex of V' either belongs to K' or to L'. By construction, $K' \cap L' = \emptyset$ since every vertex in $K' \setminus \{v_1\}$ is a neighbour of v_1 and every vertex in $L' \setminus \{v_2\}$ is a neighbour of v_2 and no vertex in V' is adjacent to both v_1 and v_2 which are themselves distinct. For a contradiction, suppose K' is not a clique. Then K' contains two non-adjacent vertices t and t'. As $K' \setminus \{v_1, x_1, y_1\}$ is complete to the clique $\{v_1, x_1, y_1\}$, we find that t and t' both belong to $K' \setminus \{v_1, x_1, y_1\}$. By construction of G', we find that $\{t, t'\}$ is anti-complete to $\{u_1, u_2, x_2\}$. By the definition of a skeleton, $\{u_1, u_2\}$ is anti-complete to $\{x_2\}$. Then $\{u_1, u_2, t, t', x_2\}$ induces a $3P_1 + P_2$ in G, a contradiction. By the same arguments, L' is a clique.

In G' we first delete the edge v_1v_2. Second, for $i \in \{1, 2\}$ we replace the vertices v_i, x_i, y_i by a new vertex v_i^* that is adjacent precisely to every vertex that is a neighbour of at least one vertex of $\{v_i, x_i, y_i\}$ in G'. This transforms the graph G' into the graph $G^* = (V^*, E^*)$. Note that in G^* there is no edge between v_1^* and v_2^*. We give each vertex $z \in V^* \setminus \{v_1^*, v_2^*\}$ weight $w^*(z) = w(z)$, and for $i \in \{1, 2\}$, we set $w^*(v_i^*) = w(v_i) + w(x_i) + w(y_i)$. See Fig. 5.

The algorithm will now solve Weighted Vertex Cut on (G^*, w^*) with terminals v_1^* and v_2^*; recall that this can be done in polynomial time by standard network flow techniques. Let S^* be the output. Then $G^* - S^*$ has two distinct connected components on vertex sets K^* and L^*, respectively, with $v_1^* \in K^*$ and $v_2^* \in L^*$. We set $K = (K^* \setminus \{v_1^*\}) \cup \{v_1, x_1, y_1\}$ and $L = (L^* \setminus \{v_2^*\}) \cup \{v_2, x_2, y_2\}$ and note that $G' - S^*$ contains $G[K]$ and $G[L]$ as distinct connected components.

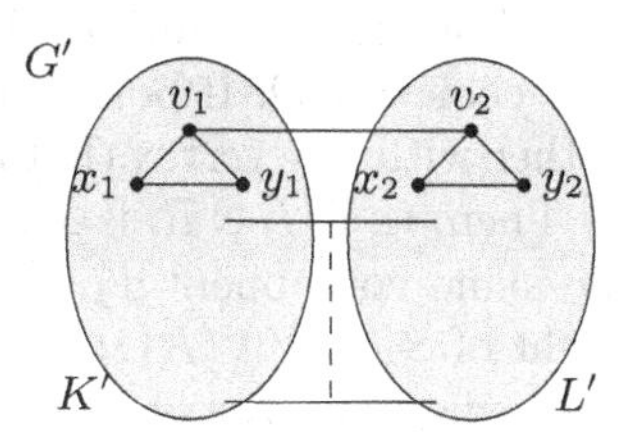

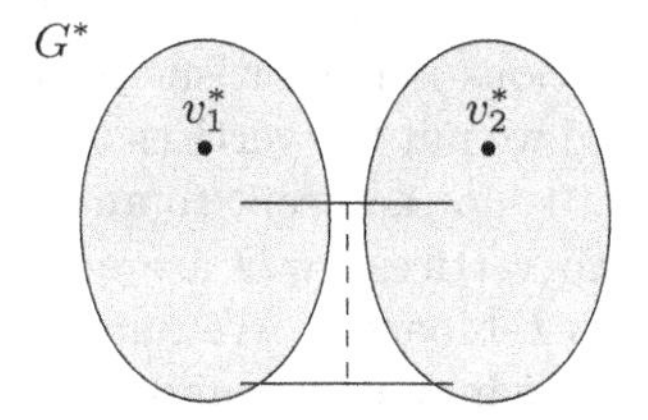

Fig. 5. The graph G' and G^* in the proof of Lemma 7.

As K is a subset of the clique K' and L is a subset of the clique L' and $V' = K' \cup L'$, we find that $G[K]$ and $G[L]$ are the only two connected components of $G' - S'$, and moreover that K and L are cliques. As no vertex of $(K \cup L) \setminus \{v_1, v_2\}$ is adjacent to u_1 or u_2, this means that $S = V \setminus (\{u_1, u_2\} \cup K \cup L)$ is a 2-clique solution for G. Moreover, as S^* is an optimal solution of Weighted Vertex Cut on instance (G^*, w^*) with terminals v_1^* and v_2^*, we find that S has minimum weight over all 2-clique solutions with skeleton $(u_1, u_2, v_1, v_2, x_1, y_1, x_2, y_2)$.

From all the $\mathcal{O}(n^8)$ 2-clique solutions computed in this way, we pick one with minimum weight; we found this 2-clique solution in polynomial time. □

The Algorithm. We are now ready to prove the main result of the section.

Theorem 4. Weighted Subset Odd Cycle Transversal *is polynomial-time solvable for* $(3P_1 + P_2)$*-free graphs.*

Proof. Let G be a $(3P_1 + P_2)$-free graph with a vertex weighting w, and let $T \subseteq V(G)$. We describe a polynomial-time algorithm for the optimization version of the problem on input (G, T, w) using the approach of Sect. 3. So, in Step 1, we compute a neutral solution of minimum weight, i.e., a minimum weight odd cycle transversal, using polynomial time due to Lemma 1 (take $s = 4$). We then compute, in Step 2, a T-full solution by setting $S_T = T$. It remains to compute a mixed solution S_T of minimum weight (Step 3) and compare its weight with the two solutions found above (Step 4). By Lemma 2 we can distinguish between two cases: $G[O]$ is connected or $G[O]$ consists of two connected components.

Case 1. $G[O]$ is connected.
We first compute in polynomial time a 2-clique solution of minimum weight by using Lemma 7. In the remainder of Case 1, we will compute a mixed solution S_T of minimum weight with connected $G[O]$ that is not a 2-clique solution. By Lemma 4 and the fact that $G[R]$ is bipartite by definition, we find that $|R| \leq 8$. We consider all $\mathcal{O}(n^8)$ possibilities for R. We discard a choice for R if $G[R]$ is not bipartite. If $G[R]$ is bipartite, we compute a solution S_T of minimum weight such that B_T contains R. Let $F_1, \ldots, F_p$ be the components of $G[R]$. By definition, $p \geq 1$. By Lemma 4 $p \leq 4$. By Lemma 6, O has at most $p \leq 4$ connectors.

We now consider all $\mathcal{O}(n^4)$ possible choices for a set D of at most four connectors. For each set D, we first check that $G[D \cup R]$ is T-bipartite and that there are no two vertices in D with a neighbour in the same F_i; if one of these conditions is not satisfied, we discard our choice of D. If both conditions are satisfied we put the vertices of D in O, together with any vertex that is not in T and that is not adjacent to any vertex of R. Then, as $G[D \cup R]$ is T-bipartite and no two vertices in D are adjacent to the same component F_i, the graph $G[R \cup O]$ is T-bipartite. We remember the weight of $S_T = V \setminus (R \cup O)$. In doing the above, we may have computed a set O that is disconnected or that contains even vertices. So we might compute some solutions more than once. However, we can compute each solution in polynomial time, and the total number of solutions we compute in Case 1 is $\mathcal{O}(n^8) \cdot \mathcal{O}(n^4) = \mathcal{O}(n^{12})$, which is polynomial as well. Out of all the 2-clique solutions and other mixed solutions we found, we pick a solution $S_T = V_T \setminus (R \cup O)$ with minimum weight as the output for Case 1.

Case 1. $G[O]$ consists of two connected components D_1 and D_2.
By Lemma 3, R is a clique of size at most 2. We consider all possible $\mathcal{O}(n^2)$ options for R. Each time R is a clique, we proceed as follows. By Lemma 5, both D_1 and D_2 have at most one connector. We consider all $\mathcal{O}(n^2)$ ways of choosing at most one connector from each of them. If we choose two, they must be non-adjacent. We discard the choice if the subgraph of G induced by R and

the chosen connector(s) is not T-bipartite. Otherwise we continue. If we chose at most one connector v, we let O consist of v and all vertices that do not belong to T and that do not have a neighbour in R. Then $G[R \cup O]$ is T-bipartite and we store $S_T = V \setminus (R \cup O)$. Note that O might not induce two connected components consisting of odd vertices, so we may duplicate some work. However, $R \cup O$ induces a T-bipartite graph and we found O in polynomial time, and this is what is relevant (together with the fact that we only use polynomial time).

When the algorithm chooses two (non-adjacent) connectors v and v' we do as follows. We remove any vertex from T and any neighbour of R other than v and v'. Let (G', w') be the resulting weighted graph (w' is the restriction of w to $V(G')$). We solve Weighted Vertex Cut in polynomial time on G', w' with v and v' as terminals. Let S be the output. We let $O = V(G') - S$. Then $G[O]$ has two connected components (as $G[R \cup \{v, v'\}]$ is T-bipartite, this implies that $G[R \cup O]$ is T-bipartite) but $G[O]$ might contain even vertices. However, what is relevant is that $G[R \cup O]$ is T-bipartite, and that we found O in polynomial time. We remember the solution $S_T = V \setminus (R \cup O)$. In the end we remember from all the solutions we computed one with minimum weight as the output for Case 2. The number of solutions is $\mathcal{O}(n^2) \cdot \mathcal{O}(n^2) = \mathcal{O}(n^4)$ and we found each solution in polynomial time so processing Case 2 takes polynomial time.

Correctness of our algorithm follows from the correctness of Cases 1 and 2, which describe all possible mixed solutions due to Lemma 2. As processing Cases 1 and 2 takes polynomial time, we compute a mixed solution of minimum weight in polynomial time. Computing a non-mixed solution of minimum weight takes polynomial time as deduced already. Hence, the running time is polynomial. □

The Proof of Theorems 2 and Theorem 3

We omit the proofs that Weighted Subset Odd Cycle Transversal is polynomial-time solvable for P_4-free graphs and (P_1+P_3)-free graphs. The reduction in [13] for Weighted Subset Feedback Vertex Set for $5P_1$-free graphs yields NP-completeness for $5P_1$-free graphs. Theorem 2 follows from Theorem 4, the above results and the result of [4] that even Odd Cycle Transversal is NP-complete on H-free graphs if H has a cycle or a claw.

We omit the proofs that Weighted Feedback Vertex Set is polynomial-time solvable for sP_2-free graphs for every $s \geq 1$, $(3P_1+P_2)$-free graphs and (P_1+P_3)-free graphs. The problem is polynomial-time solvable for P_4-free graphs [2]. Theorem 3 now follows from the above results and the results that Feedback Vertex Set is NP-complete on H-free graphs if H has a cycle [14] or a claw [10].

5 Conclusions

We determined the complexity of Weighted Subset Odd Cycle Transversal and Weighted Subset Feedback Vertex Set on H-free graphs except when $H \in \{2P_1+P_3, P_1+P_4, 2P_1+P_4\}$. We believe that the case $H = 2P_1+P_3$ is polynomial-time solvable for both problems using our methodology and our algorithms for $H = P_1 + P_3$ as a subroutine. The other two cases are open even for Odd Cycle Transversal and Feedback Vertex Set. For these cases we first need to be able to determine the complexity of finding a maximum induced disjoint union of stars in a $(P_1 + P_4)$-free graph. We refer to Table 1 for other unresolved cases in our framework and note again that our results demonstrate that the classifications of Weighed Subset Odd Cycle Transversal and Subset Odd Cycle Transversal do not coincide for H-free graphs.

References

1. Abrishami, T., Chudnovsky, M., Pilipczuk, M., Rzążewski, P., Seymour, P.: Induced subgraphs of bounded treewidth and the container method. In: Proceedings of the SODA, pp. 1948–1964 (2021)
2. Bergougnoux, B., Papadopoulos, C., Telle, J.A.: Node multiway cut and subset feedback vertex set on graphs of bounded mim-width. In: Adler, I., Müller, H. (eds.) WG 2020. LNCS, vol. 12301, pp. 388–400. Springer, Heidelberg (2020). https://doi.org/10.1007/978-3-030-60440-0_31
3. Brettell, N., Johnson, M., Paesani, G., Paulusma, D.: Computing subset transversals in H-free graphs. In: Adler, I., Müller, H. (eds.) WG 2020. LNCS, vol. 12301, pp. 187–199. Springer, Heidelberg (2020). https://doi.org/10.1007/978-3-030-60440-0_15
4. Chiarelli, N., Hartinger, T.R., Johnson, M., Milanič, M., Paulusma, D.: Minimum connected transversals in graphs: New hardness results and tractable cases using the price of connectivity. Theoret. Comput. Sci. **705**, 75–83 (2018)
5. Dabrowski, K.K., Feghali, C., Johnson, M., Paesani, G., Paulusma, D., Rzążewski, P.: On cycle transversals and their connected variants in the absence of a small linear forest. Algorithmica **82**, 2841–2866 (2020)
6. Fomin, F.V., Heggernes, P., Kratsch, D., Papadopoulos, C., Villanger, Y.: Enumerating minimal subset feedback vertex sets. Algorithmica **69**, 216–231 (2014)
7. Grzesik, A., Klimošová, T., Pilipczuk, M., Pilipczuk, M.: Polynomial-time algorithm for maximum weight independent set on P_6-free graphs. In: Proceedings of the SODA, pp. 1257–1271 (2019)
8. Johnson, M., Paesani, G., Paulusma, D.: Connected Vertex Cover for $(sP_1 + P_5)$-free graphs. Algorithmica **82**, 20–40 (2020)
9. Lozin, V., Malyshev, D., Mosca, R., Zamaraev, V.: Independent domination versus weighted independent domination. Inf. Process. Lett. **156**, 105914 (2020)
10. Munaro, A.: On line graphs of subcubic triangle-free graphs. Discret. Math. **340**, 1210–1226 (2017)
11. Paesani, G., Paulusma, D., Rzążewski, P.: Feedback vertex set and even cycle transversal for H-free graphs: finding large block graphs. CoRR abs/2105.02736 (2021)
12. Papadopoulos, C., Tzimas, S.: Polynomial-time algorithms for the subset feedback vertex set problem on interval graphs and permutation graphs. Discret. Appl. Math. **258**, 204–221 (2019)
13. Papadopoulos, C., Tzimas, S.: Subset feedback vertex set on graphs of bounded independent set size. Theoret. Comput. Sci. **814**, 177–188 (2020)
14. Poljak, S.: A note on stable sets and colorings of graphs. Comment. Math. Univ. Carol. **15**, 307–309 (1974)

Computing the Fréchet Distance Between Uncertain Curves in One Dimension

Kevin Buchin[1], Maarten Löffler[2], Tim Ophelders[1,2], Aleksandr Popov[1](✉), Jérôme Urhausen[2], and Kevin Verbeek[1]

[1] Department of Mathematics and Computer Science, TU Eindhoven, Eindhoven, The Netherlands
{k.a.buchin,a.popov,k.a.b.verbeek}@tue.nl
[2] Department of Information and Computing Sciences, Utrecht University, Utrecht, The Netherlands
{m.loffler,t.a.e.ophelders,j.e.urhausen}@uu.nl

Abstract. We consider the problem of computing the Fréchet distance between two curves for which the exact locations of the vertices are unknown. Each vertex may be placed in a given *uncertainty region* for that vertex, and the objective is to place vertices so as to minimise the Fréchet distance. This problem was recently shown to be NP-hard in 2D, and it is unclear how to compute an optimal vertex placement at all.

We give a polynomial-time algorithm for 1D curves with intervals as uncertainty regions. In contrast, we show that the problem is NP-hard in 1D in the case that vertices are placed to maximise the Fréchet distance.

We also study the weak Fréchet distance between uncertain curves. While finding the optimal placement of vertices seems more difficult than for the regular Fréchet distance—and indeed we can easily prove that the problem is NP-hard in 2D—the optimal placement of vertices in 1D can be computed in polynomial time. Finally, we investigate the discrete weak Fréchet distance, for which, somewhat surprisingly, the problem is NP-hard already in 1D.

Keywords: Curves · Uncertainty · Fréchet distance · 1D · Hardness · Weak Fréchet distance

1 Introduction

The *Fréchet distance* is a popular distance measure for curves. Its computational complexity has drawn considerable attention in computational geometry [2,5,7,8,11,18,22]. The Fréchet distance between two (polygonal) curves is often illustrated using a person and a dog: imagine a person is walking along one curve, having the dog, which walks on the other curve, on a leash. The person and the dog may change their speed independently but may not walk backwards. The Fréchet distance corresponds to the minimum leash length needed with which the person and the dog can walk from start to end on their respective curve.

A. Lubiw et al. (Eds.): WADS 2021, LNCS 12808, pp. 243–257, 2021.
https://doi.org/10.1007/978-3-030-83508-8_18

The Fréchet distance and its variants have found many applications, for instance, in the context of protein alignment [23], handwriting recognition [30], map matching [6] and construction [3,9], and trajectory similarity and clustering [12,21]. In most applications, we obtain the curves by a sequence of measurements, and these are inherently imprecise. However, it is often reasonable to assume that the true location is within a certain radius of the measurement, or that it stays within an *uncertainty region*. Think of the person and the dog, except now each is given a sequence of regions they have to visit. More specifically, they need to visit one location per region and move on a straight line between locations without going backwards. Then minimising the leash length corresponds to the following problem. Each curve is given by a sequence of uncertainty regions; minimise the Fréchet distance over all possible choices of locations in the regions. This is the *lower bound* problem for the Fréchet distance on uncertain curves.

Similar problems involving uncertainty have drawn more and more attention in the past few years in computational geometry. Most results are on uncertain point sets, where we often aim to minimise or maximise some quantity stemming from the point set, but also perform visibility queries in polygons or find Delaunay triangulations [1,14,19,24–29]. More recently there have also been several results on curves with uncertainty [4,13,17,20].

The earliest results for a variant of the problem we consider do not concern the Fréchet distance as such, but its variant the discrete Fréchet distance, where we restrict our attention to the vertices of the curves. Ahn et al. [4] show a polynomial-time algorithm that decides whether the lower bound discrete Fréchet distance is below a certain threshold, for two curves with uncertainty regions modelled as circles in constant dimension. The lower bound Fréchet distance with uncertainty regions modelled as point sets admits a simple dynamic program [13]. However, as has recently been shown, the decision problem for the continuous Fréchet distance is NP-hard already in two dimensions with vertical line segments as uncertainty regions and one precise and one uncertain curve [13]; it is not clear how to compute the lower bound at all with any uncertainty model that is not discrete. With the 2D problem being NP-hard, we turn our attention to one-dimensional curves. We present an efficient algorithm for computing the lower bound Fréchet distance with imprecision modelled as intervals. In the full version [15], we generalise this to a framework applicable in higher dimensions and restricted settings; it may not give polynomial-time solutions in many settings.

Next to the discrete Fréchet distance, the most common variant of the Fréchet distance is the weak Fréchet distance [5]. In the person–dog analogy, this variant allows backtracking on the paths. The weak Fréchet distance (for certain curves) has interesting properties in 1D [10,16,22]: it can be computed in linear time in 1D, while in 2D it cannot be computed significantly faster than quadratic time under the strong exponential-time hypothesis. To our knowledge, the weak Fréchet distance has not been studied in the uncertain setting before. We give a polynomial-time algorithm that solves the lower bound problem in 1D. In contrast to that, we show that the problem is NP-hard in 2D, and that discrete weak Fréchet distance is NP-hard already in 1D. Table 1 summarises these results.

Table 1. Complexity results for the lower bound problems for uncertain curves.

	Fréchet distance		Weak Fréchet distance	
	Discrete	Continuous	Discrete	Continuous
1D	Polynomial [4]	Polynomial	NP-hard	Polynomial
2D	Polynomial [4]	NP-hard [13]	NP-hard	NP-hard

The table provides an interesting insight. First of all, it appears that for continuous distances the dimension matters, whereas for the discrete ones the results are the same both in 1D and 2D. Moreover, it may be surprising that discretising the problem has a different effect: for the Fréchet distance it makes it easier, while for the weak Fréchet distance the problem becomes harder. We discuss the polynomial-time algorithm for the Fréchet distance in 1D in Sect. 3. We give the algorithm for the weak Fréchet distance in 1D in Sect. 5.1 and show NP-hardness for the weak (discrete) Fréchet distance in Sect. 5.2.

Finally, we also turn our attention to the problem of maximising the Fréchet distance, or finding the upper bound. It has been shown that the problem is NP-hard in 2D for several uncertainty models, including discrete point sets, both for the discrete and continuous Fréchet distance [13]. We strengthen that result by presenting a similar construction that already shows NP-hardness in 1D. The proof is discussed in Sect. 4.

2 Preliminaries

Denote $[n] \equiv \{1, 2, \ldots, n\}$. Consider a *sequence* of points $\pi = \langle p_1, p_2, \ldots, p_n \rangle$. We also use π to denote a *polygonal curve,* defined by the sequence by linearly interpolating between the points and seen as a continuous function: $\pi(i + \alpha) = (1 - \alpha)p_i + \alpha p_{i+1}$ for $i \in [n - 1]$ and $\alpha \in [0, 1]$. The *length* of such a curve is the number of its vertices, $|\pi| = n$. Denote the concatenation of two sequences π and σ by $\pi \sqcup \sigma$. Denote a subcurve from vertex i to j of π as $\pi[i : j] = p_i \sqcup p_{i+1} \sqcup \ldots \sqcup p_j$. Occasionally we use the notation $\langle \pi(i) \mid i \in I \rangle_{i=1}^{n}$ to denote a curve built on a subsequence of vertices of π, where vertices are only taken if they are in set I. For example, setting $I = \{1, 3, 4\}$, $n = 5$, $\pi = \langle p_1, p_2, \ldots, p_5 \rangle$ means $\langle \pi(i) \mid i \in I \rangle_{i=1}^{n} = \langle p_1, p_3, p_4 \rangle$.

Denote the Fréchet distance between two polygonal curves π and σ by $d_{\mathrm{F}}(\pi, \sigma)$, the discrete Fréchet distance by $d_{\mathrm{dF}}(\pi, \sigma)$, and the weak Fréchet distance by $d_{\mathrm{wF}}(\pi, \sigma)$. Recall the definition of Fréchet distance for polygonal curves of lengths m and n. It is based on *parametrisations* (non-decreasing surjections) α and β with $\alpha \colon [0, 1] \to [1, m]$, $\beta \colon [0, 1] \to [1, n]$. Parametrisations establish a *matching.* Denote the cost of a matching $\mu = (\alpha, \beta)$ as $\mathrm{cost}_\mu(\pi, \sigma) = \max_{t \in [0,1]} \|\pi \circ \alpha(t) - \sigma \circ \beta(t)\|$. Then we can define Fréchet distance between polygonal curves π and σ as the infimum of $\mathrm{cost}_\mu(\pi, \sigma)$ over all matchings μ. The discrete and weak Fréchet distance are defined similarly, using discrete and weak matchings, respectively. The discrete matching is restricted to vertices, and the weak matching is a pair of continuous surjections, i.e. a path $(\alpha, \beta): [0,1]^2 \to [1,m] \times [1,n]$, with $\alpha(0) = 1, \alpha(1) = m$ and $\beta(0) = 1, \beta(1) = n$.

An *uncertain* point in one dimension is a set of real numbers $u \subseteq \mathbb{R}$. The intuition is that only one point from this set represents the true location of the point; however, we do not know which one. A *realisation* p of such a point is one of the points from u. In this paper, we consider two special cases of uncertain points. An *indecisive* point is a finite set of numbers $u = \{x_1, \dots, x_\ell\}$. An *imprecise* point is a closed interval $u = [x_1, x_2]$. Note that a precise point is a special case of both indecisive and imprecise points.

Define an *uncertain curve* as a sequence of uncertain points $\mathcal{U} = \langle u_1, \dots, u_n \rangle$. A *realisation* $\pi \Subset \mathcal{U}$ of an uncertain curve is a polygonal curve $\pi = \langle p_1, \dots, p_n \rangle$, where each p_i is a realisation of the uncertain point u_i. For uncertain curves $\mathcal{U}$ and $\mathcal{V}$, define the lower bound and upper bound Fréchet distance. The discrete and weak Fréchet distance are defined similarly.

$$d_{\mathrm{F}}^{\min}(\mathcal{U}, \mathcal{V}) = \min_{\pi \Subset \mathcal{U}, \sigma \Subset \mathcal{V}} d_{\mathrm{F}}(\pi, \sigma), \qquad d_{\mathrm{F}}^{\max}(\mathcal{U}, \mathcal{V}) = \max_{\pi \Subset \mathcal{U}, \sigma \Subset \mathcal{V}} d_{\mathrm{F}}(\pi, \sigma).$$

3 Lower Bound Fréchet Distance in One Dimension

Problem 1. Given two uncertain curves $\mathcal{U} = \langle u_1, \dots, u_m \rangle$ and $\mathcal{V} = \langle v_1, \dots, v_n \rangle$ in $\mathbb{R}$ for some $m, n \in \mathbb{N}^+$ with uncertainty regions modelled as intervals and a threshold $\delta > 0$, decide if $d_{\mathrm{F}}^{\min}(\mathcal{U}, \mathcal{V}) \le \delta$.

We propose an efficient algorithm that solves this problem. As has been shown previously [13], the problem is NP-hard in 2D for vertical line segments as uncertainty regions, but admits a simple dynamic program for indecisive points in 2D. In the full version [15], we generalise this approach to higher dimensions and other uncertainty regions; however, the instantiations of that approach may not result in polynomial-time algorithms in many settings.

Consider the space $\mathbb{R} \times \mathbb{R}$ of the coordinates of the two curves in 1D; we want to keep track of pairs of points in uncertainty regions of the curves that are reachable, and use dynamic programming to go through the curves.

Formal Definition. Denote $\mathbb{R}^{\le 0} = \{x \in \mathbb{R} \mid x \le 0\}$ and $\mathbb{R}^{\ge 0} = \{x \in \mathbb{R} \mid x \ge 0\}$. We are interested in what is feasible within the *interval free space,* which in this space turns out to be a band around the line $y = x$ of width 2δ in L_1-distance called $\mathcal{F}_\delta$. For notational convenience, define the following regions (see Fig. 1):

$$\mathcal{F}_\delta = \{(x, y) \in \mathbb{R}^2 \mid |x - y| \le \delta\}, \qquad I_i = (u_i \times \mathbb{R}) \cap \mathcal{F}_\delta, \qquad J_j = (\mathbb{R} \times v_j) \cap \mathcal{F}_\delta.$$

We use dynamic programming, similarly to the standard free-space diagram for the Fréchet distance; however, we propagate reachable subsets of uncertainty regions on the two curves. The propagation in the interval-free-space diagram consists of starting anywhere within the current region and going in restricted directions, since we need to distinguish between going in the positive and the

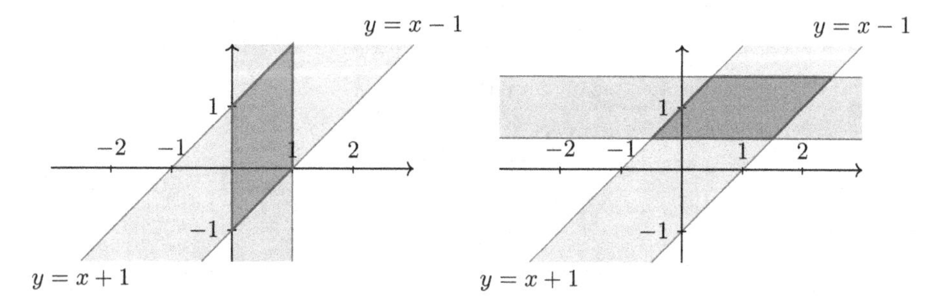

Fig. 1. On the left, the filled region is $I_i = (u_i \times \mathbb{R}) \cap \mathcal{F}_\delta$ for $u_i = [0, 1]$. On the right, the filled region is $J_j = (\mathbb{R} \times v_j) \cap \mathcal{F}_\delta$ for $v_j = [0.5, 1.5]$. In both cases $\delta = 1$.

negative x-direction along both curves. We introduce the notation for restricting the directions in the form of quadrants, half-planes, and slabs:

$$Q_{LD} = \mathbb{R}^{\leq 0} \times \mathbb{R}^{\leq 0}, \; Q_{LU} = \mathbb{R}^{\leq 0} \times \mathbb{R}^{\geq 0}, \; Q_{RD} = \mathbb{R}^{\geq 0} \times \mathbb{R}^{\leq 0}, \; Q_{RU} = \mathbb{R}^{\geq 0} \times \mathbb{R}^{\geq 0},$$

$$H_L = \mathbb{R}^{\leq 0} \times \mathbb{R}, \qquad H_R = \mathbb{R}^{\geq 0} \times \mathbb{R}, \qquad H_D = \mathbb{R} \times \mathbb{R}^{\leq 0}, \qquad H_U = \mathbb{R} \times \mathbb{R}^{\geq 0}.$$

$$S_L = \mathbb{R}^{\leq 0} \times \{0\}, \qquad S_R = \mathbb{R}^{\geq 0} \times \{0\}, \qquad S_D = \{0\} \times \mathbb{R}^{\leq 0}, \qquad S_U = \{0\} \times \mathbb{R}^{\geq 0}.$$

We introduce notation for propagating from a region by taking the appropriate Minkowski sum, denoted with $\oplus$. For $a, b \in \{L, R, U, D\}$ and a region X,

$$X^a = X \oplus H_a \,, \qquad X^{ab} = X \oplus Q_{ab} \,, \qquad X^{a0} = X \oplus S_a \,.$$

Now we can discuss the propagation. We start with the base case, where we compute the feasible combinations for the boundaries of the cells of a regular free-space diagram corresponding to the first vertex on one of the curves. For the sake of better intuition we do not use $(0, 0)$ as the base case here. So, we fix our position to the first vertex on $\mathcal{U}$ and see how far we can go along $\mathcal{V}$; and the other way around. As we are bound to the same vertex on $\mathcal{U}$, as we go along $\mathcal{V}$, we keep restricting the feasible realisations of u_1. Thus, we cut off unreachable parts of the interval as we propagate along the other curve. We do not care about the direction we were going in after we cross a vertex on the curve where we move. So, if we stay at u_1 and we cross over v_j, then we are free to go both in the negative and the positive direction of the x-axis to reach a realisation of v_{j+1}. We get the following expressions, where $U_{i,j}$ denotes the propagation upwards from the pair of vertices u_i and v_j and propagation down, left, and right is defined similarly:

$$U_{1,1} = (I_1 \cap J_1)^{U0} \cap \mathcal{F}_\delta \,, \qquad D_{1,1} = (I_1 \cap J_1)^{D0} \cap \mathcal{F}_\delta \,,$$

$$R_{1,1} = (I_1 \cap J_1)^{R0} \cap \mathcal{F}_\delta \,, \qquad L_{1,1} = (I_1 \cap J_1)^{L0} \cap \mathcal{F}_\delta \,,$$

$$U_{1,j+1} = ((U_{1,j} \cup D_{1,j}) \cap J_{j+1})^{U0} \cap \mathcal{F}_\delta \,, \qquad D_{1,j+1} = ((U_{1,j} \cup D_{1,j}) \cap J_{j+1})^{D0} \cap \mathcal{F}_\delta \,,$$

$$R_{i+1,1} = ((R_{i,1} \cup L_{i,1}) \cap I_{i+1})^{R0} \cap \mathcal{F}_\delta \,, \qquad L_{i+1,1} = ((R_{i,1} \cup L_{i,1}) \cap I_{i+1})^{L0} \cap \mathcal{F}_\delta \,.$$

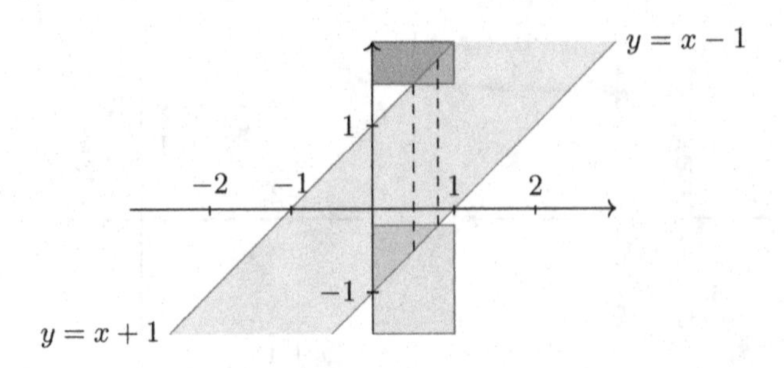

Fig. 2. An interval-free-space diagram for $u_i = [0, 1]$, $v_j = [-1.5, -0.2]$, and $v_{j+1} = [1.5, 2]$ with $\delta = 1$. Note that the feasible realisations for u_i are $[0.5, 0.8]$.

Once the boundary regions are computed, we can proceed with propagation:

$$U_{i+1,j} = (U^U_{i,j} \cup R^{RU}_{i,j} \cup L^{LU}_{i,j}) \cap I_{i+1}\,, \qquad D_{i+1,j} = (D^D_{i,j} \cup R^{RD}_{i,j} \cup L^{LD}_{i,j}) \cap I_{i+1}\,,$$

$$R_{i,j+1} = (R^R_{i,j} \cup U^{RU}_{i,j} \cup D^{RD}_{i,j}) \cap J_{j+1}\,, \qquad L_{i,j+1} = (L^L_{i,j} \cup U^{LU}_{i,j} \cup D^{LD}_{i,j}) \cap J_{j+1}\,.$$

To solve the decision problem, check if the last vertex combination is feasible:

$$((R_{m-1,n} \cup L_{m-1,n}) \cap I_m) \cup ((U_{m,n-1} \cup D_{m,n-1}) \cap J_n) \neq \emptyset\,.$$

Intuition. If the consecutive regions are always disjoint, we do not need to consider the possible directions: we always know (in 1D) where the next region is, and thus what direction we take. However, if the regions may overlap, it may be that for different realisations of a curve a segment goes in the positive or in the negative direction. The propagation we compute is based on the parameter space where we look at whether we have reached a certain vertex on each curve yet, inspired by the traditional free-space diagram. It may be that we pass by several vertices on, say, $\mathcal{V}$ while moving along a single segment on $\mathcal{U}$. The direction we choose on $\mathcal{U}$ needs to be kept consistent as we compute the next regions, otherwise we might include realisations that are invalid as feasible solutions. Therefore, we need to keep track of the chosen direction, reflected by the separate sets U, D and R, L. Otherwise, these regions in 1D are simply the feasible pairs of realisations of the last vertices on the prefixes of the curves.

It is helpful to think of the approach in terms of interval-free-space diagrams. Consider a combination of specific vertices on the two curves, say, u_i and v_j, and suppose that we want to stay at u_i but move to v_{j+1}. Which realisations of u_i, v_j, and v_{j+1} can we pick that allow this move to stay within the 2δ-band?

Suppose the x-coordinate of the diagram corresponds to the x-coordinate of $\mathcal{U}$. Then we may pick a realisation for u_i anywhere in the vertical slab corresponding to the uncertainty interval for u_i, namely, in the slab $u_i \times \mathbb{R}$. The fixed realisation for u_i would then yield a vertical line. Now suppose the y-coordinate of the diagram corresponds to the x-coordinate of $\mathcal{V}$. For v_j, picking a realisation corresponds to picking a horizontal line from the slab $\mathbb{R} \times v_j$; for v_{j+1}, it corresponds to picking a horizontal line from $\mathbb{R} \times v_{j+1}$. Picking a realisation for the pair (u_i, v_j) thus corresponds to a point in $u_i \times v_j$.

We may only maintain the matching as long the distance between the matched points is at most δ. For a fixed point on $\mathcal{U}$, this corresponds to a 2δ window for the coordinates along $\mathcal{V}$. So, the allowed matchings are contained within the band defined by $y = x \pm \delta$, and when we pick the realisations for (u_i, v_j), we only pick points from $u_i \times v_j$ for which $|y - x| \leq \delta$ holds.

As we consider the propagation to v_{j+1}, note that we may not move within u_i, so the allowed realisations for the pair (u_i, v_{j+1}) are limited. In particular, we can find that region by taking the subset of $u_i \times v_{j+1}$ for which $|y - x| \leq \delta$ holds and restricting the x-coordinate further to be feasible for the pair (u_i, v_j). See Fig. 2 for an illustration of this. In this figure, we know that v_{j+1} lies above v_j; if we did not know that, we would have to attempt propagation both upwards and downwards. For the second curve, the same holds.

We analyse the complexity of the propagated regions in the full version [15]. We conclude that their complexity is constant.

Theorem 1. *We can solve the decision problem for lower bound Fréchet distance on imprecise curves of lengths m and n in 1D in time $\Theta(mn)$.*

4 Upper Bound Fréchet Distance

We now turn our attention to the upper bound Fréchet distance. The problem is known to be NP-hard in 2D in all variants we consider [13]; we show that this remains true even in 1D. Define the following problems.

Problem 2. Upper Bound (Discrete) Fréchet: Given two uncertain trajectories $\mathcal{U}$ and $\mathcal{V}$ in 1D and a threshold $\delta > 0$, check if $d_{\mathrm{F}}^{\max}(\mathcal{U}, \mathcal{V}) \leq \delta$ ($d_{\mathrm{dF}}^{\max}(\mathcal{U}, \mathcal{V}) \leq \delta$).

We show NP-hardness by a reduction from CNF-SAT. The construction we use is similar to that used in 2D; however, in 2D the desired alignment of subcurves is achieved by having one of the curves be close enough to $(0, 0)$ at all times. Here making a curve close to 0 will not work, so we need to add extra gadgets instead that can 'eat up' the alignment of the subcurves that we do not care about. The proof can be found in the full version [15].

Theorem 2. *The problem Upper Bound (Discrete) Fréchet is NP-hard for the indecisive and for the imprecise model.*

5 Weak Fréchet Distance

In this section, we investigate the weak Fréchet distance for uncertain curves. In general, since weak matchings can revisit parts of the curve, the dynamic program for the regular Fréchet distance cannot easily be adapted, as it relies on the fact that only the realisation of the last few vertices is tracked. When computing the weak Fréchet distance for uncertain curves, one cannot simply forget the realisations of previous vertices, as the matching might revisit them. Surprisingly, we can show that for the continuous weak Fréchet distance between uncertain one-dimensional curves, we can still obtain a polynomial-time dynamic program, as shown in Sect. 5.1. One may expect that the discrete weak Fréchet distance for uncertain curves in 1D is also solvable in polynomial time; however, in Sect. 5.2 we show that this problem is NP-hard. We also show that computing the continuous weak Fréchet distance is NP-hard for uncertain curves in 2D.

5.1 Algorithm for Continuous Setting

We first introduce some definitions. Consider polygonal one-dimensional curves $\pi\colon [1, m] \to \mathbb{R}$ and $\sigma\colon [1, n] \to \mathbb{R}$ with vertices at the integer parameters. Let π^{-1} denote the reversal of a polygonal curve π. Denote by $\pi|_{[a,b]}$ the restriction of π to the domain $[a, b]$. For integer values of a and b, note that $\pi|_{[a,b]} \equiv \pi[a : b]$. Finally, define the *image* of a curve as the set of points in $\mathbb{R}$ that belong to the curve, $\mathrm{Im}(\pi) \equiv \{\pi(x) \mid x \in [1, m]\}$ for $\pi\colon [1, m] \to \mathbb{R}$. For any polygonal curve π, define the *growing curve* $\overrightarrow{\pi}$ of π as the sequence of local minima and maxima of the sequence $\langle \pi(i) \mid \pi(i) \notin \mathrm{Im}(\pi|_{[1,i)}) \rangle_{i=1}^{m}$. Thus, the vertices of a growing curve alternate between local minima and maxima, the subsequence of local maxima is strictly increasing, and the subsequence of local minima is strictly decreasing.

It has been shown that for precise one-dimensional curves, the weak Fréchet distance can be computed in linear time [16]. For uncertain curves, it is unclear how to use that linear-time algorithm; however, we can apply some of the underlying ideas. A *relaxed matching* between π and σ is defined by parametrisations $\alpha\colon [0, 1] \to [1, m]$ and $\beta\colon [0, 1] \to [1, n]$ with $\alpha(0) = 1$, $\alpha(1) = x \in [m - 1, m]$ and $\beta(0) = 1$, $\beta(1) = y \in [n - 1, n]$. Observe that the final points of parametrisations have to be on the last segments of the curves, but not necessarily at the endpoints of those segments. Moreover, define a relaxed matching (α, β) to be *cell-monotone* if for all $t \leq t'$, we have $\min(\lfloor \alpha(t) \rfloor, m - 1) \leq \alpha(t')$ and $\min(\lfloor \beta(t) \rfloor, n - 1) \leq \beta(t')$. In other words, once we pass by a vertex to the next segment on a curve, we do not allow going back to the previous segment; backtracking within a segment is allowed. Let $rm(\pi, \sigma)$ be the minimum matching cost over all cell-monotone relaxed matchings: $rm(\pi, \sigma) = \inf_{\text{cell-monotone relaxed matching } \mu} \mathrm{cost}_\mu(\pi, \sigma)$. It has been shown that $d_{\mathrm{wF}}(\pi, \sigma) = \max\left(rm(\overrightarrow{\pi}, \overrightarrow{\sigma}), rm(\overrightarrow{\pi^{-1}}, \overrightarrow{\sigma^{-1}})\right)$ for precise curves [16]. Let $rm(\pi, \sigma)[i, j] \equiv rm(\pi[1 : i], \sigma[1 : j])$. Then the value of $rm(\pi, \sigma)$ can be computed in quadratic time as $rm(\pi, \sigma)[m, n]$ using the following dynamic program:

$$
\begin{aligned}
rm(\pi,\sigma)[0,\cdot] &= rm(\pi,\sigma)[\cdot,0] = \infty\,,\\
rm(\pi,\sigma)[1,1] &= |\pi(1)-\sigma(1)|\,,\ \text{ and for } i>0 \text{ or } j>0,\\
rm(\pi,\sigma)[i+1,j+1] &= \min\begin{cases} \max\left(rm(\pi,\sigma)[i,j+1], d(\pi(i), \mathrm{Im}(\sigma[j:j+1]))\right),\\ \max\left(rm(\pi,\sigma)[i+1,j], d(\sigma(j), \mathrm{Im}(\pi[i:i+1]))\right). \end{cases}
\end{aligned}
$$

If π is a growing curve, we have $\mathrm{Im}(\pi[i, i+1]) = \mathrm{Im}(\pi[1 : i+1])$, so the following dynamic program is equivalent if π and σ are growing curves:

$$\begin{aligned}
r(\pi,\sigma)[0,\cdot] &= r(\pi,\sigma)[\cdot,0] = \infty\,,\\
r(\pi,\sigma)[1,1] &= |\pi(1)-\sigma(1)|\,,\ \text{and for } i>0 \text{ or } j>0,\\
r(\pi,\sigma)[i+1,j+1] &= \min\begin{cases}\max\big(r(\pi,\sigma)[i,j+1], d\big(\pi(i), \mathrm{Im}(\sigma[1:j+1])\big)\big)\,,\\ \max\big(r(\pi,\sigma)[i+1,j], d\big(\sigma(j), \mathrm{Im}(\pi[1:i+1])\big)\big)\,.\end{cases}
\end{aligned}$$

Let $r(\pi, \sigma) := r(\pi, \sigma)[m, n]$ when executing the dynamic program above for curves $\pi\colon [1, m] \to \mathbb{R}$ and $\sigma\colon [1, n] \to \mathbb{R}$. We have $rm(\overrightarrow{\pi}, \overrightarrow{\sigma}) = r(\overrightarrow{\pi}, \overrightarrow{\sigma})$. Moreover, observe that the final result of computing r is the same whether we apply it to the original or the growing curves. In other words, $r(\pi, \sigma) = r(\overrightarrow{\pi}, \overrightarrow{\sigma})$, so

$$\begin{aligned}
d_{\mathrm{wF}}(\pi,\sigma) &= \max\big(rm(\overrightarrow{\pi},\overrightarrow{\sigma}), rm(\overrightarrow{\pi^{-1}},\overrightarrow{\sigma^{-1}})\big) = \max\big(r(\overrightarrow{\pi},\overrightarrow{\sigma}), r(\overrightarrow{\pi^{-1}},\overrightarrow{\sigma^{-1}})\big)\\
&= \max\big(r(\pi,\sigma), r(\pi^{-1},\sigma^{-1})\big)\,.
\end{aligned}$$

With regard to computing the minimum weak Fréchet distance over realisations of uncertain curves, this roughly means that we only need to keep track of the image of the prefix (and he suffix) of π and σ. To formalise this, we split up the computation over the prefix and the suffix. Let $i_{\min}, i_{\max} \in [m]$, $j_{\min}, j_{\max} \in [n]$, $[x_{\min}, x_{\max}] \subseteq \mathbb{R}$, and $[y_{\min}, y_{\max}] \subseteq \mathbb{R}$. Abbreviate the pairs $I := (i_{\min}, i_{\max})$, $J := (j_{\min}, j_{\max})$ and the intervals $X := [x_{\min}, x_{\max}]$, $Y := [y_{\min}, y_{\max}]$, and call a realisation π of an uncertain curve *I-respecting* if $\pi(i_{\min})$ is a global minimum of π and $\pi(i_{\max})$ is a global maximum of π. Moreover, say that π is *(I, X)-respecting* if additionally $\pi(i_{\min}) = x_{\min}$ and $\pi(i_{\max}) = x_{\max}$. Let $\pi' \Subset \mathcal{U}_I$ and $\pi'' \Subset \mathcal{U}_I^X$ denote some I- and (I, X)-respecting realisations of an uncertain curve $\mathcal{U}$, respectively. Consider the minimum weak Fréchet distance between (I, X)- and (J, Y)-respecting realisations $\pi \Subset \mathcal{U}_I^X$ and $\sigma \Subset \mathcal{V}_J^Y$:

$$d_{\mathrm{wF}}^{\min}(\mathcal{U}_I^X, \mathcal{V}_J^Y) \equiv \min_{\pi \Subset \mathcal{U}_I^X, \sigma \Subset \mathcal{V}_J^Y} d_{\mathrm{wF}}(\pi,\sigma) = \min_{\pi \Subset \mathcal{U}_I^X, \sigma \Subset \mathcal{V}_J^Y} \max\big(r(\pi,\sigma), r(\pi^{-1},\sigma^{-1})\big)\,.$$

Lemma 1. *Among (I, X) - and (J, Y)-respecting realisations, the prefix and the suffix are independent:*

$$d_{\mathrm{wF}}^{\min}(\mathcal{U}_I^X, \mathcal{V}_J^Y) = \max\begin{cases}\min_{\pi \Subset \mathcal{U}_I^X, \sigma \Subset \mathcal{V}_J^Y} r(\pi,\sigma)\,,\\ \min_{\pi' \Subset \mathcal{U}_I^X, \sigma' \Subset \mathcal{V}_J^Y} r(\pi'^{-1},\sigma'^{-1})\,.\end{cases}$$

The missing details are in the full version [15]. The remainder of this section is guided by the following observations based on Lemma 1.

1. If we can compute $\min_{\pi \Subset \mathcal{U}_I^X, \sigma \Subset \mathcal{V}_J^Y} r(\pi, \sigma)$, we can compute $d_{\mathrm{wF}}^{\min}(\mathcal{U}_I^X, \mathcal{V}_J^Y)$.
2. To compute $d_{\mathrm{wF}}^{\min}(\mathcal{U}_I, \mathcal{V}_J)$, we must find an optimal pair of images X and Y for π and σ.
3. We can find $d_{\mathrm{wF}}^{\min}(\mathcal{U}, \mathcal{V})$ by computing $d_{\mathrm{wF}}^{\min}(\mathcal{U}_I, \mathcal{V}_J)$ for all $O(m^2n^2)$ values for (I, J).

Instead of computing $\min_{\pi \in \mathcal{U}_I^X, \sigma \in \mathcal{V}_J^Y} r(\pi, \sigma)$ for a specific value of (X, Y), we compute the function $(X, Y) \mapsto \min_{\pi \in \mathcal{U}_I^X, \sigma \in \mathcal{V}_J^Y} r(\pi, \sigma)$ using a dynamic program that effectively simulates the dynamic program $r(\pi, \sigma)$ for all I- and J-respecting realisations simultaneously. So let

$$R_{I,J}[i,j](x,y,X,Y) := \inf_{\substack{\pi \in \mathcal{U}_I, \mathrm{Im}(\pi[1:i])=X, \pi(i)=x \\ \sigma \in \mathcal{V}_J, \mathrm{Im}(\sigma[1:j])=Y, \sigma(j)=y}} r(\pi,\sigma)[i,j], \quad \text{then}$$

$$R_{I,J}[m,n](x,y,X,Y) = \inf_{\substack{\pi \in \mathcal{U}_I^X, \pi(m)=x \\ \sigma \in \mathcal{V}_J^Y, \sigma(n)=y}} r(\pi,\sigma).$$

We derive

$$R_{I,J}[0,\cdot](x,y,X,Y) = R_{I,J}[\cdot,0](x,y,X,Y) = \infty,$$

$$R_{I,J}[1,1](x,y,X,Y) = \inf_{\substack{\pi \in \mathcal{U}_I, \{x\}=X, \pi(1)=x \\ \sigma \in \mathcal{V}_J, \{y\}=Y, \sigma(1)=y}} |\pi(1) - \sigma(1)|, \text{ and for } (i,j) \neq (1,1)$$

$$\begin{aligned}
&R_{I,J}[i,j](x,y,X,Y) \\
&= \inf_{\substack{\pi \in \mathcal{U}_I, \mathrm{Im}(\pi[1:i])=X, \pi(i)=x \\ \sigma \in \mathcal{V}_J, \mathrm{Im}(\sigma[1:j])=Y, \sigma(j)=y}} \min \begin{cases} \max\{r(\pi,\sigma)[i-1,j], d(\pi(i-1), Y)\}, \\ \max\{r(\pi,\sigma)[i,j-1], d(\sigma(j-1), X)\} \end{cases} \\
&= \min \begin{cases} \inf_{\substack{\pi \in \mathcal{U}_I, \mathrm{Im}(\pi[1:i])=X, \pi(i)=x \\ \sigma \in \mathcal{V}_J, \mathrm{Im}(\sigma[1:j])=Y, \sigma(j)=y \\ \pi(i-1)=x'}} \max\{r(\pi,\sigma)[i-1,j], d(x', Y)\}, \\ \inf_{\substack{\pi \in \mathcal{U}_I, \mathrm{Im}(\pi[1:i])=X, \pi(i)=x \\ \sigma \in \mathcal{V}_J, \mathrm{Im}(\sigma[1:j])=Y, \sigma(j)=y \\ \sigma(j-1)=y'}} \max\{r(\pi,\sigma)[i,j-1], d(y', X)\} \end{cases} \\
&= \min \begin{cases} \inf_{\substack{\pi \in \mathcal{U}_I, \mathrm{Im}(\pi[1:i])=X, \pi(i)=x \\ \mathrm{Im}(\pi[1:i-1])=X', \pi(i-1)=x'}} \max\{R_{I,J}[i-1,j](x', y, X', Y), d(x', Y)\}, \\ \inf_{\substack{\sigma \in \mathcal{V}_J, \mathrm{Im}(\sigma[1:j])=Y, \sigma(j)=y \\ \mathrm{Im}(\sigma[1:j-1])=Y', \sigma(j-1)=y'}} \max\{R_{I,J}[i,j-1](x, y', X, Y'), d(y', X)\}, \end{cases}
\end{aligned}$$

where the conditions on x', y', X', and Y' can be checked purely in terms of $\mathcal{U}_I$ and $\mathcal{V}_J$, so the recurrence does not depend on any particular π or σ. This yields a dynamic program that constructs the function $R_{I,J}[i,j]$ based on the functions $R_{I,J}[i-1,j]$ and $R_{I,J}[i,j-1]$.

Theorem 3. *The continuous weak Fréchet distance between uncertain one-dimensional curves can be computed in polynomial time.*

Proof. We use the recurrence above, with parameters I, J, i, j, x, y, X, and Y. The first four are easy to handle, since $i \in [m]$, $j \in [n]$, $I \in [m]^2$, and $J \in [n]^2$. The other parameters are continuous. X can be represented by $x_{\min}$ and $x_{\max}$, Y by $y_{\min}$ and $y_{\max}$. To prove that we can solve the recurrence in polynomial time, it is sufficient to prove that we can restrict the computation to a polynomial number of different $x_{\min}$, $x_{\max}$, $y_{\min}$, $y_{\max}$, x and y.

We assume that each of the u_i and v_j is given as a set of intervals. This includes the cases of uncertain curves with imprecise vertices (where each of these is just one interval) and with indecisive vertices (where each interval is just a point; but in this case we get by definition only a polynomial number of different values for the parameters).

Consider the realisations $\pi = \langle p_1, \ldots, p_m \rangle$ and $\sigma = \langle q_1, \ldots, q_n \rangle$ of the curves that attain the lower bound weak Fréchet distance $d_{\mathrm{wF}}^{\min}(\mathcal{U}, \mathcal{V}) =: \delta$. In these realisations, we need to have a sequence of vertices $r_1 \leq r_2 \leq \cdots \leq r_\ell$ with the r_k alternately from the set of p_i and the set of q_j such that r_1 is at a right interval endpoint, r_ℓ is at a left interval endpoint, and $r_{k+1} - r_k = \delta$. Since $1 \leq \ell \leq m+n$, this implies that there are only $O(N^2 \cdot (m+n))$ candidates for δ, where N is the total number of interval endpoints. We can compute these candidates in time $O(N^2 \cdot (m+n))$.

Now assume that we have chosen π and σ such that none of the p_i or q_j can be increased (i.e. moved to the right) without increasing the weak Fréchet distance. Then for every p_i (and likewise q_j) there is a sequence $r_1 \leq r_2 \leq \cdots \leq r_\ell = p_i$, where r_1 is the endpoint of an interval and $r_{k+1} - r_k = \delta$. There are $O(N)$ possibilities for r_1, $O(m+n)$ possibilities for ℓ, and $O(N^2 \cdot (m+n))$ possibilities for δ, thus the total number of positions to consider for p_i is polynomial. □

5.2 Hardness of Discrete Setting

In this section, we prove that minimising the discrete weak Fréchet distance is NP-hard, already in one-dimensional space. We show this both when for indecisive and imprecise points. In the constructions in this section, the lower bound Fréchet distance is never smaller than 1. The goal is to determine whether it is equal to 1 or greater than 1.

Indecisive Points. We reduce from 3-SAT. Consider an instance with n variables and m clauses. We assign each variable a unique *height:* variable x_i gets assigned height $10i + 5$. We use slightly higher heights ($10i + 6$ and $10i + 7$) to interact with the positive state of a variable, and slightly lower heights to interact with the negative state. We construct two uncertain curves, one which represents the variables and one which represents the clauses. The first curve, $\mathcal{U}$, consists of $n + 2$ vertices. The first and last vertex are certain points, both at height 0. The remaining vertices are uncertain points, with two possible heights each:

$$\mathcal{U} = \langle 0, \{14, 16\}, \{24, 26\}, \ldots, \{10n + 4, 10n + 6\}, 0 \rangle .$$

The second curve, $\mathcal{V}$, consists of $nm+n+m+2$ vertices. For a clause $c_j = \ell_a \vee \ell_b \vee \ell_c$, let C_j be the set $\{10a + 3/7, 10b + 3/7, 10c + 3/7\}$, where for each literal we choose $+7$ if $\ell_i = x_i$ or $+3$ if $\ell_i = \neg x_i$. Let S be the set $S = \{15, 25, \ldots, 10n+5\}$ of 'neutral' variable heights. Then $\mathcal{V}$ is the curve that starts and ends at 0, has a vertex for each C_j, and has sufficiently many copies of S between them:

$$\mathcal{V} = \langle 0, S, \ldots, S, C_1, S, \ldots, S, C_2, S, \ldots, S, \ldots\ldots, C_m, 0 \rangle .$$

Consider the free-space diagram, with a 'spot' (i, j) corresponding to a pair of vertices u_i and v_j. The discrete weak Fréchet distance is equal to 1 if and only

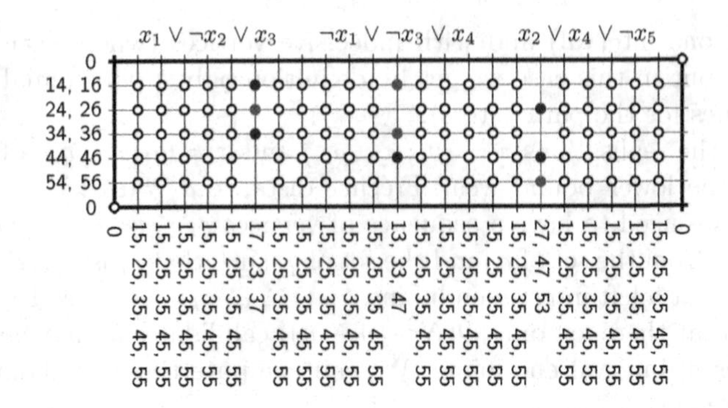

Fig. 3. An example with five variables and three clauses. White dots are always accessible, no matter the state of the variables (however, note that only one white dot per column can be used). Red/blue dots are accessible only if the corresponding variable is set to False / True. Spots without a dot are never accessible. (Color figure online)

if there is an assignment to each uncertain vertex such that the there is a path from the bottom left to the top right of the diagram that uses only accessible spots, where a spot is accessible if the assigned heights of the corresponding row and column are within 1. Figure 3 shows an example.

We can only cross a column corresponding to clause c_j if at least one of the corresponding literals is set to true. The remaining columns can always be crossed at any row. The repetition is necessary: although all spots are in principle reachable, only one spot in each column can be reachable at a time. If we have at least n columns between each pair of clauses, this will always be possible.

Theorem 4. *Given two uncertain curves $\mathcal{U}$ and $\mathcal{V}$, each given by a sequence of values and sets of values in $\mathbb{R}$, the problem of choosing a realisation of $\mathcal{U}$ and $\mathcal{V}$ minimising the weak discrete Fréchet distance between $\mathcal{U}$ and $\mathcal{V}$ is NP-hard.*

Imprecise Points and Higher Dimensions. The construction above relies on the ability to select arbitrary sets of values as uncertainty regions. We strengthen the proof in two ways: we restrict the uncertainty regions to intervals and we use uncertainty in only one curve. We then extend this result to continuous weak Fréchet distance in 2D. These results are discussed further in the full version [15].

Theorem 5. *Given an uncertain curve $\mathcal{U}$, given by a sequence of values and intervals in $\mathbb{R}$, and a certain curve $\mathcal{V}$, given by a sequence of values in $\mathbb{R}$, the problem of choosing a realisation of $\mathcal{U}$ minimising the weak discrete Fréchet distance between $\mathcal{U}$ and $\mathcal{V}$ is NP-hard.*

Corollary 1. *Given an uncertain curve $\mathcal{U}$, given by a sequence of points and regions in $\mathbb{R}^2$, and a certain curve $\mathcal{V}$, given by a sequence of points in $\mathbb{R}^2$, the problem of choosing a realisation of $\mathcal{U}$ minimising the weak Fréchet distance between $\mathcal{U}$ and $\mathcal{V}$ is NP-hard.*

Acknowledgements. Research on the topic of this paper was initiated at the 5th Workshop on Applied Geometric Algorithms (AGA 2020) in Langbroek, Netherlands. Maarten Löffler is partially supported by the Dutch Research Council (NWO) under project no. 614.001.504 and no. 628.011.005. Aleksandr Popov is supported by the Dutch Research Council (NWO) under project no. 612.001.801. Jérôme Urhausen is supported by the Dutch Research Council (NWO) under project no. 612.001.651.

References

1. Abellanas, M., et al.: Smallest color-spanning objects. In: auf der Heide, F.M. (ed.) Algorithms – ESA 2001. LNCS, vol. 2161, pp. 278–289. Springer, Berlin (2001). https://doi.org/10.1007/3-540-44676-1_23
2. Agarwal, P.K., Avraham, R.B., Kaplan, H., Sharir, M.: Computing the discrete Fréchet distance in subquadratic time. SIAM J. Comput. **43**(2), 429–449 (2014). https://doi.org/10.1137/130920526
3. Ahmed, M., Wenk, C.: Constructing street networks from GPS trajectories. In: Epstein, L., Ferragina, P. (eds.) Algorithms – ESA 2012. LNCS, vol. 7501, pp. 60–71. Springer, Berlin (2012). https://doi.org/10.1007/978-3-642-33090-2_7
4. Ahn, H.K., Knauer, C., Scherfenberg, M., Schlipf, L., Vigneron, A.: Computing the discrete Fréchet distance with imprecise input. Int. J. Comput. Geom. Appl. **22**(1), 27–44 (2012). https://doi.org/10.1142/S0218195912600023
5. Alt, H., Godau, M.: Computing the Fréchet distance between two polygonal curves. Int. J. Comput. Geom. Appl. **5**(1), 75–91 (1995). https://doi.org/10.1142/S0218195995000064
6. Brakatsoulas, S., Pfoser, D., Salas, R., Wenk, C.: On map-matching vehicle tracking data. In: Proceedings of the 31st International Conference on Very Large Data Bases, pp. 853–864. ACM, New York (2005). https://doi.org/10.5555/1083592.1083691
7. Bringmann, K.: Why walking the dog takes time: Fréchet distance has no strongly subquadratic algorithms unless SETH fails. In: Proceedings of the 55th Annual IEEE Symposium on Foundations of Computer Science (FOCS 2014), pp. 661–670. IEEE, Piscataway, NJ, USA (2014). https://doi.org/10.1109/FOCS.2014.76
8. Bringmann, K., Mulzer, W.: Approximability of the discrete Fréchet distance. J. Comput. Geom. **7**(2), 46–76 (2016). https://doi.org/10.20382/jocg.v7i2a4
9. Buchin, K., et al.: Clustering trajectories for map construction. In: Proceedings of the 25th International Conference on Advances in Geographic Information Systems (SIGSPATIAL '17), pp. 14:1–14:10. ACM, New York (2017). https://doi.org/10.1145/3139958.3139964
10. Buchin, K., Buchin, M., Knauer, C., Rote, G., Wenk, C.: How difficult is it to walk the dog? (2007). https://page.mi.fu-berlin.de/rote/Papers/pdf/How+difficult+is+it+to+walk+the+dog.pdf, presented at EuroCG 2007, Graz, Austria
11. Buchin, K., Buchin, M., Meulemans, W., Mulzer, W.: Four Soviets walk the dog: improved bounds for computing the Fréchet distance. Discret. Comput. Geom. **58**(1), 180–216 (2017). https://doi.org/10.1007/s00454-017-9878-7
12. Buchin, K., Driemel, A., van de L'Isle, N., Nusser, A.: Klcluster: center-based clustering of trajectories. In: Proceedings of the 27th International Conference on Advances in Geographic Information Systems (SIGSPATIAL '19), pp. 496–499. ACM, New York (2019). https://doi.org/10.1145/3347146.3359111

13. Buchin, K., Fan, C., Löffler, M., Popov, A., Raichel, B., Roeloffzen, M.: Fréchet distance for uncertain curves. In: 47th International Colloquium on Automata, Languages, and Programming. LIPIcs, vol. 168, pp. 20:1–20:20. Schloss Dagstuhl – Leibniz-Zentrum für Informatik, Dagstuhl, Germany (2020). https://doi.org/10.4230/LIPIcs.ICALP.2020.20
14. Buchin, K., Löffler, M., Morin, P., Mulzer, W.: Preprocessing imprecise points for Delaunay triangulation: simplified and extended. Algorithmica **61**(3), 674–693 (2011). https://doi.org/10.1007/s00453-010-9430-0
15. Buchin, K., Löffler, M., Ophelders, T., Popov, A., Urhausen, J., Verbeek, K.: Computing the Fréchet distance between uncertain curves in one dimension. arXiv preprint (2021). https://arxiv.org/abs/2105.09922
16. Buchin, K., Ophelders, T., Speckmann, B.: SETH says: weak Fréchet distance is faster, but only if it is continuous and in one dimension. In: Proceedings of the 30th Annual ACM-SIAM Symposium on Discrete Algorithms (SODA 2019), pp. 2887–2901. SIAM, Philadelphia, PA, USA (2019). https://doi.org/10.5555/3310435.3310614
17. Buchin, M., Sijben, S.: Discrete Fréchet distance for uncertain points (2016). http://www.eurocg2016.usi.ch/sites/default/files/paper_72.pdf, presented at EuroCG 2016, Lugano, Switzerland
18. Driemel, A., Har-Peled, S., Wenk, C.: Approximating the Fréchet distance for realistic curves in near linear time. Discret. Comput. Geom. **48**(1), 94–127 (2012). https://doi.org/10.1007/s00454-012-9402-z
19. Fan, C., Luo, J., Zhu, B.: Tight approximation bounds for connectivity with a color-spanning set. In: Cai, L., Cheng, S.W., Lam, T.W. (eds.) Algorithms and Computation (ISAAC 2013). LNCS, vol. 8283, pp. 590–600. Springer, Berlin (2013). https://doi.org/10.1007/978-3-642-45030-3_55
20. Fan, C., Zhu, B.: Complexity and algorithms for the discrete Fréchet distance upper bound with imprecise input. arXiv preprint (2018). https://arxiv.org/abs/1509.02576v2
21. Gudmundsson, J., Wolle, T.: Football analysis using spatio-temporal tools. Comput. Environ. Urban Syst. **47**, 16–27 (2014). https://doi.org/10.1016/j.compenvurbsys.2013.09.004
22. Har-Peled, S., Raichel, B.: The Fréchet distance revisited and extended. ACM Trans. Algorithms **10**(1), 3:1–3:22 (2014). https://doi.org/10.1145/2532646
23. Jiang, M., Xu, Y., Zhu, B.: Protein structure: structure alignment with discrete Fréchet distance. J. Bioinform. Comput. Biol. **6**(1), 51–64 (2008). https://doi.org/10.1142/s0219720008003278
24. Knauer, C., Löffler, M., Scherfenberg, M., Wolle, T.: The directed Hausdorff distance between imprecise point sets. Theor. Comput. Sci. **412**(32), 4173–4186 (2011). https://doi.org/10.1016/j.tcs.2011.01.039
25. van Kreveld, M., Löffler, M., Mitchell, J.S.B.: Preprocessing imprecise points and splitting triangulations. SIAM J. Comput. **39**(7), 2990–3000 (2010). https://doi.org/10.1137/090753620
26. Löffler, M.: Data imprecision in computational geometry. Ph.D. thesis. Universiteit Utrecht (2009). https://dspace.library.uu.nl/bitstream/handle/1874/36022/loffler.pdf
27. Löffler, M., van Kreveld, M.: Largest and smallest tours and convex hulls for imprecise points. In: Arge, L., Freivalds, R. (eds.) Algorithm Theory. LNCS, vol. 4059, pp. 375–387. Springer, Berlin (2006). https://doi.org/10.1007/11785293_35

28. Löffler, M., Mulzer, W.: Unions of onions: preprocessing imprecise points for fast onion decomposition. J. Comput. Geom. **5**(1), 1–13 (2014). https://doi.org/10.20382/jocg.v5i1a1
29. Löffler, M., Snoeyink, J.S.: Delaunay triangulations of imprecise points in linear time after preprocessing. Comput. Geom. Theory Appl. **43**(3), 234–242 (2010). https://doi.org/10.1016/j.comgeo.2008.12.007
30. Zheng, J., Gao, X., Zhan, E., Huang, Z.: Algorithm of on-line handwriting signature verification based on discrete Fréchet distance. In: Kang, L., Cai, Z., Yan, X., Liu, Y. (eds.) International Symposium on Intelligence Computation and Applications. LNCS, vol. 5370, pp. 461–469. Springer, Berlin (2008). https://doi.org/10.1007/978-3-540-92137-0_5

Finding a Largest-Area Triangle in a Terrain in Near-Linear Time

Sergio Cabello[1], Arun Kumar Das[2](✉), Sandip Das[2], and Joydeep Mukherjee[3]

[1] Faculty of Mathematics and Physics, Institute of Mathematics, Physics and Mechanics, University of Ljubljana, Ljubljana, Slovenia
[2] Advanced Computing and Microelectronics Unit, Indian Statistical Institute, Kolkata, India
[3] Department of Computer Science, Ramakrishna Mission Vivekananda Educational and Research Institute, Howrah, India

Abstract. A terrain is an x-monotone polygon whose lower boundary is a single line segment. We present an algorithm to find in a terrain a triangle of largest area in $O(n \log n)$ time, where n is the number of vertices defining the terrain. The best previous algorithm for this problem has a running time of $O(n^2)$.

Keywords: Terrain · Inclusion problem · Geometric optimisation · Hereditary segment tree

1 Introduction

An *inclusion problem* asks to find a geometric object inside a given polygon that is optimal with respect to a certain parameter of interest. This parameter can be the area, the perimeter or any other measure of the inner object that plays a role in the application at hand. Several variants of the inclusion problem come up depending on the parameter to optimize, the constraints imposed in the sought object, as well as the assumptions we can make about the containing polygon. For example, computing a largest-area or largest-perimeter convex polygon inside a given polygon is quite well studied [5,12,14]. A significant amount of work has also been done on computing largest-area triangle inside a given polygon [3,6,10,17]. In the last few years, there have been new efficient algorithms for the problems of finding a largest-area triangle [15,16], a largest-area or a largest-perimeter rectangle [4], and a largest-area quadrilateral [18] inside a given convex polygon. In this paper, we propose a deterministic $O(n \log n)$-time algorithm to find a largest-area triangle inside a given terrain, which improves the best known running time of $O(n^2)$, presented in [9]. These problems find applications in stock cutting [7], robot motion planning [19], occlusion culling [14] and many other domains of facility location and operational research.

S. Cabello—Supported by the Slovenian Research Agency (P1-0297, J1-9109, J1-1693, J1-2452).

A. Lubiw et al. (Eds.): WADS 2021, LNCS 12808, pp. 258–270, 2021.
https://doi.org/10.1007/978-3-030-83508-8_19

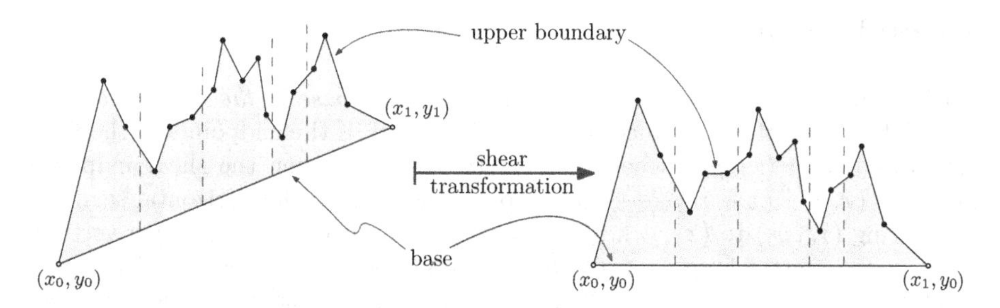

Fig. 1. Two terrains. The right one is obtained from the left one by a shear transformation to make the base horizontal

A polygon P is *x-monotone* if it has no vertical edge and each vertical line intersects P in an interval, which may be empty. An x-monotone polygon has a unique vertex with locally minimum x-coordinate, that is, a vertex whose two adjacent vertices have larger x-coordinate; see for example [2, Lemma 3.4]. Similarly, it has a unique vertex with locally maximum x-coordinate. If we split the boundary of an x-monotone polygon at the unique vertices with maximum and minimum x-coordinate, we get the *upper boundary* and the *lower boundary* of the polygon. Each vertical line intersects each of those boundaries at most once.

A *terrain* is an x-monotone polygon whose lower boundary is a single line segment, called the *base* of the terrain. The upper boundary of the terrain is connecting the endpoints of the base and lies above the base: each vertical ray from the base upwards intersects the upper boundary at exactly one point. Figure 1 shows two examples.

In this work, we focus on the problem of finding inside a terrain a triangle of largest area. We will show that when the terrain has n vertices, such a largest-area triangle can be computed in $O(n \log n)$ time. This is an improvement over the algorithm of Das et al. [9], which has a running time of $O(n^2)$. It should be noted that we compute a single triangle with largest area, even if there are more optimal solutions.

To obtain our new algorithm we build on the approach and geometric insights of [9]. More precisely, in that work, there is a single type of optimal solution that takes $O(n^2)$ time, while all the other cases can be handled in $O(n \log n)$ time. We show that the remaining case also can be solved in $O(n \log n)$ time combining shortest path trees in polygons [13], hereditary segment trees [8], search for row maxima in monotone matrices [1], and additional geometric insights.

Our new time bound, $O(n \log n)$, is a significant improvement over the best previous result. Nevertheless, the problem could be solvable in linear time. Note that the problem cannot be solved in sublinear time because we need to scan all the vertices of the polygon: any vertex of the terrain that is not scanned could be arbitrarily high and be the top vertex of a triangle with arbitrarily large area. We leave closing this gap between $O(n \log n)$ and $\Omega(n)$ as an interesting open problem for future research.

2 Preliminaries

Without loss of generality, we will assume that *the base of the terrain is horizontal*; the general case reduces to this one. Indeed, if the endpoints of the base are (x_0, y_0) and (x_1, y_1), where it must be $x_0 \neq x_1$, then the shear mapping $(x, y) \mapsto (x, y - (x - x_0)\frac{y_1 - y_0}{x_1 - x_0})$ transforms the base to the horizontal segment connecting (x_0, y_0) to (x_1, y_0). Since the mapping also transforms each vertical segment to a vertical segment, the terrain gets mapped to a terrain with a horizontal base; see Fig. 1. Since the area of any measurable region of the plane is not changed with this affine transformation, because the determinant of the Jacobian matrix is 1, it suffices to find the triangle of largest area in the resulting polygon.

For simplicity, we will assume that *no three vertices in the terrain are collinear.* This property is invariant under shear transformations. The assumption can be lifted using simulation of simplicity [11]. More precisely, we can assume that each vertex $v_i = (x_i, y_i)$ is replaced by a vertex $v'_i = (x_i, y_i + \epsilon^i)$ for a sufficiently small $\epsilon > 0$. These transformations break all collinearities if ϵ is sufficiently small. The replacement is not actually performed, but simulated. More precisely, whenever the vertices v_i, v_j and v_k are collinear, to decide their relative position after the replacements we have to check which one would be the sign of the determinant

$$\begin{vmatrix} 1 & x_i & y_i + \epsilon^i \\ 1 & x_j & y_j + \epsilon^j \\ 1 & x_k & y_k + \epsilon^k \end{vmatrix} = \begin{vmatrix} 1 & x_i & y_i \\ 1 & x_j & y_j \\ 1 & x_k & y_k \end{vmatrix} + \begin{vmatrix} 1 & x_i & \epsilon^i \\ 1 & x_j & \epsilon^j \\ 1 & x_k & \epsilon^k \end{vmatrix} = \begin{vmatrix} 1 & x_i & \epsilon^i \\ 1 & x_j & \epsilon^j \\ 1 & x_k & \epsilon^k \end{vmatrix}.$$

For example, if $i < j$ and $i < k$, then $\epsilon^i \gg \epsilon^j$ and $\epsilon^i \gg \epsilon^k$, which means that, if ϵ is positive and sufficiently small, the determinant has the sign of $\begin{vmatrix} 1 & x_j \\ 1 & x_k \end{vmatrix} = x_k - x_j$. The other cases are similar.

A vertex of a terrain is *convex* if the internal angle between the edges incident to this vertex is less than 180°. If the angle is greater than 180°, then the vertex is *reflex.* Angles of 180° do not occur because of our assumption of no 3 collinear points. The endpoints of the base of the terrain are called *base vertices.* The one with smallest x-coordinate is the *left base vertex* and is denoted by B_ℓ; the one with largest x-coordinate is the *right base vertex* and is denoted by B_r. The base vertices are convex.

3 Previous Geometric Observations

In this section, we state several observations and properties given in [9], without repeating their proofs here. The first one talks about the structure of an optimal solution.

A triangle contained in the terrain with an edge on the base of the terrain is a *grounded triangle.* For a grounded triangle, the vertex not contained in the base of the terrain is the *apex,* and the edges incident to the apex are the *left side* and the *right side*; the right side is incident to the vertex of the base with larger x-coordinate.

Lemma 1 (Lemmas 1 and 2, Corollary 1 in [9]). *In each terrain there is a largest area triangle T satisfying all of the following properties:*

(*a*) *the triangle T is grounded;*
(*b*) *the apex of T lies on the boundary of the terrain* or *each of the left and right sides of T contains two vertices of the terrain.*

Note that property (b) splits into two cases. An option is that the apex of the grounded triangle is on the boundary of the terrain. The other option is that each of the edges incident to the apex contains two vertices of the terrain. The first case is already solved in $O(n \log n)$ time.

Lemma 2 (Implicit in [9]; see the paragraph before Theorem 1). *Given a terrain with n vertices, we can find in $O(n \log n)$ time the grounded triangle with largest area that has its apex on the boundary of the terrain.*

The key insight to obtain Lemma 2 is to decompose the upper boundary of the terrain into $O(n)$ pieces with the following property: for any two points p, p' in the same piece, the largest grounded triangles with apex at p and with apex at p' have the same vertices of the terrain on the left and right sides. We refer to [9] for further details.

It remains the case when the apex is *not* contained on the boundary of the terrain. This means that each side of the optimal triangle contains two vertices of the terrain. There are two options: either both vertices contained in a side are reflex vertices, or one vertex is reflex and the other is a vertex of the base.

Recall that B_ℓ is the left endpoint of the base of the terrain. Consider the visibility graph of the vertices of the terrain and let T_ℓ be the shortest path tree from the vertex B_ℓ in the visibility graph. We regard T_ℓ as a geometric object, that is, a set of segments connecting vertices of the terrain. We orient the edges in T_ℓ away from the root, consistent with the direction that the shortest path from B_ℓ would follow them; see Fig. 2.

Consider an (oriented) edge $p \to q$ of T_ℓ; the point p is closer to B_ℓ than q is; it may be that $p = B_\ell$. When q is a reflex vertex, the *forward prolongation* of $p \to q$ is the segment obtained by extending the directed segment $p \to q$ until it reaches the boundary of the terrain. (The interior of the segment pq is not part of the prolongation.) Each point on the forward prolongation is further from B_ℓ than q is. The *backward prolongation* of $p \to q$ is the extension of $p \to q$ from p in the direction $q \to p$ until it reaches the boundary of the terrain. (A forward prolongation would be empty, is q is a convex vertex. The backward prolongation is empty if $p = B_\ell$.)

Let L be the set of non-zero-length forward prolongations of segments $p \to q$ of T_ℓ with q a reflex vertex. See Fig. 2. A similar construction is done to obtain a shortest-path tree T_r from the right endpoint B_r of the base of the terrain, the prolongations of its edges, and the set R of forward prolongations for the edges of T_r.

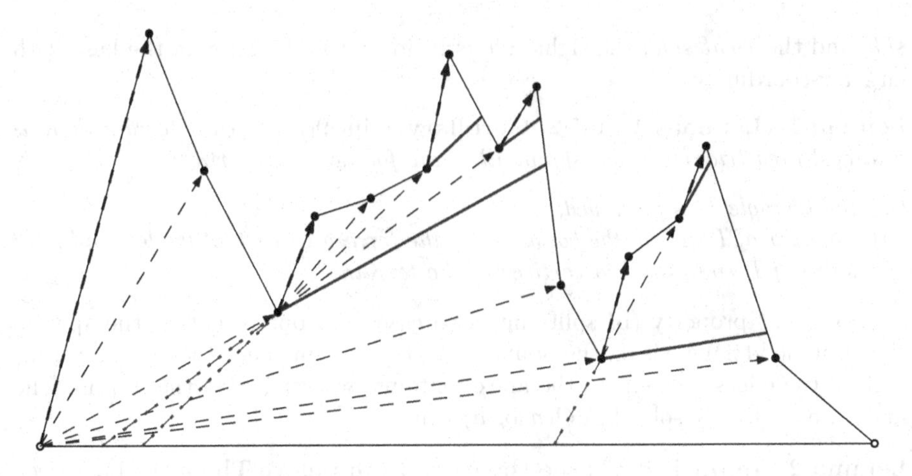

Fig. 2. The tree T_ℓ with blue dashed arcs. The set L, of forward prolongations of the edges of $E(T_\ell)$, is in solid, thick red. In dashed-dotted purple is the set of backward prolongations for the edges defining L. (Color figure online)

Using that the terrain is an x-monotone polygon and the lower boundary is a single segment, one obtains the following properties.

Lemma 3 (Lemmas 3, 4 and 5 in [9]). *The backward prolongation of each edge of $E(T_\ell) \cup E(T_r)$ has an endpoint on the base of the terrain; it may be an endpoint of the base. The segments in $E(T_\ell)$ have positive slope and the segments in $E(T_r)$ have negative slope.*

If the apex of the grounded triangle with largest area is not *on the boundary of the terrain, then there is an edge s_ℓ of L and an edge s_r of R such that: the left side of the triangle is collinear with s_ℓ, the right side of the triangle is collinear with s_r, and the apex of the triangle is the intersection $s_\ell \cap s_r$.*

4 New Algorithm

We are now going to describe the new algorithm. In fact, we describe the missing piece in the previous approach of [9]. Because of Lemma 1, it suffices to search the grounded triangle of largest area. We have two cases to consider: the apex may be on the boundary of the terrain or not. The first case can be handled using Lemma 2. To approach the second case, we use Lemma 3: in such a case the apex of the triangle belongs to $A = \{\ell \cap r \mid \ell \in L, r \in R\}$. We refer to A as the *set of candidate apices*.

We start providing a simple property for L and R.

Lemma 4. *The edges of L are pairwise interior-disjoint and can be computed in $O(n)$ time. The same holds for R.*

Proof. Consider the forward prolongation $qt \in L$ of the oriented edge $p \to q$ of T_ℓ. The shortest path from B_ℓ to any point on qt consists of the shortest

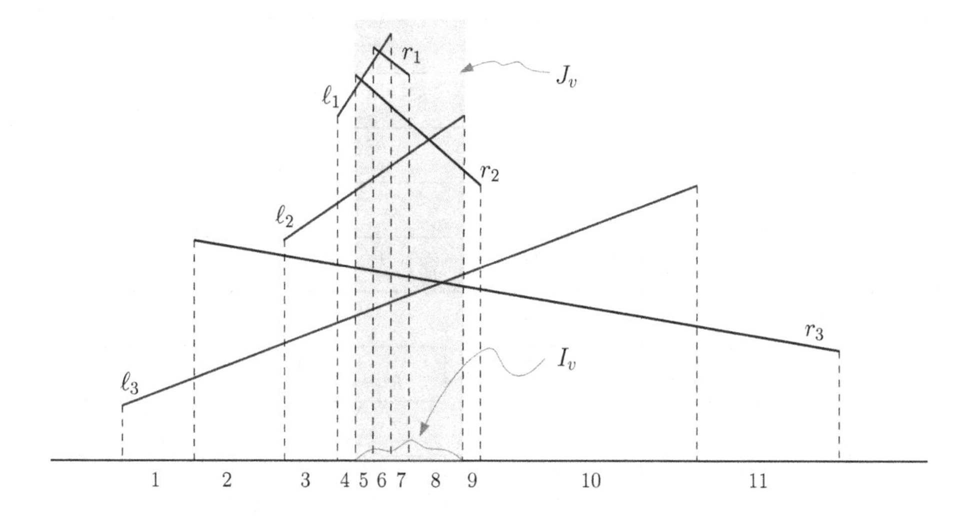

Fig. 3. Atomic intervals.

path from B_ℓ to q followed by a portion of qt. It follows that the edges of L are contained in shortest paths from B_ℓ and thus they are pairwise disjoint. (They cannot overlap because of our assumption on general position.)

Guibas et al. [13] show how to compute in $O(n)$ time the shortest path tree T_ℓ from B_ℓ and the forward extensions L. This is the extended algorithm discussed after their Theorem 2.1, where they decompose the polygon into regions such that the shortest path to any point in the region goes through the same vertices of the polygon. □

We use Lemma 4 to compute L and R in linear time. Note that $L \cup R$ has $O(n)$ segments.

We use a *hereditary segment tree*, introduced by Chazelle et al. [8], as follows. We decompose the x-axis into intervals using the x-coordinates of the endpoints of the segments in $L \cup R$. We disregard the two unbounded intervals: the leftmost and the rightmost. The resulting intervals are called the *atomic intervals*. See Fig. 3 for an example where the atomic intervals are marked as $1, 2, \ldots, 11$. We make a height-balanced binary tree $\mathcal{T}$ such that the i-th leaf represents the i-th atomic interval from left to right; see Fig. 4. For each node v of the tree $\mathcal{T}$, we define the interval I_v as the union of all the intervals stored in the leaves of $\mathcal{T}$ below v. Alternatively, for each internal node v, the interval I_v is the union of the intervals represented by its two children. In the two-dimensional setting, v represents the vertical strip bounded by the vertical lines passing through the end points of I_v. Let us denote this strip by J_v. In Fig. 3, J_v is shaded in grey for the highlighted node in Fig. 4.

Consider a node v of $\mathcal{T}$ and denote by w its parent. We maintain in v four lists of segments: L_v, R_v, L_v^h and R_v^h. The list L_v contains all the segments $\ell \in L$ such that the x-projection of ℓ contains I_v but does not contain I_w. Similarly, R_v contains the segments $r \in R$ whose projection onto the x-axis contains I_v but

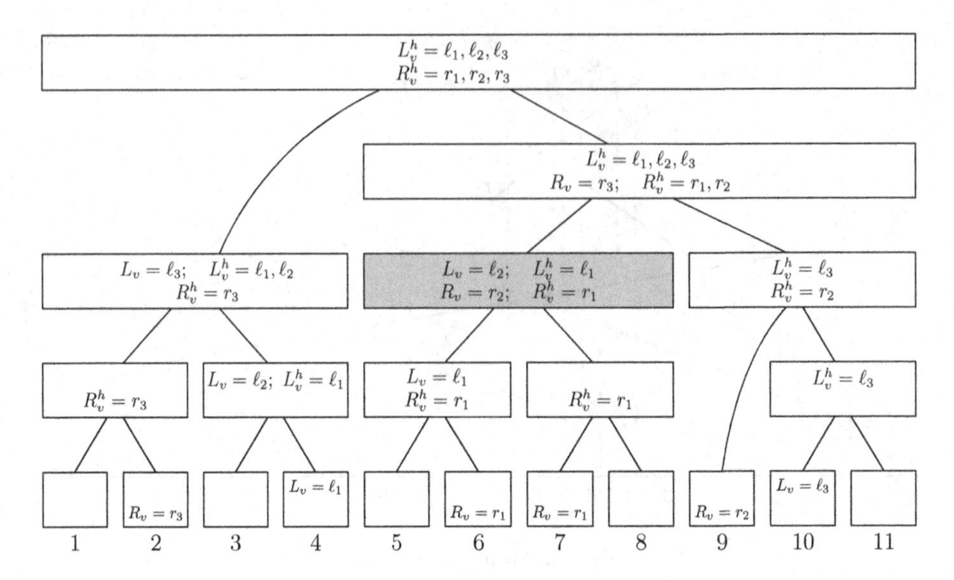

Fig. 4. Example of hereditary segment tree for Fig. 3. All the lists that are not indicated are empty.

does not contain I_w. We call L_v and R_v the *standard* lists. The list L_v^h contains the members of L_u for all proper descendants u of v in $\mathcal{T}$, that is, all descendants of v excluding v itself. Similarly, R_v^h contains the members of R_u for all proper descendants u of v in $\mathcal{T}$. We call L_v^h and R_v^h the *hereditary* lists. We put only one copy of a segment in a hereditary list of a node, even if it is stored in more than one of its descendants. See Fig. 4 for an example. All the members of the standard lists of each node are stored inside the relevant node. The members of L are colored as red and the members of R are colored as blue.

Chazelle et al. [8] noted that

$$\sum_v (|L_v| + |R_v| + |L_v^h| + |R_v^h|) \ = \ O(n \log n). \tag{1}$$

Indeed, each single segment s of $L \cup R$ appears in $O(\log n)$ standard lists, namely in at most two nodes at each level. Moreover, the nodes that contain s in their standard lists have $O(\log n)$ ancestors in total, namely the search nodes on the search path to the extreme atomic intervals contained in projection of s. It follows that s appears in $O(\log n)$ hereditary lists.

For each node v of $\mathcal{T}$ we define the intersections

$$\begin{aligned} A_v \ = \ & \{\ell \cap r \mid \ell \in L_v,\, r \in R_v,\, x(\ell \cap r) \in I_v\} \cup \\ & \{\ell \cap r \mid \ell \in L_v^h,\, r \in R_v,\, x(\ell \cap r) \in I_v\} \cup \\ & \{\ell \cap r \mid \ell \in L_v,\, r \in R_v^h,\, x(\ell \cap r) \in I_v\}. \end{aligned}$$

The set A_v is the set of candidate apices defined by the node v.

Lemma 5. *The set of candidate apices, A, is the (disjoint) union of the sets A_v, where v iterates over the nodes of $\mathcal{T}$.*

Proof. Consider a pair of intersecting segments $\ell \in L$ and $r \in R$, and let u be the leaf of $\mathcal{T}$ such that the x-coordinate of $\ell \cap r$ is contained in I_u. We walk from u upwards along the tree until the first node v with the property that $\ell \in L_v \cup L_v^h$ and $r \in R_v \cup R_v^h$ is reached. It cannot be that $\ell \in L_v^h$ and $r \in R_v^h$ because otherwise both ℓ and r would be in the lists of the descendant of v towards the leaf u. Moreover, the intersection point $\ell \cap r$ has its x-coordinate in I_v because $x(\ell \cap r) \in I_u \subseteq I_v$. It follows that $\ell \cap r \in A_v$. □

We have to find the best apex in A. Since $A = \bigcup_v A_v$ because of Lemma 5, it suffices to find the best apex in A_v for each v. For this we consider each v separately and look at the interaction between the lists L_v and R_v, the lists L_v^h and R_v, and the lists R_v^h and L_v.

4.1 Interaction Between Two Standard Lists

Consider a fixed node v and its standard lists L_v and R_v. The x-projection of each segment in $L_v \cup R_v$ is a superset of the interval I_v, and thus no endpoint of such a segment lies in the interior J_v.

Since the segments in L_v are pairwise interior-disjoint (Lemma 4) and they cross the vertical strip J_v from left to right, we can sort them with respect to their y-order within the vertical strip J_v. We sort them in decreasing y-order. Henceforth, we regard L_v as a sorted list. Thus, L_v contains $\ell_1, \dots, \ell_{|L_v|}$ and, whenever $1 \leq i < j \leq |L_v|$, the segment ℓ_i is above ℓ_j. We do the same for R_v, also by decreasing y-coordinate. Thus, R_v is a sorted list $r_1, \dots, r_{|R_v|}$ and, whenever $1 \leq i < j \leq |R_v|$, the segment r_i is above r_j.

Because of Lemma 3, each segment s of $L \cup R$ can be prolonged inside the terrain until it hits the base of the terrain. Indeed, such a prolongation contains an edge of $E(T_\ell) \cup E(T_r)$ by definition. Let $b(s)$ be the point where the prolongation of s intersects the base of the terrain.

Lemma 6. *If $1 \leq i < j \leq |L_v|$, then $b(\ell_i)$ lies to the right of $b(\ell_j)$. If $1 \leq i < j \leq |R_v|$, then $b(r_i)$ lies to the left of $b(r_j)$.*

Proof. Let s_i be the longest segment that contains ℓ_i and is contained in the terrain; let s_j be the longest segment that contains ℓ_j that is contained in the terrain. Thus $b(\ell_i)$ is an endpoint of s_i and $b(\ell_j)$ is an endpoint of s_j. Assume, for the sake of reaching a contradiction, that $b(\ell_i)$ lies to the left of $b(\ell_j)$. This means s_i and s_j are disjoint, and thus s_i is completely above s_j for any x-coordinate that they share. Then s_j cannot go through any vertex of the terrain to the left of J_v, as such a vertex would be below s_i, which is contained in the terrain. By construction of the hereditary segment tree I_v is a proper subset of the x-projections of ℓ_j and none of the end points of ℓ_j belongs to interior of J_v. This means the left end point of the ℓ_j should be to the left of J_v. Hence we arrive at the contradiction.

The argument for segments of R is similar. □

Once L_v and R_v are sorted, we can detect in $O(|L_v| + |R_v|)$ time which segments of L_v do not cross any segment of R_v inside J_v. Indeed, we can merge the lists to obtain the order π_ℓ of $L_v \cup R_v$ along the left boundary of J_v and the order π_r along the right boundary of J_v. Then we note that ℓ_i does not intersect any segment of R_v inside J_v if and only if the ranking of ℓ_i is the same in π_ℓ and in π_r. We remove from L_v the segments that do not cross any segment of R_v inside J_v. To avoid introducing additional notation, we keep denoting to the resulting list as L_v.

Within the same running time $O(|L_v| + |R_v|)$ time we can find for each $\ell_i \in L_v$ an index $\psi(i)$ such that ℓ_i and $r_{\psi(i)}$ intersect inside J_v. Indeed, if ℓ_i crosses some segment of R_v inside J_v, then it must cross one of the segments of R_v that is closest to ℓ_i in the order π_ℓ (the predecessor or the successor from R_v). These two candidates for all ℓ_i can be computed with a scan of the order π_ℓ.

Consider the $|L_v| \times |R_v|$ matrix $M = (M[i,j])_{i,j}$ defined as follows. If ℓ_i and r_j intersect in J_v, then $M[i,j]$ is the area of the grounded triangle with apex $\ell_i \cap r_j$ and sides containing ℓ_i and r_j. If ℓ_i and r_j do not intersect in J_v, and $j < \psi(i)$, then $M[i,j] = j\varepsilon$ for and infinitesimal $\varepsilon > 0$. In the remaining case, when ℓ_i and r_j do not intersect in J_v but $\psi(i) < j$, then $M[i,j] = -j\varepsilon$ for the same infinitesimal $\varepsilon > 0$. Thus, a generic row of M, when we walk it from left to right, has small positive increasing values until it reaches values defined by the area of triangles, and then it starts taking small negative values that decrease. The matrix M is not constructed explicitly, but we work with it implicitly. Given a pair of indices (i,j), we can compute $M[i,j]$ in constant time.

Note that within each row of M the non-infinitesimal elements are contiguous. Indeed, Whether a segment of L_v and a segment of R_v intersect in J_v depends only on the orders π_ℓ and π_r along the boundaries of J_v, which is the same as the order along the lists L_v and R_v. It also follows that the entries of M defined as areas of triangles form a staircase such that in lower rows it moves towards the right.

The following property shows that M is totally monotone. In fact, the lemma restates the definition of totally monotone matrix.

Lemma 7. *Consider indices i, i', j, j' such that $1 \le i < i' \le |L_v|$ and $1 \le j < j' \le |R_v|$. If $M[i',j] > M[i',j']$, then $M[i,j] > M[i,j']$.*

Proof. The cases when $s_{i'}$ and ℓ_j do not intersect or s_i and $\ell_{j'}$ do not intersect are treated by a case analysis. For example, $s_{i'}$ and ℓ_j do not intersect, then $M[i',j] > M[i',j']$ can only occur when $\psi(i') < j < j'$. In such a case ℓ_i cannot intersect r_j neither $r_{j'}$ and we must have also $\psi(i) < j < j'$. Thus $M[i,j] = -j\varepsilon > -j'\varepsilon = M[i,j']$.

The argument when s_i and $\ell_{j'}$ do not intersect is similar, but using the contrapositive. If $M[i,j] \le M[i,j']$, then both j, j' must be to the left of $\psi(i)$ and s_i dos not intersect r_j nor $r_{j'}$. In such a case $\ell_{i'}$ cannot cross s_j nor $s_{j'}$.

It remains the interesting case, when $\ell_{'i}$ and r_j intersect and also ℓ_i and $r_{j'}$ intersect. Using that ℓ_i is above $\ell_{i'}$, that r_j is above $r_{j'}$, that $\ell_{i'}$ intersects r_j, and that ℓ_i intersects $r_{j'}$, we conclude that ℓ_i also intersects r_j and that $\ell_{i'}$ also intersects $r_{j'}$. For this we just have to observe the relative order of the endpoints of the segments restricted to the boundaries of J_v.

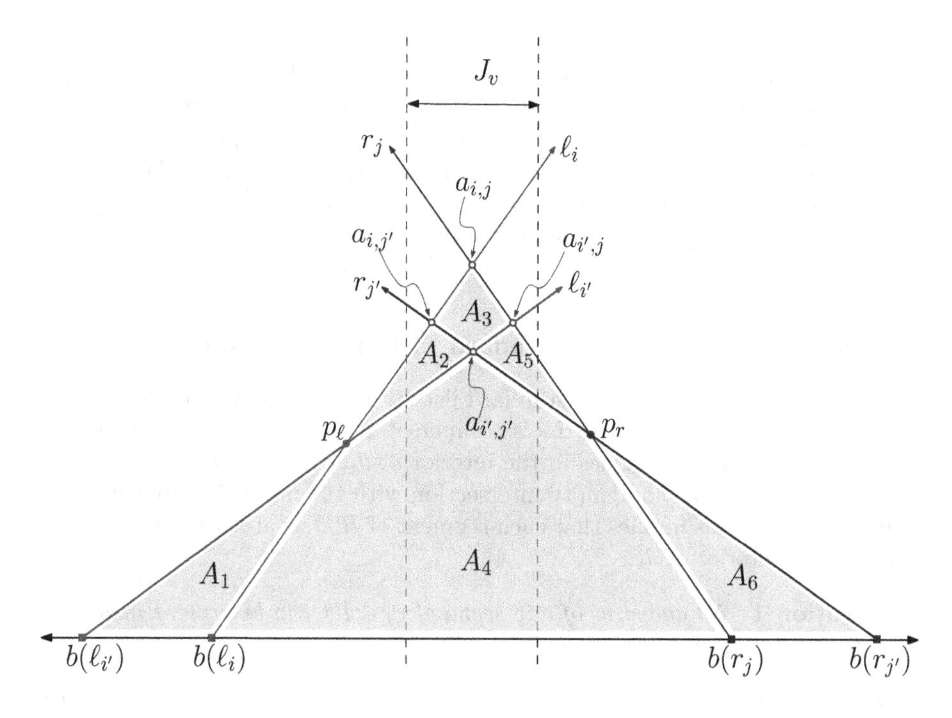

Fig. 5. Scenario in the proof of Lemma 7.

Next we use elementary geometry, as follows. See Fig. 5. Because of Lemma 6, the extensions of ℓ_i and $\ell_{i'}$ inside the terrain intersect in a point to the left of J_v, which we denote by p_ℓ. Similarly, the extensions of r_j and $r_{j'}$ intersect in a point p_r to the right of J_v.

For each $(\alpha, \beta) \in \{i, i'\} \times \{j, j'\}$, let $a_{\alpha,\beta}$ be the intersection point of ℓ_α and $r_{\varphi(\beta)}$. Thus, we have defined four points, namely $a_{i,j}, a_{i,j'}, a_{i',j}, a_{i',j'}$. We have argued before that these four points indeed exist, and they lie in J_v. We also define the triangle $T_{\alpha,\beta}$ as the grounded triangle with sides containing ℓ_α and $r_{\varphi(\beta)}$ (and thus apex $a_{\alpha,\beta}$).

We define the following areas

$$\begin{aligned}
A_1 &= \text{area}\big(\triangle(b(\ell_{i'}), b(\ell_i), p_\ell,)\big), & A_2 &= \text{area}\big(\triangle(p_\ell, a_{i',j'}, a_{i,j'})\big) \\
A_3 &= \text{area}\big(\diamond(a_{i,j'}, a_{i',j'}, a_{i',j}, a_{i,j})\big), & A_4 &= \text{area}\big(\pentagon(b(\ell_i), b(r_j), p_r, a_{i',j'}, p_\ell)\big) \\
A_5 &= \text{area}\big(\triangle(a_{i',j'}, p_r, a_{i',j})\big), & A_6 &= \text{area}\big(\triangle(b(r_j), b(r_{j'}), p_r)\big).
\end{aligned}$$

The condition $M[i', j] > M[i', j']$ translates into

$$A_1 + A_4 + A_5 \; = \; M[i', j] \; > \; M[i', j'] \; = \; A_1 + A_4 + A_6,$$

which implies that $A_5 > A_6$. We then have

$$M[i,j] \;=\; A_2 + A_3 + A_4 + A_5 \;>\; A_2 + A_3 + A_4 + A_6 \;>\; M[i,j']$$

as we wanted to show. □

For each index i with $1 \leq i \leq |L_v|$, let $\varphi(i)$ be the smallest index of columns where the maximum in the ith arrow of M is attained. Thus, $M_{i,\varphi(i)} = \max\{M_{i,j} \mid 1 \leq j \leq |R_v|\}$. Since M is totally monotone, we can compute the values $\varphi(i)$ for all $1 \leq i \leq |L_v|$ using the SWAMK algorithm of Aggarwal et al. [1]. This step takes $O(|L_v| + |R_v|)$ time.

We return the maximum among the values $M[i, \varphi(i)]$. In total we have spent $O(|L_v| + |R_v|)$ time, assuming that L_v and R_v were in sorted form.

4.2 Interaction Between a Standard List and a Hereditary List

Consider now a fixed node v, its standard list L_v and its hereditary list R_v^h. The x-projection of each segment in L_v is a superset of the interval I_v, and thus no endpoint of such a segment lies in the interior of J_v. However, the x-projection of a segment in R_v^h has non empty intersection with the interval I_v, but it is not a superset of I_v. This implies that each segment of R_v^h has at least one of its end points in the interior of J_v.

Observation 1. *No endpoint of any segment $r_j \in R_v^h$ can be present inside the strip J_v and below any segment of L_v.*

Proof. The vertical upwards ray from the endpoint is outside the terrain, because the endpoint is on the boundary of the terrain, while the segments of L_v are contained in the terrain. □

We only consider those segments of R_v^h which have their right endpoint on the exterior or on the right boundary of J_v and this right endpoint lies below the right endpoint of the topmost member of L_v in J_v. These are the members who participate in forming a feasible grounded triangle by interacting with the members of L_v.

Observation 2. *Let p be the point where the top most member of L_v intersects the right boundary of J_v. If $r'_i \in R_v^h$ intersects the right boundary of J_v below p then r'_i must have its left end point above the top most member of L_v.*

Proof. If any of such left end point q is present below any segment of L_v then the perpendicular through that point on the terrain base intersects the members of L_v, who are lying above q, outside the terrain. This is a contradiction as lies on the boundary of the terrain. □

Like before, we assume that the members of L_v are sorted by decreasing y-order. We also assume that the relevant elements of R_v^h are sorted by y-coordinate along their intersection with the right boundary of J_v.

Using Observations 1 and 2 and an argument similar to Lemma 7 we can establish that these sorted members satisfy a totally monotone property. Indeed, the segments of R_v^h do not cross the whole J_v but they cannot finish with an endpoint in J_v below an element of L_v. This suffices to argue that the crossings used in the proof of Lemma 7 exist. Thus finding a largest area grounded triangle formed by the members of the sorted lists L_v and R_v^h needs $O(|L_v| + |R_v^h|)$ amount of time. We can also handle the interaction between R_v and L_v^h in a similar fashion. This finishes the description of the interaction between a standard and a hereditary list in a node v.

4.3 Putting Things Together

Because of Lemma 5, by handling the interactions between the standard lists at each node v of $\mathcal{T}$ and between the standard list and each of the hereditary lists, we find an optimal triangle whose apex lies in A.

To get the lists sorted at each node, we can use the same technique that Chazelle et al. [8] used to improve their running time. We define a partial order in the segments of L: a segment ℓ is a predecessor of ℓ' if they share some x-coordinate and ℓ is above ℓ' at the common x-coordinate, or if they do not share any x-coordinate and ℓ is to the left of ℓ'. One can see that this definition is transitive and can be extended to a total order. This partial order can be computed with a sweep line algorithm and extended to a total order using a topological sort. Once this total order is computed at the root, it can be passed to its descendants in time proportional to the lists. (We compute this for L and for R separately.)

Theorem 1. *A largest area triangle inside a terrain with n vertices can be found in $O(n \log n)$ time.*

Proof. The computation of the total order extending the above-below relation takes $O(n \log n)$ for L and for R. After this, we can pass the sorted lists to each child in time proportional to the size of the lists. Thus, we spend additional $O(|L_v| + |R_v| + |L_v^h| + R_v^h|)$ time per node v of the hereditary tree to get the sorted lists.

Once the lists at each node v of the hereditary tree are sorted, we spend $O(|L_v| + |R_v| + |L_v^h| + R_v^h|)$ time to handle the apices of A_v, as explained above. Using (1), the total time over all nodes together is $O(n \log n)$. □

References

1. Aggarwal, A., Klawe, M., Moran, S., Shor, P., Wilber, R.: Geometric applications of a matrix-searching algorithm. Algorithmica **2**, 195–208 (1987)
2. de Berg, M., Cheong, O., van Kreveld, M.J., Overmars, M.H.: Computational Geometry: Algorithms and Applications, 3rd edn. Springer, Heidelberg (2008)
3. Boyce, J.E., Dobkin, D.P., Drysdale, III, R.L., Guibas, L.J.: Finding extremal polygons. In: STOC, pp. 282–289. ACM (1982)

4. Cabello, S., Cheong, O., Knauer, C., Schlipf, L.: Finding largest rectangles in convex polygons. Comput. Geom. Theory Appl. **51**(C), 67–74 (2016)
5. Cabello, S., Cibulka, J., Kynčl, J., Saumell, M., Valtr, P.: Peeling potatoes near-optimally in near-linear time. SIAM J. Comput. **46**(5), 1574–1602 (2017)
6. Chandran, S., Mount, D.: A parallel algorithm for enclosed and enclosing triangles. Int. J. Comput. Geometry Appl. **2**, 191–214 (1992)
7. Chang, J.S., Yap, C.K.: A polynomial solution for the potato-peeling problem. Discrete Comput. Geometry **1**(2), 155–182 (1986)
8. Chazelle, B., Edelsbrunner, H., Guibas, L., Sharir, M.: Algorithms for bichromatic line-segment problems polyhedral terrains. Algorithmica **11**, 116–132 (1994)
9. Das, A.K., Das, S., Mukherjee, J.: Largest triangle inside a terrain. Theoret. Comput. Sci. **858**, 90–99 (2021)
10. Dobkin, D.P., Snyder, L.: On a general method for maximizing and minimizing among certain geometric problems. In: SFCS, pp. 9–17 (1979)
11. Edelsbrunner, H., Mücke, E.P.: Simulation of simplicity: a technique to cope with degenerate cases in geometric algorithms. ACM Trans. Graph. **9**(1), 66–104 (1990)
12. Goodman, J.E.: On the largest convex polygon contained in a non-convex n-gon, or how to peel a potato. Geom. Dedicata. **11**(1), 99–106 (1981)
13. Guibas, L.J., Hershberger, J., Leven, D., Sharir, M., Tarjan, R.E.: Linear-time algorithms for visibility and shortest path problems inside triangulated simple polygons. Algorithmica **2**, 209–233 (1987)
14. Hall-Holt, O., Katz, M.J., Kumar, P., Mitchell, J.S.B., Sityon, A.: Finding large sticks and potatoes in polygons. In: SODA, pp. 474–483 (2006)
15. van der Hoog, I., Keikha, V., Löffler, M., Mohades, A., Urhausen, J.: Maximum-area triangle in a convex polygon, revisited. Inf. Process. Lett. **161**, 105943 (2020)
16. Kallus, Y.: A linear-time algorithm for the maximum-area inscribed triangle in a convex polygon (2017, preprint). https://arxiv.org/abs/1706.03049
17. Melissaratos, E., Souvaine, D.: Shortest paths help solve geometric optimization problems in planar regions. SIAM J. Comput. **21**(4), 601–638 (1992)
18. Rote, G.: The largest contained quadrilateral and the smallest enclosing parallelogram of a convex polygon (2019, preprint). https://arxiv.org/abs/1905.11203
19. Toth, C.D., O'Rourke, J., Goodman, J.E.: Handbook of Discrete and Computational Geometry. CRC Press, Boca Raton (2017)

Planar Drawings with Few Slopes of Halin Graphs and Nested Pseudotrees

Steven Chaplick[1], Giordano Da Lozzo[2](✉), Emilio Di Giacomo[3], Giuseppe Liotta[3], and Fabrizio Montecchiani[3]

[1] Maastricht University, Maastricht, The Netherlands
[2] Roma Tre University, Rome, Italy
giordano.dalozzo@uniroma3.it
[3] University of Perugia, Perugia, Italy

Abstract. The *planar slope number* $\text{psn}(G)$ of a planar graph G is the minimum number of edge slopes in a planar straight-line drawing of G. It is known that $\text{psn}(G) \in O(c^{\Delta})$ for every planar graph G of degree Δ. This upper bound has been improved to $O(\Delta^5)$ if G has treewidth three, and to $O(\Delta)$ if G has treewidth two. In this paper we prove $\text{psn}(G) \in \Theta(\Delta)$ when G is a Halin graph, and thus has treewidth three. Furthermore, we present the first polynomial upper bound on the planar slope number for a family of graphs having treewidth four. Namely we show that $O(\Delta^2)$ slopes suffice for nested pseudotrees.

1 Introduction

Minimizing the number of slopes used by the edge segments of a straight-line graph drawing is a well-studied problem, which has received notable attention since its introduction by Wade and Chu [22]. A break-through result by Keszegh, Pach and Pálvölgyi [18] states that every planar graph of maximum degree Δ admits a planar straight-line drawing using at most $2^{O(\Delta)}$ slopes. That is, the *planar slope number* of planar graphs is bounded by a function of Δ, which answers a question of Dujmović et al. [14]. In contrast, the slope number of non-planar graphs has been shown to be unbounded (with respect to Δ) even for $\Delta = 5$ [3,21]. Besides the above mentioned upper bound, Keszegh et al. [18] also prove a lower bound of $3\Delta - 6$, leaving as an open problem to reduce the large gap between upper and lower bounds on the planar slope number of planar graphs.

The open problem by Keszegh et al. motivated a great research effort to establish improvements for subclasses of planar graphs. Jelínek et al. [17] study planar partial 3-trees and show that their planar slope number is at most $O(\Delta^5)$. Di Giacomo et al. [12] study a subclass of planar partial 3-trees (those admitting an outer 1-planar drawing) and present an $O(\Delta^2)$ upper bound for the planar

This work began at the Graph and Network Visualization Workshop 2019. Research by GDL was partially supported by MIUR Project "AHeAD" under PRIN 20174LF3T8, by H2020-MSCA-RISE project 734922 – "CONNECT", and by Roma Tre University Azione 4 Project "GeoView".

A. Lubiw et al. (Eds.): WADS 2021, LNCS 12808, pp. 271–285, 2021.
https://doi.org/10.1007/978-3-030-83508-8_20

slope number of these graphs. Lenhart et al. [20] prove that the planar slope number of a partial 2-tree is at most 2Δ (and some partial 2-trees require at least Δ slopes). Knauer et al. [19] focus on outerplanar graphs (a subclass of partial 2-trees) and establish a tight bound of $\Delta - 1$ for the (outer)planar slope number of this graph class. Finally, Di Giacomo et al. [13] prove that the planar slope number of planar graphs of maximum degree three is four.

An algorithmic strategy to tackle the study of the planar slope number problem can be based on a *peeling-into-levels* approach. This approach has been successfully used to address the planar slope number problem for planar 3-trees [17], as well as to solve several other algorithmic problems on (near) planar graphs, including determining their pagenumber [4,5, 15,23], computing their girth [6], and constructing radial drawings [11]. In the peeling-into-levels approach the vertices of a plane graph are partitioned into levels, based on their distance from the outer face. The vertices in each level induce an outerplane graph and two consecutive levels form a 2-outerplane graph. One key ingredient is an algorithm that deals with a 2-outerplane graph with possible constraints on one of the two levels. Another ingredient is an algorithm to extend a partial solution by introducing the vertices of new levels, while taking into account the constraints defined in the already-considered levels.

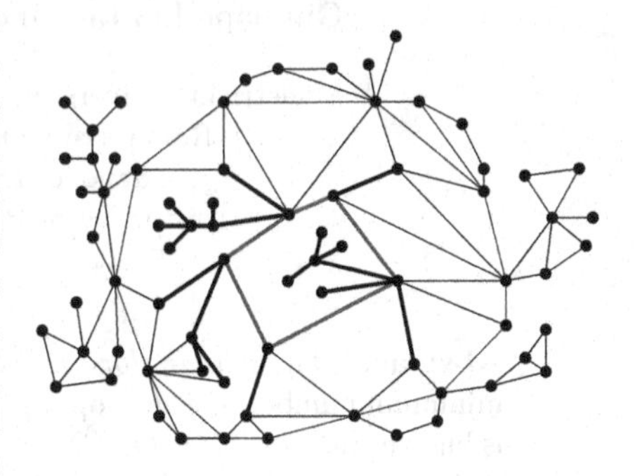

Fig. 1. A nested pseudotree: the edges of its pseudotree are bold and the cycle of its pseudotree is red. The vertices along and inside the blue cycle induce a cycle-pseudotree, defined in Sect. 4.

In an attempt to exploit the peeling-into-levels approach to prove a polynomial upper bound on the planar slope number of general planar graphs, one must be able to show a polynomial bound on the planar slope number of 2-outerplanar graphs. In this paper we take a first step in this direction by focusing on a meaningful subfamily of 2-outerplanar graphs, namely the nested pseudotrees. A *nested pseudotree* is a graph with a planar embedding such that when removing the vertices of the external face one is left with a pseudotree, that is, a connected graph with at most one cycle. See Fig. 1 for an example. The family of nested pseudotrees generalizes the well studied 2-outerplanar simply nested graphs and properly includes the Halin graphs [16], the cycle-trees [9], and the cycle-cycles [9]. Simply nested graphs were first introduced by Cimikowski [8], who proved that the inner-triangulated ones are Hamiltonian, and have been extensively studied in various contexts, such as universal point sets [1,2], square-contact representations [9], and clustered planarity [10]. Generally, nested pseudotrees have treewidth four and, as such, the best prior upper bound on their planar slope number is the one by Keszegh et al., which is exponential in Δ. Halin graphs and cycle-trees have instead treewidth three, and therefore the previously known upper bound for these graphs is $O(\Delta^5)$, as shown by Jelínek et al. [17].

We prove significantly better upper bounds for all the above mentioned graph classes. Our main results are the following.

Theorem 1. *Every degree-Δ Halin graph has planar slope number $O(\Delta)$.*

Theorem 2. *Every degree-Δ nested pseudotree has planar slope number $O(\Delta^2)$.*

The proofs of Theorems 1 and 2 are constructive and are based on a unified approach. The problem is easily reduced to the study of 2-connected instances. We then use inductive techniques to further reduce the problem to the study of triconnected instances, which are eventually treated by means of a suitable data structure called SPQ-tree [9]. Statements marked with ($\star$) are proven in [7].

2 Preliminaries

Notation. Let G be a graph. The *degree* $\deg_G(v)$ of a vertex v of G is the number of neighbors of v in G. The *degree* $\Delta(G)$ of G is $\max_{v \in G} \deg_G(v)$. When clear from the context, we omit the specification of G in the above notation and say that G is a *degree-Δ* graph. Let a, b, and c be points in $\mathbb{R}^2$; we denote by $\overline{ab}$ the straight-line segment whose endpoints are a and b, and by $\blacktriangle(abc)$ the triangle whose corners are a, b, and c.

Nested Pseudotrees. A planar drawing of a graph is *outerplanar* if all the vertices are incident to the outer face, and 2-*outerplanar* if removing the vertices of the outer face yields an outerplanar graph. A graph is 2-*outerplanar* (*outerplanar*) if it admits a 2-outerplanar drawing (outerplanar drawing). In a 2-outerplanar drawing, vertices incident to the outer face are called *external*, and all other vertices are *internal*. A 2-*outerplane* graph is a 2-outerplanar graph with a planar embedding inherited from a 2-outerplanar drawing. A 2-outerplane graph is *simply nested* if its external vertices induce a chordless cycle and its internal vertices induce either a chordless cycle or a tree. As in [9], we refer to a 2-outerplanar simply nested graph whose internal vertices induce a chordless cycle or a tree as a *cycle-cycle* or a *cycle-tree*, respectively. A *Halin graph* is a 3-connected topological graph G such that, by removing the edges incident to the outer face, one gets a tree whose internal vertices have degree at least 3 and whose leaves are incident to the outerface of G. Observe that, Halin graphs are a subfamily of the cycle-trees. A *pseudotree* is a connected graph containing at most one cycle. A *nested pseudotree* is a topological graph such that removing the vertices on the outer face yields a non-empty pseudotree. Note that the external vertices of a nested pseudotree need not induce a chordless cycle.

Theorem 3 ($\star$). *Nested pseudotrees have treewidth at most* 4, *which is tight.*

Planar Slope Number. The *slope* of a line ℓ is the smallest angle $\alpha \in [0, \pi)$ such that ℓ can be made horizontal by a clockwise rotation by α. The *slope* of a segment is the slope of the line containing it. Let G be a planar graph and let Γ be a planar straight-line drawing of G. The *planar slope number* $\text{psn}(\Gamma)$ of Γ is the number of distinct slopes used by the edges of G in Γ. The *planar slope number* $\text{psn}(G)$ of G is the minimum $\text{psn}(\Gamma)$ over all planar straight-line drawings Γ of G. If G has degree Δ, then clearly $\text{psn}(G) \geq \lceil \Delta/2 \rceil$.

Geometric Definitions. Consider a planar straight-line drawing Γ of a path $\pi = (u_1, \ldots, u_k)$ directed from u_1 to u_k, and let $x(u)$ and $y(u)$ denote the x- and y-coordinate of a vertex u in Γ, respectively. We say that π is $\nearrow$*-monotone*, if $y(u_{i+1}) \geq y(u_i)$ and $x(u_{i+1}) > x(u_i)$, for $i = 1, \ldots, k-1$. Similarly, we say that it is $\nwarrow$*-monotone*, if $y(u_{i+1}) \geq y(u_i)$ and $x(u_{i+1}) < x(u_i)$, for $i = 1, \ldots, k-1$.

Let Γ be a planar straight-line drawing of a graph G and let e be an edge of G. We say that an isosceles triangle T is *nice for* e if its base coincides with the drawing of e in Γ and T does not intersect any edge of Γ, except e.

Theorem 4 ([20]). *Let $0 < \beta < \frac{\pi}{2}$. Let $\blacktriangle(abc)$ be any isosceles triangle whose base $\overline{bc}$ is horizontal, whose apex a lies above $\overline{bc}$, and such that the interior angles at b and c are equal to β. There exists a set $\mathcal{L}(\beta, \Delta)$ of $O(\Delta)$ slopes such that any degree-Δ partial 2-tree G admits a planar straight-line drawing inside $\blacktriangle(abc)$ using the slopes in $\mathcal{L}(\beta, \Delta)$ in which any given edge (u, v) of G is drawn such that $c \equiv v$ and $b \equiv u$.*

3 Cycle-Trees and Proof of Theorem 1

In this section, we consider cycle-trees and prove that their planar slope number is $O(\Delta^2)$ in general and $O(\Delta)$ for Halin graphs. A degree-2 vertex v of a cycle-tree G whose neighbors are x and y is *contractible* if (x, y) is not an edge of G, and if deleting v and adding the edge (x, y) yields a cycle-tree; this operations is the *contraction* of v. A cycle-tree G is *irreducible* if it contains no contractible vertex. We prove the following.

Lemma 1 ($\star$). *For every degree-Δ cycle-tree G and irreducible cycle-tree G' obtained from G by any sequence of contractions,* $\text{psn}(G) \leq \text{psn}(G')$.

By Lemma 1, without loss of generality, the considered cycle-trees will have no contractible vertices. Furthermore, if the outer face of an irreducible 2-connected cycle-tree G of degree Δ has size $k \geq 3$, then the number of edges of G is $O(k\,\Delta)$, which implies that $\text{psn}(G) \in O(\Delta)$ if k is constant. This observation allows us to assume $k > 3$ for 2-connected instances, which will simplify the description.

3.1 3-Connected Instances

A *path-tree* is a plane graph G that can be augmented to a cycle-tree G' by adding an edge $e = (u, v)$ to its outer face. Suppose that, in a clockwise walk along the outer face of G', edge e is traversed from u to v; then u is the *leftmost path-vertex* and v is the *rightmost path-vertex* of G. All vertices in the outer face

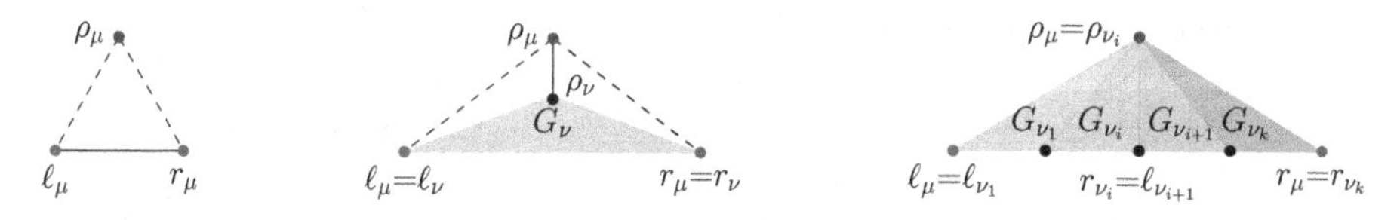

Fig. 2. Path-trees associated with a Q-node (left), an S-node (middle), and a P-node (right). Dashed edges may or may not exist. Shaded triangles represent smaller path-trees G_{ν_i} rooted at ρ_i, with leftmost path-vertex ℓ_{ν_i} and rightmost path-vertex r_{ν_i}.

of G' are *path-vertices*, while the other vertices are *tree-vertices*. Let f be the internal face of G' that contains edge e. The path induced by the path-vertices of G is the *path of* G. Analogously, the tree induced by the tree-vertices of G is the *tree of* G. The path-tree G can be *rooted* at any tree-vertex ρ on the boundary of f; then vertex ρ becomes the *root* of G. If G is rooted at ρ, then the tree of G is also rooted at ρ. A rooted path-tree with root ρ, leftmost path-vertex ℓ, and rightmost path-vertex r is *almost-3-connected* if it becomes 3-connected by adding the edges (ρ, ℓ), (ρ, r), and (ℓ, r), if missing.

SPQ-Decomposition of Path-Trees. Let G be an almost-3-connected path-tree rooted at ρ, with leftmost path-vertex ℓ and rightmost path-vertex r. The *SPQ-decomposition* of G [9] constructs a tree $\mathcal{T}$, called the *SPQ-tree* of G, whose nodes are of three different kinds: *S-*, *P-*, and *Q-nodes*. Each node μ of $\mathcal{T}$ is associated with an almost-3-connected rooted path-tree G_μ, called the *pertinent graph* of μ. To avoid special cases, we extend the definition of path-trees so to include graphs whose path is a single edge (ℓ, r) and whose tree consists of a single vertex ρ, possibly not adjacent to ℓ or r. As a consequence, we also extend the definition of almost-3-connected path-trees to graphs such that adding (ρ, ℓ), (ρ, r), and (ℓ, r), if missing, yields a 3-cycle.

Q-NODE: The pertinent graph G_μ of a *Q-node* μ is an almost-3-connected rooted path-tree consisting of three vertices: one tree-vertex ρ_μ and two path-vertices ℓ_μ and r_μ. Vertices ρ_μ, ℓ_μ, and r_μ are the root, the leftmost path-vertex, and the rightmost path-vertex of G_μ, respectively. G_μ always has edge (ℓ_μ, r_μ), while (ρ_μ, ℓ_μ) and (ρ_μ, r_μ) may not exist; see Fig. 2(left).

S-NODE: The pertinent graph G_μ of an *S-node* μ is an almost-3-connected rooted path-tree consisting of a root ρ_μ adjacent to the root ρ_ν of one almost-3-connected rooted path-tree G_ν, and possibly to the leftmost path-vertex ℓ_ν and to the rightmost path-vertex r_ν of G_ν. The node ν whose pertinent graph is G_ν is the unique child of μ in $\mathcal{T}$. The leftmost and the rightmost path-vertices of G_μ are ℓ_ν and r_ν, respectively; see Fig. 2(middle).

P-NODE: The pertinent graph G_μ of a *P-node* μ is an almost-3-connected rooted path-tree obtained from almost-3-connected rooted path-trees $G_{\nu_1}, \dots, G_{\nu_k}$, with $k > 1$, as follows. First, the roots of $G_{\nu_1}, \dots, G_{\nu_k}$ are identified into the

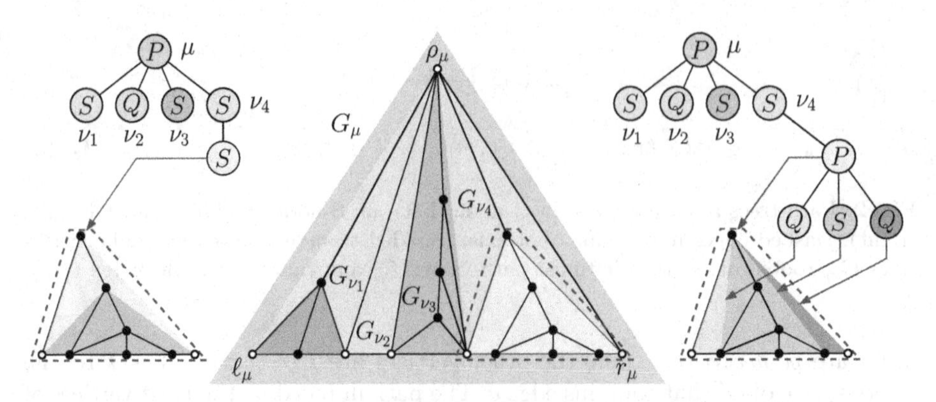

Fig. 3. Two alternative partial SPQ-trees of the almost-3-connected path-tree in the center: the child of node ν_4 is an S-node on the left and a P-node on the right.

root ρ_μ of G_μ. Second, the leftmost path-vertex of G_{ν_i} is identified with the rightmost path-vertex of $G_{\nu_{i-1}}$, for $i = 2, \ldots, k$. The nodes $\nu_1, \ldots, \nu_k$ whose pertinent graphs are $G_{\nu_1}, \ldots, G_{\nu_k}$, respectively, are the children of μ in $\mathcal{T}$, and the left-to-right order in which they appear in $\mathcal{T}$ is $\nu_1, \ldots, \nu_k$. The leftmost and the rightmost path-vertices of G_μ are ℓ_{ν_1} and r_{ν_k}, respectively; see Fig. 2(right).

The SPQ-tree $\mathcal{T}$ of G is such that: **(i)** Q-nodes are leaves of $\mathcal{T}$. **(ii)** If the pertinent graph of an S-node μ contains neither (ρ_μ, ℓ_μ) nor (ρ_μ, r_μ), then the parent of μ is a P-node. **(iii)** Every P-node has at most $2\Delta+1$ children. Figure 3 provides two alternative SPQ-trees of the same graph.

For simplicity, we assume that the pertinent graphs of the children of a P-node μ are induced subgraphs of G_μ. This implies that if G_μ contains an edge (ρ_μ, v), where v is a path-vertex, then such an edge belongs to every child of μ whose pertinent graph contains v.

Property 1. Any SPQ-tree $\mathcal{T}$ can be modified so that each child of every P-node is either an S- or a Q-node.

Let μ be a node of $\mathcal{T}$. The *left path* of μ is the path directed from ℓ_μ to ρ_μ, consisting of edges belonging to the outer face of G_μ, and not containing r_μ. The definition of the *right path* of μ is symmetric. Observe that, if μ is a Q-node whose pertinent graph G_μ does not contain the edge (ρ_μ, ℓ_μ), then the left path of μ is the empty path. Similarly, the right path of μ is the empty path if G_μ does not contain the edge (ρ_μ, r_μ).

Lemma 2 ([9]). *Every almost-3-connected rooted path-tree admits an SPQ-tree.*

The cornerstone of our contribution is a construction for almost-3-connected rooted path-trees using $O(\Delta^2)$ slopes. We start by defining the slope set.

Slope Set. Let G be an almost-3-connected path-tree and let $\mathcal{T}$ be an SPQ-tree of G. For any node μ of $\mathcal{T}$ and for any path-vertex v in G_μ we let $\delta_\mu(v) = \deg_{G_\mu}(v)$ and we let δ^* be the maximum $\delta_\mu(v)$ over all nodes μ and path-vertices

v. Consider the equilateral triangle $\blacktriangle(abc)$ with vertices a, b, and c in counter-clockwise order; refer to Fig. 4(a). Assume that the side $\overline{bc}$ is horizontal, and that a lies above $\overline{bc}$. Let $b = u_0, u_1, \ldots, u_{2\Delta+1} = c$ be the $2\Delta + 2$ equispaced points along $\overline{bc}$. We define the following slope sets:

Black Slope: The slope 0, i.e., the slope of an horizontal line.

Orange Slopes: The *i-th orange slope* O_i is the slope of $\overline{au_i}$, with $1 \leq i \leq 2\Delta$.

Blue Slopes: The *i-th blue slope* B_i is the slope of $\overline{av_i}$, where v_i is the vertex of the equilateral triangle inside $\blacktriangle(abc)$ with vertices v_i, u_i, and u_{i+1}, with $0 \leq i \leq 2\Delta$.

Magenta Slopes: We have two sets of magenta slopes:

- *Left-magenta slopes:* The *i-th l-magenta slope* M_i^l is $\frac{i\pi}{3\delta^*}$, with $1 \leq i \leq \delta^* - 1$. For convenience, we let $M_{\delta^*}^l = B_0$ and consider B_0 to be also left-magenta.
- *Right-magenta slopes:* The *i-th r-magenta* slope M_i^r is $\pi - M_i^l$, with $1 \leq i \leq \delta^* - 1$. Again, we let $M_{\delta^*}^r = B_{2\Delta}$ and consider $B_{2\Delta}$ to be also right-magenta.

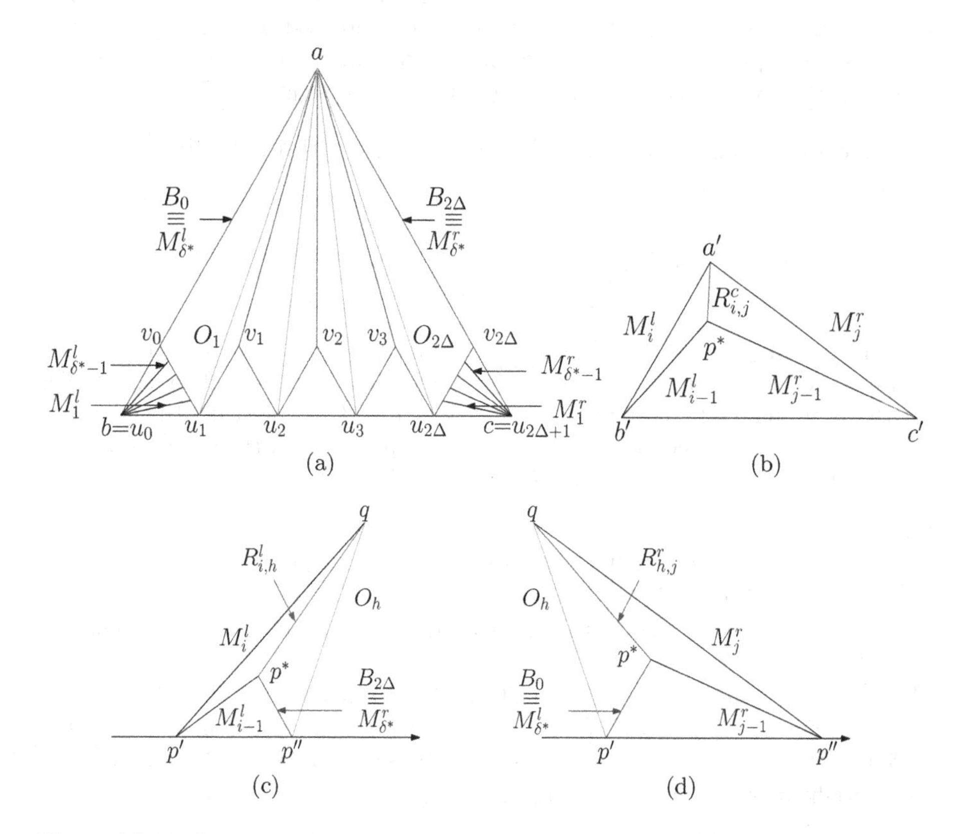

Fig. 4. (a) Black, orange, blue, left- and right-magenta slopes; (b) c-red slope $R_{i,j}^c$; (c) l-red slope $R_{i,h}^l$; and (d) r-red slope $R_{h,j}^r$. (Color figure online)

Red Slopes: Let M_i^l be a left-magenta slope, with $2 \leq i \leq \delta^*$, and let M_j^r be a right-magenta slope, with $2 \leq j \leq \delta^*$. Also, let $1 \leq h \leq 2\Delta$. We have:

- *Central-red slopes:* Let $\blacktriangle(a'b'c')$ be a triangle such that the slope of $\overline{b'c'}$ is the black slope, the slope of $\overline{c'a'}$ is M_j^r, and the slope of $\overline{a'b'}$ is M_i^l. Let p^* be the intersection point between the line with slope M_{i-1}^l passing through b' and the line with slope M_{j-1}^l passing through c'. The *c-red slope* $R_{i,j}^c$ is the slope of the segment $\overline{a'p^*}$; see Fig. 4(b).
- *Left-red slopes:* Let q be a point above the x-axis. Let p' be the intersection point between the line with slope M_i^l passing through q and the x-axis. Also, let p'' be the intersection point between the line with slope O_h passing through q and the x-axis. Further, let p^* be the intersection point between the line with slope M_{i-1}^l passing through p' and the line with slope $B_{2\Delta}$ passing through p''. The *l-red slope* $R_{i,h}^l$ is the slope of the segment $\overline{qp^*}$; see Fig. 4(c).
- *Right-red slopes:* Let q be a point above the x-axis. Let p' be the intersection point between the line with slope O_h passing through q and the x-axis. Also, let p'' be the intersection point between the line with slope M_j^r passing through q and the x-axis. Further, let p^* be the intersection point between the line with slope B_0 passing through p' and the line with slope M_{j-1}^r passing through p''. The *r-red slope* $R_{h,j}^r$ is the slope of the segment $\overline{qp^*}$; see Fig. 4(d).

Let $\mathcal{S}$ be the union of these slope sets together with the black slope. Note that,

$$\begin{aligned}|\mathcal{S}| &= 1+2\Delta+2\Delta+1+2(\delta^*-1)+(\delta^*-1)^2+4\Delta(\delta^*-1)\\ &= \delta^{*2}+4\Delta\delta^*+1\leq 5\Delta^2-1 \qquad (1)\end{aligned}$$

Construction. In what follows we assume that G is rooted at ρ, with leftmost path-vertex ℓ and rightmost path-vertex r and that $\mathcal{T}$ satisfies Property 1. We say that a triangle $\blacktriangle(a_\mu b_\mu c_\mu)$ is *good for* a node μ of $\mathcal{T}$, if it satisfies the following properties. First, the side $\overline{b_\mu c_\mu}$ has the black slope. Second, the slopes s_l and s_r of the sides $\overline{a_\mu b_\mu}$ and $\overline{a_\mu c_\mu}$, respectively, are such that:

G.1 If $s_l = O_i$ and $s_r = O_j$ are orange, then $j = i + 1$.
G.2 If μ is an S- or a Q-node, then s_l is either (i) orange or (ii) a left-magenta slope such that $s_l \geq M_{\delta_\mu(\ell_\mu)}^l$;
G.3 If μ is an S- or a Q-node, then s_r is either (i) orange or (ii) a right-magenta slope such that $s_r \leq M_{\delta_\mu(r_\mu)}^r$;
G.4 If μ is an S-node whose pertinent graph contains neither the edge (ρ_μ, ℓ_μ) nor the edge (ρ_μ, r_μ), then at least one among s_l and s_r is an orange slope;
G.5 If μ is a P-node, s_l is a left-magenta slope such that $s_l \geq M_{\delta_\mu(\ell_\mu)}^l$ and s_r is a right-magenta slope such that $s_r \leq M_{\delta_\mu(r_\mu)}^r$.

Let μ be a node of $\mathcal{T}$ and let $\blacktriangle(a_\mu b_\mu c_\mu)$ be a good triangle for μ. Let s_l and s_r be the slopes of $\overline{a_\mu b_\mu}$ and $\overline{a_\mu c_\mu}$, respectively. We will recursively construct a planar straight-line drawing Γ_μ of G_μ with the following *geometric properties*:

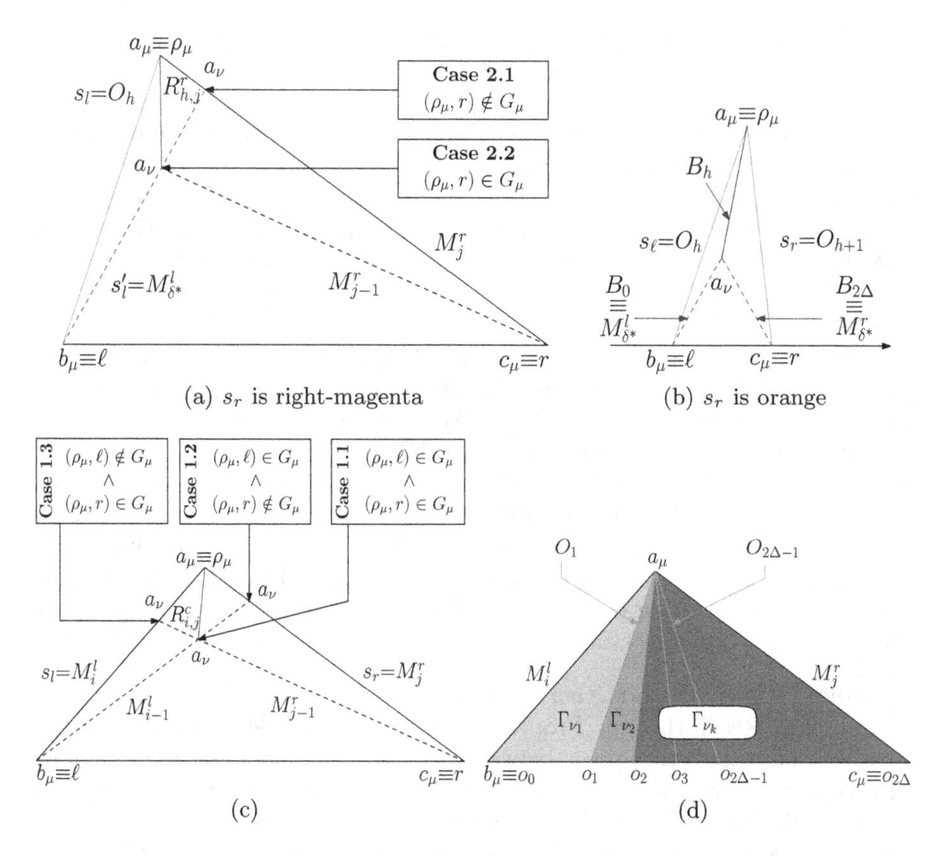

Fig. 5. (a)-(c) Construction of a good triangle for the child of an S-node: (a)-(b) s_l is orange; (c) s_l and s_r are magenta. (d) Construction of good triangles for the children of a P-node with $k = 3$ children. The triangle of each child has a distinct opacity. (Color figure online)

P.1 Γ_μ uses the slopes in $\mathcal{S}$.

P.2 The convex hull of Γ_μ is the given triangle $\blacktriangle(a_\mu b_\mu c_\mu)$, and the vertices ρ_μ, ℓ_μ, and r_μ are mapped to the points a_μ, b_μ, and c_μ, respectively.

P.3 If s_l (resp. s_r) is left-magenta (resp. right-magenta), then the left path (resp. right path) is $\nearrow$-monotone (resp. $\nwarrow$-monotone); if s_l (resp. s_r) is orange, then the left path (resp. right path) is $\nearrow$-monotone (resp. $\nwarrow$-monotone) except, possibly, for the edge incident to ρ_μ.

We remark Property P.3 is not needed to compute a drawing of a cycle-tree, but it will turn out to be fundamental to handle nested pseudotrees.

We describe how to construct Γ_μ in a given good triangle $\blacktriangle(a_\mu b_\mu c_\mu)$ for μ, based on the type of μ. The proof that the construction satisfies Properties P.1, P.2, and P.3 can be found in [7]. When μ is the root of $\mathcal{T}$, the algorithm yields a planar straight-line drawing Γ of G using the slopes in $\mathcal{S}$.

Q-NODES. If μ is a Q-node, we obtain Γ_μ by placing ρ_μ, ℓ_μ, and r_μ at the points a_μ, b_μ, and c_μ, respectively.

S-NODES. If μ is an S-node, then the construction of Γ_μ depends on the degree of ρ_μ in G_μ. Let ν be the unique child of μ. For convenience, we let $\ell = \ell_\mu = \ell_\nu$ and $r = r_\mu = r_\nu$. We first recursively build a drawing Γ_ν of G_ν in a triangle $\blacktriangle(a_\nu b_\nu c_\nu)$ that is good for ν, where a_ν is appropriately placed in the interior of $\blacktriangle(a_\mu b_\mu c_\mu)$ while $b_\nu = b_\mu$ and $c_\nu = c_\mu$. Then, Γ_μ is obtained from Γ_ν by simply placing ρ_μ at a_μ, and by drawing the edges incident to ρ_μ as straight-line segments.

Note that, ρ_μ is adjacent to the root ρ_ν of G_ν, and to either ℓ, or r, or both. In order to define the point a_ν, we now choose the slopes s'_l and s'_r of the segments $\overline{a_\nu b_\nu}$ and $\overline{a_\nu c_\nu}$, respectively, as follows. We start with s'_l. Since μ is an S-node, s_l is either orange or a left-magenta slope M^l_i. If s_l is orange, then $s'_l = M^l_{\delta*}$. See Fig. 5(a) and Fig. 5(b). If $s_l = M^l_i$ and (ρ_μ, ℓ) belongs to G_μ, we have that $s'_l = M^l_{i-1}$. Notice that, by Property G.2 $i \geq \delta_\mu(\ell)$, and since ℓ is incident at least to (ρ_μ, ℓ) and to an edge of the path of G, we have $i \geq 2$. If $s_l = M^l_i$ and (ρ_μ, ℓ) does not belong to G_μ, we have that $s'_l = s_l = M^l_i$. See Fig. 5(c). The choice of s'_r is symmetric, based on the existence of (ρ_μ, r). Notice that, by Property G.4, if both s_l and s_r are magenta, then one between (ρ_μ, ℓ) and (ρ_μ, r) exists.

P-NODES. If μ is a P-node, then let $\nu_1, \nu_2, \ldots, \nu_k$, with $2 \leq k \leq 2\Delta + 1$ be the children of μ. Recall that, by Property 1, no ν_i is a P-node. Refer to Fig. 5(d). Let o_i be the intersection point between $\overline{b_\mu c_\mu}$ and the line passing through a_μ with slope O_i, for $i = 1, \ldots, 2\Delta - 1$. For convenience, we let $o_0 = b_\mu$ and $o_{2\Delta} = c_\mu$. We recursively build a drawing Γ_{ν_i} of G_{ν_i}, with $i = 1, \ldots, k-1$, in the triangle $\blacktriangle(a_\mu o_{i-1} o_i)$, which is good for ν_i, and a drawing Γ_{ν_k} of G_{ν_k} in the triangle $\blacktriangle(a_\mu o_{k-1} o_{2\Delta})$, which is good for ν_k. Γ_μ is the union of the Γ_{ν_i}'s.

Lemma 3. *For any almost-3-connected path-tree G and any triangle $\blacktriangle(abc)$ that is good for the root of an SPQ-tree of G, the graph G admits a planar straight-line drawing inside $\blacktriangle(abc)$ that satisfies Properties* P.1, P.2, *and* P.3.

3-Connected Cycle-Trees. Let G be a degree-Δ 3-connected cycle-tree. We show how to exploit Lemma 3 to draw G using $O(|\mathcal{S}|) = O(\Delta^2)$ slopes. Similarly to path-trees, we call *cycle-vertices* the vertices on the outer boundary of G and *tree-vertices* the remaining vertices of G. Let ℓ, v, and r be three cycle-vertices that appear in this clockwise order along the outer face of G; refer to Fig. 6. Remove v and its incident edges from G. Denote by G^- the resulting topological graph. Let π be the graph formed by the edges that belong to the outer face of G^- and do not belong to the outer face of G.

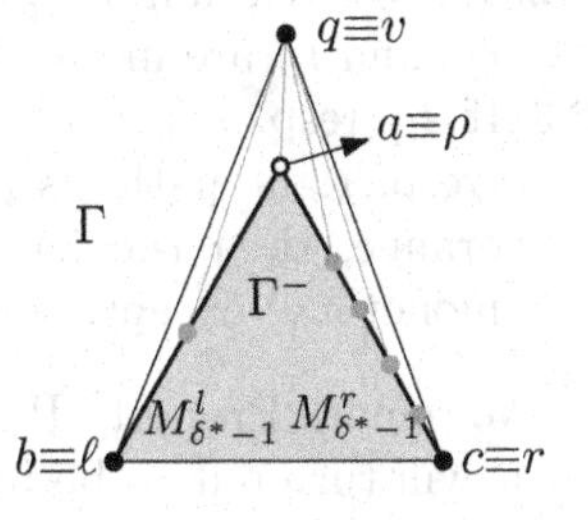

Fig. 6. How to draw a 3-connected cycle-tree.

Since G is 3-connected, we have that G^- is at least 2-connected and that π is a path connecting ℓ and r that contains at least one tree-vertex different from v. Let ρ be *any* such vertex encountered when traversing π from ℓ to r. Moreover, the only degree-2 vertices of G^-, if any, belong to π. Let G^* be the graph obtained from G^- by replacing each degree-2 vertex of π different from ℓ, ρ, and r, if any, with an edge connecting its endpoints. We have that G^* is an almost-3-connected path-tree rooted at ρ, with leftmost path-vertex ℓ and rightmost path-vertex r.

Lemma 4. *Every degree-Δ 3-connected cycle-tree G has* $\text{psn}(G) \in O(|\mathcal{S}|)$.

Proof. If the outer boundary of G has 3 vertices, the total number of edges of G is $O(\Delta)$ and hence $\text{psn}(G) \in O(\Delta) \subseteq O(|\mathcal{S}|)$. So assume that the outer boundary of G has more than 3 vertices.

Let $\mathcal{T}$ be the SPQ-tree of G^* and let $\blacktriangle(abc)$ be an equilateral triangle. Note that an equilateral triangle is good for the root of $\mathcal{T}$, regardless of its type. Let Γ^* be the planar straight-line drawing of G^* inside $\blacktriangle(abc)$, obtained by applying Lemma 3. We prove that there exists a planar straight-line drawing Γ of G such that $\text{psn}(\Gamma) \leq \text{psn}(\Gamma^*) + \Delta$, which implies the statement because $\text{psn}(\Gamma^*) \in O(|\mathcal{S}|) = O(\Delta^2)$ by Lemma 3. Note that, the slopes s_ℓ and s_r of $\overline{ab}$ and $\overline{ac}$ are the largest l-magenta slope $M^l_{\delta^*}$ and the smallest r-magenta slope $M^r_{\delta^*}$, respectively. Moreover, since the drawing Γ^* inside $\blacktriangle(abc)$ has been obtained by applying Lemma 3, we have that Γ^* satisfies Property P.3. We construct a planar straight-line drawing Γ of G as follows; refer to Fig. 6. First, we obtain a planar straight-line drawing Γ^- of G^- from Γ^*, by subdividing the edges that stemmed from the contraction operations (which yielded G^* from G^-). Clearly, $\text{psn}(\Gamma^-) = \text{psn}(\Gamma^*)$. Γ^- exhibits the following useful property: By Property P.3 of Γ^*, we have that the subpath of π from ℓ to ρ is $\nearrow$-monotone and that the subpath of π from r to ρ is $\nwarrow$-monotone. Second, we select a point q vertically above ρ such that all the straight-line segments connecting q to each of the vertices of π do not cross Γ^-. The existence of such a point is guaranteed by the above property. Finally, we obtain Γ from Γ^- by placing v at point q, and by drawing its incident edges as straight-line segments. Since v has at most degree Δ, we have that $\text{psn}(\Gamma) \leq \text{psn}(\Gamma^-) + \Delta$.

Proof of Theorem 1. Halin graphs are 3-connected cycle-trees with $\delta^*=3$ because each path-vertex has two incident edges that are incident to the outer face and it is a leaf when these two edges are removed. From Sect. 1 we have $|\mathcal{S}| = 12\Delta + 10$. Thus Lemma 4 implies Theorem 1. □

3.2 2-Connected and 1-Connected Instances

We can extend the result of Lemma 4 to 2- and 1-connected cycle-trees. To this aim we define the concept of (w, z)-flag and of c-flag of G.

Let c be a cut-vertex of G. By removing c from G, we obtain $k \geq 2$ connected subgraphs $H_1, H_2, \ldots, H_k$. The subgraph C_i of G induced by $V(H_i) \cup \{c\}$ is a component of G with respect to c ($1 \leq i \leq k$). One of such components, say C_1, contains all the cycle-vertices of G. The union of all components different from C_1 is called the *c-flag* of G. We say that a cut-vertex c' is *dominated* by c if c' belongs to the c-flag of G. A cut-vertex is *dominant* when it is not dominated by another cut-vertex. Let $\{w, z\}$ be a 2-cut of G, where w is a tree-vertex and z is a cycle-vertex. By removing w and z from G, we obtain $k \geq 2$ connected subgraphs $H_1, H_2, \ldots, H_k$. The subgraph C_i of G induced by $V(H_i) \cup \{w, z\}$ is a component of G with respect to $\{w, z\}$ ($1 \leq i \leq k$). One of such components, say C_1, contains all the cycle-vertices of G. The union of all components different from C_1 is called the *(w, z)-flag* of G. We say that a 2-cut $\{w', z\}$, with $w' \neq w$, is *dominated* by $\{w, z\}$ if w' belongs to the (w, z)-flag of G. A 2-cut $\{w, z\}$ is *dominant* when no other 2-cut dominates it.

Let G_1 be the graph obtained from G by removing, for each dominant cut-vertex c, all vertices of the c-flag of G except c. We call G_1 the *1-frame graph* of G. Let G_2 be the graph obtained from G_1 by removing, for each dominant 2-cut $\{w, z\}$, all vertices of the (w, z)-flag of G_1 except w and z, and by adding the edge (w, z), called the *virtual edge* of $\{w, z\}$, if it does not already exist in G. We call G_2 the *2-frame graph* of G.

We draw the 2-frame by means of Lemma 4 and then we add back the (w, z)-flags and the c-flag. Notice that, removing the vertex z from a (w, z)-flag yields a tree, and therefore each (w, z)-flag is a partial 2-tree. It follows that the (w, z)-flags and the c-flags (which are trees) can be drawn by exploiting Theorem 4 (see [7] for details).

Theorem 5. *Every degree-Δ cycle-tree has planar slope number $O(\Delta^2)$.*

4 Nested Pseudotrees

To prove Theorem 2, we first consider nested-pseudotrees whose external boundary is a chordless cycle. We call such graphs *cycle-pseudotrees* (see Fig. 1). The omitted details of this section can be found in [7].

Cycle-Pseudotrees. Let H be a degree-Δ cycle-pseudotree with pseudotree P and let $e = (u, v)$ be a given edge of the cycle of P, which we call the *reference edge*. Let G be the graph obtained by removing e from H. Clearly, G is a cycle-tree, and u and v are tree-vertices of G. Let G_1 be the 1-frame of G and let G_2 be the 2-frame of G_1. A vertex x *belongs* to the c-flag of G (for some cut-vertex c) if x is a tree-vertex of the flag different from c; analogously x belongs to the (w, z)-flag of G (for some 2-cut $\{w, z\}$), if x is a tree-vertex of the flag and $x \neq w$.

We distinguish cases based on the endpoints of e. For each case, we will show how to obtain a planar straight-line drawing Γ of H using $O(\Delta^2)$ slopes. All the flags considered below arise from dominant cut-vertices and dominant 2-cuts.

Case A. There exists a cut-vertex c of G such that u belongs to the c-flag. We distinguish four subcases: **A.1** v belongs to the c-flag, **A.2** v belongs to the c'-flag of some cut-vertex $c' \neq c$ of G, **A.3** v belongs to the (w, z)-flag of a 2-cut $\{w, z\}$ of G_1, or **A.4** v belongs to G_2.

If Case A does not apply, then we may assume that neither u nor v belong to a c-flag of any cut-vertex c of G. In particular, both u and v belong to G_1.

Case B. There exists a 2-cut $\{w, z\}$ of G_1 such that u belongs to the (w, z)-flag. We distinguish three subcases: **B.1** v belongs to the (w, z)-flag, **B.2** v belongs to the (w', z')-flag of some 2-cut $\{w', z'\} \neq \{w, z\}$ of G_1, **B.3** v belongs to G_2.

Case C. If Case A and Case B do not apply, then both u and v belong to G_2.

We obtain Γ recursively. Each of the cases yields either a smaller instance to which a different case applies or it is a base case (i.e., **A.1**, **B.1**, and **C**) in which we use Theorem 5 to obtain a planar straight-line drawing using $O(\Delta^2)$ slopes. Crucially in all cases the depth of the recursion is constant and each recursive call increases the number of slopes by $O(\Delta)$. This leads to the following.

Theorem 6. *Every degree-Δ cycle-pseudotree has planar slope number $O(\Delta^2)$.*

Proof of Theorem 2. A nested pseudotree G is a cycle-pseudotree together with a (possibly empty) set of partial two-trees hanging from 2-cuts formed by edges of the chordless cycle C containing the psuedotree or from cut-vertices of C. We first draw the cycle-psudeotree via Theorem 6 and then attach the hanging partial 2-trees as drawn with Theorem 4. This results in the use of $O(\Delta^2)$ slopes.

5 Conclusions and Open Problems

In this paper we proved a quadratic upper bound on the planar slope number of nested pseudotrees. In the special case of Halin graphs we have an asymptotically tight $O(\Delta)$ bound. Our proofs are constructive and yield polynomial-time algorithms. Still the coordinates of the vertices may use a super-linear number of bits. It remains open whether the same upper bounds on the slope number can be achieved if the vertices must lie on an integer grid of polynomial size. Also it would be interesting to establish whether the upper bound of Theorem 2 is tight and whether it also applies to nested pseudoforests. Finally, is there a subexponential upper bound on the planar slope number of 2-outerplanar graphs? This question is interesting even for 2-connected graphs.

References

1. Angelini, P., et al.: Small universal point sets for k-outerplanar graphs. Discrete Comput. Geometry **60**(2), 430–470 (2018). https://doi.org/10.1007/s00454-018-0009-x
2. Angelini, P., Di Battista, G., Kaufmann, M., Mchedlidze, T., Roselli, V., Squarcella, C.: Small point sets for simply-nested planar graphs. In: van Kreveld, M., Speckmann, B. (eds.) GD 2011. LNCS, vol. 7034, pp. 75–85. Springer, Heidelberg (2011). https://doi.org/10.1007/978-3-642-25878-7_8
3. Barát, J., Matousek, J., Wood, D.R.: Bounded-degree graphs have arbitrarily large geometric thickness. Electr. J. Comb. **13**(1), 1–14 (2006)

4. Bekos, M.A., Bruckdorfer, T., Kaufmann, M., Raftopoulou, C.N.: The book thickness of 1-planar graphs is constant. Algorithmica **79**(2), 444–465 (2016). https://doi.org/10.1007/s00453-016-0203-2
5. Bekos, M.A., Da Lozzo, G., Griesbach, S., Gronemann, M., Montecchiani, F., Raftopoulou, C.N.: Book embeddings of nonplanar graphs with small faces in few pages. In: Cabello, S., Chen, D.Z. (eds.) 36th International Symposium on Computational Geometry, SoCG 2020. LIPIcs, vol. 164, pp. 16:1–16:17. Schloss Dagstuhl - Leibniz-Zentrum für Informatik (2020). https://doi.org/10.4230/LIPIcs.SoCG.2020.16
6. Chang, H., Lu, H.: Computing the girth of a planar graph in linear time. SIAM J. Comput. **42**(3), 1077–1094 (2013). https://doi.org/10.1137/110832033
7. Chaplick, S., Da Lozzo, G., Di Giacomo, E., Liotta, G., Montecchiani, F.: Planar drawings with few slopes of halin graphs and nested pseudotrees. CoRR abs/2105.08124 (2021). https://arxiv.org/abs/2105.08124
8. Cimikowski, R.J.: Finding hamiltonian cycles in certain planar graphs. Inf. Process. Lett. **35**(5), 249–254 (1990). https://doi.org/10.1016/0020-0190(90)90053-Z
9. Da Lozzo, G., Devanny, W.E., Eppstein, D., Johnson, T.: Square-contact representations of partial 2-trees and triconnected simply-nested graphs. In: Okamoto, Y., Tokuyama, T. (eds.) ISAAC 2017. LIPIcs, vol. 92, pp. 24:1–24:14. Schloss Dagstuhl - Leibniz-Zentrum für Informatik (2017). https://doi.org/10.4230/LIPIcs.ISAAC.2017.24
10. Da Lozzo, G., Eppstein, D., Goodrich, M.T., Gupta, S.: Subexponential-time and FPT algorithms for embedded flat clustered planarity. In: Brandstädt, A., Köhler, E., Meer, K. (eds.) WG 2018. LNCS, vol. 11159, pp. 111–124. Springer, Cham (2018). https://doi.org/10.1007/978-3-030-00256-5_10
11. Di Giacomo, E., Didimo, W., Liotta, G., Meijer, H.: Computing radial drawings on the minimum number of circles. J. Graph Algorithms Appl. **9**(3), 365–389 (2005). https://doi.org/10.7155/jgaa.00114
12. Di Giacomo, E., Liotta, G., Montecchiani, F.: Drawing outer 1-planar graphs with few slopes. J. Graph Algorithms Appl. **19**(2), 707–741 (2015)
13. Di Giacomo, E., Liotta, G., Montecchiani, F.: Drawing subcubic planar graphs with four slopes and optimal angular resolution. Theor. Comput. Sci. **714**, 51–73 (2018). https://doi.org/10.1016/j.tcs.2017.12.004
14. Dujmović, V., Eppstein, D., Suderman, M., Wood, D.R.: Drawings of planar graphs with few slopes and segments. Comput. Geom. **38**(3), 194–212 (2007). https://doi.org/10.1016/j.comgeo.2006.09.002
15. Dujmovic, V., Frati, F.: Stack and queue layouts via layered separators. J. Graph Algorithms Appl. **22**(1), 89–99 (2018). https://doi.org/10.7155/jgaa.00454
16. Halin, R.: Studies on minimally n-connected graphs. In: Combinatorial Mathematics and its Applications (Proc. Conf., Oxford, 1969), pp. 129–136. Academic Press, London (1971)
17. Jelínek, V., Jelínková, E., Kratochvíl, J., Lidický, B., Tesar, M., Vyskocil, T.: The planar slope number of planar partial 3-trees of bounded degree. Graphs Comb. **29**(4), 981–1005 (2013)
18. Keszegh, B., Pach, J., Pálvölgyi, D.: Drawing planar graphs of bounded degree with few slopes. SIAM J. Discrete Math. **27**(2), 1171–1183 (2013)
19. Knauer, K.B., Micek, P., Walczak, B.: Outerplanar graph drawings with few slopes. Comput. Geom. **47**(5), 614–624 (2014)
20. Lenhart, W., Liotta, G., Mondal, D., Nishat, R.I.: Planar and plane slope number of partial 2-trees. In: Wismath, S., Wolff, A. (eds.) GD 2013. LNCS, vol. 8242, pp. 412–423. Springer, Cham (2013). https://doi.org/10.1007/978-3-319-03841-4_36

21. Pach, J., Pálvölgyi, D.: Bounded-degree graphs can have arbitrarily large slope numbers. Electr. J. Comb. **13**(1) (2006)
22. Wade, G.A., Chu, J.H.: Drawability of complete graphs using a minimal slope set. Comput. J. **37**(2), 139–142 (1994)
23. Yannakakis, M.: Embedding planar graphs in four pages. J. Comput. Syst. Sci. **38**(1), 36–67 (1989). https://doi.org/10.1016/0022-0000(89)90032-9

An APTAS for Bin Packing with Clique-Graph Conflicts

Ilan Doron-Arad, Ariel Kulik, and Hadas Shachnai(✉)

Computer Science Department, Technion, 3200003 Haifa, Israel
{idoron-arad,kulik,hadas}@cs.technion.ac.il

Abstract. We study the following variant of the classic *bin packing* problem. Given a set of items of various sizes, partitioned into groups, find a packing of the items in a minimum number of identical (unit-size) bins, such that no two items of the same group are assigned to the same bin. This problem, known as *bin packing with clique-graph conflicts*, has natural applications in storing file replicas, security in cloud computing and signal distribution.

Our main result is an *asymptotic polynomial time approximation scheme (APTAS)* for the problem, improving upon the best known ratio of 2. As a key tool, we apply a novel *Shift & Swap* technique which generalizes the classic linear shifting technique to scenarios allowing conflicts between items. The major challenge of packing *small* items using only a small number of extra bins is tackled through an intricate combination of enumeration and a greedy-based approach that utilizes the rounded solution of a *linear program*.

1 Introduction

In the classic *bin packing (BP)* problem, we seek a packing of items of various sizes into a minimum number of unit-size bins. This fundamental problem arises in a wide variety of contexts and has been studied extensively since the early 1970's. In some common scenarios, the input is partitioned into *disjoint groups*, such that items in the same group are *conflicting* and therefore cannot be packed together. For example, television and radio stations often assign a set of programs to their channels. Each program falls into a genre such as comedy, documentary or sports on TV, or various musical genres on radio. To maintain a diverse daily schedule of programs, the station would like to avoid broadcasting two programs of the same genre in one channel. Thus, we have a set of items (programs) partitioned into groups (genres) that need to be packed into a set of bins (channels), such that items belonging to the same group cannot be packed together.

We consider this natural variant of the classic bin packing problem that we call *group bin packing (GBP)*. Formally, the input is a set of N items $I = \{1, \ldots, N\}$ with corresponding sizes $s_1, ..., s_N \in (0, 1]$, partitioned into n disjoint groups $G_1, ..., G_n$, i.e., $I = G_1 \cup G_2 \cup \ldots \cup G_n$. The items need to be packed in

A. Lubiw et al. (Eds.): WADS 2021, LNCS 12808, pp. 286–299, 2021.
https://doi.org/10.1007/978-3-030-83508-8_21

unit-size bins. A packing is *feasible* if the total size of items in each bin does not exceed the bin capacity, and no two items from the same group are packed in the same bin. We seek a feasible packing of all items in a minimum number of unit-size bins. We give in [6] some natural applications of GBP.

Group bin packing can be viewed as a special case of *bin packing with conflicts (BPC)*, in which the input is a set of items I, each having size in $(0, 1]$, along with a conflict graph $G = (V, E)$. An item $i \in I$ is represented by a vertex $i \in V$, and there is an edge $(i, j) \in E$ if items i and j cannot be packed in the same bin. The goal is to pack the items in a minimum number of unit-size bins such that items assigned to each bin form an *independent set* in G.

Indeed, GBP is the special case where the conflict graph is a union of cliques. Thus, GBP is also known as *bin packing with clique-graph conflicts* (see Sect. 1.2).

1.1 Contribution and Techniques

Our main result (in Sect. 3) is an APTAS for the group bin packing problem, improving upon the best known ratio of 2 [1].[1]

Existing algorithms for BPC often rely on initial *coloring* of the instance. This enables to apply in later steps known techniques for bin packing, considering each color class (i.e., a subset of non-conflicting items) separately. In contrast, our approach uses a refined packing of the original instance while eliminating conflicts, thus generalizing techniques for classic BP.

Our first technical contribution is an enhancement of the *linear shifting* technique of [8]. This enables our scheme to enumerate in polynomial time over packings of relatively large items, while guaranteeing that these packings respect the group constraints. Our *Shift & Swap* technique considers the set of large items that are associated with different groups (satisfying certain properties) as a classic BP instance, i.e., the group constraints are initially relaxed. Then the scheme applies to these items the linear shifting technique of [8]. In the process, items of the same group may be packed in the same bin. Our *Swapping* algorithm resolves all conflicts, with no increase in the total number of bins used (see Sects. 3.1 and 3.2).

A common approach used for deriving APTASs for BP is to pack in a bounded number of extra bins a set of discarded small items of total size $O(\varepsilon)OPT$, where $OPT = OPT(I)$ is the minimum number of bins required for packing the given instance I, and $\varepsilon \in (0, 1)$ is the accuracy parameter of the scheme. As shown in [6], this approach may fail for GBP, e.g., when the discarded items belong to the same group. Our second contribution is an algorithm that overcomes this hurdle. The crux is to find a set of small items of total size $O(\varepsilon)OPT$ containing $O(\varepsilon)OPT$ items from each group. This would enable to pack these items in a small number of extra bins. Furthermore, the remaining small items should be

[1] We note that 2 is the best known absolute as well as *asymptotic* approximation ratio for the problem (see Sect. 1.2). We give formal definitions of absolute/asymptotic ratios in Sect. 2.

feasibly assigned to partially packed OPT bins. Our algorithm identifies such sets of small items through an intricate combination of enumeration and a greedy-based approach that utilizes the rounded solution of a *linear program.*

1.2 Related Work

The classic bin packing problem is known to be NP-hard. Furthermore, it cannot be approximated within a ratio better than $\frac{3}{2}$, unless P = NP. This ratio is achieved by the simple First-Fit Decreasing algorithm [21]. The paper [8] presents an APTAS for bin packing, which uses at most $(1+\varepsilon)OPT+1$ bins, for any fixed $\varepsilon \in (0, 1/2)$. The paper [16] gives an approximation algorithm that uses at most $OPT + O(\log^2(OPT))$ bins. The additive factor was improved in [20] to $O(\log OPT \cdot \log\log OPT)$. For comprehensive surveys of known results for BP see, e.g., [3,4].

The problem of *bin packing with conflicts (BPC)* was introduced in [15]. As BPC includes as a special case the classic *graph coloring* problem, it cannot be approximated within factor $N^{1-\varepsilon}$ for an input of N items, for all $\varepsilon > 0$, unless $P = NP$ [22]. Thus, most of the research work focused on obtaining approximation algorithms for BPC on classes of conflict graphs that can be optimally colored in polynomial time. Epstein and Levin [7] presented sophisticated algorithms for two such classes, namely, a $\frac{5}{2}$-approximation for BPC with a *perfect* conflict graph,[2] and $\frac{7}{4}$-approximation for a *bipartite* conflict graph.

The hardness of approximation of GBP (with respect to absolute approximation ratio) follows from the hardness of BP, which is the special case of GBP where the conflict graph is an independent set. A 2.7-approximation algorithm for general instances follows from a result of [15]. Oh and Son [19] showed that a simple algorithm based on First-Fit outputs a packing of any GBP instance I in $1.7OPT + 2.19v_{max}$ bins, where $v_{max} = \max_{1\leq j\leq n} |G_j|$. The paper [18] shows that some special cases of the problem are solvable in polynomial time. The best known ratio for GBP is 2 due to [1].

Jansen [12] presented an *asymptotic fully polynomial time approximation scheme (AFPTAS)* for BPC on d-inductive conflict graphs,[3] where $d \geq 1$ is some constant. The scheme of [12] uses for packing a given instance I at most $(1 + \varepsilon)OPT + O(d/\varepsilon^2)$ bins. This implies that GBP admits an AFPTAS on instances where the maximum clique size is some constant d. Thus, the existence of an asymptotic approximation scheme for general instances remained open.

Das and Wiese [5] introduced the problem of makespan minimization with bag constraints. In this generalization of the classic makespan minimization problem, each job belongs to a *bag*. The goal is to schedule the jobs on a set of m identical machines, for some $m \geq 1$, such that no two jobs in the same bag are assigned to the same machine, and the makespan is minimized. For the classic

[2] For the subclass of interval graphs the paper [7] gives a $\frac{7}{3}$-approximation algorithm.
[3] A graph G is *d-inductive* if the vertices of G can be numbered such that each vertex is connected by an edge to at most d lower numbered vertices.

problem of makespan minimization with no bag constraints, there are known PTAS [11,17] as well as EPTAS [2,10,13,14]. Das and Wiese [5] developed a PTAS for the problem with bag constraints. Later, Grage et al. [9] obtained an EPTAS.

Due to space constraints, some of our results and formal proofs are given in the full version of the paper [6].

2 Preliminaries: Scheduling with Bag Constraints

Our scheme is inspired by the elaborate framework of Das and Wiese [5] for makespan minimization with bag constraints. For completeness, we give below an overview of the scheme of [5]. Given a set of jobs I partitioned into bags and m identical machines, let $p_\ell > 0$ be the processing time of job $\ell \in I$. The instance is scaled such that the optimal makespan is 1. The jobs and bags are then classified using the next lemma.

Lemma 2.1. *For any instance I and $\varepsilon \in (0,1)$, there is an integer $k \in \{1, ..., \lceil \frac{1}{\varepsilon^2} \rceil\}$ such that $\sum_{\ell \in I:\, p_\ell \in [\varepsilon^{k+1}, \varepsilon^k)} p_\ell \leq \varepsilon^2 m$.*

A job ℓ is *small* if $p_\ell < \varepsilon^{k+1}$, *medium* if $p_\ell \in [\varepsilon^{k+1}, \varepsilon^k)$ and *large* if $p_\ell \geq \varepsilon^k$, where k is the value found in Lemma 2.1. A bag is *large* if the number of large and medium jobs it contains is at least εm, and *small* otherwise.

The scheme of [5] initially enumerates over *slot patterns* for packing large and medium jobs from large bags optimally in polynomial time. The enumeration is enhanced by using dynamic programming and a flow network to schedule also the large jobs from small bags. The medium jobs in each small bag are scheduled across the m machines almost evenly, causing only small increase to the makespan. The small jobs are partitioned among *machine groups* with the same processing time and containing jobs from the same subset of large bags. Then, a greedy approach is used with respect to the bags to schedule the jobs within each machine group, such that the overall makespan is at most $1 + O(\varepsilon)$.

Our scheme classifies the items and groups similar to the classification of jobs and bags in [5]. We then apply enumeration over patterns to pack the large and medium items. Thus, Lemmas 3.2, 3.6 and 3.8 in this paper are adaptations of results obtained in [5]. However, the remaining components of our scheme are different. One crucial difference is our use of a Shift & Swap technique to round the sizes of large and medium items. Indeed, rounding the item sizes using the approach of [5] may cause overflow in the bins, requiring a large number of extra bins to accommodate the excess items. Furthermore, packing the *small* items using $O(\varepsilon)OPT$ extra bins requires new ideas (see Sect. 3).

We use standard definitions of approximation ratio and *asymptotic* approximation ratio. Given a minimization problem Π, let $\mathcal{A}$ be a polynomial-time algorithm for Π. For an instance I of Π, denote by $OPT(I)$ and $\mathcal{A}(I)$ the values of an optimal solution and the solution returned by $\mathcal{A}$ for I, respectively. We say that $\mathcal{A}$ is a ρ-approximation algorithm for Π, for some $\rho \geq 1$, if $\mathcal{A}(I) \leq \rho \cdot OPT(I)$ for any instance I of Π. $\mathcal{A}$ is an *asymptotic* ρ-approximation for Π if there is a constant $c \in \mathbb{R}$ such that $\mathcal{A}(I) \leq \rho \cdot OPT(I) + c$ for any instance I of Π. An APTAS for Π is a family of algorithms $(A_\varepsilon)_{\varepsilon>0}$ such that A_ε is a polynomial-time asymptotic $(1+\varepsilon)$-approximation for each $\varepsilon > 0$. When clear from the context, we use $OPT = OPT(I)$.

3 An APTAS for GBP

In this section we present an APTAS for GBP. Let OPT be the optimal number of bins for an instance I. Our scheme uses as a subroutine a BalancedColoring algorithm proposed in [1] for the *group packing* problem (see the details in [6]). Let $S(I)$ be the total size of items in I, i.e., $S(I) = \sum_{\ell \in [N]} s_\ell$. Recall that v_{max} is the maximum cardinality of any group. The next lemma follows from a result of [1].

Lemma 3.1. *Let I be an instance of GBP. Then BalancedColoring packs I in at most* $\max\{2S(I), S(I) + v_{max}\}$ *bins.*

By the above, given an instance I of GBP, we can guess OPT in polynomial time, by iterating over all integer values in $[1, \max\{2S(I), S(I)+v_{max}]$ and taking the minimal number of bins for which a feasible solution exists.

Similar to Lemma 2.1, we can find a value of k, $1 \leq k \leq \lceil \frac{1}{\varepsilon^2} \rceil$, satisfying $\sum_{\ell \in I:\, s_\ell \in [\varepsilon^{k+1}, \varepsilon^k)} s_\ell \leq \varepsilon^2 \cdot OPT$. Now, we classify item ℓ as *small* if $s_\ell < \varepsilon^{k+1}$, *medium* if $s_\ell \in [\varepsilon^{k+1}, \varepsilon^k)$ and *large* otherwise. A group is *large* if the number of large and medium items of that group is at least $\varepsilon^{k+2} \cdot OPT$, and *small* otherwise. Given an instance I of GBP and a constant $\varepsilon \in (0,1)$, we also assume that $OPT > \frac{3}{\varepsilon^{k+2}}$ (otherwise, the conflict graph is d-inductive, where d is a constant, and the problem admits an AFPTAS [12]).

Lemma 3.2. *There are at most $\frac{1}{\varepsilon^{2k+3}}$ large groups.*

3.1 Rounding of Large and Medium Items

We start by reducing the number of distinct sizes for the large and medium items. Recall that in the linear shifting technique we are given a BP instance of N items and a parameter $Q \in (0, N]$. The items are sorted in non-increasing order by sizes and then partitioned into classes. Each class (except maybe the last one) contains $\max\{Q, 1\}$ items. The items in class 1 (i.e., largest items) are discarded (the discarded items are handled in a later stage of the algorithm). The sizes of items in each class are then rounded up to the maximum size of an item in this class. For more details see, e.g., [8].

We apply linear shifting to the large and medium items in each large group with parameter $Q = \lfloor \varepsilon^{2k+4} \cdot OPT \rfloor$. Let I, I' be the instance before and after the shifting over large groups, respectively.

Lemma 3.3. $OPT(I') \leq OPT(I)$.

Lemma 3.4. *Given a feasible packing of I' in OPT bins, we can find a feasible packing of I in $(1+O(\varepsilon))OPT$ bins.*

Next, we round the sizes of large items in small groups. As the number of these groups may be large, we use the following *Shift & Swap* technique. We merge all of the large items in small groups into a single group, to which we apply linear shifting with parameter $Q = \lfloor 2\varepsilon \cdot OPT \rfloor$. In addition to items in class 1, which are discarded due to linear shifting, we also discard the items in the last size class; these items are packed in a new set of bins (see the proof of Lemma 3.15 in [6]).

Lemma 3.5. *After rounding, there are at most $O(1)$ distinct sizes of large and medium items from large groups, and large items from small groups.*

Relaxing the *feasibility* requirement for the packing of rounded large items from small groups, the statements of Lemma 3.3 and Lemma 3.4 hold for these items as well. To obtain a feasible packing of these items, we apply a Swapping subroutine which resolves the possible conflicts caused while packing the items.

Our scheme packs in each step a subset of items, using OPT bins, while discarding some items. The discarded items are packed later in a set of $O(\varepsilon) \cdot OPT + 1$ extra bins. In Sect. 3.2 we pack the large and medium items using enumeration over patterns followed by our Swapping algorithm to resolve conflicts. Section 3.3 presents an algorithm for packing the small items by combining recursive enumeration (for relatively "large" items) with a greedy-based algorithm that utilizes the rounded solution of a linear program (for relatively "small" items). In Sect. 3.4 we show that the components of our scheme combine together to an APTAS for GBP.

3.2 Large and Medium Items

The large items and medium items from large groups are packed in the bins using *slot patterns*. Let $G_{i_1}, \ldots, G_{i_L}$ be the large groups, and let 'u' be a label representing all the small groups. Given the modified instance I', a slot is a pair (s_ℓ, j), where s_ℓ is the rounded size of a large or medium item $\ell \in I'$ and $j \in \{i_1, \ldots, i_L\} \cup \{u\}$. A *pattern* is a multiset $\{t_1, \ldots, t_\beta\}$ for some $1 \leq \beta \leq \lfloor \frac{1}{\varepsilon^{k+1}} \rfloor$, where t_i is a slot for each $i \in [\beta]$.[4]

Lemma 3.6. *By using enumeration over patterns, we find a pattern for each bin for the large and medium items, such that these patterns correspond to an optimal solution. The running time is $O(N^{O(1)})$.*

Given slot patterns corresponding to an optimal solution, large and medium items from large groups can be packed optimally, since they are identified both by a label and a size. On the other hand, large items from small groups are

[4] Recall that the number of medium/large items that fit in a single bin is at most $\lfloor \frac{1}{\varepsilon^{k+1}} \rfloor$.

identified solely by their sizes. A greedy packing of these items, relating only to their corresponding patterns, may result in conflicts (i.e., two large items of the same small group are packed in the same bin). Therefore, we incorporate a process of swapping items of the same (rounded) size between their hosting bins, until there are no conflicts.

Given an item ℓ that conflicts with another item in bin b, for an item y in bin c such that $s_\ell = s_y$, *swap*(ℓ, y) is *bad* if it causes a conflict (either because y conflicts with an item in bin b, ℓ conflicts with an item in bin c, or $c = b$); otherwise, *swap*(ℓ, y) is *good*. We now describe our algorithm for packing the large items from small groups.

Let ζ be the given slot patterns for OPT bins. Initially, the items are packed by these patterns, where items from small groups are packed ignoring the group constraints. This can be done simply by placing an arbitrary item of size s from some small group in each slot (s, u). If ζ corresponds to an optimal solution, we meet the capacity constraint of each bin. However, this may result with conflicting items in some bins. Suppose there is a conflict in bin b. Then for one of the conflicting items, ℓ, we find a good *swap*(ℓ, y) with item y in a different bin, such that $s_y = s_\ell$. We repeat this process until there are no conflicts. We give the pseudocode of Swapping in Algorithm 1.

Algorithm 1. *Swapping*($\zeta, G_1, \ldots, G_n$)

1: Pack the large and medium items from large groups in slots corresponding to their sizes and by labels.
2: Pack large items from small groups in slots corresponding to their sizes.
3: **while** there is an item ℓ involved in a conflict **do**
4: Find a good *swap*(ℓ, y) and resolve the conflict.

Theorem 3.7. *Given a packing of large and medium items by slot patterns corresponding to an optimal solution, Algorithm 1 resolves all conflicts in polynomial time.*

We use the Swapping algorithm for each possible guess of patterns to obtain a feasible packing of the large items and medium items from large groups in OPT bins.

Now, we discard the medium items from small groups and pack them later in a new set of bins with other discarded items. This requires only a small number of extra bins (see the proof of Lemma 3.15 in [6]).

3.3 Small Items

Up to this point, all large items and the medium items from large groups are feasibly packed in OPT bins. We proceed to pack the small items. Let I_0, B be the set of unpacked items and the set of OPT partially packed bins, respectively.

The packing of the small items is done in four phases: an *optimal phase*, an *eviction phase*, a *partition phase* and a *greedy phase*.

The optimal phase is an iterative process consisting of a constant number of iterations. In each iteration, a subset of bins is packed with a subset of items whose (rounded) sizes are large relative to the free space in each of these bins. As these items belong to a *small* collection of groups among $G_1, \ldots, G_n$, they can be selected using enumeration. Thus, we obtain a packing of these items which corresponds to an optimal solution. For packing the remaining items, we want each item to be small relative to the free space in its assigned bin. To this end, in the eviction phase we discard from some bins items of non-negligible size (a single item from each bin). Then, in the partition phase, the unpacked items are partitioned into a constant number of sets satisfying certain properties, which guarantee that these items can be feasibly packed in the available free space in the bins. Finally, in the greedy phase, the items in each set are packed in their allotted subset of bins greedily, achieving a feasible packing of all items, except for a small number of items from each group, of small total size. The pseudocode of our algorithm for packing the small items is given in Algorithm 4.

The Optimal Phase: For any $b \in B$, denote by f_b^0 the free capacity in bin b, i.e., $f_b^0 = 1 - \sum_{\ell \in b} s_\ell$. We say that item ℓ is *b-negligible* if $s_\ell \leq \varepsilon^2 f_b^0$, and ℓ is *b-non-negligible* otherwise. We start by classifying the bins into two disjoint sets. Let $E_0 = \{b \in B | \ 0 < f_b^0 < \varepsilon\}$ and $D_0 = B \setminus E_0$.

We now partition B into *types*. Each type contains bins having the same total size of packed large/medium items; also, the items packed in each bin type belong to the same set of large groups, and the same number of slots is allocated in these bins to items from small groups. Formally, for each pattern p we denote by t_p the subset of bins packed with p.[5] Let T denote the set of bin types. Then $|T| = |P|$, where P is the set of all patterns. The *cardinality* of type $t \in T$ is the number of bins of this type. We use for the optimal phase algorithm *RecursiveEnum* (see the pseudocode in Algorithm 2).

Lemma 3.8. *There are* $O(1)$ *types before Step 1 of Algorithm 2.*

Once we have the classification of bins, each type t of cardinality smaller than $1/\varepsilon^4$ is padded with empty bins so that $|t| \geq 1/\varepsilon^4$. An item ℓ is *t-negligible* if ℓ is b-negligible for all bins b of type t (all bins in the same type have the same free capacity), and *t-non-negligible* otherwise. Denote by I_t' the large/medium items that are packed in the bins of type t, and let $I_t(g)$ be the set of small items that are packed in t in some solution g (in addition to I_t'). For any $1 \leq i \leq n$, a group G_i is *$t(g)$-significant* if $I_t(g)$ contains at least $\varepsilon^4|t|$ t-non-negligible items from G_i, and G_i is *$t(g)$-insignificant* otherwise.

RecursiveEnum proceeds in iterations. In the first iteration, it *guesses* for each type $t \subseteq E_0$ a subset of the items $I_t(g_{opt}) \subseteq I_0$, where g_{opt} corresponds to an optimal solution for completing the packing of t. Specifically, *RecursiveEnum* initially guesses $L(t, g_{opt})$ groups that are $t(g_{opt})$-significant: $G_{i_1}, \ldots, G_{i_{L(t,g_{opt})}}$.

[5] For the definition of patterns see Sect. 3.2.

For each $G_{i_j}, j \in \{1, \ldots, L(t, g_{opt})\}$, the algorithm guesses which items of G_{i_j} are added to $I_{t(g_{opt})}$. Since the number of guesses might be exponential, we apply to G_{i_j} linear shifting as follows. Guess $\lceil \frac{1}{\varepsilon^3} \rceil$ *representatives* in G_{i_j}, of sizes $s_{\ell_1} \leq s_{\ell_2} \leq \ldots s_{\ell_{\lceil 1/\varepsilon^3 \rceil}}$. The kth representative is the largest item in size class k, $1 \leq k \leq \lceil \frac{1}{\varepsilon^3} \rceil$ for the linear shifting of G_{i_j} in type t. Using the parameter $Q^t_{i_j} = \varepsilon^3 |t|$, the item sizes in class k are rounded up to s_{ℓ_k}, for $1 \leq k \leq \lceil \frac{1}{\varepsilon^3} \rceil$. Given a correct guess of the representatives, the actual items in size class k are selected at the end of algorithm *RecursiveEnum* (in Step 16). Denote the chosen items from G_{i_j} to bins of type t by $G^t_{i_j}$.

We now extend the definition of patterns for each type t. A slot is a pair (s_ℓ, j), where s_ℓ is the (rounded) size of a t-non-negligible item $\ell \in I_t(g_{opt})$, and there is a label for each $t(g_{opt})$-significant group $G_{i_j}, j \in \{1, \ldots, L(t, g_{opt})\}$.[6] A *t-pattern* is a multiset $\{q_1, \ldots, q_{\beta_t}\}$ containing at most $\lfloor \frac{1}{\varepsilon^2} \rfloor$ elements, where q_i is a slot for each $i \in \{1, \ldots, \beta_t\}$. Now, for each type $t \in T$ we use enumeration over patterns for assigning $G^t_{i_1}, \ldots, G^t_{i_{L(t,g_{opt})}}$ to bins in t. This completes the first iteration, and the algorithm proceeds recursively.

We now update D_0, E_0 for the next iteration by removing from E_0 bins b that have a considerably large free capacity with respect to f^0_b. For each $b \in B$, let f^1_b be the capacity available in b after iteration 1. Then $E_1 = \{b \in E_0 | \, 0 < f^1_b < \varepsilon f^0_b\}$ and $D_1 = B \setminus E_1$.

Now, each type $t \in T$ is partitioned into *sub-types* that differ by the packing of $I_t(g_{opt})$ in the first iteration. The set of types T is updated to contain these sub-types. At this point, a recursive call to *RecursiveEnum* computes for each bin type $t \subseteq E_1$ a guessing and a packing of its t-non-negligible items.[7] We repeat this recursive process $\alpha = \frac{1}{\varepsilon} + 5 = O(1)$ times.

Let G^t_i be the subset of items (of rounded sizes) assigned from G_i to bins of type t at the end of *RecursiveEnum*, for $1 \leq i \leq n$ and $t \in T$. Recall that the algorithm did not select specific items in G^t_i; that is, we only have their rounded sizes and the number of items in each size class. The algorithm proceeds to pack items from G_i in all types t for which G_i was $t(g_{opt})$-significant in some iteration. Let T_{G_i} be the set of these types. The algorithm considers first the type $t \in T_{G_i}$ for which the class C of largest size items contains the item of maximal size, where the maximum is taken over all types $t \in T_{G_i}$. The algorithm packs in bins of type t the Q^t_i largest remaining items in G_i in the slots allocated to items in C; it then proceeds similarly to the remaining size classes in types $t \in T_{G_i}$ and the remaining items in G^t_i.

Lemma 3.9. *The following hold for RecursiveEnum: (i) the running time is polynomial; (ii) the increase in the number of bins in Step 7 is at most εOPT; (iii) In Step 11 we discard at most εOPT items from each group of total size at most εOPT. (iv) One of the guesses in Steps 8, 11 corresponds to an optimal solution.*

[6] Note that we do not need a label for the $t(g_{opt})$-insignificant groups, because their items are packed separately.

[7] An item is b-non-negligible w.r.t f^1_b in this iteration, or w.r.t f^h_b in iteration $h+1, h \in \{0, \ldots, \alpha - 1\}$.

Algorithm 2. *RecursiveEnum*(I_0, B)

1: Let f_b^0 be the remaining free capacity in bin $b \in B$.
2: Let $E_0 = \{b \in B | 0 < f_b^0 < \varepsilon\}$ and $D_0 = B \setminus E_0$.
3: Denote by T the collection of bin types.
4: **for** $h = 0, \ldots, \alpha$ **do**
5:     **for** all types $t \subseteq E_h$ **do**
6:         **if** $|t| < \frac{1}{\varepsilon^4}$ **then**
7:             increase the cardinality of t to $\frac{1}{\varepsilon^4}$.
8:         Guess $t(g_{opt})$-significant groups: $G_{i_1}, \ldots, G_{i_{L(t,g_{opt})}}$
9:         **for** $j = 1, \ldots, L(t, g_{opt})$ **do**
10:             Guess the number of items from G_{i_j} to be added to bins of type t.
11:             Guess a representative for each size class of t-non-negligible items of G_{i_j} for linear shifting.
12:         Guess $|t|$ t-patterns for bins in t using the sizes after linear shifting of $G_{i_1}, \ldots, G_{i_{L(t,g_{opt})}}$.
13:         Replace type t in T by all of the sub-types of t.
14:     Let f_b^{h+1} be the remaining free capacity in bin $b \in B$.
15:     Let $E_{h+1} = \{b \in E_h | 0 < f_b^{h+1} < \varepsilon f_b^h\}$ and $D_{h+1} = B \setminus E_{h+1}$.
16: Complete the packing of all size classes by assigning items greedily.

The Eviction Phase: One of the guesses in the optimal phase corresponds to an optimal solution. For simplicity, henceforth assume that we have this guess. Recall that E_α is the set of all bins b for which $0 < f_b^\alpha < \varepsilon f_b^{\alpha-1}$. In Step 3 of *PackSmallItems* (Algorithm 4) we evict an item from each $b \in E_\alpha$ such that the available capacity of b increases to at least $\frac{f_b^\alpha}{\varepsilon}$. This is done greedily: consider the bins in E_α one by one in arbitrary order. From each bin discard a small item $\ell \in G_i$, for some G_i, $1 \leq i \leq n$, such that the following hold: (i) $s_\ell \geq \frac{f_b^\alpha}{\varepsilon}$, and (ii) less than εOPT items were discarded from G_i in this phase. Since α is large enough, this phase can be completed successfully, as shown below. Let $T = \{t_1, \ldots, t_\mu, t'\}$ be the types after the optimal phase, where t' is a new type such that $|t'| = \varepsilon OPT$. Bins of type t' are empty, i.e., each bin b of type t' has free space 1. Denote by $f(t)$ the free space in each bin b of type t after the eviction phase, and let I_L be the large items from small groups (already packed in the bins).

Lemma 3.10. *After Step 4 of PackSmallItems there exists a partition of I_α into types $I_{t_1}, \ldots, I_{t_\mu}, I_{t'}$, for which the following hold. For each $t \in T$, (i) $|G_j^t| = |G_j \cap I_t| \leq |t| - |(I_t' \setminus I_L) \cap G_j|$, for all $1 \leq j \leq n$. (ii) for any $\ell \in I_t : s_\ell \leq \varepsilon f(t)$, and (iii) $S(I_t) \leq f(t)|t|$.*

We explain the conditions of the lemma below.

The Partition Phase: Let T be the set of types after Step 4 of Algorithm 4, and I_α the remaining unpacked items.[8] We seek a partition of I_α into subsets

[8] Recall that we consider only items that were not discarded in previous steps, as discarded items are packed in a separate set of bins.

associated with bin types such that the items assigned to each type t are relatively tiny; also, the total size and the cardinality of the set of items assigned to t allow to feasibly pack these items in bins of this type. This is done by proving that a polytope representing the conditions in Lemma 3.10 has vertices at points which are integral up to a constant number of coordinates. Each such coordinate, $x_{\ell,t}$, corresponds to a fractional selection of some item $\ell \in I_\alpha$ to type $t \in T$. We use G_j to denote the subset of remaining items in G_j, $1 \leq j \leq n$.

Formally, we define a polytope P as the set of all points $x \in [0,1]^{I_\alpha \times T}$ which satisfy the following constraints.

$$\begin{aligned}
&\forall \ell \in I_\alpha, t \in T \text{ s.t. } s_\ell > \varepsilon f(t) &&: \; x_{\ell,t} = 0 \\
&\forall t \in T &&: \; \sum_{\ell \in I_\alpha} x_{\ell,t} s_\ell \leq f(t)|t| \\
&\forall \ell \in I_\alpha &&: \; \sum_{t \in T} x_{\ell,t} = 1 \\
&\forall 1 \leq j \leq n, t \in T &&: \; \sum_{\ell \in G_j} x_{\ell,t} \leq |t| - |(I'_t \setminus I_L) \cap G_j|
\end{aligned}$$

The first constraint refers to condition (*ii*) in Lemma 3.10, which implies that items assigned to type t need to be tiny w.r.t the free space in the bins of this type. The second constraint reflects condition (*iii*) in the lemma, which guarantees that the items in I_t can be feasibly packed in the bins of type t. The third constraint ensures that overall each item $\ell \in I_\alpha$ is (fractionally) assigned exactly once.

The last constraint reflects condition (*i*) in Lemma 3.10. Overall, we want to have at most $|t|$ items of G_j assigned to bins of type t. Recall that these bins may already contain large/medium items from G_j packed in previous steps. While large/medium items from *large* groups are packed optimally, the packing of large items from *small* groups, i.e., I_L, is not necessarily optimal. In particular, the items in I_L packed by our scheme in bins of type t may not appear in these bins in the optimal solution g_{opt} to which our packing corresponds. Thus, we exclude these items and only require that the number of items assigned from G_j to bins of type t is bounded by $|t| - |(I'_t \setminus I_L) \cap G_j|$.

Theorem 3.11. *Let $x \in P$ be a vertex of P. Then,*

$$|\{\ell \in I_\alpha \mid \exists t \in T : \; x_{\ell,t} \in (0,1)\}| = O(1).$$

By Theorem 3.11, we can find a feasible partition (with respect to the constraints of the polytope) by finding a vertex of the polytope, and then discarding the $O(1)$ fractional items. These items can be packed in $O(1)$ extra bins. By Lemma 3.10 we have that $P \neq \emptyset$; thus, a vertex of P exists and the partition can be found in polynomial time.

The Greedy Phase: In this phase we pack the remaining items using algorithm *GreedyPack* (see the pseudocode in Algorithm 3). Let $G_1^t, \dots, G_n^t$ be the items in I_t from each group, and let $S(I_t)$ be the total size of these items, i.e., $S(I_t) = \sum_{j=1}^{n} \sum_{\ell \in G_j^t} s_\ell$.

Algorithm 3. $GreedyPack(I_t = \{G_{i_1}^t, \dots, G_{i_H}^t\}, t = \{b_1, \dots, b_{|t|}\}$

```
1: for j = 1, ..., H do
2:     Sort G^t_{i_j} in a non-increasing order by sizes.
3: Let y_{i_j} be the largest remaining item in G^t_{i_j}, j = 1, ..., H.
4: for each bin b ∈ t do
5:     Add to bin b the items y_{i_1}, ..., y_{i_H}.
6:     while total size of items packed in bin b > 1 do
7:         Select a group G^t_{i_j} ∈ {G^t_{i_1}, ..., G^t_{i_H}} such that y_{i_j} is not last in G^t_{i_j}.
8:         if cannot complete last step then
9:             return failure
10:        Return y_{i_j} to G^t_{i_j}.
11:        Let y'_{i_j} be the next largest item in G^t_{i_j}.
12:        Add y'_{i_j} to bin b.
13:    for j = 1, ..., H do
14:        if G^t_{i_j} has a large item in bin b then
15:            discard the small item.
```

We now describe the packing of the remaining items in I_t in bins of type t. First, we add $2\varepsilon|t|$ extra bins to t. The extra bins are empty and thus have capacity 1; however, we assume that they have capacity $f(t) \le 1$. This increases the overall number of bins in the solution by $2\varepsilon OPT$. Consider the items in each group in non-increasing order by sizes. For each bin $b \in t$ in an arbitrary order, GreedyPack assigns to b the largest remaining item in each group $G_1^t, \dots, G_n^t$. If an overflow occurs, replace an item from some group G_j^t by the next item in G_j^t. This is repeated until there is no overflow in b. W.l.o.g., we may assume that $|G_j| = OPT$ for all $1 \le j \le n$; thus, b contains one item from each group (otherwise, we can add to G_j dummy items of size 0, with no increase to the number of bins in an optimal solution).

Recall that the large items from small groups are packed using the Swapping algorithm, that yields a feasible packing. Yet, it does not guarantee that the small items can be added without causing conflicts. Hence, GreedyPack may output a packing in which a small and large item from the same small group are packed in the same bin. Such conflicts are resolved by discarding the small item in each.

Lemma 3.12. *The total size of items discarded in GreedyPack in Step 15 due to conflicts is at most εOPT, and at most $\varepsilon^{k+2} \cdot OPT$ items are discarded from each group.*

Proof. The number of items discarded from each group is at most $\varepsilon^{k+2} \cdot OPT$, since all groups are small. Assume that the total size of these items is strictly larger than εOPT. Since each discarded item is *coupled* with a large conflicting item from the same group, whose size is at least $1/\varepsilon$ times larger (recall that the medium items are discarded), this implies that the total size of large conflicting items is greater than OPT. Contradiction. □

Algorithm 4. $PackSmallItems(I_0, B)$

1: **for** each guess of $RecursiveEnum(I_0, B)$ **do**
2: **for** $b \in E_\alpha$ **do**
3: evict from b the largest item ℓ satisfying: ℓ is small, and less than εOPT items where evicted from G_i, where $\ell \in G_i$.
4: Add to T a new type t' consisting of εOPT empty bins.
5: Compute a feasible partition of I_α into the types in T.
6: **for** $t \in T$ **do**
7: Add $2\varepsilon|t|$ extra bins to t.
8: Assign I_t to bins of type t using $GreedyPack(I_t, t)$.

Lemma 3.13. *For any $t \in T$, given a parameter $0 < \delta < \frac{1}{2}$ and a set of items I_t such that (i) $|G_j^t| \leq |t| - |(I_t' \setminus I_L) \cap G_j|$; (ii) for all $\ell \in I_t : s_\ell \leq \delta f(t)$, and (iii) $S(I_t) \leq (1 - \delta) f(t)|t|$, GreedyPack finds a feasible packing of I_t in bins of type t.*

Lemma 3.14. *Algorithm 4 assigns in Step 8 to OPT bins all items except for $O(\varepsilon)OPT$ items from each group, of total size $O(\varepsilon)OPT$.*

3.4 Putting It All Together

It remains to show that the items discarded throughout the execution of the scheme can be packed in a small number of extra bins.

Lemma 3.15. *The medium items from small groups and all discarded items can be packed in $O(\varepsilon) \cdot OPT$ extra bins.*

We summarize in the next result.

Theorem 3.16. *There is an APTAS for the group bin packing problem.*

References

1. Adany, R., et al.: All-or-nothing generalized assignment with application to scheduling advertising campaigns. ACM Trans. Algorithms (TALG) **12**(3), 1–25 (2016)

2. Alon, N., Azar, Y., Woeginger, G.J., Yadid, T.: Approximation schemes for scheduling on parallel machines. J. Sched. **1**(1), 55–66 (1998)
3. Christensen, H.I., Khan, A., Pokutta, S., Tetali, P.: Approximation and online algorithms for multidimensional bin packing: a survey. Comput. Sci. Rev. **24**, 63–79 (2017)
4. Coffman, E.G., Csirik, J., Galambos, G., Martello, S., Vigo, D.: Bin packing approximation algorithms: survey and classification. In: Handbook of Combinatorial Optimization, pp. 455–531 (2013)
5. Das, S., Wiese, A.: On minimizing the makespan when some jobs cannot be assigned on the same machine. In: 25th Annual European Symposium on Algorithms, ESA, pp. 31:1–31:14 (2017)
6. Doron-Arad, I., Kulik, A., Shachnai, H.: An APTAS for bin packing with clique-graph conflicts. arXiv preprint arXiv:2011.04273 (2020)
7. Epstein, L., Levin, A.: On bin packing with conflicts. SIAM J. Optim. **19**(3), 1270–1298 (2008)
8. Fernandez de la Vega, W., Lueker, G.S.: Bin packing can be solved within $1 + \varepsilon$ in linear time. Combinatorica **1**, 349–355 (1981)
9. Grage, K., Jansen, K., Klein, K.M.: An EPTAS for machine scheduling with bag-constraints. In: The 31st ACM Symposium on Parallelism in Algorithms and Architectures, pp. 135–144 (2019)
10. Hochbaum, D.S. (ed.): Approximation Algorithms for NP-Hard Problems. PWS Publishing Co., USA (1996)
11. Hochbaum, D.S., Shmoys, D.B.: Using dual approximation algorithms for scheduling problems theoretical and practical results. J. ACM **34**(1), 144–162 (1987)
12. Jansen, K.: An approximation scheme for bin packing with conflicts. J. Comb. Optim. **3**(4), 363–377 (1999)
13. Jansen, K.: An EPTAS for scheduling jobs on uniform processors: using an MILP relaxation with a constant number of integral variables. SIAM J. Discret. Math. **24**(2), 457–485 (2010)
14. Jansen, K., Klein, K., Verschae, J.: Closing the gap for makespan scheduling via sparsification techniques. In: 43rd International Colloquium on Automata, Languages, and Programming (ICALP), pp. 72:1–72:13 (2016)
15. Jansen, K., Öhring, S.R.: Approximation algorithms for time constrained scheduling. Inf. Comput. **132**(2), 85–108 (1997)
16. Karmarkar, N., Karp, R.M.: An efficient approximation scheme for the one-dimensional bin-packing problem. In: 23rd Annual Symposium on Foundations of Computer Science, pp. 312–320. IEEE (1982)
17. Leung, J.Y.: Bin packing with restricted piece sizes. Inf. Process. Lett. **31**(3), 145–149 (1989)
18. McCloskey, B., Shankar, A.: Approaches to bin packing with clique-graph conflicts. Computer Science Division, University of California (2005)
19. Oh, Y., Son, S.: On a constrained bin-packing problem. Technical report CS-95-14 (1995)
20. Rothvoß, T.: Approximating bin packing within O(log OPT * log log OPT) bins. In: 54th Annual IEEE Symposium on Foundations of Computer Science, pp. 20–29. IEEE Computer Society (2013)
21. Simchi-Levi, D.: New worst-case results for the bin-packing problem. Naval Res. Logist. (NRL) **41**(4), 579–585 (1994)
22. Zuckerman, D.: Linear degree extractors and the inapproximability of max clique and chromatic number. Theory Comput. **3**(1), 103–128 (2007)

Fast Deterministic Algorithms for Computing All Eccentricities in (Hyperbolic) Helly Graphs

Feodor F. Dragan[1(✉)], Guillaume Ducoffe[2], and Heather M. Guarnera[3]

[1] Computer Science Department, Kent State University, Kent, USA
dragan@cs.kent.edu

[2] National Institute for Research and Development in Informatics and University of Bucharest, Bucureşti, Romania
guillaume.ducoffe@ici.ro

[3] Department of Mathematical and Computational Sciences, The College of Wooster, Wooster, USA
hguarnera@wooster.edu

Abstract. A graph is Helly if every family of pairwise intersecting balls has a nonempty common intersection. The class of Helly graphs is the discrete analogue of the class of hyperconvex metric spaces. It is also known that every graph isometrically embeds into a Helly graph, making the latter an important class of graphs in Metric Graph Theory. We study diameter, radius and all eccentricity computations within the Helly graphs. Under plausible complexity assumptions, neither the diameter nor the radius can be computed in truly subquadratic time on general graphs. In contrast to these negative results, it was recently shown that the radius and the diameter of an n-vertex m-edge Helly graph G can be computed with high probability in $\tilde{\mathcal{O}}(m\sqrt{n})$ time (*i.e.*, subquadratic in $n+m$). In this paper, we improve that result by presenting a deterministic $\mathcal{O}(m\sqrt{n})$ time algorithm which computes not only the radius and the diameter but also all vertex eccentricities in a Helly graph. Furthermore, we give a parameterized linear-time algorithm for this problem on Helly graphs, with the parameter being the Gromov hyperbolicity δ. More specifically, we show that the radius and a central vertex of an m-edge δ-hyperbolic Helly graph G can be computed in $\mathcal{O}(\delta m)$ time and that all vertex eccentricities in G can be computed in $\mathcal{O}(\delta^2 m)$ time. To show this more general result, we heavily use our new structural properties obtained for Helly graphs.

1 Introduction

Given an undirected unweighted graph $G = (V, E)$, the distance $d_G(u, v)$ between two vertices u and v is the minimum number of edges on any path connecting u

This work was supported by project PN 19 37 04 01 "New solutions for complex problems in current ICT research fields based on modelling and optimization", funded by the Romanian Core Program of the Ministry of Research and Innovation (MCI), 2019–2022.

A. Lubiw et al. (Eds.): WADS 2021, LNCS 12808, pp. 300–314, 2021.
https://doi.org/10.1007/978-3-030-83508-8_22

and v in G. The eccentricity $e_G(u)$ of a vertex u is the maximum distance from u to any other vertex. The radius and the diameter of G, denoted by $rad(G)$ and $diam(G)$, are the smallest and the largest eccentricities of vertices in G, respectively. A vertex with eccentricity equal to $rad(G)$ is called a central vertex of G. We are interested in the fundamental problems of finding a central vertex and of computing the diameter and the radius of a graph. The problem of finding a central vertex of a graph is one of the most famous facility location problems in Operation Research and in Location Science. The diameter and radius of a graph play an important role in the design and analysis of networks in a variety of networking environments like social networks, communication networks, electric power grids, and transportation networks. A naive algorithm which runs a BFS from each vertex to compute its eccentricity and then (in order to compute the radius, the diameter and a central vertex) picks the smallest and the largest eccentricities and a vertex with smallest eccentricity has running time $\mathcal{O}(nm)$ on an n-vertex m-edge graph. Interestingly, this naive algorithm is conditionally optimal for general graphs as well as for some restricted families of graphs [1,6,18,54] since, under plausible complexity assumptions, neither the diameter nor the radius can be computed in truly subquadratic time (i.e., in $\mathcal{O}(n^a m^b)$, for some positive a, b such that $a + b < 2$) on those graphs. Already for split graphs (a subclass of chordal graphs), computing the diameter is roughly equivalent to Disjoint Sets, *a.k.a.*, the monochromatic Orthogonal Vector problem [15]. Under the Strong Exponential-Time Hypothesis (SETH), we cannot solve Disjoint Sets in truly subquadratic time, and so neither we can compute the diameter of split graphs in truly subquadratic time [6].

In a quest to break this quadratic barrier (in the size $n + m$ of the input), there has been a long line of work presenting more efficient algorithms for computing the diameter and/or the radius on some special graph classes, by exploiting their geometric and tree-like representations and/or some forbidden pattern (*e.g.*, excluding a minor, or a family of induced subgraphs). For example, although the diameter of a split graph can unlikely be computed in subquadratic time, there is an elegant linear-time algorithm for computing the radius and a central vertex of a chordal graph [16]. Efficient algorithms for computing the diameter and/or the radius or finding a central vertex are also known for interval graphs [34,53], AT-free graphs [37], directed path graphs [19], distance-hereditary graphs [21,26,29,33], strongly chordal graphs [23], dually chordal graphs [7,25], chordal bipartite graphs [40], outerplanar graphs [43], planar graphs [11,45], graphs with bounded clique-width [21,39], graphs with bounded tree-width [1,9,42] and, more generally, H-minor free graphs and graphs of bounded (distance) VC-dimension [42].

We here study the *Helly graphs* as a broad generalization of dually chordal graphs which in turn contain all interval graphs, directed path graphs and strongly chordal graphs. Recall that a graph is Helly if every family of pairwise intersecting balls has a non-empty common intersection. This latter property on the balls will be simply referred to as the Helly property in what follows. Helly graphs have unbounded tree-width and unbounded clique-width, they do not exclude any fixed minor and they cannot be characterized via some forbidden structures. They are sometimes called absolute retracts or disk-Helly graphs by opposition to other Helly-type properties on graphs [22]. The Helly graphs are well studied in Metric Graph Theory. *E.g.*, see the survey [3] and the papers cited therein. This is partly because every graph is an isometric subgraph of some Helly graph, thereby making of the latter the discrete equivalent of hyperconvex metric spaces [36,47]. A minimal by inclusion Helly graph H which contains a given graph G as an isometric subgraph is unique and called the injective hull [47] or the tight span [36] of G. Polynomial-time recognition algorithms for the Helly graphs were presented in [4,23,51]. Several structural properties of these graphs were also identified (see [3] and the references cited therein). The dually chordal graphs are exactly the Helly graphs in which the intersection graph of balls is chordal, and they were studied independently from the general Helly graphs [7,8,25,35]. As we already mentioned it [7,25,35], the diameter, the radius and a central vertex of a dually chordal graph can be found in linear time, that is optimal. However, it was open until recently whether there are truly subquadratic-time algorithms for these problems on general Helly graphs. First such algorithms were recently presented in [41] for computing both the radius and the diameter and in [38] for finding a central vertex. Those algorithms are randomized and run, with high probability, in $\tilde{\mathcal{O}}(m\sqrt{n})$ time on a given n-vertex m-edge Helly graph (*i.e.*, subquadratic in $n+m$). They make use of the Helly property and of the unimodality of the eccentricity function in Helly graphs [24]: every vertex of locally minimum eccentricity is a central vertex. In [41], a linear-time algorithm for computing all eccentricities in C_4-free Helly graphs was also presented. The C_4-free Helly graphs are exactly the Helly graphs whose balls are convex. They properly include strongly chordal graphs as well as bridged Helly graphs and hereditary Helly graphs [41].

Our Contribution. We improve those results from [41] and [38] by presenting a deterministic $\mathcal{O}(m\sqrt{n})$ time algorithm which computes not only the radius and the diameter but also all vertex eccentricities in an n-vertex m-edge Helly graph. Being able to efficiently compute all vertex eccentricities is of great importance. For example, in the analysis of social networks (e.g., citation networks or recommendation networks), biological systems (e.g., protein interaction networks), computer networks (e.g., the Internet or peer-to-peer networks), transportation networks (e.g., public transportation or road networks), etc., the eccentricity $e_G(v)$ of a vertex v is used to measure its importance in the network: the *eccentricity centrality index* of v [50] is defined as $\frac{1}{e_G(v)}$.

We complete this above result with a parameterized *linear-time* algorithm for computing all vertex eccentricities in Helly graphs, with the parameter being the Gromov hyperbolicity δ, as defined by the following four point condition. The hyperbolicity of a graph G [46] is the smallest half-integer $\delta \geq 0$ such that, for any four vertices u, v, w, x, the two largest of the three distance sums $d(u,v) + d(w,x)$, $d(u,w) + d(v,x)$, $d(u,x) + d(v,w)$ differ by at most 2δ. In this case we say that G is δ-hyperbolic. As the tree-width of a graph measures its combinatorial tree-likeness, so does the hyperbolicity of a graph measure its metric tree-likeness. In other words, the smaller the hyperbolicity δ of G is, the closer G is to a tree metrically. The hyperbolicity of an n-vertex graph can be computed in polynomial-time (e.g., in $\mathcal{O}(n^{3.69})$ time [44]), however it is unlikely that it can be done in subquadratic time [6,20,44]. A 2-approximation of hyperbolicity can be computed in $\mathcal{O}(n^{2.69})$ time [44] and an 8-approximation can be computed in $\mathcal{O}(n^2)$ time [12] (assuming that the input is the distance matrix of the graph). Graph hyperbolicity has attracted attention recently due to the empirical evidence that it takes small values in many real-world networks, such as biological networks, social networks, Internet application networks, and collaboration networks, to name a few (see, e.g., [2,5,48,52]). Furthermore, many special graph classes (e.g., interval graphs, chordal graphs, dually chordal graphs, AT-free graphs, weakly chordal graphs and many others) have constant hyperbolicity [2,10,17,28,32,49,55]. In fact, the dually chordal graphs and the C_4-free Helly graphs are known to be proper subclasses of the 1-hyperbolic Helly graphs (this follows from results in [8,28]). Notice also that any graph is δ-hyperbolic for some $\delta \leq diam(G)/2$.

We show that the radius and a central vertex of an m-edge Helly graph G with hyperbolicity δ can be computed in $\mathcal{O}(\delta m)$ time and that all vertex eccentricities in G can be computed in $\mathcal{O}(\delta^2 m \log \delta)$ time, even if δ is not known to us. If either δ or a constant approximation of it is known, then the running time of our algorithm can be lowered to $\mathcal{O}(\delta^2 m)$. Thus, for Helly graphs with constant hyperbolicity, all vertex eccentricities can be computed in linear time. As a byproduct, we get a linear time algorithm for computing all eccentricities in C_4-free Helly graphs as well as in dually chordal graphs, generalizing known results from [7,25,41]. Previously, for dually chordal graphs, it was only known that a central vertex can be found in linear time [7,25]. Notice that the diameter problem can unlikely be solved in truly subquadratic time in general 1-hyperbolic graphs and that the radius problem can unlikely be solved in truly subquadratic time in general 2-hyperbolic graphs [18]. For general δ-hyperbolic graphs, there are only additive $\mathcal{O}(\delta)$-approximations of the diameter and the radius, that can be computed in linear time [17,30,31].

To show our more general results, additionally to the unimodality of the eccentricity function in Helly graphs, we rely on new structural properties obtained for this class. It turns out that the hyperbolicity of a Helly graph G is governed by the size of a largest isometric rectilinear grid in G. As a consequence, the hyperbolicity of an n-vertex Helly graph is at most $\sqrt{n}+1$ and the diameter of the center of G is at most $2\sqrt{n}+3$. These properties, along with others, play a crucial role in efficient computations of all eccentricities in Helly graphs. We also give new characterizations of the Helly graphs. Among others, we show that the Helly property for balls of equal radii implies the Helly property for balls with variable radii. It would be interesting to know whether a similar result holds for all (discrete) metric spaces. We are not aware of such a general result.

Notations. Recall that $d_G(u, v)$ denotes the distance between vertices u and v in $G = (V, E)$. Let $n = |V|$ be the number of vertices and $m = |E|$ be the number of edges in G. The ball of radius r and center v is defined as $\{u \in V : d_G(u,v) \leq r\}$, and denoted by $N^r_G[v]$. Sometimes, $N^r_G[v]$ is called the r-neighborhood of v. In particular, $N_G[v] := N^1_G[v]$ and $N_G(v) := N_G[v] \setminus \{v\}$ denote the closed and open neighbourhoods of a vertex v, respectively. More generally, for any vertex-subset S and a vertex u, we define $d_G(u,S) := \min_{v \in S} d_G(u,v)$, $N^r_G[S] := \bigcup_{v \in S} N^r_G[v]$, $N_G[S] := N^1_G[S]$ and $N_G(S) := N_G[S] \setminus S$. The metric projection of a vertex u on S, denoted by $Pr_G(u,S)$, is defined as $\{v \in S : d_G(u,v) = d_G(u,S)\}$. The metric interval $I_G(u,v)$ between u and v is $\{w \in V : d_G(u,w) + d_G(w,v) = d_G(u,v)\}$. For any $k \leq d_G(u,v)$, we can also define the slice $L(u,k,v) := \{w \in I_G(u,v) : d_G(u,w) = k\}$. Recall that the eccentricity of a vertex u is defined as $\max_{v \in V} d_G(u,v)$ and denoted by $e_G(u)$. Note that we will omit the subscript if the graph G is clear from the context. The radius and the diameter of a graph G are denoted by $rad(G)$ and $diam(G)$, respectively. A vertex c is called central in G if $e_G(c) = rad(G)$. The set of all central vertices of G is denoted by $C(G) := \{v \in V : e_G(v) = rad(G)\}$ and called the center of G. The eccentricity function $e_G(v)$ of a graph G is said to be *unimodal*, if for every non-central vertex v of G there is a neighbor $u \in N_G(v)$ such that $e_G(u) < e_G(v)$ (that is, every local minimum of the eccentricity function is a global minimum). Recall also that a vertex set $S \subseteq V$ is called convex in G if, for every vertices $x, y \in S$, all shortest paths connecting them are contained in S (i.e., $I_G(x,y) \subseteq S$). For $\beta \geq 0$, we say that S is *β-pseudoconvex* [30] if, for every vertices $x, y \in S$, any vertex $z \in I_G(x,y) \setminus S$ satisfies $\min\{d_G(z,x), d_G(z,y)\} \leq \beta$. A subgraph H of G is called isometric (or distance-preserving) if, for every vertices x, y of H, $d_G(x,y) = d_H(x,y)$.

2 Helly Graphs and Their Hyperbolicity

Here we demonstrate that for Helly graphs, having a constant hyperbolicity is equivalent to the following properties: having β-pseudoconvexity of balls with a constant β, or having the diameter of the center bounded by a constant for all subsets of vertices, or not having a large $(\gamma \times \gamma)$ rectilinear grid as an isometric subgraph. These results generalize some known results from [13,14,17,28,30].

First we give new characterizations of Helly graphs through a formula for the eccentricity function and relations between diameter and radius for all subsets of vertices. For this we need to generalize our basic notations. Define for any set $M \subseteq V$ and any vertex $v \in V$ the eccentricity of v in G with respect to M as $e_M(v) = \max_{u \in M} d_G(u,v)$. Let $diam_M(G) = \max_{v \in M} e_M(v)$, $rad_M(G) = \min_{v \in V} e_M(v)$, $C_M(G) = \{v \in V : e_M(v) = rad_M(G)\}$. When $M = V$, these agree with earlier definitions.

Theorem 1. *For a graph G the following statements are equivalent:*

(1) G is Helly;
(2) the eccentricity function $e_M(\cdot)$ is unimodal for every set $M \subseteq V$;
(3) $e_M(v) = d_G(v, C_M(G)) + rad_M(G)$ holds for every set $M \subseteq V$ and every vertex $v \in V$;

(4) $2rad_M(G) - 1 \leq diam_M(G) \leq 2rad_M(G)$ *holds for every set* $M \subseteq V$*;*
(5) $rad_M(G) = \lfloor \frac{diam_M(G)+1}{2} \rfloor$ *holds for every set* $M \subseteq V$*.*

Proof of this theorem and of all other statements of this section can be found in full version of this paper [27]. The equivalence between (1) and (5) can be rephrased as follows.

Corollary 1. *For every graph* $G = (V, E)$*, the family of all balls* $\{N_G^r[v] : v \in V, r \in \mathbb{N}\}$ *of* G *has the Helly property if and only if the family of* k*-neighborhoods* $\{N_G^k[v] : v \in V\}$ *of* G *has the Helly property for every natural number* k*.*

That is, the Helly property for balls of equal radii implies the Helly property for balls with variable radii. It would be interesting to know whether a similar result holds for all (discrete) metric spaces. We are not aware of such a general result and did not find its analog in the literature.

We will also need the following lemma from [23].

Lemma 1 [23]. *For every Helly graph* $G = (V, E)$ *and every set* $M \subseteq V$*, the graph induced by the center* $C_M(G)$ *is Helly and it is an isometric (and hence connected) subgraph of* G*.*

Given this lemma, it will be convenient to denote by $C_M(G)$ not only the set of central vertices but also the subgraph of G induced by this set. Then, $diam(C_M(G))$ denotes the diameter of this graph ($diam(C_M(G)) = diam_{C_M(G)}(G)$ by this isometricity).

Let $\delta(G)$ be the smallest half-integer $\delta \geq 0$ such that G is δ-hyperbolic. Let $\gamma(G)$ be the largest integer $\gamma \geq 0$ such that G has a $(\gamma \times \gamma)$ rectilinear grid as an isometric subgraph. Let $\beta(G)$ be the smallest integer $\beta \geq 0$ such that all balls in G are β-pseudoconvex. Finally, let $\kappa(G)$ be the smallest integer $\kappa \geq 0$ such that $diam(C_M(G)) \leq \kappa$ for every set $M \subseteq V$.

Theorem 2. *For every Helly graph* G*, a constant bound on one parameter from* $\{\delta(G), \gamma(G), \beta(G), \kappa(G)\}$ *implies a constant bound on all others.*

The following corollaries of Theorem 2 will play an important role in efficient computations of all eccentricities of a Helly graph. Corollary 2 gives a sublinear bound on the hyperbolicity of an n-vertex Helly graph. Corollary 3 gives a sublinear bound on the diameter of the center of an n-vertex Helly graph.

Corollary 2. *The hyperbolicity of an* n*-vertex Helly graph* G *is at most* $\sqrt{n}+1$*.*

Corollary 3. *For any Helly graph* G*,* $diam(C(G)) \leq 2\delta(G) + 1 \leq 2\sqrt{n} + 3$*.*

3 All Eccentricities in Helly Graphs

It is known that the radius (see [41]) and a central vertex (see [38]) of an n-vertex m-edge Helly graph can be computed in $\tilde{\mathcal{O}}(m\sqrt{n})$-time with high probability. In this section, we improve those results by presenting a deterministic $\mathcal{O}(m\sqrt{n})$

time algorithm which computes not only the radius and a central vertex but also all vertex eccentricities in a Helly graph. To show this more general result, we heavily make use of our new structural results from Sect. 2. In particular, the fact that both the hyperbolicity of a Helly graph G and the diameter of its center $C(G)$ are upper bounded by $\mathcal{O}(\sqrt{n})$ will be very handy. The following results from [38,41] and [17,30,31] will be also very useful.

Lemma 2 [41]. *Let G be an m-edge Helly graph and k be a natural number. One can compute the set of all vertices of G of eccentricity at most k, and their respective eccentricities, in $\mathcal{O}(km)$ time.*

Lemma 3 [38]. *Let G be an m-edge Helly graph and v be an arbitrary vertex. There is an $\mathcal{O}(m)$-time algorithm which either certifies that v is a central vertex of G or finds a neighbor u of v such that $e(u) < e(v)$.*

Lemma 4 [17,30,31]. *Let G be an arbitrary m-edge graph and δ be its hyperbolicity. There is an $\mathcal{O}(\delta m)$-time algorithm which finds in G a vertex c with eccentricity at most $rad(G) + 2\delta$. The algorithm does not need to know the value of δ in order to work correctly.*

First, by combining Lemmas 3 and 4, we show that a central vertex of a Helly graph G can be computed in $\mathcal{O}(\delta m)$ time, where δ is the hyperbolicity of G.

Lemma 5. *If G is an m-edge Helly graph, then one can compute a central vertex and the radius of G in $\mathcal{O}(\delta m)$ time, where δ is the hyperbolicity of G.*

Proof. We use Lemma 4 in order to find, in $\mathcal{O}(\delta m)$ time, a vertex c of G with eccentricity $e(c) \leq rad(G) + 2\delta$. Then we apply Lemma 3 at most 2δ times in order to descend from c to a central vertex c^*. It takes $\mathcal{O}(\delta m)$ time. □

Combining this with Corollary 2, we get.

Corollary 4. *For any n-vertex m-edge Helly graph G, a central vertex and the radius of G can be computed in $\mathcal{O}(m\sqrt{n})$ time.*

We are now ready to prove our main result of this section.

Theorem 3. *All vertex eccentricities in an n-vertex m-edge Helly graph G can be computed in total $\mathcal{O}(m\sqrt{n})$ time.*

Proof. Our goal is to compute $e(v)$ for every $v \in V$. For that, we first find a central vertex c and compute the radius $rad(G)$ of G, which takes $\mathcal{O}(m\sqrt{n})$ time by Corollary 4. If $rad(G) \leq 5\sqrt{n} + 6$ (the choice of this number will be clear later), then $diam(G) \leq 2rad(G) \leq 10\sqrt{n} + 12$ and we are done by Lemma 2 (applied for $k = 10\sqrt{n}+12$); it takes in this case total time $\mathcal{O}(m\sqrt{n})$ to compute all eccentricities in G. Thus, from now on, we assume $rad(G) > 5\sqrt{n} + 6$. By Theorem 1(3), for every $v \in V$, $e(v) = d(v, C(G)) + rad(G)$ holds. Thus, in order to compute all the eccentricities, it is sufficient to compute $C(G)$. For a central vertex $c \in C(G)$ found earlier, let $S = N_G^{2\sqrt{n}+3}[c]$. By Corollary 3, $C(G) \subseteq S$.

In what follows, let $r = rad(G)$. Consider the BFS layers $L_i(S) = \{v \in V : d(v,S) = i\}$. Note that if $i \leq r - 4\sqrt{n} - 6 \leq r - diam_S(G)$, then all the vertices of $L_i(S)$ are at distance at most r from all the vertices in S. As a result, in order to compute $C(G)$, it is sufficient to consider the layers $L_i(S)$, for $i > r - 4\sqrt{n} - 6$. Set $A = \bigcup_{i > r - 4\sqrt{n} - 6} L_i(S)$. Since for every $v \notin S$, $d(v,c) = d(v,S) + 2\sqrt{n} + 3 \leq r$, we deduce that there are at most $(r - 2\sqrt{n} - 3) - (r - 4\sqrt{n} - 6) = 2\sqrt{n} + 3$ nonempty layers in A.

We will need to consider the "critical band" of all the layers $L_i(S)$, for $1 \leq i \leq r - 4\sqrt{n} - 6$ (all the layers between S and A). We claim that there are at least $\sqrt{n}$ layers in this band. Indeed, under the above assumption, $r > 5\sqrt{n} + 6$. Then, the number of layers is exactly $e(c) - 2\sqrt{n} - 3 > 3\sqrt{n} + 3$, minus at most $2\sqrt{n} + 3$ layers most distant from c (layers in A). Overall, there are at least $\sqrt{n}$ layers in the critical band, as claimed. Then, one layer in the critical band, call it L, contains at most $n/\sqrt{n} = \sqrt{n}$ vertices.

Claim 1. *For every $a \in A$, there exists a "distant gate" $a^* \in Pr(a,L)$ with the following property: $N^r[a] \cap S = N^{r-d(a,L)}[a^*] \cap S$.*

In order to prove the claim, set $p = d(a,L)$ and $q = d(a,c) \leq r$. Let us consider a family of balls $\mathcal{F} = \{N^p[a],\ N^{q-p}[c]\} \cup \{N^{r-p}[s]\ :\ s \in N^r[a] \cap (S \setminus c)\}$. We stress that $N^p[a] \cap N^{q-p}[c] = Pr(a,L)$. Then, in order to prove the existence of a distant gate, it suffices to prove that the balls in $\mathcal{F}$ intersect; indeed, if it is the case then we may choose for a^* any vertex in the common intersection of the balls in $\mathcal{F}$. Clearly, $N^p[a] \cap N^{q-p}[c] \neq \emptyset$ and, in the same way, $N^p[a] \cap N^{r-p}[s] \neq \emptyset$ for each $s \in N^r[a] \cap (S \setminus c)$. Furthermore, since L is in the critical band, $d(c,L) > 2\sqrt{n} + 3$, and therefore we have for each $s, s' \in S$:

$$2(r-p) \geq 2(q-p) = 2d(c,L) > diam_S(G) \geq d(s,s').$$

In the same way $(q-p) + (r-p) \geq 2(q-p) > diam_S(G) \geq d(s,c)$. The latter proves that the balls in $\mathcal{F}$ intersect. This concludes the proof of Claim 1.

We finally explain how to compute these distant gates, and how to use this information in order to compute $S \cap C(G)$. Specifically:

- We make a BFS from every $u \in L$. it takes $\mathcal{O}(m|L|) = \mathcal{O}(m\sqrt{n})$ time. Doing so, we can compute $\forall a \in A,\ Pr(a,L)$, in total $\mathcal{O}(|A||L|) = \mathcal{O}(n\sqrt{n})$ time.
- Since A contains at most $\mathcal{O}(\sqrt{n})$ nonempty layers, then the number of pairwise distinct distances $d(a,L)$, for $a \in A$, is also in $\mathcal{O}(\sqrt{n})$. Call the set of all these distances I_A. Then, $\forall u \in L$, and $\forall i \in I_A$, we also compute $p(u,i) = |N_G^{r-i}[u] \cap S|$. For that, we consider the vertices $u \in L$ sequentially. Recall that we computed a BFS tree rooted at u. In particular, we can order the vertices of S by increasing distance to u. It takes $\mathcal{O}(n)$ time. Similarly, we can order I_A in $\mathcal{O}(\sqrt{n}\log n) = o(n)$ time. In order to compute all the values $p(u,i)$, it suffices to scan in parallel these two ordered lists. The running time is $\mathcal{O}(n)$ for every fixed $u \in L$, and so the total running time is $\mathcal{O}(n|L|) = \mathcal{O}(n\sqrt{n})$.

- Now, in order to compute a distant gate a^*, for $a \in A$, we proceed as follows. Let $i = d(a, L)$. We scan $Pr(a, L)$ and we store a vertex a^* maximizing $p(a^*, i)$. It takes $\mathcal{O}(|A||L|) = \mathcal{O}(n\sqrt{n})$ time. On the way, $\forall u \in L$, let $q(u)$ be the maximum i such that $a^* \equiv u$ is the distant gate of some vertex $a \in A$, such that $d(a, L) = i$ (possibly, $q(u) = 0$ if u was not chosen as the distant gate of any vertex).
- Let $s \in S$ be arbitrary. For having $s \in S \cap C(G)$, it is necessary and sufficient to have $s \in N^r[a] \cap S, \forall a \in A$. Equivalently, $\forall u \in L$, one must have $d(s, u) \leq r - q(u)$. This can be checked in time $\mathcal{O}(|L|)$ per vertex in S, and so, in total $\mathcal{O}(n\sqrt{n})$ time.

□

4 Eccentricities in Helly Graphs with Small Hyperbolicity

In the previous section we showed that a central vertex of a Helly graph G can be computed in $\mathcal{O}(\delta m)$ time, where δ is the hyperbolicity of G. This nice result, combined with the property that all Helly graphs have hyperbolicity $\mathcal{O}(\sqrt{n})$ (Corollary 2), was key to the design of our $\mathcal{O}(m\sqrt{n})$-time algorithm for computing all vertex eccentricities. Next, we deepen the connection between hyperbolicity and fast eccentricity computation within Helly graphs.

As we have mentioned earlier, many graph classes (e.g., interval graphs, chordal graphs, dually chordal graphs, AT-free graphs, weakly chordal graphs and many others) have constant hyperbolicity. In particular, the dually chordal graphs and the C_4-free Helly graphs (superclasses of the interval graphs and of the strongly chordal graphs) are proper subclasses of the 1-hyperbolic Helly graphs. This raises the question whether all vertex eccentricities can be computed in linear time in a Helly graph G if its hyperbolicity δ is a constant.

We prove in what follows that it is indeed the case, which is the main result of this section. The following result could also be considered as a parameterized algorithm on Helly graphs with δ as the parameter.

Theorem 4. *If G is an m-edge Helly graph of hyperbolicity δ, then the eccentricity of all vertices of G can be computed in $\mathcal{O}(\delta^2 m \log \delta)$ time. The algorithm does not need to know the value of δ in order to work correctly. If δ (or a constant approximation of it) is known, then the running time is $\mathcal{O}(\delta^2 m)$.*

As a byproduct, we get a linear time algorithm for computing all vertex eccentricities in C_4-free Helly graphs as well as in dually chordal graphs, generalizing known results from [7,25,41]. We recall that for dually chordal graphs, until this paper it was only known that a central vertex of such a graph can be found in linear time [7,25].

The remainder of this section is devoted to proving Theorem 4. For that, the following result is proved in Subsect. 4.1:

Lemma 6. *Let G be an m-edge Helly graph, c be a central vertex of G and k be a natural number. There is an $\mathcal{O}(k^2 m)$-time algorithm which computes $C(G) \cap N^k[c]$.*

Proof (Proof of Theorem 4 assuming Lemma 6). Since, by Theorem 1(3), $e(v) = d(v, C(G)) + rad(G)$ holds for every $v \in V$, as before, in order to compute all the eccentricities, it is sufficient to compute $C(G)$. We first find a central vertex c and compute the radius $rad(G)$ of G. This takes $\mathcal{O}(\delta m)$ time by Lemma 5.

By Corollary 3, we know that $diam(C(G)) \leq 2\delta + 1$. Therefore, $C(G) \subseteq N^{2\delta+1}[c]$. If δ is known to us, we fix $k := 2\delta + 1$ (if only a constant approximation $\delta' \geq \delta$ of δ is known, we set $k = 2\delta' + 1$). Then, we are done applying Lemma 6. Otherwise, we work sequentially with $k = 2, 3, 4, 5, 8, 9, \ldots, 2^p, 2^p + 1, 2^{p+1}, 2^{p+1}+1, \ldots$, and we stop after finding the smallest integer (power of 2) k such that $C(G) \cap N^k[c] = C(G) \cap N^{k+1}[c]$. Indeed, by the isometricity (and hence connectedness) of $C(G)$ in G (see Lemma 1), the set $C(G) \cap N^k[c]$ will contain all central vertices of G, i.e., $C(G) \cap N^k[c] = C(G)$. The latter will happen for some $k < 2(2\delta + 1)$ after at most $\mathcal{O}(\log \delta)$ probes. Overall, since we need to apply Lemma 6 at most $\mathcal{O}(\log \delta)$ times, for some values $k < 2(2\delta + 1)$, the total running time is $\mathcal{O}(\delta^2 m \log \delta)$. If δ (or a constant approximation of it is known), then we call Lemma 6 only once, and therefore the running time goes down to $\mathcal{O}(\delta^2 m)$. □

4.1 Proof of Lemma 6

In what follows, G is a Helly graph, k is an integer and $r = rad(G)$. Let $S_k = N^k[c]$. If $r \leq 2k$, we can compute all central vertices in $\mathcal{O}(km)$ time (see Lemma 2). Thus from now on, $r > 2k$. As $diam_{S_k}(G) \leq 2k$, to find all central vertices in S_k (i.e., the set $C(G) \cap S_k$), we will need to consider only the vertices at distance $> r - 2k$ from S_k.

Let $i < 2k$ be fixed (we need to consider all possible i between k and $2k-1$ sequentially). Let $A_{k,i} = L_{r-i}(S_k)$ (where we recall that $L_{r-i}(S_k) = \{v \in V : d(v, S_k) = r - i\}$). We want to compute $S_{k,i} := \{s \in S_k : A_{k,i} \subseteq N^r[s]\}$. Indeed, $C(G) \cap S_k = \bigcap_{i=k}^{2k-1} S_{k,i}$. The computation of $S_{k,i}$ (for k, i fixed) works by phases. We describe below the two main phases of the process.

First Phase of the Algorithm. To give the intuition of our approach, we will need the following simple claim. For a vertex $v \in V$ and an integer j, let $L(v, j, S_k) := \{u \in V : d(v, S_k) = d(v, u) + d(u, S_k) \text{ and } d(v, u) = j\}$.

Claim 2. *Let $B \subseteq A_{k,i}$ be such that $\bigcap\{L(b, j, S_k) : b \in B\} \neq \emptyset$, for some $0 \leq j < r - i$. Then, for every $s \in S_k$, $\max_{b \in B} d(s, b) \leq r$ if and only if $d(s, \bigcap\{L(b, j, S_k) : b \in B\}) \leq r - j$.*

Proof. If $d(s, \bigcap\{L(b, j, S_k) : b \in B\}) \leq r - j$, then $\max_{b \in B} d(s, b) \leq r$. Conversely, let us assume $\max_{b \in B} d(s, b) \leq r$. Set $\mathcal{F} = \{N_G^{r-j}[s], N_G^{r+k-(i+j)}[c]\} \cup \{N_G^j[b] : b \in B\}$. We prove that the balls in $\mathcal{F}$ intersect. For each $b, b' \in B$, $N_G^j[b] \cap N_G^j[b'] \supseteq \bigcap\{L(b, j, S_k) : b \in B\} \neq \emptyset$. Since we assume $\max_{b \in B} d(s, b) \leq r$, $N_G^j[b] \cap N_G^{r-j}[s] \neq \emptyset$. Furthermore, as for each $b \in B$ we have $d(b, c) = d(b, S_k) + k = r - i + k$, we obtain $N_G^{r-i+k-j}[c] \cap N_G^j[b] = L(b, j, S_k) \neq \emptyset$. Finally, since we have $j < r - i$, $(r - i + k - j) + (r - j) > k + i \geq k \geq d(s, c)$.

Therefore, $N_G^{r+k-(i+j)}[c] \cap N_G^{r-j}[s] \neq \emptyset$. It follows from the above that the balls in $\mathcal{F}$ pairwise intersect. By the Helly property, there exists a vertex y in the common intersection of all the balls in $\mathcal{F}$. As for each $b \in B$, $y \in N_G^{r-i+k-j}[c] \cap N_G^j[b] = L(b, j, S_k)$, we deduce that $y \in \bigcap\{L(b, j, S_k) : b \in B\}$. Finally, we have $d(s, \bigcap\{L(b, j, S_k) : b \in B\}) \leq d(s, y) \leq r - j$. □

We are now ready to present the first phase of our algorithm (for k, i fixed). It is divided into $r - i$ steps: from $j = 0$ to $j = r - i - 1$. At step j, for $0 \leq j < r - i$, the intermediate output is a collection of disjoint subsets $V_j^1, V_j^2, ..., V_j^{p_j}$ of the layer $L_{r-i-j}(S_k)$. These disjoint subsets are in one-to-one correspondence with some partition $B_1, B_2, ..., B_{p_j}$ of $A_{k,i}$. Specifically, the algorithm ensures that: $\forall 1 \leq t \leq p_j$, $V_j^t = \bigcap\{L(b, j, S_k) : b \in B_t\} \neq \emptyset$. Doing so, by the above Claim 2, for any $s \in S_k$ we have $\max_{z \in A_{k,i}} d(s, z) \leq r \iff \max_{1 \leq t \leq p_j} d(s, V_j^t) \leq r - j$.

Initially, for $j = 0$, every set B_t is a singleton. Furthermore, $B_t = V_0^t$. Then, we show how to partition $L_{r-i-(j+1)}(S_k)$ from $V_j^1, V_j^2, ..., V_j^{p_j}$ in total $\mathcal{O}(\sum_{x \in L_{r-i-j}(S_k)} |N_G(x)|)$ time. Note that in doing so we get a total running time in $\mathcal{O}(m)$ for that phase. For that, let us define $W_j^t = N(V_j^t) \cap L_{r-i-(j+1)}(S_k)$. Since the subsets V_j^t are pairwise disjoint, the construction of the W_j^t's takes total $\mathcal{O}(\sum_{x \in L_{r-i-j}(S_k)} |N_G(x)|)$ time. Furthermore:

Claim 3. $W_j^t = \bigcap\{L(b, j+1, S_k) : b \in B_t\}$.

Proof. We only need to prove that we have $\bigcap\{L(b, j+1, S_k) : b \in B_t\} \subseteq W_j^t$ (the other inclusion being trivial by construction). For that, let $x \in \bigcap\{L(b, j+1, S_k) : b \in B_t\}$ be arbitrary. Recall that we have, for each $b \in B_t$, $d(b, c) = k + d(b, S_k) = r - i + k$. In particular, $x \in L(b, j+1, S_k) = L(b, j+1, c)$. It implies that the balls in $\{N_G[x], N_G^{r-i+k-j}[c]\} \cup \{N_G^j[b] : b \in B_t\}$ pairwise intersect. By the Helly property, x has a neighbour in $N_G^{r-i+k-j}[c] \cap (\bigcap\{N^j[b] : b \in B_t\}) = \bigcap\{L(b, j, S_k) : b \in B_t\} = V_j^t$. Since $x \in L_{r-i-(j+1)}(S_k)$, we get that $x \in W_j^t$. □

Finally, in order to compute the new sets $V_{j+1}^{t'}$, we proceed as follows. Let $\mathcal{W} = \{W_j^t : 1 \leq t \leq p_j\}$. While $\mathcal{W} \neq \emptyset$, we select some vertex $x \in L_{r-i-(j+1)}(S_k)$ maximizing $\#\{t : x \in W_j^t\}$. Then, we create a new set $\bigcap_{t: x \in W_j^t} W_j^t$, and we remove $\{W_j^t : x \in W_j^t\}$ from $\mathcal{W}$. Note that, by the above Claim 3, $\bigcap_{t: x \in W_j^t} W_j^t = \bigcap_{t: x \in W_j^t} \bigcap\{L(b, j+1, S_k) : b \in B_t\} = \bigcap\{L(b, j+1, S_k) : b \in \bigcup_{t: x \in W_j^t} B_t\}$. Furthermore, by maximality of vertex x, $\bigcap_{t: x \in W_j^t} W_j^t$ is disjoint from the subsets in $\{W_j^t : x \notin W_j^t\}$. The latter ensures that all the new sets we create are pairwise disjoint. In order to implement this above process efficiently, we store each $x \in L_{r-i-(j+1)}(S_k)$ in a list indexed by $\#\{t : x \in W_j^t\}$. Then, we traverse these lists by decreasing index. We keep, for each $x \in L_{r-i-(j+1)}(S_k)$, a pointer to its current position in order to dynamically change its list throughout the process. See also the proof of Lemma 2 in [41]. The running time is proportional to $\sum\{|W_j^t| : 1 \leq t \leq p_j\} = \mathcal{O}(\sum_{x \in L_{r-i-j}(S_k)} |N_G(x)|)$.

Second Phase of the Algorithm. Let $C_1, C_2, ..., C_p$ denote the sets $V_{r-i-1}^1, ..., V_{r-i-1}^{p_{r-i-1}}$ (i.e., those obtained at the end of the first phase of our algorithm). Note that $C_1, C_2, ..., C_p$ are subsets of $L_1(S_k)$ $(= N_G(S_k))$. At this point, it is not possible anymore to follow the shortest-paths between $A_{k,i}$ and S_k. Then, let $X = A_{k,i} \cup \{c\}$. Set $\alpha(c) = k + i + 2$ and $\alpha(a) = r$ for each $a \in A_{k,i}$. We define the set $Y = \{y : \forall x \in X, d(y,x) \leq \alpha(x)\}$. Observe that $S_{k,i} = Y \cap S_k$ (recall that $S_{k,i}$ was defined as $\{s \in S_k : A_{k,i} \subseteq N^r[s]\}$). Therefore, in order to compute $S_{k,i}$, it suffices to compute Y.

For that, we proceed in $i + 2$ steps. At step ℓ, for $0 \leq \ell \leq i + 1$, we maintain a family of nonempty pairwise disjoint sets $Z_\ell^1, Z_\ell^2, ..., Z_\ell^{q_\ell}$ and a covering $X_\ell^1, X_\ell^2, \ldots, X_\ell^{q_\ell}$ of X such that the following is true: for every $1 \leq t \leq q_\ell$, $Z_\ell^t = \bigcap_{x \in X_\ell^t} N_G^{\alpha(x)-(i+1)+\ell}[x]$. Doing so, after $i+2$ steps, the set Y is nonempty if and only if $q_{i+1} = 1$ (the above partition is reduced to one group). Furthermore, if it is the case, $Y = Z_{i+1}^1$.

Initially, for $\ell = 0$, we start from $Z_0^1 = C_1, ..., Z_0^p = C_p$, and then the corresponding covering is $\forall 1 \leq t \leq p$, $X_1^t = B_t \cup \{c\}$ (with $B_1, B_2, ..., B_p$ being the partition of $A_{k,i}$ after the first phase of our algorithm). – Note that this is only a covering, and not a partition, because the vertex c is contained in all the groups. – For going from ℓ to $\ell + 1$, we proceed as we did during the first phase. Specifically, for every t, let $U_\ell^t = N_G[Z_\ell^t]$. Since the sets Z_ℓ^t are pairwise disjoint, the computation of all the intermediate sets U_ℓ^t takes total $\mathcal{O}(m)$ time.

Claim 4. $U_\ell^t = \bigcap_{x \in X_\ell^t} N_G^{\alpha(x)-(i+1)+(\ell+1)}[x]$.

The proof is similar to that of Claim 3. Finally, in order to compute the new sets $Z_{\ell+1}^{t'}$, let $\mathcal{U} = \{U_\ell^t : 1 \leq t \leq q_\ell\}$. While $\mathcal{U} \neq \emptyset$, we select some vertex $u \in V$ maximizing $\#\{t : u \in U_\ell^t\}$. Then, we create a new set $\bigcap_{t: u \in U_\ell^t} U_\ell^t$, and we remove $\{U_\ell^t : u \in U_\ell^t\}$ from $\mathcal{U}$. The running time is proportional to $\sum\{|U_\ell^t| : 1 \leq t \leq q_\ell\} = \mathcal{O}(m)$.

Complexity Analysis. Overall, the first phase runs in $\mathcal{O}(m)$ time, and the second phase runs in $\mathcal{O}(im) = \mathcal{O}(km)$ time. Since it applies for k, i fixed, the total running time of the algorithm of Lemma 6 (for k fixed) is in $\mathcal{O}(k^2m)$. □

References

1. Abboud, A., Vassilevska Williams, V., Wang, J.: Approximation and fixed parameter subquadratic algorithms for radius and diameter in sparse graphs. In: SODA, pp. 377–391. SIAM (2016)
2. Abu-Ata, M., Dragan, F.F.: Metric tree-like structures in real-world networks: an empirical study. Networks **67**(1), 49–68 (2016)
3. Bandelt, H.-J., Chepoi, V.: Metric graph theory and geometry: a survey. Contemp. Math. **453**, 49–86 (2008)
4. Bandelt, H.-J., Pesch, E.: Dismantling absolute retracts of reflexive graphs. Eur. J. Comb. **10**(3), 211–220 (1989)
5. Borassi, M., Coudert, D., Crescenzi, P., Marino, A.: On computing the hyperbolicity of real-world graphs. In: Bansal, N., Finocchi, I. (eds.) ESA 2015. LNCS, vol. 9294, pp. 215–226. Springer, Heidelberg (2015). https://doi.org/10.1007/978-3-662-48350-3_19

6. Borassi, M., Crescenzi, P., Habib, M.: Into the square: on the complexity of some quadratic-time solvable problems. Electron. Notes TCS **322**, 51–67 (2016)
7. Brandstädt, A., Chepoi, V., Dragan, F.F.: The algorithmic use of hypertree structure and maximum neighbourhood orderings. DAM **82**(1–3), 43–77 (1998)
8. Brandstädt, A., Dragan, F.F., Chepoi, V., Voloshin, V.: Dually chordal graphs. SIDMA **11**(3), 437–455 (1998)
9. Bringmann, K., Husfeldt, T., Magnusson, M.: Multivariate analysis of orthogonal range searching and graph distances parameterized by treewidth. In: IPEC (2018)
10. Brinkmann, G., Koolen, J., Moulton, V.: On the hyperbolicity of chordal graphs. Ann. Comb. **5**(1), 61–69 (2001)
11. Cabello, S.: Subquadratic algorithms for the diameter and the sum of pairwise distances in planar graphs. ACM TALG **15**(2), 21 (2018)
12. Chalopin, J., Chepoi, V., Dragan, F.F., Ducoffe, G., Mohammed, A., Vaxès, Y.: Fast approximation and exact computation of negative curvature parameters of graphs. In: SoCG 2018, pp. 22:1–22:15 (2018)
13. Chalopin, J., Chepoi, V., Genevois, A., Hirai, H., Osajda, D.: Helly groups (2020)
14. Chalopin, J., Chepoi, V., Hirai, H., Osajda, D.: Weakly modular graphs and nonpositive curvature. Mem. Amer. Math. Soc. 159 (2020)
15. Chepoi, V., Dragan, F.F.: Disjoint sets problem (1992)
16. Chepoi, V., Dragan, F.: A linear-time algorithm for finding a central vertex of a chordal graph. In: van Leeuwen, J. (ed.) ESA 1994. LNCS, vol. 855, pp. 159–170. Springer, Heidelberg (1994). https://doi.org/10.1007/BFb0049406
17. Chepoi, V., Dragan, F.F., Estellon, B., Habib, M., Vaxès, Y.: Diameters, centers, and approximating trees of δ-hyperbolic geodesic spaces and graphs. In: SoCG 2008, pp. 59–68. ACM (2008)
18. Chepoi, V., Dragan, F.F.F., Habib, M., Vaxès, Y., Alrasheed, H.: Fast approximation of eccentricities and distances in hyperbolic graphs. J. Graph Algorithms Appl. **23**(2), 393–433 (2019)
19. Corneil, D., Dragan, F.F., Habib, M., Paul, C.: Diameter determination on restricted graph families. DAM **113**(2–3), 143–166 (2001)
20. Coudert, D., Ducoffe, G.: Recognition of c_4-free and 1/2-hyperbolic graphs. SIDMA **28**(3), 1601–1617 (2014)
21. Coudert, D., Ducoffe, G., Popa, A.: Fully polynomial FPT algorithms for some classes of bounded clique-width graphs. ACM TALG **15**(3) (2019)
22. Dourado, M., Protti, F., Szwarcfiter, J.: Complexity aspects of the Helly property: graphs and hypergraphs. EJC **1000**, 17-12 (2009)
23. Dragan, F.F.: Centers of Graphs and the Helly Property. Ph.D. thesis, Moldava State University, Chişinău (1989). (in Russian)
24. Dragan, F.F.: Conditions for coincidence of local and global minima for eccentricity function on graphs and the Helly property. Stud. Appl. Math. Inf. Sci. 49–56 (1990). (in Russian)
25. Dragan, F.F.: HT-graphs: centers, connected r-domination and Steiner trees. Comput. Sci. J. Moldova (Kishinev) **1**(2), 64–83 (1993)
26. Dragan, F.F.: Dominating cliques in distance-hereditary graphs. In: Schmidt, E.M., Skyum, S. (eds.) SWAT 1994. LNCS, vol. 824, pp. 370–381. Springer, Heidelberg (1994). https://doi.org/10.1007/3-540-58218-5_34
27. Dragan, F.F., Ducoffe, G., Guarnera, H.M.: Fast deterministic algorithms for computing all eccentricities in (hyperbolic) Helly graphs. CoRR, arXiv:2102.08349 (2021)
28. Dragan, F.F., Guarnera, H.M.: Obstructions to a small hyperbolicity in Helly graphs. Discret. Math. **342**(2), 326–338 (2019)

29. Dragan, F.F., Guarnera, H.M.: Eccentricity function in distance-hereditary graphs. Theor. Comput. Sci. **833**, 26–40 (2020)
30. Dragan, F.F., Guarnera, H.M.: Eccentricity terrain of δ-hyperbolic graphs. J. Comput. Syst. Sci. **112**, 50–65 (2020)
31. Dragan, F.F., Habib, M., Viennot, L.: Revisiting radius, diameter, and all eccentricity computation in graphs through certificates. CoRR, arXiv:1803.04660 (2018)
32. Dragan, F.F., Mohammed, A.: Slimness of graphs. DMTCS **21**(3) (2019)
33. Dragan, F.F., Nicolai, F.: LexBFS-orderings of distance-hereditary graphs with application to the diametral pair problem. DAM **98**(3), 191–207 (2000)
34. Dragan, F.F., Nicolai, F., Brandstädt, A.: LexBFS-orderings and powers of graphs. In: d'Amore, F., Franciosa, P.G., Marchetti-Spaccamela, A. (eds.) WG 1996. LNCS, vol. 1197, pp. 166–180. Springer, Heidelberg (1997). https://doi.org/10.1007/3-540-62559-3_15
35. Dragan, F.F., Prisakaru, K., Chepoi, V.: The location problem on graphs and the Helly problem. Diskret. Mat. **4**(4), 67–73 (1992)
36. Dress, A.: Trees, tight extensions of metric spaces, and the cohomological dimension of certain groups: a note on combinatorial properties of metric spaces. Adv. Math. **53**(3), 321–402 (1984)
37. Ducoffe, G.: Around the diameter of AT-free graphs. CoRR, arXiv:2010.15814 (2020)
38. Ducoffe, G.: Distance problems within Helly graphs and k-Helly graphs. CoRR, arXiv:2011.00001 (2020)
39. Ducoffe, G.: Optimal diameter computation within bounded clique-width graphs. CoRR, arXiv:2011.08448 (2020)
40. Ducoffe, G.: Beyond Helly graphs: the diameter problem on absolute retracts. CoRR, arXiv:2101.03574 (2021)
41. Ducoffe, G., Dragan, F.F.: A story of diameter, radius, and (almost) Helly property. Networks, to appear
42. Ducoffe, G., Habib, M., Viennot, L.: Diameter computation on H-minor free graphs and graphs of bounded (distance) VC-dimension. In: SODA, pp. 1905–1922. SIAM (2020)
43. Farley, A., Proskurowski, A.: Computation of the center and diameter of outerplanar graphs. DAM **2**(3), 185–191 (1980)
44. Fournier, H., Ismail, A., Vigneron, A.: Computing the Gromov hyperbolicity of a discrete metric space. IPL **115**(6–8), 576–579 (2015)
45. Gawrychowski, P., Kaplan, H., Mozes, S., Sharir, M., Weimann, O.: Voronoi diagrams on planar graphs, and computing the diameter in deterministic $\tilde{O}(n^{5/3})$ time. In: SODA, pp. 495–514. SIAM (2018)
46. Gromov, M.: Hyperbolic groups. In: Gersten, S.M. (ed.) Essays in Group Theory. MSRI, vol. 8, pp. 75–263. Springer, New York (1987). https://doi.org/10.1007/978-1-4613-9586-7_3
47. Isbell, J.: Six theorems about injective metric spaces. Commentarii Mathematici Helvetici **39**(1), 65–76 (1964)
48. Kennedy, W.S., Saniee, I., Narayan, O.: On the hyperbolicity of large-scale networks and its estimation. In: BigData 2016, pp. 3344–3351. IEEE (2016)
49. Koolen, J.H., Moulton, V.: Hyperbolic bridged graphs. Eur. J. Comb. **23**(6), 683–699 (2002)
50. Koschützki, D., Lehmann, K.A., Peeters, L., Richter, S., Tenfelde-Podehl, D., Zlotowski, O.: Centrality indices. In: Brandes, U., Erlebach, T. (eds.) Network Analysis. LNCS, vol. 3418, pp. 16–61. Springer, Heidelberg (2005). https://doi.org/10.1007/978-3-540-31955-9_3

51. Lin, M., Szwarcfiter, J.: Faster recognition of clique-Helly and hereditary clique-Helly graphs. Inf. Process. Lett. **103**(1), 40–43 (2007)
52. Narayan, O., Saniee, I.: Large-scale curvature of networks. Phys. Rev. E, **84**(6), 066108 (2011)
53. Olariu, S.: A simple linear-time algorithm for computing the center of an interval graph. Int. J. Comput. Math. **34**(3–4), 121–128 (1990)
54. Roditty, L., Vassilevska Williams, V.: Fast approximation algorithms for the diameter and radius of sparse graphs. In: STOC, pp. 515–524. ACM (2013)
55. Wu, Y., Zhang, C.: Hyperbolicity and chordality of a graph. Electr. J. Comb. **18**(1), Paper #P43 (2011)

ANN for Time Series Under the Fréchet Distance

Anne Driemel and Ioannis Psarros(✉)

Hausdorff Center for Mathematics, University of Bonn, Bonn, Germany
driemel@cs.uni-bonn.de, ipsarros@uni-bonn.de

Abstract. We study approximate-near-neighbor data structures for time series under the continuous Fréchet distance. For an attainable approximation factor $c > 1$ and a query radius r, an approximate-near-neighbor data structure can be used to preprocess n curves in $\mathbb{R}$ (aka time series), each of complexity m, to answer queries with a curve of complexity k by either returning a curve that lies within Fréchet distance cr, or answering that there exists no curve in the input within distance r. In both cases, the answer is correct. Our first data structure achieves a $(5+\epsilon)$ approximation factor, uses space in $n \cdot \mathcal{O}\left(\epsilon^{-1}\right)^k + \mathcal{O}(nm)$ and has query time in $\mathcal{O}(k)$. Our second data structure achieves a $(2+\epsilon)$ approximation factor, uses space in $n \cdot \mathcal{O}\left(\frac{m}{k\epsilon}\right)^k + \mathcal{O}(nm)$ and has query time in $\mathcal{O}\left(k \cdot 2^k\right)$. Our third positive result is a probabilistic data structure based on locality-sensitive hashing, which achieves space in $\mathcal{O}(n \log n + nm)$ and query time in $\mathcal{O}(k \log n)$, and which answers queries with an approximation factor in $\mathcal{O}(k)$. All of our data structures make use of the concept of signatures, which were originally introduced for the problem of clustering time series under the Fréchet distance. In addition, we show lower bounds for this problem. Consider any data structure which achieves an approximation factor less than 2 and which supports curves of arc length up to L and answers the query using only a constant number of probes. We show that under reasonable assumptions on the word size any such data structure needs space in $L^{\Omega(k)}$.

Keywords: Data structures · Approximate nearest neighbor · Fréchet distance

1 Introduction

For a long time, Indyk's result on approximate nearest neighbor algorithms for the discrete Fréchet distance of 2002 [20] was the only result known for proximity searching under the Fréchet distance. However, recently there has been a raised

A full version of this paper can be found on arXiv [9]. We thank Karl Bringmann and André Nusser for useful discussions on the topic of this paper. Special thanks go to the anonymous reviewer who pointed out an error in an earlier version of the manuscript, and to Andrea Cremer for careful reading.

A. Lubiw et al. (Eds.): WADS 2021, LNCS 12808, pp. 315–328, 2021.
https://doi.org/10.1007/978-3-030-83508-8_23

interest in this area and several new results have been published [1–3,6,8,10,11,13,15,16,24,25]. An intuitive definition of the Fréchet distance uses the metaphor of a person walking a dog. Imagine the dog walker being restricted to follow the path defined by the first curve while the dog is restricted to the second curve. In this analogy, the Fréchet distance is the shortest length of a dog leash that makes a dog walk feasible. Despite the many results in this area and despite the popularity of the Fréchet distance it is still an open problem how to build efficient data structures for it. Known results either suffer from a large approximation factor or high complexity bounds with dependency on the arclength of the curve, or only support a very restricted set of queries. Before we discuss previous work in more detail, we give a formal definition of the problem we study.

Definition 1 (Fréchet distance). *Given two curves $\pi, \tau : [0,1] \mapsto \mathbb{R}$, their Fréchet distance is:*

$$\mathrm{d}_F(\pi,\tau) = \min_{\substack{f:[0,1]\mapsto[0,1] \\ g:[0,1]\mapsto[0,1]}} \max_{\alpha\in[0,1]} \|\pi(f(\alpha)) - \tau(g(\alpha))\|_2,$$

where f and g range over all continuous, non-decreasing functions with $f(0) = g(0) = 0$, and $f(1) = g(1) = 1$.

Definition 2 (c-ANN problem). *The input consists of n curves Π in $\mathbb{R}^d$. Given a distance threshold $r > 0$, an approximation factor $c > 1$, preprocess Π into a data structure such that for any query τ, the data structure reports as follows:*

- *if $\exists \pi \in \Pi$ s.t. $\mathrm{d}_F(\pi,\tau) \leq r$, then it returns $\pi' \in \Pi$ s.t. $\mathrm{d}_F(\pi',\tau) \leq cr$,*
- *if $\forall \pi \in \Pi$, $\mathrm{d}_F(\pi,\tau) \geq cr$ then it returns "no",*
- *otherwise, it either returns a curve $\pi \in \Pi$ s.t. $\mathrm{d}_F(\pi,\tau) \leq cr$, or "no".*

1.1 Previous Work

Most previous results on data structures for ANN search of curves, concern the *discrete* Fréchet distance. This is a simplification of the distance measure that only takes into account the vertices of the curves. The first non-trivial ANN-data structure for the discrete Fréchet distance from 2002 by Indyk [20] achieved approximation factor $\mathcal{O}((\log m + \log\log n)^{t-1})$, where m is the maximum length of a sequence, and $t > 1$ is a trade-off parameter. More recently, in 2017, Driemel and Silvestri [10] showed that locality-sensitive hashing can be applied and obtained a data structure of near-linear size which achieves approximation factor $\mathcal{O}(k)$, where k is the length of the query sequence. They show how to improve the approximation factor to $\mathcal{O}(d^{3/2})$ at the expense of additional space usage (now exponential in k), and a follow-up result by Emiris and Psarros [11] achieves a $(1+\epsilon)$ approximation, at the expense of further increasing space usage. Recently, Filtser et al. [13] showed how to build a $(1+\epsilon)$-approximate data structure using space in $n \cdot \mathcal{O}(1/\epsilon)^{kd}$ and with query time in $\mathcal{O}(kd)$.

These results are relevant in our setting, since the continuous Fréchet distance can naively be approximated using the discrete Fréchet distance. However, to the best of our knowledge, all known such methods introduce a dependency on the arclength of the curves (resp. the maximum length of an edge), either in the complexity bounds or in the approximation factor. It is not at all obvious how to avoid this when approximating the continuous with the discrete Fréchet distance.

For the continuous Fréchet distance, a recent result by Mirzanezhad [24] can be described as follows. The main ingredient of this data structure is the discretization of the space of query curves with a grid, achieving an approximation factor of $1+\epsilon$. Alas, the space required for each input curve is high, namely roughly D^{dk}, where D is the diameter of the set of vertices of the input, d is the dimension of the input space and k is the complexity of the query.

Interestingly, there are some data structures for the related problem of range searching, which are especially tailored to the case of the continuous Fréchet distance and which do not have a dependency on the arclength.

The subset of input curves, that lie within the search radius of the query curve is called the range of the query. A range query should return all input curves inside the range, or a statistic thereof. Driemel and Afshani [1] consider the exact range searching problem for polygonal curves under the Fréchet distance. For n curves of complexity m in $\mathbb{R}^2$, their data structure uses space in $\mathcal{O}\left(n(\log\log n)^{\mathcal{O}(m^2)}\right)$ and the query time costs $\mathcal{O}\left(\sqrt{n}\log^{\mathcal{O}(m^2)} n\right)$, assuming that the complexity of the query curves is at most $\log^{\mathcal{O}(1)} n$. They also show lower bounds in the pointer model of computation that match the number of log factors used in the upper bounds asymptotically. The new lower bounds that we show in this paper also hold for the case of range searching (more specifically, range emptiness queries), but we assume a different computational model, namely the cell-probe model. While the lower bound of Afshani and Driemel only holds in the case of exact range reporting and uses curves in the plane, our new lower bound also holds in the case of approximation and is meaningful from $d \geq 1$.

1.2 Known Techniques

Our techniques are based on a number of different techniques that were previously used only for the discrete Fréchet distance. In this section we give an overview of these techniques and highlight the main challenges that distinguish the discrete Fréchet distance from the continuous Fréchet distance.

The locality-sensitive hashing scheme proposed by Driemel and Silvestri [10] achieves linear space and query time in $\mathcal{O}(k)$, with an approximation factor of $\mathcal{O}(k)$ for the *discrete* Fréchet distance. The data structure is based on snapping vertices to a randomly shifted grid and then removing consecutive duplicates in the sequence of grid points produced by snapping. Any two near curves produce the same sequence of grid points with constant probability while any two curves, which are sufficiently far away from each other, produce two non-equal sequences of grid points with certainty. The main argument used in the analysis of this scheme involves the optimal discrete matching of the vertices of the two curves. This analysis is not directly applicable to the continuous Fréchet distance as the optimal matching is not always realized at the vertices of the curves.

There are several ANN data structures with fast query time and small approximation factor which store a set of representative query candidates together with precomputed answers for these queries so that a query can be answered approximately with a lookup table. One example of this is the $(1+\epsilon)$-ANN data structure for the ℓ_p norms [17], which employs a grid and stores all those grid points which are near to some data point, and a pointer to the data point that they represent. The side-length of the grid controls the approximation factor and using hashing for storing precomputed solutions leads to an efficient query time. A similar approach was used by Filtser et al. [13] for the $(1+\epsilon)$-ANN problem under the *discrete* Fréchet distance. The algorithm discretizes the query space with a canonical grid and stores representative point sequences on this grid.

There are several challenges when trying to apply the same approach to the ANN problem under the continuous Fréchet distance. Computing good representatives in this case is more intricate: two curves may be near but some of their vertices may be far from any other vertex on the other curve. Hence, picking representative curves which are defined by vertices in the proximity of the vertices of the data curve is not sufficient. In case the input consists of curves with bounded arclength only, one can enumerate all curves which are defined by grid points and lie within a given Fréchet distance. However, this results in a large dependency on the arclength. One of the main questions that we attempt to answer in our paper is whether efficient ANN data structures for the continuous Fréchet distance are possible without such dependency on the arclength.

1.3 Preliminaries

For any $x \in \mathbb{R}$, $|x|$ denotes the absolute value of x. For any positive integer n, $[n]$ denotes the set $\{1,\ldots,n\}$. Throughout this paper, a *curve* is a continuous function $[0,1] \mapsto \mathbb{R}$ and we may refer to such a curve as a *time series*. We can define a curve π as $\pi := \langle x_1,\ldots,x_m\rangle$, which means that π is obtained by linearly interpolating $x_1,\ldots,x_m$. The *vertices* of $\pi : [0,1] \mapsto \mathbb{R}$ are those points which are local extrema in π. For any curve π, $\mathcal{V}(\pi)$ denotes the sequence of vertices of π. The number of vertices $|\mathcal{V}(\pi)|$ is called the *complexity* of π and it is also denoted by $|\pi|$. For any two points x, y, $\overline{xy}$ denotes the directed line segment connecting x with y in the direction from x to y. The segment defined by two consecutive vertices is called an *edge*. For any two $0 \leq p_a < p_b \leq 1$ and any curve π, we denote by $\pi[p_a, p_b]$ the subcurve of π starting at $\pi(p_a)$ and ending at $\pi(p_b)$. For any two curves π_1, π_2, with vertices $x_1,\ldots,x_k$ and $x_k,\ldots,x_m$ respectively, $\pi_1 \oplus \pi_2$ denotes the curve $\langle x_1,\ldots,x_k,\ldots x_m\rangle$, that is the concatenation of π_1 and π_2. We define the *arclength* $\lambda(\pi)$ of a curve π as the total sum of lengths of the edges of π. We refer to a pair of continuous, non-decreasing functions $f : [0,1] \mapsto [0,1]$, $g : [0,1] \mapsto [0,1]$ such that $f(0) = g(0)$, $f(1) = g(1)$, as a *matching*. If a matching $\phi = (f,g)$ of two curves π, τ satisfies

$\max_{\alpha \in [0,1]} \|\pi(f(\alpha)) - \tau(g(\alpha))\| \leq \delta$, then we say that ϕ is a δ-matching of π and τ. Given two curves $\pi : [0,1] \to \mathbb{R}$, $\tau : [0,1] \to \mathbb{R}$. The δ-free space is the subset of the parametric space $[0,1]^2$ defined as $\{(x,y) \in [0,1]^2 \mid |\pi(x) - \tau(y)| \leq \delta\}$.

Our data structures make use of a *dictionary* data structure. A dictionary stores a set of (key, value) pairs and when presented with a key, returns the associated value. Assume we have to store n (key,value) pairs, where the keys come from a universe U^k. Perfect hashing provides us with a dictionary using $O(n)$ space and $O(k)$ query time which can be constructed in $O(n)$ expected time [14]. During look-up, we compute the hash function in $\mathcal{O}(k)$ time, we access the corresponding bucket in the hashtable in $\mathcal{O}(1)$ time and check if the key stored there is equal to the query in $\mathcal{O}(k)$ time.

All of our data structures can operate in the *word-RAM model.* For the sake of simplicity, we state our results in the *real-RAM model,* assuming the availability of a floor function operation in constant time. Our lower bounds are for the *cell-probe model.* The cell-probe model of computation counts the number of memory accesses (cell probes) to the data structure which are performed by a query. Given a universe of data and a universe of queries, a cell-probe data structure with performance parameters s, t, w, is a structure which consists of s memory cells, each able to store w bits, and any query can be answered by accessing t memory cells. Our lower bound concerns approximate distance oracles. A Fréchet distance oracle is a data structure which, given one input curve π, a distance threshold r, and an approximation factor $c > 0$, reports for any query curve τ as follows:(i) if $\mathrm{d}_F(\pi, \tau) \leq r$ then the answer is "yes", (ii) if $\mathrm{d}_F(\pi, \tau) > cr$ then the answer is "no", (iii) otherwise the answer can be either "yes" or "no".

1.4 Our Contributions

We study the c-ANN problem for time series under the continuous Fréchet distance. Our first result is a data structure that achieves approximation factor $5 + \epsilon$ for any $\epsilon > 0$. The data structure is described in Sect. 2 and leads to the following theorem.

Theorem 1. *Let $\epsilon \in (0,1]$. There is a data structure for the $(5+\epsilon)$-ANN problem, which stores n time series of complexity m and supports query time series of complexity k, uses space in $n \cdot \mathcal{O}\left(\frac{1}{\epsilon}\right)^k + \mathcal{O}(nm)$, needs $\mathcal{O}(nm) \cdot \mathcal{O}\left(\frac{1}{\epsilon}\right)^k$ expected preprocessing time and answers a query in $\mathcal{O}(k)$ time.*

To achieve this result, we generate a discrete approximation of the set of all possible non-empty queries. To this end, we employ the concept of signatures, previously introduced in [7]. The signature of a time series provides us with a selection of the local extrema of the function graph, which we use to approximate the set of queries.

We extend these ideas to improve the approximation factor to $(2+\epsilon)$, albeit with an increase in space and query time. In particular, we generate all curves with vertices that lie in the vicinity of the vertices of the input curves. We combine this with a careful analysis of the involved matchings and a more elaborate query algorithm. The resulting data structure can be found in Sect. 3 and leads to the following theorem.

Theorem 2. *Let $\epsilon \in (0, 1]$. There is a data structure for the $(2+\epsilon)$-ANN problem, which stores n time series of complexity m and supports query time series of complexity k, uses space in $n \cdot \mathcal{O}\left(\frac{m}{k\epsilon}\right)^k$, needs $\mathcal{O}(nm) \cdot \mathcal{O}\left(\frac{m}{k\epsilon}\right)^k$ expected preprocessing time and answers a query in $\mathcal{O}(k \cdot 2^k)$ time.*

Our third result is a data structure that uses space in $\mathcal{O}(n \log n + nm)$ and has query time in $\mathcal{O}(k \log n)$. This improvement in the space complexity comes with a sacrifice in the approximation factor achieved by the data structure, which is now in $\mathcal{O}(k)$.

Theorem 3. *There is a data structure for the $(24k+1)$-ANN problem, which stores n time series of complexity m and supports queries with time series of complexity k, uses space in $\mathcal{O}(n \log n + nm)$, needs $\mathcal{O}(nm \log n)$ expected preprocessing time and answers a query in $\mathcal{O}(k \log n)$ time. For a fixed query, the preprocessing succeeds with probability at least $1 - 1/\operatorname{poly}(n)$.*

To achieve this result, we combine the notion of signatures with the ideas of the locality-sensitive scheme that was previously used [10] for the discrete Fréchet distance. In the discrete case, it is sufficient to snap the vertices of the curves to a grid of well-chosen resolution and to remove repetitions of grid points along the curve to obtain a hash index with good probability. In the continuous case, we first compute a signature, which filters the salient points of the curve, and only then apply the grid snapping to this signature to obtain the hash index. The resulting data structure is surprisingly simple. The description of the data structure can be found in Sect. 4.

Finally, we give a lower bound in the cell-probe model of computation, which seems to indicate that for data structures that achieve approximation factor better than 2 and that use a constant number of probes per query, a dependency on the arc-length of the curve is necessary.

Theorem 4. *Consider any Fréchet distance oracle with approximation factor $2-\gamma$, for any $\gamma \in (0, 1]$, distance threshold $r = 1$, in the cell-probe model, which supports time series as follows: it stores any polygonal curve in $\mathbb{R}$ of arclength at most L, for $L \geq 6$, it supports queries of arclength up to L and complexity k, where $k \leq L/6$, and it achieves performance parameters t, w, s. There exist*

$$w_0 = \Omega\left(\frac{L^{1-\epsilon}}{t}\right), \qquad s_0 = 2^{\Omega\left(\frac{k \log(L/k)}{t}\right)}$$

such that if $w < w_0$ then $s \geq s_0$, for any constant $\epsilon > 0$.

To achieve this result we observe that a technique first introduced by Miltersen [23] can be applied here. Miltersen shows that lower bounds for communication problems can be translated into lower bounds for cell-probe data structures. In particular, we use a reduction from the lopsided disjointness problem (see Sect. 5). The proof of the theorem can be found in the full version [9].

In addition, we extend these lower bound results to the case of the discrete Fréchet distance (in the full version [9]). Here, our reduction is more intricate. We adapt a reduction by Bringmann and Mulzer [5], which was used for showing lower bounds for computing the Fréchet distance. Our results show that an exponential dependence on k for the space is necessary when the number of probes is constant (such as in [13]).

1.5 Signatures

A crucial ingredient to our algorithms is the notion of *signatures* which was first introduced in [7]. We define signatures as follows.

Definition 3 (δ-signatures). *A curve $\sigma : [0,1] \mapsto \mathbb{R}$ is a δ-signature of $\tau : [0,1] \mapsto \mathbb{R}$ if it is a curve defined by a series of values $0 = t_1 < \cdots < t_\ell = 1$ as the linear interpolation of $\tau(t_i)$ in the order of the index i, and satisfies the following properties. For $1 \leq i \leq \ell - 1$ the following conditions hold:*

i) (non-degeneracy) if $i \in [2, \ell-1]$ then $\tau(t_i) \notin \overline{\tau(t_{i-1}), \tau(t_{i+1})}$,
ii) (direction-preserving) if $\tau(t_i) < \tau(t_{i+1})$ for $t < t' \in [t_i, t_{i+1}]$: $\tau(t) - \tau(t') \leq 2\delta$ and if $\tau(t_i) > \tau(t_{i+1})$ for $t < t' \in [t_i, t_{i+1}]$: $\tau(t') - \tau(t) \leq 2\delta$,
iii) (minimum edge length) if $i \in [2, \ell-2]$ then $|\tau(t_{i+1}) - \tau(t_i)| > 2\delta$, and if $i \in \{1, \ell-1\}$ then $|\tau(t_{i+1}) - \tau(t_i)| > \delta$,
iv) (range) for $t \in [t_i, t_{i+1}]$: if $i \in [2, \ell-2]$ then $\tau(t) \in \overline{\tau(t_i)\tau(t_{i+1})}$, and if $i = 1$ and $\ell > 2$ then $\tau(t) \in \overline{\tau(t_i)\tau(t_{i+1})} \cup \overline{(\tau(t_i)-\delta)(\tau(t_i)+\delta)}$, and if $i = \ell - 1$ and $\ell > 2$ then $\tau(t) \in \overline{\tau(t_{i-1})\tau(t_i)} \cup \overline{(\tau(t_i)-\delta)(\tau(t_i)+\delta)}$, and if $i = 1$ and $\ell = 2$ then $\tau(t) \in \overline{\tau(t_1)\tau(t_2)} \cup \overline{(\tau(t_1)-\delta)(\tau(t_1)+\delta)} \cup \overline{(\tau(t_2)-\delta)(\tau(t_2)+\delta)}$.

For any $\delta > 0$ and any curve $\pi : [0,1] \mapsto \mathbb{R}$ of complexity m, a δ-signature of π can be computed in $\mathcal{O}(m)$ time [7]. We now state some basic results about signatures.

Lemma 1 (Lemma 3.1 [7]). *It holds for any δ-signature σ of τ: $d_F(\sigma, \tau) \leq \delta$.*

Lemma 2 (Lemma 3.2 [7]). *Let σ with vertices $v_1, \ldots, v_\ell$, be a δ-signature of π with vertices $u_1, \ldots, u_m$. Let $r_i = [v_i - \delta, v_i + \delta]$, for $1 \leq i \leq \ell$, be ranges centered at the vertices of σ ordered along σ. It holds for any time series τ if $d_F(\pi, \tau) \leq \delta$, then τ has a vertex in each range r_i, and such that these vertices appear on τ in the order of i.*

2 A Constant-Factor Approximation for Time Series

In this section, we describe the data structure for Theorem 1. The data structure achieves approximation factor $(5 + \epsilon)$. The full proof as well as pseudocode of the algorithms can be found in the full version of our paper [9].

The Data Structure. The input consists of a set Π of n curves in $\mathbb{R}$, and the approximation error $\epsilon > 0$. To simplify our exposition, we assume that the distance threshold r is equal to 1 (otherwise, we scale the input uniformly). To solve the problem for a different value of r, the input set can be uniformly scaled. Let $\mathcal{G}_w := \{i \cdot w \mid i \in \mathbb{Z}\}$ be the regular grid with side-length $w := \epsilon/2$. Let $\mathcal{H}$ be a dictionary, which is initially empty. For each input curve $\pi \in \Pi$, we compute its 1-signature σ_π, with vertices $\mathcal{V}(\sigma_\pi) = v_1, \ldots, v_\ell$, and for each $v_i \in \mathcal{V}(\sigma_\pi)$ we define the range $r_i := [v_i - 2 - w, v_i + 2 + w]$. We enumerate all curves with at most k vertices, chosen from the sets $r_1 \cap G_w, r_2 \cap G_w, \ldots$, and satisfying the order of i, and we store them in a set $\mathcal{C}'$. Next, we compute the set $\mathcal{C}(\pi) := \{\sigma \in \mathcal{C}' \mid \mathrm{d}_F(\sigma, \pi) \leq 3\}$. We store $\mathcal{C}(\pi)$ in $\mathcal{H}$ as follows: for each $\sigma \in \mathcal{C}(\pi)$, we use as key the sequence of its vertices $\mathcal{V}(\sigma)$: if $\mathcal{V}(\sigma)$ is not already stored in $\mathcal{H}$, then we insert the pair $(\mathcal{V}(\sigma),\pi)$ into $\mathcal{H}$.

The total space required is $\mathcal{O}(n \cdot \max_{\pi \in \Pi} |\mathcal{C}(\pi)|)$.

Our intuition is the following. We would like the set $\mathcal{C}(\pi)$ to contain all those curves that correspond to 2-signatures of query curves that have π as an approximate near neighbor in the set Π. So when presented with a query we can simply compute its 2-signature and do a lookup in $\mathcal{H}$. However, the set of all possible 2-signatures with non-empty query is infinite. Therefore, we snap the vertices to a grid to obtain a discrete set of bounded size.

The Query Algorithm. When presented with a query curve τ, we first compute a 2-signature σ_τ, and then we compute a key by snapping the vertices to the same grid $\mathcal{G}_w$. Snapping to $\mathcal{G}_w$ is implemented as follows: if $\mathcal{V}(\sigma_\tau) = v_1, \ldots, v_\ell$ then $\sigma'_\tau := \langle g_w(v_1), \ldots, g_w(v_\ell) \rangle$, where for any $x \in \mathbb{R}$, $g_w(x)$ is the nearest point of x in $\mathcal{G}_w$. We perform a lookup in $\mathcal{H}$ with the key $\mathcal{V}(\sigma'_\tau)$ and return the result: if $\mathcal{V}(\sigma'_\tau)$ is stored in $\mathcal{H}$ then we return the associated curve, otherwise we return "no".

Lemma 3. *Let τ be a query curve of complexity k. If the query algorithm returns an input curve $\pi' \in \Pi$, then $\mathrm{d}_F(\pi', \tau) \leq 5 + \epsilon$. If the query algorithm returns "no", then there is no $\pi \in \Pi$ such that $\mathrm{d}_F(\pi, \tau) \leq 1$.*

Proof. Let π be any input curve in Π and let σ_π be the 1-signature of π. Let τ be a query curve, let σ_τ be its 2-signature and let σ'_τ be as defined in the query algorithm. First suppose that $\mathrm{d}_F(\pi, \tau) \leq 1$. By the triangle inequality and Lemma 1, $\mathrm{d}_F(\pi, \sigma_\tau) \leq 3 + w$. Let $u_1, \ldots, u_{\ell'}$ be the vertices of σ_τ and define for each $i \in [\ell']$, $r'_i := [u_i - 2, u_i + 2]$. By Lemma 2, σ_π has a vertex in each range r'_i and these vertices appear on σ_π in the order of i. This guarantees that the vertices of σ'_τ lie in the ranges $r_1, \ldots, r_\ell$ and it will be considered during preprocessing. Hence, σ'_τ will be generated when preprocessing π. This implies that $\mathcal{V}(\sigma'_\tau)$ is stored in $\mathcal{H}$. It is possible that σ'_τ was also generated and stored for a different input curve, say $\pi' \neq \pi$ with $\mathrm{d}_F(\pi', \sigma'_\tau) \leq 3$. We claim that $\mathrm{d}_F(\pi', \tau) \leq 5 + 2w$. Indeed, we have by the triangle inequality

$$\mathrm{d}_F(\pi', \tau) \leq \mathrm{d}_F(\pi', \sigma'_\tau) + \mathrm{d}_F(\sigma'_\tau, \sigma_\tau) + \mathrm{d}_F(\sigma_\tau, \tau) \leq 5 + 2w.$$

This proves that any curve returned by the query algorithm has Fréchet distance at most $5 + 2w = 5 + \epsilon$ to the query curve, and if the query algorithm returns "no", then there is no input curve within Fréchet distance 1 to the query curve. □

By Lemma 3 the data structure returns a correct result. To show Theorem 1, it remains to analyze the complexity. We sketch the analysis of the candidate set which is generated during preprocessing. Indeed, we will show now that $|\mathcal{C}'| \leq \mathcal{O}\left(\frac{1}{\epsilon}\right)^k$. Notice that if there exists a curve with k vertices which is within distance 1 from π then $\ell \leq k$, by Lemma 2. Recall that the curves in $|\mathcal{C}'|$ have vertices in the ranges $r_i \cap \mathcal{G}_w$ and the vertices respect the order of i. If we fix the choices of $t_1, \ldots, t_\ell$, where each t_i denotes the number of vertices in $r_i \cap \mathcal{G}_w$, we can produce at most $\prod_{i=1}^{\ell} |r_i \cap \mathcal{G}_w|^{t_i}$ distinct sequences of vertices of length $\sum_{i=1}^{\ell} t_i$ and hence at most $\prod_{i=1}^{\ell} |r_i \cap \mathcal{G}_w|^{t_i}$ curves of length at most $\sum_{i=1}^{\ell} t_i$. Hence,

$$|\mathcal{C}'| \leq \sum_{\substack{t_1+\ldots+t_\ell=k \\ \forall i:\ t_i \geq 0 \\ t_1 \geq 1, t_\ell \geq 1}} \prod_{i=1}^{\ell} \left(\frac{4}{\epsilon}+2\right)^{t_i} \leq \sum_{\substack{t_1+\ldots+t_\ell=k \\ \forall i:\ t_i \geq 0}} \left(\frac{4}{\epsilon}+2\right)^k$$
$$\leq \binom{k+\ell-1}{k} \cdot \left(\frac{4}{\epsilon}+2\right)^k \leq (2\mathrm{e})^k \cdot \left(\frac{4}{\epsilon}+2\right)^k = \mathcal{O}\left(\frac{1}{\epsilon}\right)^k.$$

3 Improving the Approximation Factor to $(2 + \epsilon)$

In this section, we describe the data structure for Theorem 2. The full proof of this theorem as well as pseudocode can be found in the full version of our paper [9]. We build upon the ideas developed in Sect. 2. The key to circumventing the larger approximation factor resulting from the use of the triangle inequality seems to be a careful construction of matchings. For this we define the notion of a δ-tight matching for two curves. Figure 1 illustrates the approach.

Definition 4. (δ-tight matching). *Given two curves π and τ, consider a monotone path λ through the parametric space of π and τ consisting of two types of segments:*

(i) a segment contained in the 0-free space (corresponding to identical subcurves of π and τ),
(ii) a horizontal line segment contained in the δ-free space (corresponding to a point on π and a subcurve on τ).

If λ exists, we say λ is a tight matching *of width δ from π to τ.*

Lemma 4. *Let $X = \overline{ab} \subset \mathbb{R}$ be a line segment and let τ and π be curves with $[a, b] \subseteq [\tau(0), \tau(1)]$ and $[a, b] \subseteq [\pi(0), \pi(1)]$. If $\mathrm{d}_F(X, \tau) = \delta_1$ and $\mathrm{d}_F(X, \pi) = \delta_2$, then $\mathrm{d}_F(\tau, \pi) \leq \max(\delta_1, \delta_2)$.*

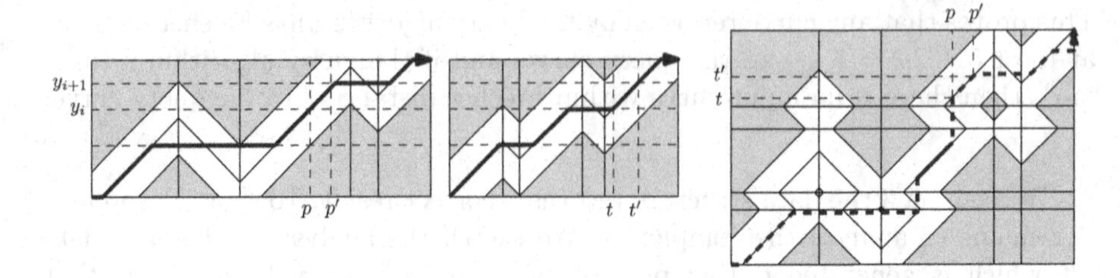

Fig. 1. Example of the path constructed in the proof of Lemma 4. The left figure shows a tight matching from X to π. The middle figure shows a tight matching from X to τ. Diagonal edges of the 0-free space of these can be transferred to the diagram on the right, which is the free space diagram of π and τ. The final path results from connecting these diagonal segments using horizontal and vertical line segments.

Theorem 5. *Let τ be a curve with vertices $\tau(t_1), \ldots, \tau(t_m)$, and let σ_τ be a δ-signature of τ with vertices $\tau(t_{s_1}), \ldots, \tau(t_{s_\ell})$. Let τ' be a curve obtained by deleting any subset of vertices of τ which are not in σ_τ, i.e. $\tau' = \langle \tau(t'_1), \ldots, \tau(t'_k) \rangle$, where $\{t_{s_1}, \ldots t_{s_\ell}\} \subseteq \{t'_1, \ldots, t'_k\} \subseteq \{t_1, \ldots, t_m\}$. Then $\mathrm{d}_F(\tau, \tau') \leq \delta$.*

The Data Structure. The input consists of a set Π of n curves in $\mathbb{R}$, and the approximation error $\epsilon > 0$. As before, we assume that the distance threshold is $r := 1$ (otherwise we can uniformly scale the input). To discretize the query space, we use the regular grid $\mathcal{G}_w := \{i \cdot w \mid i \in \mathbb{Z}\}$, where $w := \epsilon/2$. Let $\mathcal{H}$ be a dictionary which is initially empty. For each input curve $\pi \in \Pi$, with vertices $\mathcal{V}(\pi) = v_1, \ldots, v_m$, we set $r_i = [v_i - 4 - w, v_i + 4 + w]$, for $i \in [m]$, and we compute a set $\mathcal{C}' := \mathcal{C}'(\pi)$ which contains all curves with at most k vertices such that each vertex belongs to some $r_i \cap \mathcal{G}_w$ and the vertices are ordered in the order of i. More formally,

$$\mathcal{C}' = \{\langle u_1, \ldots, u_\ell \rangle \mid \ell \leq k, \exists i_1, \ldots, i_\ell \text{ s.t. } i_1 \leq \cdots \leq i_\ell \text{ and } \forall j \in [\ell]\ u_j \in r_{i_j} \cap \mathcal{G}_w\}.$$

Next, we filter $\mathcal{C}'$ to obtain the set $\mathcal{C}(\pi) := \{\sigma \in \mathcal{C}' \mid \mathrm{d}_F(\sigma, \pi) \leq 1 + w\}$.

We store $\mathcal{C}(\pi)$ in $\mathcal{H}$ as follows: for each $\sigma \in \mathcal{C}(\pi)$, we use as key the sequence of its vertices $\mathcal{V}(\sigma)$: if $\mathcal{V}(\sigma)$ is not already stored in $\mathcal{H}$, then we insert the pair $(\mathcal{V}(\sigma), \pi)$ into $\mathcal{H}$. The total space required is $\mathcal{O}(n \cdot \max_{\pi \in \Pi} |\mathcal{C}(\pi)|)$.

The Query Algorithm. For a query curve τ, the algorithm `query`(τ) first computes the 1-signature of τ, namely σ, and then enumerates all possible curves τ_{key} which are produced from τ by deleting vertices that are not in σ. For each possible τ_{key}, we compute $\tilde{\tau}_{key} := \langle g_w(v_1), \ldots, g_w(v_\ell) \rangle$, where for any $x \in \mathbb{R}$, $g_w(x)$ is the nearest point of x in $\mathcal{G}_w$. For each $\tilde{\tau}_{key}$ we perform a lookup in $\mathcal{H}$, with key $\mathcal{V}(\tilde{\tau}_{key})$: if $\mathcal{V}(\tilde{\tau}_{key})$ is stored in $\mathcal{H}$ then we return the associated curve. If there is no $\tilde{\tau}_{key}$ such that $\mathcal{V}(\tilde{\tau}_{key})$ is stored in $\mathcal{H}$ then the algorithm returns "no".

4 An $\mathcal{O}(k)$-ANN Data Structure with Near-Linear Space

In this section we give the data structure for Theorem 3. The full proof can be found in the full version of our paper [9]. The data structure has approximation factor of order $\mathcal{O}(k)$, but it uses space in $\mathcal{O}(n \log n + nm)$ and query time in $\mathcal{O}(k \log n)$. Our main ingredient is a properly-tuned randomly shifted grid: Let $w > 0$ be a fixed parameter and z chosen uniformly at random from the set $[0, w]$. The function $g_{w,z}(x) = \lfloor w^{-1}(x - z) \rfloor$ induces a random partition of the line.

The Data Structure. The input consists of a set Π of n curves in $\mathbb{R}$. As before, we assume that the distance threshold is $r := 1$. Let $w = 48k$. We build $s = O(\log n)$ dictionaries $\mathcal{H}_1, \ldots, \mathcal{H}_s$ which are initially empty. For each $i \in [s]$, we sample z_i uniformly and independently at random from $[0, w]$. For each input curve $\pi \in \Pi$, we compute its 1-signature σ_π, with vertices $\mathcal{V}(\sigma_\pi) = v_1, \ldots, v_\ell$, and for each $i \in [s]$ we compute the curve $\sigma'_{\pi|i} = \langle g_{w,z_i}(v_1), \ldots, g_{w,z_i}(v_\ell) \rangle$. For each $\pi \in \Pi$, such that $|V(\sigma_\pi)| \leq k$, we use as key in $\mathcal{H}_i$ the sequence of vertices $\mathcal{V}(\sigma'_{\pi|i})$: if $\mathcal{V}(\sigma'_{\pi|i})$ is not already stored in $\mathcal{H}_i$, then we insert the pair $(\mathcal{V}(\sigma'_{\pi|i}), \pi)$.

The Query Algorithm. When presented with a query curve τ, with vertices $u_1, \ldots, u_k$, we compute for each $i \in [s]$, the curve $\tau'_i = \langle g_{w,z_i}(u_1), \ldots, g_{w,z_i}(u_k) \rangle$. Then, for each $i \in [s]$, we perform a lookup in $\mathcal{H}_i$ with the key $\mathcal{V}(\tau_i{}')$ and return the result: if $\exists i \in [s]$ such that $\mathcal{V}(\tau_i{}')$ is stored in $\mathcal{H}_i$ then we return the curve associated with it. Otherwise we return "no". (Recall that $\mathcal{V}(\tau_i{}')$ only retains the maxima and minima of the sequence $g_{w,z_i}(u_1), \ldots, g_{w,z_i}(u_k)$.)

5 A Lower Bound in the Cell-Probe Model

Our lower bound of Theorem 4 works by reducing the lopsided set disjointness problem to the problem of approximating the Fréchet distance of two curves in $\mathbb{R}$. (A similar reduction appears in [22], which however works for curves in $\mathbb{R}^2$.) The full proof is diverted to the full version [9].

First consider an instance of the set disjointness problem: Alice has a set $A = \{\alpha_1, \ldots, \alpha_k\} \subset [U]$ and Bob has a set $B = \{\beta_1, \ldots, \beta_m\} \subset [U]$, where U is the size of the universe. We now describe our main gadgets which will be used to define one curve of complexity $\mathcal{O}(k)$ for A and one curve of complexity $\mathcal{O}(U - m)$ for B. For each $i \in [U]$:

- If $i \in A$ then $x_{2i-1} := 4i + 4$, $x_{2i} := 4i$,
- If $i \notin A$ then $x_{2i-1} := 4i$, $x_{2i} := 4i$,
- If $i \in B$ then $y_{2i-1} := 4i$, $y_{2i} := 4i$,
- If $i \notin B$ then $y_{2i-1} := 4i + 3$, $y_{2i} := 4i + 1$,

We now define $\tilde{x} := \langle 0, x_1, \ldots, x_{2U}, 4U + 5 \rangle$ and $\tilde{y} := \langle 0, y_1, \ldots, y_{2U}, 4U + 5 \rangle$. Notice that the number of vertices of $\tilde{x}$ is $2k + 2$, and the number of vertices of $\tilde{y}$ is $2(U - m) + 2$, because we only take into account vertices which are local extremes. The arclength of any of $\tilde{x}$, $\tilde{y}$ is at most $12U + 2$.

Theorem 6. *If $A \cap B = \emptyset$ then $\mathrm{d}_F(\tilde{x}, \tilde{y}) \leq 1$. If $A \cap B \neq \emptyset$ then $\mathrm{d}_F(\tilde{x}, \tilde{y}) \geq 2$.*

6 Conclusions

We have described and analyzed a simple $(5+\epsilon)$ -ANN data structure. Focusing on improving the approximation factor, while compromising other performance parameters, we presented a $(2+\epsilon)$-ANN data structure for time series under the continuous Fréchet distance. In doing so, we have presented the new technique of constructing so-called *tight matchings*, which may be of independent interest. In addition, we have also presented a $\mathcal{O}(k)$-ANN randomized data structure for time series under the Fréchet distance, with near-linear space usage and query time in $\mathcal{O}(k \log n)$. We also showed lower bounds in the cell-probe model, which indicate that an approximation better than 2 cannot be achieved, unless we allow space usage depending on the arclength of the time series or allow superconstant number of probes. Our bounds are not tight. In particular, they leave open the possibility of a data structure with approximation factor $(2+\epsilon)$, with space usage in $n \cdot \mathcal{O}(\epsilon^{-1})^k$, and which answers any query using only a constant number of probes.[1] Moreover, it is possible that even an approximation factor of $(1+\epsilon)$ can be achieved with space and query time similar to Theorem 1.

Apart from these improvements, several open questions remain, we discuss two main research directions:

1. Are there data structures with similar guarantees for the ANN problem under the continuous Fréchet distance for curves in the plane (or higher dimensions)? Our approach uses signatures, which are tailored to the 1-dimensional setting. A related concept for curves in higher dimensions is the *curve simplification*. It is an open problem if it is possible to apply simplifications in place of signatures to obtain similar results.
2. The lower bounds presented in this paper are only meaningful when the number of probes is constant. Can we find lower bounds for the setting that query time is polynomial in k and m, and logarithmic in n?

One of the aspects that make our results and these open questions interesting is that known generic approaches designed for general classes of metric spaces cannot be applied. There exist several data structures which operate on general metric spaces with bounded doubling dimension (see e.g. [4,18,21]). However, the doubling dimension of the metric space defined over the space of time series with the continuous Fréchet distance is unbounded [7]. Another aspect that makes our problem difficult, is that the Fréchet distance does not exhibit a norm structure. In this sense it is very similar to the well-known Hausdorff distance for sets, which is equally challenging from the point of view of data structures (see also the discussion in [12,19]). We hope that answering the above research questions will lead to new techniques for handling such distance measures.

[1] In fact, an earlier version of this manuscript claimed such a result, but it contained a flaw.

References

1. Afshani, P., Driemel, A.: On the complexity of range searching among curves. In: Proceedings of the 28th Annual ACM-SIAM Symposium on Discrete Algorithms, SODA 2018, pp. 898–917 (2018). https://doi.org/10.1137/1.9781611975031.58
2. Aronov, B., Filtser, O., Horton, M., Katz, M.J., Sheikhan, K.: Efficient nearest-neighbor query and clustering of planar curves. In: Friggstad, Z., Sack, J.-R., Salavatipour, M.R. (eds.) WADS 2019. LNCS, vol. 11646, pp. 28–42. Springer, Cham (2019). https://doi.org/10.1007/978-3-030-24766-9_3
3. de Berg, M., Gudmundsson, J., Mehrabi, A.D.: A dynamic data structure for approximate proximity queries in trajectory data. In: Proceedings of the 25th ACM SIGSPATIAL International Conference on Advances in Geographic Information Systems, SIGSPATIAL 2017 (2017). https://doi.org/10.1145/3139958.3140023
4. Beygelzimer, A., Kakade, S.M., Langford, J.: Cover trees for nearest neighbor. In: Proceedings of the 23rd International Conference (ICML) on Machine Learning, pp. 97–104 (2006). https://doi.org/10.1145/1143844.1143857
5. Bringmann, K., Mulzer, W.: Approximability of the discrete Fréchet distance. JoCG **7**(2), 46–76 (2016). https://doi.org/10.20382/jocg.v7i2a4
6. De Berg, M., Cook, A.F., Gudmundsson, J.: Fast Fréchet queries. Comput. Geom. **46**(6), 747–755 (2013)
7. Driemel, A., Krivosija, A., Sohler, C.: Clustering time series under the Fréchet distance. In: Proceedings of the 27th Annual ACM-SIAM Symposium on Discrete Algorithms, SODA, pp. 766–785 (2016). https://doi.org/10.1137/1.9781611974331.ch55
8. Driemel, A., Phillips, J.M., Psarros, I.: The VC dimension of metric balls under Fréchet and Hausdorff distances. In: Proceedings of the 35th International Symposium on Computational Geometry, pp. 28:2–28:16 (2019)
9. Driemel, A., Psarros, I.: $(2+\epsilon)$-ANN for time series under the Fréchet distance. CoRR abs/2008.09406 (2020). https://arxiv.org/abs/2008.09406
10. Driemel, A., Silvestri, F.: Locally-sensitive hashing of curves. In: Proceedings of 33rd International Symposium on Computational Geometry, pp. 37:1–37:16 (2017)
11. Emiris, I.Z., Psarros, I.: Products of Euclidean metrics and applications to proximity questions among curves. In: Proceedings of 34th International Symposium on Computational Geometry (SoCG), LIPIcs, vol. 99, pp. 37:1–37:13 (2018)
12. Farach-Colton, M., Indyk, P.: Approximate nearest neighbor algorithms for Hausdorff metrics via embeddings. In: 40th Annual Symposium on Foundations of Computer Science, FOCS 1999, New York, NY, USA, 17–18 October 1999, pp. 171–180 (1999). https://doi.org/10.1109/SFFCS.1999.814589
13. Filtser, A., Filtser, O., Katz, M.J.: Approximate nearest neighbor for curves - simple, efficient, and deterministic. In: 47th International Colloquium on Automata, Languages, and Programming, ICALP 2020, pp. 48:1–48:19 (2020). https://doi.org/10.4230/LIPIcs.ICALP.2020.48
14. Fredman, M.L., Komlós, J., Szemerédi, E.: Storing a sparse table with $O(1)$ worst case access time. J. ACM **31**(3), 538–544 (1984)
15. Gudmundsson, J., Horton, M., Pfeifer, J., Seybold, M.P.: A practical index structure supporting Fréchet proximity queries among trajectories (2020)
16. Gudmundsson, J., Smid, M.: Fast algorithms for approximate Fréchet matching queries in geometric trees. Comput. Geom. **48**(6), 479–494 (2015). https://doi.org/10.1016/j.comgeo.2015.02.003

17. Har-Peled, S., Indyk, P., Motwani, R.: Approximate nearest neighbor: towards removing the curse of dimensionality. Theory Comput. **8**(1), 321–350 (2012). https://doi.org/10.4086/toc.2012.v008a014
18. Har-Peled, S., Mendel, M.: Fast construction of nets in low-dimensional metrics and their applications. SIAM J. Comput. **35**(5), 1148–1184 (2006). https://doi.org/10.1137/S0097539704446281
19. Indyk, P.: On approximate nearest neighbors in non-Euclidean spaces. In: 39th Annual Symposium on Foundations of Computer Science, FOCS 1998, pp. 148–155 (1998). https://doi.org/10.1109/SFCS.1998.743438
20. Indyk, P.: Approximate nearest neighbor algorithms for Fréchet distance via product metrics. In: Symposium on Computational Geometry, pp. 102–106 (2002)
21. Krauthgamer, R., Lee, J.R.: Navigating nets: simple algorithms for proximity search. In: Proceedings of the Fifteenth Annual ACM-SIAM Symposium on Discrete Algorithms, SODA 2004, pp. 798–807 (2004). http://dl.acm.org/citation.cfm?id=982792.982913
22. Meintrup, S., Munteanu, A., Rohde, D.: Random projections and sampling algorithms for clustering of high-dimensional polygonal curves. NeurIPS **2019**, 12807–12817 (2019)
23. Miltersen, P.B.: Lower bounds for union-split-find related problems on random access machines. In: Proceedings of the Twenty-sixth Annual ACM Symposium on Theory of Computing, STOC 1994, pp. 625–634. ACM (1994). https://doi.org/10.1145/195058.195415
24. Mirzanezhad, M.: On the approximate nearest neighbor queries among curves under the Fréchet distance. CoRR abs/2004.08444 (2020). https://arxiv.org/abs/2004.08444
25. Werner, M., Oliver, D.: ACM SIGSPATIAL GIS Cup 2017: range queries under Fréchet distance. SIGSPATIAL Spec. **10**(1), 24–27 (2018). https://doi.org/10.1145/3231541.3231549

Strictly In-Place Algorithms for Permuting and Inverting Permutations

Bartłomiej Dudek, Paweł Gawrychowski(✉), and Karol Pokorski

Institute of Computer Science, University of Wrocław, Wrocław, Poland
{bartlomiej.dudek,gawry,pokorski}@cs.uni.wroc.pl

Abstract. We revisit the problem of permuting an array of length n according to a given permutation in place, that is, using only a small number of bits of extra storage. Fich, Munro and Poblete [FOCS 1990, SICOMP 1995] obtained an elegant $\mathcal{O}(n \log n)$-time algorithm using only $\mathcal{O}(\log^2 n)$ bits of extra space for this basic problem by designing a procedure that scans the permutation and outputs exactly one element from each of its cycles. However, in the strict sense in place should be understood as using only an asymptotically optimal $\mathcal{O}(\log n)$ bits of extra space, or storing a constant number of indices. The problem of permuting in this version is, in fact, a well-known interview question, with the expected solution being a quadratic-time algorithm. Surprisingly, no faster algorithm seems to be known in the literature.

Our first contribution is a strictly in-place generalisation of the method of Fich et al. that works in $\mathcal{O}_\varepsilon(n^{1+\varepsilon})$ time, for any $\varepsilon > 0$. Then, we build on this generalisation to obtain a strictly in-place algorithm for inverting a given permutation on n elements working in the same complexity. This is a significant improvement on a recent result of Guśpiel [arXiv 2019], who designed an $\mathcal{O}(n^{1.5})$-time algorithm.

1 Introduction

Permutations are often used as building blocks in combinatorial algorithms operating on more complex objects. This brings the need for being able to efficiently operate on permutations. One of the most fundamental operations is rearranging an array $A[1..n]$ according to a permutation π. This can be used, for example, to transpose a rectangular array [24, Ex. 1.3.3-12]. Denoting by a_i the value stored in $A[i]$, the goal is to make every $A[i] = a_{\pi^{-1}(i)}$. This is trivial if we can allocate a temporary array $B[1..n]$ and, after setting $B[\pi(i)] \leftarrow A[i]$ for every i, copy $B[1..n]$ to $A[1..n]$. Alternatively, one can iterate over the cycles of π and rearrange the values on each cycle. Then there is no need for allocating a temporary array as long as we can recognise the elements of π in already processed cycles. This is easy if we can overwrite π, say by setting $\pi(i) \leftarrow i$ after having processed $\pi(i)$. However, we might want to use the same π later, and thus cannot overwrite its elements. In such a case, assuming that every $\pi(i)$ can store at least one extra bit, we could mark the processed elements by temporarily setting $\pi(i) \leftarrow -\pi(i)$, and after having rearranged the array restoring the original π. Even though

A. Lubiw et al. (Eds.): WADS 2021, LNCS 12808, pp. 329–342, 2021.
https://doi.org/10.1007/978-3-030-83508-8_24

this reduces the extra space to just one bit per element, this might be still too much, and $\pi(i)$ might be not stored explicitly but computed on-the-fly, as in the example of transposing a rectangular array. This motivates the challenge of designing an efficient algorithm that only assumes access to π through an oracle and uses a small number of bits of extra storage. This is in fact a known interview puzzle [1, Sec. 6.9]. The expected solution is a quadratic-time algorithm that identifies the cycles of π by iterating over $i = 1, 2, \ldots, n$ and checking if i is the smallest on its cycle in π. Having identified such i, we permute the values $A[i], A[\pi(i)], A[\pi^2(i)], \ldots$. This uses only a constant number of auxiliary variables, or $\mathcal{O}(\log n)$ bits of additional space, which is asymptotically optimal as we need to be able to specify an index consisting of $\lceil \log n \rceil$ bits. However, its worst-case running time is quadratic. Designing a faster solution is nontrivial, but Fich, Munro and Poblete obtained an elegant $\mathcal{O}(n \log n)$-time algorithm using only $\mathcal{O}(\log^2 n)$ bits of extra space [13]. Their approach is also based on identifying the cycles of π. This is implemented by scanning the elements $i = 1, 2, \ldots, n$ while testing if the current i is the leader on its cycle, designating exactly one element on every cycle to be its leader. We call this a cycle leaders procedure. Due to the unidirectional nature of the input, the test must be implemented by considering the elements $i, \pi(i), \pi^2(i), \ldots$ until we can conclude if i is the leader of its cycle. The main contribution of Fich et al. is an appropriate definition of a leader that allows to implement such a test while storing only $\mathcal{O}(\log n)$ indices and making the total number of accesses to π only $\mathcal{O}(n \log n)$. They also show an algorithm running in $\mathcal{O}(n^2/b)$ time and $b + \mathcal{O}(\log n)$ bits of space for arbitrary $b \leq n$.

A procedure that transforms the input using only a small number of bits of extra storage is usually referred to as an in-place algorithm. The allowed extra space depends on the problem, but in the most strict form, this is $\mathcal{O}(\log n)$ bits where n is the size of the input. We call such a procedure strictly in-place. This is related to the well-studied complexity class L capturing decision problems solvable by a deterministic Turing machine with $\mathcal{O}(\log n)$ bits of additional writable space with read-only access to the input. There is a large body of work concerned with time-space tradeoffs assuming read-only random access to the input. Example problems include: sorting and selection [2,4–7,17,25–27,29], constructing the convex hull [8], multiple pattern matching [14] or constructing the sparse suffix array [3,18]. This raises the question: is there a deterministic subquadratic strictly in-place algorithm for permuting an array? We provide an affirmative answer to this question by designing, for every $\varepsilon > 0$, a strictly in-place algorithm for this problem that works in $\mathcal{O}_\varepsilon(n^{1+\varepsilon})$ time.[1]

Previous and Related Work. Fich et al. [13] designed a cycle leaders algorithm that works in $\mathcal{O}(n \log n)$ time and uses $\mathcal{O}(\log^2 n)$ bits of extra space. Given oracle access to both π and π^{-1}, they also show a simpler algorithm that needs only $\mathcal{O}(\log n)$ bits of extra space and the same time. Both algorithms have interesting connections to leader election in, respectively, unidirectional [9,28] and bidirec-

[1] We write $\mathcal{O}_\varepsilon(f(n))$ to emphasise that the hidden constant depends on ε.

tional [16,21] rings. While in certain scenarios, such as π being specified with an explicit formula, one can assume access to both π and π^{-1}, in the general case this is known to significantly increase the necessary space [19]. For a random input, traversing the cycle from i until we encounter a smaller element takes in total $\mathcal{O}(n \log n)$ average time and uses only $\mathcal{O}(\log n)$ bits of space [23]. Using hashing, one can also design an algorithm using expected $\mathcal{O}(n \log n)$ time and $\mathcal{O}(\log n)$ bits of space without any assumption on π (interestingly, a different application of randomisation is known to help in leader election in anonymous unidirectional rings [22]). Better cycle leaders algorithms are known for some specific permutations, such as the perfect shuffle [12].

An interesting related question is that of inverting a given permutation π on n elements. For example, it allows us to work with both the suffix array and the inverse suffix array without explicitly storing both of them [15]. The goal of this problem is to replace π with its inverse π^{-1} efficiently while using a small amount of extra space. El-Zein, Munro and Robertson [11] solve this in $\mathcal{O}(n \log n)$ time using only $\mathcal{O}(\log^2 n)$ bits of extra space. The high-level idea of their procedure is to identify cycle leaders and invert every cycle of π at its leader i. The difficulty in such an approach is that the leader i' of the inverted cycle might be encountered again, forcing the cycle to be restored to its original state. El-Zein et al. deal with this hurdle by temporarily lifting the restriction that π is a permutation. Very recently, Guśpiel [20] designed a strictly in-place algorithm for this problem that works in $\mathcal{O}(n^{1.5})$ time. We stress that his approach does not provide a subquadratic strictly in-place solution for identifying cycle leaders nor permuting an array. See Table 1 with the summary of previous work.

Table 1. Comparison of different algorithms for in-place permuting and inverting a permutation. $(f(n), g(n))$ denotes that the algorithm runs in $\mathcal{O}(f(n))$ time and uses $\mathcal{O}(g(n))$ bits of space.

<table>
<tr><th></th><th>Permuting</th><th>Inverting</th></tr>
<tr><td rowspan="2">Trivial</td><td colspan="2">(n, n)</td></tr>
<tr><td colspan="2">$(n^2, \log n)$</td></tr>
<tr><td>With hashing[2]</td><td>$(n \log n, \log n)$</td><td>-</td></tr>
<tr><td rowspan="2">Fich et al. [13]</td><td>$(n^2/b, b + \log n)$</td><td>-</td></tr>
<tr><td>$(n \log n, \log^2 n)$</td><td>-</td></tr>
<tr><td>El-Zein et al. [11]</td><td>-</td><td>$(n \log n, \log^2 n)$</td></tr>
<tr><td>Guśpiel [20]</td><td>-</td><td>$(n^{1.5}, \log n)$</td></tr>
<tr><td>This work[3]</td><td colspan="2">$(n^{1+\varepsilon}, \log n)$</td></tr>
</table>

Our Contribution. Building on the approach of Fich et al. we design, for every $\varepsilon > 0$, a cycle leaders algorithm that works in $\mathcal{O}_\varepsilon(n^{1+\varepsilon})$ time and uses $\mathcal{O}_\varepsilon(\log n)$

[2] This approach runs in expected $\mathcal{O}(n \log n)$ time.

[3] The constant in time and space complexity depends on ε.

bits of extra space. This implies a strictly in-place algorithm for permuting an array in $\mathcal{O}_\varepsilon(n^{1+\varepsilon})$ time. In other words, we show that by increasing the number of auxiliary variables to a larger constant we can make the exponent in the running time arbitrarily close to 1. Then, we apply our improved cycle leaders algorithm to obtain a solution for inverting a given permutation in the same time and space. This significantly improves on the recent result of Guśpiel [20].

Techniques and Roadmap. The main high-level idea in Fich et al. is to work with local minima, defined as the elements $i \in E_1 = [n]$ of $\pi_1 = \pi$ such that $i < \pi^{-1}(i) \wedge i < \pi(i)$. This is applied iteratively by defining a new permutation π_2 on the set E_2 of local minima, and repeating the same construction on π_2. After at most $t \leq \lfloor \log n \rfloor$ iterations, there is only the smallest element m remaining and the leader is chosen as the unique i on the cycle such that $\pi_t \circ \ldots \circ \pi_1(i) = m$. Checking if i is a leader is done by introducing the so-called *elbows*. For completeness, we provide a description of the algorithm in the full version of the paper [10]. Compared to the original version, we introduce new notation and change some implementation details to make the subsequent modifications easier to state. The crucial point is that the extra space used by the algorithm is bounded by a constant number of words per iteration in the above definition. The natural approach for decreasing the space to $\mathcal{O}(\log n)$ is to modify the definition of the local minimum to decrease the number of iterations to a constant. To this end, we work with b-local minima defined as elements less than all of their b successors and predecessors, for $b = \lceil n^\varepsilon \rceil$ where ε is a sufficiently small constant. This decreases the number of iterations to $\log n / \log b = 1/\varepsilon = \mathcal{O}_\varepsilon(1)$. There is, however, a nontrivial technical difficulty when trying to work with this idea. In the original version, one can check $\pi_r(x) \in E_r$ should be "promoted" to E_{r+1} by explicitly maintaining $x, \pi_r(x)$ and $\pi_r(\pi_r(x))$ and simply comparing $\pi_r(x)$ with x and $\pi_r(\pi_r(x))$. For larger values of b, this translates into explicitly maintaining $x, \pi_r(x), \pi_r(\pi_r(x)), \ldots, \pi_r^{2b}(x)$ to check if $\pi_r^b(x) \in E_{r+1}$, which of course takes too much space. We overcome this difficulty by designing and analysing a recursive pointer. This gives us our cycle leaders algorithm described in Sect. 3.

In a way similar to El-Zein et al. [11], we use our cycle leaders algorithm to design a solution to the permutation inversion problem. The high-level idea is to identify the leader i of a cycle, and then invert the cycle by traversing it from i. We need to somehow guarantee that the cycle is not inverted again, but do not have enough extra space to store i. El-Zein et al. mark the already inverted cycles that otherwise would be again inverted in the future by converting them to paths, that is, changing some $\pi(x)$ to undefined. This is then gradually repaired to a cycle, which requires a nontrivial interleave of four different scans. Our starting point is a simplification of their algorithm described in the full version of the paper [10] based on encoding slightly more (but still very little) information about the cycle. In Sect. 4 we further extend this method to work with b-local minima. Due to space constraints, some details and proofs are deferred to the full version of the paper [10].

Algorithm 1. A general framework of *left-to-right* algorithms.

1: **for** $i = 1..n$ **do** PROCESS(i)

Algorithm 2. A naive cycle leaders algorithm.

1: **function** PROCESS(i)
2: **if** $i = \text{MINR}(i, i)$ **then** Report that i is a leader.

2 Preliminaries

$[n]$ denotes the set of integers $\{1, 2, \ldots, n\}$ and for a function f and a nonnegative integer k we define $f^k(x)$ to be x when $k = 0$ and $f(f^{k-1}(x))$ otherwise. If f is a permutation then f^{-1} denotes its inverse and we define $f^{-k}(x)$ to be $f^{-k}(x) = f^{-1}(f^{-k+1}(x))$ for $k > 0$. Throughout the paper log denotes $\log_2$.

In the cycle leaders problem, we assume that the permutation π on $[n]$ is given through an oracle that returns any $\pi(i)$ in constant time. The goal is to identify exactly one element on each cycle of π as a leader. All of our algorithms follow the same left-to-right scheme: we consider the elements $i = 1, 2, \ldots, n$ in this order and test if the current i is the leader of its cycle by considering the elements $i, \pi(i), \pi^2(i), \ldots$ until we can determine if i is a leader.

Whenever we refer to a range $x \ldots y$, we mean $x, \pi(x), \pi(\pi(x)), \ldots, y$. We will also consider ranges $x \ldots y$ longer than a full cycle, but in such cases there will be always an middle point z (clear from the context) such that $x \ldots y$ consists of two ranges $x \ldots z$ and $z \ldots y$, each shorter than a full cycle. For example, in the range $\pi^{-k}(x) \ldots \pi^k(x)$ there are elements from $\pi^{-k}(x) \ldots x$ and from $x \ldots \pi^k(x)$. We use $\ell(i, i')$ to denote $\min\{k > 0 : \pi^k(i) = i'\}$. $\text{MINR}(a, b)$ naively finds the minimum between a and b on the same cycle of π, that is, $\min\{a, \pi(a), \pi^2(a), \ldots, b\}$. If $a = b$, we assume it computes the minimum of the full cycle of a. We also create a ternary MINR as follows: $\text{MINR}(x, y, z) = \min(\text{MINR}(x, y), \text{MINR}(y, z))$. Using this notation, the naive cycle leaders algorithm is presented in Algorithm 2.

When inverting π, we assume constant-time random access to the input. Due to the final goal being replacing every $\pi(i)$ with $\pi^{-1}(i)$, we allow temporarily overwriting $\pi(i)$ with any value from $[n]$ as long as after the algorithm terminates the input is overwritten as required. Additional space used by our algorithms consists of a number of auxiliary variables called words, each capable of storing a single integer from $[n]$. We assume that basic operations on such variables take constant time. We assume that the value of n is known to the algorithm.

3 Leader Election in Smaller Space

We extend the algorithm of Fich et al. [13] to obtain, for any $\varepsilon > 0$, a solution to the cycle leaders problem in $\mathcal{O}_\varepsilon(n^{1+\varepsilon})$ time using $\mathcal{O}_\varepsilon(\log n)$ bits of extra space.

Let $E_1 = [n]$ and $\pi_1 = \pi$. We denote by $b = \lceil n^\varepsilon \rceil$ the size of the neighborhood considered while determining local minima and declare an element of a permutation to be a b-local minimum if it is strictly smaller than all of its b successors and predecessors. Then, E_r is the set of all b-local minima encountered following π_{r-1} in E_{r-1}, that is $E_r = \{i \in E_{r-1} : i < \pi_{r-1}^k(i) \text{ for all } k \in \{-b, \ldots, b\} \setminus \{0\}\}$. We say that an element is on level r if it belongs to E_r. We define $\pi_r : E_r \to E_r$ as follows: $\pi_r(e)$ is the first element in π after e that belongs to E_r, formally: $\pi_r(e) = \pi_{r-1}^k(e)$ where $\pi_{r-1}^k(e) \in E_r$ and $\forall_{0<k'<k} \pi_{r-1}^{k'} \notin E_r$. For $b = 1$ this is exactly the definition used by Fich et al. [13].

Definition 1. *A path of π is a partial function obtained from a cycle C of π by replacing $\pi(x)$ with $\perp$, for any $x \in C$.*

$\perp$ should be understood as an undefined element. For a path, π_r is undefined for the last element from E_r, and similarly π_r^{-1} is undefined for the first element from E_r. When deciding if an element is a b-local minimum on a path, we disregard comparisons with such undefined elements. Furthermore, we assume that $\pi_r(\perp) = \perp$ and $\pi_r^{-1}(\perp) = \perp$ for every r.

For each level r, only at most $\frac{1}{b+1}$ of its elements can belong to the level $r+1$. For a cycle or a path C, let t be the largest number such that $|E_t \cap C| > b$. We set $t = 0$ for $|C| \le b$ and observe that $t < \frac{1}{\epsilon}$ for every C. We note that if C is a cycle then $E_{t+2} \cap C = \emptyset$, but when C is a path $|E_{t+2} \cap C| = 1$. Because all algorithms presented in this paper follow the framework given in Algorithm 1 and during the execution of Process(i), we only consider the elements that can be reached from i, we restrict our considerations to just one cycle or path and we are going to omit the "$\cap C$" part everywhere later.

Definition 2. *A* b-staircase *of size r from i is a sequence of elements $(i = i_1, i_2, \ldots, i_{r+1} = m = j_{r+1}, j_r, j_{r-1}, \ldots, j_1 = i')$ such that $i_k, j_k \in E_k$, for $k \in [r+1]$ and $i_{k+1} = \pi_k^b(i_k)$, $j_k = \pi_k^b(j_{k+1})$ for $k \in [r]$. Elements i, m and i' are called* the start, the middle *and* the end *of the b-staircase, respectively. The part of the b-staircase from the start to the middle is called its* left part *and the part from the middle to the end is called its* right part.

Definition 3. *An* almost b-staircase *of size r from i is a sequence of elements $(i = i_1, i_2, \ldots, i_{r+1} = m = j_{r+1}, j_r, j_{r-1}, \ldots, j_1 = i')$, such that $i_k, j_k \in E_k$, for $k \in [r]$ and $i_{k+1} = \pi_k^b(i_k)$ and $j_k = \pi_k^b(j_{k+1})$ for $k \in [r]$.*

The difference between the two above definitions is that for b-staircase we require $m \in E_{r+1}$, but for almost almost b-staircase it suffices that $m \in E_r$.

Definition 4. *A b-staircase or an almost b-staircase of size r is called* proper *if $|E_r| > b$.*

Definition 5. *A b-staircase of size r from i is called the* best b-staircase from i *if there is no proper almost b-staircase of size $r+1$ from i (possibly there is no best b-staircase from i).*

Definition 6. *An element i is the leader of its cycle if the best b-staircase from i exists and its middle m is the minimum on the cycle.*

Lemma 1. *There is exactly one leader on any cycle of a permutation.*

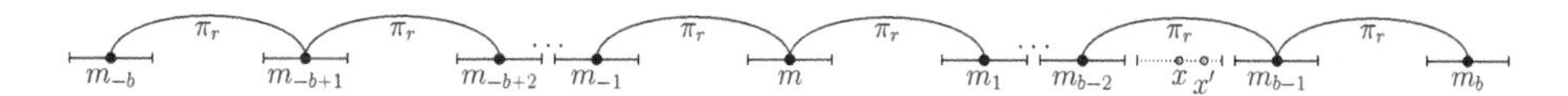

Fig. 1. Filled nodes represent elements $m_k = \pi_r^k(m)$ for $k \in \{-b, \ldots, b\}$ and each segment represents the range around m_k. All the elements in the ranges (different than m) are larger than m by induction hypothesis.

Proof. Let m be the minimum on the cycle and t be the largest size of a proper almost b-staircase on the cycle. Consider the almost b-staircase B of size t from $i = \pi_1^{-b}(\ldots(\pi_{t-1}^{-b}(\pi_t^{-b}(m)))\ldots)$. By the choice of t, there is no proper almost b-staircase of size $t+1$ from i. As m is the minimum on the cycle and $|E_t| > b$, B is in fact a b-staircase. Thus, B is the best b-staircase from i and i is the leader. Furthermore, as m and t are both uniquely defined, there are no other leaders on the cycle. □

A b-staircase may be longer than a full cycle if it visits some elements twice (before and after reaching its middle) but all elements occur at most once in each part. We allow computing $\text{MinR}(x, y)$ for $x = \perp$ or $y = \perp$. In such a case, the minimum is computed from the beginning or to the end of the path, respectively.

Lemma 2. *Consider $m \in E_r$ on a cycle or a path where $|E_r| > b$. Then m is a b-local minimum on level r in π if and only if $m = \text{MinR}(\pi_r^{-b}(m), m, \pi_r^b(m))$.*

Proof. ($\Leftarrow$) Trivial. ($\Rightarrow$) Proof by induction on r. The case $r = 1$ is immediate.

For $r > 1$, assume that m is a b-local minimum on level r. We consider the elements in $\pi_r^{-b}(m) \ldots \pi_r^b(m)$, appropriately truncated if we are on a path. Among these elements, only $m_k = \pi_r^k(m) \neq \perp$ for $k \in \{-b, \ldots, b\}$ are b-local minima on level $r-1$. All of them are also larger than m (if $k \neq 0$). By the induction hypothesis, for all $k \in \{-b, \ldots, b\}$ (such that $m_k \neq \perp$) any element between $\pi_{r-1}^{-b}(m_k)$ and $\pi_{r-1}^b(m_k)$ on the cycle/path is larger than or equal to m_k and hence also larger than m. We call the above ranges the *ranges around* m_k and they are represented as segments in Fig. 1.

Consider an element x in $\pi_r^{-b}(m) \ldots \pi_r^b(m)$ which is not in the range around m_k for any k. In Fig. 1, this corresponds to x in a gap between the ranges. Clearly, $x \notin E_r$, so there is an $\ell < r$ such that $x \in E_\ell \setminus E_{\ell+1}$. Hence x is larger than some $x' \in E_{\ell'} \setminus E_{\ell'+1}$ between $\pi_\ell^{-b}(x)$ and $\pi_\ell^b(x)$ for $\ell < \ell' \leq r$. If there is k such that x' is in the range around m_k, then $x' > m_k > m$ and the lemma follows. Otherwise, observe that x' belongs to the same gap as x, so we can apply the same reasoning for x' instead of x. Because the considered values are always decreasing and each gap contains only finite number of elements, finally we obtain a value in one of the ranges around m_k and conclude that x is larger than one of the m_k and also than m. Hence, m is the smallest element in $\pi_r^{-b}(m) \ldots \pi_r^b(m)$. □

The lemma enables us to check if an element m is a b-local minimum on level r without knowing which of the elements in $\pi_r^{-b}(m) \ldots \pi_r^b(m)$ belong to the level r.

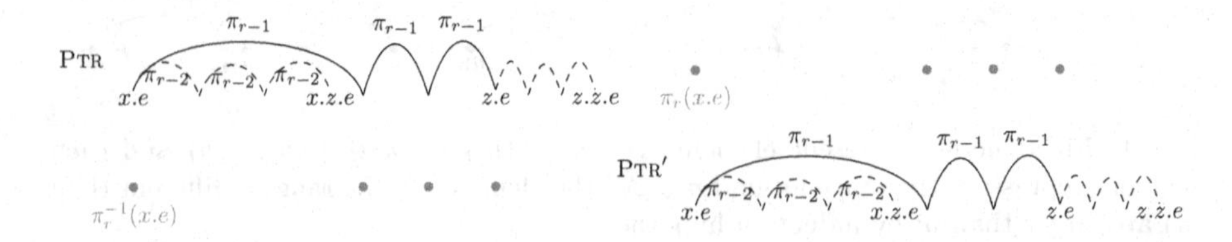

Fig. 2. A sketch of PTR structure for $b = 3$. On the bottom there is PTR′, the result of executing the ADVANCE method on PTR.The gray dots are not part of pointer, they only show the alignment between PTR and PTR′.

Fich et al. designed procedure NEXT(r) which computes, for an element on E_r, its successor on π_r. We start with extending their idea to work with larger values of the parameter b. For $b = \lceil n^\epsilon \rceil$, the number of elements to compare with is no longer constant. We define a recursive pointer PTR with the following fields:

- r – the level of the pointer,
- e – an element of E_r pointed at by the pointer,
- x – PTR of level $r - 1$ pointing to e (or NULL if $r = 1$),
- y and z – PTRs of level $r - 1$ both pointing to $\pi_{r-1}^b(e)$ (or NULL if $r = 1$).

PTR has the ADVANCE method which moves the pointer from e to $\pi_r(e)$ and updates x, y and z accordingly. See Fig. 2. Intuitively, PTR points to a single element e, but it also contains a number (exponential in its level r) of different pointers pointing to different elements. While updating PTR we will ensure that the above properties of all the other pointers are also satisfied.

Before and after calling ADVANCE, y is equal to z, but during the execution they are different. In fact, it is enough to only store e, x and z in every PTR and keep a local y reused between every call to ADVANCE method.

We now analyse PTR :: ADVANCE implemented in Algorithm 3 for a pointer of level r. At the end of the function, e is updated to point to the element $x.e \in E_r$. As we have to also deal with paths, we add the check for $\perp$ in line 3. For $r \geq 2$, z is moved forward b times along π_{r-1}, so after line 6, y and z point to $\pi_{r-1}^b(x.e)$ and $\pi_{r-1}^{2b}(x.e)$, respectively. In line 7 we compare $y.e$ with all elements between $\pi_{r-1}^{-b}(y.e)$ and $\pi_{r-1}^b(y.e)$ and all three pointers are simultaneously advanced to the next elements along π_{r-1} until y points to a b-local minimum on level r. By Lemma 2 this is equivalent to checking if $y.e$ is a b-local minimum on level $r - 1$ as long as $|E_{r-1}| > b$.

In Algorithm 4 we implement BESTbSTAIRCASE(i) that constructs the best b-staircase from i returning PTR structure pointing to the middle of the staircase. Before running the main loop, the algorithm checks if $|E_1| > b$, to make sure it is allowed to execute ADVANCE method on p. We start with a b-staircase of size 0. Each iteration of the main loop starts with p of level r representing the right part of the b-staircase of size $r - 1$ from i. The invariant $|E_r| > b$ is preserved in the main loop. During each iteration, p is subject to change and the updated p

Algorithm 3. Implementation of the ADVANCE method in the PTR structure. It moves the PTR object onto the next element on the same level.

```
1: function PTR :: ADVANCE
2:     if this.r = 1 then
3:         if π(this.e) =⊥ then abort
4:         this.e ← π(this.e)
5:         return
6:     for j = 1..b do this.z.ADVANCE()
7:     while this.y ≠ MINR(this.x.e, this.y.e, this.z.e) do
8:         this.x.ADVANCE()
9:         this.y.ADVANCE()
10:        this.z.ADVANCE()
11:    this.x ← this.y
12:    this.e ← this.x.e
13:    this.y ← this.z
```

may not represent a (proper) b-staircase, so at the beginning of the iteration, it is stored in g. Then, ADVANCE method is called $2b$ times on p and enumerates the elements from the set $S = \{\pi_r^k(e) : k \in \{1, \ldots, 2b\}\}$. The pointer m_p is stored after the b-th ADVANCE so it points to $\pi_r^b(e)$, that is the middle of the almost staircase of size r. If $m_p.e$ is not the smallest in S, then the best b-staircase from i does not exist and we terminate in line 16. However, if any of $\pi_r^k(e)$ for $k \in \{1, \ldots, b\}$ is equal to e, then $|S| \leq b$, so there is no proper almost staircase of size r from i (both left and right part would self-overlap) and we return g as the best staircase. The same happens for paths in case any ADVANCE call aborts. Only when $|E_r| > b$ and the new middle m_e is less than all b pairwise-distinct left and b pairwise-distinct right neighbours on E_r, the algorithm extends the b-staircase to size r and proceeds to the next iteration with p of level $r+1$.

We are ready to provide an algorithm for the cycle leaders problem. We proceed as in Algorithm 1, but we alter the PROCESS function to work with the new definition of leader, see Algorithm 5. BESTbSTAIRCASE(i) returns PTR representing the right part of the best b-staircase from i. Its middle can be read from the field e of the returned PTR.

Theorem 1. *For every $\varepsilon > 0$, there exists an algorithm for reporting leaders of a permutation π on n elements in $\mathcal{O}_\varepsilon(n^{1+\epsilon})$ time using $\mathcal{O}_\varepsilon(\log n)$ additional bits of space.*

Proof. We analyse Algorithm 1 with the implementation of PROCESS provided in Algorithm 5. Each PTR of level r has three PTRs on level $r-1$, so there are $\sum_{k=0}^{t} 3^k = \mathcal{O}(3^t)$ pointers in total. As $t \leq \frac{1}{\epsilon}$, our algorithm uses $\mathcal{O}(3^{\frac{1}{\epsilon}} \cdot \log n) = \mathcal{O}_\varepsilon(\log n)$ extra bits of space.

The total time for reporting leaders is dominated by the time spent in PTR :: ADVANCE. We sum up (for all created pointers) just the time inside ADVANCE call of each PTR without the recursive calls. The time for recursive calls will be accounted to the descendant pointers. Inside ADVANCE, the time is dominated

Algorithm 4. Constructs the best b-staircase from i (if any).

1: **function** BESTbSTAIRCASE(i)
2: $p \leftarrow$ PTR($e \leftarrow i,\ r \leftarrow 1,\ x \leftarrow$ NULL, $y \leftarrow$ NULL, $z \leftarrow$ NULL)
3: $x \leftarrow i$
4: **for** $j = 1..b$ **do**
5: $x \leftarrow \pi(x)$
6: **if** $x \in \{\perp, i\}$ **then return** p
7: **while** TRUE **do**
8: $g \leftarrow p$
9: $m_e \leftarrow p.e$
10: **for** $j = 1..2b$ **do**
11: **try** p.ADVANCE() **catch** */abort/* **return** g
12: **if** $j \leq b$ **and** $p.e = g.e$ **then return** g
13: **if** $j = b$ **then** $m_p \leftarrow p$
14: **if** $p.e < m_e$ **then** $m_e \leftarrow p.e$
15: **if** $m_p.e = m_e$ **then** $p \leftarrow$ PTR($e \leftarrow m_p.e, r \leftarrow p.r + 1, x \leftarrow m_p, y \leftarrow p, z \leftarrow p$)
16: **else return** NULL

Algorithm 5 Modified function for checking if i is the leader of its cycle.

1: **function** PROCESS(i)
2: $p \leftarrow$ BESTbSTAIRCASE(i)
3: **if** $p \neq$ NULL **and** MINR(i, i) $= p.e$ **then** Report that i is the leader

by MINR executions. As each advanced position is covered by at most $2b + 1$ MINRs, if a pointer p proceeds k steps along π (starting with $p.x.e = s$ and ending with $p.z.e = \pi^k(s)$) during ADVANCE, the total time spent on executions of MINR for the call is $\mathcal{O}(kb)$.

Now we count the total length of traversals for all created pointers. This, multiplied by $\mathcal{O}(b)$, is the final complexity of the algorithm. Recall that for every element, we check if it is possible to create the best b-staircase from it by constructing proper almost b-staircases and validating if these are b-staircases. Thus, if it is possible to construct a b-staircase of size r and not $r+1$ from i, every pointer traverses (in the worst possible case) all elements of the almost b-staircase of size $r + 1$ from i. We charge the traversed range to the pair (r, m), where $m \in E_{r+1} \setminus E_{r+2}$ is the middle of the b-staircase of size r from i. Observe that i is between $\pi_{r+1}^{-1}(m)$ and m and the execution of BESTbSTAIRCASE(i) finishes before reaching $\pi_{r+1}^{2b+1}(m)$, see Fig. 3. Thus, for each r, every element belongs to $\mathcal{O}(b)$ ranges charged to the elements from $E_{r+1} \setminus E_{r+2}$. The total length of all such ranges is therefore $\mathcal{O}(nb)$. Recall that there are $\mathcal{O}(3^r) = \mathcal{O}(3^t) = \mathcal{O}_\epsilon(1)$ pointers on each level r, so summing over all $r \leq t$ we obtain that the running time of Algorithm 5 is $\mathcal{O}(\frac{1}{\epsilon} \cdot 3^t \cdot nb^2) = \mathcal{O}_\epsilon(n^{1+2\epsilon})$. By adjusting ϵ, the running time becomes $\mathcal{O}_\epsilon(n^{1+\epsilon})$. □

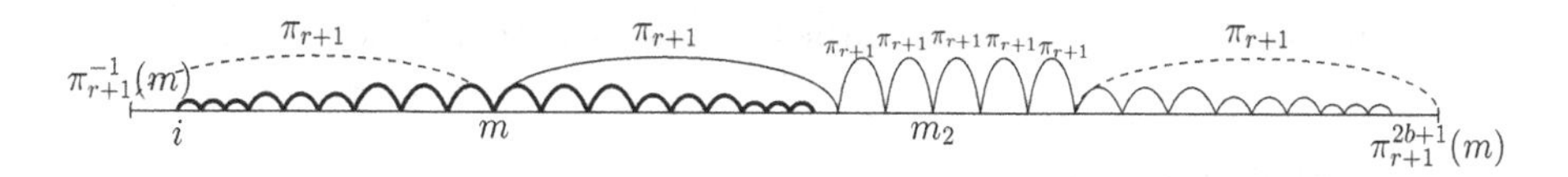

Fig. 3. The maximum potential traversal range for a middle m of a b-staircase ($b = 3$) of size r. The b-staircase of size r with the middle element m is drawn with a bold line and the almost b-staircase of size $r + 1$ with new middle m_2 is drawn with a solid line.

4 Inverting Permutations in Smaller Space

In this section, we extend the results of Sect. 3 to obtain an $\mathcal{O}(n^{1+\epsilon})$-time algorithm for inverting permutations using $\mathcal{O}_\varepsilon(\log n)$ bits by extending the algorithm given by El-Zein et al. [11]. We follow the same natural idea of inverting each cycle upon reaching its leader. However, there are some additional technical complications. Due to space constraints, we defer some of the details to the full version of the paper [10], where we also describe a simpler version of the algorithm of El-Zein et al. [11].

We proceed from left to right as described in Sect. 3. After having detected that i is a leader, we invert its cycle. However, this might result in creating a cycle with leader $i' > i$ that would be inverted again in the future. This problematic situation can be detected during the scan, as after inverting a cycle, the middle of the best b-staircase remains the same and the only best b-staircase on the inverted cycle is exactly the reverse of the previous best b-staircase. Thus, we can compute the end i' of the largest b-staircase constructed from i with GetEnd method of Ptr returning the end of the staircase in the following way: $p.\text{GetEnd}() = p.z.\text{GetEnd}()$ if $p.z \neq$ `NULL` and $p.\text{GetEnd}() = p.e$ otherwise.

Definition 7. *A cycle C of π is* hard *if for the best b-staircase $(i, \dots, m, \dots, i')$ we have $i' > i$. Otherwise, C is* easy.

To deal with hard cycles, we need the notion of *cutting before* y, where $y = \pi(x)$. This operation is implemented by setting $\pi(x) \leftarrow \perp$, where $\perp$ is called a null, to transform a cycle into a path P with y as its first element and x as the last element. For a hard cycle, we will cut the inverted cycle before the new leader. As in [11], in the description of our algorithm we will use multiple types of nulls to encode some additional information, and then simulate these types using the following lemma.

Lemma 3 ([11]). *For any $k \in [n]$, it is possible to simulate an extension of the range of stored values of π from $[n]$ to $[n] \cup \{\perp_1, \dots, \perp_k\}$ with each value from $[n]$ occurring at most c times in π using $\mathcal{O}(ck \log n)$ bits with $\mathcal{O}(c)$-time overhead.*

During the execution of the algorithm, π consists of cycles and components, where a component is either a path or a proper sigma.

Definition 8. Sigma *is a function $\pi : C \to C$, such that there is at least one element $y \in C$ with $\{\pi^k(y) : k \in \{0, 1, 2, \dots\}\} = C$. We call sigma* proper *if it is not a cycle.*

Depending on the number of times k an element that looks like a possible leader was found for a cycle, we can be in one of the following states:

- not inverted cycle (for $k = 0$),
- inverted path (for $k = 1$ and only if the cycle was hard),
- inverted sigma (for easy cycles if $k = 1$ and for hard cycles if $k \geq 2$).

We use the name *component* for both paths and proper sigmas. For a path we call the element without the predecessor the *start* of the path and the element pointing to null the *end* of the path. Similarly, in a proper sigma we call element y_1 without predecessor the *start* of the sigma and among the two elements x_1, x_2 such that $\pi(x_1) = \pi(x_2) = y_2$ satisfying $\ell(y_1, x_1) < \ell(y_1, x_2)$, x_2 is called the *end* and y_2 the *intersection* of the sigma. A part of sigma before the intersection (from y_1 to x_1) is called the *tail* and the rest (a cycle with y_2 and x_2) is the *loop* of sigma. *Fixing* a component consists in changing the successor of its end to its start as to create a cycle.

Definition 9. *The* leader *of a component is the element that would be the leader in its fixed component.*

Definition 10. *The* rank *of element i on a path or on a tail of a sigma in π is the size of the best b-staircase from i. If there is no best b-staircase from i then its rank is undefined.*

Definition 11. *An element i is* outstanding *if has the largest rank among the elements on its component.*

During the algorithm, we maintain the following invariants: (1) the start of a component is its leader (and is outstanding), (2) the intersection of a sigma is the only element from which there is a best staircase on the loop of the sigma (and is already processed). We want to proceed differently depending on whether i is on a cycle, a path or the tail of a sigma. Distinguishing between the three cases would be too time consuming, but turns out to not be necessary as in every case we first compute BESTbSTAIRCASE(i) and terminate if the computation failed or aborted. If the computation succeeds then in fact we have enough time to also locate the element a which is either predecessor of i (for i on the cycle) or the end of the component (for i on a path or the tail of a proper sigma).

When the leader i of a cycle is found for the first time, the algorithm finds the end i' of the corresponding b-staircase and inverts the cycle, so i' becomes the leader of the inverted cycle. If $i' > i$, then the cycle is hard, and we change it into a path by cutting before i'. The rank of i' on the path is stored in the type of the null.

The path remains in this state at least until we consider one of its outstanding elements. If the algorithm finds an outstanding element i, it creates a temporary, larger sigma, by pointing the end of the sigma to i, then checks if there is the best b-staircase from i on the loop of sigma. If so, the change is in effect or is reverted otherwise. Because of this check, we have no other elements than i with the best staircase on the loop of sigma.

The crux of the time analysis is there are only $\mathcal{O}(b)$ outstanding elements on every path, so every sigma will be enlarged only at most that many times. By appropriately adjusting the parameters we arrive at out final theorem.

Theorem 2. *For every $\varepsilon > 0$, there exists an algorithm for inverting a permutation π on n elements in $\mathcal{O}_\varepsilon(n^{1+\epsilon})$ time using $\mathcal{O}_\varepsilon(\log n)$ additional bits of space.*

Acknowledgments. B. Dudek was supported by the National Science Centre, Poland, under grant number 2017/27/N/ST6/02719.

References

1. Aziz, A., Lee, T.H., Prakash, A.: Elements of Programming Interviews in Java: The Insiders' Guide. CreateSpace Independent Publishing Platform, USA (2015)
2. Beame, P.: A general sequential time-space tradeoff for finding unique elements. SIAM J. Comput. **20**(2), 270–277 (1991)
3. Birenzwige, O., Golan, S., Porat, E.: Locally consistent parsing for text indexing in small space. In: 31st SODA, pp. 607–626. SIAM (2020)
4. Borodin, A., Cook, S.A.: A time-space tradeoff for sorting on a general sequential model of computation. SIAM J. Comput. **11**(2), 287–297 (1982)
5. Borodin, A., Fischer, M.J., Kirkpatrick, D.G., Lynch, N.A., Tompa, M.: A time-space tradeoff for sorting on non-oblivious machines. J. Comput. Syst. Sci. **22**(3), 351–364 (1981)
6. Chan, T.M.: Comparison-based time-space lower bounds for selection. ACM Trans. Algorithms **6**(2), 1–16 (2010)
7. Chan, T.M., Munro, J.I., Raman, V.: Finding median in read-only memory on integer input. Theor. Comput. Sci. **583**, 51–56 (2015)
8. Darwish, O., Elmasry, A.: Optimal time-space tradeoff for the 2d convex-hull problem. In: Schulz, A.S., Wagner, D. (eds.) ESA 2014. LNCS, vol. 8737, pp. 284–295. Springer, Heidelberg (2014). https://doi.org/10.1007/978-3-662-44777-2_24
9. Dolev, D., Klawe, M.M., Rodeh, M.: An $O(n \log n)$ unidirectional distributed algorithm for extrema finding in a circle. J. Algorithms **3**(3), 245–260 (1982)
10. Dudek, B., Gawrychowski, P., Pokorski, K.: Strictly in-place algorithms for permuting and inverting permutations. CoRR abs/2101.03978 (2021)
11. El-Zein, H., Munro, J.I., Robertson, M.: Raising permutations to powers in place. In: 27th ISAAC. LIPIcs, vol. 64, pp. 1–12. Schloss Dagstuhl - Leibniz-Zentrum fuer Informatik (2016)
12. Ellis, J.A., Krahn, T., Fan, H.: Computing the cycles in the perfect shuffle permutation. Inf. Process. Lett. **75**(5), 217–224 (2000)
13. Fich, F.E., Munro, J.I., Poblete, P.V.: Permuting in place. SIAM J. Comput. **24**(2), 266–278 (1995)
14. Fischer, J., Gagie, T., Gawrychowski, P., Kociumaka, T.: Approximating LZ77 via small-space multiple-pattern matching. In: Bansal, N., Finocchi, I. (eds.) ESA 2015. LNCS, vol. 9294, pp. 533–544. Springer, Heidelberg (2015). https://doi.org/10.1007/978-3-662-48350-3_45
15. Fischer, J., I, T., Köppl, D., Sadakane, K.: Lempel-ziv factorization powered by space efficient suffix trees. Algorithmica **80**(7), 2048–2081 (2018)

16. Franklin, W.R.: On an improved algorithm for decentralized extrema finding in circular configurations of processors. Commun. ACM **25**(5), 336–337 (1982)
17. Frederickson, G.N.: Upper bounds for time-space trade-offs in sorting and selection. J. Comput. Syst. Sci. **34**(1), 19–26 (1987)
18. Gawrychowski, P., Kociumaka, T.: Sparse suffix tree construction in optimal time and space. In: 28th SODA, pp. 425–439. SIAM (2017)
19. Golynski, A.: Cell probe lower bounds for succinct data structures. In: 20th SODA, pp. 625–634. SIAM (2009)
20. Guśpiel, G.: An in-place, subquadratic algorithm for permutation inversion. CoRR abs/1901.01926 (2019)
21. Hirschberg, D.S., Sinclair, J.B.: Decentralized extrema-finding in circular configurations of processors. Commun. ACM **23**(11), 627–628 (1980)
22. Itai, A., Rodeh, M.: Symmetry breaking in distributed networks. Inf. Comput. **88**(1), 60–87 (1990)
23. Knuth, D.E.: Mathematical analysis of algorithms. In: IFIP Congress (1), pp. 19–27 (1971)
24. Knuth, D.E.: The art of computer programming, Volume I: Fundamental Algorithms, 3rd edn. Addison-Wesley (1997)
25. Munro, J.I., Paterson, M.: Selection and sorting with limited storage. Theor. Comput. Sci. **12**, 315–323 (1980)
26. Munro, J.I., Raman, V.: Selection from read-only memory and sorting with minimum data movement. Theor. Comput. Sci. **165**(2), 311–323 (1996)
27. Pagter, J., Rauhe, T.: Optimal time-space trade-offs for sorting. In: 39th FOCS, pp. 264–268. IEEE Computer Society (1998)
28. Peterson, G.L.: An $O(n \log n)$ unidirectional algorithm for the circular extrema problem. ACM Trans. Program. Lang. Syst. **4**(4), 758–762 (1982)
29. Raman, V., Ramnath, S.: Improved upper bounds for time-space trade-offs for selection. Nord. J. Comput. **6**(2), 162–180 (1999)

A Stronger Lower Bound on Parametric Minimum Spanning Trees

David Eppstein(✉)

Computer Science Department, University of California, Irvine, Irvine, USA
eppstein@uci.edu

Abstract. We prove that, for an undirected graph with n vertices and m edges, each labeled with a linear function of a parameter λ, the number of different minimum spanning trees obtained as the parameter varies can be $\Omega(m \log n)$.

1 Introduction

In the *parametric minimum spanning tree problem* [16], the input is a graph G whose edges are labeled with linear functions of a parameter λ. For any value of λ, one can obtain a spanning tree T_λ as the minimum spanning tree of the weight functions, evaluated at λ. Varying λ continuously from $-\infty$ to ∞ produces in this way a discrete sequence of trees, each of which is minimum within some range of values of λ. How many different spanning trees can belong to this sequence, for a worst case graph, and how can we construct them all efficiently? Known bounds are that the number of trees in a graph with n vertices and m edges can be $\Omega\big(m\alpha(n)\big)$ (where α is the inverse Ackermann function) [9] and is always $O(mn^{1/3})$ [7]; both bounds date from the 1990s and, although far apart, have not been improved since. The sequence of trees can be constructed in time $O(mn \log n)$ [13] or in time $O(n^{2/3} \log^{O(1)} n)$ per tree [1]; faster algorithms are also known for planar graphs [12] or for related optimization problems that construct only a single tree in the parametric sequence [6,19]. In this paper we improve the 25-year-old lower bound on the number of parametric minimum spanning trees from $\Omega\big(m\alpha(n)\big)$ to $\Omega(m \log n)$.

A broad class of applications of this problem involves bicriterion optimization, where each edge of a graph has two real weights of different types (say, investment cost and eventual profit) and one wishes to find a tree optimizing a nonlinear combination of the sums of these two weights (such as the ratio of total profit to total investment cost, the return on the investment). Each spanning tree of G may be represented by a planar point whose Cartesian coordinates are the sums of its two kinds of weights, giving an exponentially large cloud of points, one per tree. The convex hull of this point cloud has as its vertices the parametric minimum spanning trees (and maximum spanning trees) for linear weight functions obtained from the pair of weight values on each edge by using these values as coefficients. (Essentially, this construction of weight functions from pairs of weights is

A. Lubiw et al. (Eds.): WADS 2021, LNCS 12808, pp. 343–356, 2021.
https://doi.org/10.1007/978-3-030-83508-8_25

a form of projective duality transforming points into lines, and the equivalence between the convex hull of the points representing trees into the lower envelope of lines representing their total weight is a standard reflection of that projective duality.) Any bicriterion optimization problem that can be expressed as maximizing a quasiconvex function (or minimizing a quasiconcave function) of the two kinds of total weight automatically has its optimum at a convex hull vertex, and can be solved by constructing the sequence of parametric minimum spanning trees and evaluating the combination of weights for each one [18]. Other combinatorial optimization problems that have been considered from the same parametric and bicriterion point of view include shortest paths [3–5, 11], optimal subtrees of rooted trees [2], minimum-weight bases of matroids [9], minimum-weight closures of directed graphs [10], and the knapsack problem [8, 15, 17].

The main idea behind our new lower bound is a recursive construction of a family of graphs (more specifically, 2-trees), formed by repeated replacement of edges by triangles (Fig. 1). We also determine the parametric weight functions of these graphs by a separate recursive construction (Fig. 3). However, this only produces an $\Omega(n \log n)$ lower bound, because for a graph constructed in this way with n vertices, the number of edges is $2n - 3$, only a constant factor larger than the number of vertices. To obtain our claimed $\Omega(m \log n)$ lower bound we use an additional packing argument, in which we find a dense graph containing many copies of our sparse lower bound construction, each contributing its own subsequence of parametric minimum spanning trees to the total.

2 Background and Preliminaries

The *minimum spanning tree* of a connected undirected graph with real-valued edge weights is a tree formed as a subgraph of the given graph, having the minimum possible total edge weight. As outlined by Tarjan [22], standard methods for constructing minimum spanning trees are based on two rules, stated most simply for the case when all edge weights are distinct. The *cut rule* concerns cuts in the graph, partitions of the vertices into two subsets; an edge *spans* a cut when its two endpoints are in different subsets. The cut rule states that (for distinct edge weights) the minimum-weight edge spanning any given cut in a graph belongs to its unique minimum spanning tree. The *cycle rule*, on the other hand, states that (again for distinct edge weights) the maximum-weight edge in any cycle of the graph does not belong to its unique spanning tree. One consequence of these rules is that the minimum spanning tree depends only on the sorted ordering of the edge weights, rather than on more detailed properties of their numeric values.

An input to the *parametric minimum spanning tree* problem consists of an undirected connected graph whose edges are labeled with linear functions of a parameter λ rather than with real numbers. For any value of λ, plugging λ into these functions produces a system of real weights for the edges, and therefore a minimum spanning tree T_λ. Different values of λ may produce different trees, and the task is either to obtain a complete description of which tree is minimum for each possible value of λ or, in some versions of the problem, to find a value λ and its tree optimizing another objective function.

If we plot the graphs of the linear functions of a parametric minimum spanning tree instance, as lines in the (λ, weight) plane, then the geometric properties of this *arrangement of lines* are closely related to the combinatorial properties of the parametric minimum spanning tree problem. If no two edges have the same weight function, then all edge weights will be distinct except at a finite set of values of λ, the λ-coordinates of points where two lines in the arrangement cross. As λ varies continuously, the sorted ordering of the weights will remain unchanged except when λ passes through one of these crossing points, where the set of lines involved in any crossing will reverse their weight order. It follows from these considerations that the sequence of parametric minimum spanning trees is finite, and that these trees change only at certain *breakpoints* which are necessarily the λ-coordinates of crossings of lines. In particular, m lines have $O(m^2)$ crossings and there can be only $O(m^2)$ distinct trees in the sequence of parametric minimum spanning trees. However, a stronger bound, $O(mn^{1/3})$, is known [7].

The worst-case instances of the parametric minimum spanning tree problem, the ones with the most trees for their numbers of edges and vertices, have distinct edge weight functions whose arrangement of lines has only *simple crossings*, crossings of exactly two lines. For, in any other instance, perturbing the edge weight functions by a small amount will preserve the ordering of weights away from the crossings of its lines, and therefore will preserve its sequence of trees away from these crossings, while only possibly increasing the number of breakpoints near perturbed crossings of multiple lines, which become multiple simple crossings. For an instance in which the lines have only simple crossings, the only possible change to the minimum spanning tree at a breakpoint is a *swap*, a change to the tree in which one edge (corresponding to one of the two crossing lines at a simple crossing) is removed, and the other edge (corresponding to the other of the two crossing lines) is added in its place. For details on this correspondence between the geometry of line arrangements and the sequence of parametric minimum spanning trees, and generalizations of this correspondence to other matroids than the matroid of spanning trees, see our previous paper on this topic [9].

3 Replacing Edges by Triangles

A *2-tree* is a graph obtained from the two-vertex one-edge graph K_2 by repeatedly adding new degree-two vertices, adjacent to pairs of adjacent earlier vertices. Equivalently, they are obtained by repeatedly replacing edges by triangles. These graphs are planar and include the maximal outerplanar graphs [20]; their subgraphs are the *partial 2-trees*, graphs of treewidth ≤ 2 [23]. The graphs we use in our lower bound are a special case of this construction where we apply this edge replacement process simultaneously to all edges in a smaller graph of the same type. We define the first graph T_0 in our sequence of graphs to be the graph K_2, and then for all $i > 0$ we define T_i to be the graph obtained by replacing all edges of T_{i-1} by triangles. It seems natural to call these *complete*

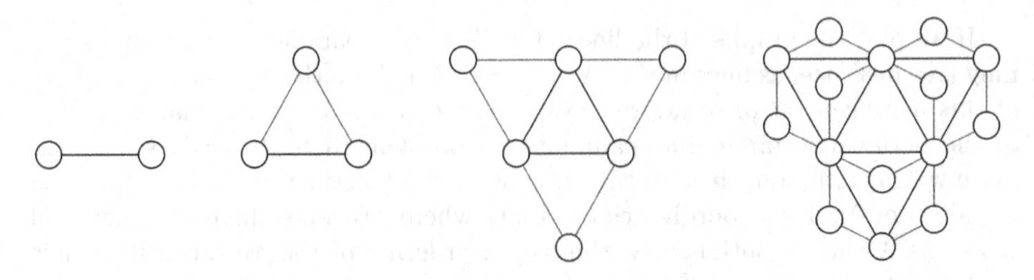

Fig. 1. Recursively constructing a family of 2-trees T_i (here, $i = 0, 1, 2, 3$ in left-to-right order) by repeatedly replacing every edge of T_{i-1} by a triangle.

2-trees, by analogy to complete trees (whose leaves are repeatedly replaced by stars for a given number of levels) but we have been unable to find this usage in the literature. The graphs T_i for $i \leq 3$ are depicted in Fig. 1.

Lemma 1. *T_i has 3^i edges and $(3^i + 3)/2$ vertices.*

Proof. The bound on the number of edges follows from the fact that each replacement of edges by triangles triples the number of edges. The bound on the number of vertices follows easily by induction on i, using the observations that each edge of T_{i-1} leads to a newly added vertex in T_i and that $(3^{i-1} + 3)/2 + 3^{i-1} = (3^i + 3)/2$. □

What happens when we replace an edge by a triangle in a parametric spanning tree problem? For a non-parametric minimum spanning tree, the answer is given by the following lemma.

Lemma 2. *Let graph G contain edge pq, and replace this edge by a triangle pqr to form a larger graph G^+. Suppose that the edges in G^+ have distinct edge weights, and use these weights to assign weights to the edges in G, with the following exception: in G, give edge pq the weight of the* bottleneck *edge in triangle pqr (the maximum-weight edge on path from p to q in the minimum spanning tree of the graph of the triangle) instead of the weight of pq. Then, the minimum spanning tree of $G+$ has the same set of edge weights as the minimum spanning tree of G, together with the minimum weight of a non-bottleneck edge in triangle pqr.*

Proof. If pq is the heaviest edge in pqr then the path from p to q in the minimum spanning tree of pqr passes through r, the bottleneck edge is the heavier of the two edges on this path, and the minimum non-bottleneck edge is the lighter of its two edges. Otherwise, pq is the bottleneck edge and again the minimum non-bottleneck edge is the lighter of the two remaining edges incident to r. Applying the cut rule to the cut separating r from the rest of the graph shows that the minimum non-bottleneck edge is an edge of the minimum spanning tree of G^+. Since we did not include its edge weight in the weights for G, its weight is not included in the set of edge weights of the minimum spanning tree for G.

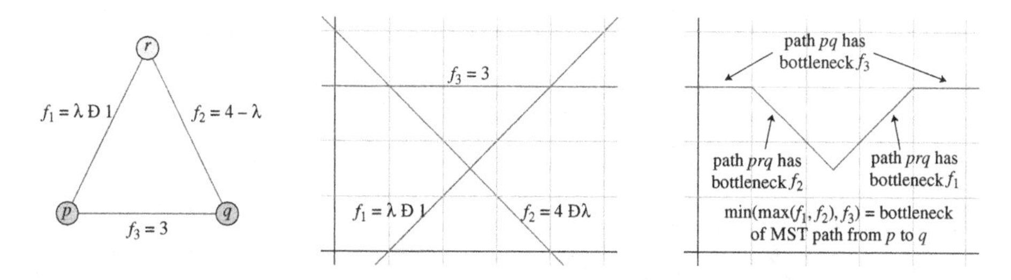

Fig. 2. A parametric spanning tree problem on a single triangle pqr, and the graph of the bottleneck edge weight on the path from p to q in the parametric spanning tree, as a function of the parameter λ. (Color figure online)

Contracting this minimum non-bottleneck edge in G^+ produces a multigraph with two copies of edge pq, the lighter of which is the bottleneck edge. Therefore, if we keep only the lighter of the two edges, we obtain the weighting on G as a contraction of a minimum spanning tree edge in G^+. This contraction preserves the set of remaining minimum spanning tree weights, as the lemma states. □

It follows that in the parametric case, replacing an edge pq by a triangle pqr, with linear parametric weights on each triangle edge, causes that edge to behave as if it has a nonlinear piecewise linear weight function attached to it, the function mapping the parameter λ to the bottleneck weight from p to q in triangle pqr. Figure 2 shows an example of three parametric weights on a triangle pqr and this bottleneck weight function, with the weights chosen so that the function has three breakpoints. Clearly, we can perturb these three weight functions within small neighborhoods of their coefficients, and obtain a qualitatively similar bottleneck weight function.

4 Weighted 2-Trees

We now describe how to assign parametric weights to the edges of T_i to obtain our $\Omega(n \log n)$ lower bound. As a base case, we may use any linear function as the weight of the single edge of T_0; it can have only one spanning tree, regardless of this choice. For T_i, with $i > 0$, we perform the following steps to assign its weights:

- Construct the weight functions for the edges of T_{i-1}, recursively.
- Apply a linear transformation to the parameter of these weight functions (the same transformation for each edge) so that, in the arrangement of lines representing the graphs of these weight functions, all crossings occur in the interval $[0, 1]$ of λ-coordinates. Additionally, scale these weight functions by a sufficiently small factor ϵ so that, within this interval, they are close enough to the λ-axis, for a meaning of "close enough" to be specified below.
- Construct T_i by replacing each edge pq in T_{i-1} by a triangle pqr, with a new vertex for each triangle. Color the three edges of each triangle red, blue, and

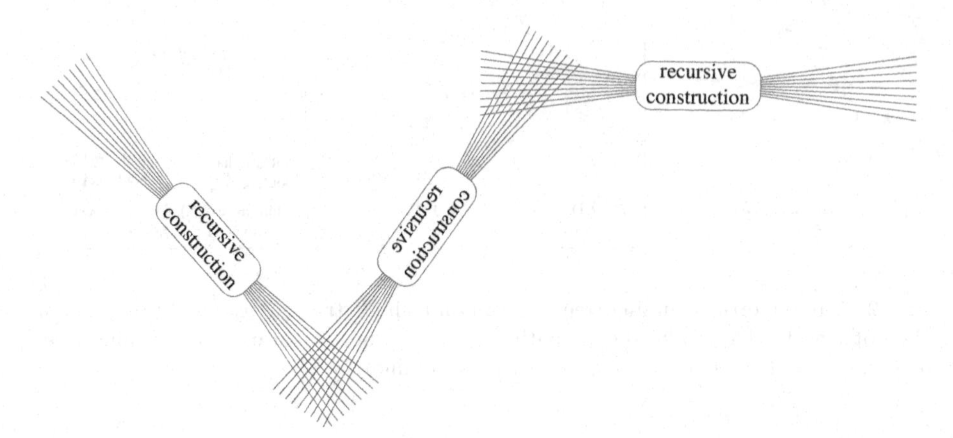

Fig. 3. Recursive construction for the parametric weight functions of the graphs T_i, shown here as an arrangement of lines in a plane whose horizontal coordinate is the parameter λ and whose vertical coordinate is the edge weight at that parameter value. The reversed text in the central recursive construction indicates that the construction is reversed left-to-right relative to the other two copies.

green, as in Fig. 2(left), with pq colored green and the other two edges colored red and blue (choosing arbitrarily which one to color red and which one to color blue).

- Give each edge of T_i a transformed copy of the weight function of the corresponding edge of T_{i-1}, transformed as follows:
  - For a green edge pq, corresponding to an edge of T_{i-1} with weight function $f(\lambda)$, give pq the weight function $f(\lambda - 4.5) + 3$. This transformation shifts the part of the weight function where the crossings with other green edges occur to be close to the right green segment of Fig. 2(right).
  - For a red edge pr, corresponding to an edge pq of T_{i-1} with weight function $f(\lambda)$, give pr the weight function $f(3.75 - \lambda) + \lambda - 1$. This transformation shifts the part of the weight function where the crossings with other red edges occur to be close to the red segment of Fig. 2(right), and (by negating λ in the argument to f) reverses the ordering of the crossings within that region.
  - For a blue edge qr, corresponding to an edge pq of T_{i-1} with weight function $f(\lambda)$, give qr the weight function $f(\lambda - 1.25) + 4 - \lambda$. This transformation shifts the part of the weight function where the crossings with other red edges occur to be close to the blue segment of Fig. 2(right).
- Perturb all of the weight functions, if necessary, so that all crossings of two weight functions have different λ-coordinates, without changing the left-to-right ordering of the crossings between any one weight function and the rest of them.

This construction is depicted schematically, in the (λ, weight) plane, in Fig. 3. We are now ready to define what it means for the weight scaling factor ϵ to be small enough, so that the scaled weight functions are "close enough" to the λ axis: as shown in the figure, the left-to-right ordering of the crossings of the lines graphing the weight functions should be:

1. All crossings of blue with green lines
2. All crossings of two blue lines, in one copy of the recursive construction
3. All crossings of blue with red lines
4. All crossings of two red lines, in a second (reversed) copy of the recursive construction
5. All crossings of red with green lines
6. All crossings of two green lines, in the third copy of the recursive construction

Our construction automatically places all monochromatic crossings into disjoint unit-length intervals with these orderings. The bichromatic crossings of Fig. 2 are separated from these unit-length intervals by a horizontal distance of at least 0.25, and sufficiently small values of ϵ will cause the bichromatic crossings of T_i to be close to the positions of the crossings with the same color in Fig. 2. Therefore, by choosing ϵ small enough, we can ensure that the crossing ordering described above is obtained. Figure 4 depicts this construction for T_2.

We observe that, within each of the unit-length intervals containing a copy of the recursive construction, the bottleneck edges for each triangle pqr in the construction of T_i are exactly the ones of the color for that copy of the recursive construction, and that within these intervals, the minimum non-bottleneck edge in each triangle does not change. Therefore, by Lemma 2, the changes in the sequence of parametric minimum spanning trees within these intervals exactly correspond to the changes in the trees of T_{i-1} from the recursive construction.

Lemma 3. *For weights constructed as above, the number of distinct parametric minimum spanning trees for T_i is at least as large as*

$$N(i) = \frac{i3^i}{2} + \frac{3^i + 3}{4}.$$

Proof. We prove by induction on i that the number of trees is at least as large as the solution to the recurrence

$$N(i) = 3N(i-1) + \frac{3^i - 3}{2}.$$

To prove this, it is easier to count the number of breakpoints, values of λ at which the tree structure changes; the number of trees is the number of breakpoints plus one. In each copy of the recursive construction, this number of breakpoints is exactly $N(i-1) - 1$, so the total number of breakpoints appearing in these three copies is $3N(i-1) - 3$.

Additional breakpoints happen within the ranges of values for λ at which (in the (λ, weight) plane) pairs of lines of two different colors cross. Because of the reversal of the red copy of the recursive construction, the minimum spanning trees immediately to the left and right of these regions of bichromatic crossings correspond to the same trees in T_{i-1}: the bottleneck edges that are included in these

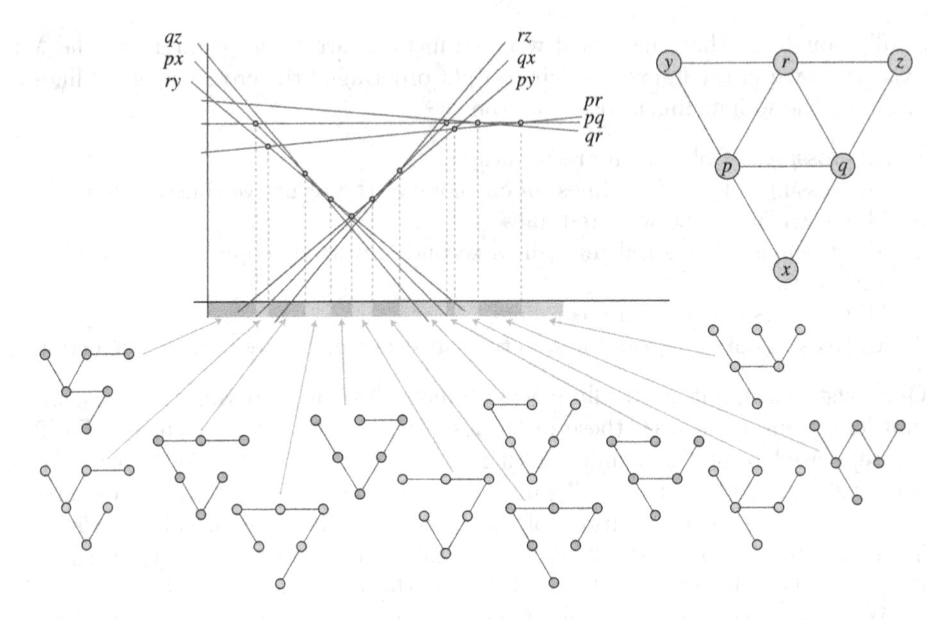

Fig. 4. T_2 (upper right) as parametrically weighted in our construction, with the graphs of each weight function shown as lines in the (λ, w) plane (upper right), and the resulting sequence of 12 parametric minimum spanning trees (bottom). The marked yellow crossings of pairs of lines correspond to breakpoints in the sequence of trees. (Color figure online)

minimum spanning trees come from the same triangles, but with different colors. In the regions where the green lines cross lines of other colors, the minimum non-bottleneck edge in each triangle does not change, so each green bottleneck edge in the minimum spanning tree must be exchanged for a red or blue one. Each change to a tree within this crossing region removes a single edge from the minimum spanning tree and replaces it with another single edge, the two edges whose two lines cross at the λ-coordinate of that change. Therefore, no matter what sequence of changes is performed, to exchange all green bottleneck edges for all red or blue ones requires a number of crossings equal to the number of edges in the minimum spanning tree of T_{i-1}, which is $(3^{i-1} + 1)/2$ by Lemma 1. We get this number of breakpoints at the region where the green and blue lines cross, and the same number at the region where the red and green lines cross.

The analysis of the number of breakpoints at the region where the blue and red lines cross is similar, but slightly different. Immediately to the left and right of this region, the bottleneck edge in each triangle and the minimum non-bottleneck edge in the triangle are red and blue, but in a different order to the left and to the right. Therefore, in triangles where the bottleneck edge is part of the minimum spanning tree (as is always the minimum non-bottleneck edge), nothing changes. However, in triangles where the bottleneck edge is *not* part of the minimum spanning tree, there is a change, to the minimum non-bottleneck edge, from before this crossing region to after it. These triangles correspond to edges of T_{i-1} which do not belong to the minimum spanning tree (for the parameter values in this range), of which there are $(3^{i-1} - 1)/2$ by Lemma 1. By the same argument as before, the crossing region must contain at least this many breakpoints.

Adding together the $3N(i-1) - 3$ breakpoints from the recursive copies, the $(3^{i-1} + 1)/2$ breakpoints for the green–red and green–blue crossing regions, the $(3^{i-1} - 1)/2$ breakpoints for the red–blue crossing region, and $+1$ to convert numbers of breakpoints to numbers of distinct trees, and simplifying, gives the right hand side of the recurrence. A straightforward induction shows that the solution to the recurrence is the formula given in the statement of the lemma. □

For $i = 0, 1, 2, \ldots$ the number of trees given by this formula is

$$1, 3, 12, 48, 183, 669, 2370, 8202, 27885, 93495 \ldots$$

For instance, T_1 has three trees with the weighting given in Fig. 2: the bottleneck function shown in the figure has four linear pieces, but the red and blue pieces both correspond to the same tree, with a different edge on the path pqr as the bottleneck edge. Figure 4 shows the 12 trees for T_2.

5 Packing into Dense Graphs

The lower bound obtained from Lemma 3 applies only to sparse graphs, where the numbers of vertices and edges are within constant factors of each other. However, we want a bound that applies more generally, for graphs with significantly more edges than vertices. The other direction, for graphs with significantly fewer edges than vertices, is less interesting. To achieve many fewer edges than vertices, it is necessary to allow disconnected graphs, and consider minimum spanning forests instead of minimum spanning trees; but with these modifications one can obtain a lower bound simply by adding isolated vertices to the construction of Lemma 3.

To achieve many more edges than vertices, we use the following construction for packing many instances of a sparse lower bound graph into a single denser graph. It does not require any detailed knowledge of the structure of the sparse graph.

Lemma 4. *Let G be a parametrically weighted graph with N vertices and M edges, whose sequence of parametric minimum spanning trees has length T, and let k be a positive integer satisfying $k \leq M$. Then there is a parametrically weighted graph H with $N + 3M$ vertices and $(2k+2)M$ edges whose sequence of parametric minimum spanning trees has length at least $2kT$.*

Proof. We construct H from G in the following steps, illustrated in Fig. 5.

- Number the edges of G as $e_0, e_2, \ldots e_{M-1}$ arbitrarily.
- Subdivide each edge e_i of G, connecting two vertices u and v, into a four-edge path u–a_i–b_i–c_i–v. (It is arbitrary which vertex of this path we call a_i and which we call c_i.)

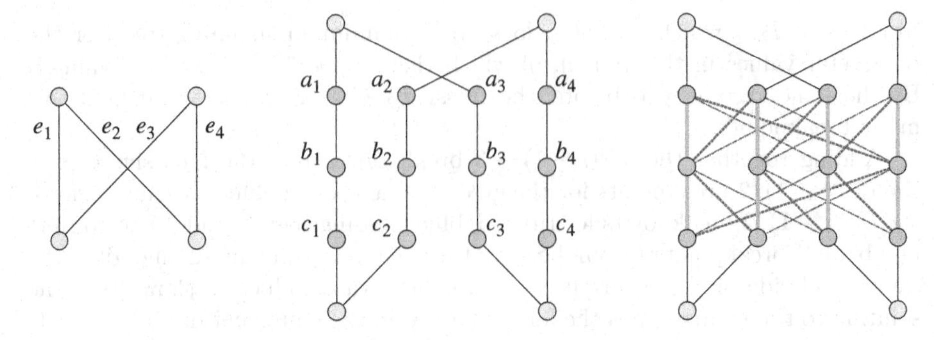

Fig. 5. The construction of Lemma 4, applied to a graph G with four vertices and four edges (left), with the parameter $k = 3$. The central graph is a subdivision of each edge of this graph into a four-edge path, with vertices labeled as shown, and the graph on the right is the final construction H, with the colors and textures of edges indicating the partition of its edges into four subgraphs H_0 (thin black edges), H_1 (thick yellow edges), H_2 (dotted blue edges), and H_3 (dashed red edges). (Color figure online)

- Add additional edges from b_i to a_j and c_j, for each i and each $j = i+1, i+2, \dots i+k-1 \bmod M$.

Given this construction, we define subgraphs H_j as follows:

- H_0 consists of all edges connecting vertices of G to new vertices a_i or c_i.
- H_j consists of all edges from b_i to a_{i+j-1} or c_{i+j-1}, for all i, with indexes taken modulo m.

Then, for $i = 1, 2, \dots k$, the graph $H_0 \cup H_i$ is isomorphic to a subdivision of G, with $H_0 \cup H_1$ being the subdivision we used to construct H and the others obtained in the same way but with permuted connections.

As in Lemma 3, we flatten the arrangement of lines for the weighting of G so that its crossings all lie within a small neighborhood of the unit interval of the λ-axis, without changing its sequence of parametric minimum spanning trees. We then apply linear transformations to the system of weights for the edges in each copy H_j with $j > 0$, as detailed below, while using small-enough weights for all edges in H_0 so that these edges belong to all minimum spanning trees for parameters in the range covered by the transformed unit intervals shown in Fig. 6. More specifically, for each $j > 0$ we use one transformed copy of the weights in G for the a–b edges in H_j, and a second transformed copy for the b–c edges, arranged so that the transformed unit intervals containing the crossings within each copy project to disjoint intervals of the λ-axis, and so that all crossings of the a–b edges appear above all lines for the b–c edges and vice versa. Therefore, in the graph $H_0 \cup H_j$, the parametric trees in the parameter range where the a–b edges cross each other consist of all b–c edges (because those have smaller weight than the a–b edges in each path) together with a subset of the a–b edges corresponding to a spanning tree of G. Because we copied and transformed the weights of G for the a–b edges in this parameter range, we obtain

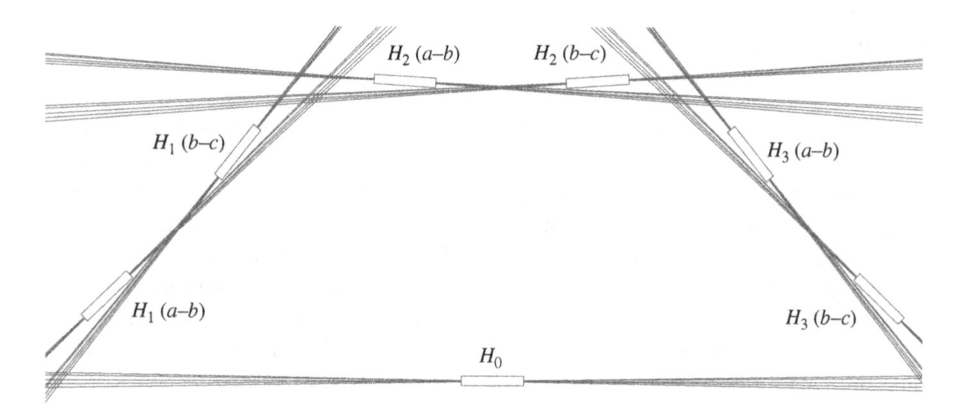

Fig. 6. An arrangement of lines for the weight functions of Lemma 4 with $k = 3$. The small rectangles indicate transformed neighborhoods of the unit λ-interval, containing all crossings of the bundle of lines associated with each subgraph.

T distinct trees of this type. To arrange the a–b and b–c parameter weights for H_i in this fashion, we transform them so that the a–b weights lie near the line $w = 3 - \lambda$, with crossings in the range $\lambda \in [1, 2]$, and so that the b–c weights lie near the line $w = \lambda - 3$, with crossings in the range $\lambda \in [4, 5]$. Then, we transform and flatten these combined weights of H_i, so that they again lie near the λ-axis with all crossings of edges of either type in the range $[0, 1]$.

We arrange the sets of lines associated with H_1, H_2, etc., so that the lines from each H_j pass above the crossings for each other H_j', $j \neq j'$, and so that the range of parameters within which H_j has the lowest lines contains the two subranges where its a–b lines cross and where its b–c lines cross, again as shown in the figure. We may do this by finding a convex-downward polygonal chain with k sides (for instance the upper part of a regular $2k$-gon), in which all sides project to a range of λ-coordinates of more than unit length, and by transforming the weights of each H_i so that the unit interval of the λ-axis, near which all crossings of these weights occur, is transformed to the interior of one of the sides of this polygonal chain. Figure 6 shows the weights for three subgraphs H_1, H_2, and H_3, transformed in this way so that they are near the upper three sides of a hexagon. The weights for H_0 can be chosen to be near a horizontal line, below all crossings of the other weight functions, as also shown in the figure.

Therefore, within these subranges, the parametric minimum spanning trees for all of H will be the same as the trees for $H_0 \cup H_j$, because $H_0 \cup H_j$ spans H and has lower edge weights than any of the remaining edges. With this arrangement, we get $2kT$ distinct parametric minimum spanning trees, $2T$ for each H_j with $j > 0$, as well as additional trees that are not counted in the lemma. □

With this, we are ready to prove our main result:

Theorem 1. *There exists a constant C such that the following is true. Let n and m be integers with $n > 0$ and $2n - 3 \leq m \leq \binom{m}{2}$. Then there exists a* *parametrically weighted graph with n vertices and m edges, with at least $Cm \log n$ parametric minimum spanning trees.*

Proof. Let $G = T_i$, $N = (3^i + 3)/2$, and $M = 3^i$, with i chosen as large as possible so that $N + 3M \leq n$ and $4M \leq m$, and choose k as large as possible so that $(2k+2)M \leq m$; then $N = \Theta(n)$ and $M = \Theta(m/n)$. Apply Lemma 3 to give weights to G so that it has $\Omega(n \log n)$ parametric minimum spanning trees, and apply Lemma 4 to construct a parametrically weighted graph H with $N + 3M$ vertices and $(2k+2)M$ edges that has $\Omega(m \log n)$ parametric minimum spanning trees. If necessary, add leaf vertices to H to increase its number of vertices to n, and then add high-weight edges to increase its number of edges to m without affecting this sequence of parametric spanning trees. □

6 Conclusions

We have shown that the number of parametric minimum spanning trees can be $\Omega(m \log n)$ in the worst case, improving a 25-year-old $\Omega(m\alpha(n))$ lower bound. Because of the structure of the graphs used in our lower bound construction, the new lower bound applies as well to the special cases of planar graphs and of bounded-treewidth graphs, both of which can have $\Omega(n \log n)$ parametric minimum spanning trees. However, our new lower bound is still far from the $O(mn^{1/3})$ upper bound, so there is plenty of room for additional improvement.

Another related question concerns the *parametric bottleneck shortest path problem*, a parametric version of the problem of finding a path between two specified vertices that minimizes the maximum edge weight on the path. In the non-parametric version of the problem, a minimum spanning tree path is an optimal path, although faster algorithms are possible and the problem is also of interest in the case of directed graphs [14]. The same problem is also known in the equivalent maximin form as the widest path problem, where an optimal solution can be found as a maximum spanning tree path [21]. The parametric versions of these problems differ somewhat: a breakpoint in the piecewise linear parametric minimum spanning tree function (the function mapping the parameter value λ to the weight of its minimum spanning tree) might not be a breakpoint in the bottleneck shortest path problem (the maximum weight of an edge on the bottleneck shortest path problem) or vice versa. However, the bottleneck breakpoints that look locally like the minimum of two linear functions do correspond to breakpoints of the minimum spanning tree problem. For this reason, any asymptotic lower bound on the parametric bottleneck shortest path problem would also be a lower bound for parametric minimum spanning trees, and any asymptotic upper bound on the parametric minimum spanning tree problem (including the known $O(mn^{1/3})$ bound) is also an upper bound on parametric bottleneck shortest paths. In fact, our previous $\Omega(m\alpha(n))$ lower bound also applies to parametric bottleneck shortest paths, but our new $\Omega(m \log n)$ bound does not. Can we strengthen the $\Omega(m\alpha(n))$ bound for this problem?

References

1. Agarwal, P.K., Eppstein, D., Guibas, L.J., Henzinger, M.R.: Parametric and kinetic minimum spanning trees. In: Proceedings of the 39th IEEE Symposium on Foundations of Computer Science (FOCS 1998), pp. 596–605 (1998). https://doi.org/10.1109/SFCS.1998.743510
2. Carlson, J., Eppstein, D.: The weighted maximum-mean subtree and other bicriterion subtree problems. In: Arge, L., Freivalds, R. (eds.) SWAT 2006. LNCS, vol. 4059, pp. 400–410. Springer, Heidelberg (2006). https://doi.org/10.1007/11785293_37
3. Carstensen, P.J.: Parametric cost shortest path problems. Unpublished Bellcore memo (1984)
4. Castelli, L., Labbé, M., Violin, A.: Network pricing problem with unit toll. Networks **69**(1), 83–93 (2017). https://doi.org/10.1002/net.21701
5. Chakraborty, S., Fischer, E., Lachish, O., Yuster, R.: Two-phase algorithms for the parametric shortest path problem. In: Marion, J.-Y., Schwentick, T. (eds.) Proceedings of the 27th International Symposium on Theoretical Aspects of Computer Science (STACS 2010), Volume 5 of LIPIcs, pp. 167–178. Schloss Dagstuhl - Leibniz-Zentrum für Informatik (2010). https://doi.org/10.4230/LIPIcs.STACS.2010.2452
6. Chan, T.M.: Finding the shortest bottleneck edge in a parametric minimum spanning tree. In: Proceedings of the 16th ACM-SIAM Symposium on Discrete Algorithms (SODA 2005), pp. 917–918. SIAM (2005). https://dl.acm.org/citation.cfm?id=1070432.1070561
7. Dey, T.K.: Improved bounds for planar k-sets and related problems. Discrete Comput. Geom. **19**(3), 373–382 (1998). https://doi.org/10.1007/PL00009354
8. Eben-Chaime, M.: Parametric solution for linear bicriteria knapsack models. Manag. Sci. **42**(11), 1565–1575 (1996). https://doi.org/10.1287/mnsc.42.11.1565
9. Eppstein, D.: Geometric lower bounds for parametric matroid optimization. Discrete Comput. Geom. **20**(4), 463–476 (1998). https://doi.org/10.1007/PL00009396
10. Eppstein, D.: The parametric closure problem. ACM Trans. Algorithms **14**(1), A2:1–A2:22 (2018). https://doi.org/10.1145/3147212
11. Erickson, J.: Maximum flows and parametric shortest paths in planar graphs. In: Charikar, M. (ed.) Proceedings of the 21st ACM-SIAM Symposium on Discrete Algorithms (SODA 2010), pp. 794–804. SIAM (2010). https://doi.org/10.1137/1.9781611973075.65
12. Fernández-Baca, D., Slutzki, G.: Linear-time algorithms for parametric minimum spanning tree problems on planar graphs. Theor. Comput. Sci. **181**(1), 57–74 (1997). https://doi.org/10.1016/S0304-3975(96)00262-9
13. Fernández-Baca, D., Slutzki, G., Eppstein, D.: Using sparsification for parametric minimum spanning tree problems. Nordic J. Comput. **3**(4), 352–366 (1996)
14. Gabow, H.N., Tarjan, R.E.: Algorithms for two bottleneck optimization problems. J. Algorithms **9**(3), 411–417 (1988). https://doi.org/10.1016/0196-6774(88)90031-4
15. Giudici, A., Halffmann, P., Ruzika, S., Thielen, C.: Approximation schemes for the parametric knapsack problem. Inf. Process. Lett. **120**, 11–15 (2017). https://doi.org/10.1016/j.ipl.2016.12.003
16. Gusfield, D.: Bounds for the parametric minimum spanning tree problem. In: Proceedings of the West Coast Conference on Combinatorics, Graph Theory and Computing (Humboldt State University, Arcata, California, 1979), Volume 26 of Congress Number, Winnipeg, Manitoba, pp. 173–181. Utilitas Math (1980)

17. Holzhauser, M., Krumke, S.O.: An FPTAS for the parametric knapsack problem. Inf. Process. Lett. **126**, 43–47 (2017). https://doi.org/10.1016/j.ipl.2017.06.006
18. Katoh, N.: Bicriteria network optimization problems. IEICE Trans. Fundam. Electron. Commun. Comput. Sci. **E75**, A:321–A:329 (1992)
19. Katoh, N., Tokuyama, T.: Notes on computing peaks in k-levels and parametric spanning trees. In: Souvaine, D.L. (ed.) Proceedings of the 17th Symposium on Computational Geometry (SoCG 2001), pp. 241–248. ACM (2001). https://doi.org/10.1145/378583.378675
20. Mitchell, S.L.: Linear algorithms to recognize outerplanar and maximal outerplanar graphs. Inf. Process. Lett. **9**(5), 229–232 (1979). https://doi.org/10.1016/0020-0190(79)90075-9
21. Pollack, M.: The maximum capacity route through a network. Oper. Res. **8**, 733–736 (1960). https://doi.org/10.1287/opre.8.5.733
22. Tarjan, R.E.: Data Structures and Network Algorithms, Volume 44 of CBMS-NSF Regional Conference Series in Applied Mathematics. Society for Industrial and Applied Mathematics (1983). https://doi.org/10.1137/1.9781611970265
23. Wald, J.A., Colbourn, C.J.: Steiner trees, partial 2-trees, and minimum IFI networks. Networks **13**(2), 159–167 (1983). https://doi.org/10.1002/net.3230130202

Online Bin Packing of Squares and Cubes

Leah Epstein[1(✉)] and Loay Mualem[2]

[1] Department of Mathematics, University of Haifa, Haifa, Israel
[2] Department of Computer science, University of Haifa, Haifa, Israel

Abstract. In the d-dimensional online bin packing problem, hypercubes of positive sizes no larger than 1 are presented one by one to be assigned to positions in d-dimensional unit cube bins. In this work, we provide improved upper bounds on the asymptotic competitive ratio for square and cube bin packing problems, where our bounds do not exceed 2.0885 and 2.5735 for square and cube packing, respectively. To achieve these results, we adapt and improve a previously designed harmonic-type algorithm, and apply a different method for defining weight functions. We detect deficiencies in the state-of-the-art results by providing counter-examples to the current best algorithms and the analysis, where the claimed bounds were 2.1187 for square packing and 2.6161 for cube packing.

1 Introduction

Bin Packing (BP) has been the cornerstone of approximation algorithms and has been extensively studied since the early 1970's. This problem and its variants are important problems with numerous classic applications, such as machine scheduling, cutting stock problems, and storage allocation. Recent applications include also cloud storage.

Bin packing was first introduced and investigated by Ullman in 1951 [38] (see also [2,4,5,10,19,27–30,36,39]). In the classic or standard one-dimensional bin packing problem, we are given a list $L = \{i_1, i_2, \ldots, i_n\}$ of items, and item sizes $S = \{s_1, s_2, \ldots, s_n\}$, where $s_j \in (0, 1]$ is the size of i_j for any $1 \leq j \leq n$. The goal is to pack these items into the minimum number of bins for this input. More precisely, for a subset of items B, we let $|B| = \sum_{i_j \in B} s_j$, and the goal is to partition L into a set of subsets $B = \{b_1, b_2, b_3, \ldots, b_\ell\}$, where $1 \leq \ell \leq n$, such that $|b_k| \leq 1$ holds for $k = 1, \ldots, \ell$, where ℓ is minimized.

A bin packing algorithm is called *online* if it is given the items from L one at a time, and it must assign each item into a bin immediately upon arrival. A newly arriving item is packed according to the packing and sizes of items that have already been presented before its arrival. There is no information about subsequent items, and removing an already packed item from its position is not allowed. As opposed to online algorithms, offline algorithms for bin packing have complete knowledge about the list of items. An offline algorithm simply maps L into a set of bins (in a valid way), and the ordering of the items in L plays

A. Lubiw et al. (Eds.): WADS 2021, LNCS 12808, pp. 357–370, 2021.
https://doi.org/10.1007/978-3-030-83508-8_26

no role. The offline problem is known to be NP-hard [23]; thus, research for this variant has concentrated on the study and development of fast algorithms that can produce near-optimal solutions for the problem in polynomial time. That is, extensive research has gone into developing approximation algorithms for this problem. These algorithms have proven performance for any possible input, and process the input items in polynomial time. See [19,30,35] for such work.

Online algorithms are analyzed via the (absolute or asymptotic) competitive ratio. This is the worst-case cost ratio between outputs of an online algorithm and those of an optimal offline algorithm (for the same inputs). There is vast research on online variants as well [2,4,5,10,27–29,36,39].

We define the competitive ratio more precisely. Given an input list L, let $ALG(L)$ be the cost (number of bins used) obtained by applying algorithm ALG on the input L. Let OPT be an optimal offline algorithm, that uses the minimum number of bins for packing the items, and let $OPT(L)$ denote the number of bins that OPT uses for a given input L. The algorithm is *absolutely r-competitive* if for any input $ALG(L) \leq r \cdot OPT(L)$ and *asymptotically* r-competitive if there exists a constant C such that for any input $ALG(L) \leq r \cdot OPT(L) + C$. The asymptotic competitive ratio for ALG is the infimum r such that ALG is asymptotically r-competitive. Since the last measure is the common one for bin packing, we only discuss this measure in this text, and sometimes omit the words asymptotic and asymptotically. For offline problems, the approximation ratio is defined analogously.

In this work, we deal with *online bin packing of cubes*, and we improve the asymptotic competitive ratio for the d-dimensional bin packing problem of cubes for $d = 2$ and $d = 3$. In the $d = 2$, cubes are in fact squares. The case $d = 1$ is sometimes seen as the classic variant of bin packing. We define the more general case of box packing as follows. The input consists of a list L of items, where each item is a d-dimensional box, and in each dimension, the side length of an item does not exceed 1. The output is a packing of all input items of L into d-dimensional hyper-cube bins. The goal is to minimize the number of used bins. A packing is an assignment of positions in bins to all items such that the following two requirements hold. No two items in a bin overlap with each other (except for their boundaries), and the sides of item are parallel to sides of bins. Note that we do not exclude the option of rotation, also called non-oriented box packing, though we deal here with the asymptotic competitive ratio for squares and cubes, where rotation is meaningless.

Previous Results. Recall that bin packing is an NP-hard problem, and a large part of the research in this field of study focused on finding (asymptotic) approximation bounds. The offline problem has asymptotic approximation schemes (where an approximation scheme is a family of asymptotic approximation algorithms with approximation ratio $1 + \varepsilon$ for any $\varepsilon > 0$) [19,30].

The online bin packing problem was first introduced by Ullman [38]. Johnson [28] showed that a greedy algorithm called Next Fit (NF) has an asymptotic (and an absolute) competitive ratio of 2. It was also shown by Johnson et al. [29] that another greedy algorithm called First Fit has an asymptotic competitive ratio of $\frac{17}{10}$ [29]. Lee and Lee [32] presented the Harmonic algorithm. This algorithm uses bounded space (a constant number of bins can receive items at each time), and it achieves an asymptotic competitive ratio of approximately 1.69103 for large values of its parameter. They also developed the Refined Harmonic algorithm, which has an asymptotic competitive ratio that does not exceed 1.63597. Shortly afterwards, Ramanan et al. [34] introduced Modified Harmonic and Modified Harmonic 2, and showed that these algorithms have asymptotic competitive ratios not exceeding approximately 1.61562 and 1.61217, respectively. The upper bound was improved further later [27,36]. The best upper bound on the asymptotic competitive ratio known so far is 1.57829 by Balogh et al. [2].

As for lower bounds, Yao showed that no online algorithm has an competitive ratio smaller than 1.5 [41]. Later, Brown [8] and Liang [31] improved this lower bound to 1.53635, and Van Vliet [39] improved this lower bound known to 1.54014. Balogh et al. [5] improved this lower bound to 1.54037. The tightest lower bound known so far is 1.54278 By Balogh et al. [4].

Here, we study the d-dimensional bin packing of cubes for $d = 2, 3$, and improve the existing bounds for this problem. In what follows, we discuss previous work for that variant. The hyper-cube online packing problem was studied by Coppersmith and Raghavan [11] who showed upper bounds of 2.6875 and 6.25 on the asymptotic competitive ratios for online square packing and online cube packing, respectively. Seiden and van Stee [37] improved the upper bound for square packing to $395/162 \approx 2.438272$, Miyazawa and Wakabayashi [33] improved the upper bound for cube packing to 3.954, where the algorithm was based on that of [11]. Epstein and van Stee proved an upper bound of 2.24437 for square packing and an upper bound of 2.9421 for online cube packing [17]. These algorithms are similar to the one-dimensional modifications of harmonic algorithms. Han et al. [26] gave upper bounds of 2.1187 and 2.6161 for the asymptotic competitive ratios for square packing and cube packing, respectively. We note that these last bounds are not valid for the algorithms as they were defined and their analysis, as we show in this work, by providing counter-examples to the action of the algorithm for $d = 2$, and by explaining why the analysis does not hold in general. There is also an earlier version of that work [25] but there are flaws in that analysis as well. As for lower bounds on the competitive ratio, there has been some work on that direction as well [3,7,17,37,40], and the current best lower bound is approximately 1.75154 [3].

The offline variant of the square and cube bin packing also have asymptotic approximation schemes [6]. In addition, there is work for more general variants of online rectangle and box bin packing with or without rotation, and vector packing. See [1,7,9,11–15,18,20–22,24,40,40]. Naturally, the bounds for the more general case are larger.

Our Contribution. For decades, bounds for asymptotic competitive ratios of bin packing problems have been extensively studied. In this work, we present improved results for d-dimensional bin packing problem. We provide a new harmonic-type algorithm for d-dimensional bin packing problem. The key components of our algorithm are classification of the items, and an extension of the framework suggested by [32] with respect to the one-dimensional bin packing problem, similar to [36]. Our algorithm is specified by a general structure that is based on that of [26] and a new set of parameters. However, for the analysis, we do not use the analysis as in [26,36], but we define a new weighting technique for the d-dimensional bin packing problem. A related method was used in the past for standard bin packing [2], that is, for the one-dimensional case, but no such method was defined for variants in multiple dimensions. Here we show that it allows one to improve the bounds for another bin packing problem. To emphasize the effectiveness of our new suggested algorithm, we established an improved asymptotic competitive ratio for the cases $d = 2, 3$. We obtain tighter upper bounds for the asymptotic competitive ratio for the online square and cube packing. Specifically, the algorithms have asymptotic competitive ratios of at most 2.0885 and 2.5735, respectively. This is to be compared to the currently known bounds of 2.1187 for square packing and 2.6161 for cube packing by [26] (which are unfortunately incorrect, but it might be possible to prove slightly inferior bounds for these algorithms using our method of analysis).

Additionally, we present a counter example for the previous upper bound claimed by [26] for $d = 2$, showing that it is higher than 2.12. We also explain why their analysis is incorrect and cannot yield a bound below 2.24 for square packing (though we believe that an upper bound of approximately 2.14 can be shown for their algorithm for $d = 2$ using our method of analysis). Our analysis is based on introducing weight functions and bounding the asymptotic competitive ratio by showing that the total weight of bins of the algorithm is equal to the total weight (up to an additive constant) while bounding the total weight of any bin of an optimal offline solution from above [32,36]. For obtaining the upper bounds on weights, we use computer-assisted proofs.

In the full version of this work [16], we present our algorithm for the online d-dimensional bin packing problem of squares and cubes, which is based on previously known algorithms. In Sect. 3, we present our weighting functions and present their analysis for our algorithm and for optimal solutions. Another section that appears only in the full version of the paper contains the specific parameters for our algorithms, which lead to the improved bounds. In the full version of the paper, we also show the counter example for the algorithm of [26], and omitted explanations and proofs can also be found in the full version.

2 Algorithm Extended Harmonic (EH)

Now, we define our algorithm Extended Harmonic (EH) for hyper-cube packing. For any $M \geq 110$, Let N be fixed positive integer and let $t_i \in [0,1]$ for every $i \in \{1,\ldots,N+1\}$ such that $t_i \geq t_{i+1}$, $t_1 = 1$, $t_{N+1} = 1/M$. Let the interval I_i be $(t_{i+1}, t_i]$ for every $i \in \{1,\ldots,N\}$. An item t is categorized as type i if its size is in I_i, i.e., $s(t) \in (t_{i+1}, t_i]$. The algorithm is split into two main components, where we categorize all items into *small* and *large*, small items are packed by AssignSmall [18] (see the full version of this work [16]), and large items are packed by EH which we defined in this section.

For every $i \in \{1, \ldots, N\}$, each item of type i, is either colored red or blue. We then define two sets of counters $\{e_j\}_{j=1}^N$ and $\{n_j\}_{j=1}^N$ such that each counter is initialized with zero, e_i denotes the number of red colored items of type i while n_i denotes the number of items of type i. In addition, for every $i \in \{1, \ldots, N\}$, we define α_i to be an approximate fraction of the red items of type i with respect to n_i, that is $0 \leq \alpha_i \leq 1$ for all i. The invariant $e_i \approx \alpha_i \cdot n_i$ will be maintained throughout the whole process of the algorithm (in the sense that $|e_i - \alpha_i \cdot n_i| = O(1)$). In addition to using e_j, n_j, and α_j for $j = 1, 2, \ldots, N$, there are auxiliary values calculated based on item types. The maximum number items of type i that can be packed in one bin will be based on a parameter β_i for every $i \in \{1, \ldots, N\}$. This parameter will be used for blue items, since for them the maximum number will be packed (except for at most one bin for every type). The amount of unused (free) space in bins filled with β_i^d items from interval I_i will be based on a value denoted by δ_i (this definition of δ_i will be slightly modified later). This value is defined according to the maximum size of any item of type i, which is t_i. This algorithm exploits this free space to pack red items of other types. Thus, $\delta_i = 1 - \beta_i \cdot t_i$. Note that this is the space in one dimension, while, for example, for $d = 2$ the space is an L-shaped area whose width is δ_i. We sometimes decide not to use the entire space of δ_i for red items. For simplicity of the algorithm and its analysis, we define the set $D = \{\Delta_0 = 0, \Delta_1, \ldots, \Delta_k\}$ to describe the set of spaces into which red items can be placed, such that $\Delta_k < 1/2$, $\Delta_i \leq \Delta_{i+1}$, and $\Delta_1 > 0$ for every i. The set may contain all values of the form δ_i or just some of them. Let $\phi : \{1, \ldots, N\} \rightarrow \{0, \ldots, k\}$ denote a mapping function from item types to their corresponding index Δ_j, and for any $i \in \{1, \ldots, N\}$ denote by $\Delta_{\phi(i)}$, the amount of space used to hold red items in a bin which holds blue items of type i. We require that the function ϕ satisfies $\Delta_{\phi(i)} \leq \delta_i$. If $\phi(i) = 0$ holds, then no red items are accepted in bins filled with β_i items. For example, if $\Delta_1 = 0.28$, $\Delta_2 = 0.3$, $\Delta_3 = 0.32$ and $\delta_i = 0.31$, we can choose $\phi(i) = 2$. We could also choose $\phi(i) = 1$, but usually largest j is chosen such that $\Delta_j \leq \delta_i$, in order to save space. To ensure that for every red item there may potentially exist a bin to pack it, we require that $\alpha_i = 0$ for every $i \in \{1, \ldots, N\}$ such that $t_i > \Delta_k$.

We follow some of the literature of this type of algorithms, and use γ_i to denote the maximum number of red items of type i (for every $i \in \{1 \ldots, N\}$) that can be packed in the bin, where $\gamma_i = 0$ if $t_i > \Delta_k$ and $\gamma_i = \max\{1, \{\Delta_1/t_i\}\}$ otherwise. This value is the number of items that can fit in one dimension. For example, if $t_i = \frac{1}{30}$ and $\Delta_1 = 0.21$, then $\gamma_i = 6$, but in the case $t_i = 0.22$, and if $\Delta_k = 0.3$, we will have $\gamma_i = 1$, which means that there will be just one red item of type i next to blue items in each dimension.

To generalize the usage of γ_i to d-dimensional bin packing, we define θ_i which denotes the maximum number of red items of type i (for every $i \in \{1, \ldots, N\}$) that can be packed in a single d-dimensional bin as follows: $\theta_i = \beta_i^d - (\beta_i - \gamma_i)^d$. For example, if $d = 2$, $\beta_i = 5$ and $\gamma_i = 2$, we get $\theta_i = 16$. Thus, a cube with β_i items of type i packed in each dimension is created, and a smaller cube with $\beta_i - \gamma_i$ items in each dimension is removed to make space for other items. The definition $\theta_i = 0$ for $t_i > \Delta_k$ means that there is no place at any bin for red items of type i since the blue items are too large. As mentioned in the preceding paragraph, we require that $\alpha_i = 0$ in these cases, i.e., all the items from interval I_i are colored blue and there are no red items from interval I_i. It is possible that other values of α_i will also be equal to zero.

For simplicity, we redefine the values δ_i to be exactly the $\Delta_{\phi(i)}$ values (by possibly reducing some of these values). Thus, a red item of type j can be packed with blue items of type i if and only if $t_j \leq \delta_i$. The main ingredients for our algorithm are as follows. A pair of integers N and k, such that N denotes number of intervals, and k denotes the number of different sizes of spaces for red items. Rational numbers $t_1 = 1 > t_2 > \cdots > t_N > t_{N+1} = 0$, which denote the intervals boundaries, i.e., the ith interval is $(t_{i+1}, t_i]$. Rational numbers $\alpha_1, \ldots, \alpha_N, \in [0,1]$, where for every $i \in \{1, \ldots, N\}$, α_i denotes the fraction of red items from the whole set of items in the ith interval. Parameters $0 < \Delta_1 < \Delta_2 < \cdots < \Delta_k < 1/2$, which denote set of spaces into which red items can be placed. A function $\phi : \{1, \ldots, N\} \rightarrow \{0. \ldots, k\}$, which denotes a mapping function from item types to their corresponding indexes of spaces for red items. It always holds that $\Delta_{\phi(i)} \geq 1 - \beta_i \cdot t_i$. For simplicity we denote $\Delta_{\phi(i)}$ by δ_i. An item x of size $s(x)$ has a type $\tau(x)$ where $\tau(x) = j \quad \Leftrightarrow \quad s(x) \in I_j$. A table describing bin types can be found in the full version of the paper [16]. Note that not all bin types have the required number of items. Bins that have a smaller number of items (less than β_i^d for blue items of type i, or less than θ_j for red items of type j) is called indeterminate.

An Overview of the Code of Algorithm Extended Harmonic. In what follows, we give an overview of this algorithm, which is our main algorithm. The algorithm is defined for any dimension $d \geq 2$, and we will use this algorithm for the cases $d = 2, 3$. We present the pseudo-code for our algorithm in the full version of the paper [16]. The algorithm colors each incoming item as blue or red. The coloring is based on the number of items of the same type that already arrived, such that the percentage of red items will be correct. Specifically, for every type i, n_i will be the total number of items of this type at each time, and e_i will be the number of items of type i whose color is red. Recall that the algorithm maintains the property $e_i = \alpha_i \cdot n_i$ approximately (since the numbers of items e_i and n_i are integers, while $\alpha_i \cdot n_i$ is not necessarily an integer). This is done by testing the ratio between e_i and the new value of n_i after an item of type i arrives.

Types of bins are marked by pairs of indexes of types, where the first one is the type of blue items for this bin, and the second one is the type of red items. A type that was not decided yet appears as a question mark. Thus, a bin of type $(?, j)$ is a bin that already has at least one red item of type j and no blue items. For an item of type i whose color is red, the algorithm checks whether there exists an already existing bin that can be used to accommodate the new item. This has to be a bin that requires at least one additional item of type i that is red. This may be a bin of type $(?, i)$ or (j, i) for some type $j \neq i$, where its pre-determined number of items of type i was not packed yet. We see such a bin as *open*. If there is no such bin, it will check whether there is a bin with blue items but no red items, where red items of type i can be accepted, and if indeed this is possible, such a bin is selected. If there is no open bin to pack the new item (as a red item), then the algorithm opens a new bin of type $(?, i)$ and packs the new item into it. Thus, the algorithm will pack the new item in the first open bin from the following ordered list of bins. First, a bin of type $(?, i)$ or (j, i) with less than θ_i red items in the bin. Then, a bin of type $(j, ?)$ such that $\delta_j \geq \gamma_i \cdot t_i$, and finally a bin of $(?, i)$ (a new bin). The crucial part of this ordering that the algorithm avoids making new decisions as much as possible. The new item is packed into an open bin for red items of type i if this is possible. If not, the algorithm still tries to use an existing bin, in order to avoid a situation where there are bins of types $(?, i)$ and $(j, ?)$ which could be combined. Only if there is no other option, a new bin is introduced. In this case one can deduce that the current status of the output is such that all spaces that could receive red items of i are already exhausted. Note that this can still change throughout the execution of the algorithm and we analyze only the final output.

For a new item of type i whose color is blue, the case where this type cannot receive red items in its bins is easy. The item is either packed into a bin that does not have its full number of items, or if there is no such bin (which can also be called open), a new bin is opened. Such bins are denoted by type (i). If this type can receive red items into the packing of its blue items, the algorithm checks whether there exists a bin that already received at least one blue item of type i, but it did not receive its full number of blue items, where this can be a bin of type $(i, ?)$ or (i, j) for some $j \neq i$. If there is no such bin, once again the algorithm prefers an existing bin with red items, and only if no such bin can accept blue items of type i, a new bin of type $(i, ?)$ is opened. Thus, the algorithm will pack the new item into the first open bin from the following ordered list of bins. First, a bin of type $(i, ?)$ or (i, j) with less than β_i blue items in the bin. Then, a bin of type $(?, j)$ such that $\delta_j \geq \gamma_j \cdot t_j$, and finally, a bin of type $(i, ?)$ (a new bin). Note that the algorithmic approach is almost identical to those of [26,36]. The algorithm runs one copy of AssignSmall and packs every new small items with this algorithm.

3 Weighting Functions and Results

In what follows, we describe the weighting technique and present the specific weight functions which we use in our algorithm. As it was done in the past [2,32,36], we split the different inputs into cases, based on a classification of the output. We will define one weight function for every case, and the different weight functions are independent in the sense that every case will be analyzed separately. Obviously, all weight functions are based on the parameters of the algorithm, and those are common to all cases.

We assume here that L consists of large items (only). Small items are packed separately, and the weight function used for them is not different from those used in the past. Specifically, the weight of a small item is $\frac{(M+1)^d}{(M^d-1)}$ times its area or volume. In this section we only find the relation between the cost of the algorithm for large items and the total weight. Obviously, when we consider optimal solutions, and we find the relation of weights to their costs, we will consider small items as well, adding the weights of small items as well.

In the past [26,36], two weight functions were designed for the cases in the analysis with bin types that all of them have both red items and blue items. In the analysis, the two functions were compared in the sense that the better one was finally used. The intuition for the two functions was that either the cost of these bins is calculated as a part of the weights of the items that are blue, or it is taken into account in the weights of the items that are red. Informally, while these bins had both blue and red items, in this kind of analysis, the cost of the bins is either paid for by blue items or by red items. The core of the technique which we use for our weight functions is the partitioning the cost of bins of type (i, j) between red and blue items. This can be done in the cases described above, when there are no bins of types $(i, ?)$ and $(?, j)$. For applying the method used here for the design of weight functions, we use a parameter w $(0 \leq w \leq 1)$, where w is the share of the blue items in bins where the cost is split, and $1 - w$ is the share of red items. The value w is not necessarily the same for all the cases, and it is typically different (any value can be used for any case and will lead to a correct proof, be we use values that allow us to prove upper bounds that are as tight as possible). The approach of previous work with two weight functions can be seen as the special case where the choice of w had to be out of $\{0, 1\}$, while we allow w to be a rational number in $[0, 1]$ and usually it is not an integer.

First, we define weighting functions for items such that the number of bins used by our algorithm is bounded by the total weight of the input sequence. Every weight function will be used for one case, where cases are defined later. For a weight function U, for any set X of items, we let $U(X) = \sum_{p \in X} U(p)$.

The following lemma is similar to Lemma 2.2 of [36].

Lemma 1. *The number of all indeterminate bins is $O(1)$, where the constant is independent of the input size.*

Given the last lemma, we assume that no such bins exist in the output. Let B_i and R_i be the number of bins containing blue items and red items, respectively, for type i (bins with both blue and red items are counted in two such values). Let λ_i be the number of items of type i in L. The algorithm keeps the proportion of red items out of all items for a given type almost exactly, up to a constant number of items for every type. The next lemma was proved for the one-dimensional case [36], and it holds for multiple dimensions since it deals with numbers of items, and not with sizes or possible ways to pack items. Since there are no red items for types $1, 2, \ldots, 17$, we let $R_i = 0$ for these types.

The next lemma is also similar to Lemma 2.2 of [36] (see also [26]).

Lemma 2. $B_i = \frac{1-\alpha_i}{\beta_i^d} \cdot \lambda_i + O(1)$, *and* $R_i = \frac{\alpha_i}{\theta_i} \cdot \lambda_i + O(1)$.

Let Y denote number of bins of type (i, j) for all values of i and j, i.e., the number of bins which have both red and blue items. The next property holds due to the double counting of such bins.

Lemma 3. $A(L) \leq \sum_i B_i + \sum_i R_i - Y$.

Let q be the maximum index $i \leq 17$ such that there is at least one bin at termination that satisfies the following condition: the bin is of the type $(i, ?)$ if $i \notin \{2, 3, \ldots, 8\}$ and the bin is of the type $(i, ?)$ or $(20 + i, ?)$ for $2 \leq i \leq 8$. If there is no such i, we let $q = 1$. The motivation is to find whether there are bins with only blue items that are ready to receive red items. If there are such bins, we are interested in the largest value δ_g such that there is a bin of type $(g, ?)$. Let e be the maximum index $j \geq 18$ such that there is at least one bin of the type $(?, j)$ at termination, and if there is no such j, we let $e = 0$. There will be no red items for type 18, and therefore in the case where $e > 0$, where have $e \geq 19$.

Lemma 4. *If $2 \leq q \leq 9$, it holds that $e \leq 37 - q$. If $10 \leq q \leq 16$, it holds that $e \leq 35 - q$.*

The next lemma holds by definition.

Lemma 5. *Assume that $q \in \{2, 3, \ldots, 16\}$. For any $i \in \{q + 1, \ldots, 17\}$, there are no bins of type $(i, ?)$, and for any $j \geq e + 1$ there are no bins of type $(?, j)$.*

Definition 1. *Let $0 \leq w \leq 1$ be a parameter used for the analysis, as explained above. Let $q \in \{2, \ldots, 16\}, e \in \{19, \ldots, 151\}$. Define the weight of an item p of size x to be*

$$V_{e,q}(p) = \begin{cases} 1, & \text{if } x \in I_i, \text{ for } i = 1, \ldots, q \\ w, & \text{if } x \in I_i, \text{ for } i = q + 1, \ldots, 17 \\ \frac{\alpha_i}{\theta_i} + \frac{1-\alpha_i}{\beta_i^d}, & \text{if } x \in I_i, \text{ for } i = 18, \ldots, e \\ \frac{(1-w)\cdot\alpha_i}{\theta_i} + \frac{1-\alpha_i}{\beta_i^d}, & \text{if } x \in I_i, \text{ for } i = e + 1, \ldots, 151. \end{cases}$$

Lemma 6. *Let $q \in \{2, \ldots, 16\}, e \in \{19, \ldots, 151\}$, and let $V_{e,q}(p)$ be as in Definition 1 such that e satisfies Lemma 4 as its maximum value ($e = 37 - q$ if $q \leq 9$, and $e = 35 - q$ otherwise). Then, $A(L) \leq \sum_{p \in I_i} V_{e,q}(p) + O(1)$.*

Next, we define weighting functions for large items such that

$$A(L) \leq \max_{1 \leq i \leq 17} W_i(L) + O(1).$$

We split our proof into 17 cases such that in each case we will use different weighting functions. Among 15 of these cases, i.e., cases $2, 3, \ldots, 16$, we will define the weighting function using Definition 1 with respect to e, q.

Handling Case 1*:* This is the case where $q = 1$. In this case it holds that all bins with blue items of sizes above $\frac{1}{3}$ that can be combined with red items were indeed combined with them.

In what follows, we define the weight of an item p of size x in this case.

$$W_1(p) = \begin{cases} \frac{1-\alpha_i}{\beta_i^d}, & \text{if } x \in I_i, \text{ for } i = 1, 18 \\ 0, & \text{if } x \in I_i, \text{ for } i = 2, \ldots, 17 \\ \frac{\alpha_i}{\theta_i}, & \text{if } x \in I_i, \text{ for } i = 22, \ldots, 28 \\ \frac{\alpha_i}{\theta_i} + \frac{1-\alpha_i}{\beta_i^d}, & \text{if } x \in I_i, \text{ for } i = 19, \ldots, 21, 29, \ldots, 151. \end{cases}$$

The definition of this case implies that bin types $(2,?), \ldots, (17,?)$ and bin types $(22,?), \ldots, (28,?)$ do not exist. Hence, $Y \geq \sum_{i=1}^{17} B_i + \sum_{i=22}^{28} B_i$. We use the property $\alpha_i = 0$ for $1 \leq i \leq 18$, and get

$$\begin{aligned} A(L) \leq \sum_{i=1}^{N} B_i + \sum_{i=1}^{N} R_i - Y &= \sum_{i=1}^{151} B_i + \sum_{i=1}^{151} R_i - \sum_{i=2}^{17} B_i - \sum_{i=22}^{28} B_i \\ &= \sum_{i=1,18,19,20,21,29,\ldots,151} (B_i + R_i) + \sum_{i=2,\ldots,17,22\ldots,28} R_i \\ &= \sum_{i=1,18,19,20,21,29,\ldots,151} \left(\frac{1-\alpha_i}{\beta_i^d} \cdot \lambda_i + \frac{\alpha_i}{\theta_i} \cdot \lambda_i\right) \\ &+ \sum_{i=2,\ldots,17,22\ldots,28} \frac{\alpha_i}{\theta_i} \cdot \lambda_i + O(1) = \sum_{p \in I_i} W_1(p) + O(1). \end{aligned}$$

Handling Cases **2, 3, . . . , 16***:* a table in the full version of this work [16], contains the cases which rely on using both e and q. The definitions are based on our discussion above. Since $t_e = 1 - t_{q+1}$ always holds, substituting the values e, q and $v_{e,q}$, into Lemma 6, yields that $A(L) \leq \sum_{p \in I_i} V_{e,q}(p) + O(1)$.

Note that in these weight functions we did not take into account the fact that the definition of q considers also items of sizes in $(\frac{1}{3}, \frac{1}{2}]$ as blue items that can receive red items in their bins. The relevant cases are easy in the sense that the asymptotic competitive ratios for them are small even without reducing these weights (cases $2, \ldots, 7$), and reducing these weights will not change the competitive ratio of the algorithm.

Handling Case **17***:* In this case $q = 17$. Any red item could have been combined into a blue bin of the form $(17,?)$, and thus, there are no $(?, j)$ bins at all. In what follows, we define the weight of an item p of type $i \leq 151$ in this case.

$$W_{17}(p) = \frac{1-\alpha_i}{\beta_i^d}.$$

Since the number of bins type $(?, j)$ is zero for any j, all the red items are packed in bins which include blue items. i.e., the only type of bins that may exist are $(i, j), (i), (i, ?)$, which means that there are blue items packed into every bin. Hence, we get that $Y = \sum_i R_i$. Which yields

$$A(L) \le \sum_i B_i + \sum_i R_i - Y = \sum_i B_i + \sum_i R_i - \sum_i R_i$$
$$= \sum_{i \in 1,\ldots,151} B_i = \sum_{p \in I_i} W_{17}(p) + O(1),$$

where first inequality holds by Lemma 3, and the last equality holds by definition of W_{17}, B_i and Lemma 2. by the analysis above we get that

Lemma 7. $A(L) \le \max_{1 \le i \le 17} W_i(L) + O(1)$.

3.1 Upper Bounds on the Asymptotic Competitive Ratio

In this section, we provide the α_i parameters for square and cube packing, respectively. We also provide upper bounds on the asymptotic competitive ratio for each case in Table 1.

For each $j \in \{1, 2, \ldots, 17\}$, we use the following integer program for obtaining an upper bound on the asymptotic competitive ratio,

$$\text{maximize} \quad f_j(X) = \sum_{i=1}^{151} w_i \cdot x_i + \frac{112^d}{111^d - 1}\left(1 - \sum_{i=1}^{151} x_i \cdot t_{i+1}^d\right)$$

subject to

$$\sum_{i=1}^{151} x_i \cdot t_{i+1}^d \le 1 \qquad (1)$$

$$\sum_{i=1}^{151} \lfloor (t_{i+1} \cdot (u+1)) \rfloor^d \cdot x_i \le u^d \qquad \forall u \in \{1, \ldots, 220\} \qquad (2)$$

$$x_i \ge 0 \ \text{ and } \ x_i \in \mathbb{Z} \qquad \forall i \in \{1, \ldots, 151\}$$

Here X is a feasible set of items which fit into a single bin (of an optimal solution), x_i is the number of items type i in X, and w_i is the weight of an item of type i, defined in the previous part of the section by the function W_i. The value $1 - \sum_{i=1}^{111} x_i \cdot t_{i+1}^d$ is an upper bound on the total volume (or area) of all the small items in X, and $\frac{112^d}{111^d-1} \cdot \left(1 - \sum_{i=1}^{111} x_i \cdot t_{i+1}^d\right)$ is an upper bound of the total weight of all the small items in X.

The second type of constraints is based on a simple property that for an integer $u \ge 1$, no bin can contain more than u^d items of size above $\frac{1}{u+1}$ (see for example Claim 2.1 of [14]). For every item type, the constraint takes into account the number of independent items of size above $\frac{1}{u+1}$ it can be split into. An item of type i has a side above $\frac{1}{t+1}$, so every side can be split into $\lfloor \frac{t_{i+1}}{1/(u+1)} \rfloor$ parts. For example, an item of side above $\frac{1}{2}$ can be split into three items of sides above $\frac{1}{6}$ in every dimension.

In order to obtain a slightly better result, we added two constraints of a different form to the integer program for the case $d = 2$, as follows.

The first constraint is:

$$\sum_{i=1}^{16} 21 \cdot x_i + \sum_{i=17}^{28} 11 \cdot x_i + \sum_{i=29}^{38} x_i \leq 57. \quad (3)$$

The second constraint is:

$$\sum_{i=1}^{16} 80 \cdot x_i + \sum_{i=17}^{28} 30 \cdot x_i + \sum_{i=29}^{37} 10 \cdot x_i + x_{38} \leq 190. \quad (4)$$

Lemma 8. *Conditions* (3) *and* (4) *hold for every valid bin of an optimal solution (for $d = 2$).*

The next theorem states our main result.

Theorem 1. *The asymptotic performance ratio of Algorithm EH for square packing is at most* 2.0885*, while for cube packing is at most* 2.5735*.*

Proof. We set the parameters α_i according to the corresponding table (for square packing and for cube packing). For each case we applied a simple integer program solver in order to find the worst case bound. We obtain the results for square and cube packing, as described in Table 1. Hence, we get that $A(L) \leq 2.5735 \cdot OPT(L)+O(1)$ for cube packing, and $A(L) \leq 2.0885 \cdot OPT(L)+O(1)$ for square packing. □

Table 1. Square and cube packing: upper bounds on the total weights for each case.

	Square packing	Cube packing
Case 1	2.088447879968511	2.5731896581108735
Case 2	1.9438375658626355	2.45464218336544
Case 3	2.0109397168059324	2.475823071455533
Case 4	1.9607242494316246	2.455719344199358
Case 5	1.9942453743436321	2.5115525001235937
Case 6	1.9875046382360564	2.5339175799806912
Case 7	1.9554146240072456	2.5016302664189443
Case 8	1.9441281429162531	2.493821911539605
Case 9	2.0884478982863968	2.5734762658161277
Case 10	2.0884277288254993	2.5593413871191126
Case 11	2.088445077308426	2.5567398601707696
Case 12	2.0876840226666538	2.557631911023032
Case 13	2.0847781920964583	2.5498950440578287
Case 14	2.07732977965866	2.5226265870712448
Case 15	2.0656430335436333	2.527717407098689
Case 16	2.0437751234561317	2.5385458044738085
Case 17	2.088086287477056	2.5718658072279847

References

1. Azar, Y., Cohen, I.R., Kamara, S., Shepherd, F.B.: Tight bounds for online vector bin packing. In: Proceedings of the 45th ACM Symposium on Theory of Computing (STOC 2013), pp. 961–970 (2013)
2. Balogh, J., Békési, J., Dósa, G., Epstein, L., Levin, A.: A new and improved algorithm for online bin packing. In: Proceedings of the 26th European Symposium on Algorithms (ESA 2018), pp. 5:1–5:14 (2018)
3. Balogh, J., Békési, J., Dósa, G., Epstein, L., Levin, A.: Lower bounds for several online variants of bin packing. Theory Comput. Syst. **63**(8), 1757–1780 (2019). https://doi.org/10.1007/s00224-019-09915-1
4. Balogh, J., Békési, J., Dósa, G., Epstein, L., Levin, A.: A new lower bound for classic online bin packing. CoRR, abs/1807.05554 (2018). Also in Proceedings of the WAOA 2019
5. Balogh, J., Békési, J., Galambos, G.: New lower bounds for certain classes of bin packing algorithms. Theory Comput. Sys. **440–441**, 1–13 (2012)
6. Bansal, N., Correa, J.R., Kenyon, C., Sviridenko, M.: Bin packing in multiple dimensions: inapproximability results and approximation schemes. Math. Oper. Res. **31**(1), 31–49 (2006)
7. Blitz, D., Heydrich, S., van Stee, R., van Vliet, A., Woeginger, G.J.: Improved lower bounds for online hypercube and rectangle packing. CoRR, abs/1607.01229v2 (2016)
8. Brown, D.J.: A lower bound for on-line one-dimensional bin packing algorithms. Coordinated Science Laboratory report no. R-864 (UILU-ENG 78–2257) (1979)
9. Christensen, H.I., Khan, A., Pokutta, S., Tetali, P.: Multidimensional bin packing and other related problems: a survey. Comput. Sci. Rev. **24**, 63–79 (2017)
10. Coffman Jr., E.G., Garey, M., Johnson, D.S.: Approximation algorithms for bin packing: a survey. In: Hochbaum, D. (ed.) Approximation Algorithms for NP-Hard Problems, pp. 46–93. PWS Publishing Co., Boston (1996)
11. Coppersmith, D., Raghavan, P.: Multidimensional online bin packing: algorithms and worst case analysis. Oper. Res. Lett. **8**, 17–20 (1989)
12. Csirik, J., Frenk, J.B.G., Labbe, M.: Two-dimensional rectangle packing: on-line methods and results. Discrete Appl. Math. **45**(3), 197–204 (1993)
13. Csirik, J., van Vliet, A.: An on-line algorithm for multidimensional bin packing. Oper. Res. Lett. **13**(3), 149–158 (1993)
14. Epstein, L.: Two-dimensional online bin packing with rotation. Theor. Comput. Sci. **411**(31–33), 2899–2911 (2010)
15. Epstein, L.: A lower bound for online rectangle packing. J. Comb. Optim. **38**(3), 846–866 (2019). https://doi.org/10.1007/s10878-019-00423-z
16. Epstein, L., Mualem, L.: Online bin packing of squares and cubes. CoRR, abs/2105.08763 (2021)
17. Epstein, L., van Stee, R.: Online square and cube packing. Acta Informatica **41**(9), 595–606 (2005). https://doi.org/10.1007/s00236-005-0169-z
18. Epstein, L., van Stee, R.: Optimal online algorithms for multidimensional packing problems. SIAM J. Comput. **35**(2), 431–448 (2005)
19. Fernandez de la Vega, W., Lueker, G.S.: Bin packing can be solved within $1+\varepsilon$ in linear time. Combinatorica **1**(4), 349–355 (1981). https://doi.org/10.1007/BF02579456
20. Fujita, S., Hada, T.: Two-dimensional on-line bin packing problem with rotatable items. Theor. Comput. Sci. **289**(2), 939–952 (2002)

21. Galambos, G.: A 1.6 lower-bound for the two-dimensional on-line rectangle bin-packing. Acta Cybernet. **10**(1–2), 21–24 (1991)
22. Galambos, G., van Vliet, A.: Lower bounds for 1-, 2- and 3-dimensional on-line bin packing algorithms. Computing **52**(3), 281–297 (1994). https://doi.org/10.1007/BF02246509
23. Garey, M.R., Johnson, D.S.: Computers and Intractability: A Guide to the Theory of of NP-Completeness. Freeman and Company, San Francisco (1979)
24. Han, X., Chin, F.Y., Ting, H.-F., Zhang, G., Zhang, Y.: A new upper bound 2.5545 on 2D online bin packing. ACM Trans. Algorithms **7**(4) (2011). Article 50
25. Han, X., Ye, D., Zhou, Y.: Improved online hypercube packing. CoRR, abs/cs/0607045 (2016). Also in Proceedings of the WAOA 2006
26. Han, X., Ye, D., Zhou, Y.: A note on online hypercube packing. CEJOR **18**(2), 221–239 (2010). https://doi.org/10.1007/s10100-009-0109-z
27. Heydrich, S., van Stee, R.: Beating the harmonic lower bound for online bin packing. In: Proceedings of the 43rd International Colloquium on Automata, Languages, and Programming (ICALP 2016), pp. 41:1–41:14 (2016)
28. Johnson, D.S.: Fast algorithms for bin packing. J. Comput. Syst. Sci. **8**(3), 272–314 (1974)
29. Johnson, D.S., Demers, A., Ullman, J., Garey, M., Graham, R.: Worst-case performance bounds for simple one-dimensional packing algorithms. SIAM J. Comput. **3**(4), 299–325 (1974)
30. Karmarkar, N., Karp, R.M.: An efficient approximation scheme for the one-dimensional bin-packing problem. In: Proceedings of the 23rd Annual Symposium on Foundations of Computer Science (FOCS 1982), pp. 312–320 (1982)
31. Liang, F.M.: A lower bound for on-line bin packing. Inf. Process. Lett. **10**(2), 76–79 (1980)
32. Lee, C.C., Lee, D.T.: A simple online bin packing algorithm. J. ACM **32**(3), 562–572 (1985)
33. Miyazawa, F.K., Wakabayashi, Y.: Cube packing. Theor. Comput. Sci. **297**(1–3), 355–366 (2003)
34. Ramanan, P., Brown, D.J., Lee, C.C., Lee, D.T.: On-line bin packing in linear time. J. Algorithms **10**(3), 305–326 (1989)
35. Rothvoss, T.: Better bin packing approximations via discrepancy theory. SIAM J. Comput. **45**(3), 930–946 (2016)
36. Seiden, S.S.: On the online bin packing problem. J. ACM **49**(5), 640–671 (2002)
37. Seiden, S.S., van Stee, R.: New bounds for multidimensional packing. Algorithmica **36**(3), 261–293 (2003). https://doi.org/10.1007/s00453-003-1016-7
38. Ullman, J.D.: The performance of a memory allocation algorithm. Technical report 100, Princeton University, Princeton, NJ (1971)
39. van Vliet, A.: An improved lower bound for online bin packing algorithms. Inf. Process. Lett. **43**(5), 227–284 (1992)
40. van Vliet, A.: Lower and upper bounds for online bin packing and scheduling heuristics. Ph.D. thesis, Erasmus University, Rotterdam, The Netherlands (1995)
41. Yao, A.C.C.: New algorithms for bin packing. J. ACM **27**(2), 207–227 (1980)

Exploration of k-Edge-Deficient Temporal Graphs

Thomas Erlebach and Jakob T. Spooner(✉)

School of Informatics, University of Leicester, Leicester, England
{te17,jts21}@le.ac.uk

Abstract. An *always-connected temporal graph* $\mathcal{G} = \langle G_1, ..., G_L \rangle$ with underlying graph $G = (V, E)$ is a sequence of graphs $G_t \subseteq G$ such that $V(G_t) = V$ and G_t is connected for all t. This paper considers the property of *k-edge-deficiency* for temporal graphs; such graphs satisfy $G_t = (V, E - X_t)$ for all t, where $X_t \subseteq E$ and $|X_t| \leq k$. We study the TEMPORAL EXPLORATION problem (compute a temporal walk that visits all vertices $v \in V$ at least once and finishes as early as possible) restricted to always-connected, k-edge-deficient temporal graphs and give constructive proofs that show that k-edge-deficient and 1-edge-deficient temporal graphs can be explored in $O(kn \log n)$ and $O(n)$ timesteps, respectively. We also give a lower-bound construction of an infinite family of always-connected k-edge-deficient temporal graphs for which any exploration schedule requires at least $\Omega(n \log k)$ timesteps.

Keywords: Graph algorithms · Temporal graphs · Graph exploration

1 Introduction

Given a simple, connected, undirected graph G and a *start vertex* $s \in V(G)$, the task of *exploring* G, i.e., computing a sequence of consecutively crossed edges $e \in E(G)$ that begins at s and visits every vertex $v \in V(G)$ at least once, is both natural and well-understood. A closely related problem was initially considered by Shannon [19], who designed a mechanical maze-solving machine which implemented a depth first search-type technique in order to locate, within a given maze, a prespecified goal. This 'searching' problem is indeed related to graph exploration: if our task is to simply complete an exploration of G, then a solution can be straightforwardly found by performing a DFS starting from s and stopping once all vertices have been visited at least once – clearly this requires $\Theta(n)$ edge-traversals in total.

The graph exploration problem in the context of temporal graphs (i.e., graphs whose edge set can change over time) has also received significant attention in recent years. This problem, known as TEMPORAL EXPLORATION (TEXP), but restricted to *k-edge-deficient* temporal graphs (which we define formally later) is the focus of this paper. Given a temporal graph $\mathcal{G}$, the problem asks that we compute a temporal walk, starting at some prespecified vertex $s \in V(\mathcal{G})$,

A. Lubiw et al. (Eds.): WADS 2021, LNCS 12808, pp. 371–384, 2021.
https://doi.org/10.1007/978-3-030-83508-8_27

that makes at most a single edge-traversal in each timestep, and that visits all vertices at least once by the earliest time possible. We formally define the problem and temporal graph model in Sect. 2, but refer the interested reader to [5,16] for more on temporal graphs in general, or [6,17] for more on TEXP. In the most general setting, TEXP makes no assumptions about the input temporal graph, aside from the assumption that the input temporal graph is connected in each timestep (i.e., *always-connected*), which ensures exploration is always possible. This allows an arbitrary number of edges from the underlying graph to be missing in each timestep, and thus the graphs in different timesteps can differ substantially, which leads to pessimistic bounds on the worst-case exploration time. It is therefore interesting to study the question whether better exploration times can be guaranteed if the number of missing edges in each time step is small. To study this question, we also consider always-connected temporal graphs but, in contrast to previous work, we consider k-edge-deficient temporal graphs, whose structure in each step is 'close' to that of its underlying graph, in the sense that at most k edges are missing. Such graphs were defined by Gotoh et al., in [11], where they were considered in a distributed setting. We assume that the temporal structure of an input temporal graph is known in full to an algorithm prior to it computing a solution, as opposed to a setting in which the structure of the graph in each step is revealed online and over time.

Contribution. We introduce the temporal graph property of *k-edge-deficiency*, and consider TEMPORAL EXPLORATION on always-connected temporal graphs that are k-edge-deficient for some $k \in \mathbb{N}$. We define the property formally in Sect. 2, but essentially these are temporal graphs $\mathcal{G}$ with *underlying graph* G, such that, during each timestep t of $\mathcal{G}$'s lifetime, there are at most k edges $e \in E$ in the underlying graph that are untraversable in (or 'missing' from) $\mathcal{G}$. Let $n = |V(G)|$. In Sect. 3 we prove for arbitrary $k \in \mathbb{N}$ that k-edge-deficient always-connected temporal graphs can be explored in $O(kn \log n)$ timesteps. In Sect. 4 we additionally show that 1-edge-deficient graphs can always be explored in $O(n)$ timesteps, giving a recursive exploration algorithm that exploits a number of existing structural/algorithmic results originating from traditional graph theory. Finally, in Sect. 5, we sketch a modification of an existing $\Omega(n \log n)$ lower bound on the number of timesteps required to explore always-connected temporal graphs with planar underlying graph of maximum degree ≤ 4, presented in [6], that allows us to obtain an $\Omega(n \log k)$ bound on the worst-case time required to explore arbitrary always-connected k-edge-deficient temporal graphs.

Related Work. Brodén et al. [3] consider the TEMPORAL TRAVELLING SALESPERSON PROBLEM on a complete graph with n vertices, with edge costs that can differ between 1 and 2 in each timestep. They show that when an edge's cost changes at most k times over the input graph's lifetime, the problem is NP-complete, but provide a $(2 - \frac{2}{3k})$-approximation; for the same problem, Michail and Spirakis [17] prove APX-hardness and provide a $(1.7 + \epsilon)$-approximation. Bui-Xuan et al. [4] propose multiple objectives for optimisation when computing temporal walks/paths: e.g., *fastest* (fewest steps used) and *foremost* (arriving at the destination at the earliest time possible). The decision version of the

TEMPORAL EXPLORATION problem, which asks whether or not a given temporal graph admits a temporal walk that visits all vertices at least once, is also considered in [17]. They show that the problem is NP-complete when no restrictions are placed on the input; they also propose considering the problem under the *always-connected* assumption, which ensures that exploration is possible provided the lifetime of the input graph is sufficiently long [17]. Erlebach et al. [6] further consider the optimisation variant of the TEMPORAL EXPLORATION problem under the always-connected assumption. They prove an $\Omega(n^2)$ lower bound on the time needed to explore general always-connected temporal graphs, and provide a proof that temporal graphs within this class can be explored in n^2 steps. They also prove a number of bounds on the number of timesteps required to explore various restricted temporal graph classes. Bodlaender and van der Zanden [2] examine TEXP when restricted to graphs of pathwidth at most 2 in each timestep, showing the problem to be NP-complete even under these limiting restrictions. In [14] and [13], Ilcinkas et al. respectively consider TEXP restricted to temporal graphs with underlying cycle or cactus graphs. Akrida et al. [1] consider RETURN-TO-BASE TEXP in which a candidate solution must return to the vertex from which it initially departed. In [7], Erlebach et al. prove an $O(dn^{1.75})$ bound on the number of time steps required to explore any temporal graph with degree bounded by d in each step, a considerable improvement over the previously best known $O(\frac{n^2 \log d}{\log n})$ bound [8]. In [9], a *non-strict* variant of TEXP is studied – here, a computed walk may make an unlimited number of edge-traversals in each given timestep. Notions of strict/non-strict paths which respectively allow for a single edge/unlimited number of edge(s) to be crossed in any timestep have been considered before, notably by Kempe et al. in [15] and Zschoche et al. in [20]. In this paper, we only consider strict temporal walks. Gotoh et al. in [12] consider TEXP on temporal graphs with underlying cycle under the so-called (H, S)-view, in which only the availability of edges at most H hops away for at most the next S timesteps is known to an algorithm. Casteigts et al. examined the fixed-parameter tractability of the problem of finding temporal paths between a source and destination that wait no longer than Δ consecutive timesteps at any vertex they visit. Temporal graph exploration has also been studied in a distributed setting: in [11], Gotoh et al. consider a variant in which a collection of cooperating mobile agents construct a map of a temporal graph. In the same paper, they defined the class of k-edge-deficient graphs (under a different name), proving bounds on the number of cooperating agents required to ensure that exploration is possible under a variety of different distributed settings.

2 Preliminaries

We denote by $[n]$ the set $\{1, ..., n\}$. Let $G = (V, E)$ and G' be simple, undirected graphs. We write $G' \subseteq G$ if G' is a (not necessarily induced) subgraph of G. $|V|$ is the *order* of G; $|E|$ is G's *size*. If $X \subseteq V$ is a subset of G's vertex set, we denote by $G - X$ the subgraph of G induced by $V(G) - X$.

Definition 2.1 (Temporal graph). *A temporal graph $\mathcal{G} = \langle G_1, G_2, ..., G_L \rangle$ with underlying graph $G = (V, E)$, order $n = |V|$ and lifetime L is an ordered sequence of subgraphs $G_t = (V, E_t)$ of G, indexed by the timesteps $t \in [L]$. In particular, we have that $V(G_t) = V = V(\mathcal{G})$ and $E_t \subseteq E$ for all $t \in [L]$.*

Let $\mathcal{G} = \langle G_1, G_2, ..., G_L \rangle$ be an arbitrary temporal graph. An edge $e \in E$ that satisfies $e \in E_t$ is *present* during timestep t. If $e \in E$ satisfies $e \notin E_t$, we say that e is *missing* in timestep t. A temporal graph $\mathcal{G} = \langle G_1, G_2, ..., G_L \rangle$ is said to be *always-connected* if it is such that G_t is connected for all $t \in [L]$.

Definition 2.2 (Temporal walk). *A temporal walk $\mathcal{W}$ in a temporal graph $\mathcal{G}$ is an alternating sequence of vertices and edge-time pairs,*

$$\mathcal{W} = v_1, (e_1, t_1), v_2, ..., v_{k-1}, (e_{k-1}, t_{k-1}), v_k.$$

Each edge-time pair (e_j, t_j) denotes the traversal of edge $e_j = \{v_j, v_{j+1}\}$ at timestep t_j, which implies that $e_j \in E_{t_j}$. We require that $t_0 < t_1 < ... < t_{k-1}$, i.e., that the timesteps at which the consecutive edges of $\mathcal{W}$ are traversed are strictly increasing. We say that the walk starts at vertex v_0, and for all $i \in [k]$, we say that $\mathcal{W}$ visits *$v_i \in V(G)$.*

$\mathcal{W}$ is an *exploration schedule* of $\mathcal{G}$ with start vertex $s \in V(\mathcal{G})$ if $\mathcal{W}$ is a temporal walk in $\mathcal{G}$ that starts at s and visits all vertices $v \in V(G)$. Let $\mathcal{W}$ be an exploration schedule in a temporal graph $\mathcal{G}$ with underlying graph G. We denote by $a(\mathcal{W})$ the timestep at which $\mathcal{W}$ first visits the n-th unique vertex $v \in V(G)$; this is the *arrival time* of $\mathcal{W}$. If $\mathcal{W}$ satisfies $a(\mathcal{W}) \leq a(\mathcal{W}')$ for any other exploration schedule $\mathcal{W}'$ with the same start vertex in $\mathcal{G}$, then we say that $\mathcal{W}$ is *foremost.*

Definition 2.3 (Temporal Exploration). *An instance of the* Temporal Exploration *(TEXP) problem is given as a pair $(\mathcal{G}, s)$, where $\mathcal{G} = \langle G_1, G_2, ..., G_L \rangle$ is an arbitrary temporal graph on n vertices with lifetime $L \geq |V(\mathcal{G})|^2 = n^2$, and $s \in V(\mathcal{G})$ is a start vertex. The problem asks that we compute an exploration schedule $\mathcal{W}$ such that $\mathcal{W}$ is foremost and starts at vertex s. It is assumed that G_t ($t \in [L]$) is known to an algorithm prior to it computing a solution.*

It was proven in [6] that arbitrary always-connected temporal graphs admit at least one exploration schedule $\mathcal{W}$ such that $\alpha(\mathcal{W}) \leq n^2$. Hence having $L \geq |V(\mathcal{G})|^2$ ensures that an exploration schedule exists.

Definition 2.4 (k-edge-deficient). *Let $\mathcal{G} = \langle G_1, ..., G_L \rangle$ be a temporal graph with underlying graph $G = (V, E)$ and order $n = |V|$. Then $\mathcal{G}$ is k-edge-deficient (for $k \in \mathbb{N}$) if, for all $t \in [L]$, we have $G_t = (V, E - X_t)$ for some $X_t \subseteq E$ with $|X_t| \leq k$.*

When constructing a walk in a k-edge-deficient temporal graph $\mathcal{G}$, we may speak of an agent *following* a walk W in the underlying graph G. By this, we mean that the agent traverses in $\mathcal{G}$ the edges in the same order as they are traversed by

W, and does this whenever it is possible to do so, i.e., whenever the next edge e traversed by W is present in the current timestep t. If that edge is not present, the agent is *blocked on e in step t*. For always-connected k-edge-deficient temporal graphs we require that $G_t = (V, E - X_t)$ is connected for all $t \in [L]$. We consider only always-connected, k-edge-deficient temporal graphs with finite lifetime $L \geq n^2$ – as such, any temporal graph we refer to (unless stated otherwise) is assumed to hold these properties. The following lemma from [6] will be useful.

Lemma 2.5 (Reachability lemma; Erlebach et al. [6]). *Let $\mathcal{G}$ be an arbitrary always-connected temporal graph with vertex set V and lifetime L. Then an agent situated at any vertex $u \in V$ at any time $t \leq L - n$ can reach any other vertex $v \in V$ in at most $|V| - 1 = n - 1$ steps, i.e., by time step $t + n - 1$.*

3 $O(kn \log n)$-Time Exploration of k-Edge-Deficient Temporal Graphs

We present an algorithm that proceeds in rounds. In each round, it considers a forest consisting of $k+1$ edge-disjoint subtrees of a spanning tree of the underlying graph and ensures that all edges of one of these trees can be traversed in the round. The following lemma allows us to split a tree T into a pair of edge-disjoint subtrees (whose union covers $E(T)$) in a balanced way:

Lemma 3.1. *Let T be a tree with $m \geq 2$ edges. Then one can compute two edge-disjoint subtrees T' and T'' such that $|E(T')|, |E(T'')| \in [m/3, 2m/3]$, and such that $E(T') \cup E(T'') = E(T)$.*

Say that a set S of edge-disjoint subtrees $T' \subseteq F$ is a *subtree-cover* of a forest F if, for every $e \in E(F)$ we have $e \in E(T')$ for some $T' \in S$. Call such a subtree-cover S *balanced* if it satisfies the additional property that the tree of largest size in S contains at most three times the number of edges contained by the smallest. By applying Lemma 3.1 to the largest tree in a balanced sub-tree cover, we can show the following lemma:

Lemma 3.2. *Let S be a balanced subtree-cover of some forest F such that $|S| = x$ and $|E(F)| \geq x + 1$ hold. Then one can obtain a balanced subtree cover S' of F such that $|S'| = x + 1$.*

Theorem 3.3. *Let $\mathcal{G} = \langle G_1, ..., G_L \rangle$ be an always-connected, k-edge-deficient temporal graph (for some $k \in \mathbb{N}$) with underlying graph G, and let $|V(G)| = n$. Then, for any start vertex s, there is an exploration schedule $\mathcal{W}$ of $\mathcal{G}$ with $a(\mathcal{W}) = O(kn \log n)$. Moreover, such a schedule can be computed in polynomial time.*

Proof. For $k \geq n-1$ the result clearly holds as every always-connected temporal graph can be explored in $\leq n(n-1)$ time steps (by repeated application of Lemma 2.5 [6]), so we assume $k < n - 1$ for the rest of the proof.

Compute an arbitrary spanning tree T of G, and let $m = |E(T)|$ – assume w.l.o.g. that $m > k + 1$, otherwise G can be explored in $O(kn)$ steps via $O(k)$

applications of Lemma 2.5. Let $S = \{T\}$ and note that S is a balanced subtree-cover of T. Now apply Lemma 3.2 to S k times to obtain a balanced subtree-cover S^* of size $k+1$ (possible since $k \leq n-2$). Let F denote a forest containing all subtrees induced by edges of T that may not yet have been traversed, initially $F = T$.

We now specify our algorithm in terms of an agent that explores the graph in consecutive *rounds*. We denote by t the first step of a given round, and by v the vertex at which the agent is positioned at the beginning of timestep t. Let $m' = \sum_{T_i \in S^*} |E(T_i)|$. At the beginning of the first round $t = 1$, $v = s$, $F = \{T\}$, S^* is a balanced subtree-cover of F (with size $k+1$), and $m' = m$. While F contains more than $k+1$ edges, execute a round as follows: Consider the graph from step $t+n$ onward, and place a single virtual agent at an arbitrary vertex v_i in each of the $k+1$ subtrees $T_i \in S^*$. For each $i \in [k+1]$, compute an Euler tour of T_i starting from vertex v_i, then let the agents follow the Euler tours of their respective trees for the following $6m'$ steps. Since there are $k+1$ virtual agents following tours in edge-disjoint subtrees, and since $\mathcal{G}$ is k-edge-deficient, it follows that there are no edges missing from at least one subtree $T' \in S^*$ in every step. Let T_{i^*} be the subtree that had no edges missing during the largest number of steps in the considered $6m'$-step period. Then T_{i^*} had no edge missing for $\geq \frac{6m'}{k+1}$ steps. Since $|S^*| = k+1$, the smallest tree in S^* cannot contain $> \frac{m'}{k+1}$ edges, so because S^* is balanced the largest tree in S^* contains $\leq \frac{3m'}{k+1}$ edges. Therefore, the $\geq \frac{6m'}{k+1}$ steps in which the virtual agent positioned in T_{i^*} is able to traverse an edge are enough for that agent to complete their Euler tour of T_{i^*} and arrive back at v_{i^*}. Using the steps in the interval $[t, t+n-1]$, move the real agent, using Lemma 2.5, from v to the vertex v_{i^*} at which the virtual agent began their tour of T_{i^*}. Let W^* be the tour followed by the virtual agent positioned in T_{i^*}; from step $t+n$ to step $t' = t+n+6m'-1$, let the real agent complete W^*. Once completed, check if $> k+1$ edges remain untraversed; if so, consider the set $S' = S^* - \{T_{i^*}\}$ and note that $|S'| = k$. Observe that S' is balanced since S^* was balanced and removing a tree cannot violate this property. Since we have $S' = S^* - \{T_{i^*}\}$, and since S^* covered T, we have that S' covers the forest F' obtained from F by removing the edges of T_{i^*}. Apply Lemma 3.2 to S' to obtain a balanced subtree-cover S'' of F' such that $|S''| = k+1$ – note that doing so is valid since $|E(F')| > k+1 = |S'|+1$, as is required by Lemma 3.2. Now, set $S^* = S''$, $F = F'$, $v = v_{i^*}$ and $t = t'+1$ and start the next round as above. Once a round is completed and at most $k+1$ edges remain, stop and use $O(n)$ steps to explore up to $2k+2$ remaining unexplored vertices one by one using Lemma 2.5.

Note that every vertex v in $V(T) = V(G)$ either (1) belongs to an edge of T that was traversed by the algorithm, or (2) was visited via an application of Lemma 2.5. Hence, the computed walk is an exploration schedule and it remains only to bound its arrival time. In each round, a subtree containing at least a $\frac{1}{3(k+1)}$ fraction of the edges of F is traversed in its entirety. To see this, observe that $|S^*| = k+1$, so the largest tree in S^* must contain $\geq \frac{m'}{k+1}$ edges; because

S^* is balanced, it follows that all trees in S^* have size $\geq \frac{m'}{3(k+1)}$. Hence, after x rounds, the total number of edges in T that have not yet been removed from F is $\leq m(1 - \frac{1}{3(k+1)})^x$. Thus, after $x = 3(k+1)\ln(\frac{m}{k+1}) = O(k \log m) = O(k \log n)$ (recall that $m = |E(T)| = n-1$) rounds there are $\leq m(1-\frac{1}{3(k+1)})^{3(k+1)\ln(\frac{m}{k+1})} \leq k+1$ unexplored edges remaining in F. As each round takes $n+6m' \leq n+6m = O(n)$ steps, the total number of steps after $O(k \log n)$ rounds is $O(kn \log m) = O(kn \log n)$. A further at most $(2k+2)n$ steps are needed to explore up to $2k+2$ remaining unvisited vertices. Hence, the entire exploration takes $O(kn \log n) + (2k+2)n = O(kn \log n)$, as required.

Finally, it is easy to see that the algorithm for determining the exploration schedule can be implemented to run in polynomial time. □

4 Linear-Time Exploration of 1-Edge-Deficient Temporal Graphs

A graph $G = (V, E)$ is *k-vertex-connected* (or simply k-connected) if, for any subset $X \subseteq V(G)$ such that $|X| < k$, the subgraph of G induced by $V - X$ is connected. Let $G = (V, E)$ be a connected graph. An edge $e \in E$ is a *bridge* if $G' = (V, E - \{e\})$ is disconnected. A graph $G = (V, E)$ is *2-edge-connected* if it is connected and does not contain a bridge. A 2-edge-connected component (abbreviated 2-ecc) of a graph G is a vertex-maximal induced subgraph $C \subseteq G$ such that C is 2-edge-connected. Note that a 2-ecc can also be a single vertex. We say that a spanning subgraph G'' of G *preserves 2-edge-connectivity* if it contains all bridges of G and, for every 2-ecc C of G, the subgraph of G'' induced by $V(C)$ is 2-edge-connected. In order to show that every connected graph G has a spanning subgraph that preserves 2-edge-connectivity and has only a linear number of edges, we make use of the following result by Nagamochi and Ibaraki.

Theorem 4.1 (Nagamochi and Ibaraki, [18]). *Every k-connected graph $G = (V, E)$ admits a k-connected spanning subgraph $G' = (V', E')$ such that $|E'| \leq k|V|$. Moreover, G' can be computed in $O(|E|)$-time.*

By applying Theorem 4.1 to each biconnected component of a given connected graph G, we can show the following:

Lemma 4.2. *Let G be an arbitrary connected graph and let $\mathcal{C}$ be the set of all 2-eccs of G. Then, G admits a spanning subgraph G^* such that (1) the vertices of each 2-ecc $C \in \mathcal{C}$ form a 2-ecc C^* in G^* with $|E(C^*)| \leq 5|V(C^*)|$; (2) $|E(G^*)| \leq 5|V(G)|$; and (3) $V(G^*) = V(G)$.*

If $\mathcal{G}$ is a 1-edge-deficient, always-connected temporal graph with underlying graph G and G^* is a spanning subgraph of G that preserves 2-edge-connectivity, then the temporal graph $\mathcal{G}^*$ with underlying graph G^* that is obtained from $\mathcal{G}$ by removing all edges that are not in G^* is also always-connected and 1-edge-deficient. This also implies that every cycle C of G^* induces a connected subgraph in every timestep of $\mathcal{G}^*$.

A *circuit* C in a graph G is a closed walk in G that does not repeat edges. In 1-deficient temporal graphs, a circuit behaves like an always-connected temporal graph with underlying cycle, as at most one edge of the circuit can be missing in each step. Thus, we get the following theorem, which was shown in [6] for always-connected cycles.

Theorem 4.3. (Erlebach, Hoffmann and Kammer, [6]). *For every 1-edge deficient temporal graph $\mathcal{G}$ with underlying circuit C, there exists a start vertex from which the graph can be explored in at most $|E(C)| - 1$ steps.*

The following theorem by Fan allows us to reduce the exploration of a 2-ecc to the exploration of at most three circuits.

Theorem 4.4. (Fan, [10]). *The edges of any 2-edge-connected graph $G = (V, E)$ can be covered by at most 3 circuits. Moreover, such a cover can be computed in $O(|E| \cdot |V|)$-time.*

The edges which belong to no 2-ecc of an arbitrary connected graph G are precisely the bridges of G. Hence, one can represent the structure of G as a tree T, called the *2-ecc tree* of G, by identifying each 2-ecc with a vertex, and joining two vertices by an edge in T if and only if their corresponding 2-eccs are connected by a bridge in G. In the proof of Theorem 4.6, we will therefore re-use standard terminology for trees: We choose a 2-ecc C as the *root component.* If C' and C'' are 2-eccs such that C' lies on the path from C to C'' in T, then C'' is a *descendant* of C'. If C' and C'' correspond to neighbouring nodes in T and C'' is a descendant of C', then C'' is a *child* of C' and C' is the *parent* of C''. The *subtree* rooted at a 2-ecc C' consists of all 2-eccs that are descendants of C', and the subgraph of G consisting of all those 2-eccs and the bridge edges between them is said to *correspond* to that child subtree. For any child C' of the root C of the 2-ecc tree, we call the subgraph of G that corresponds to the subtree rooted at C' a *child subgraph.*

Lemma 4.5. *Let G be an arbitrary connected graph on n vertices. Then, there is a 2-ecc C^* of G such that rooting the 2-ecc tree of G at C^* ensures that the child subgraphs (i.e., the subgraphs of G corresponding to the subtrees rooted at children of C^*) each contain at most $n/2$ vertices.*

Proof. Consider the tree T obtained by identifying each 2-edge-connected component C of G with a vertex v_C. Root T at an arbitrary node $v_{C'}$, then process the vertices in a bottom up manner, labelling a vertex v_C with the integer $x_C = |\{u \in V(G) : u \in V(C')$ for a descendant C' of C in $T\}|$. Select a vertex v_{C^*} such that $x_{C^*} \geq n/2$ and such that v_{C^*} has largest depth in T amongst all such vertices. If v_{C^*} is already the root of T, we are done. Otherwise, let $v_{C'}$ be the parent of v_{C^*} and reroot T at v_{C^*} to form a 2-ecc tree T^*, in which $v_{C'}$ is a child of v_{C^*}. We have that for every child $v_C \neq v_{C'}$ of v_{C^*} in T^* we have $x_C < n/2$, because otherwise the algorithm would have picked v_C rather than v_{C^*}. Furthermore, we have $x_{C^*} \geq n/2$, and so the total number of vertices in all components C'' such that $v_{C''}$ is a descendant of $v_{C'}$ in T^* must be $\leq n/2$. □

Theorem 4.6. *Let $\mathcal{G} = \langle G_1, ..., G_L \rangle$ be an always-connected, 1-edge-deficient temporal graph with arbitrary underlying graph G, and let $|V(G)| = n$. Then, for any start vertex s, there is an exploration schedule $\mathcal{W}$ of $\mathcal{G}$ with $a(\mathcal{W}) = O(n)$. Moreover, such a schedule can be computed in polynomial time.*

Proof. Apply Lemma 4.2 to G in order to obtain a spanning subgraph $G^* \subseteq G$ (with $|E(G^*)| \leq 5n$) such that each 2-ecc C of G forms a 2-ecc C^* in G^* with $|E(C^*)| \leq 5|V(C^*)|$. Apply Lemma 4.5 to G^* to obtain a 2-ecc tree T of G^* with a root component C_1 such that the child subgraphs $G_i \subseteq G^*$ satisfy $|V(G_i)| \leq n/2$. Let k denote the number of 2-eccs in G^*. Let $T(n,k)$ denote the maximum number of timesteps required to explore an arbitrary 1-edge-deficient, always-connected temporal graph on n vertices whose underlying graph has k 2-eccs, at most $5n$ edges, and is such that every 2-ecc C^* satisfies $|E(C^*)| \leq 5|V(C^*)|$, starting from an arbitrary vertex s in the graph at timestep 1. We now specify our exploration algorithm and prove by induction on k that $T(n,k) \leq 164n$.

Base Case (Arbitrary n, $k = 1$): G^* consists of a single 2-ecc C_1; without loss of generality assume that $|V(C_1)| \geq 3$. Apply Theorem 4.4 to C_1, obtaining a circuit cover $\{X_1, ..., X_c\}$ of C_1 containing c circuits, where $1 \leq c \leq 3$. Consider now the following 3 time intervals, noting that $|E(X_i)| \leq |E(C_1)| \leq 5n$ for all $i \in [3]$: $I_1 = [n+1, 6n], I_2 = [7n+1, 12n]$ and $I_3 = [13n+1, 18n]$. During the steps of I_i apply Theorem 4.3 to X_i to determine a vertex $v_i \in X_i$ such that an exploration schedule of X_i using at most $|E(X_i)| - 1 \leq 5n - 1$ timesteps begins at v_i at the first step of I_i. Beginning at the start vertex $s \in V(G)$ in timestep 1, employ Lemma 2.5 to move in at most n steps to vertex v_1, wait until the first step of interval I_1, then follow the walk obtained by the application of Theorem 4.3 during interval I_1. If $c > 1$, repeat these steps for all remaining circuits X_i in the computed circuit cover of C_1. Once Theorem 4.3 has been applied to X_c, notice that, for all $i \in [c]$, all vertices of X_i have been visited. Since $\{X_1, ..., X_c\}$ covers all edges of C_1 (and also all edges of G^* since G^* consists only of C_1), it follows that all vertices of G^* have been visited at least once. The number of timesteps taken to achieve this is at most $c(n+5n) \leq 18n$.

Inductive step (Arbitrary n, $k > 1$): Assume that $T(n,j) \leq 164n$ for all $j < k$ and consider the root component C_1 of G^*. We now distinguish two cases:

Case 1: $|C_1| \geq 2$. Apply Theorem 4.4 to C_1 and obtain a circuit cover $X^* = \{X_1, ..., X_c\}$ of C_1 containing c circuits, where $1 \leq c \leq 3$. Let $V' = \{v \in V(C_1) : v \in e \text{ for some bridge } e\}$. Construct a function $\alpha : V' \to X^*$ by arbitrarily mapping each vertex $v \in V'$ to some circuit $X_i \in X^*$ such that $v \in X_i$. Recall that we root the 2-ecc tree T of G^* at C_1. For each child C_i of C_1 in T, we denote by G_i the child subgraph of G^* corresponding to the subtree of T rooted at C_i. Let $\mathsf{Br} = \{e \in E(G^*) : e \text{ is a bridge and } e \cap V' \neq \emptyset\}$ and, for any $v \in V'$, let $\beta(v) = \{G_i : \{v,u\} \in \mathsf{Br} \text{ for some } u \in G_i\}$. For $i \in [3]$, let $F_i = \bigcup_{\{v \in V' : \alpha(v) = X_i\}} \beta(v)$.

Let $G_{X_i} \subseteq G^*$ be the subgraph of G^* induced by $V(X_i \cup F_i)$ $(i \in [c])$. For each $i \in [c]$, we construct a closed walk in G_{X_i} that will be followed (in opposite directions) by two virtual agents. Both agents start at some arbitrary vertex

$s_i \in V(X_i)$ and follow the walk in opposite directions whenever possible, i.e., whenever they are not blocked on the next edge they need to cross. Starting at some timestep t_i, let the agents do the following: Move along the edges of X_i, one in the clockwise direction (agent CW) and the other in the counter-clockwise direction (agent CCW). Whenever either agent reaches for the first time a vertex $v \in V'$ such that $\alpha(v) = X_i$ the agent descends into each $G_j \in \beta(v)$ via the bridge connecting it and vertex $v \in X_i$, and explores G_j via a depth-first search. The only exception is the vertex s_i: If $s_i \in V'$ and $\alpha(s_i) = X_i$, then agent CW descends into each $G_j \in \beta(s_i)$ immediately at the start of the walk (before traversing any edge of X_i), while agent CCW does so only when it returns to s_i after having traversed all edges of X_i. Agent CW processes the subgraphs in $\beta(v)$ in increasing order of their indices, whilst agent CCW processes them in decreasing order of their indices. Once an agent has explored all subgraphs $G_j \in \beta(v)$, then that agent attempts to cross the next edge in X_i. Both agents continue this until the first timestep in which both agents are blocked on the same edge e. If every edge of G^* were to be present in every timestep, it would take each agent at most $\mathsf{Exp}(X_i) = |E(X_i)| + \sum_{G_j \in F_i} 2|V(G_j)|$ steps to carry out their respective walks in G_{X_i}: 1 step to traverse each of the edges of X_i, $2V(G_j)| - 2$ steps spent exploring G_j via a DFS, and 2 steps spent traversing the bridge edges connecting X_i and each $G_j \in F_i$. Since G^* is 1-edge-deficient, it is possible for the agents to both be blocked on the same edge during the same timestep. We distinguish three subcases as follows. Recall that t_i denotes the timestep in which the exploration of G_{X_i} begins. We use t'_i to denote an upper bound on the timestep by which the exploration of G_{X_i} (possibly except one subgraph, see below for details) is completed by at least one of the two agents.

Case 1.1: If the agents are never blocked on the same edge e during any step t in $[t_i, t'_i]$ for $t'_i = t_i + 2\mathsf{Exp}(X_i)$, then, in each timestep $t \in [t_i, t'_i]$, at least one of the two agents is able to cross the next edge of their respective walk. In this case, we have that by the end of timestep t'_i, the agent that was blocked on an edge in the least number of timesteps $t \in [t_i, t'_i]$ will have not been blocked in $\geq \mathsf{Exp}(X_i)$ timesteps and, as such, will have completed their exploration of G_{X_i}.

It remains to consider the situation that the agents are blocked on the same edge of G^* during some timestep in $[t_i, t'_i]$, where $t'_i = t_i + 3\mathsf{Exp}(X_i)$ in Case 1.2 and $t'_i = t_i + 4\mathsf{Exp}(X_i)$ in Case 1.3. Consider the timestep t in which the agents are first both blocked on the same edge e.

Case 1.2: $e \in X_i$. Check whether or not e is present during any step $t' \in [t+1, t+|E(X_i)|]$. If yes, wait until that step, then let both agents cross e. If not, let both agents apply Lemma 2.5 in X_1, using at most $|E(X_1)| - 1$ timesteps to move to the opposite endpoint of e, then continue attempting to traverse the next edge of their walk whenever possible. Notice that, during any step $t' \in [t_i, t-1]$, at least one of the two agents was able to cross the next edge in their respective walk, since t is the first timestep in which both agents are blocked on the same edge. When the agents are blocked on e during step t, they either wait at their current vertex for at most $|E(X_i)| - 1$ steps until e is present again, or spend $\leq |E(X_i)| - 1$ steps reaching the opposite endpoint of e by applying Lemma 2.5 in X_i. In either case, it takes at most $|E(X_i)| - 1$ steps for them to reach the opposite endpoint of e. At this point, observe that the vertices $x \in V(X_i)$ and the $G_j \in F_i$ that remain to be explored/processed by agent CW are exactly those that have already been explored/processed by agent CCW (and vice versa). Hence, it follows that the two agents will not be blocked on the same edge again for the remainder of their walks. In all remaining steps, since the sets of vertices unexplored by the walks of the two agents are disjoint, we again have that at least one of the two agents will be able to cross the next edge of their respective walk in all steps $t' \in [t + |E(X_i)|, t'_i]$. Concluding, during the entire time interval $[t_i, t'_i]$, there are $\leq |E(X_i)|$ steps in which neither of the agents can cross the next edge of their respective walk, and by step $t'_i \leq t_i + 2\mathsf{Exp}(X_i) + |E(X_i)| \leq t_i + 3\mathsf{Exp}(X_i)$, it is ensured that the agent who was blocked during the least number of steps since the start of step t_i has completed their exploration of G_{X_i} in at most $3\mathsf{Exp}(X_i)$ steps.

Case 1.3: $e \in G_j$ for some $G_j \in F_i$. Let $b = \{v, u\}$, where $\{v, u\} \in \mathsf{Br}$, $v \in X_i$ satisfies $\alpha(v) = X_i$, and $u \in V(G_j)$. Consider the timestep $t \in [t_i, t'_i]$, during which the two agents are first blocked on e. Let $t_1^*, t_2^* \in [t_i, t'_i]$ denote respectively the timesteps at which the first agent (say agent A_1) and second agent (agent A_2) traverse the edge b from v toward u – clearly we have $t_1^* \leq t_2^* < t$, since $e \in E(G_j)$ and any vertex in $V(G_j)$ can only be reached from X_i by traversing b. We now retrospectively alter the walks of both agents: First, change the walk of A_1 so that, during the interval $[t_1^*, t_2^* - 1]$, A_1 waits at vertex v. Now, change the walks of both A_1 and A_2 during the steps $[t_2^*, t'_i]$, so that they both do not process subgraph G_j, but continue their exploration of X_i and all $G_{j'} \in F_i$ such that $G_{j'} \neq G_j$. We claim that $t_2^* \leq t_i + 2\mathsf{Exp}(X_i)$. To see this, observe that $t \leq t_i + 2\mathsf{Exp}(X_i)$ since, if $t > t_i + 2\mathsf{Exp}(X_i)$, the two agents will not have been blocked on the same edge during any of the steps $[t_i, t_i + 2\mathsf{Exp}(X_i)]$, and so the agent who was blocked on an edge in the least amount of steps during this interval would have traversed an edge of their walk in $\geq \mathsf{Exp}(X_i)$ timesteps – enough to have finished the entire exploration of G_{X_i}. Hence we have $t_2^* \leq t < t_i + 2\mathsf{Exp}(X_i)$, as required. Both agents can then continue following their respective walks during the interval $[t_2^*, t'_i]$ without the possibility of being blocked on the same edge again; by our earlier reasoning this requires of the agent that is blocked during the least number of steps in this period another $\leq 2\mathsf{Exp}(X_i)$ steps. Concluding, one of the two agents will have visited all vertices in $G_{X_i} \setminus V(G_j)$ by the end of step $t_2^* + 2\mathsf{Exp}(X_i) \leq t_i + 4\mathsf{Exp}(X_i)$.

In all three subcases, one of the two agents has explored all vertices of G_{X_i}, except possibly those of a single subgraph G_j of G_{X_i}, in at most $4\mathsf{Exp}(X_i)$ timesteps, and we will let the real agent follow that agent's walk.

After processing all c circuits X_i in this way, there will be at most c subgraphs that have not yet been explored. We next reduce those unexplored subgraphs to at most one: While there are two or more unexplored subgraphs, we repeatedly (1) choose a circuit X in C_1 that contains two vertices of V' that have a bridge to an unexplored subgraph (note that a circuit X such that $|E(X)| \leq 2\,V(C_1)|$

must exist), and then (2) process X and the two unexplored subgraphs in the same way as we processed X_i for $1 \leq i \leq c$ above.

After this, there will be at most a single subgraph G_j corresponding to a child subtree rooted at a child of C_1 in the 2-ecc tree that is not yet explored. That subgraph has at most $n/2$ vertices (by choice of C_1) and has at most $k-1$ 2-eccs (because it does not contain the 2-ecc C_1). We now apply the inductive hypothesis to explore G_j recursively in at most $164 \cdot n/2 = 82n$ steps.

To bound the overall number of timesteps, we assume that $c = 3$, that 3 subgraphs remain unexplored after processing G_{X_i} for $i \in [3]$, that two iterations of the procedure for reducing the number of unexplored subgraphs are needed, and that a recursive call needs to be made to explore the final unexplored subgraph. We omit the details, but one can straightforwardly show (via a case analysis) that this is the worst case for the total number of steps needed to complete the exploration.

The whole exploration then consists of the following parts: At most n steps to move from s to a vertex v_1 in X_1; at most $4\mathsf{Exp}(X_1)$ steps to explore G_{X_1} apart from at most one child subgraph G_j. Another at most $n + 4\mathsf{Exp}(X_2)$ steps to do the same for G_{X_2} (where the first n steps allow the agent to move from the vertex where the exploration of G_{X_1} ends to a vertex in X_2), and another at most $n + 4\mathsf{Exp}(X_3)$ steps to do the same for G_{X_3}. Then, at most twice: n steps to move to a vertex in a circuit X (recall that $|E(X)| \leq 2|V(C_0)|$) and $4\mathsf{Exp}(X)$ steps to explore it and at least one of the two subgraphs attached to it. Finally, $\leq n$ steps are needed to move to a vertex in the last unexplored subgraph G_j, and another $\leq 82n$ steps are required to explore that subgraph recursively.

As $\mathsf{Exp}(X_i) = |E(X_i)| + \sum_{G_j \in F_i} 2|V(G_j)|$, we have $\sum_{i=1}^{3} \mathsf{Exp}(X_i) \leq 3|E(C_1)| + \sum_{i=1}^{3} \sum_{G_j \in F_i} 2\,V(G_j)| \leq 15\,V(C_1)| + 2\sum_{i=1}^{3} \sum_{G_j \in F_i} |V(G_j)| \leq 15n$. Furthermore, for any circuit X in C_1 with two subgraphs G_1 and G_2 attached via bridges, we have $\mathsf{Exp}(X) \leq 2|V(C_1)| + 2|V(G_1)| + 2|V(G_2)| \leq 2n$. Thus, the total exploration time is at most $6n + 4 \cdot 15n + 8 \cdot 2n + 82n = 82n + 82n = 164n$.

Case 2: $|C_1| = 1$. In this case we apply a similar technique to that used in Case 1, but this case is simpler as the root component consists of a single vertex and all child subgraphs are attached to that same vertex via bridges.

Finally, we remark that all steps in the construction of the exploration schedule can be implemented in polynomial time. □

5 Lower Bound

To complement the upper bounds from Sects. 3 and 4, we also present a lower bound on the worst-case exploration time of k-edge-deficient temporal graphs.

Theorem 5.1. *For arbitrarily large n and every k with $2 \leq k \leq \frac{n}{2} - 1$, there is a k-edge-deficient temporal graph with n vertices for which an optimal exploration takes $\Omega(n \log k)$ steps.*

The theorem can be shown by adapting the construction of a lower bound of $\Omega(n \log n)$ on the exploration time of temporal graphs with underlying planar graphs of maximum degree 4 from [6, Theorem 2]. That construction has a time-varying part (in which $n/2$ edges are missing in each step) and a fixed part (a static path of $n/2$ edges). By reducing the size of the time-varying part and increasing the size of the static part, we obtain Theorem 5.1

6 Conclusion

We have shown that always-connected k-edge-deficient temporal graphs admit an exploration schedule $\mathcal{W}$ with arrival time $O(kn \log n)$; if $k = 1$, the arrival time improves to $O(n)$. The provided proofs are both constructive, yielding polynomial-time algorithms for computing such exploration schedules. As $n-1$ steps are necessary to explore any graph, our results also yield $O(k \log n)$ and $O(1)$-approximation algorithms for TEXP for the $k \in \mathbb{N}$ and $k = 1$ cases, respectively, as well as an $O(\log n)$-approximation if $k = O(1)$. Furthermore, we gave an infinite family of k-edge-deficient temporal graphs that require $\Omega(n \log k)$ timesteps to be explored. It would be interesting to close the gap between the lower and upper bounds. In particular, an interesting question is whether always-connected k-edge-deficient graphs for $k = O(1)$ can be explored in $O(n)$ steps.

References

1. Akrida, E.C., Mertzios, G.B., Spirakis, P.G.: The temporal explorer who returns to the base. In: Heggernes, P. (ed.) CIAC 2019. LNCS, vol. 11485, pp. 13–24. Springer, Cham (2019). https://doi.org/10.1007/978-3-030-17402-6_2
2. Bodlaender, H.L., van der Zanden, T.C.: On exploring always-connected temporal graphs of small pathwidth. Inf. Process. Lett. **142**, 68–71 (2019). https://doi.org/10.1016/j.ipl.2018.10.016
3. Brodén, B., Hammar, M., Nilsson, B.J.: Online and offline algorithms for the time-dependent TSP with time zones. Algorithmica **39**(4), 299–319 (2004). https://doi.org/10.1007/s00453-004-1088-z
4. Bui-Xuan, B., Ferreira, A., Jarry, A.: Computing shortest, fastest, and foremost journeys in dynamic networks. Int. J. Found. Comput. Sci. **14**(2), 267–285 (2003). https://doi.org/10.1142/S0129054103001728
5. Casteigts, A., Flocchini, P., Quattrociocchi, W., Santoro, N.: Time-varying graphs and dynamic networks. Int. J. Parallel Emergent Distrib. Syst. **27**(5), 387–408 (2012). https://doi.org/10.1080/17445760.2012.668546
6. Erlebach, T., Hoffmann, M., Kammer, F.: On temporal graph exploration. J. Comput. Syst. Sci. **119**, 1–18 (2021). https://doi.org/10.1016/j.jcss.2021.01.005
7. Erlebach, T., Kammer, F., Luo, K., Sajenko, A., Spooner, J.T.: Two moves per time step make a difference. In: Baier, C., Chatzigiannakis, I., Flocchini, P., Leonardi, S. (eds.) 46th International Colloquium on Automata, Languages, and Programming (ICALP 2019). Leibniz International Proceedings in Informatics (LIPIcs), vol. 132, pp. 141:1–141:14. Schloss Dagstuhl-Leibniz-Zentrum fuer Informatik, Dagstuhl (2019). https://doi.org/10.4230/LIPIcs.ICALP.2019.141

8. Erlebach, T., Spooner, J.T.: Faster exploration of degree-bounded temporal graphs. In: Potapov, I., Spirakis, P., Worrell, J. (eds.) 43rd International Symposium on Mathematical Foundations of Computer Science (MFCS 2018). Leibniz International Proceedings in Informatics (LIPIcs), vol. 117, pp. 36:1–36:13. Schloss Dagstuhl-Leibniz-Zentrum fuer Informatik, Dagstuhl (2018). https://doi.org/10.4230/LIPIcs.MFCS.2018.36
9. Erlebach, T., Spooner, J.T.: Non-strict temporal exploration. In: Richa, A.W., Scheideler, C. (eds.) SIROCCO 2020. LNCS, vol. 12156, pp. 129–145. Springer, Cham (2020). https://doi.org/10.1007/978-3-030-54921-3_8
10. Fan, G.: Covering graphs by cycles. SIAM J. Discrete Math. **5**(4), 491–496 (1992). https://doi.org/10.1137/0405039
11. Gotoh, T., Flocchini, P., Masuzawa, T., Santoro, N.: Tight bounds on distributed exploration of temporal graphs. In: Felber, P., Friedman, R., Gilbert, S., Miller, A. (eds.) 23rd International Conference on Principles of Distributed Systems (OPODIS 2019). Leibniz International Proceedings in Informatics (LIPIcs), vol. 153, pp. 22:1–22:16. Schloss Dagstuhl-Leibniz-Zentrum fuer Informatik, Dagstuhl (2020). https://doi.org/10.4230/LIPIcs.OPODIS.2019.22
12. Gotoh, T., Sudo, Y., Ooshita, F., Masuzawa, T.: Dynamic ring exploration with (H, S) view. Algorithms **13**(6) (2020). https://doi.org/10.3390/a13060141
13. Ilcinkas, D., Klasing, R., Wade, A.M.: Exploration of constantly connected dynamic graphs based on cactuses. In: Halldórsson, M.M. (ed.) SIROCCO 2014. LNCS, vol. 8576, pp. 250–262. Springer, Cham (2014). https://doi.org/10.1007/978-3-319-09620-9_20
14. Ilcinkas, D., Wade, A.M.: Exploration of the T-interval-connected dynamic graphs: the case of the ring. In: Moscibroda, T., Rescigno, A.A. (eds.) SIROCCO 2013. LNCS, vol. 8179, pp. 13–23. Springer, Cham (2013). https://doi.org/10.1007/978-3-319-03578-9_2
15. Kempe, D., Kleinberg, J.M., Kumar, A.: Connectivity and inference problems for temporal networks. J. Comput. Syst. Sci. **64**(4), 820–842 (2002). https://doi.org/10.1006/jcss.2002.1829
16. Michail, O.: An introduction to temporal graphs: an algorithmic perspective. In: Zaroliagis, C., Pantziou, G., Kontogiannis, S. (eds.) Algorithms, Probability, Networks, and Games. LNCS, vol. 9295, pp. 308–343. Springer, Cham (2015). https://doi.org/10.1007/978-3-319-24024-4_18
17. Michail, O., Spirakis, P.G.: Traveling salesman problems in temporal graphs. Theor. Comput. Sci. **634**, 1–23 (2016). https://doi.org/10.1016/j.tcs.2016.04.006
18. Nagamochi, H., Ibaraki, T.: A linear-time algorithm for finding a sparse k-connected spanning subgraph of a k-connected graph. Algorithmica **7**(5 & 6), 583–596 (1992). https://doi.org/10.1007/BF01758778
19. Shannon, C.E.: Presentation of a maze-solving machine. In: Sloane, N.J.A., Wyner, A.D. (eds.) Claude Elwood Shannon - Collected Papers, pp. 681–687. IEEE Press (1993)
20. Zschoche, P., Fluschnik, T., Molter, H., Niedermeier, R.: The complexity of finding small separators in temporal graphs. J. Comput. Syst. Sci. **107**, 72–92 (2020). https://doi.org/10.1016/j.jcss.2019.07.006

Parameterized Complexity of Categorical Clustering with Size Constraints

Fedor V. Fomin, Petr A. Golovach(✉), and Nidhi Purohit

Department of Informatics, University of Bergen, Bergen, Norway
{Fedor.Fomin,Petr.Golovach,Nidhi.Purohit}@uib.no

Abstract. In the Categorical Clustering problem, we are given a set of vectors (matrix) $\mathbf{A} = \{\mathbf{a}_1, \ldots, \mathbf{a}_n\}$ over Σ^m, where Σ is a finite alphabet, and integers k and B. The task is to partition $\mathbf{A}$ into k clusters such that the median objective of the clustering in the Hamming norm is at most B. That is, we seek a partition $\{I_1, \ldots, I_k\}$ of $\{1, \ldots, n\}$ and vectors $\mathbf{c}_1, \ldots, \mathbf{c}_k \in \Sigma^m$ such that $\sum_{i=1}^{k} \sum_{j \in I_i} d_H(\mathbf{c}_i, \mathbf{a}_j) \leq B$, where $d_H(\mathbf{a}, \mathbf{b})$ is the Hamming distance between vectors $\mathbf{a}$ and $\mathbf{b}$. Fomin, Golovach, and Panolan [ICALP 2018] proved that the problem is fixed-parameter tractable (for binary case $\Sigma = \{0, 1\}$) by giving an algorithm that solves the problem in time $2^{\mathcal{O}(B \log B)} \cdot (mn)^{\mathcal{O}(1)}$. We extend this algorithmic result to a popular capacitated clustering model, where in addition the sizes of the clusters should satisfy certain constraints. More precisely, in Capacitated Clustering, in addition, we are given two non-negative integers p and q, and seek a clustering with $p \leq |I_i| \leq q$ for all $i \in \{1, \ldots, k\}$. Our main theorem is that Capacitated Clustering is solvable in time $2^{\mathcal{O}(B \log B)} |\Sigma|^B \cdot (mn)^{\mathcal{O}(1)}$. The theorem not only extends the previous algorithmic results to a significantly more general model, it also implies algorithms for several other variants of Categorical Clustering with constraints on cluster sizes.

Keywords: Categorical clustering · Capacitated clustering · Parameterized complexity

1 Introduction

While many problems in machine learning concerns numerical data, there is a large class of problems about learning from categorical data. The term categorical data refers to the type of data whose values are discrete and belong to a specific finite set of categories. It could be text, some numeric values, or even unstructured data like images. The most popular clustering objectives for numerical data are k-means and k-median, that are based on distances in the ℓ_1

The research leading to these results have been supported by the Research Council of Norway via the project "MULTIVAL" (grant no. 263317) and the European Research Council (ERC) via grant LOPPRE, reference 819416.

A. Lubiw et al. (Eds.): WADS 2021, LNCS 12808, pp. 385–398, 2021.
https://doi.org/10.1007/978-3-030-83508-8_28

and ℓ_2-norm. For categorical data, other meters, like Hamming distance, could be much more useful.

We study the parameterized complexity of clustering problems with constraints on the sizes of the clusters. The need for clustering with constraints comes from various application. The survey of Banerjee and Ghosh [5] contains a number of examples of clustering with balancing constraints in Direct Marketing [35], Category Management [29], Clustering of Documents [3,25], and Energy Aware Sensor Networks [21,22] among others. However, introducing constraints on the sizes of clustering usual makes clustering tasks much more computationally challenging.

In this paper we focus on categorical data clustering, where data features admit a fixed number of possible values. We work with vectors from Σ^m, where Σ is a finite alphabet. The most commonly used similarity measure for categorical data is the Hamming distance. For two vectors $\mathbf{a}, \mathbf{b} \in \Sigma^m$ or, equivalently, for two strings of length m over Σ, we use $d_H(\mathbf{a}, \mathbf{b})$, to denote the *Hamming distance* between $\mathbf{a}$ and $\mathbf{b}$, that is, the number of indices $i \in \{1, \ldots, m\}$ where the i-th elements of $\mathbf{a}$ and $\mathbf{b}$ differ. The task of the vanilla CATEGORICAL CLUSTERING problem is, given an $m \times n$ matrix $\mathbf{A}$ with columns $(\mathbf{a}_1, \ldots, \mathbf{a}_n)$ over a finite alphabet Σ, a positive integer k, and a nonnegative integer B, decide whether there is a partition $\{I_1, \ldots, I_k\}$ of $\{1, \ldots, n\}$ and vectors $\mathbf{c}_1, \ldots, \mathbf{c}_k \in \Sigma^m$ such that $\sum_{i=1}^{k} \sum_{j \in I_i} d_H(\mathbf{c}_i, \mathbf{a}_j) \le B$. The sets $I_1, \ldots, I_k$ are called *clusters* and the vectors $\mathbf{c}_1, \ldots, \mathbf{c}_k$ are *medians* (or *centers*)[1]. We consider the generalization of the problem, where the size of each cluster should be within a given interval:

CAPACITATED CLUSTERING

Input: An $m \times n$ matrix $\mathbf{A}$ with columns $(\mathbf{a}_1, \ldots, \mathbf{a}_n)$ over a finite alphabet Σ, a positive integer k, a nonnegative integer B, and positive integers p and q such that $p \le q$.

Task: Decide whether there is a partition $\{I_1, \ldots, I_k\}$ of $\{1, \ldots, n\}$, where $p \le |I_i| \le q$, and vectors $\mathbf{c}_1, \ldots, \mathbf{c}_k \in \Sigma^m$ such that $\sum_{i=1}^{k} \sum_{j \in I_i} d_H(\mathbf{c}_i, \mathbf{a}_j) \le B$.

Parameterized algorithms for the vanilla variant of CAPACITATED CLUSTERING (without constraints on the sizes of clusters) were given by Fomin, Golovach and Panolan in [17]. One of the main results of their paper is the theorem providing an algorithm of running time $2^{\mathcal{O}(B \log B)} \cdot (nm)^{\mathcal{O}(1)}$ for vanilla clustering over binary field. In other words, the problem is fixed-parameter tractable (FPT) parameterized by B. The main question that we address in this paper is whether clustering constraints impact the problem's parameterized complexity.

Our Results. Our main result is that CAPACITATED CLUSTERING is FPT when parameterized by the budget B and the alphabet size. More precisely, we show

Theorem 1. CAPACITATED CLUSTERING *can be solved in* $2^{\mathcal{O}(B \log B)} \cdot |\Sigma|^B \cdot (mn)^{\mathcal{O}(1)}$ *time.*

[1] Some authors call $\mathbf{c}_1, \ldots, \mathbf{c}_k$ *means* in the case of Hamming distances.

Fomin, Golovach and Panolan [17, Theorem 1] proved that CATEGORICAL CLUSTERING for binary matrices is FPT when parameterized by the budget B. Theorem 1 generalizes this result. Interestingly, for approximation algorithms, introducing clustering constraints makes the problem much more computationally challenging. However, from parameterized complexity perspective, adding constraints does not change the complexity of the problem.

We also observe that CAPACITATED CLUSTERING is NP-complete even for binary matrices, $k = 2$ and $p = q = \frac{n}{k}$. Theorem 1 can be used to establish fixed-parameter tractability of several other variants of constrained clustering discussed in the literature. In some applications, it is natural to require that the sizes of clusters should be approximately equal, see e.g. [33]. We consider variants of CATEGORICAL CLUSTERING, where the input contains additional parameters besides a matrix $\mathbf{A} = (a_1, \ldots, a_n)$ and integers k and B, and the task is to find clusters $I_1, \ldots, I_k$ and medians $\mathbf{c}_1, \ldots, \mathbf{c}_k \in \Sigma^m$ such that $\sum_{i=1}^{k} \sum_{j \in I_i} d_H(\mathbf{c}_i, \mathbf{a}_j) \leq B$ and the sizes of the clusters satisfy special balance properties.

- In BALANCED CLUSTERING, we are additionally given a nonnegative integer δ and it should hold that $||I_i| - |I_j|| \leq \delta$ for all $i, j \in \{1, \ldots, k\}$, that is, the sizes of clusters can differ by at most δ.
- In FACTOR-BALANCED CLUSTERING, we are given a real $\alpha \geq 1$ and it is required that $|I_i| \leq \alpha |I_j|$ for all $i, j \in \{1, \ldots, k\}$, that is, the ratio of the clusters sizes is upper bounded by α.

By making use of Theorem 1, we prove that BALANCED CLUSTERING and FACTOR-BALANCED CLUSTERING are solvable in time $2^{\mathcal{O}(B \log B)} |\Sigma|^B \cdot (mn)^{\mathcal{O}(1)}$.

We conclude by discussing kernelization for these problems. In particular, we show that BALANCED CLUSTERING admits a polynomial kernel under the combined parameterization by k, B and δ. We also observe that neither of considered problems has a polynomial kernel when parameterized by B only, unless coNP $\subseteq$ NP/poly, even for the binary case.

High-Level Overview of the Proof of Theorem 1. The algorithm for the vanilla problem of Fomin et al. [17], as well as the algorithm of Fomin, Golovach and Simonov for clustering in ℓ_p-norm [19], use the result of Marx [27] about enumeration of subhypergraphs with certain properties of a given hypergraph of a special type. Basically, these algorithms can be seen as an intricate reduction of a clustering instance to a hypergraph of special type and then calling the result of Marx as a black box. In the context of the categorical clustering problems, a similar reduction implies that all potential medians can be listed in $2^{\mathcal{O}(B \log B)} |\Sigma|^B \cdot (mn)^{\mathcal{O}(1)}$ time.

However, this strategy does not work to prove Theorem 1. Here the difficulties are due to the constraints on sizes of clusters. The algorithm for CATEGORICAL CLUSTERING in [17] uses an observation that identical columns $\mathbf{a}_i$ and $\mathbf{a}_j$ of $\mathbf{A}$ can be clustered together. That is, $i, j \in I_h$ for a cluster I_h of an optimal solution. Hence, a solution can be seen as a partition of the family of *initial* clusters, i.e., inclusion maximal sets of indices $J \subseteq \{1, \ldots, n\}$ such that the columns $\mathbf{a}_i$ for $i \in J$ are the same. Since the number of initial clusters that are part of *composite* clusters of a solution, that is, clusters including at least two initial clusters, is at most $2B$ in any yes-instance, the color coding technique of Alon, Yuster and Zwick [2] allows to highlight initial clusters that may be included in a single composite cluster of a solution. This way, the initial problem is reduced to selecting a single composite cluster of minimum cost that contains a given number of initial clusters. To solve this problem, the result of Marx [27] about enumeration of subhypergraphs becomes handy.

This scheme does not work for Capacitated Clustering, because it may happen that splitting of an initial cluster between clusters of a solution is inevitable due to size constraints. This makes it impossible to select composite clusters independently from each other and destroys the approach from [17,19].

The main insight that allows to overcome the above issues is the very specific structure of possible splitting of initial clusters. For a clustering $\mathcal{I} = \{I_1, \ldots, I_k\}$ and the partition $\mathcal{J}$ of the column indices into initial clusters, we look at the structure of the intersection graph $G(\mathcal{I}, \mathcal{J})$ defined by the two partitions of $\{1, \ldots, n\}$. The crucial fact we prove here is that there is an optimal solution such that this intersection graph is a forest. It can be seen that $G(\mathcal{I}, \mathcal{J})$ has at most $3B$ vertices in connected components with at least three vertices for such a solution. This allows to guess the structure of $G(\mathcal{I}, \mathcal{J})$, that is, guess a forest F isomorphic to $G(\mathcal{I}, \mathcal{J})$, by using the brute force. Then for a given F, we find a solution $\mathcal{I}$ with $G(\mathcal{I}, \mathcal{J})$ isomorphic to F by combining dynamic programming with color coding and enumeration of subhypergraphs of Marx.

Due to space constraints, the proofs are either omitted or sketched in this extended abstract. The details are available in the full version [18].

Related Work. Clustering is one of the most common procedures in unsupervised machine learning. Capacitated Clustering is the variant of the popular k-median clustering with the Hamming norm. In many applications of clustering, constraints come naturally. For example, the lower bound on the size of a cluster ensures certain anonymity of data and is often required for data privacy [32]. There is a rich literature on approximation algorithms for various versions of capacitated clustering [1,6–10,12,14,24,26,33]. However, to the best of our knowledge, no parameterized algorithms for categorical clustering with constraints on the sizes of clusters, were known prior to our work.

Several approximations and parameterized algorithms are known for the vanilla case of Categorical Clustering without constraints can be found in the literature. For binary field, Categorical Clustering was introduced by Kleinberg, Papadimitriou, and Raghavan [23] as one of the examples of segmentation problems. The problem appears under different names in the literature [11,28]. Feige proved in [15] that the problem is NP-complete for every $k \geq 2$. We use several ideas from Feige's construction for our lower bounds. Ostrovsky and Rabani [30] gave a randomized PTAS for binary Categorical Clustering which was recently improved to EPTAS in [16] and [4]. Fomin, Golovach and Simonov in [19] studied k-clustering with various distance norms in Categorical Clustering. One of their results is that clustering with Hamming-distance

(ℓ_0-distance) (but unbounded size of the alphabet Σ) is W[1]-hard parameterized by $m + B$. The following paper about binary variant of CATEGORICAL CLUSTERING is highly relevant to this paper. Fomin, Golovach and Panolan [17] gave two parameterized algorithms for binary case of CATEGORICAL CLUSTERING with running time $2^{\mathcal{O}(B \log B)} \cdot (nm)^{\mathcal{O}(1)}$ and $2^{\mathcal{O}(\sqrt{kB \log (k+B) \log k})} \cdot (nm)^{\mathcal{O}(1)}$.

2 Preliminaries

In this section we introduce the terminology used throughout the paper and obtain some auxiliary results.

We refer to the book of Cygan et al. [13] for the detailed introduction to Parameterized Complexity. The input of a parameterized problem contains an integer value k that is referred as a *parameter*. A parameterized problem is *fixed-parameter tractable* (FPT) if there is an algorithm solving it in $f(k) \cdot |I|^{\mathcal{O}(1)}$ time, where I is an input, k is a parameter, and $f(\cdot)$ is a computable function; the parameterized complexity class FPT is composed by fixed-parameter tractable problems.

All matrices and vectors considered in this paper are assumed to be over a finite alphabet Σ and we say that a matrix (vector) is *binary* if $\Sigma = \{0, 1\}$. Therefore, to simplify notation, we omit Σ in the notation whenever it does not create confusion. We use m and n to denote the number of rows and columns, respectively, of input matrices if it does not create confusion. We write $\mathbf{A} = (\mathbf{a}_1, \ldots, \mathbf{a}_n)$ to denote that $\mathbf{A}$ is a matrix with n columns $\mathbf{a}_1, \ldots, \mathbf{a}_n$. For a partition $\mathcal{I} = \{I_1, \ldots, I_k\}$ of $\{1, \ldots, n\}$, we say that $\{I_1, \ldots, I_k\}$ is a *k-clustering* for $\mathbf{A}$. For an inclusion maximal $J \subseteq \{1, \ldots, n\}$ such that the columns $\mathbf{a}_i$ are identical for all $i \in J$, we say that J is an *initial cluster*. We say that a cluster I_i of $\mathcal{I}$ is *simple* if $I_i \subseteq J$ for some initial cluster J and I_i is *composite*, otherwise, that is, if I_i contains some $h, j \in \{1, \ldots, n\}$ such that $\mathbf{a}_h$ and $\mathbf{a}_j$ are distinct. For a vector $\mathbf{a} \in \Sigma^m$, we use $\mathbf{a}[i]$ to denote the i-th element of the vector for $i \in \{1, \ldots, m\}$. Thus, for two vectors $\mathbf{a}, \mathbf{b} \in \Sigma^m$, $d_H(\mathbf{a}, \mathbf{b}) = |\{i \in \{1, \ldots, m\} \mid \mathbf{a}[i] \neq \mathbf{b}[i]\}|$. Let a_{ij} for $i \in \{1, \ldots, m\}$ and $j \in \{1, \ldots, n\}$ be the elements of $\mathbf{A}$. For $I \subseteq \{1, \ldots, m\}$ and $J \subseteq \{1, \ldots, n\}$, we denote by $\mathbf{A}[I, J]$ the $|I| \times |J|$-submatrix of $\mathbf{A}$ with the elements a_{ij} where $i \in I$ and $j \in J$.

Formally, for CATEGORICAL CLUSTERING and its variants, a solution is formed by clusters $I_1, \ldots, I_k$ together with the corresponding medians $\mathbf{c}_1, \ldots, \mathbf{c}_k$. However, given clusters $I_1, \ldots, I_k$, optimal medians $\mathbf{c}_1, \ldots, \mathbf{c}_k$ can be computed by the easy *majority rule*. Let $\mathbf{A} = (\mathbf{a}_1, \ldots, \mathbf{a}_n)$ and let $\{I_1, \ldots, I_k\}$ be an k-clustering. For every $i \in \{1, \ldots, k\}$, we compute $\mathbf{c}_i \in \Sigma^m$ as follows. For each $j \in \{1, \ldots, m\}$, we consider the multiset $R_{ij} = \{\mathbf{a}_h[j] \mid h \in I_i\}$ of elements of Σ. For each $s \in R_{ij}$, we compute the number of its occurrences in the multiset and find an element s^* that occurs most often (ties are broken arbitrarily). Then we set $\mathbf{c}_i[j] = s^*$. It is straightforward to verify that for every $\mathbf{c} \in \Sigma^m$, $\sum_{h \in I_i} d_H(\mathbf{c}_i, \mathbf{a}_h) \leq \sum_{h \in I_i} d_H(\mathbf{c}, \mathbf{a}_h)$. Therefore, the choice of $\mathbf{c}_i$ is optimal. This gives the following observation.

Observation 1. *Given a matrix* $\mathbf{A} = (\mathbf{a}_1, \ldots, \mathbf{a}_n)$ *and a* k*-clustering* $\{I_1, \ldots, I_k\}$, *a family of vectors* $\mathbf{c}_1, \ldots, \mathbf{c}_k \in \Sigma^m$ *such that* $\sum_{i=1}^{k} \sum_{j \in I_i} d_H(\mathbf{c}_i, \mathbf{a}_j)$ *is minimum can be computed in polynomial time by the majority rule.*

For a k-clustering $\{I_1, \ldots, I_k\}$, we define the *cost* $\mathsf{cost}(I_1, \ldots, I_k)$ as the minimum value of $\sum_{i=1}^{k} \sum_{j \in I_i} d_H(\mathbf{c}_i, \mathbf{a}_j)$ over all k-tuples of vectors $\mathbf{c}_1, \ldots, \mathbf{c}_k \in \Sigma^m$. By Observation 1, we have that $\mathsf{cost}(I_1, \ldots, I_k)$ can be computed in polynomial time. Then the task of Categorical Clustering and its variants is reduced to finding a k-clustering of cost at most B (with the respective constraints of the cluster sizes). Thus, we may refer to a k-clustering as a solution without specifying medians.

Observe that given vectors $\mathbf{c}_1, \ldots, \mathbf{c}_k$, we can find an k-clustering $\{I_1, \ldots, I_k\}$ that minimizes $\sum_{i=1}^{k} \sum_{j \in I_i} d_H(\mathbf{c}_i, \mathbf{a}_j)$ by the greedy procedure: for each $i \in \{1, \ldots, n\}$, we find $j \in \{1, \ldots, k\}$ such that $d_H(\mathbf{c}_j, \mathbf{a}_i)$ is minimum (ties are broken arbitrarily) and place i in the cluster I_j. However, the constructed k-clustering does not respect the size constraints of our problems. Still, given vectors $\mathbf{c}_1, \ldots, \mathbf{c}_k$, we can decide in polynomial time whether an instance of Capacitated Clustering has a solution with the medians $\mathbf{c}_1, \ldots, \mathbf{c}_k$ using a reduction to the classical Minimum Weight Perfect Matching problem.

Lemma 1 ($*$[2])**.** *Let* $\mathbf{c}_1, \ldots, \mathbf{c}_k \in \Sigma^m$. *For an instance of* Capacitated Clustering, *it can be decided in polynomial time whether the instance has a solution with the family of medians* $\{\mathbf{c}_1, \ldots, \mathbf{c}_k\}$.

By Lemma 1, we have that solving our problems can be reduced to finding a family of medians $\{\mathbf{c}_1, \ldots, \mathbf{c}_k\}$ (notice that some medians may be the same).

Since we are interested in the parameterized complexity of clustering problems, in the last part of this section, we argue that Capacitated Clustering is NP-hard for very restricted instances. In [15], Feige proved that the problem is NP-complete for $k = 2$ and binary matrices, that is, for the case $\Sigma = \{0, 1\}$. This result immediately implies that Capacitated Clustering is also NP-complete for $k = 2$ and binary matrices. However, we would like to underline that Capacitated Clustering is NP-hard even if $p = q$. For this, we use some details of the hardness proof of Feige [15].

Theorem 2 ($*$)**.** *For every fixed integer constant* $c \geq 0$, Capacitated Clustering *is* NP*-complete for* $k = 2$, *binary matrices and* $q - p \leq c$.

3 FPT Algorithm for Parameterization by B

In this section, we show that Capacitated Clustering is FPT when parameterized by B and $|\Sigma|$. Our main result is Theorem 1 that we restate here.

Theorem 1. Capacitated Clustering *can be solved in* $2^{\mathcal{O}(B \log B)} \cdot |\Sigma|^B \cdot (mn)^{\mathcal{O}(1)}$ *time.*

---

[2] The proofs labeled ($*$) are omitted in this extended abstract.

Note that this result is tight in the sense that it is unlikely that the dependence on the alphabet size could be made polynomial. It was shown in [19], that CATEGORICAL CLUSTERING is W[1]-hard when parameterized by k and the number of rows m of the input matrix if $\Sigma = \mathbb{Z}$, i.e., for an infinite alphabet. However, it is straightforward to see that this result holds for $\Sigma = \{0, \ldots, n-1\}$, because our measure is the Hamming distance. This immediately leads to the following proposition.

Proposition 1. CAPACITATED CLUSTERING *is* W[1]*-hard when parameterized by* B *and* m.

Now we sketch the proof of Theorem 1.

Sketch of the Proof of Theorem 1

The proof is constructive and we sketch the algorithm and its analysis. Let $(\mathbf{A}, \Sigma, k, B, p, q)$ be an instance of CAPACITATED CLUSTERING with $\mathbf{A} = (\mathbf{a}_1, \ldots, \mathbf{a}_n)$. First, we compute the partition $\mathcal{J} = \{J_1, \ldots, J_s\}$ of $\{1, \ldots, n\}$ into initial clusters.

It can be shown that if $(\mathbf{A}, \Sigma, k, B, p, q)$ is a yes-instance, then there is a solution $\mathcal{I} = \{I_1, \ldots, I_k\}$ such that the intersection graph $G(\mathcal{I}, \mathcal{J})$ of the initial clusters and the clusters of the solution is a forest. We call such a solution (or k-clustering) *acyclic*. Thus to solve the problem, it is sufficient to check whether the instance has an acyclic solution.

We observe that any k-clustering for $\mathbf{A}$ of cost at most B has at most B composite clusters. For each t from 0 to $\min\{k, B\}$, we want to verify whether there is a solution $\mathcal{I} = \{I_1, \ldots, I_k\}$ with exactly t composite clusters. If we find such a solution, then we return that $\mathbf{A}$ is a yes-instance and stop. Otherwise, if we have no solution for every value of t, we report that $(\mathbf{A}, \Sigma, k, B, p, q)$ is a no-instance. In what follows, we sketch how to verify the existence of a solution with t composite clusters for a fixed nonnegative $t \leq \min\{k, B\}$.

We consider the special case $t = 0$ separately. If $t = 0$, then a solution $\mathcal{I}$ has no composite cluster, that is, the clusters of the solution form partitions of the initial clusters, and we can solve the problem directly.

From now, we assume that $t \geq 1$. Note that we also can assume that $B \geq 1$, because for $B = 0$, no cluster of a solution can be composite.

Observe there can be at most $2B$ initial clusters with nonempty intersections with the composite clusters of a solution $\mathcal{I}$. Since $G(\mathcal{I}, \mathcal{J})$ is a forest, it is easy to observe that at least $t + 1$ initial clusters have nonempty intersections with the composite clusters. We consider $\ell = t + 1, \ldots, 2B$, and for each ℓ, we check whether there is a solution $\mathcal{I} = \{I_1, \ldots, I_k\}$ such that exactly ℓ initial clusters have nonempty intersections with the composite clusters of $\mathcal{I}$. If we find such a solution, then we return the yes-answer and stop. Otherwise, if we have no solution for all the values of ℓ, we report that $(\mathbf{A}, \Sigma, k, B, p, q)$ is a no-instance. From now, we assume that positive $t + 1 \leq \ell \leq 2B$ is given.

We use the deep result of Marx [27] to construct the set $\mathcal{M} = \mathcal{M}(\mathbf{A}, B)$ of potential medians (see [19] for a similar construction). This set has size

$2^{\mathcal{O}(B\log B)}|\Sigma|^B\cdot(mn)^{\mathcal{O}(1)}$ and can be computed in $2^{\mathcal{O}(B\log B)}|\Sigma|^B\cdot(mn)^{\mathcal{O}(1)}$ time. For a k-clustering $\mathcal{I}=\{I_1,\ldots,I_k\}$, we define the *minimum cost (with respect to* $\mathcal{M}$*)*, as $\min\{\sum_{i=1}^{k}\sum_{j\in I_i} d_H(\mathbf{c}_i,\mathbf{a}_j) \mid \mathbf{c}_1,\ldots,\mathbf{c}_k\in\mathcal{M}\}$. If $(\mathbf{A},\Sigma,k,B,p,q)$ is a yes-instance, then it has a solution such that the medians are in $\mathcal{M}$. Throughout this section, whenever we say that $\mathcal{I}$ is a clustering of minimum cost, we mean that the cost is minimum with respect to $\mathcal{M}$.

We use the *color coding* technique of Alon, Yuster and Zwick [2] (see [13, Chapter 5] for the detailed introduction). For simplicity, we sketch a Monte Carlo algorithm with false negatives. This algorithm can be derandomized by standard tools [2] (see also [13, Chapter 5]). The main idea is to highlight the initial clusters with nonempty intersections with clusters of a potential solution. We color the initial clusters by ℓ colors uniformly at random. We say that a k-clustering $\mathcal{I}=\{I_1,\ldots,I_k\}$ of cost at most B is a *colorful* solution if the initial clusters with nonempty intersections with the clusters of $\mathcal{I}$ have distinct colors. The algorithm exploits the property that with probability at least e^{-2B}, a yes-instance admits a colorful solution.

Our next task is to explain how to verify that there is a colorful solution for a given random coloring $\psi\colon\mathcal{J}\to\{1,\ldots,\ell\}$.

Recall that we are looking for an acyclic solution $\mathcal{I}=\{I_1,\ldots,I_k\}$, that is, $G(\mathcal{I},\mathcal{J})$ is required to be a forest. Let $\mathcal{I}$ be such a k-clustering. Let $\mathcal{I}'\subseteq\mathcal{I}$ be the set of composite clusters and let $\mathcal{J}'\subseteq\mathcal{J}$ be the set of initial clusters having nonempty intersections with the composite clusters. The main idea behind our algorithm for finding a colorful solution is to guess the structure of the forest $G(\mathcal{I}',\mathcal{J}')=G(\mathcal{I},\mathcal{J})[\mathcal{I}'\cup\mathcal{J}']$ and then do dynamic programming over it. Now we sketch the main ideas behind a simplified version of our dynamic programming algorithm. A more accurate version with all the details is given in the full version of the paper. Recall that $|\mathcal{I}'|=t$ and $|\mathcal{J}'|=\ell$ by our assumptions. Note also that the leaves of $G(\mathcal{I}',\mathcal{J}')$ are initial clusters and every connected component of this forest contains at least three vertices.

We consider all forests F on $t+\ell$ vertices such that (i) each connected component of F has at least three vertices, and (ii) F admits a bipartition (U,W) of its vertex set with $|U|=t$ and $|W|=\ell$ such that the leaves of F are in W, and we consider all possible bijective mappings $\alpha\colon W\to\{1,\ldots,\ell\}$. Since $t\leq B$ and $\ell\leq 2B$, the number of possible forests F is $2^{\mathcal{O}(B)}$ [31] and they can be listed in $2^{\mathcal{O}(B)}$ time (see, e.g., [34]). Note that since the leaves of F required to be in W, the bipartition (U,W) is unique. Clearly, the total number of mappings α is $\ell!=2^{\mathcal{O}(B\log B)}$.

For a given forest F and mapping α, we say that an acyclic k-clustering $\{I_1,\ldots,I_k\}$ for $\mathbf{A}$ is *feasible* if the following holds:

(i) $p\leq|I_i|\leq q$ for $i\in\{1,\ldots,k\}$,
(ii) the set $\mathcal{I}'\subseteq\mathcal{I}$ of composite clusters has size t and the set $\mathcal{J}'\subseteq\mathcal{J}$ of initial clusters having nonempty intersections with the composite clusters has size ℓ,
(iii) the initial clusters in $\mathcal{J}'$ are colored by distinct colors by ψ, and

(iv) $G(\mathcal{I}', \mathcal{J}')$ is isomorphic to F with an isomorphism φ that bijectively maps $\mathcal{I}'$ to U and $\mathcal{J}'$ to W in such a way that $\psi(J) = \alpha(\varphi(J))$ for $J \in \mathcal{J}'$.

The problem of finding a colorful solution boils down to checking whether there are F and α such that there is a feasible k-clustering of cost at most B. We do the check by considering all the forests F and bijections α. If we find that there is a feasible k-clustering of cost at most k for one of the choices, we stop and return the yes-answer. Otherwise, we conclude that there is no colorful solution.

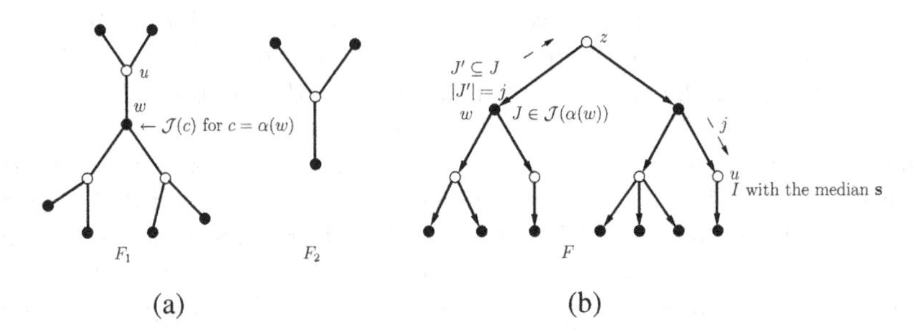

Fig. 1. An example of F (a) and the dynamic programming scheme (b); the vertices of U are shown in white and the vertices of W are black.

Assume that a forest F with the bipartition (U, W) and a bijective mapping $\alpha \colon W \to \{1, \ldots, \ell\}$ are given (see Fig. 1(a) for an example). The mapping α assigns to each vertex $w \in W$ the color $\alpha(w)$. For a color $c \in \{1, \ldots, \ell\}$, denote by $\mathcal{J}(c)$ the set of initial clusters of color c. With every vertex $w \in W$, we associate the set of initial clusters $\mathcal{J}(c)$ for $c = \alpha(w)$. To construct a feasible k-clustering, we have to select a cluster $J \in \mathcal{J}(c)$ and then split it between the composite clusters corresponding to the neighbors of w in F. Notice that some parts of J may form simple clusters of a solution, and the initial clusters from $\mathcal{J}(c)\setminus\{J\}$ are used to form simple clusters. For every vertex $u \in U$, we are constructing a composite cluster from the parts of initial clusters selected for its neighbors. Note that the median of this composite cluster is selected from $\mathcal{M}$.

Denote by $F_1, \ldots, F_f$ the connected components of F. Notice that the clustering for each component can be done independently, because the colors of initial clusters associated with the vertices of distinct components are disjoint. For every $i \in \{1, \ldots, f\}$ and a positive $h \leq k$, we define $\omega_i(h)$ to be the minimum cost of a feasible h-clusterings for the matrix $\mathbf{A}_i$ formed by the columns of $\mathbf{A}$ with the indices from the initial clusters of colors $\alpha(W \cap V(F_i))$. It is easy to see that $\mathbf{A}$ has a feasible k-clustering of cost at most B if and only if there are positive integers $h_1, \ldots, h_f$ such that $h_1 + \cdots + h_f = k$ and $\omega_1(h_1) + \cdots + \omega_f(h_f) \leq B$. The existence of such integers $h_1, \ldots, h_f$ can be checked in polynomial time by a straightforward dynamic programming algorithm.

By these arguments, we have to compute the tables of values of $\omega_i(h)$. For this, we use the fact that $F_1, \ldots, F_f$ are trees and this allows us to use dynamic programming over these trees. We explain the algorithm under the simplifying assumption that F itself is a tree (see Fig. 1 (b)) and $h = k$. Since the computations for distinct components are independent, this assumption can be made without loss of generality. We select a vertex $z \in U$ as a root of F. This selection defines a parent-child relation on the set of vertices. For a vertex $x \in V(F)$, we denote by F_x the subtree of F induced by the descendants of x (including the vertex itself). For every $x \in V(T)$, we compute the tables of auxiliary values depending on whether $x \in U$ or $x \in W$. To define these values, denote by $C(X) \subseteq \{1,\ldots,n\}$ the set of indices in the initial clusters with their colors in $X \subseteq \{1,\ldots,\ell\}$.

Let $w \in W$ and $c = \alpha(w)$. For every nonnegative integer $h \leq k$, every $J \in \mathcal{J}(c)$, and every nonnegative integer $j \leq |J|$, we define $\omega_w^{(1)}(h,J,j)$ as the minimum cost of an h-clustering for the matrix $\mathbf{A}_w$ that is feasible with respect to F_w with the property that the cluster J is selected for w, where $\mathbf{A}_w = \mathbf{A}[\{1,\ldots,m\}, C(\alpha(W \cap V(F_w))\setminus J']$ for an arbitrary $J' \subseteq J$ of size j. In words, the definition means that we cluster columns of $\mathbf{A}$ with their indices from the initial clusters with the colors corresponding to the vertices of $W \cap F_w$ except j columns with the indices in $J' \subseteq J$ of size j that are expected to be included in the composite cluster corresponding to the parent of w. Notice that for a leaf, we can have $h = 0$ if a unique initial cluster J of color c has size j. It is assumed that $\omega_w^{(1)}(h,J,j) = +\infty$ if there is no h-clustering satisfying the constraints.

Let now $u \in U$. For every positive integer $h \leq k$, every nonnegative integer $j \leq q$, and every $\mathbf{s} \in \mathcal{M}$, we define $\omega_u^{(2)}(h,j,\mathbf{s})$ as the minimum cost of an h-clustering for the matrix $\mathbf{A}_u = \mathbf{A}[\{1,\ldots,m\}, C(\alpha(W \cap V(F_u))]$ that has a selected cluster I corresponding to u that has $\mathbf{s}$ as its median, and the clustering satisfies the requirements of a feasible clustering with respect to F_u, except the size conditions for I that are changed to $q - j \leq |I| \leq p - j$. The idea behind this definition is that we do clustering for $\mathbf{A}_u$ expecting to complement the special cluster I with the fixed median $\mathbf{s}$ by j elements of the initial cluster corresponding to the parent of u (unless u is the root and $j = 0$ in this case). As above, $\omega_u^{(2)}(h,j,\mathbf{s}) = +\infty$ if there is no clustering with the required properties.

We compute the tables of values of the functions $\omega_w^{(1)}(\cdot,\cdot,\cdot)$ and $\omega_u^{(2)}(\cdot,\cdot,\cdot)$ bottom-up starting from the leaves. Observe that a feasible k-clustering for $\mathbf{A}$ of cost at most B exists if and only if $\min_{\mathbf{s}\in\mathcal{M}} \omega_z^{(2)}(k,0,\mathbf{s}) \leq B$. Hence, by computing the table for the root, we solve the problem.

If $w \in W$ is a leaf, then the table of values of $\omega_w^{(1)}(\cdot,\cdot,\cdot)$ can be computed in polynomial time directly. Let w be a internal vertex of F and assume that the tables of values of $\omega_u^{(2)}(\cdot,\cdot,\cdot)$ are constructed for all children u of w. Then the computation of $\omega_w^{(1)}(h,J,j)$ is based on the following observation. Assume that u is a child of w and we expect that (i) j' elements of J should be included in the composite cluster I corresponding to u and (ii) $\mathbf{A}_u$ gives exactly h' clusters. Then the contribution of j' elements of J and $\mathbf{A}_u$ to the total cost is $\min\{\omega_u^{(2)}(h',j',\mathbf{s}) + j'd_H(\mathbf{s},\mathbf{a}_i) \mid \mathbf{s} \in \mathcal{M}\}$ for an arbitrary $i \in J$. This means that the contribution depends only on j' and h'. This implies that $\omega_w^{(1)}(h,J,j)$ can be computed in $2^{\mathcal{O}(B\log B)}|\Sigma|^B \cdot (mn)^{\mathcal{O}(1)}$ time by the dynamic programming algorithm over the children of w.

Let $u \in U$ and assume that the tables of values of $\omega_w^{(1)}(\cdot,\cdot,\cdot)$ are constructed for all children $w \in W$ of u. To compute $\omega_u^{(2)}(h, j, \mathbf{s})$, we make the following observation for a child w of u. If (i) exactly j' elements of an initial cluster J corresponding to w are included in the composite cluster I corresponding to u, and (ii) $\mathbf{A}_w$ gives exactly h' clusters, then the contribution of j' elements of J and $\mathbf{A}_w$ to the total cost is $\min\{\omega_w^{(1)}(h', J, j') + j'd_H(\mathbf{s}, \mathbf{a}_i) \mid J \in \mathcal{J}(c),\ i \in J\}$. Thus, the contribution depends only on j' and h'. Then again, we can do dynamic programming over the children of u and compute $\omega_u^{(2)}(h, j, \mathbf{s})$ in polynomial time.

Summarizing, we conclude that the tables of values of $\omega_w^{(1)}(\cdot,\cdot,\cdot)$ and $\omega_u^{(2)}(\cdot,\cdot,\cdot)$ can be computed in $2^{\mathcal{O}(B \log B)}|\Sigma|^B \cdot (mn)^{\mathcal{O}(1)}$ time. Therefore, we can decide whether there is a feasible k-clustering for $\mathbf{A}$ of cost at most B in this time.

Combining this with the previous arguments about dealing with the case when F is disconnected, we conclude that it can be decided in $2^{\mathcal{O}(B \log B)}|\Sigma|^B \cdot (mn)^{\mathcal{O}(1)}$ time whether there is a feasible k-clustering for $\mathbf{A}$ of cost at most B for every given F and α. Therefore, given a coloring $\psi\colon \mathcal{J} \to \{1, \dots, \ell\}$, it can be decided in $2^{\mathcal{O}(B \log B)}|\Sigma|^B \cdot (mn)^{\mathcal{O}(1)}$ time whether $(\mathbf{A}, \Sigma, k, B, p, q)$ has an acyclic colorful solution with t composite clusters such that exactly ℓ initial clusters have nonempty intersections with the composite clusters of the solution.

Recall that with probability at least e^{-2B}, a yes-instance admits a colorful solution. This implies that if we try e^{2B} random colorings, then the probability that we fail to find a colorful solution for a yes-instance for each of the colorings is at most $e^{-1} < 1$. This leads to the randomized algorithm with running time $e^{2B} \cdot 2^{\mathcal{O}(B \log B)}|\Sigma|^B \cdot (mn)^{\mathcal{O}(1)}$, that is, $2^{\mathcal{O}(B \log B)}|\Sigma|^B \cdot (mn)^{\mathcal{O}(1)}$. This concludes the sketch of the proof.

4 Clustering with Size Constraints

In this section, we discuss other variants of Categorical Clustering with cluster size constraints: Balanced Clustering and Factor-Balanced Clustering. We also discuss the special case of Capacitated Clustering for $p = q = n/k$ and refer to this problem as Equal Clustering.

Recall that by Theorem 2, Capacitated Clustering is NP-complete for $k = 2$ and $p = q = n/2$, that is, Equal Clustering is NP-complete for $k = 2$. Using the same arguments as in the proof of Theorem 2, we can show the following more general claim.

Theorem 3. *For every fixed $\alpha \geq 1$ ($\delta \geq 0$, respectively),* Factor-Balanced Clustering *(*Balanced Clustering*, respectively) is* NP*-complete for $k = 2$ and binary matrices.*

From the positive side, we observe that Balanced Clustering and Factor-Balanced Clustering admit Turing reductions to Capacitated Clustering, that is, Capacitated Clustering is the most general among the considered problems. For this, we make the following straightforward observation.

Observation 2. *An instance* $(\mathbf{A}, \Sigma, k, B, \delta)$ *of* BALANCED CLUSTERING *(an instance* $(\mathbf{A}, \Sigma, k, B, \alpha)$ *of* FACTOR-BALANCED CLUSTERING, *respectively) is a yes-instance if and only if there is a nonnegative integer* p *such that* $\frac{n}{k} - \delta \leq p \leq \frac{n}{k}$ $(\frac{n}{\alpha k} \leq p \leq \frac{n}{k}$, *respectively) and for* $q = p + \delta$ $(q = \alpha p$, *respectively),* $(\mathbf{A}, \Sigma, k, B, p, q)$ *is a yes-instance of* CAPACITATED CLUSTERING.

This allows us to obtain the following corollary of Theorem 1.

Corollary 1. BALANCED CLUSTERING *and* FACTOR-BALANCED CLUSTERING *are solvable in time* $2^{\mathcal{O}(B \log B)}|\Sigma|^B \cdot (mn)^{\mathcal{O}(1)}$.

5 Conclusion

We proved that CAPACITATED CLUSTERING can be solved in $2^{\mathcal{O}(B \log B)}|\Sigma|^B \cdot (mn)^{\mathcal{O}(1)}$ time. This also implies that the same holds for BALANCED CLUSTERING and FACTOR-BALANCED CLUSTERING. The natural question is whether it is possible to improve the dependence on B? We do not know the answer to this question even for the special case of EQUAL CLUSTERING.

Another important direction of research in the investigation of kernelization for clustering problems with size constraints (we refer to the recent book Fomin et al. on kernelization [20] for basic definitions). In [17, Theorem 3], Fomin, Golovach and Panolan proved that CATEGORICAL CLUSTERING does not admit a polynomial kernel when parmeterized by B, unless $\mathsf{NP} \subseteq \mathsf{coNP/poly}$. This immediately implies the following proposition.

Proposition 2. CAPACITATED CLUSTERING *(*BALANCED CLUSTERING *and* FACTOR-BALANCED CLUSTERING, *respectively) has no polynomial kernel when paramterized by* B, *unless* $\mathsf{NP} \subseteq \mathsf{coNP/poly}$, *even if* $\Sigma = \{0, 1\}$.

Also by Theorems 2 and 3 the problems are already NP-hard for k. Thus, for kernelization, we have to consider more restrictive parameterizations. Up to now, we have only partial results. In particular, we can show BALANCED CLUSTERING admits a polynomial kernel when parameterzied by B, k and δ.

Theorem 4 (*). BALANCED CLUSTERING *admits a kernel, where the output matrix has* $\mathcal{O}(B(B + k))$ *rows and* $O(k(B + \delta k))$ *columns, and is a matrix over an alphabet of size at most* $B + k$.

Theorem 4 leads to the question whether FACTOR-BALANCED CLUSTERING admits a polynomial kernel when parameterized by k and B with the assumption that α is a fixed constant. A more general question is whether there are polynomial kernel for CAPACITATED CLUSTERING, BALANCED CLUSTERING and FACTOR-BALANCED CLUSTERING parameterized by k and B. Notice that CATEGORICAL CLUSTERING has a polynomial kernel for this parmeterization [17, Theorem 2]. Another direction of research is to investigate kernels of other types. Are there polynomial *Turing kernels* and do these problem admit polynomial *lossy kernels*, that is, approximative kernels? (We refer to the book [20] for the definition of the notions.)

References

1. Aggarwal, G., et al.: Achieving anonymity via clustering. ACM Trans. Algorithms **6**(3), 1–19 (2010)
2. Alon, N., Yuster, R., Zwick, U.: Color-coding. J. ACM **42**(4), 844–856 (1995). https://doi.org/10.1145/210332.210337
3. Baeza-Yates, R., Ribeiro-Neto, B.: Modern Information Retrieval, vol. 463. ACM Press, New York (1999)
4. Ban, F., Bhattiprolu, V., Bringmann, K., Kolev, P., Lee, E., Woodruff, D.P.: A PTAS for ℓ_p-low rank approximation. In: 30th Annual ACM-SIAM Symposium on Discrete Algorithms, SODA 2019, pp. 747–766. SIAM (2019). https://doi.org/10.1137/1.9781611975482.47
5. Banerjee, A., Ghosh, J.: Clustering with balancing constraints. In: Constrained Clustering: Advances in Algorithms, Theory, and Applications, pp. 171–200. CRC Press (2008)
6. Byrka, J., Fleszar, K., Rybicki, B., Spoerhase, J.: Bi-factor approximation algorithms for hard capacitated k-median problems. In: 26th Annual ACM-SIAM Symposium on Discrete Algorithms, SODA 2015, pp. 722–736. SIAM (2015)
7. Byrka, J., Rybicki, B., Uniyal, S.: An approximation algorithm for uniform capacitated k-median problem with $1+\epsilon$ capacity violation. In: Louveaux, Q., Skutella, M. (eds.) IPCO 2016. LNCS, vol. 9682, pp. 262–274. Springer, Cham (2016). https://doi.org/10.1007/978-3-319-33461-5_22
8. Charikar, M., Guha, S., Tardos, É., Shmoys, D.B.: A constant-factor approximation algorithm for the k-median problem. J. Comput. Syst. Sci. **65**(1), 129–149 (2002)
9. Chen, D.Z., Li, J., Liang, H., Wang, H.: Matroid and knapsack center problems. Algorithmica **75**(1), 27–52 (2016). https://doi.org/10.1007/s00453-015-0010-1
10. Chuzhoy, J., Rabani, Y.: Approximating k-median with non-uniform capacities. In: 16th Annual ACM-SIAM Symposium on Discrete Algorithms, SODA 2005, pp. 952–958. SIAM (2005)
11. Cilibrasi, R., van Iersel, L., Kelk, S., Tromp, J.: The complexity of the single individual SNP haplotyping problem. Algorithmica **49**(1), 13–36 (2007). https://doi.org/10.1007/s00453-007-0029-z
12. Cohen-Addad, V., Li, J.: On the fixed-parameter tractability of capacitated clustering. In: 46th International Colloquium on Automata, Languages, and Programming, ICALP 2019. LIPIcs, vol. 132, pp. 41:1–41:14. Schloss Dagstuhl - Leibniz-Zentrum für Informatik (2019)
13. Cygan, M., et al.: Parameterized Algorithms. Springer, Heidelberg (2015). https://doi.org/10.1007/978-3-319-21275-3
14. Demirci, H.G., Li, S.: Constant approximation for capacitated k-median with $(1+\varepsilon)$-capacity violation. In: 43rd International Colloquium on Automata, Languages, and Programming, ICALP 2016. LIPIcs, vol. 55, pp. 73:1–73:14. Schloss Dagstuhl - Leibniz-Zentrum für Informatik (2016)
15. Feige, U.: NP-hardness of hypercube 2-segmentation. CoRR abs/1411.0821 (2014). http://arxiv.org/abs/1411.0821
16. Fomin, F.V., Golovach, P.A., Lokshtanov, D., Panolan, F., Saurabh, S.: Approximation schemes for low-rank binary matrix approximation problems. ACM Trans. Algorithms **16**(1), 12:1–12:39 (2020). https://doi.org/10.1145/3365653
17. Fomin, F.V., Golovach, P.A., Panolan, F.: Parameterized low-rank binary matrix approximation. Data Min. Knowl. Discov. **34**(2), 478–532 (2020). https://doi.org/10.1007/s10618-019-00669-5

18. Fomin, F.V., Golovach, P.A., Purohit, N.: Parameterized complexity of categorical clustering with size constraints. CoRR 2104.07974 (2021). https://arxiv.org/abs/2104.07974
19. Fomin, F.V., Golovach, P.A., Simonov, K.: Parameterized k-clustering: tractability island. J. Comput. Syst. Sci. **117**, 50–74 (2021). https://doi.org/10.1016/j.jcss.2020.10.005
20. Fomin, F.V., Lokshtanov, D., Saurabh, S., Zehavi, M.: Kernelization: Theory of Parameterized Preprocessing. Cambridge University Press, Cambridge (2019)
21. Ghiasi, S., Srivastava, A., Yang, X., Sarrafzadeh, M.: Optimal energy aware clustering in sensor networks. Sensors **2**(7), 258–269 (2002)
22. Gupta, G., Younis, M.: Load-balanced clustering of wireless sensor networks. In: IEEE International Conference on Communications (ICC), vol. 3, pp. 1848–1852. IEEE (2003)
23. Kleinberg, J., Papadimitriou, C., Raghavan, P.: Segmentation problems. J. ACM **51**(2), 263–280 (2004). https://doi.org/10.1145/972639.972644
24. Li, S.: On uniform capacitated k-median beyond the natural LP relaxation. ACM Trans. Algorithms **13**(2), 22:1–22:18 (2017)
25. Lynch, P.J., Horton, S., Horton, S.: Web Style Guide: Basic Design Principles for Creating Web Sites. Universities Press (1999)
26. Malinen, M.I., Fränti, P.: Balanced K-means for clustering. In: Fränti, P., Brown, G., Loog, M., Escolano, F., Pelillo, M. (eds.) S+SSPR 2014. LNCS, vol. 8621, pp. 32–41. Springer, Heidelberg (2014). https://doi.org/10.1007/978-3-662-44415-3_4
27. Marx, D.: Closest substring problems with small distances. SIAM J. Comput. **38**(4), 1382–1410 (2008). https://doi.org/10.1137/060673898
28. Miettinen, P., Mielikäinen, T., Gionis, A., Das, G., Mannila, H.: The discrete basis problem. IEEE Trans. Knowl. Data Eng. **20**(10), 1348–1362 (2008). https://doi.org/10.1109/TKDE.2008.53
29. Nielsen, A.: Category management: positioning your organization to win, Chicago (1992)
30. Ostrovsky, R., Rabani, Y.: Polynomial-time approximation schemes for geometric min-sum median clustering. J. ACM **49**(2), 139–156 (2002). https://doi.org/10.1145/506147.506149
31. Otter, R.: The number of trees. Ann. Math. **49**(3), 583–599 (1948). https://doi.org/10.2307/1969046
32. Rösner, C., Schmidt, M.: Privacy preserving clustering with constraints. In: 45th International Colloquium on Automata, Languages, and Programming (ICALP 2018). LIPIcs, vol. 107, pp. 96:1–96:14. Schloss Dagstuhl-Leibniz-Zentrum fuer Informatik (2018)
33. Vallejo-Huanga, D., Morillo, P., Ferri, C.: Semi-supervised clustering algorithms for grouping scientific articles. In: International Conference on Computational Science (ICCS) (2017). Procedia Comput. Sci. **108**, 325–334. https://doi.org/10.1016/j.procs.2017.05.206
34. Wright, R.A., Richmond, L.B., Odlyzko, A.M., McKay, B.D.: Constant time generation of free trees. SIAM J. Comput. **15**(2), 540–548 (1986). https://doi.org/10.1137/0215039
35. Yang, Y., Padmanabhan, B.: Segmenting customer transactions using a pattern-based clustering approach. In: Proceedings of the 3rd IEEE International Conference on Data Mining (ICDM), pp. 411–418. IEEE Computer Society (2003). https://ieeexplore.ieee.org/xpl/conhome/8854/proceeding

Graph Pricing with Limited Supply

Zachary Friggstad(✉) and Maryam Mahboub

Department of Computing Science, University of Alberta, Edmonton, Canada
{zacharyf,mahboub}@ualberta.ca

Abstract. We study approximation algorithms for graph pricing with vertex capacities yet without the traditional envy-free constraint. Specifically, we have a set of items V and a set of customers X where each customer $i \in X$ has a budget b_i and is interested in a bundle of items $S_i \subseteq V$ with $|S_i| \leq 2$. However, there is a limited supply of each item: we only have μ_v copies of item v to sell for each $v \in V$. We should assign a price $p(v)$ to each $v \in V$ and choose a subset $Y \subseteq X$ of customers so that each $i \in Y$ can afford their bundle ($p(S_i) \leq b_i$) and at most μ_v chosen customers have item v in their bundle for each item $v \in V$. Each customer $i \in Y$ pays $p(S_i)$ for the bundle they purchased: our goal is to do this in a way that maximizes revenue. Such pricing problems have been studied from the perspective of *envy-freeness* where we also must ensure that $p(S_i) \geq b_i$ for each $i \notin Y$. However, the version where we simply allocate items to customers after setting prices and do not worry about the envy-free condition has received less attention.

Our main result is an 8-approximation for the capacitated case via local search and a 7.8096-approximation in simple graphs with uniform vertex capacities. The latter is obtained by combing a more involved analysis of a multi-swap local search algorithm for constant capacities and an LP-rounding algorithm for larger capacities. If all capacities are bounded by a constant C, we further show a multi-swap local search algorithm yields a $\left(4 \cdot \frac{2C-1}{C} + \epsilon\right)$-approximation. We show the analysis of the locality gaps of our algorithms is tight, at one point using an interesting construction based on regular, high-girth graphs. We also give a $(4 + \epsilon)$-approximation in simple graphs through LP rounding when all capacities are very large as a function of ϵ.

Keywords: Graph pricing · Capacitated pricing · Approximation algorithms · Local search · Linear programming

1 Introduction

Choosing prices to sell items in order to maximize revenue is a complicated task even in environments where one can be certain of customer behaviour. Indeed,

This research was undertaken, in part, thanks to funding from the Canada Research Chairs program and an NSERC Discovery Grant.

A. Lubiw et al. (Eds.): WADS 2021, LNCS 12808, pp. 399–413, 2021.
https://doi.org/10.1007/978-3-030-83508-8_29

many so-called *pricing problems* have been studied in combinatorial optimization. One popular setting is this: a collection of items V is available to be sold where we have $\mu_v \in \mathbb{Z}_{\geq 0} \cup \{\infty\}$ copies of item $v \in V$. Additionally, we are given a collection of customers X where each $i \in X$ has some budget $b_i \geq 0$. In the *single-minded* setting, each customer $i \in X$ is interested in a bundle $S_i \subseteq V$. We must assign prices $p : V \to \mathbb{R}_{\geq 0}$ to the items and sell them to some customers $Y \subseteq X$ while respecting two constraints:

- **Affordability**: $p(S_i) := \sum_{v \in S_i} p(v) \leq b_i$ for $i \in Y$, and
- **Supply/Capacity Constraints**: $|\{i \in Y : v \in S_i\}| \leq \mu_v$ for $v \in V$.

That is, each customer that purchases their bundle can afford it and no item is oversold. Such a solution (p, Y) is said to be **feasible**, and the goal is to find a feasible (p, Y) maximizing revenue, i.e. $\sum_{i \in Y} p(S_i)$.

Much attention has been given to the **envy-free** setting, where a feasible solution must additionally satisfy the property $p(S_i) \geq b_i$ for $i \notin Y$ or to the **unlimited supply** setting where $\mu_v = \infty$ for each $v \in V$. Observe that in the unlimited supply setting, any pricing yields an envy-free solution by simply choosing the customers that can afford the price. However, the problem still remains **APX**-hard in this relaxed setting and, further, is hard to approximate within a factor better than 4 unless the Unique Games conjecture fails to hold, see the related works section.

The single-minded, envy-free pricing (SMEFP) problem with limited supply was studied by Cheung and Swamy [5]. Somewhat informally, they show the following. If there is an LP-based α-approximation to the problem of choosing the best customers Y when given prices p (without regard to the envy-free condition), then there is an $O(\alpha \cdot \log C)$-approximation to SMEFP where $C = \max_{v \in V} \mu_v$. In the special case where $|S_i|$ is bounded by a constant for each $i \in X$, this yields a logarithmic approximation for SMEFP.

We study single-minded pricing problems yet *without* the envy-free constraint. This is a natural variant of pricing problems where customer satisfaction is less of a concern than overall revenue generation. To the best of our knowledge, it seems that pricing problems without the envy-free condition like this have received virtually no attention so far except in simpler cases of unlimited supply where envy-freeness is a superfluous constraint, *i.e.* any solution can be trivially be made envy-free without losing revenue.

More specifically we mainly consider the case when $|S_i| = 2$ for each customer i. That is, the set of customers can be thought of as edges E in a graph $G = (V, E)$ with vertex capacities and, perhaps, parallel edges. We show that without the envy-free condition, the problem admits a constant-factor approximation. In fact, this is relatively easy to obtain via randomized rounding (with alterations) of an LP relaxation. Our focus is on obtaining smaller constants by considering a more intricate *local search* approximation algorithm.

1.1 Our Results

We use shorthand notation like $e = uv \in E$ when we want to consider an edge $e \in E$ in some graph $G = (V, E)$ and also want name the endpoints u, v of e. This allows us to name distinct customers (i.e. e) who are interested in the same bundle of items (*i.e.* $\{u, v\}$). We focus on the following problem.

Definition 1. *Let $G = (V, E)$ be a graph with vertex capacities $\mu_v \in \mathbb{Z}_{\geq 0} \cup \{\infty\}$ where each $e = uv \in E$ has a budget $b_e \geq 0$ and is interested in the bundle of vertices $\{u, v\}$. In* CAPACITATED GRAPH PRICING, *we want to find a pricing $p : V \to \mathbb{R}_{\geq 0}$ and $F \subseteq E$ with (p, F) feasible while maximizing revenue $\sum_{e=uv \in F} p(u) + p(v)$.*

All of our algorithmic results extend in a simple way to the case where each customer is interested in a bundle of size *at most* 2, but it is slightly simpler to describe the algorithms and their analysis for the case where each customer wants precisely two different items. Unless otherwise stated, all graphs may have parallel edges. We use the term *simple graph* to indicate it does not have parallel edges.

To obtain approximations for CAPACITATED GRAPH PRICING, we use a reduction from Balcan and Blum [2] to reduce the problem to bipartite graphs where all items on one side will be priced 0. Specifically, we consider the following problem.

Definition 2. *In* L-SIDED PRICING, *we are given a* CAPACITATED GRAPH PRICING *instance in a bipartite graph $(L \cup R, E)$. A feasible solution (p, F) must also have $p(v) = 0$ for $v \in R$.*

Though they only focused on uncapacitated pricing problems, the reduction in [2] extends to CAPACITATED GRAPH PRICING without modification. We also remark that the reduction would not be valid if one were looking for envy-free solutions for CAPACITATED GRAPH PRICING. The proof of the following appears in Appendix A for completeness.

Lemma 1 (Modified version of Balcan and Blum [2]). *A 4α-approximation* CAPACITATED GRAPH PRICING *exists if there is an α-approximation for* L-SIDED PRICING.

Approximation Algorithms

We develop approximation algorithms for L-SIDED PRICING in order to approximate CAPACITATED GRAPH PRICING. It is possible to get a 4-approximation for L-SIDED PRICING through straightforward rounding of a natural linear programming relaxation (briefly summarized at then end of Sect. 4), thus leading to a 16-approximation for CAPACITATED GRAPH PRICING overall. We consider an alternative approach to get a better approximation guarantee.

Theorem 1. L-SIDED PRICING *has a polynomial-time 2-approximation.*

This is fairly simple to obtain using a local search procedure that iteratively tries to change the price of one item at a time until we cannot get a more profitable allocation by doing so. We think it nicely highlights a direction for designing approximations for pricing+packing problems. We also consider a multi-swap variant of this algorithm that tries to change the prices of $O(1)$ items at a time.

Theorem 2. *For any constants $C \geq 2, \epsilon > 0$, there is a poly-time $(\frac{2C-1}{C} + \epsilon)$-approximation for* L-SIDED PRICING *if $\mu_v \leq C$ for all $v \in L$.*

Note, Theorem 2 does not require bounds on capacities for nodes in R. Observe if $C = 1$ then both CAPACITATED GRAPH PRICING and L-SIDED PRICING reduce to maximum-weight matching; we can easily set prices to match the full budget of all edges in any matching.

On the other hand, intuitively it should be easier to get approximations for large capacities as L-SIDED PRICING with unbounded capacities can be solve in polynomial time [2]. We confirm this in simple graphs using LP rounding.

Theorem 3. *For $\epsilon > 0$, the integrality gap of LP relaxation* (LP-Pricing) *is $1 - 2\epsilon$ in simple graphs when $\mu_v \geq 3\ln(1/\epsilon)/\epsilon^2 + 1$ for all $v \in R$.*

The LP relaxation referenced in the theorem statement is a bit long and can be found in Sect. 4.

The assumption of simplicity is required for us to apply Chernoff bounds to ensure certain random events are independent. By combining Theorems 2 and 3 along with additional scrutiny in the dependence of μ on ϵ, we obtain slightly improved bounds in simple graphs with uniform capacities (i.e. the capacities need not be large or small, just as long as they are uniform).

Theorem 4. *Instances of* L-SIDED PRICING *in simple graphs where all vertices in $L \cup R$ have the same capacity μ admit a randomized, polynomial-time* 1.952381*-approximation, yielding a* 7.8096*-approximation for* CAPACITATED GRAPH PRICING.

A Structural Result

To prove Theorem 2, we develop a result about covering directed graphs by directed balls in a uniform way. This may be of independent interest in other settings, so we state it here in the introduction.

Let $H = (L, F)$ be a directed graph. For any $u \in L$ and $r \geq 0$ consider the "directed ball" $B^+(u, r) = \{v \in L : d_H(u, v) \leq r\}$ of nodes reachable from u in at most r steps. Similarly, let $\partial B^+(u, r) = \{v \in L : d(u, v) = r\}$ be nodes v such that the shortest $u - v$ path in H has length exactly r (the *boundary* of $B^+(u, r)$). We prove the following covering result for directed graphs.

Theorem 5. *Let $H = (L, F)$ be a directed graph where the indegree of each node is at most C and let $d \in \mathbb{Z}_{\geq 0}$. There is a "weighting" of directed balls $\tau : L \times \{0, 1, \ldots, d\} \to \mathbb{Z}_{\geq 0}$ with the following properties. For any $v \in V$,*

$$\sum_{u \in L, 0 \leq r \leq d : v \in B^+(u,r)} \tau(u, r) = \frac{C^{d+1} - 1}{C - 1} \quad \text{and} \quad \sum_{u \in L, 0 \leq r \leq d : v \in \partial B^+(u,r)} \tau(u, r) = C^d.$$

Furthermore, $\tau(u, r) \leq C^{d-r}$ *for each* $u \in V$ *and* $0 \leq r \leq d$.

That is, each $v \in L$ lies in these balls with weight precisely $\frac{C^{d+1}-1}{C-1}$ and appears on the boundary of the balls with weight precisely C^d. The bound on $\tau(u, r)$ at the end of the statement is required to ensure the local search algorithm used to prove Theorem 2 runs in polynomial time.

Lower Bounds
As shown in [2], L-SIDED PRICING with unbounded capacities can be solved in polynomial time with a simple greedy algorithm. We show L-SIDED PRICING remains **APX**-hard even in graphs with small capacities, so we cannot expect to get a matching 4-approximation for CAPACITATED GRAPH PRICING using a reduction to L-SIDED PRICING.

Theorem 6. L-SIDED PRICING *is* **APX**-*hard, even if all capacities are at most 4 and all customers have a budget of 1 or 2.*

We also show the analysis of both local search algorithms are tight. Here, we say the locality gap of an instance is the maximum ratio of the optimum solution value and the value of a locally-optimum solution. The locality gap of a local search algorithm is the supremum of the locality gaps over all instances of the problem.

Theorem 7. *For any* $\epsilon > 0$*, the locality gap of the single-swap algorithm (Algorithm 1) is at least* $2 - \epsilon$ *even in instances where all capacities are 1.*

Theorem 8. *For any* $C \geq 2, \rho \geq 1, \epsilon > 0$*, the locality gap of the* ρ*-swap algorithm (Algorithm 3) on* L-SIDED PRICING *instances having maximum capacity at most* C *is at least* $\left(\frac{2C-1}{C} - \epsilon\right)$.

The first construction is quite simple, but the second construction for the multi-swap analysis is much more involved. As a starting point for the construction, we require simple graphs of constant degree but arbitrarily large girth. Such graphs were shown to exist by Sachs [14].

1.2 Related Work

The basic model of pricing problems of this sort were introduced by Guruswami *et al.* [10]. Among other results, they given an $O(\log n + \log m)$-approximation for the case of single-minded pricing without item capacities if we have n items and m customers. Here, the bundle S_i for each customer i may be any subset of items (not just size 2). This was later improved by Briest and Krysta to an $O(\log D + \log k)$-approximation where each set has size at most k and each item appears in at most D sets [3]. These logarithmic guarantees are essentially tight: Chalermsook *et al.* show for any constant $\epsilon > 0$ there is no $O(\log^{1-\epsilon}(m+n))$-approximation unless $\mathbf{NP} \subseteq \mathbf{DTIME}(n^{\mathrm{polylog(n)}})$ [4].

In the case with no capacities, Balcan and Blum give an $O(k)$-approximation for the case where all customers are interested in a set of size at most k, which specializes to a 4-approximation in the case $k = 2$ [2]. Amazingly, this may also be tight: building on work by Khadekar *et al.* [11], Lee showed that there is no $(4 - \epsilon)$-approximation when $k = 2$ for any constant $\epsilon > 0$ unless the Unique Games conjecture fails [12].

Cheung and Swamy studied the envy-free variant of capacitated pricing problems [5]. As mentioned earlier, they show that LP-based approximations that choose the maximum-profit set of customers for given prices translate to approximation algorithms for envy-free pricing with capacities while losing an $O(\log \mu_{\max})$-factor. In particular, for envy-free CAPACITATED GRAPH PRICING they present an $O(\log \mu_{\max})$-approximation.

Other variants of envy-free pricing problems have been studied, we do not attempt to comprehensively survey all such models and just sample a few to discuss. For example, it could be that each customer is interested in acquiring just a single item from their subset (rather than all items). This was also studied in [10] and follow-up work (*e.g.* [6]). Other directions have considered more restricted subsets of items in single-minded pricing, for example the customers may be interested in the edges of a subpath of a large path (the **highway problem**) or subpaths of a tree (the **tollbooth problem**). See [9] and [8] for definitions and state-of-the-art approximations for these problems.

1.3 Organization

We present our algorithmic contributions in the body of this paper: Sect. 2 introduces notation and discusses the reduction from CAPACITATED GRAPH PRICING to L-SIDED PRICING. Section 3 gives the local search algorithms to prove Theorems 1 and 2 and, in doing so, also proves the structural result in Theorem 5. The randomized LP rounding algorithm which proves Theorem 3 and, ultimately, the proof of Theorem 4 is sketched in Sect. 4. Due to space constraints, the full proofs of Theorem 3 and 4 and all of our lower bound results are deferred to the full version of this paper.

2 Preliminaries

We consider graphs that may have parallel edges unless we explicitly specify otherwise. We do not consider loops. It is easy to extend our algorithms to cases where some customers may only be interested a singleton bundle while ensuring the same approximation guarantees that we present here, this is discussed in the full version of our paper.

For a set of nodes S in a graph $G = (V, E)$, we let $N(S)$ denote all nodes not in S that are neighbours of some node in S. For $u \in V$ we let $\delta_G(u)$ be all edges having u as an endpoint. Often the subscript G is omitted when it is clear from the context. For a subset of edges B, we let $\delta_B(u) = \delta(u) \cap B$ when the graph G is clear from the context.

Again, we sometimes refer to an edge e by uv where u, v are the endpoints of e. For brevity, we may use notation like $e = uv \in E$ when we want to consider an edge $e \in E$ but also want to name the endpoints u, v of e as well. The reason for using this notation rather than simply saying $uv \in E$ is that our local search algorithms do work for graphs with parallel edges (i.e. customers interested in identical bundles), so e would be one particular customer and u, v would name the items that e is interested in.

Given a function $f : T \to \mathbb{R}$ on some finite set T, for any $S \subseteq T$ we let $f(S)$ denote $\sum_{x \in S} f(x)$ (in particular, $f(\emptyset) = 0$). Similarly, if $p : V \to \mathbb{R}_{\geq 0}$ is a pricing of the vertices of a graph $G = (V, E)$, for an edge $e = uv \in E$ we let $p(e)$ denote $p(u) + p(v)$. For two pricings $p, p' : V \to \mathbb{R}_{\geq 0}$ of the nodes of a graph, we let $\mathtt{HW}(p, p') = |\{v \in V : p(v) \neq p'(v)\}|$.

Finally, consider an instance $G = (L \cup R, E)$ of L-Sided Pricing where edges have budgets b_e and vertices have capacities μ_v. For any pricing p of L, let $\mathtt{val}(p) = \max_{F \subseteq E \text{ and (p,F) feasible}} \sum_{e \in F} p(e)$ be the maximum profit of a feasible solution with prices p. Note that $\mathtt{val}(p)$ can be computed in polynomial time as it is merely asking for a maximum-weight μ-matching solution using only edges $e = uv$ with $p(e) \leq b_e$ with the weight of such an edge being $p(e)$.

3 Local-Search Algorithms

We consider local-search algorithms for L-Sided Pricing. Recall we are given a bipartite graph $G = (L \cup R, E)$ where each $v \in L \cup R$ has a capacity $\mu_v \geq 0$, each $e \in E$ has a budget b_e, and we are restricted to setting $p(v) = 0$ for each $v \in R$. It is clear that there is an optimal solution p such that for each $u \in L$ we have $p(u) = b_e$ for some $e \in \delta(u)$. Otherwise we could increase $p(u)$ to the next budget of an edge touching u (or decrease, if $p(u)$ exceeds all budgets of edges touching u) while not decreasing the value of the solution. Thus, for $u \in L$ we define $P_u = \{b_e : e \in \delta(u)\}$, the different budgets of customers interested in u.

We run a local-search approximation based on this observation. Here, a vector p over L is a **pricing** if $p(u) \in P_u$ for each $u \in L$. The local-search algorithm iteratively tries to improve a pricing by changing the price of only one vertex until no such improvement is possible. The full algorithm is presented in Algorithm 1. Because a price $p(u)$ is chosen from P_u for each $u \in L$, it is clear that an iteration can be executed in polynomial time.

Algorithm 1. Single-Swap Algorithm for L-Sided Pricing.

let p be any pricing
while $\mathtt{val}(p') > \mathtt{val}(p)$ for some pricing p' with $\mathtt{HW}(p, p') = 1$ **do**
  $p \leftarrow \arg\max\{\mathtt{val}(p') : p' \text{ a pricing with } \mathtt{HW}(p, p') = 1\}$
return p

Call a pricing p *locally optimal* if it cannot be improved by changing the price for any $u \in L$, note Algorithm 1 returns a locally-optimal pricing. As is common in local search, we analyze the quality of a locally-optimal solution. In the next subsection we show $\mathtt{val}(p) \geq \mathtt{val}(p^*)/2$ for any locally-optimal pricing p where p^* is an optimal pricing for the L-Sided Pricing instance.

The main concern is then the efficiency of the algorithm. Clearly each iteration can be executed in polynomial time but the number of iterations is not apparently bounded. In fact, with some approximation algorithms based on local search it is **PLS**-complete to find a locally-optimal solution [1]. To cope with this problem, we use a more recent observation from [7] that essentially shows after a polynomial number of iterations of Algorithm 1, the resulting pricing p, while not necessarily locally-optimal, still has value at least $\texttt{val}(p^*)/2$. The straightforward details of this adaptation are deferred to the full version of this paper. This completes the proof of Theorem 1.

3.1 Single-Swap Analysis

We fix p^* to be some particular optimal pricing.

Lemma 2. *For any locally-optimal pricing p, $\texttt{val}(p) \geq \texttt{val}(p^*)/2$.*

Proof. Let $B \subseteq E$ be the edges that are bought in the local optimum solution, and $B^* \subseteq E$ the edges that are bought in the global optimum solution. Thus, $\texttt{val}(p) = \sum_{u \in L} p(u) \cdot |\delta_B(u)|$ and $p(e) \leq b_e$ for each $e \in \delta_B(u)$.

For each $u \in L$, consider the local search step that changes the price of u from $p(u)$ to $p^*(u)$. That is, consider p^u where $p^u(u) = p^*(u)$ and $p^u(u') = p(u')$ for $u' \in L - \{u\}$. We refer to this swap as the $p \to p^u$ swap. For brevity, let $\Delta_u := \texttt{val}(p^u) - \texttt{val}(p)$ and note $\Delta_u \leq 0$ because p is a local optimum. We provide a lower bound on Δ_u in a way that relates part of the global optimum with part of the local optimum.

First, construct a subset $B' \subseteq B^*$ and an injective mapping $\sigma : B' \to B$ iteratively as follows in Algorithm 2. Intuitively, it greedily pairs some edges in B^* with edges in B sharing the same endpoint in R until no more pairs can be made. After this pairing, for each $v \in R$ we either have $\delta_{B^*}(v) \subseteq B'$ or $\delta_B(v) \subseteq \sigma(B')$ (or both).

Algorithm 2. Constructing B' and σ.

$B' := \emptyset$
for each $e^* = uv \in B^*$ where $v \in R$ **do**
  if there is some $e \in \delta_B(v)$ such that no $e' \in B'$ has $\sigma(e') = e$ **then**
    set $B' := B' \cup \{e^*\}$ and $\sigma(e^*) := e$

Now we bound Δ_u. One possible matching with the modified prices p^u is $B^u := B \cup \delta_{B^*}(u) - \delta_B(u) - \{\sigma(e) : e \in \delta_{B'}(u)\}$. A simple inspection of the definition of B' and σ shows this is feasible. That is, it alters B by swapping $\delta_B(u)$ for $\delta_{B^*}(u)$ and removes edges paired, via σ, with $\delta_{B^*}(u)$ to make room across nodes in R for these new edges. It could be that some edges in $\delta_{B^*}(u)$ are not paired by σ but this indicates their right-endpoints already have enough room to accommodate these edges without removing other edges from B. So, B^u respects the vertex capacities.

Now, Δ_u represents the cost change when using the maximum value matching with the new profits. This can be bounded as follows, based on the fact that B^u is a feasible solution under prices p^u:

$$0 \geq \Delta_u \geq p^*(u) \cdot |\delta_{B^*}(u)| - p(u) \cdot |\delta_B(u)| - \sum_{e' \in \delta_{B'}(u)} p(\sigma(e')).$$

Summing over all $u \in L$ and noting each $e \in B$ has its corresponding term appearing in the last sum for at most one $u \in L$ because σ' is one-to-one shows $0 \geq \texttt{val}(p^*) - 2 \cdot \texttt{val}(p)$.

3.2 An Improved Multi-swap Algorithm for Bounded Capacities

Here we consider the restriction of L-Sided Pricing to instances where $\mu_u \leq C$ for each $u \in L$ for some fixed constant $C \geq 2$. We do not require capacities of $v \in R$ to be bounded.

Let $d \geq 1$ be a fixed integer: larger d will result in better approximation guarantees with a slower, but still polynomial-time, algorithm. The multi-swap algorithm we consider is given in Algorithm 3. Let $\rho = 1 + C + C^2 + \ldots + C^d = \frac{C^{d+1}-1}{C-1}$. An iteration runs in polynomial time because ρ is a constant.

Algorithm 3. Multi-Swap Algorithm For L-Sided Pricing.

let p be any pricing
while $\texttt{val}(p') > \texttt{val}(p)$ for some pricing p' with $\texttt{HW}(p, p') \leq \rho$ **do**
  $p \leftarrow \arg\max\{\texttt{val}(p') : p'$ a pricing with $\texttt{HW}(p, p') \leq \rho\}$
return p

As before, call a pricing p *locally optimal* if $\texttt{val}(p') \leq \texttt{val}(p)$ for any pricing p' with $\texttt{HW}(p, p') \leq \rho$. Recall P_u for $u \in L$ is the set of distinct budgets of the edges incident to u and that, in L-Sided Pricing, we can assume any pricing p has $p(u) \in P_u$ for all $u \in L$. So, as C and d are constants, a single iteration can be executed in polynomial time by trying all subsets $S \subseteq L$ of bounded size and, for each of those, trying all $\prod_{u \in S}(|P_u| - 1) \leq |E|^{O(1)}$ ways to change the prices of all $u \in S$. We prove the following.

Lemma 3. *Let p be a locally-optimal solution and p^* a global optimum solution. Then $\texttt{val}(p) \geq \frac{C - C^{-d}}{2C - 1 - C^{-d}} \cdot \texttt{val}(p^*)$.*

We use the same trick as in the single-swap case to ensure polynomial running time: after a polynomial number of iterations of the algorithm we have $\texttt{val}(p)$ being at least that of the guarantee from Lemma 3. Again, this detail will appear in the full version of this paper. Theorem 2 follows by choosing large enough d.

We will soon prove Theorem 5 stated in Sect. 1.1. For now, we show how to complete the local search analysis using this result.

Proof (Proof of Lemma 3). Let p^* denote an optimal pricing, $B \subseteq E$ the edges bought in the local optimum p, and $B^* \subseteq E$ the edges bought under p^*. Let

$\sigma : B' \to B$ be a pairing constructed in the same way as in the single swap analysis (using Algorithm 2) where $B' \subseteq B^*$.

To describe the swaps used in the analysis, first consider the following auxiliary directed graph $H = (L, F)$ whose nodes are the same as the left side of this L-Sided Pricing instance and whose edges are given as follows. For any $e^* = uv \in B'$, let $w \in L$ be the left-endpoint of $\sigma(e^*)$. Add a directed edge from u to w in F.

Observe that both the indegree and outdegree of a vertex in H is at most C by this construction, so Theorem 5 applies. Let $\tau : L \times \{0, 1, \ldots, d\}$ be the given weighting of directed balls in H. These weights will be used to combine inequalities generated by the test swaps below.

Test Swaps
For any $u \in L$ and any $0 \leq i \leq d$, consider the prices $p^{u,i}$ defined by

$$p^{u,i}(v) = \begin{cases} p^*(v) & \texttt{if } d_H(u,v) \leq i \\ p(v) & \texttt{otherwise} \end{cases}$$

Note $\texttt{HW}(p, p^{u,i}) = |B^+(u,i)| \leq C^0 + C^1 + \ldots + C^i \leq \rho$ because the out-degree of each vertex is at most C, so $p \to p^{u,i}$ is a valid test swap. Let $\Delta_{u,i} = \texttt{val}(p^{u,i}) - \texttt{val}(p)$ and note $\Delta_{u,i} \leq 0$ by local optimality. We bound the difference by explicitly describing a feasible set of edges $B^{u,i}$, namely:

$$B^{u,i} = B \cup \delta_{B^*}(B^+(u,i)) - \delta_B(B^+(u,i)) - \sigma(\delta_{B'}(\partial B^+(u,i))).$$

That is, add all edges from B^* touching a vertex in the directed ball $B^+(u,i)$ and remove all edge from B that either touch $B^+(u,i)$ or are paired (via σ) with an edge in B' that touches $\partial B^+(u,i)$. It is again easy to check that $(p^{u,i}, B^{u,i})$ is a feasible solution: across $u \in L$ we simply exchanged edges in B touching U for edges in B^* touching u and we ensured any new $e^* \in B'$ has $\sigma(e^*)$ removed to make room for e^* across its right-endpoint. Observe for any $e^* \in \delta_{B'}(B^+(u,i-1))$ that $\sigma(e^*)$ is already removed when $\delta_B(B^+(u,i))$ is removed from B, which is why the last part of the definition of $B^{u,i}$ only uses the boundary $\partial B^+(u,i)$ instead of all of $B^+(u,i)$ to remove the remaining edges of B that are paired with $\delta_{B'}(B^+(u,i))$.

Weighting the inequalities by $\tau(u,i)$,

$$\begin{aligned} 0 \geq \tau(u,i) \cdot \Delta_{u,i} &\geq \tau(u,i) \cdot \left(\sum_{e \in B^{u,i}} p^{u,i}(e) - \sum_{e \in B} p(e) \right) \\ &= \tau(u,i) \cdot \sum_{e \in B^* \cap B^{u,i}} p^*(e) - \tau(u,i) \cdot \sum_{e \in B - B^{u,i}} p(e). \end{aligned} \qquad (1)$$

To finish, consider the contribution of each edge in B^* and B to this bound if we sum over all $u \in L, 0 \leq i \leq d$. Observe an edge $e = vw \in B^*$ is "swapped in" in this analysis for the swap $p \to p^{u,i}$ if and only if $v \in B^+(u,i)$. So by Theorem 5, the total contribution of $p^*(e)$ to $\sum_{u,i} \tau(u,i) \cdot \Delta_{u,i}$ is precisely $\frac{C^{d+1}-1}{C-1}$.

On the other hand, an edge $e = vw \in B$ is "swapped out" in this analysis for the swap $p \to p^{u,i}$ if and only if $v \in B^+(u,i)$ or $\sigma^{-1}(e) \in \partial B^+(u,i)$ (if e is indeed paired by σ). Again by Theorem 5, the total τ-weight of the first event is exactly $\frac{C^{d+1}-1}{C-1}$ and, if $\sigma^{-1}(e)$ is defined, the total τ-weight of the second event is exactly C^d. Thus,

$$0 \geq \sum_{u \in L 0 \leq i \leq d} \tau(u,i) \cdot \Delta(u,i) \geq \frac{C^{d+1}-1}{C-1} \cdot \mathtt{val}(p^*) - \left(\frac{C^{d+1}-1}{C-1} + C^d\right) \cdot \mathtt{val}(p),$$

which proves Theorem 3.

3.3 Proof of Theorem 5

Inductively define $\tau(u,i)$ for $u \in L$ and $0 \leq i \leq d$ as follows:

$$\tau(u,i) = \begin{cases} 1 & \text{if } i = d, \\ C^{d-i} - \sum_{j=i+1}^{d} \sum_{v \in L : d_H(v,u) = j-i} \tau(v,j) & \text{otherwise, i.e. } i < d. \end{cases}$$

The inspiration behind this construction is that in general we would have $d_H(u,v) = i$ for only *at most* C^i nodes u. So we consider smaller directed balls to make up this deficiency. If we think that the distance i requirement for each $v \in V$ is exactly C^i, then for each $u \in L$ the ball $B^+(u,j)$ contributes to the distance $d - j + d_H(u,v)$ requirement for each $v \in B^+(u,j)$.

The recurrence above ensures the total contribution to the distance i requirement for each v by all directed balls is exactly C^i. We formalize this idea and show the τ values are nonnegative in Lemma 4 below.

Lemma 4. *For each $u \in L, 0 \leq i \leq d$ we have* $\sum_{j=i}^{d} \sum_{v \in L : d_H(v,u) = j-i} \tau(v,j) = C^{d-i}$ *and $0 \leq \tau(u,i) \leq C^{d-i}$.*

Proof. The equality is by construction and the observation that $d_H(v,u) = 0$ if and only if $v = u$. The inequalities are proven inductively with the base case $i = d$ being given. Now suppose for $i < d$ we know $0 \leq \tau(u,j) \leq C^{d-j}$ for any $i < j \leq d$ and any $u \in L$. By the recurrence for $\tau(u,i)$ and because $\tau(v,j) \geq 0$ for any $i < j \leq d$ and $v \in V$, we see $\tau(u,i) \leq C^{d-i}$.

Next, we prove $\tau(u,i) \geq 0$ for each $u \in L$. For any $i < j \leq d$ and any $v \in L$ with $d_H(v,u) = j - i$, there is some $w \in L$ such that $d_H(v,w) = i - j - 1$ and $d_H(w,u) = 1$. That is, consider a shortest $v - u$ path P in H, as $i < j$, we have $v \neq u$ so the second-last node on this path is a node w whose distance to u is 1 (it could be $w = v$, if $j - i = 1$).

From this and using the equality from the first part of the theorem statement, we bound the double sum in the recurrence defining $\tau(u,i)$ by

$$\sum_{j=i+1}^{d} \sum_{v \in L : d_H(v,u)=j-i} \tau(v,j) \leq \sum_{w:d_H(w,u)=1} \sum_{j=i+1}^{d} \sum_{v \in L : d_H(v,w)=j-(i+1)} \tau(v,j)$$
$$= \sum_{w:d_H(w,u)=1} C^{d-(i+1)} \leq C^{d-i}.$$

The last bound follows as each $v \in L$ has indegree at most C in H. Thus, from the recurrence again, we see $\tau(u,i) \geq 0$.

Lemma 4 finishes the proof of Theorem 5 as follows. The first bullet point in Theorem 5 follows by summing over all $0 \leq i \leq d$. The second point follows by fixing $i = 0$.

4 LP-Based Approximations

So far, our focus has been on approximations based on local search. Here, we consider linear programming relaxations for L-Sided Pricing. Recall for each $u \in L$ that $P_u = \{b_e : e \in \delta(u)\}$ is a set of possible prices for vertex u and that there is an optimal solution that selects $p(u)$ from P_u for each $u \in L$.

For $u \in L$ and $p \in P_u$, we let $y_{u,p}$ be a variable indicating we select price p for u. Similarly, for each $e = uv \in E$ and $p \in P_u$ we let $x_{e,p}$ be a variable indicating edge e is selected and vertex u is assigned price p (so e buys their bundle at price p). The following relaxation provides an upper bound on the optimal solution to the given instance of the L-Sided Pricing.

$$\textbf{maximize}: \quad \sum_{e=uv} \sum_{p \in P_u} p \cdot x_{e,p} \qquad \textbf{(LP-Pricing)}$$

$$\begin{aligned} \textbf{subject to}: \quad & \sum_{p \in P_u} y_{u,p} = 1 && \forall\, u \in L \\ & \sum_{e \in \delta(u)} x_{e,p} \leq y_{u,p} \cdot \mu_u && \forall\, u \in L, p \in P_u \\ & \sum_{e=uv \in \delta(v)} \sum_{p \in P_u} x_{e,p} \leq \mu_v && \forall\, v \in R \\ & x_{e,p} \leq y_{u,p} && \forall\, u \in L, e \in \delta(u), p \in P_u \\ & x_{e,p} = 0 && \forall\, e = uv, p \in P_u \text{ s.t. } p > b_e \\ & x, y \geq 0 \end{aligned} \qquad (2)$$

A solution to L-Sided Pricing naturally corresponds to an integer solution, so the optimum LP value provides an upper bound on the optimum solution value.

Theorem 3 is proven through a simple rounding procedure. Independently for each $u \in L$, a single price $p'(u)$ for u is chosen at random from the distribution that places probability $y_{u,p}$ on each $p \in P_u$. As noted earlier, we can then compute an optimal set of customers by solving the corresponding maximum-profit μ-matching problem.

For the sake of space, we simply sketch the analysis. To bound the profit of the μ-matching, we construct a fractional μ-matching for these prices by assigning a fractional weight of x'_e of an customer $e \in \delta(u)$ with $p'(u) \leq b_e$ to $(1-\epsilon) \cdot \frac{x_{e,p'(u)}}{y_{u,p'(u)}}$. Constraint (2) ensures this fractional μ-matching x' does not violate the capacity of any $u \in L$. For $v \in R$, simplicity of the graph allows us to use Chernoff bounds. That is, the capacity constraint for v is violated by x' with low probability: at most $\exp(-\mu(v) \cdot \epsilon^2/3)$. Since $\mu(v)$ is sufficiently large, this is at most ϵ. If this rare event does occur, namely v's capacity constraint is violated, we simply reset x'_e to 0 for each $e \in \delta(v)$. The expected profit of the final fractional μ-matching is then at least $1 - 2 \cdot \epsilon$ times the optimum value of (LP-Pricing). By integrality of the μ-matching polytope, the expected profit from the final selection of customers obtained by solving the μ-matching problem is also at least this quantity.

Finally, to get the 4-approximation based on rounding this LP in the case where graphs are not simple, we use the same rounding algorithm and initially construct $x'_e := \frac{1}{2} \cdot \frac{x_{e,p'(u)}}{y_{u,p'(u)}}$ for $e \in \delta(u)$. By Markov's inequality, the probability any $v \in R$ has its capacity constraint violated by x' is at most $\frac{1}{2}$, in which case we reset $x'_e := 0$ for each $e \in \delta(v)$. The expected profit of this fractional μ-matching is at least $\frac{1}{4}$ of the optimum value of (LP-Pricing). Of course, this is inferior to our local search procedure but it does demonstrate the integrality gap remains bounded by a constant even if the graph is not simple.

4.1 Proof Sketch for Theorem 4

Here we combine the results from Theorem 2 and 3 to provide an improvement over the 2-approximation for the instances with uniform capacities. We begin with a more refined analysis of the randomized rounding procedure. We used simpler Chernoff bounds in the proof of Theorem 3 in order to present the dependence on ϵ in a simpler way. But since we are interested in optimal constants at this point, we analyze a tighter Chernoff bound. Our analysis may still not be optimal for our approach, it could be that one can get even better constants using more refined scrutiny of the randomized rounding algorithm for small C. Though, the constants we chose are optimal for our analysis technique.

Lemma 5. *The randomized rounding procedure produces a solution for* L-SIDED PRICING *with expected profit at least* $0.516 \cdot OPT_{LP}$ *in simple graphs where* $\mu_v \geq 22$ *for all* $v \in R$.

Proof (Proof of Theorem 4). If $C \leq 21$, use the multiswap local search algorithm to get a solution with profit $\geq \left(\frac{21}{41} - \epsilon\right) \cdot OPT$ for L-SIDED PRICING. If $C \geq 22$, use the randomized rounding procedure to get a solution whose cost is at least

$0.516 \cdot OPT$. For small enough ϵ, $0.516 > \frac{21}{41} - \epsilon$, so in either case we get profit at least $\left(\frac{21}{41} - \epsilon\right) \cdot OPT$. In terms of approximation guarantees, this yields an approximation guarantee of at most 1.952381 (again, for small enough ϵ). Using Lemma 1, we get a 7.8096-approximation for Capacitated Graph Pricing.

A Reduction to L-Pricing

Proof (Proof of Lemma 1). Let $G = (V, E)$ be an instance of Capacitated Graph Pricing with capacities μ and budgets b. Randomly form L by including each vertex independently with probability 1/2 and set $R = V - L$. Discard all edges with both endpoints in the same set of the partition. Consider an optimum pricing p^* for G and corresponding set of customers $F^* \subseteq E$. Let F' be the restriction of F^* to edges e with endpoints in each of L and R and consider prices p' where $p'(u) = p^*(u)$ for $u \in L$ and $p'(v) = 0$ for $v \in R$. One can easily check $\mathbf{E}\left[\sum_{e \in F'} p'(e)\right] = \frac{1}{4} \cdot \sum_{e \in F^*} p^*(e)$.

This can be efficiently derandomized because we only require pairwise independence of the events $u \in L$ for various $u \in V$, see [13] for details behind this technique.

References

1. Alekseeva, E., Kochetov, Y., Alexsandr, P.: Complexity of local search for the p-median problem. Eur. J. Oper. Res. **191**(3), 736–752 (2008)
2. Balcan, M.F., Blum, A.: Approximation algorithms and online mechanisms for item pricing. Theory Comput. **3**(9), 179–195 (2007)
3. Briest, P., Krysta, P.: Single-minded unlimited supply setting pricing on sparse instances. In: Proceedings of SODA, pp. 1093–1102 (2006)
4. Chalermsook, P., Chuzhoy, J., Kannan, S., Khanna, S.: Improved hardness results for profit maximization pricing problems with unlimited supply. In: Gupta, A., Jansen, K., Rolim, J., Servedio, R. (eds.) APPROX/RANDOM -2012. LNCS, vol. 7408, pp. 73–84. Springer, Heidelberg (2012). https://doi.org/10.1007/978-3-642-32512-0_7
5. Cheung, M., Swamy, C.: Approximation algorithms for single-minded envy-free profit-maximization problems with limited supply. In: Proceedings of FOCS, pp. 35–44 (2008)
6. Elbassioni, K., Fouz, M., Swamy, C.: Approximation algorithms for non-single-minded profit-maximization problems with limited supply. In: Saberi, A. (ed.) WINE 2010. LNCS, vol. 6484, pp. 462–472. Springer, Heidelberg (2010). https://doi.org/10.1007/978-3-642-17572-5_39
7. Friggstad, Z., Khodamoradi, K., Salavatipour, M.R.: Exact algorithms and lower bounds for stable instances of euclidean k-means. In: Proceedings of SODA, pp. 2958–2972 (2019)
8. Gamzu, I., Segev, D.: A sublogarithmic approximation for highway and tollbooth pricing. In: Abramsky, S., Gavoille, C., Kirchner, C., Meyer auf der Heide, F., Spirakis, P.G. (eds.) ICALP 2010. LNCS, vol. 6198, pp. 582–593. Springer, Heidelberg (2010). https://doi.org/10.1007/978-3-642-14165-2_49

9. Grandoni, F., Rothvoß, T.: Pricing on paths: a PTAS for the highway problem. In: Proceedings of SODA, pp. 675–684 (2011)
10. Guruswami, V., Hartline, J.D., Karlin, A.R., Kempe, D., Kenyon, C., McSherry, F.: On profit-maximizing envy-free pricing. In: Proceedings of the Sixteenth Annual ACM-SIAM Symposium on Discrete Algorithms, SODA 2005, Vancouver, BC, Canada, 23–25 January 2005, pp. 1164–1173 (2005)
11. Khandekar, R., Kimbrel, T., Makarychev, K., Sviridenko, M.: On hardness of pricing items for single-minded bidders. In: Dinur, I., Jansen, K., Naor, J., Rolim, J. (eds.) APPROX/RANDOM -2009. LNCS, vol. 5687, pp. 202–216. Springer, Heidelberg (2009). https://doi.org/10.1007/978-3-642-03685-9_16
12. Lee, E.: Hardness of graph pricing through generalized Max-Dicut. In: Proceedings of the Forty-seventh Annual ACM Symposium on Theory of Computing, pp. 391–399 (2015)
13. Motwani, R., Raghavan, P.: Randomized Algorithms. Cambridge University Press, Cambridge (1995)
14. Sachs, H.: Regular graphs with given girth and restricted circuits. J. Lond. Math. Soc. **1**(1), 423–429 (1963)

Fair Correlation Clustering with Global and Local Guarantees

Zachary Friggstad and Ramin Mousavi(✉)

Department of Computing Science, University of Alberta, Edmonton, AB, Canada
{zacharyf,mousavih}@ualberta.ca

Abstract. CORRELATION CLUSTERING is a model that aims to group items according to similarity and dissimilarity measures. In general, for a given set of items V we are given weights $w_{u,v}$ between items $u, v \in V$ indicating how similar they are: these weights can be negative, which indicates dissimilarity between the items. The objective is to partition the items V into groups to minimize the total weight of pairs u, v with $w_{u,v} < 0$ that are put in the same group plus the total weight of pairs u, v with $w_{u,v} > 0$ that are put in different groups (i.e. *violated* edges). In general, CORRELATION CLUSTERING is at least as hard to approximate as the MULTICUT problem but the important *unweighted complete* case where $w_{u,v} \in \{-1, +1\}$ for every distinct $u, v \in V$ admits constant-factor approximations.

More recently, attention has been drawn to *fair* clustering where items come with labels and clusters are further required to maintain proportional representation of the labels. Specifically, we consider the case of FAIR CORRELATION CLUSTERING where each item in V is either *red* or *blue* and each cluster should receive an equal number of red and blue points. In this setting, Ahmadi et al. (2020) show that an α-approximation for standard correlation clustering without the fairness constraint yields an $O(\alpha)$-approximation for FAIR CORRELATION CLUSTERING.

Our main results are twofold. First, we give an improved constant-factor approximation for FAIR CORRELATION CLUSTERING in unweighted settings. In this case, Ahmadi et al. give a 10.18-approximation. Our algorithm gives an improved 6.18-approximation. Further, we describe an alternative approach that seems to yield a 5.5-approximation: the analysis involves a computer-assisted verification of a bound. Second, we give the first constant-factor approximation where the objective is to minimize the maximum number of violated edges incident to any single vertex: FAIR CORRELATION CLUSTERING with local guarantees.

We also consider extensions to instances where each cluster should have a b-to-1 proportion of the two labels and give an $O(b^2)$-approximation for FAIR CORRELATION CLUSTERING with local guarantees.

Z. Friggstad—Research supported by an NSERC Discovery Grant and Discovery Accelerator Supplement Award.

A. Lubiw et al. (Eds.): WADS 2021, LNCS 12808, pp. 414–427, 2021.
https://doi.org/10.1007/978-3-030-83508-8_30

Keywords: Correlation clustering · Fair clustering · Local objective · Global objective

1 Introduction

In **Correlation Clustering** on unweighted complete graphs, we are given a graph $G = (V, E)$, and a partitioning of edges $E = E^- \sqcup E^+$. The goal is to find a partitioning $\mathcal{C} \subseteq 2^V$ (clustering) of V such that the total number of disagreement (unhappy) edges is minimized. That is, for a clustering $\mathcal{C}$ of V, we say $e = (u, v) \in E^+$ is unhappy if u and v belong to different parts of $\mathcal{C}$. Similarly, $e = (u, v) \in E^-$ is unhappy if u and v belong to the same part of $\mathcal{C}$. This definition generalizes to **uniformly-weighted Correlation Clustering**. Given a value c, each edge $e = (u, v) \in E$ has two weights $w^+_{u,v} \geq 0$ and $w^-_{u,v} \geq 0$ such that $w^+_{u,v} + w^-_{u,v} = c$. If u and v are in the same cluster, then the unhappiness of e is $w^-_{u,v}$ and if u and v lie in different clusters then the unhappiness of e is $w^+_{u,v}$. The objective is to minimize the total unhappiness of edges.

From theory side, correlation clustering is related to some fundamental problems in combinatorial optimization like MINIMUM $s - t$ CUT, MULTIWAY CUT, and MULTICUT problems [9]. And from the practical side, it has applications in areas like image segmenting, databases, and statistics [14,20], and [15].

Following a series of work [4,6,10], Chawla et al. [11] designed an approximation algorithm with guarantee 2.06 for the uniformly-weighted instances of CORRELATION CLUSTERING on complete graphs. One can consider a further generalization of CORRELATION CLUSTERING where the two weights of an edge are arbitrary. Charikar et al. [10] give an $O(\log n)$-approximation for this generalization, but also show it is (asymptotically) at least as hard to approximate as the MINIMUM MULTICUT problem, so any improvement would be a breakthrough.

Recently, the notion of fair clustering has been considered. Specifically, each vertex/data point in V also comes with a label $\sigma(v)$ from some given set of labels L. The goal is to find a clustering such that the proportion of each label type in each cluster equals (or is close to) the proportion of that label type in the entire input. That is, for each $\ell \in L$ if $\frac{1}{b_\ell} \cdot |V|$ many vertices in V have label ℓ then each cluster C should also have (exactly or approximately) $\frac{1}{b_\ell} \cdot |C|$ vertices with label ℓ. For example, such a model can be used to ensure proportional representation of demographics in each cluster.

This concept of fair clustering was first introduced by Chierichetti et al. [13]. Since this initial work, researchers have designed approximation algorithms for fair variants of fundamental problems like k-MEDIAN, k-MEANS, and k-CENTER [5,7,8,12,17]. Fair clustering has even been considered in k-MEANS clustering in the streaming model [19].

Ahmadi et al. [1] studied FAIR CORRELATION CLUSTERING, which is the fair version of CORRELATION CLUSTERING where exactly half of the vertices are labelled blue and half are labelled red. So the objective is to find a partitioning of the vertices such that the number of blue vertices and the number of red vertices are the same in each cluster. Their result for Fair Correlation Clustering is that an α-approximation for Correlation Clustering can be used as a black-box to obtain a $(3\alpha+4)$-approximation algorithm for Fair Correlation Clustering. Putting $\alpha = 2.06$, their algorithm is a 10.18-approximation algorithm. They also show that Fair Correlation Clustering is NP-complete.

More generally, Ahmadi et al. [1] also consider the case where we have k colors $c_1, \ldots, c_k$ and we are additionally given ratios $\frac{1}{b_2}, \ldots, \frac{1}{b_k}$. A feasible solution is a partitioning such that in each cluster the ratio of the number of c_1 color vertices to the number of c_i color vertices is $\frac{1}{b_i}$ for all $2 \le i \le k$. Note, the case discussed above is $k = 2$ (two colors) and $b_2 = 1$ (1-to-1 ratio in each cluster). They give an $O(\max_{i=1}^{k}\{b_i\} \cdot k^2)$. Recently, [3] considered a wide range of fairness concepts for correlation clustering and, similar to [1], they also use black-box reductions from fair version to the classical version of Correlation Clustering.

A different objective for Correlation Clustering was proposed first by Puleo and Milenkovic [18], where they consider partitioning V in a way that minimizes the maximum number of unhappy edges incident to a vertex: i.e. to minimize $\max_{v \in V} |\{e \in \delta(v) : e \text{ unhappy with } \mathcal{C}\}|$. This problem in literature sometimes is referred to as Min-Max Correlation Clustering or Correlation Clustering with Local Guarantees as we are more concerned with minimizing the maximum dissatisfaction incident to any single vertex. Like Correlation Clustering, this problem also generalizes natural graph cut problems, e.g., min-max $s-t$ cut, min-max multiway cut, and min-max multicut, please see [9] for more details and definitions of these problems. This variant of Correlation Clustering has received considerable attention as well: following a series of work [2,9,18], the state of the art for this problem is a 5-approximation algorithm given by Kalhan, Makarychev, and Zhou [16].

We also consider Local Fair Correlation Clustering, the fair variant of Correlation Clustering with local guarantees. Here, we are given a complete unweighted graph $G = (V, E)$ where each edge has label either $+$ or $-$. In addition, every vertex is assigned one of the k colors $c_1, \ldots, c_k$ and we are additionally given ratios $\frac{1}{b_2}, \ldots, \frac{1}{b_k}$. A feasible partitioning is a partitioning such that in each cluster the ratio of c_1 color vertices to c_i color vertices is $\frac{1}{b_i}$ for all $2 \le i \le k$. Again, the objective is to find a clustering that minimizes the Correlation Clustering with Local Guarantees objective.

1.1 Our Results

Unless stated otherwise, in each of our results exactly half of the vertices are blue and half of the vertices are red, and the goal is to find a clustering solution where each cluster has an equal number of red and blue vertices. Our first result is an improved approximation for Fair Correlation Clustering in unweighted complete graphs.

Theorem 1. *There is a 6.18-approximation algorithm for* Fair Correlation Clustering *on unweighted complete graphs.*

This is an improvement over the guarantee of 10.18 by Ahmadi et al. [1]. Until our paper, the approximation algorithms for FAIR CORRELATION CLUSTERING are obtained by using previous approximations for CORRELATION CLUSTERING as a black-box [1,3]. For our improvements, we leverage the fact that previous approximations rely on linear programming (LP) relaxations.

We start by considering a weighted bipartite matching instance between the red and blue vertices where the weights of the edges (u, v) are meant to reflect that some unhappy edges will be unavoidable if we choose to match u with v, so there will be an optimum matching that is comparable to OPT, the optimal FAIR CORRELATION CLUSTERING solution cost. We then contract all edges in a minimum-cost matching and obtain an instance of uniformly-weighted CORRELATION CLUSTERING with $c = 4$. Next, we show that an optimal LP solution x^* to the CORRELATION CLUSTERING instance that underlies the given FAIR CORRELATION CLUSTERING instance (i.e. if we ignore the fair requirement) can be converted to an LP solution $\overline{x}$ for this new uniformly-weighted CORRELATION CLUSTERING instance whose cost can be charged to the cost of x^* and the cost of the matching, i.e. $O(OPT)$. Then we use the LP rounding algorithm in [11] to get a CORRELATION CLUSTERING solution in this contracted graph, which is then a FAIR CORRELATION CLUSTERING solution in the original graph.

We show that it is possible to improve the approximation factor in Theorem 1 to 5.5. For the analysis, we delve into the LP-rounding algorithm from [4] for CORRELATION CLUSTERING. We point out that the analysis of 2.5-approximation factor for CORRELATION CLUSTERING on complete graphs has a lemma, in particular Lemma 18 in [4], that its verification requires a tedious calculations that is left out of the paper. This lemma will translate to our setting but its proof requires even more case analysis and each case involves a more complicated constrained optimization problem, see Sect. 2.3 for more details on this conjecture.

Conjecture 1. There is a 5.5-approximation algorithm for FAIR CORRELATION CLUSTERING on unweighted complete graphs.

Our second class of results pertain to LOCAL FAIR CORRELATION CLUSTERING.

Theorem 2. *There is a 80-approximation algorithm for* LOCAL FAIR CORRELATION CLUSTERING *on unweighted complete graphs.*

While this constant is a bit high, to the best of our knowledge no prior results were known for this problem. Our algorithm here behaves in a way that is similar to our algorithm for Theorem 1. Namely, we first construct a matching between the red and blue points where the weight of an edge (u, v) reflects the number of edges in G incident to u or v that will inevitably become unhappy if u and v are grouped together. We also show how an LP solution for the unfair version maps to an LP solution for the resulting instance in the contracted graph and use the LP-based rounding algorithm in [16] to get our final solution. However, placing a bound on the resulting LP solution encounters new challenges in this min-max setting. Namely, it could be that after contracting matched vertices u and v to a single node, the resulting LP solution has the contracted node $\{u, v\}$ being incident to a much larger amount of unhappy edges than either u or v saw in the original LP solution. We articulate how this could be the case, and show that there must then be another vertex w in G that was nearly as unhappy as the contracted node $\{u, v\}$ so the unhappiness of $\{u, v\}$ can still be bounded by $O(OPT)$.

Finally, we provide a generalization of our result for LOCAL FAIR CORRELATION CLUSTERING to instances with more colors and non-uniform requirements on the ratios of colors.

Theorem 3. *Consider an instance of* LOCAL FAIR CORRELATION CLUSTERING *where there are* k *colors and the ratios are* $\frac{1}{b_2}, \ldots, \frac{1}{b_k}$. *Let* $B := \sum_{i=2}^{k} b_i$. *Then, there is a* $(7 \cdot B^2 + 43 \cdot B + 30) \cdot \mathrm{OPT}$*-approximation algorithm for this problem.*

1.2 Organization

The proof of Theorem 1 appears in Sect. 2 followed by the proof of Theorem 2 in Sect. 3. For the sake of space and to provide clearer exposition of our main ideas, the proof of most of the lemmas, proof of Theorem 3, and a detailed discussion of Conjecture 1 are deferred to the full version of the paper.

2 Fair Correlation Clustering

In this section, the FAIR CORRELATION CLUSTERING instances we consider are presented as unweighted complete graph $G = (V, E)$, a partitioning of edges $E = E^- \sqcup E^+$, and a partitioning of vertices $V = V_R \sqcup V_B$ (i.e. the *red* and *blue* vertices). The *label* of an edge $e \in E$ is $-$ if $e \in E^-$ or $+$ if $e \in E^+$. The goal is to find a partitioning $\mathcal{C} \subseteq 2^V$ (clustering) of V such that each cluster has the same number of red and blue vertices, and the total number of *unhappy* edges is minimized. Given a clustering of V, we say $e = (u, v) \in E^+$ is unhappy if u and v belong to different clusters. Similarly $e = (u, v) \in E^-$ is unhappy if u and v belong to the same cluster.

We have the following LP_{cc} for classical fair correlation clustering. The variable $x_{u,v}$ indicates whether u and v are in the same cluster or not (0 if they are in the same cluster and 1 otherwise). The constraints ensure that x is a metric.

$$\begin{aligned} \min \quad & \sum_{(u,v)\in E^+} x_{u,v} + \sum_{(u,v)\in E^-} (1 - x_{u,v}) & & \\ \text{s.t.} \quad & x_{u,v} + x_{v,w} \geq x_{u,w} & \forall u, v \in V & \\ & x_{u,v} = x_{v,u} & \forall u, v \in V & \qquad (LP_{cc}) \\ & x_{u,u} = 0 & \forall u \in V & \\ & 0 \leq x_{u,v} \leq 1 & \forall (u,v) \in E & \end{aligned}$$

2.1 6.18-Approximation Algorithm

Consider an instance of FAIR CORRELATION CLUSTERING. Let OPT denote the optimal value of a clustering. Note that the optimal value of LP_{cc} is at most OPT. For three vertices u, v, and w such that $u \in V_R$ and $v \in V_B$, we say $\overset{\triangle}{uvw}$ is a ***bad triangle with the base*** (u, v) if the labels of (u, w) and (v, w) are different.

As in [1], we compute a minimum-cost matching of V_R with V_B in an auxiliary bipartite graph whose edge weights approximately reflect the number of edges that will be unhappy in any clustering that includes a matched pair in the same cluster. In this auxiliary graph, for each pair of vertices u, v where $u \in V_R$ and $v \in V_B$ we add an edge (u, v) and we set

$$w_{u,v} = \begin{cases} (\#\text{ of bad triangles with base } (u,v)) & \text{if } (u,v) \in E^+ \\ 1 + (\#\text{ of bad triangles with base } (u,v)) & \text{if } (u,v) \in E^- \end{cases}$$

We have the following upper bound on the cost of the minimum cost perfect matching in the auxiliary graph. For brevity, let $w(M) = \sum_{(u,v)\in M} w_{u,v}$ denote the cost of a matching.

Lemma 1 (Lemma 1 in [1]). *Let M be a minimum-cost perfect matching in the auxiliary graph. Then $w(M) \leq 2 \cdot \text{OPT}$.*

We note that our use of M differs in a key way from their algorithm. We first compute a matching and then use this to define an auxiliary CORRELATION CLUSTERING instance obtained by contracting the endpoints of matching edges into a single node. We then apply a CORRELATION CLUSTERING approximation on this graph, but our analysis does not use CORRELATION CLUSTERING approximations as a black-box: we need to expand on details that are unique to our setting to complete the proof. By way of contrast, their algorithm uses CORRELATION CLUSTERING approximations as a black-box on the subgraph induced by V_R and, independently, computes the minimum-cost matching in this auxiliary graph and places each $v \in V_B$ in the same cluster as its matched counterpart in V_R.

So, given a minimum-cost matching M we build an instance of uniformly-weighted CORRELATION CLUSTERING. The graph, which we call the *contracted graph*, is $G_{contracted} = (V_{super}, E_{super})$. Here, V_{super} is the set of vertices obtained by contracting the matching edges (u, v) into a single node $\{u, v\}$, we call these *super nodes*. For each super node $i = \{u, v\}$ and $j = \{u', v'\}$ we have an edge (i, j) in E_{super} and call these edges *super edges*. There are four edges in G *associated* with the super edge (i, j), namely, (u, u'), (u, v'), (v, u'), and (v, v'). Let $w^+_{i,j}$ be the number of $+$ edges among these four edges, and let $w^-_{i,j}$ be the number of $-$ edges among these four edges, see Fig. 1. Note $w^+_{i,j} + w^-_{i,j} = 4$.

Let x^* be an optimal solution to LP_{cc}. Let $i = \{u, v\}$ and $j = \{u', v'\}$ be two super nodes in $G_{contracted}$. We define a feasible solution $\overline{x}$ to LP_{wcc}, the standard LP for the weighted CORRELATION CLUSTERING, on the auxiliary graph by taking the average of the four edges associated to (i, j), more precisely, $\overline{x}_{i,j} :=$

$\frac{x^*_{u,u'}+x^*_{u,v'}+x^*_{v,u'}+x^*_{v,v'}}{4}$. It is easy to see that $\overline{x}$ satisfies the triangle inequalities in $G_{contracted}$ and thus it is a feasible solution for LP_{wcc}.

$$\begin{aligned} \min \quad & \sum_{(i,j)\in E_{super}} w^+_{i,j}x_{i,j} + w^-_{i,j}(1-x_{i,j}) & & \\ \text{s.t. } & x_{i,j} + x_{j,k} \geq x_{i,k} & \forall i,j,k \in V_{super} & \\ & x_{i,j} = x_{j,i} & \forall i,j \in V_{super} & \quad (LP_{wcc}) \\ & x_{i,i} = 0 & \forall i \in V_{super} & \\ & 1 \leq x_{i,j} \leq 0 & \forall (i,j) \in E_{super} & \end{aligned}$$

Finally, we use the LP-based CORRELATION CLUSTERING approximation from [11] on $G_{contracted}$ to get a clustering $\mathcal{C}'$. The output of our algorithm is obtained by replacing each super node in the clustering $\mathcal{C}'$ by both of the original nodes it represents. Algorithm 1 summarizes these steps.

Algorithm 1: Algorithm for FAIR CORRELATION CLUSTERING

1: Compute a minimum-cost perfect matching M on the auxiliary graph and form $G_{contracted}$ as above.
2: Compute an optimal solution x^* of LP_{cc} and set $\overline{x}$ as above for each super edge (i,j) of $G_{contracted}$.
3: Run the LP-based 2.06-approximation from [11] for CORRELATION CLUSTERING on $G_{contracted}$ using LP solution $\overline{x}$ to get a clustering $\mathcal{C}'$ (see Theorem 4 below).
4: Return the clustering $\mathcal{C} = \{\phi(C') : C' \in \mathcal{C}'\}$ where $\phi(C') = \cup_{i=\{u,v\}\in C'}\{u,v\}$.

2.2 Analysis of 6.18-Approximation Algorithm

For a partitioning $\mathcal{C}$ of the vertices in a CORRELATION CLUSTERING or FAIR CORRELATION CLUSTERING instance, let Cost($\mathcal{C}$) denote the cost of this partitioning (i.e. the total number, or weight, of unhappy edges). Let $\mathcal{C}$ be the output of Algorithm 1 and $\mathcal{C}'$ the clustering of $G_{contracted}$ from Step 3. We first relate Cost($\mathcal{C}$) and Cost($\mathcal{C}'$).

For each super node $i = \{u,v\}$, if $(u,v) \in E^-$, then (u,v) contributes 1 unit to M. Let M^- denote the sum of all such contributions for all $(i,j) \in E_{super}$. Let $M_{i,j}$ be the number of bad triangles "involved" in super edge (i,j), i.e., if $w^+_{i,j} = 1,\ 2,\ or\ 3$ then $M_{i,j} = 2$ and $M_{i,j} = 0$ otherwise. This is because when $w^+_{i,j} \neq 0$ and $w^+_{i,j} \neq 4$, then there are exactly 2 bad triangles whose base lies in $\{(u,v),(u',v')\}$ that contribute to $w_{u,v}$ or $w_{u',v'}$. For example, in Fig. 1, $\overset{\triangle}{uu'v}$ and $\overset{\triangle}{u'uv'}$ are two bad triangles that contribute one unit to $w_{u,v}$ and one unit to $w_{u'v'}$, respectively. In this way, we have $w(M) = M^- + \sum_{(i,j)\in E_{super}} M_{i,j}$.

Lemma 2. Cost($\mathcal{C}$) = Cost($\mathcal{C}'$) + M^-.

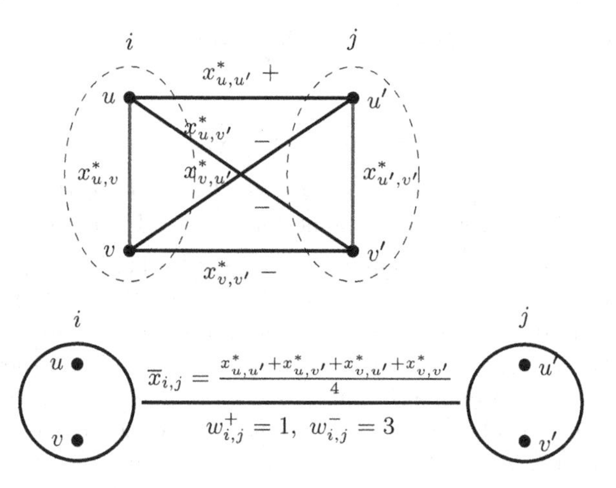

Fig. 1. In the top picture, the red edges are in the matching M. x^* is an optimal solution for LP_{cc}. i and j are the super nodes obtained by contracting (u, v) and (u', v') respectively. The bottom picture shows the resulting graph after contraction. Edges $(u, u'), (u, v'), (v, u')$, and (v, v') are edges in G associated to the super edge (i, j).

We proceed to bound $\text{Cost}(\mathcal{C}')$. Let $value(x^*)$ denotes the objective value of the LP_{cc} under solution x^*. Note that $value(x^*) \leq \text{OPT}$ where OPT is the optimal value for the Fair Correlation Clustering on G.

Lemma 3. *Let $M_{i,j}$ be the contribution of super edge (i, j) to the matching (so $M_{i,j}$ is either* 0 *or* 2*). Then,* $\text{Cost}(\mathcal{C}') \leq 2.06 \cdot (value(x^*) + \sum_{(i,j)\in E_{super}} M_{i,j})$.

This is proven below. For now, we show how to complete the analysis.

Proof (of Theorem 1*).* By Lemmas 2 and 3, we have $\text{Cost}(\mathcal{C}) \leq 2.06 \cdot \big(value(x^*) + w(M) - M^-\big) + M^-$ (recall $w(M) = M^- + \sum_{(i,j)\in E_{super}} M_{i,j}$). Then by Lemma 1 we see $\text{Cost}(\mathcal{C}) \leq 2.06 \cdot (\text{OPT} + 2 \cdot \text{OPT}) = 6.18 \cdot \text{OPT}$. □

So it remains to prove Lemma 3. We use the following result in [11].

Theorem 4 (Theorem 1 & 20 in [11], paraphrased). *Given a feasible solution $\overline{x}$ to LP_{wcc} for a uniformly-weighted* Correlation Clustering *instance, there is a polynomial-time algorithm that outputs a partitioning $\mathcal{C}'$ such that* $\text{Cost}(\mathcal{C}') \leq 2.06 \cdot value(\overline{x})$.

Remark 1. In [11], their result are stated for instances when $w^+_{i,j} + w^-_{i,j} = 1$ even if the weights are not integers, by scaling weights this holds when $w^+_{i,j} + w^-_{i,j} = c$ for any fixed c.

The proof of Lemma 3 follows immediately from the following bound and Theorem 4.

Lemma 4. *value*$(\overline{x}) \leq$ *value*$(x^*) + \sum_{(i,j)\in E_{super}} M_{i,j}$.

2.3 Towards a 5.5-Approximation Algorithm

We modify Algorithm 1 by introducing a small change in the auxiliary graph and using a particular rounding scheme for the LP. First, we add 0.6 to $w_{u,v}$ (instead of 1) for each bad triangle $\overset{\triangle}{uvz}$ where $z \in V$. Therefore, for a super edge (i,j) such that $w^+_{i,j} = 1,\ 2,\ or\ 3$, we have $M_{i,j} = 1.2$ (instead of 2). Then, we can show that a minimum-cost perfect matching in the auxiliary graph has value at most $1.2 \cdot \mathrm{OPT}$ with the same reasoning as in Lemma 1. The other difference is that we use the 2.5-approximation algorithm [4] in line 3 in Algorithm 1 instead of 2.06-approximation algorithm. For the analysis, we modify the analysis of 2.5-approximation for Correlation Clustering [4] to be suitable in our setting. See the full version of the paper for detailed description of the algorithm and the analysis. One of the lemmas in the analysis requires case analysis (5^3 cases, one could be smart and reduce the number of cases using symmetries but still the number of remained cases is very large) and each case requires finding the minimum of a degree 3 multivariate polynomial (with 12 variables) subject to linear constraints.

So we use a MATLAB program to verify the bound holds in each case (the cases are easy to enumerate). The MATLAB program requires us to try different starting points to find minima: we tested with a variety of starting points and the desired bound held each time. But since we cannot exhaustively search all the starting points, we state the 5.5-approximation algorithm as a conjecture for now and are working towards completing a more rigorous proof for the full version of this work.

3 Local Fair Correlation Clustering

We now consider Local Fair Correlation Clustering. The input is identical to Fair Correlation Clustering and a feasible solution is a fair partitioning $\mathcal{C}$ of $V = V_R \sqcup V_B$. But now we define $\mathrm{Cost}(\mathcal{C})$ differently, here: $\mathrm{Cost}(\mathcal{C}) = \max_{v \in V} |\{e \in \delta(v) : e \text{ unhappy with } \mathcal{C}\}|$. That is, the unhappiness of a vertex v is the number of unhappy edges incident to v. The cost of a solution is then the maximum unhappiness over all nodes. We use a natural LP relaxation for this problem, in it we have variables $x_{u,v}$ for $(u,v) \in E$ indicating whether u and v are in the different clusters or not and variables D_u for $u \in V$ counting the number of unhappy edges incident to u.

$$\begin{aligned}
\min\ & \max_{u \in V} D_u \\
\text{s.t.}\ & \sum_{v:(u,v)\in E^+} x_{u,v} + \sum_{v:(u,v)\in E^-} (1 - x_{u,v}) = D_u \quad \forall u \in V \\
& x_{u,w} \leq x_{u,v} + x_{v,w} \qquad \forall u,v,w \in V \\
& x_{u,v} = x_{v,u} \qquad \forall u,v \in V \\
& 0 \leq x \leq 1
\end{aligned} \qquad (LP_{local})$$

To turn this into a proper linear program, we can have the objective function minimize a new variable D subject to $D \geq D_u$ for each $u \in V$. As before, we rely on known integrality gap bounds for this LP relaxation in our algorithm.

Theorem 5 (Theorem 5.1 in [16]). *Given a feasible solution $\overline{x}$ to LP_{local}, there is a rounding procedure that outputs a partitioning $\mathcal{C}'$ such that* $\text{Cost}(\mathcal{C}') \leq 5 \cdot value(\overline{x})$.

Denote by OPT the optimal value of the LOCAL FAIR CORRELATION CLUSTERING instance. Our algorithm starts by considering the same auxiliary bipartite graph between V_R and V_B as in Sect. 2.

Lemma 5. *There is a perfect matching in the auxiliary graph such that* $\max_{(u,v)\in M} w_{u,v} \leq 2 \cdot \text{OPT}$.

We can guess the value of OPT and delete all the edges in the auxiliary graph that have weight more than twice of our guess, and then find a perfect matching M. So from now on we assume we have a matching M such that $w_{u,v} \leq 2 \cdot \text{OPT}$ for all $(u, v) \in M$.

Given a matching M and graph G, we build an instance of CORRELATION CLUSTERING WITH LOCAL GUARANTEES on a complete graph called *contracted graph* $\text{G}_{contracted} = (V_{super}, E_{super})$ similar to before. Here, V_{super} is the set of vertices obtained by contracting the matching edges, we call these *super nodes* and for each pair of distinct super nodes $S_{u,v} = \{u, v\}$ and $S_{u',v'} = \{u', v'\}$. we add an edge $(S_{u,v}, S_{u',v'})$ between them to E_{super} and call these edges *super edges.* However, instead of assigning weights to super edges we simply label them with $-$ or $+$. There are four edges in G *associated* with the super edge $(S_{u,v}, S_{u',v'})$, namely (u, u'), (u, v'), (v, u'), and (v, v'). The label of $(S_{u,v}, S_{u',v'})$ is the majority of the labels of these four edges of G, using $+$ if there is no majority label[1]. See Fig. 2 for an illustration.

Given an optimal solution x^* to LP_{local} on G, we build a solution $\overline{x}$ for LP_{local} on $\text{G}_{contracted}$ as follows: Let $\overline{x}_{S_{u,v},S_{u',v'}}$ be the average of the four edges in G associated to $(S_{u,v}, S_{u',v'})$, i.e., $\overline{x}_{S_{u,v},S_{u',v'}} := \frac{x^*_{u,u'}+x^*_{u,v'}+x^*_{v,u'}+x^*_{v,v'}}{4}$. It is easy to see that $\overline{x}$ satisfies the triangle inequalities and thus it is feasible for LP_{local} on $\text{G}_{contracted}$.

Our algorithm then uses Theorem 5 to round this LP solution to get a FAIR CORRELATION CLUSTERING solution in $\text{G}_{contracted}$. These super nodes in each

[1] In this case it does not matter if we use $+$ or $-$.

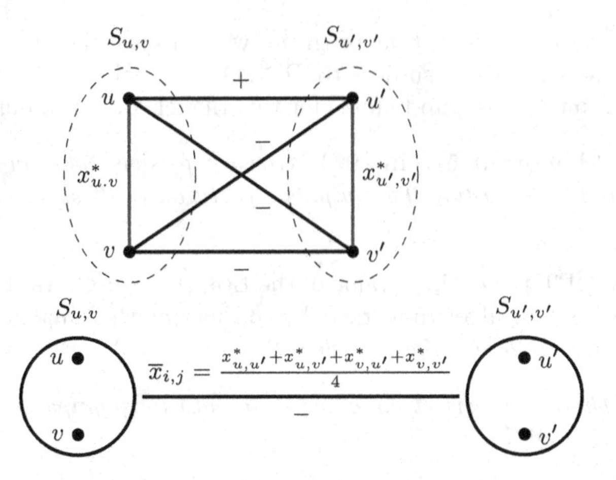

Fig. 2. In the top picture, the red edges are in the matching M. x^* is an optimal solution for LP_{local}. $S_{u,v}$ and $S_{u',v'}$ are the super nodes obtained by contracting (u,v) and (u',v'), respectively. The bottom picture shows the resulting graph after contraction. Edges $(u,u'), (u,v'), (v,u')$, and (v,v') are edges in G associated to the super edge $(S_{u,v}, S_{u',v'})$. Its label is $-$ because the majority label among the labels of the four edges in G associated to $(S_{u,v}, S_{u',v'})$ is $-$.

cluster are then expanded so that we get a clustering in the original instance G. Algorithm 2 summarizes these steps.

Algorithm 2: Algorithm for Local Fair Correlation Clustering

1: Compute a perfect matching M that minimizes $\max_{(u,v)\in M} w_{u,v}$.
2: Compute an optimal solution x^* of LP_{local} and set $\overline{x}$ as above for each super edge (i,j) of $G_{contracted}$.
3: Run the LP-based 5-approximation algorithm of [16] on $(G_{contracted}, \overline{x})$ to get a clustering $\mathcal{C}'$.
4: Return the clustering $\mathcal{C} = \{\phi(C') : C' \in \mathcal{C}'\}$ where $\phi(C') = \cup_{i=\{u,v\}\in C'}\{u,v\}$.

3.1 Analysis of Algorithm 2

We let $\mathcal{C}$ be the output of Algorithm 2 and $\mathcal{C}'$ the clustering found in step 3. The main step in our analysis is bounding the value of $\overline{x}$ constructed in line 2. For this, we need to construct new arguments that are specific to our setting: edges of E_{super} are classified depending on both how they were labelled and also on how the LP weight is distributed across their underlying edges of G.

Let $\overline{u} \in V$ be a vertex that is incident to a maximum number of unhappy edges with respect to $\mathcal{C}$, we need to bound the number of such unhappy edges incident to $\overline{u}$. Without loss of generality, say $\overline{u} \in V_{red}$ and let $\overline{v} \in V_B$ be the node that $\overline{u}$ is matched with in M. We group super edges that contain $\overline{u}$, i.e.,

$(S_{\overline{u},\overline{v}}, S_{u',v'})$ for all $(u', v') \in M$ into four types. Below, $0 < \epsilon \leq 1/2$ is a quantity that we will fix later.

So, consider a super edge $e' = (S_{\overline{u},\overline{v}}, S_{u',v'})$. We assign it a type according to the following cases. For brevity, for an edge $(u, v) \in E$ (in the underlying graph) we let $\ell(u, v) \in \{-, +\}$ denote the label of the edge.

- **Type 1:** $\ell(\overline{u}, u') = \ell(\overline{u}, v') = \ell(\overline{v}, u') = \ell(\overline{v}, v')$. That is, all the four edges associated with e' have the same label.
- **Type 2:** Both properties below are satisfied.
  (i) $\ell(\overline{u}, u') = \ell(\overline{v}, u')$ and $\ell(\overline{u}, v') = \ell(\overline{v}, v')$, yet $\ell(\overline{u}, u') \neq \ell(\overline{u}, v')$
  (ii) $(\overline{u}, u')$ is a $+$ edge and either $x^*_{\overline{u},u'} \geq \epsilon$ or $x^*_{\overline{u},v'} \leq 1 - \epsilon$ **or** $(\overline{u}, u')$ is a $-$ edge and either $x^*_{\overline{u},u'} \leq 1 - \epsilon$ or $x^*_{\overline{u},v'} \geq \epsilon$.
- **Type 3:** Only property (i) of type 2 edges is satisfied.
- **Type 4:** All super edges that are not of types 1, 2, or 3.

In the following we bound the total number of super edges involving $\overline{u}$ of type 2, 3, and 4 by $O(1) \cdot \mathrm{OPT}$, this will even include such edges that are happy with $\mathcal{C}$. Then, we show that the number of type 1 edges involving $\overline{u}$ that are unhappy with $\mathcal{C}$ is at most $O(1) \cdot \mathrm{OPT}$. Recall $D_{\overline{u}}$ is the fractional number of unhappy edges incident to $\overline{u}$ with respect to solution x^*, see LP_{local}.

Lemma 6. *Let k_2 be the number of edges of type 2. Then, $k_2 \leq \frac{1}{\epsilon} \cdot \mathrm{OPT}$.*

Lemma 7. *Let k_3 be the number of edges of type 3. Then, we have $\mathrm{OPT} \geq \frac{1-2\cdot\epsilon}{4(1-\epsilon)} \cdot k_3$.*

If we set $\epsilon = \frac{1}{4}$, then we have the following upper bound on the total number of super edges of type 2 and 3:

Corollary 1. $k_2 + k_3 \leq 10 \cdot \mathrm{OPT}$.

Lemma 8. *Let k_4 be the number of edges of type 4. Then, $k_4 \leq 2 \cdot \mathrm{OPT}$.*

Note that the number of unhappy edges incident to $\overline{u}$ with respect to $\mathcal{C}$ that comes from super edges of type 2 and 3 is exactly equal to the number of these super edges. Because, for each such super edge, there are two associated edges in G incident to $\overline{u}$ with opposite labels. So no matter how we cluster, exactly one of these two edges will be unhappy. Also the number of unhappy edges incident to $\overline{u}$ that comes from super edges of type 4 is at most twice the number of such super edges, since for each such super edge, there are two edges incident to $\overline{u}$ in G and we can make both of them unhappy at worst case. In summary, we have the following fact.

Corollary 2. *The number of unhappy edges incident to $\overline{u}$ with respect to $\mathcal{C}$ that comes from super edges of type 2, 3, and 4 is at most $k_2 + k_3 + 2 \cdot k_4 \leq 10 \cdot \mathrm{OPT} + 4 \cdot \mathrm{OPT} = 14 \cdot \mathrm{OPT}$.*

Define $\mathrm{Disagree}_{\mathrm{type1}}(\overline{u})$ to be the number of unhappy edges incident to $\overline{u}$ with respect to $\mathcal{C}$ that come from super edges of type 1.

Lemma 9. $\text{Disagree}_{\text{type1}}(\overline{u}) \leq 65 \cdot \text{OPT}$.

Now we can state the final result.

Proof (of Theorem 2). Combining Corollary 2 and Lemma 9 we have the total disagreement on $\overline{u}$ with respect to $\mathcal{C}$ without considering edge $(\overline{u}, \overline{v})$ is at most $79 \cdot \text{OPT}$ and if $(\overline{u}, \overline{v})$ is an unhappy edge then a the total disagreement is at most $79 \cdot \text{OPT} + 1$.

Since we picked $\overline{u}$ to be a vertex with maximum unhappy edges incident to it, this is a 80-approximation for LOCAL FAIR CORRELATION CLUSTERING assuming $\text{OPT} \geq 1$ (see the remark below regarding this assumption). □

Remark 2. We assumed $\text{OPT} \geq 1$. If $\text{OPT} = 0$, then contract each + edge. This rise to a natural clustering and each cluster must be fair too.

This approach generalizes to the setting with multiple colors in different ratios. See the full version of the paper for the proof of Theorem 3.

References

1. Ahmadi, S., Galhotra, S., Saha, B., Schwartz, R.: Fair correlation clustering. arXiv preprint arXiv:2002.03508 (2020)
2. Ahmadi, S., Khuller, S., Saha, B.: Min-Max correlation clustering via MultiCut. In: Lodi, A., Nagarajan, V. (eds.) IPCO 2019. LNCS, vol. 11480, pp. 13–26. Springer, Cham (2019). https://doi.org/10.1007/978-3-030-17953-3_2
3. Ahmadian, S., Epasto, A., Kumar, R., Mahdian, M.: Fair correlation clustering. In: International Conference on Artificial Intelligence and Statistics. pp. 4195–4205. PMLR (2020)
4. Ailon, N., Charikar, M., Newman, A.: Aggregating inconsistent information: ranking and clustering. J. ACM **55**(5), 1–27 (2008)
5. Backurs, A., Indyk, P., Onak, K., Schieber, B., Vakilian, A., Wagner, T.: Scalable fair clustering. In: International Conference on Machine Learning, pp. 405–413 (2019)
6. Bansal, N., Blum, A., Chawla, S.: Correlation clustering. Mach. Learn. **56**(1–3), 89–113 (2004)
7. Bera, S., Chakrabarty, D., Flores, N., Negahbani, M.: Fair algorithms for clustering. In: Advances in Neural Information Processing Systems, pp. 4954–4965 (2019)
8. Bercea, I.O., et al.: On the cost of essentially fair clusterings. arXiv preprint arXiv:1811.10319 (2018)
9. Charikar, M., Gupta, N., Schwartz, R.: Local guarantees in graph cuts and clustering. In: Eisenbrand, F., Koenemann, J. (eds.) IPCO 2017. LNCS, vol. 10328, pp. 136–147. Springer, Cham (2017). https://doi.org/10.1007/978-3-319-59250-3_12
10. Charikar, M., Guruswami, V., Wirth, A.: Clustering with qualitative information. J. Comput. Syst. Sci. **71**(3), 360–383 (2005)
11. Chawla, S., Makarychev, K., Schramm, T., Yaroslavtsev, G.: Near optimal LP rounding algorithm for correlation clustering on complete and complete k-partite graphs. In: Proceedings of the Forty-Seventh Annual ACM Symposium on Theory of Computing, pp. 219–228 (2015)

12. Chen, X., Fain, B., Lyu, L., Munagala, K.: Proportionally fair clustering. In: International Conference on Machine Learning, pp. 1032–1041 (2019)
13. Chierichetti, F., Kumar, R., Lattanzi, S., Vassilvitskii, S.: Fair clustering through fairlets. In: Advances in Neural Information Processing Systems, pp. 5029–5037 (2017)
14. Dwork, C., Kumar, R., Naor, M., Sivakumar, D.: Rank aggregation methods for the web. In: Proceedings of the 10th International Conference on World Wide Web, pp. 613–622 (2001)
15. Fagin, R., Kumar, R., Sivakumar, D.: Efficient similarity search and classification via rank aggregation. In: Proceedings of the 2003 ACM SIGMOD International Conference on Management of Data, pp. 301–312 (2003)
16. Kalhan, S., Makarychev, K., Zhou, T.: Correlation clustering with local objectives. In: Advances in Neural Information Processing Systems, pp. 9346–9355 (2019)
17. Kleindessner, M., Samadi, S., Awasthi, P., Morgenstern, J.: Guarantees for spectral clustering with fairness constraints. In: International Conference on Machine Learning, pp. 3458–3467 (2019)
18. Puleo, G., Milenkovic, O.: Correlation clustering and biclustering with locally bounded errors. In: International Conference on Machine Learning, pp. 869–877. PMLR (2016)
19. Schmidt, M., Schwiegelshohn, C., Sohler, C.: Fair coresets and streaming algorithms for fair k-means. In: Bampis, E., Megow, N. (eds.) WAOA 2019. LNCS, vol. 11926, pp. 232–251. Springer, Cham (2020). https://doi.org/10.1007/978-3-030-39479-0_16
20. Wirth, A.: Correlation clustering. In: Sammut, C., Webb, G.I. (eds.) Encyclopedia of Machine Learning and Data Mining, pp. 280–284. Springer, Boston (2017). https://doi.org/10.1007/978-1-4899-7687-1_176

Better Distance Labeling for Unweighted Planar Graphs

Paweł Gawrychowski(✉) and Przemysław Uznański

Institute of Computer Science, University of Wrocław, Wrocław, Poland
gawry@cs.uni.wroc.pl

Abstract. A distance labeling scheme is an assignment of labels, that is, binary strings, to all nodes of a graph, so that the distance between any two nodes can be computed from their labels without any additional information about the graph. The goal is to minimize the maximum length of a label as a function of the number of nodes. A major open problem in this area is to determine the complexity of distance labeling in unweighted planar (undirected) graphs. It is known that, in such a graph on n nodes, some labels must consist of $\Omega(n^{1/3})$ bits, but the best known labeling scheme constructs labels of length $\mathcal{O}(\sqrt{n}\log n)$ [Gavoille, Peleg, Pérennes, and Raz, J. Algorithms, 2004]. For weighted planar graphs with edges of length polynomial in n, we know that labels of length $\Omega(\sqrt{n}\log n)$ are necessary [Abboud and Dahlgaard, FOCS 2016]. Surprisingly, we do not know if distance labeling for weighted planar graphs with edges of length polynomial in n is harder than distance labeling for unweighted planar graphs. We prove that this is indeed the case by designing a distance labeling scheme for unweighted planar graphs on n nodes with labels consisting of $\mathcal{O}(\sqrt{n})$ bits with a simple and (in our opinion) elegant method. We augment the construction with a mechanism that allows us to compute the distance between two nodes in only polylogarithmic time while increasing the length by $\mathcal{O}(\sqrt{n\log n})$. The previous scheme required $\Omega(\sqrt{n})$ time to answer a query in this model.

1 Introduction

An informative labeling scheme is an elegant formalization of the idea that identifiers of nodes in a network can be chosen to carry some additional information. Peleg [30] defined such a scheme for a function f defined on subsets of nodes to consist of two components: an encoder and a decoder. First, the encoder is given a description of the whole graph G and assigns a binary string to each of its nodes. The string assigned to a node is called its label. Second, the decoder is given the labels assigned to a subset of nodes W and needs to calculate $f(W)$. This must be done without any information about the graph except for the given labels and the fact that G belongs to a specific family $\mathcal{G}$. The main goal is to make the labels as short as possible, that is, to minimize the maximum length of a label assigned to a node in G. A particularly clean example of a function f that one might want to consider in this model is adjacency. Kannan et al. [22]

A. Lubiw et al. (Eds.): WADS 2021, LNCS 12808, pp. 428–441, 2021.
https://doi.org/10.1007/978-3-030-83508-8_31

observed that an adjacency labeling scheme is related (in fact, equivalent) to a so-called vertex-induced universal graph, a purely combinatorial object that has been considered already in the 60s [29]. By now, we have a rich body of work concerning not only adjacency labeling [3,4,8–10,13,31], but also flows and connectivity labeling [20,23,25], Steiner tree labeling [30] and distance labeling.

Distance Labeling. A distance labeling scheme is an assignment of labels, that is, binary strings, to all nodes of a graph G, so that the distance $\delta_G(u,v)$ between any two nodes u,v can be computed from their labels. Unless specified otherwise, we consider unweighted graphs, so $\delta_G(u,v)$ is the smallest number of edges on a path between u and v. The main goal is to make the labels as short as possible, that is, to minimize the maximum length of a label. The secondary goal is to optimize the query time, that is the time necessary to compute $\delta_G(u,v)$ given the labels of u and v. Distance labeling for general unweighted undirected graphs on n nodes was first considered by Graham and Pollak [19], who obtained labels consisting of $\mathcal{O}(n)$ bits. The decoding time was subsequently improved to $\mathcal{O}(\log\log n)$ by Gavoille et al. [16], then to $\mathcal{O}(\log^* n)$ by Weimann and Peleg [32], and finally Alstrup et al. [6] obtained $\mathcal{O}(1)$ decoding time with labels of length $\frac{\log 3}{2}n+o(n)$.[1] It is known that some labels must consist of at least $\frac{n}{2}$ bits [22,29].

Better schemes for distance labeling are known for restricted classes of graphs. As a prime example, trees admit a distance labeling scheme with labels of length $\frac{1}{4}\log^2 n + o(\log^2 n)$ bits [15], and this is known to be tight up to lower-order terms [7]. In fact, any sparse graph admits a sublinear distance labeling scheme [5] (see also [17] for a somewhat simpler construction). However, the best known upper bound is still rather far away from the best known lower bound of $\Omega(\sqrt{n})$ [16], and recently Kosowski et al. [26] showed that, for a natural class of schemes based on storing the distances to a carefully chosen set of hubs, the best achievable hub-label size and distance-label size in sparse graphs may be $\Theta(n/2^{(\log n)^c})$ for some $0 < c < 1$.

Planar Graphs. An important subclass of sparse graphs are planar graphs, for which Gavoille et al. [16] constructed a scheme with labels of length $\mathcal{O}(\sqrt{n}\log n)$. They also proved that in any such scheme some label must consist of $\Omega(n^{1/3})$ bits. In fact, their upper bound of $\mathcal{O}(\sqrt{n}\log n)$ bits is also valid for weighted planar graphs, under a natural assumption that the weights are bounded by a polynomial in n. The lower bound is based on designing a family of grid-like graphs on $k \times k$ nodes and each edge being of length $\mathcal{O}(k)$. The family consists of $2^{\Theta(k^2)}$ graphs and admits the following property: the pairwise distances of $\mathcal{O}(k)$ nodes on the boundaries uniquely determine the graph. This construction immediately implies that, for weighted planar graphs, there must be a node with label consisting of $\Omega(\sqrt{n})$ bits. However, for unweighted planar graphs, this only implies a lower bound of $\Omega(n^{1/3})$, as one needs to replace an edge of length ℓ with ℓ edges, thus increasing the size of the graph to k^3. Abboud and Dahlgaard [1] extended this construction to show that, in fact, for graphs with

[1] All logarithms are in base 2.

the length of each edge bounded by a polynomial in n, there must be a node with label consisting of $\Omega(\sqrt{n}\log n)$ bits. Interestingly, they were able to use essentially the same construction to establish a strong conditional lower bound for dynamic planar graph algorithms. Unfortunately, there has been no progress in improving the construction for unweighted planar graphs.

Abboud et al. [2] provided a reasonable explanation for the lack of progress on improving the unweighted grid-like construction. They showed that for any unweighted planar graph G with k distinguished nodes, there is an encoding consisting of $\tilde{\mathcal{O}}(\min\{k^2, \sqrt{k \cdot n}\})$ bits that allows calculating the distance between any pair of distinguished nodes. This implies that the approach based on fixing a family $\mathcal{G}$ of unweighted planar graphs, with each graph containing k distinguished nodes such that their pairwise distance uniquely determine $G \in \mathcal{G}$, cannot result in a higher lower bound than $\tilde{\mathcal{O}}(\min\{k^2, \sqrt{k \cdot n}\})/k = \tilde{\mathcal{O}}(n^{1/3})$. This indicates that we should seek a significantly different proof technique or a better upper bound. Determining the complexity of distance labeling in unweighted planar graphs remains to be a major open problem in this area. The current state of our knowledge is summarized below.

Class of planar graphs	Lower bound	Upper bound
Lengths polynomial in n	$\Omega(\sqrt{n}\log n)$	$\mathcal{O}(\sqrt{n}\log n)$
Unweighted	$\Omega(n^{1/3})$	$\mathcal{O}(\sqrt{n}\log n)$

Our Contribution. We present an improved upper bound for distance labeling of unweighted planar graphs on n nodes. We design a distance labeling scheme with labels consisting of $\mathcal{O}(\sqrt{n})$ bits. While this might be seen as "only" a logarithmic improvement, it provides a separation for distance labeling between unweighted and weighted planar graphs. Furthermore, we believe that lack of any progress on resolving the complexity of distance labeling in unweighted planar graphs in the last 16 years makes any asymptotic decrease desirable. Our method easily extends to undirected planar graphs with edges of length from $[1, W]$, allowing us to decrease the label length from $\mathcal{O}(\sqrt{n}\log(nW))$ to $\mathcal{O}(\sqrt{n}\log W)$, and (unweighted) undirected graphs with genus g, decreasing the label length from $\mathcal{O}(\sqrt{ng}\log n)$ to $\mathcal{O}(\sqrt{ng}\log g)$ for graphs of genus at most g. Decoding time in our construction for planar unweighted graphs is at least $\mathcal{O}(\sqrt{n})$ (as in the previously known scheme of Gavoille et al. [16]), but we augment it with a mechanism that computes the distance in polylogarithmic time while increasing the label length to $\mathcal{O}(\sqrt{n\log n})$.

Techniques and Roadmap. As in the previous scheme of Gavoille et al. [16], we apply a recursive separator decomposition. This scheme is presented in detail in Sect. 2. Our improvement is based on the following observation: if each separator is, in fact, a cycle, then we can shave off a factor of $\log n$ by appropriately encoding the stored distances. For a triangulated graph, one can indeed always find a balanced cycle separator, but our graph does not have to be triangulated. In some applications, the solution to this problem is to simply triangulate with edges of sufficiently large length (as to not change the distance), but we need to keep the graph unweighted. In Sect. 3, we overcome this difficulty by designing a novel method of replacing each face of the original graph G with an appropriately chosen gadget to obtain a new unweighted graph G' with every face of length at most 4. The crucial property is that, for any two nodes that exists in both G and G', their distance in G' is at least the logarithm of their distance in G. So, while inserting the gadgets may decrease the distances, we are able to lower bound this decrease. We believe that this might be of independent interest. To facilitate efficient decoding, in Sect. 4 we build on the distance oracle of Gawrychowski et al. [18]. This requires some tweaks in their point location structure to make it smaller at the expense of increasing the query time (but still keeping it polylogarithmic) and adjusting our scheme to balance the lengths of different parts of the table.

Computational Model. When discussing the decoding time we assume the Word RAM model with words of logarithmic length. A label of length ℓ is packed in $\lceil \ell / \log n \rceil$ words, and the decoder computing the distance between u and v can access in constant time any word from their labels. Standard arithmetic and Boolean operations on words are assumed to take constant time.

2 Previous Scheme

We briefly recap the scheme of Gavoille et al. [16]. Their construction is based on the notion of separators, that is, sets of nodes which can be removed from the graph so that every remaining connected component consists of at most $\frac{2}{3}n$ nodes. By the classical result of Lipton and Tarjan [27] any planar graph on n nodes has such a separator consisting of $\mathcal{O}(\sqrt{n})$ nodes. Now the whole construction for a connected graph G proceeds as follows: find a separator S of G, and let $G_1, G_2, \ldots$ be the connected components of $G \setminus S$. The label of $v \in G_i$ in G, denoted $\ell_G(v)$, is composed of i, recursively constructed $\ell_{G_i}(v)$, and the distances $\delta_G(v, u)$ for all $u \in S$ written down in the same order for every $v \in G$. A label of $v \in S$ consists of only the distances $\delta_G(v, u)$ for all $u \in S$.

The space complexity of the whole scheme is dominated by the space required to store $|S|$ distances, each consisting of $\log n$ bits, resulting in $\mathcal{O}(\sqrt{n} \log n)$ bits in total. The bound of $\mathcal{O}(\sqrt{n})$ on the size of a separator is asymptotically tight. However, the total length of the label of $v \in G$ (in bits) depends not on the size of the separator, but on the number of bits necessary to encode the distances from v to the nodes of the separator. If the separator is a simple cycle $(u_1, u_2, \ldots, u_{|S|})$ then $|\delta_G(v, u_i) - \delta_G(v, u_{i+1})| \leq 1$, for every $i = 1, 2, \ldots, |S| - 1$, and consequently writing down $\delta_G(v, u_1)$ explicitly and then storing all the differences $\delta_G(v, u_i) - \delta_G(v, u_{i+1})$ takes only $\mathcal{O}(\sqrt{n})$ bits in total. It is known that if the graph is triangulated, there always exists a simple cycle separator [28], so for such graphs labels of length $\mathcal{O}(\sqrt{n})$ are enough. We show that, in fact, for any planar graph it is possible to select a separator so that the obtained sequence of differences is compressible. This is done by inserting some gadgets into every face of the graph.

3 Improved Scheme

We use the notion of weighted separators, as introduced in [28]. Consider a planar graph, where every node has a non-negative weight and all these weights sum up to 1. Then a set of nodes is a weighted separator if after removing these nodes the total weight of every remaining connected component is at most $\frac{2}{3}$. We have the following well-known theorem (the result is in fact more general and allows assigning weights also to edges and faces, but this is not needed in our application):

Lemma 1 ([28]). *For every planar graph on n nodes having assigned non-negative weights summing up to 1, either there exists a node that is a weighted separator or there exists a simple cycle of length at most $2\sqrt{2\lfloor d/2\rfloor n}$ which is a weighted separator, where d is the maximum face size.*

Lemma 2. *Any planar graph G has a separator S, such that*

$$\sum_{i=1}^{|S|-1} (1 + \log \delta_G(u_i, u_{i+1})) = \mathcal{O}(\sqrt{n})$$

for some ordering $u_1, u_2 \ldots, u_{|S|}$ of all nodes of S.

Before proving the lemma, we first describe a family of *subdivided cycles.* A subdivided cycle on $s \geq 3$ nodes, denoted D_s, consists of a cycle $C_s = (v_1, \ldots, v_s)$ and possibly some auxiliary nodes. D_3 and D_4 are simply C_3 and C_4, respectively. For $s > 4$, we add $\lceil \frac{s}{2} \rceil$ auxiliary nodes $u_1, \ldots, u_{\lceil \frac{s}{2} \rceil}$, and connect every v_i with $u_{\lceil \frac{i}{2} \rceil}$. To complete the construction, we recursively build $D_{\lceil \frac{s}{2} \rceil}$ and identify its cycle with $(u_1, \ldots, u_{\lceil \frac{s}{2} \rceil})$. (An example of such a subdivided cycle on 10 nodes is shown in Fig. 1.) We have the following property.

Lemma 3. *For any distinct $u, v \in C_s$, $\delta_{D_s}(u, v) \geq 1 + \log \delta_{C_s}(u, v)$.*

Proof. We apply induction on s. It is easy to check that the lemma holds when $s \leq 4$, so we assume $s \geq 5$. Let us denote $\delta_{D_s}(u, v) = d'$ and $\delta_{C_s}(u, v) = d$. We proceed with another induction on d'. When $d' = 1$ then u and v must be neighbors on C_s, so $d = 1$ and the claim holds. When $d = 1$ then the claim trivially holds for any $d' \geq 1$. Now assume $d', d \geq 2$ and consider a shortest path connecting u and v in D_s. If it consists of only auxiliary nodes except for the endpoints u and v, then we consider the immediate neighbors of u and v on the path, denoted u' and v', respectively. Since u' and v' must belong to the cycle of $D_{\lceil \frac{s}{2} \rceil}$ and the distance between them in the corresponding $C_{\lceil \frac{s}{2} \rceil}$ is at least $\lfloor \frac{d}{2} \rfloor$, by the inductive assumption applied with smaller $\lceil \frac{s}{2} \rceil < s$ (and using $d \geq 2$):

$$d' \geq 2 + 1 + \log\lfloor d/2 \rfloor = 1 + \log(4\lfloor d/2 \rfloor) > 1 + \log d.$$

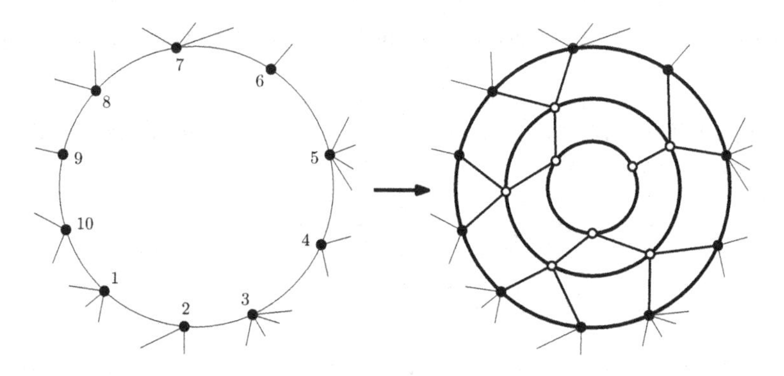

Fig. 1. A face of size 10 is transformed by replacing C_{10} with D_{10} containing 8 new auxiliary nodes.

Otherwise, let w be an intermediate node of the path that belongs to the cycle C_s. Let $\delta_{D_s}(u,w) = d'_0$ and $\delta_{C_s}(u,w) = d_0$ and $\delta_{D_s}(w,v) = d'_1$ and $\delta_{C_s}(w,v) = d_1$. Because w is an intermediate node, we can apply the inductive assumption with the same s but smaller $d'_0, d'_1 < d'$ to obtain $d'_0 \geq 1 + \log d_0$ and $d'_1 \geq 1 + \log d_1$. Then:

$$\begin{aligned} d' &= d'_0 + d'_1 \\ &\geq 1 + \log d_0 + 1 + \log d_1 && \text{by the inductive assumption} \\ &= 1 + \log(2d_0 d_1) \\ &\geq 1 + \log(d_0 + d_1) && 2xy \geq x + y \text{ for any } x, y \geq 1 \\ &\geq 1 + \log d && \text{by the triangle inequality} \end{aligned}$$

as required. □

Proof of Lemma 2. Let G' be the graph constructed from G by replacing every face (including the external face) with a subdivided cycle of appropriate size. More precisely, let $(v_1, v_2, \ldots, v_s)$ be a boundary walk of a face of G. Note that nodes v_i are not necessarily distinct. We create a subdivided cycle D_s and identify its cycle C_s with $(v_1, v_2, \ldots, v_s)$. Clearly G' is also planar and each of its faces is either a triangle or a square. Since any subdivided cycle has at most twice as many auxiliary nodes as cycle nodes and the lengths of all boundary walks sum up to twice the number of edges, which is at most $3n - 6$ (as G is planar), G' contains at most $n' = n + 4 \cdot (3n - 6) < 13n$ nodes.

We assign weights to nodes of G' so that every node also appearing in G has weight 1 and every new node has weight 0. By Lemma 1 either there exists a single node s that is a weighted separator or there exists a weighted simple cycle separator S' in G' of size at most $2\sqrt{52n}$. In the former case, s is also a weighted separator in G and there is nothing to prove. In the latter case, let $S = S' \cap G$ be a separator in G. Because S' is a simple cycle separator, $S = (u_1, u_2, \ldots, u_c)$, and u_i and u_{i+1} are incident to the same face f of G. The boundary walk of

f, consisting of s_i nodes, has been identified with the cycle C_{s_i} of a copy of the subdivided cycle D_{s_i} such that S' connects u_i and u_{i+1} either directly or by visiting some auxiliary nodes of D_{s_i}, for every $i = 1, 2, \ldots, c$ (we assume $u_{c+1} = u_1$). Let v_i and v_i' denote nodes of D_{s_i} that have been identified with u_i and u_{i+1}, respectively. Then:

$$\sum_i \delta_{D_{s_i}}(v_i, v_i') \leq |S'| = \mathcal{O}(\sqrt{n}).$$

By Lemma 3, $\delta_{D_{s_i}}(v_i, v_i') \geq 1 + \log \delta_{C_{s_i}}(v_i, v_i')$, so:

$$\sum_i (1 + \log \delta_G(u_i, u_{i+1})) \leq \sum_i (1 + \log \delta_{C_{s_i}}(v_i, v_i')) = \mathcal{O}(\sqrt{n})$$

□

Theorem 1. *Any planar graph on n nodes admits a distance labeling scheme of length $\mathcal{O}(\sqrt{n})$.* □

Proof. We proceed as in the previously known scheme of size $\mathcal{O}(\sqrt{n}\log n)$, except that in every step we apply our Lemma 2. In more detail, to construct the label of every $v \in G$ we proceed as follows. First, we find a separator $S = (u_1, u_2 \ldots, u_c)$ using Lemma 2. We have $\sum_i (1 + \log \delta_G(u_i, u_{i+1})) = \mathcal{O}(\sqrt{n})$, so in particular $c = \mathcal{O}(\sqrt{n})$ and $\sum_{i=1}^{c-1} \log \delta_G(u_i, u_{i+1}) = \mathcal{O}(\sqrt{n})$. For every $v \in G$ we encode its distances to all nodes of the separator as follows. We use Elias γ code [14] which gives a prefix-free encoding of a number x using $2\lceil \log x \rceil + 1$ bits. We first encode $\delta_G(v, u_1)$ using Elias γ code. Then we encode the differences $\delta_G(v, u_i) - \delta_G(v, u_{i-1})$, for all $i = 2, \ldots, c$, also using Elias γ code and an extra bit to denote the sign. Encodings are concatenated, and by the prefix-free property given the concatenation we can recover all the distances. The total length of the concatenation is:

$$\mathcal{O}(\sqrt{n} + \sum_{i=2}^{c} \log |\delta_G(v, u_i) - \delta_G(v, u_{i-1})|).$$

Consequently, by $|\delta_G(v, u_i) - \delta_G(v, u_{i-1})| \leq \delta_G(u_{i-1}, u_i)$ and the properties of our separator the encoding takes $\mathcal{O}(\sqrt{n})$ bits. Second, for every node we store the name of its connected component of $G \setminus S$ in $\mathcal{O}(\log n) = \mathcal{O}(\sqrt{n})$ bits. Third, we recurse on every connected component of $G \setminus S$ and append the obtained labels to the current labels. To calculate $\delta_G(u, v)$, we first compute $d = \min_{w \in S}(\delta_G(u, w) + \delta_G(w, v))$, extracting $\delta_G(u, w)$ from the label of u and $\delta_G(w, v) = \delta_G(v, w)$ from the label of v. Then, if u and v belong to the same connected component of $G \setminus S$, we proceed recursively there and return the minimum of d and the recursively found distance in the connected component. The correctness is clear: either a shortest path between u and v is fully within one of the connected components, or it visits some $w \in S$, and in such case we can recover $\delta_G(u, w) + \delta_G(w, v)$ from the stored distances. The final size of every label is $\mathcal{O}(\sqrt{n} + \sqrt{\frac{2}{3}n} + \ldots) = \mathcal{O}(\sqrt{n})$ bits. □

4 Efficient Decoding

The drawback of the scheme from Theorem 1 is its high decoding time. Computing $\delta_G(u,v)$ given $\ell_G(u)$ and $\ell_G(v)$ is done as follows. First, we iterate over $w \in S$ and consider $\delta_G(u,w)+\delta_G(v,w)$ as a possible distance, extracting $\delta_G(u,w)$ from $\ell_G(u)$ and $\delta_G(v,w)$ from $\ell_G(v)$. Then, we check if u and v belong to the same component G_i, and if so recurse on $\ell_{G_i}(u)$ and $\ell_{G_i}(v)$. Even assuming that extracting any $\delta_G(u,w)$ takes constant time, it is not clear how to avoid iterating over all $w \in S$, so we cannot hope for anything faster than $\mathcal{O}(\sqrt{n})$. In this section we show how to overcome this difficulty by applying the machinery of Voronoi diagrams on planar graphs, introduced for computing the diameter of a planar graph in subquadratic time by Cabello [11]. Our method roughly follows the approach of Gawrychowski et al. [18] (also see [12]), but we need to make sure that the information can be distributed among the labels, and carefully adjust the parameters of the whole construction. We start with presenting the necessary definitions and tools.

r-divisions. A region R of G is an edge-induced subgraph of G. An r-division of G is a collection of regions such that each edge of G is in at least one region, there are $\mathcal{O}(n/r)$ regions, each region has at most r nodes and $\mathcal{O}(\sqrt{r})$ boundary nodes that belong to more than one region. We work with a fixed planar embedding of G, and all of its subgraphs, in particular the regions, inherit this embedding. A hole of a region R is a face that is not a face of G. An r-division with few holes has the additional property that each edge belonging to two regions is on a hole in each of them, and each region has $\mathcal{O}(1)$ holes.

Lemma 4 ([24]). *For a constant s, there is a linear-time algorithm that, for any biconnected triangulated planar embedded graph G and any $r \geq s$, outputs an r-division of G with few holes.*

The above theorem additionally guarantees that each region is connected and its boundary nodes are exactly the nodes incident to its holes.

Voronoi Diagrams. Following the description in [18], let S be the nodes (called sites) incident to the external face h of an internally triangulated planar graph G. Each site $u \in S$ has a weight $\omega(u)$, and the distance between a site $u \in S$ and a node v, denoted by $d(u,v)$, is defined as $\omega(u) + \delta_G(u,v)$. The (additively) weighted Voronoi diagram of (S,ω) within G, denoted $\mathsf{VD}(S,\omega)$, is a partition of the nodes of G into pairwise disjoint sets, one set $\mathsf{Vor}(u)$ for every $u \in S$. $\mathsf{Vor}(u)$ is called the cell of u and contains all nodes of G that are closer to u than any other site u' (w.l.o.g. all distances are unique). We work with a dual representation of $\mathsf{VD}(S,\omega)$, denoted by $\mathsf{VD}^*(S,\omega)$, defined as follows. Let G^* be the dual of G, and VD_0 consist of the dual of edges (u,v) of G such that u and v belong to different cells. Then, let VD_1 be obtained from VD_0 by contracting edges incident to vertices of degree 2 one-by-one as long as possible. A vertex of VD_1 is called a Voronoi vertex, and is dual to a face f such that the nodes incident to f belong to at least three Voronoi cells. In particular, h^* is a Voronoi vertex. Finally, $\mathsf{VD}^*(S, \omega)$ is obtained from VD_1 by replacing h^* by multiple copies, one for each incident edge. The complexity of $\mathsf{VD}^*(S, \omega)$ is $\mathcal{O}(|S|)$ and, assuming that every node belongs to its cell, $\mathsf{VD}^*(S, \omega)$ is a tree. In the remaining part of the description we will assume that this is indeed the case, the general case can be handled as described in [18], or by (conceptually) modifying G so that the only nodes incident to the external face are those with nonempty cells. We will also assume that the graph is biconnected, so that the boundary walk of the external face is simple.

Point Location. The main technical contribution of [18] is a point location structure for $\mathsf{VD}(S, \omega)$ that, given a node v, finds its cell in $\mathcal{O}(\log |S|)$ time, assuming constant-time access to certain primitives. We briefly describe the required primitives and then the high-level idea of this structure, but the reader is strongly advised to consult the original description.

For any site u, let T_u be the shortest path tree rooted at u. Additionally, for each face f other than h we add an artificial node v_f whose embedding coincides with the embedding of f^*. In T_u, v_f is a leaf connected with a zero-length edge to the node y_f incident to f that is closest to u. For any node v, we need to have access to the following information:

1. $d(u, v)$,
2. in T_u, is v on the path from u to y_f, or left/right[2] of this path.

Let $s_1, s_2, \ldots, s_{|S|}$ be the boundary walk of the external face containing every site. Recall that $\mathsf{VD}^*(S, \omega)$ is a tree with no vertices of degree 2. A centroid decomposition of a tree T on n nodes is recursively defined as follows: we find a centroid $u \in T$ such that removing u from T and replacing it with copies, one for each edge incident to u, results in a set of trees, each with at most $(n+1)/2$ edges, and repeat the reasoning on each of these trees. The construction terminates when the tree has no nodes of degree 3 or more (i.e. it consists of a single edge). The point location structure consists of a centroid decomposition of $\mathsf{VD}^*(S, \omega)$. In the query, the decomposition is traversed starting from the root. In every step, we consider a centroid f^*. Assuming that the graph is triangulated, there are three nodes y_0, y_1, y_2 incident to f, where $y_i \in \mathsf{Vor}(s_{i_j})$. By accessing the information we can detect in constant time that $v \in \mathsf{Vor}(s_{i_j})$, or descend down in the centroid decomposition to find the cell of v. We need the following observation: consider all nodes for which this procedure reaches a vertex at depth k with subtree of size s in the centroid decomposition. All sites s_i such that $v \in \mathsf{Vor}(s_i)$ for a node v can be represented as $\mathcal{O}(k)$ contiguous segments of total length $\mathcal{O}(s)$. The depth of the centroid decomposition is of course $\mathcal{O}(\log |S|)$.

[2] Left/right is defined using a fixed planar embedding by considering how the path from u to v emanates from the path from u to y_f. The tree inherits the embedding from the graph, and for two nodes of a tree we can check being on the path or left/right by operating on their pre- and post-order number.

Bitvectors. Recall that in the proof of Theorem 1 we stored the differences $\delta_G(v, u_i) - \delta_G(v, u_{i-1})$, for $i = 2, 3, \ldots, c$, by concatenating their Elias γ encodings. Now we need to compute any prefix sum $\delta_G(v, u_i) - \delta_G(v, u_1)$ in constant time. The following lemma can be proved by augmenting the concatenation of all Elias γ encodings with some extra information and using a rank/select structure [21] to access the required data in constant time.

Lemma 5. *partialsums For any $\epsilon > 0$ and a sequence of integers $\Delta_1, \Delta_2, \ldots, \Delta_s$, such that $\sum_{i=1}^{j} \Delta_i \in [-n, n]$ for every j, we can construct a structure consisting of $\mathcal{O}(n^\epsilon + \sum_i (1 + \log \Delta_i))$ bits that returns $\sum_{i=1}^{j} \Delta_i$, for any j, in constant time.*

Having gathered all the technical ingredients, we are now ready to describe a modification of the proof Theorem 1 that allows us to guarantee polylogarithmic decoding time. We first describe the high-level idea, then highlight two technical difficulties and proceed with a detailed description.

We would like to apply reasoning from the proof of Theorem 1 to find a balanced Jordan curve separator $S = (u_1, u_2, \ldots, u_c)$ in G with the property that the distances in G from a node u to all nodes in S can be encoded in $\mathcal{O}(\sqrt{n})$ bits. S partitions G into the external part G_{ext} and the internal part G_{int}, and we want to augment the labels with enough information so that, given $\ell_G(v)$ and $\ell_G(v')$ for $v \in G_{ext}$ and $v' \in G_{int}$, we can compute $\delta_G(v, v')$ in polylogarithmic time. By repeating the reasoning on G_{ext} and G_{int} recursively, this allows us to compute any $\delta_G(v, v')$. The natural idea would be to define a Voronoi diagram of G_{int} by setting the weight of each node in u_i to be $\delta_G(v, u_i)$, and store its point location structure that allows us to efficiently minimize $\delta_G(v, u_i) + \delta_{G_{ext}}(u_i, v')$ (which is equal to the sought $\delta_G(v, v')$). However, this takes too much space, as the point location structure is a tree on $c = \Theta(\sqrt{n})$ nodes, and it appears that we need to store a constant number of integers consisting of $\log n$ bits for each node of this tree. To overcome this, one might try to store only the top part of the centroid decomposition corresponding to subtrees of sufficiently large size, say $\log n$. Then we can afford to store a description of this top part in $\ell_G(v)$, and it can be used to either find the nearest site, or narrow down the set of remaining sites to $\mathcal{O}(\log n)$. However, this still requires some information about v', and in particular we need its position in every T_{u_i} (there is no clear way of how to restrict the number of sites u_i for which such information needs to be stored, as v' is oblivious to v, and different nodes v might need to access different sites when traversing their top parts of the centroid decomposition). Therefore, we need a more complex approach that adds $\mathcal{O}(\sqrt{n \log n})$ bits to every label.

Before we proceed with the modified construction, we need to verify that G' obtained from G in the proof of Lemma 2 is biconnected.

Lemma 6. *If $n \geq 4$ then G' is biconnected.*

Proof. It is straightforward to verify that removing an auxiliary node cannot disconnect the graph. Consider a node u of G, and let $v_0, v_1, \ldots, v_{d-1}$ be its neighbors arranged in a clockwise order. We claim that v_i and $v_{(i+1) \bmod d}$ are still connected in G' after removing u from G'. Consider the face containing $v_i, u, v_{(i+1) \bmod d}$ as a part of the boundary walk. If the boundary walk is of size at least 5 then the artificial nodes guarantee the connectivity. Similarly when the boundary walk contains just one occurrence of u. The only remaining case is that the boundary walk is $u, v_i, u, v_{(i+1) \bmod d}$, but then there are no other edges in G and $n \leq 3$. □

The modified construction proceeds as follows. G' is biconnected but not necessarily triangulated, as there might be faces of length 4. We triangulate G' to obtain G'', and then apply Lemma 4 to obtain an r-division with $r = n/\log n$. By the properties of an r-division, there are $\mathcal{O}(\log n)$ regions. Each region R contains $\mathcal{O}(\sqrt{n/\log n})$ boundary nodes incident to $\mathcal{O}(1)$ holes. The boundary walk $(u_1, u_2, \ldots, u_c)$ of every hole h of R is a (not necessarily simple) cycle in G'', and by the construction of G'' we can find a cycle $(u'_1, u'_2, \ldots, u'_{c'})$ in G' that contains $(u_1, u_2, \ldots, u_c)$ as a subsequence, and $c' = \mathcal{O}(c) = \mathcal{O}(\sqrt{n/\log n})$. The r-division of G'' naturally induces an r-division of G', and we will refer to $(u'_1, u'_2, \ldots, u'_{c'})$ as a boundary walk of h. Note that because we have defined the r-division applying Lemma 4 to G'', some nodes u'_i might not belong to R, and we do not guarantee that all nodes incident to a hole are boundary.

The label of every node v of G consists of two asymmetric parts. Let h be a hole of a region R, and $(u'_1, u'_2, \ldots, u'_{c'})$ a boundary walk of h, where $c' = \mathcal{O}(\sqrt{n/\log n})$. Furthermore, let $(u''_1, u''_2, \ldots, u''_{c''})$ be a subsequence of $(u'_1, u'_2, \ldots, u'_{c'})$ consisting of the nodes of G. By the reasoning from the proof of Theorem 1, we have $\sum_i (1 + \log \delta_G(u''_i, u''_{i-1})) = \mathcal{O}(\sqrt{n/\log n})$. The first part of the label of v in G encodes $\delta_G(v, u''_1)$ in $\mathcal{O}(\log n)$ bits, and then the differences $\delta_G(v, u''_i) - \delta_G(v, u''_{i-1})$ using Lemma 5. This takes $\mathcal{O}(\log n + \sqrt{n/\log n})$ bits for every hole by setting $\epsilon < 1/2$, so $\mathcal{O}(\sqrt{n \log n})$ in total for all R and h. If v' is a boundary node of R incident to a hole h then we store the identity of R and h in the label of v' (there could be multiple such pairs R and h, we choose any of them), together with the position of any occurrence of v' in $(u''_1, u''_2, \ldots, u''_{c''})$. This is already enough to determine $\delta_G(v, v')$ in constant time for any boundary node v'. Otherwise, v' belongs to exactly one region R, and either v is not a boundary node and belongs to the same region R or the shortest path from v to v' goes through one of the boundary nodes of R. To take the former case into the account, we consider the connected components of the subgraph of G consisting of the non-boundary nodes of R. Each such node stores the identity of its component in $\mathcal{O}(\log n)$ bits, so that we can verify if v and v' belong to the same connected component of R and recurse there if so (that is, the whole construction is repeated on every connected component, and the resulting label is a concatenation of the labels defined in the subsequent steps of the recursion). To deal with the latter case, we need to show how to find the shortest path in G from v to v' that first goes from v to a boundary node u of R and then goes to v' without visiting any other boundary node (note that this might happen even when v and v' belong to the same connected component). We focus on this in the remaining part of the description.

Consider a region R and its hole h. We make h the external face, triangulate the non-external faces, and make the weight of every edge that does not belong to G infinite to obtain a weighted graph R'. Let $u_1, u_2, \ldots$ be the boundary nodes of R incident to h. We construct the Voronoi diagram of the obtained weighted graph R' with sites $u_1, u_2, \ldots$, setting the weight of every u_i to be its distance from v in G. Storing the centroid decomposition of this Voronoi diagram would take $\mathcal{O}(\sqrt{n/\log n})$ bits, which is too much. Instead, we store its top part obtained by stopping as soon as the size of the current subtree is less than $\log^2 n$. The size of this top part is $\mathcal{O}(\sqrt{n/\log n}/\log^2 n) = \mathcal{O}(\sqrt{n/\log^5 n})$ by the following lemma.

Lemma 7. *Consider the centroid decomposition of a bounded-degree tree T and a parameter b. The decomposition contains $\mathcal{O}(|T|/b)$ subtrees of size less than b obtained by choosing a centroid in a subtree of size at least b.*

Proof. The decomposition can be naturally interpreted as a tree $\mathcal{T}$, with every node corresponding to a subtree obtained during the process. The leaves of $\mathcal{T}$ correspond to single edges of T, and internal nodes of $\mathcal{T}$ correspond to larger subtrees. The weight $w(u)$ of $u \in \mathcal{T}$ is the number of leaves in its subtree (equal to the size of the corresponding subtree of T), and for each child v we have $w(v) \leq (w(u)+1)/2$. Because the degrees of T are bounded by a constant, it is enough to count $u \in \mathcal{T}$ such that $w(u) \geq b$, we call them heavy. There are clearly at most $(|T|+1)/b$ heavy nodes with no heavy children, as their subtrees are disjoint. This also bounds the number of heavy nodes with more than one heavy child. It remains to bound the number of heavy nodes u with exactly one heavy child v. However, such u must have some non-heavy children $v_1, v_2, \ldots$ of total weight at least $b-1$, as $w(v) \leq (w(u)+1)/2$ and $b \leq w(v)$ so $b-1 \leq w(u)-w(v)$ and $w(u) - w(v) = w(v_1) + w(v_2) + \ldots$, so there are no more than $|T|/(b-1)$ such nodes u. Overall, this is $\mathcal{O}(|T|/b)$ as claimed. □

For each leaf in the top part of the decomposition, we store $\mathcal{O}(\log n)$ contiguous segments of the sites that might be relevant. This takes $\mathcal{O}(\sqrt{n/\log^5 n} \cdot \log^2 n) = \mathcal{O}(\sqrt{n/\log n})$ bits. For every non-leaf, we have a centroid f^* used for deciding where to descend. We store the preorder number of the artificial node corresponding to f^* in the respective three shortest path trees. This takes $\mathcal{O}(\sqrt{n/\log^5 n}\log n) = \mathcal{O}(\sqrt{n/\log n})$ bits. Overall, this is $\mathcal{O}(\sqrt{n\log n})$ bits.

To use the centroid decomposition, we need to store enough information in the label of v' as to be able to compute any $\delta_{R'}(u_i, v')$ in constant time. Recall that for a non-boundary node v' we have exactly one relevant R and a constant number of Voronoi diagrams corresponding to the holes of R. Therefore, because there are only $\mathcal{O}(\sqrt{n/\log n})$ sites in every Voronoi diagram, we can afford to store every $\delta_{R'}(u_i, v')$ in binary using $\mathcal{O}(\sqrt{n\log n})$ bits overall. Additionally, v' stores its pre- and postorder number in the shortest path tree rooted at every T_{u_i}, this also takes $\mathcal{O}(\sqrt{n\log n})$ bits.

To compute $\delta_G(v, v')$, we consider every hole h of the unique region R containing v'. We first use the stored top part of the centroid decomposition, where we navigate by using the stored pre- and postorder numbers that allow us to check if v' is on the path (in which case we terminate) or left/right of the path. After reaching a leaf in the top part of the centroid decomposition, we simply consider the remaining $\mathcal{O}(\log^2 n)$ possible sites one-by-one. This takes $\mathcal{O}(\log n)$ to traverse the top part, and then $\mathcal{O}(\log^2 n)$ for a leaf. Finally, if v and v' belong to the same connected component we recurse there. Overall, the decoding time is $\mathcal{O}(\log^3 n)$.

Theorem 2. *Any planar graph on n nodes admits a distance labeling scheme of length $\mathcal{O}(\sqrt{n \log n})$ with $\mathcal{O}(\log^3 n)$ decoding time.*

References

1. Abboud, A., Dahlgaard, S.: Popular conjectures as a barrier for dynamic planar graph algorithms. In: 57th FOCS, pp. 477–486 (2016)
2. Abboud, A., Gawrychowski, P., Mozes, S., Weimann, O.: Near-optimal compression for the planar graph metric. In: 29th SODA, pp. 530–549 (2018)
3. Alon, N., Nenadov, R.: Optimal induced universal graphs for bounded-degree graphs. In: 28th SODA, pp. 1149–1157 (2017)
4. Alstrup, S., Dahlgaard, S., Knudsen, M.B.T.: Optimal induced universal graphs and adjacency labeling for trees. In: 56th FOCS, pp. 1311–1326 (2015)
5. Alstrup, S., Dahlgaard, S., Knudsen, M.B.T., Porat, E.: Sublinear distance labeling. In: 24th ESA, pp. 5:1–5:15 (2016)
6. Alstrup, S., Gavoille, C., Halvorsen, E.B., Petersen, H.: Simpler, faster and shorter labels for distances in graphs. In: 27th SODA, pp. 338–350 (2016)
7. Alstrup, S., Gørtz, I.L., Halvorsen, E.B., Porat, E.: Distance labeling schemes for trees. In: 43rd ICALP, pp. 132:1–132:16 (2016)
8. Alstrup, S., Kaplan, H., Thorup, M., Zwick, U.: Adjacency labeling schemes and induced-universal graphs. In: 47th STOC, pp. 625–634 (2015)
9. Bonamy, M., Gavoille, C., Pilipczuk, M.: Shorter labeling schemes for planar graphs. In: 30th SODA, pp. 446–462 (2020)
10. Bonichon, N., Gavoille, C., Labourel, A.: Short labels by traversal and jumping. Electron. Notes Discret. Math. **28**, 153–160 (2007)
11. Cabello, S.: Subquadratic algorithms for the diameter and the sum of pairwise distances in planar graphs. ACM Trans. Algorithms **15**(2), 21:1-21:38 (2019)
12. Charalampopoulos, P., Gawrychowski, P., Mozes, S., Weimann, O.: Almost optimal distance oracles for planar graphs. In: 51st STOC, pp. 138–151. ACM (2019)
13. Dujmovic, V., Esperet, L., Gavoille, C., Joret, G., Micek, P., Morin, P.: Adjacency labelling for planar graphs (and beyond). In: 61st FOCS, pp. 577–588. IEEE (2020)
14. Elias, P.: Universal codeword sets and representations of the integers. IEEE Trans. Inf. Theory **21**(2), 194–203 (1975)
15. Freedman, O., Gawrychowski, P., Nicholson, P.K., Weimann, O.: Optimal distance labeling schemes for trees. In: 36th PODC, pp. 185–194 (2017)
16. Gavoille, C., Peleg, D., Pérennes, S., Raz, R.: Distance labeling in graphs. J. Algorithms **53**(1), 85–112 (2004)

17. Gawrychowski, P., Kosowski, A., Uznański, P.: Sublinear-space distance labeling using hubs. In: Gavoille, C., Ilcinkas, D. (eds.) DISC 2016. LNCS, vol. 9888, pp. 230–242. Springer, Heidelberg (2016). https://doi.org/10.1007/978-3-662-53426-7_17
18. Gawrychowski, P., Mozes, S., Weimann, O., Wulff-Nilsen, C.: Better tradeoffs for exact distance oracles in planar graphs. In: 29th SODA, pp. 515–529. SIAM (2018)
19. Graham, R.L., Pollak, H.O.: On embedding graphs in squashed cubes. In: Alavi, Y., Lick, D.R., White, A.T. (eds.) Graph Theory and Applications. LNM, vol. 303, pp. 99–110. Springer, Heidelberg (1972). https://doi.org/10.1007/BFb0067362
20. Hsu, T.-H., Lu, H.-I.: An optimal labeling for node connectivity. In: Dong, Y., Du, D.-Z., Ibarra, O. (eds.) ISAAC 2009. LNCS, vol. 5878, pp. 303–310. Springer, Heidelberg (2009). https://doi.org/10.1007/978-3-642-10631-6_32
21. Jacobson, G.: Space-efficient static trees and graphs. In: 30th FOCS, pp. 549–554. IEEE Computer Society (1989)
22. Kannan, S., Naor, M., Rudich, S.: Implicit representation of graphs. SIAM J. Discret. Math. **5**(4), 596–603 (1992)
23. Katz, M., Katz, N.A., Korman, A., Peleg, D.: Labeling schemes for flow and connectivity. SIAM J. Comput. **34**(1), 23–40 (2004)
24. Klein, P.N., Mozes, S., Sommer, C.: Structured recursive separator decompositions for planar graphs in linear time. In: 45th STOC, pp. 505–514. ACM (2013)
25. Korman, A.: Labeling schemes for vertex connectivity. ACM Trans. Algorithms **6**(2), 39:1-39:10 (2010)
26. Kosowski, A., Uznański, P., Viennot, L.: Hardness of exact distance queries in sparse graphs through hub labeling. In: 38th PODC, pp. 272–279 (2019)
27. Lipton, R.J., Tarjan, R.E.: Applications of a planar separator theorem. SIAM J. Comput. **9**(3), 615–627 (1980)
28. Miller, G.L.: Finding small simple cycle separators for 2-connected planar graphs. J. Comput. Syst. Sci. **32**(3), 265–279 (1986)
29. Moon, J.W.: On Minimal n-Universal Graphs. vol. 7, pp. 32–33. Cambridge University Press, Cambridge (1965)
30. Peleg, D.: Informative labeling schemes for graphs. Theor. Comput. Sci. **340**(3), 577–593 (2005)
31. Petersen, C., Rotbart, N., Simonsen, J.G., Wulff-Nilsen, C.: Near-optimal adjacency labeling scheme for power-law graphs. In: 43rd ICALP, pp. 133:1–133:15 (2016)
32. Weimann, O., Peleg, D.: A note on exact distance labeling. Inf. Process. Lett. **111**(14), 671–673 (2011)

How to Catch Marathon Cheaters: New Approximation Algorithms for Tracking Paths

Michael T. Goodrich[1], Siddharth Gupta[2], Hadi Khodabandeh[1], and Pedro Matias[1(✉)]

[1] Department of Computer Science, University of California Irvine, Irvine, USA
{goodrich,khodabah,pmatias}@uci.edu

[2] Department of Computer Science, Ben-Gurion University of the Negev, Beersheba, Israel
siddhart@post.bgu.ac.il

Abstract. Given an undirected graph, G, and vertices, s and t in G, the *tracking paths* problem is that of finding the smallest subset of vertices in G whose intersection with any s-t path results in a unique sequence. This problem is known to be NP-complete and has applications to animal migration tracking and detecting marathon course-cutting, but its approximability is largely unknown. In this paper, we address this latter issue, giving novel algorithms having approximation ratios of $(1+\epsilon)$, $O(\lg OPT)$ and $O(\lg n)$, for H-minor-free, general, and weighted graphs, respectively. We also give a linear kernel for H-minor-free graphs.

Keywords: Graph algorithms · Approximation algorithms · Graph minor · Fixed-parameter tractability · Kernelization · Minor-free graphs · Road networks

1 Introduction

In most modern marathons, each runner is provided with a small RFID tag, which is worn on the runner's shoe or embedded in the runner's bib. RFID readers are placed throughout the course and are used to track the progress of the runners [9,36]. In spite these measures, some runners try to cheat by taking shortcuts [37]. To detect all possible course-cutting, we are interested in the combinatorial optimization problem of placing the minimum number of RFID readers in the environment of a marathon to determine every possible path from the start to the finish, including paths that deviate from the official course, just from the sequence of RFID readers that are crossed by a runner taking a given path. In addition to detecting marathon course-cutting, solutions to this optimization problem could also allow for a type of marathon where each runner

The full version of this paper is available in [26]. Our research was supported in part by NSF Grant 1815073 and by the Zuckerman STEM Leadership Program.

A. Lubiw et al. (Eds.): WADS 2021, LNCS 12808, pp. 442–456, 2021.
https://doi.org/10.1007/978-3-030-83508-8_32

could be allowed to map out their own path from the start to finish so long as their path is at least the required length.

Formally, we model a city road network [18,20,21] through which a marathon will be run as an undirected graph, $G = (V, E)$, where V is the set of road intersections and possible RFID reader locations in the city, as well as the placements of the start and finish lines, and E is the set of road segments joining two points in V without having any other elements of V in its interior. Given a start-finish pair, (s, t), of vertices in G, a ***tracking set*** for (s, t) is a subset, T, of V, such that for any s-t path[1] P in G, the sequence $\mathcal{S}^T(P)$ of vertices in T traversed by P uniquely identifies P. In other words, T is a tracking set if $\mathcal{S}^T(P) \neq \mathcal{S}^T(Q)$ for all distinct s-t paths P and Q. We formally define the optimization problem, which is called the ***tracking paths*** problem, as follows:

> TRACKING(G, s, t):
> **Input:** An undirected simple graph $G = (V, E)$ and vertices $s, t \in V$.
> **Output:** A smallest tracking set for (s, t) in G.

We denote by WEIGHTEDTRACKING the vertex-weighted version, whose goal is to find a tracking set of least total weight. Further, we denote by k-TRACKING the decision version of TRACKING, which asks whether there exists a tracking set of size at most k (for any given integer k). For conciseness, we refer to the "tracking set of G", when s and t are clear from context.

Related Work. TRACKING has been shown to be NP-Complete [3], even when the input graph is planar [19] or has bounded degree [10]. It is fixed-parameter tractable (FPT): when parameterized by the solution size (a.k.a., the natural parameter), it admits a quadratic kernel in general and a linear kernel when the graph is planar [11] (other parameterizations have been studied in [12]). Further, it admits approximation ratios of 4 [19] for planar graphs and of $2\Delta + 1$ [10] for degree-Δ graphs. Exact polynomial time algorithms exist for bounded clique-width graphs [19], as well as chordal and tournament graphs [10]. For the NP-hard variant of tracking only shortest paths between multiple start-finish pairs, there exists a $O(\sqrt{n} \lg n)$-approximation [5]. We refer the reader to the full version of the paper [26] for more details on related work.

Our Contributions. Our results are summarized below:

1. **Linear kernel for H-minor-free graphs**. Previously, we only knew of a linear kernel for planar graphs [11]. This result also immediately implies an efficient $O(1)$-approximation.
2. **$(1 + \epsilon)$-approximation for H-minor-free graphs**. Previous best was a 4-approximation for planar graphs [19].
3. **$O(\lg OPT)$-approximation for** TRACKING, where OPT denotes the cardinality of an optimal tracking set. This is the first algorithm for general graphs with a non-trivial approximation ratio. Previously, we only knew of a $O(\sqrt{n} \lg n)$-approximation for tracking shortest paths only [5].

[1] In this paper, paths do not repeat vertices. We denote a path from u to v by u-v.

4. **$O(\lg n)$-approximation for** WEIGHTEDTRACKING. This is the first approximation for weighted graphs, among all variants of TRACKING.

Preliminaries. We use standard terminology concerning graphs, approximation algorithms and kernelization, which is detailed in the full version of the paper [26]. For space considerations, content marked with a link symbol "⊛" is provided in more detail and/or proved in the full version of the paper [26].

2 Structural Properties

Definition 1 (Entry-exit subgraph). *Let (G, s, t) be an instance of* TRACKING. *An **entry-exit subgraph** is a triple (G', s', t'), where G' is a subgraph of G, and (s', t') is the **entry-exit pair** corresponding to vertices in C that satisfy the following conditions:*

1. *There exists a path s-s' from s to the **entry** vertex s'*
2. *There exists a path t'-t from the **exit** vertex t' to t*
3. *Paths s-s' and t'-t are vertex-disjoint*
4. *Path s-s' (resp. t'-t) and G' share exactly one vertex: s' (resp. t').*

Notice that the same subgraph G' of G may contain multiple entry-exit pairs.

Definition 2 (Entry-exit cycle). *An **entry-exit cycle** is an entry-exit subgraph (C, s', t'), where C is a cycle (see Fig. 1).*

We say that a vertex v ***tracks*** (C, s', t') if $v \in C \setminus \{s', t'\}$. Moreover, we say that (C, s', t') is ***tracked*** if there exists a tracker in a vertex that tracks it. A cycle C is ***tracked*** if all entry-exit cycles with entry-exit pairs in C are tracked. If C contains either (i) 3 trackers or (ii) s or t and 1 tracker in a non-entry/non-exit vertex, then it must be tracked. We say that these cycles are ***trivially tracked***.

We rely on the following alternative characterization of a tracking set, due to Banik et al. [3, Lemma 2], which establishes TRACKING as a covering problem.

Lemma 1 ([3]). *For a graph $G = (V, E)$, a subset $T \subseteq V$ is a tracking set if and only if every simple cycle C in G is tracked with respect to T.*

Reduction Rules. Let us recall some reduction rules previously used to obtain polynomial kernels [3,11] and approximation algorithms [4,10,12,19].

Rule 1. [3] If there exists an edge or vertex that does not participate in any s-t path, remove it from the graph.

Rule 2. [11] If the degree of s (or t) is 1 and $N(s) \neq \{t\}$ ($N(t) \neq \{s\}$), then remove s (t), and label the vertex adjacent to it as s (t).

Rule 3. [19] If there exist adjacent vertices $a, b \notin \{s, t\}$ such that $\deg(a) = \deg(b) = 2$, then contract the edge ab.

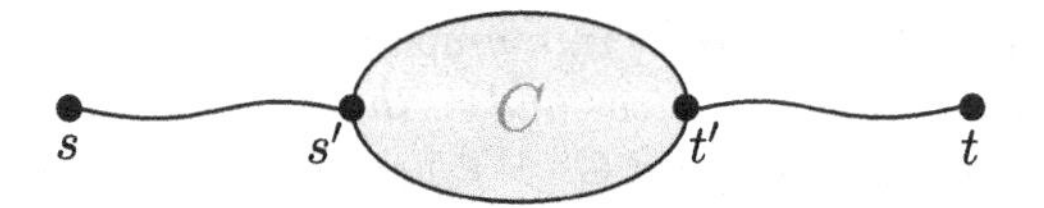

Fig. 1. Entry-exit pair illustration, with entry vertex s' and exit vertex t'.

Definition 3. *We say that an undirected graph G is* ***reduced by Rule X*** *if it cannot be further by reduced Rule X. Further, we say that G is* ***reduced*** *if it is reduced by Rules 1, 2 and 3.*

After exhaustive application of Rules 1 and 2, the graph is either a single edge, (s, t), or all its vertices have degree at least 2. Henceforth, we assume the latter, since the problem becomes trivial in the former case. Rule 3, which precludes the existence of adjacent vertices of degree 2, is used to bound the overall number of degree-2 vertices. Let us highlight a few additional useful consequences of Rule 1.

Remark 1 ([3]). Let G be a graph reduced by Rule 1. Then, every subgraph of G containing at least one edge has at least one entry-exit pair.

Remark 2 ([3]). Let G be a graph reduced by Rule 1. Then, any tracking set of G is also an FVS of G.

Remark 3. Let G be a graph reduced by Rule 1. Then the block-cut tree of G is an s-t path (see Fig. 2).

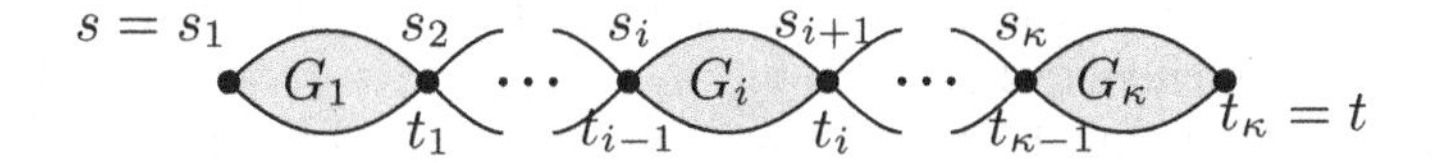

Fig. 2. The block-cut tree of a graph G reduced by Rule 1 (see Remark 4).

In other words, the latter remark says that the graph G that results from exhaustively applying Rule 1 consists of a sequence of $\kappa \geq 1$ biconnected components attached together by cut-vertices in a way that is analogous to series composition in series-parallel graphs. Thus, we can turn an instance (G, s, t) of TRACKING into one or more subproblems on biconnected graphs, (G_i, s_i, t_i), one for each biconnected component, as depicted in Fig. 2.

Remark 4. ⊛ Let G be a graph reduced by Rule 1. Then, an optimal tracking set for (G, s, t) is the ***disjoint union*** of optimal tracking sets for all (G_i, s_i, t_i).

Lower Bounds. We expand on a result by Choudhary and Raman [11], which provides a lower bound on the size of a tracking set, based on the presence of a tree-sink structure in the graph.

Definition 4 ([11]). *A* ***tree-sink***[2] *in a graph G is a pair (Tr, x), where Tr is a subtree of G with at least two vertices and x, the* ***sink****, a vertex not in Tr that is adjacent to all the leaves*[3] *of Tr in G. We use $G(Tr, x)$ to denote the subgraph induced by (Tr, x). (Notice that this definition does not preclude the adjacency between non-leaf vertices and x.)*

Lemma 2 ([11]). *Let (Tr, x) be a tree-sink in a reduced graph G, such that $|N_{Tr}(x)| = \delta$. Further let (s', t') be an entry-exit pair of $G(Tr, x)$. Then, if $x \in \{s', t'\}$, any tracking set of G contains at least $\delta - 1$ vertices in $V(Tr)$.*

The above lemma is a generalization of the lower bound given by the maximum number of vertex-disjoint paths between any two vertices [3], and it can be generalized further to obtain a more useful lower bound, established as the maximum degree among non-cut vertices (this follows from [11, Corollary 5]):

Lemma 3 ([11]). *Let G' be a subgraph of a reduced graph G and x a vertex in G', such that $G' - x$ is connected and $N_{G'}(x) = \delta$. Then, any tracking set of G contains at least $\delta - 2$ vertices in $G' - x$.*

3 *H*-Minor-Free Graphs

A graph is H***-minor-free*** if it does not contain a fixed graph H as a minor. In this section, we present a linear kernel for H-minor-free graphs and use this kernel, as well as some ideas intrinsic to its construction, to design an efficient polynomial-time approximation scheme (EPTAS). An EPTAS is a $(1 \pm \epsilon)$-approximate algorithm whose running time is $O(n^c)$ for an input of size n and a constant c independent of ϵ.

Unlike the minimum FVS problem, which also consists of covering cycles, TRACKING is not minor-closed [11] (i.e., an optimal solution for a minor of G may require more trackers than an optimal solution for G), so the powerful framework of bidimensionality [22] cannot be used to obtain either linear kernels [30] or PTASs for H-minor-free graphs [14]. Moreover, TRACKING does not possess the "local" properties required by Baker's technique to develop EPTASs for planar graphs [2], or apex-minor-free graphs [17].

Linear Kernel. The following theorem about the sparsity of H-minor-free graphs will be helpful throughout the section.

Theorem 1 (Mader [31]**).** *Any simple H-minor-free graph with n vertices has at most $\sigma_H n$ edges, where σ_H depends solely on $|V(H)|$.*

We now give the following lemma concerning a relationship between the sizes of the vertex sets in certain bipartite minor-free graphs.

[2] This is illustrated in [11], or in the full version of the paper [26].

[3] We consider a leaf in an unrooted tree to be any vertex of degree 1.

Lemma 4. ⊛ *Let $B = (U \cup V, E)$ be a simple H-minor-free bipartite graph, such that: (i) every vertex in V has degree at least 2, and (ii) there exist at most δ neighbors in common between any pair u_1, u_2 in U, i.e., $|N(u_1) \cap N(u_2)| \leq \delta$ for all $u_1, u_2 \in U$. Then $|V| \leq \delta\sigma_H|U|$.*

Next, we give a lemma which will be useful throughout the paper.

Lemma 5. ⊛ *Let F be an FVS of a reduced graph G. Then $|V(G - F)| \leq 4|X| - 5$, where X is the cut set defined by $(F, G - F)$, consisting of edges with endpoints in both F and $G - F$.*

We will use Lemmas 4 and 5 above to give, in the next lemma, a linear kernel for a biconnected reduced H-minor-free graph.

Lemma 6. *Let G be a biconnected reduced H-minor-free graph with start s and finish t. Then, G has at most $(16\sigma_H^2 + 8\sigma_H + 1)OPT - 5$ vertices and at most $(20\sigma_H^2 + 11\sigma_H)OPT - 6$ edges, where OPT denotes the size of an optimal tracking set of G.*

Proof. Let T^* be an optimal tracking set of (G, s, t), i.e., $|T^*| = OPT$. Note that $G - T^*$ is a forest, since T^* is an FVS of G. We assume that $|T^*| \geq 2$, since otherwise one could check, in polynomial time, which vertex of G belongs to T^*. We now give some claims about the structure of G:

Claim 1: Let u_1, u_2 be two vertices in T^*. There exist at most 2 trees in $G - T^*$ that are adjacent[4] to both u_1 and u_2.

Claim 2: Every tree in $G - T^*$ is adjacent to at least 2 vertices in T^*.

Claim 3: Every tree in $G - T^*$ contains at most 2 vertices adjacent to the same vertex in T^*.

The first claim follows from Lemma 3. If there existed 3 or more trees adjacent to both u_1 and u_2, then the graph G', induced by u_1, u_2 and the trees, would require at least 1 tracker in $V(G') \setminus \{u_1\}$ and 1 tracker in $V(G') \setminus \{u_2\}$, contradicting the feasibility of T^*. The last claim also follows from Lemma 3 in a similar fashion. The second claim follows from the fact that G is biconnected.

To show the bound on the size of the vertex set and the edge set of G, we construct a new graph as follows. Let us contract each tree Tr in $G - T^*$ into a ***tree vertex*** v_{Tr}. Let F be the set of all tree vertices. Note that this operation may create parallel edges between a vertex in T^* and a tree vertex, but never between two vertices in T^* or F. Furthermore, we remove any edges between vertices in T^*. The resulting graph is bipartite, with vertex set partitioned into T^* and F, and is H-minor-free (since the class of minor-free graphs is minor-closed). By Claims 1 and 2, any 2 vertices in T^* have at most 2 common neighbors, and every vertex in F is adjacent to at least 2 vertices in T^*. Hence, by Lemma 4,

$$|F| \leq 2\sigma_H|T^*|.$$

[4] In this context, a tree is adjacent to v if it includes a vertex that is adjacent to v.

As a consequence of Claim 3, there are at most 2 parallel edges between a vertex in T^* and a vertex in F. Thus, by Theorem 1, the set of edges, X, in the bipartite graph is at most

$$2 \cdot \sigma_H(|F| + |T^*|) \leq (4\sigma_H^2 + 2\sigma_H)|T^*|.$$

Notice that X is the cut set defined by $(T^*, G - T^*)$, consisting of edges with endpoints in both T^* and $G - T^*$. Hence, by Lemma 5, $|V(G - T^*)| \leq 4|X| - 5$, giving us:

$$|V(G)| \leq (16\sigma_H^2 + 8\sigma_H + 1)|T^*| - 5.$$

The edges of G consist of (a) edges in $G - T^*$ (at most $|V(G - T^*)| - 1$), (b) the cut set X, and (c) edges with both endpoints in T^* (at most $\sigma_H|T^*|$ by Theorem 1). Thus,

$$\begin{aligned}|E(G)| &\leq (4|X| - 6) + |X| + (\sigma_H|T^*|)\\ &\leq (20\sigma_H^2 + 11\sigma_H)|T^*| - 6.\end{aligned}$$

□

By Remark 4 and the application of the above lemma to each biconnected component of a reduced graph, we obtain the following.

Theorem 2. *k-*TRACKING *admits a kernel for H-minor-free graphs of size bounded by $(16\sigma_H^2 + 8\sigma_H + 1)k - 5$ vertices and $(20\sigma_H^2 + 11\sigma_H)k - 6$ edges.*

Corollary 1. TRACKING *admits a $O(1)$-approximation for H-minor-free graphs.*

Even though we develop a $(1 + \epsilon)$-approximation in the next section, the latter corollary can be more useful in practice, when running time is a concern.

EPTAS. Given the unsuitability of bidimensionality and Baker's technique discussed earlier, we shall resort to the use of balanced separators. Our algorithm relies on ***balanced separators***, sets of vertices whose removal partitions the graph into two roughly equal-sized parts. Ungar [33] first showed that every n-vertex planar graph has a balanced separator of size $O(\sqrt{n}\lg^{3/2} n)$. This was later improved by Lipton and Tarjan [28] to $\sqrt{8n}$, and Goodrich [25] showed how to compute these recursively in linear time. The Lipton-Tarjan separator theorem has been further refined (e.g., see [13,15]) and generalized to bounded-genus graphs (e.g., see [16,24]) as well as to H-minor-free graphs (e.g., see [1,32]).

Theorem 3 (Minor-free Separator Theorem [1]). *Let G be an H-minor-free graph with n vertices, where H is a simple graph with $h \geq 1$ vertices. Then a balanced separator for G of size at most $c_H^1\sqrt{n}$ can be found in $O(h^{O(1)}n^{O(1)})$ time, where c_H^1 is a positive constant depending solely on h.*

We use the Minor-free Separator Theorem recursively to decompose the graph into a set $\mathcal{R}$ of edge-disjoint subgraphs, called ***regions***. The vertices of a region $R \in \mathcal{R}$ which belong to at least one other region are called ***boundary vertices*** and the set of these vertices is denoted by $\partial(R)$. The remaining vertices of R are called ***interior vertices*** and are denote by $int(R)$.

Definition 5 (Relaxed r-division). *A* ***relaxed*** r***-division*** *of an n-vertex graph G is a decomposition of G into $\Theta(n/r)$ regions, each of which has at most r vertices, such that the total number boundary vertices is $O(n/\sqrt{r})$.*

Computing a relaxed r-division is the first step in Frederickson's algorithm [23] for constructing an r-division in a planar graph, a decomposition which additionally requires every region to have $O(\sqrt{r})$ boundary vertices (we won't need this property). Both decompositions can easily be generalized to any class of graphs that is characterized by the existence of sublinear balanced separators, which includes H-minor-free graphs.

Theorem 4 (Minor-free Separator Theorem (3) + Frederickson [23]). *There is an $O(n \lg n)$ algorithm that, given an H-minor-free graph G and a positive integer r, computes a relaxed r-division of G.*

Our strategy will be to (i) construct a relaxed r-division of a smaller graph, K, which is itself an $O(1)$-approximate tracking set, (ii) solve optimally for each region, and (iii) combine the solutions for each region into a solution for the original graph with quality comparable to that of an optimal solution. This approach has been used to obtain EPTASs for minimum FVS [6,39], maximum independent set [29] and minimum vertex cover [8]. However, and in contrast to these problems, the step of constructing a close to optimal solution from the solutions of each region is not obvious. Indeed, the difficulty of this step emerges from the very "nonlocal" structure of TRACKING, which requires special attention to the location of (s, t) in the graph, in addition to the nonlocal structure of cycles. Our EPTAS is as follows:

1. Compute a linear kernel K of G by reducing it with Rules 1, 2, 3, such that an optimal tracking set of K is $\Omega(|V(K)|)$ (see Corollary 1).
2. Compute a relaxed r-division $\mathcal{R}$ of K with $r = (2c_H^1 c_H^2 (c_H^3 + 1)/\epsilon)^2$, for any choice of $\epsilon > 0$ and constants $c_H^1, c_H^2, c_H^3 > 0$ specified later.
3. For each region R in $\mathcal{R}$, compute an optimal tracking set $OPT(R)$ for the subset of entry-cycles (with respect to (s, t)) which are completely contained in R.
4. Output $T = \bigcup_{R \in \mathcal{R}} (OPT(R) \cup \partial(R) \cup \mathcal{N}(R))$.
Here, $\mathcal{N}(R) := N_{\Pi(R)}(\partial(\Pi(R)))$ defines an appropriate neighborhood of the boundary vertices of R, where $\Pi(R)$ is the subgraph of R consisting of the union of each path in R that: (i) is not an edge, (ii) has $\partial(R)$ vertices as endpoints, and (iii) traverses no ***internal*** vertices that are in $OPT(R)$. We let $\partial(\Pi(R)) := \partial(R) \cap \Pi(R)$. See Fig. 3.

We will now give the details of the algorithm and its correctness. We refer to the Reduction Rules defined in Sect. 2. As a reminder, after exhaustive application of Rules 1 and 2, the graph is either a single edge between s and t, or all its vertices have degree at least 2. Henceforth, we will assume the latter, since a minimum tracking set is trivial in the former. Notice that none of the reduction rules introduce trackers, so there is no lifting required at the end of our algorithm, i.e., adding back any trackers introduced during the reduction.

Observation 1. *No entry-exit cycles are removed during Rules 1, 2 or 3, so a tracking set of the resulting kernel K is a tracking set of the input graph G. Therefore, any minimum tracking set of K is also a minimum tracking set of G.*

Next, we explain how to compute in polynomial time optimal tracking sets for each region in a relaxed r-division of a kernel K.

Lemma 7. ⊛ *Let $\mathcal{C}(R)$ be the set of all entry-exit cycles in G whose vertices are a subset of $V(R)$, where R is a subgraph of G. Then one can compute a minimum subset of $V(R)$ that tracks every entry-cycle of $\mathcal{C}(R)$ in $O(2^{|V(R)|} \cdot n^{O(1)})$ time.*

Let us now argue that our algorithm computes a $(1+\epsilon)$-approximate tracking set. Let $T = \bigcup_{R \in \mathcal{R}} (OPT(R) \cup \partial(R) \cup \mathcal{N}(R))$ be the output of the algorithm.

Lemma 8. ⊛ *T is a tracking set of the input graph G.*

Let us denote by OPT the size of an optimal tracking set of the input graph G. To argue that $|T| \leq (1+\epsilon)OPT$, we will need to argue that the set of trackers in the special neighborhoods defined by $\mathcal{N}(R)$, for all regions R, have small cardinalities, i.e., roughly equal to $O(\epsilon OPT)$. This is the key argument to our EPTAS, which the next lemma addresses. Its proof is not immediately obvious, since the number of neighbors of all boundary vertices could be $\Omega(OPT)$, a consequence of the quadratic gap between $|\partial(R)|$ and $|V(R)|$.

Lemma 9. ⊛ *$|\mathcal{N}(R)| \leq c_H^3 |\partial(\Pi(R))|$, where $c_H^3 \geq 9\sigma_H^2 + 3\sigma_H$.*

Proof. (Sketch) The set of untracked cycles between 2 regions R and R', which must exist in $\Pi(R) \cup \Pi(R')$, induces a forest on either region if we remove $\partial(R)$ and $\partial(R')$. Using arguments similar to those in the proof of Lemma 6, we can show that the bipartite graph with bipartition $(F, \partial(\Pi(R)))$ has the properties required by Lemma 4, but also that there exists $O(1)$ edges between a tree and a boundary vertex, where F is the set of trees in $\Pi(R) - \partial(\Pi(R))$. As a consequence, we can get an appropriate bound on the number of edges in this bipartite graph, from which the lemma follows. (See [26] for details.) □

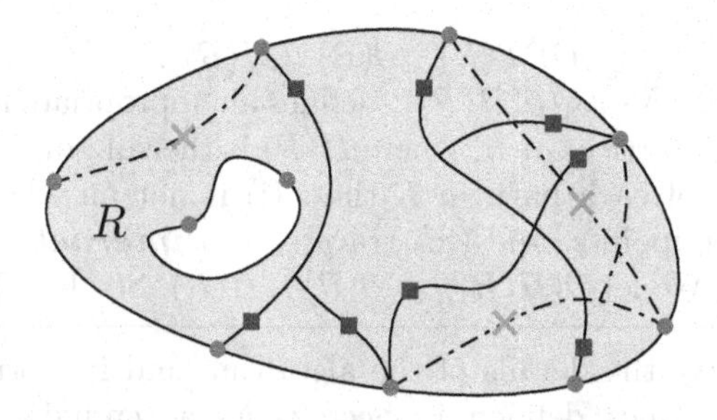

Fig. 3. Illustration of $\Pi(R)$ and of $\mathcal{N}(R)$ for a region R in a relaxed r-division $\mathcal{R}$. Vertices in $\partial(R)$ are depicted in red circles. $\Pi(R)$ consists of the union of all boundary-to-boundary paths in R (solid black), which are not edges and do not traverse $OPT(R)$ (green crosses). The dashed lines represent paths in $R - \Pi(R)$. $\mathcal{N}(R)$ is depicted in blue squares. (Color figure online)

Before proving that the output of our algorithm is a $(1+\epsilon)$-approximate tracking set, let us first recall a result from Frederickson [23, Lemma 1], which concerns the sum, for each boundary vertex b of the number of regions $\Delta(b)$ containing b in a relaxed r-division $\mathcal{R}$ of a planar graph. Even though this result was given in the context of planar graphs, it can easily be generalized to any graph whose subgraphs G' admit balanced separators of size $O(\sqrt{|V(G')|})$. We denote the set of all boundary vertices by $\partial(\mathcal{R})$. Further, let $B(\mathcal{R}) = \sum_{b\in\partial(\mathcal{R})}(\Delta(b)-1)$.

Lemma 10 **([23]).** *Let $\mathcal{R}$ be a relaxed r-division of an n-vertex graph whose subgraphs G' admit balanced separators of size at most $c\sqrt{|V(G')|}$. Then $B(\mathcal{R}) \leq c \cdot n/\sqrt{r}$, for a constant c independent of r and n.*

We will use the latter lemma to bound the overall number of trackers in the next theorem.

Theorem 5. TRACKING *admits an EPTAS for H-minor-free graphs.*

Proof. Consider the algorithm given at the beginning of the section. As a reminder, let $T = \bigcup_{R\in\mathcal{R}}(OPT(R)\cup\partial(R)\cup\mathcal{N}(R))$ be the output of the algorithm, for a relaxed r-division $\mathcal{R}$ of a kernel K of G, where $OPT(R)$ is the optimal tracking set computed with respect to entry-exit cycles in R. By Lemma 8, T is a tracking set. Next, we argue about the approximation ratio. By a union bound,

$$|T| \leq |\partial(\mathcal{R})| + \sum_{R\in\mathcal{R}}|OPT(R)| + \sum_{R\in\mathcal{R}}|\mathcal{N}(R)|.$$

Let $n' = |V(K)|$ be the number of vertices in K. Clearly, $|\partial(\mathcal{R})| \leq B(\mathcal{R})$. Moreover, we have that $\sum_{R\in\mathcal{R}}|\partial(R)| \leq 2B(\mathcal{R})$, so by Lemma 9, we have:

$$\sum_{R\in\mathcal{R}}|\mathcal{N}(R)| \leq 2c_H^3 B(\mathcal{R}).$$

Let T^* be an optimal tracking set of K, i.e., $|T^*| = OPT$ (by Observation 1). Since T^* is a tracking set, but not necessarily an optimal one, for all entry-exit cycles within any region $R\in\mathcal{R}$, we have that $|OPT(R)| \leq |T^*\cap V(R)|$. Thus,

$$\sum_{R\in\mathcal{R}}|OPT(R)| \leq OPT + B(\mathcal{R}).$$

Overall, for $r = (2c_H^1 c_H^2(c_H^3+1)/\epsilon)^2$,

$$\begin{aligned} |T| &\leq OPT + 2(c_H^3+1)B(\mathcal{R}) \\ &\leq OPT + 2c_H^1(c_H^3+1)n'/\sqrt{r} && \text{(Lemma 10, Theorem 3)} \\ &\leq OPT + 2c_H^1c_H^2(c_H^3+1)OPT/\sqrt{r} && (\text{Theorem 2}, c_H^2 \geq 16\sigma_H^2+8\sigma_H+1) \\ &= (1+\epsilon)OPT. \end{aligned}$$

Step 1 of the algorithm takes $O(n^{O(1)})$ time, since it consists of applying Rules 1, 2, 3. Step 2 can be done in $O(n \lg n)$ time [23]. Step 3 takes $O(2^r \cdot n^{O(1)})$ time, by Lemma 7. Finally, step 4 takes $O(n^{O(1)})$ time. Overall, these amount to $O(2^{O(1/\epsilon^2)} n^{O(1)})$. □

4 General Graphs

In this section, we derive an $O(\lg n)$-approximation algorithm for WEIGHTEDTRACKING on general graphs, as well as an $O(\lg OPT)$-approximation algorithm for TRACKING.

We reduce an instance (G, s, t, w') of WEIGHTEDTRACKING, for a weight function $w' : V(G) \to \mathbb{Q}$, into an instance $(\mathcal{U}, \mathcal{X}, w)$ of SETCOVER, which asks for the sub-collection of $\mathcal{X}$ of minimum total weight, whose union equals the universe $\mathcal{U}$. Here, $(\mathcal{U}, \mathcal{X})$ defines a set system, i.e., a collection $\mathcal{X}$ of subsets of a set $\mathcal{U}$, and w is the weight function $w : \mathcal{X} \to \mathbb{Q}$. It is well known that there exist greedy polynomial-time algorithms achieving approximation ratios of $(1 + \ln M)$ or of $(1 + \Delta)$ [35,38], where M is the size of the largest set in $\mathcal{X}$ and Δ is the maximum number, over all elements u in $\mathcal{U}$, of sets in $\mathcal{X}$ that contain u.

Let $\mathcal{C}$ be the set of all entry-exit cycles in our input graph G, which we assume w.l.o.g. to be reduced by Rule 1. Further, let $\mathcal{C}_F$ be the set of all entry-exit cycles in G, each of which contains at most 2 vertices from the subset $F \subseteq V$. That is, $\mathcal{C}_F := \{(C, s', t') \in \mathcal{C} : |C \cap F| \leq 2\}$. Our algorithm is as follows.

1. Compute a 2-approximate FVS F of G (see [35,38]).
2. Use the greedy algorithm of [35,38] to compute an approximate set covering, $S \subseteq V(G)$, for an instance $(\mathcal{U}, \mathcal{X}, w)$ of SETCOVER where:
   (i) the universe, $\mathcal{U}$, of elements to be covered is $\mathcal{C}_F$
   (ii) the collection of covering sets, $\mathcal{X}$, is a 1-1 correspondence with $V(G)$, where each covering set with corresponding vertex v is the subset of $\mathcal{C}_F$ which are tracked by v, that is,
   $$\mathcal{X} = \{\{(C, s', t') \in \mathcal{C}_F \mid v \text{ tracks } (C, s', t')\}\}_{v \in V(G)}.$$
   (iii) the weight function w is the weight function w' defined for WEIGHTEDTRACKING, given the 1-1 correspondence between $\mathcal{X}$ and $V(G)$.
3. Output $T = S \cup F$.

We can show that $|\mathcal{C}_F| = O(n^{O(1)})$. From the observation that every tracking set F is an FVS (see Remark 2), it follows that there are at most $O(n^{O(1)})$ entry-exit cycles not tracked by F. Thus, our claim follows (details in [26]).

Theorem 6. ⊛ WEIGHTEDTRACKING *admits an* $O(\lg n)$*-approximation.*

Unweighted Graphs. We show that the dual of the above set cover formulation has bounded VC-dimension [27,34]. This immediately improves the approximation ratio to $O(\lg OPT)$ for TRACKING (unweighted version) as a consequence of a result by Brönnimann and Goodrich [7], which establishes an approximation-ratio of $O(d \lg(dc))$ for unweighted set cover instances with dual VC-dimension d and optimal covers of size at most c.

Let $(\mathcal{U}, \mathcal{X})$ be a set system and Y a subset of $\mathcal{U}$. We say that Y is ***shattered*** if $\mathcal{X} \cap Y = 2^Y$, where $\mathcal{X} \cap Y := \{X \cap Y \mid X \in \mathcal{X}\}$. In other words, Y is shattered if the set of intersections of Y with each $X \in \mathcal{X}$ contains all the possible subsets of Y. The set system $(\mathcal{U}, \mathcal{X})$ has ***VC-dimension*** d if d is the largest integer for which there exists a subset $Y \subseteq \mathcal{U}$, of cardinality $|Y| = d$, that can be shattered.

The dual problem of an unweighted instance $(\mathcal{U}, \mathcal{X})$ of SetCover is finding a ***hitting set*** of minimum size, where a hitting set is a subset of $\mathcal{U}$ that has a non-empty intersection with every set in $\mathcal{X}$. In our case, it corresponds to finding the smallest subset of entry-exit cycles that covers every vertex, where a vertex is covered if it tracks least one entry-exit cycle in the subset. This is equivalent to an unweighted instance of SetCover with set system $(V, \mathcal{C}_F^*)$, where $V = V(G)$ and $\mathcal{C}_F^* := \{V(C) \setminus \{s', t'\} : (C, s', t') \in \mathcal{C}_F\}$ is the collection of sets, one for each entry-exit cycle, of vertices which can track that entry-exit cycle.

Lemma 11. *The set system $(V, \mathcal{C}_F^*)$ has VC-dimension at most 9.*

Proof. We show that there exists no subset $Y \subseteq V$ of size $|Y| \geq 10$ that can be shattered by $\mathcal{C}_F^*$. Since every element of $\mathcal{C}_F^*$ contains at most 2 vertices from F (by definition of $\mathcal{C}_F$), we cannot have more than 2 vertices from F in Y (since we would then require an entry-exit cycle containing at least 3 vertices in F to shatter Y). Thus, the lemma follows if we show that no subset $Y \subseteq V \setminus F$ of size $|Y| \geq 8$ can be shattered by $\mathcal{C}_F^*$. Let us assume, by contradiction, that this is possible. Then, if $Y \subseteq V \setminus F$ is to be shattered by $\mathcal{C}_F^*$, there must exist 2 entry-exit cycles (C_1, s_1', t_1') and (C_2, s_2', t_2') in $\mathcal{C}_F$, such that[5]:

- C_1 traverses all vertices of Y, say in the order $y_1, y_2, \ldots, y_{|Y|}$ (for all $y_j \in Y$),
- C_2 traverses every other vertex of Y traversed by C_1, say $Y' = \{y_2, y_4, \ldots, y_{|Y|}\}$, but not necessarily in the same order (we assume w.l.o.g. $|Y|$ is even).

Consider the graph consisting of the union of the cycles C_1, C_2. Let us contract every shared edge between C_1, C_2. Note that C_1 remains a cycle that traverses Y and C_2 remains a cycle that traverses Y' but not any vertex of $Y \setminus Y'$. So we can safely assume that C_1 and C_2 do not share any edges. Thus, the union of C_1, C_2 is a graph with $|C_1| + |C_2| - |Y|/2$ vertices and $|C_1| + |C_2|$ edges. Since both entry-exit cycles are in $\mathcal{C}_F$, each of C_1, C_2 shares at most 2 vertices with F. Let us remove such vertices, say there's $k \leq 4$ of them. The result is a graph with $|C_1| + |C_2| - |Y|/2 - k$ vertices and, at best, $|C_1| + |C_2| - 2k$ edges (the removed vertices cannot be in Y, so they have degree 2). In order for this graph to be acyclic (since F is an FVS by Remark 2, and our contractions preserve cycles) we would then require $|Y| < 8$ (since any acyclic graph with n vertices has at most $n - 1$ edges), a contradiction. □

[5] This is illustrated in the full version of the paper [26].

The above lemma, combined with the result of Brönnimann and Goodrich [7] gives us the following.

Theorem 7. TRACKING *admits an* $O(\lg OPT)$*-approximation, where* OPT *is the size of an optimal tracking set.*

References

1. Alon, N., Seymour, P.D., Thomas, R.: A separator theorem for graphs with an excluded minor and its applications. In: Ortiz, H. (ed.) STOC, pp. 293–299. ACM (1990). https://doi.org/10.1145/100216.100254
2. Baker, B.S.: Approximation algorithms for NP-complete problems on planar graphs. J. ACM **41**(1), 153–180 (1994). https://doi.org/10.1145/174644.174650
3. Banik, A., Choudhary, P., Lokshtanov, D., Raman, V., Saurabh, S.: A polynomial sized kernel for tracking paths problem. Algorithmica **82**(1), 41–63 (2019). https://doi.org/10.1007/s00453-019-00602-8
4. Banik, A., Katz, M.J., Packer, E., Simakov, M.: Tracking paths. In: Fotakis, D., Pagourtzis, A., Paschos, V.T. (eds.) CIAC 2017. LNCS, vol. 10236, pp. 67–79. Springer, Cham (2017). https://doi.org/10.1007/978-3-319-57586-5_7
5. Bilò, D., Gualà, L., Leucci, S., Proietti, G.: Tracking routes in communication networks. Theor. Comput. Sci. **844**, 1–15 (2020). https://doi.org/10.1016/j.tcs.2020.07.012
6. Borradaile, G., Le, H., Zheng, B.: Engineering a PTAS for minimum feedback vertex set in planar graphs. In: Kotsireas, I., Pardalos, P., Parsopoulos, K.E., Souravlias, D., Tsokas, A. (eds.) SEA 2019. LNCS, vol. 11544, pp. 98–113. Springer, Cham (2019). https://doi.org/10.1007/978-3-030-34029-2_7
7. Brönnimann, H., Goodrich, M.T.: Almost optimal set covers in finite VC-dimension. Discret. Comput. Geom. **14**(4), 463–479 (1995). https://doi.org/10.1007/BF02570718
8. Chiba, N., Nishizeki, T., Saito, N.: Applications of the Lipton and Tarjan's planar separator theorem. J. Inf. Process **4**(4), 203–207 (1981)
9. Chokchai, C.: Low cost and high performance UHF RFID system using Arduino based on IoT applications for marathon competition. WPMC **2018**, 15–20 (2018). https://doi.org/10.1109/WPMC.2018.8713018
10. Choudhary, P.: Polynomial time algorithms for tracking path problems. In: Gąsieniec, L., Klasing, R., Radzik, T. (eds.) IWOCA 2020. LNCS, vol. 12126, pp. 166–179. Springer, Cham (2020). https://doi.org/10.1007/978-3-030-48966-3_13
11. Choudhary, P., Raman, V.: Improved kernels for tracking path problems. CoRR abs/2001.03161 (2020). http://arxiv.org/abs/2001.03161
12. Choudhary, P., Raman, V.: Structural parameterizations of tracking paths problem. In: Cordasco, G., Gargano, L., Rescigno, A.A. (eds.) CEUR. CEUR Workshop Proceedings, vol. 2756, pp. 15–27. CEUR-WS.org (2020)
13. Chung, F.R.: Separator theorems and their applications. Universität Bonn, Institut für Ökonometrie und Operations Research (1988)
14. Demaine, E.D., Hajiaghayi, M.T.: Bidimensionality: new connections between FPT algorithms and PTASs. In: SODA, pp. 590–601. SIAM (2005)
15. Djidjev, H., Venkatesan, S.M.: Reduced constants for simple cycle graph separation. Acta Informatica **34**(3), 231–243 (1997). https://doi.org/10.1007/s002360050082

16. Djidjev, H.N.: A linear algorithm for partitioning graphs of fixed genus. Serdica. Bulgariacae mathematicae publicationes **11**(4), 369–387 (1985)
17. Eppstein, D.: Diameter and treewidth in minor-closed graph families. Algorithmica **27**(3), 275–291 (2000). https://doi.org/10.1007/s004530010020
18. Eppstein, D., Goodrich, M.T.: Studying (non-planar) road networks through an algorithmic lens. In: SIGSPATIAL. GIS, ACM (2008). https://doi.org/10.1145/1463434.1463455
19. Eppstein, D., Goodrich, M.T., Liu, J.A., Matias, P.: Tracking paths in planar graphs. In: Lu, P., Zhang, G. (eds.) ISAAC. LIPIcs, vol. 149, pp. 54:1–54:17. Schloss Dagstuhl - Leibniz-Zentrum für Informatik (2019). https://doi.org/10.4230/LIPIcs.ISAAC.2019.54
20. Eppstein, D., Gupta, S.: Crossing patterns in nonplanar road networks. In: SIGSPATIAL. GIS, ACM (2017). https://doi.org/10.1145/3139958.3139999
21. Eppstein, D., Khodabandeh, H.: On the edge crossings of the greedy spanner. CoRR abs/2002.05854 (2020). https://arxiv.org/abs/2002.05854
22. Fomin, F.V., Demaine, E.D., Hajiaghayi, M.T., Thilikos, D.M.: Bidimensionality. In: Kao, M.-Y. (ed.) Encyclopedia of Algorithms, pp. 203–207. Springer, Heidelberg (2016). https://doi.org/10.1007/978-1-4939-2864-4_47
23. Frederickson, G.N.: Fast algorithms for shortest paths in planar graphs, with applications. SIAM J. Comput. **16**(6), 1004–1022 (1987). https://doi.org/10.1137/0216064
24. Gilbert, J.R., Hutchinson, J.P., Tarjan, R.E.: A separator theorem for graphs of bounded genus. J. Algorithms **5**(3), 391–407 (1984). https://doi.org/10.1016/0196-6774(84)90019-1
25. Goodrich, M.T.: Planar separators and parallel polygon triangulation. J. Comput. Syst. Sci. **51**(3), 374–389 (1995)
26. Goodrich, M.T., Gupta, S., Khodabandeh, H., Matias, P.: How to catch marathon cheaters: new approximation algorithms for tracking paths. CoRR abs/2104.12337 (2021). https://arxiv.org/abs/2104.12337
27. Haussler, D., Welzl, E.: ϵ-nets and simplex range queries. Discret. Comput. Geom. **2**(2), 127–151 (1987). https://doi.org/10.1007/BF02187876
28. Lipton, R.J., Tarjan, R.E.: A separator theorem for planar graphs. SIAM J. Appl. Math. **36**(2), 177–189 (1979)
29. Lipton, R.J., Tarjan, R.E.: Applications of a planar separator theorem. SIAM J. Comput. **9**(3), 615–627 (1980). https://doi.org/10.1137/0209046
30. Lokshtanov, D.: Kernelization, bidimensionality and kernels. In: Kao, M.-Y. (ed.) Encyclopedia of Algorithms, pp. 1006–1011. Springer, Heidelberg (2016). https://doi.org/10.1007/978-1-4939-2864-4_526
31. Mader, W.: Homomorphiesätze für graphen. Math. Ann. **178**(2), 154–168 (1968)
32. Reed, B.A., Wood, D.R.: A linear-time algorithm to find a separator in a graph excluding a minor. ACM Trans. Algorithms **5**(4), 39:1–39:16 (2009). https://doi.org/10.1145/1597036.1597043
33. Ungar, P.: A theorem on planar graphs. J. Lond. Math. Soc. **1**(4), 256–262 (1951)
34. Vapnik, V.N., Chervonenkis, A.Y.: On the uniform convergence of relative frequencies of events to their probabilities. In: Vovk, V., Papadopoulos, H., Gammerman, A. (eds.) Measures of Complexity, pp. 11–30. Springer, Cham (2015). https://doi.org/10.1007/978-3-319-21852-6_3
35. Vazirani, V.V.: Approximation Algorithms. Springer, Heidelberg (2001)
36. Want, R.: An introduction to RFID technology. IEEE Pervasive Comput. **5**(1), 25–33 (2006). https://doi.org/10.1109/MPRV.2006.2

37. Wikipedia contributors: Marathon course-cutting (2019). https://en.wikipedia.org/wiki/Marathon_course-cutting. Accessed 16 Feb 2021
38. Williamson, D.P., Shmoys, D.B.: The Design of Approximation Algorithms. Cambridge University Press, Cambridge (2011)
39. Zheng, B.: Approximation schemes in planar graphs. Ph.D. thesis, Oregon State University (2018). https://ir.library.oregonstate.edu/concern/graduate_thesis_or_dissertations/7w62ff609. Accessed 07 Jan 2021

Algorithms for Radius-Optimally Augmenting Trees in a Metric Space

Joachim Gudmundsson and Yuan Sha(✉)

The University of Sydney, Sydney, NSW 2006, Australia
joachim.gudmundsson@sydney.edu.au, ysha3185@uni.sydney.edu.au

Abstract. Let T be a tree with n vertices embedded in a metric space. We consider the problem of adding one edge to T to minimize the radius of the resulting graph.

For the *continuous* version of the problem where a center may be a point in the interior of an edge of the graph we give a linear time algorithm. In the case when the center is restricted to lie on a vertex, the *discrete* version, we give an $O(n \log n)$ expected time algorithm.

Previously linear-time algorithms were known for the special case when the input graph is a path.

1 Introduction

Given a graph and a positive integer k. The *Radius-Optimally Augmenting Graph* (ROAG) problem asks to add k edges to graph such that the radius of the resulting graph is minimized.

Let $G = (V, E)$ be a graph with n vertices. A *point* on G can be either at a vertex of G, or in the interior of an edge of E. The eccentricity of a point c on G is the maximum shortest path length from c to all the vertices of G. The point c on G with minimum eccentricity is called the center of G, and the eccentricity of c is said to be the *radius* of G.

Two variants have been considered in the literature: the *discrete* version when the center is restricted to lie on a vertex of G and the *continuous* version where a center may be a point in the interior of an edge of G.

The discrete version of the problem cannot be approximated within a factor of $(5/3 - \varepsilon)$ in polynomial time, for any $\varepsilon > 0$, unless $P = NP$ [10]. In the same paper a cubic time algorithm is given for the case when G is a metric tree. For the special case when $k = 1$ and the input graph is a path embedded in a metric space, Johnson and Wang [11] gave a linear-time algorithm for the continuous version and recently Wang and Zhao [16] gave a linear-time algorithm for the discrete version.

In this paper we consider the case when the input graph is a tree $T = (V, E)$ embedded in a metric space and $k = 1$, that is, add one edge $e \in V^2 \setminus E$ to T such the radius of the resulting graph $T' = T \cup \{e\}$ is minimized. We refer to this problem as the *Radius-Optimally Augmenting Trees problem*, or ROAT for short. We prove the following two results:

A. Lubiw et al. (Eds.): WADS 2021, LNCS 12808, pp. 457–470, 2021.
https://doi.org/10.1007/978-3-030-83508-8_33

- a linear-time algorithm for the continuous version, and
- an $O(n \log^2 n)$ time algorithm for the discrete case, which is then improved to $O(n \log n)$ expected time by a randomization.

The ROAG problem is closely related to the problem of minimising the diameter of the augmented graph, which has received considerable attention in the community [3,6–8]. The general diameter-optimally augmentation problem of adding k edges to a graph was shown to be NP-hard [13] and even $W[2]$-hard [7,8]. Due to the hardness of the problem a number of approximation algorithms and special cases have been considered. Below we only discuss exact algorithms for restricted graphs for $k = 1$ since it is more relevant to our setting.

Most of the existing research has focused on the discrete case where only vertices are considered. In 2015 Große et al. [9] gave an $O(n \log^3 n)$ time algorithm for the case when the input graph is a metric path. Wang [14] later improved the running time to $O(n \log n)$. For metric trees Bilò [2] gave an $O(n \log n)$ time algorithm. Note that Bilò's algorithm cannot be used to solve our ROAT problem. Section 3 Fig. 2(b) gives an example where the optimal shortcut for the discrete ROAT problem is not the optimal shortcut for minimizing the diameter of the augmented graph. For arbitrary trees the problem is harder. Oh and Ahn [12] argued for an $\Omega(n^2)$ lower bound and designed an $O(n^2 \log^3 n)$ time algorithm for the problem. Very recently, Wang and Zhao [15] gave an improved $O(n^2 \log n)$ time algorithm for the problem.

Less is known for the continuous case. For paths in the Euclidean plane De Carufel et al. showed a linear-time algorithm [4], and for trees in the Euclidan plane an $O(n \log n)$ time algorithm was presented [5]. Oh and Ahn [12] also considered the continuous version in their paper and gave an $O(n^2 \log^3 n)$ time algorithm for arbitrary trees.

1.1 Our Approach

For both the discrete ROAT and continuous ROAT, we start by pre-computing a longest path P in T, which can be done in linear time. For the continuous version we then prove that there exists an optimal solution (u, v) such that both u and v lie on P. By using this observation we are able to reduce the continuous ROAT problem to the *Radius-Optimally Augmenting Monotone vertex-Weighted Paths* (ROAMWP problem) by making a vertex-weight transformation. The continuous ROAMWP problem can be solved by applying the algorithm by Johnson and Wang for paths [11] with only one minor modification. Each step of the algorithm only requires linear time.

For the discrete version the situation is slightly more complicated. For this case we observe that there exists an optimal edge such that one of the endpoints of the optimal edge must lie on the longest path and the other endpoint must coincide with the center of the augmented graph. We develop an $O(n \log^2 n)$ time algorithm, which is then improved to $O(n \log n)$ expected time by using a randomization.

2 Preliminaries

We will briefly define some of the notations that will be used throughout the paper. For a graph G embedded in a metric space, let $|uv|$ denote the distance between two vertices u and v in V and let $d_G(u,v)$ denote the length of the shortest path between u and v. The eccentricity of a point p on G, denoted as $ecc_G(p)$, is the length of the longest shortest path from p to all the vertices of G. The point c in G with the smallest eccentricity is called the center of G.

Let $P = \langle v_1, v_2, \ldots, v_n \rangle$ be a path embedded in a metric space. For any two points p and q on P, let $P(p,q)$ denote the subpath of P between p and q. For any two vertices v_i and v_j on P, let $G(i,j) = P \cup (v_i, v_j)$.

If each vertex v_k, $1 \leqslant k \leqslant n$, in P is assigned a weight $w(k)$, then P is said to be *vertex weighted.* The vertex-weighted distance from a point p to a vertex v_j is $d_P(p, v_j) + w(j)$, i.e. the path length from p to v_j plus the weight of v_j. Similarly, the vertex-weighted distance from a point p on $G(i,j)$ to a vertex v_j in $G(i,j)$ is $d_{G(i,j)}(p, v_j) + w(j)$.

For a tree T, let $T(p,q)$ denote the unique simple path between two points p and q on T.

3 An $O(n \log n)$ Expected Time Algorithm for the Discrete ROAT problem

As input we are given a tree $T = (V, E)$ embedded in a metric space. We consider the discrete version of the ROAT problem where the center of the augmented graph is restricted to lie on a vertex of V. The presented algorithm will consider two different cases: (1) the center of the augmented tree lies on a longest path P of T (Sect. 3.1), and (2) the center of the augmented tree does not lie on P (Sect. 3.2).

Given a vertex v in T, consider the problem of adding an edge to T such that the augmented tree T' minimizes the eccentricity of v. We call this problem the Eccentricity Augmenting Tree (EAT) problem with respect to v. We start with the following useful lemma.

Lemma 1. *Given a tree $T = (V, E)$ embedded in a metric space and a vertex s in V, there is an optimal solution (s, q) for the* EAT *problem with respect to s.*

Proof. Assume (u, v) is added to T such that the eccentricity of s is minimized in $G = (V, E \cup \{(u, v)\})$. Let T^* be the shortest path tree of G rooted at s, as illustrated in Fig. 1a. If (u, v) is incident to s then the lemma holds immediately. Otherwise (u, v) is not incident to s. We will prove that (u, v) can be replaced by either (s, u) or (s, v) such that the eccentricity of s in the resulting graph does not increase.

If (u, v) is an edge in T^*, there is a set U of vertices that use (u, v) on their shortest paths to s. Without loss of generality, assume u is closer to s than v, as in Fig. 1. The set U consists of all the vertices in the subtree $T^*(v)$ of T^* rooted at v. By triangle inequality, we have $|sv| \leqslant d_{T^*}(s, v)$. Replace (u, v) by (s, v).

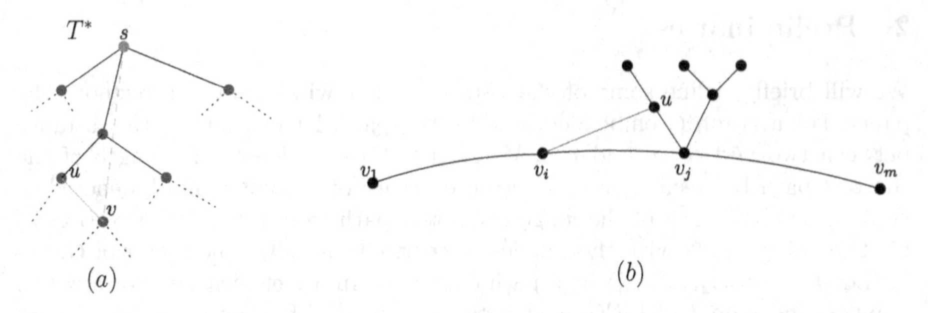

Fig. 1. (a) A shortest path tree T^* of T rooted at s. (b) Illustration for the proof of Lemma 2.

Then every vertex in U now have a path to s no longer than its path to s in T^*. Vertices not in U can still use their paths to s in T^*. So the eccentricity of s in the graph after replacement has not increased.

If (u, v) is not an edge in T^* then we can replace (u, v) by either (s, u) or (s, v). Since every vertex can still use its shortest path in T^*, the eccentricity of s does not increase. This completes the proof. $\square$

3.1 Case 1: The Center Lies on P

We start by considering the case when the center of the optimal augmented tree lies on the longest path of T. Consider a longest path $P = \langle v_1, \ldots, v_m \rangle$ of T, which can be computed in $O(n)$ time. If we remove the edges in P we obtain a forest with m trees. Let T_i, $1 \leqslant i \leqslant m$, be the tree rooted at v_i.

Next we prove that there exists an optimal edge (x, u) for the EAT problem with respect to v for which x coincides with v and u lies on P.

Lemma 2. *Given a tree T embedded in a metric space and a vertex v_i on the longest path $P = \langle v_1, \ldots, v_m \rangle$ of T, there exists an optimal solution (x, u) for the* EAT *problem with respect to v_i such that x coincides with v_i and u lies on the subpath of P between v_i and v_l where $l = \arg\max_{k \in \{1,m\}} \{d_P(v_i, v_k)\}$.*

Proof. Without loss of generality, assume $v_l = v_m$, hence, $d_P(v_i, v_m) \geqslant d_P(v_i, v_1)$, as shown in Fig. 1(b).

According to Lemma 1, x coincide with v_i. Consider the placement of u, and assume that it belongs to the tree T_j, $1 \leqslant j \leqslant m$. If $j \leqslant i$ then the eccentricity of v_i is not reduced by adding (v_i, u) since the eccentricity is decided by $d_T(v_i, v_m)$.

It remains to consider the case when $j > i$ and u does not coincide with v_j. We will show that edge (v_i, v_j) is at least as good as (v_i, u). We may assume that $|v_i u| + d_T(u, v_j) \leqslant d_P(v_i, v_j)$ since otherwise the eccentricity of v_i is again equal to $d_T(v_i, v_m)$ and we are done. Let $T^* = T \cup (v_i, u)$. For any vertex q in T_j we have

$$\begin{aligned} d_{T^*}(v_i, q) &\leqslant |v_i u| + d_T(u, v_j) + d_T(v_j, q) \\ &\leqslant |v_i u| + d_T(u, v_j) + d_P(v_j, v_m) \\ &= d_{T^*}(v_i, v_m). \end{aligned}$$

Thus the vertex farthest from v_i in $\cup_{i<j\leqslant m} T_j$ is either v_m or a vertex on T_k where $i < k < j$. By the triangle inequality, $|v_i u| + d_T(u, v_j) \geqslant |v_i v_j|$ which implies that any edge from v_i to a vertex in T_j can not be better than the edge (v_i, v_j). This proves the lemma. □

The above lemma states that if the center of the optimal augmented tree lies on P then we only need to consider shortcuts with one endpoint at this center and the other endpoint at a vertex of P. A possible approach is to compute an optimal shortcut (EAT) for every vertex on the longest path P. The vertex on the longest path with smallest eccentricity in the augmented tree (for EAT) is the center, and the computed edge is the final output.

Following the observation made in Lemma 2, one approach would be to scan P from v_1 until reaching the first v_i for which $d_P(v_i, v_m)$ is less than $d_P(v_i, v_1)$. At that point the algorithm reverse the direction and scan P from v_m. Scanning from both directions are symmetric so we will only discuss scanning from v_1 below.

For every vertex v_j along P let $w(j) = \max_{u\in T_j}\{d_T(v_j, u)\}$ be the *weight* of v_j. The vertex-weighted distance $d_P(v_i, v_j) + w(j)$ is the maximum distance from v_i to any vertex in T_j. Let $G(i, j) = P \cup (v_i, v_j)$ and let

$$\begin{aligned} S(i, j) &= \max_{i<l\leqslant j}\{d_{G(i,j)}(v_i, v_l) + w(l)\} \\ U(i, j) &= |v_i v_j| + d_P(v_j, v_m) \end{aligned}$$

Note that $S(i, j)$ is the maximum vertex-weighted distance from v_i to any vertex in $\langle v_{i+1}, \ldots, v_j\rangle$ and $U(i, j)$ is the vertex-weighted distance from v_i to v_m. From the proof of Lemma 2, we know $ecc_{G(i,j)}(v_i) = \max\{d_P(v_1, v_i), S(i, j), U(i, j)\}$ and we want to minimize $\max\{S(i, j), U(i, j)\}$.

To be able to compute these vertex-weighted distances fast, we need the following preprocessing step.

Preprocessing. In a preprocessing step, we first compute a longest path P and the weights $\{w(k)\}$ for all vertices on P from T. These steps can be done in $O(n)$ time.

Next we compute the prefix sum array $A[1\ldots m]$ in $O(m)$ time such that $A[k] = \sum_{1<l\leqslant k} |v_{l-1} v_l|$ for $k \in [2, m]$ and $A[1] = 0$. For any $1 \leqslant i < j \leqslant m$, $d_P(v_i, v_j) = A[j] - A[i]$, which can be computed in constant time. Thus $U(i, j) = |v_i v_j| + d_P(v_j, v_m)$ can be computed in constant time.

Compute two *vertex-weighted distance* arrays $B_1[1\ldots m]$ and $B_2[1\ldots m]$, where $B_1[k] = d_P(v_1, v_k) + w(k), k \in [1, m]$ and $B_2[k] = d_P(v_m, v_k) + w(k), k \in [1, m]$. B_1 and B_2 are computed in $O(m)$ time. Using the structure by Bender and Farach-Colton [1] the two arrays can be preprocessed in linear time such that range maxima queries can be answered in constant time. That is, given two indices i and j a range maxima query on B_1, or B_2, returns the maxima in the interval $B_1[i, j]$, or $B_2[i, j]$.

Let $I(i, j)$ be the maximum index k between i and j such that $d_P(v_i, v_k) \leqslant |v_i v_j| + d_P(v_j, v_k)$. The index $I(i, j)$ can be found in $O(\log m)$ time by binary search. The maximum vertex-weighted distance from v_i to a vertex v_k, $i < k \leqslant I(i, j)$, is $\max\{d_P(v_i, v_k) + w(k)\}$, which can be found in constant time by querying the range maxima query structure for B_1 (note that $\max\{d_P(v_i, v_k) + w(k)\} + d_P(v_1, v_i) = \max\{d_P(v_1, v_k) + w(k)\}$).

Similarly, the maximum weighted distance from v_i to a vertex v_k, $I(i, j) < k \leqslant j$, is $|v_i v_j| + \max\{d_P(v_j, v_k) + w(k)\}$ and can be found in constant time by querying the range maxima query structure for B_2. Thus $S(i, j)$ can be computed in $O(\log m)$ time if $I(i, j)$ is not known, and in constant time if $I(i, j)$ is known. To summarize we have the following lemma.

Lemma 3. *Given a tree T, one can preprocess T in $O(n)$ time such that given a query i, j, with $1 \leqslant i < j \leqslant m$, $d_P(v_i, v_j)$ and $U(i, j)$ can be returned in constant time and $S(i, j)$ can be returned in $O(\log m)$ time. If $I(i, j)$ is known, $S(i, j)$ can be returned in constant time.*

We proceed to discuss how to compute an optimal solution for the case when the center lies on the longest path of T.

Finding an Optimal Solution for a Center on P. From the triangle inequality we know that $S(i, j)$ is non-decreasing while $U(i, j)$ is non-increasing as j increases. Let $f(i)$ be the minimum j such that $S(i, j)$ is greater than $U(i, j)$. If we fix i then $\max\{S(i, j), U(i, j)\}$ is a unimodal function of j whose minimum value is at either $f(i)$ or $f(i) - 1$, depicted in Fig. 2(a).

Since $\max\{S(i, j), U(i, j)\}$ is a unimodal function, we can find its minimum in $O(\log m)$ time using binary search. Each binary search requires the computation of $S(i, j)$ and $U(i, j)$, which costs $O(\log m)$ time, by Lemma 3.

As a result from the above discussion we get that for any vertex v_i on the longest path, an optimal solution to the EAT problem with respect to v_i can be found in $O(\log^2 m)$ time using $O(n)$ preprocessing. However, we can improve this further. We will next show that we can compute the optimal solutions for every vertex on P in linear total time by extending the approach in [11].

The linear time algorithm in [11] relies on monotonicity arguments. The next observation summarizes the properties we need.

Observation 1. It holds that

(a) $I(i, j) \leqslant I(i, j + 1)$, for all $j \in [i + 1, m - 1]$,
(b) $I(i, j) \leqslant I(i + 1, j)$, for all $1 \leqslant i < j \leqslant m$, and
(c) $f(i) \leqslant f(i + 1)$, for all $i \in [1, m - 1]$.

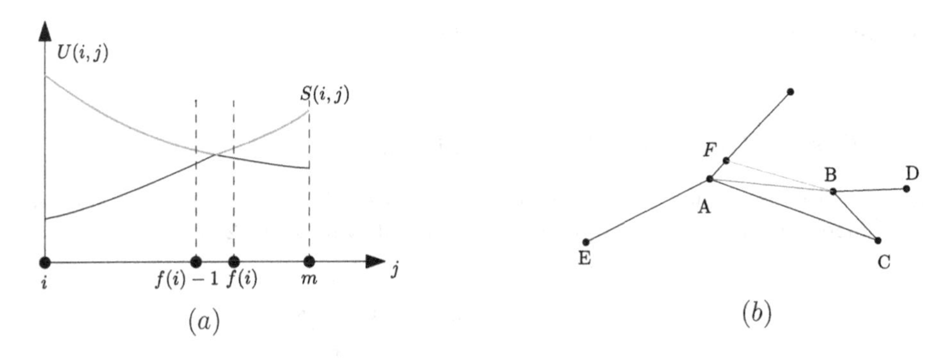

Fig. 2. (a) $S(i,j)$ and $U(i,j)$ for $j \in [i,m]$. The function $\max\{S(i,j),U(i,j)\}$ is shown in red. (b) Tree edges in black. $E-A-C-B-D$ is the longest path. The best center point on the longest path of T is A and the best shortcut is (A,B). However, F as the center with (F,B) as the shortcut is a better solution. (Color figure online)

Proof. We start with (a). By the triangle inequality, $|v_iv_j| \leqslant |v_jv_{j+1}| + |v_iv_{j+1}|$, and as a result

$$\begin{aligned} d_P(v_i, v_{I(i,j)}) &\leqslant |v_iv_{j+1}| + |v_jv_{j+1}| + d_P(v_{I(i,j)}, v_j) \\ &= |v_iv_{j+1}| + d_P(v_{I(i,j)}, v_{j+1}). \end{aligned}$$

For (b) let $k = I(i,j)$. $d_P(v_i,v_k) \leqslant |v_iv_j| + d_P(v_j,v_k)$. By triangle inequality, $|v_iv_{i+1}| \geqslant |v_iv_j| - |v_{i+1}v_j|$. Thus

$$\begin{aligned} d_P(v_{i+1}, v_k) &= d_P(v_i, v_k) - |v_iv_{i+1}| \\ &\leqslant d_P(v_j, v_k) + |v_{i+1}v_j|. \end{aligned}$$

To prove (c) we know from the definition of $f(i)$, that for index $k = f(i) - 1$, $S(i,k) \leqslant U(i,k)$. For all vertices in $P(i+1,k)$, the weighted distances from v_{i+1} either decreases by $|v_iv_{i+1}|$, if the shortest paths from v_{i+1} goes along P, or decreases by $|v_iv_j| - |v_{i+1}v_j|$, if the shortest paths from v_{i+1} use shortcut (v_{i+1}, v_j).

Since $|v_iv_j| - |v_{i+1}v_j| \leqslant |v_iv_{i+1}|$, it holds that $S(i,k) - S(i+1,k) \geqslant |v_iv_j| - |v_{i+1}v_j|$. From the definition we have $U(i,k) - U(i+1,k) = |v_iv_j| - |v_{i+1}v_j|$, hence, since $S(i,k) \leqslant U(i,k)$ we get $S(i+1,k) \leqslant U(i+1,k)$. □

Remember that our algorithm computes the optimal solution to the EAT problem for each vertex v_i, $1 \leqslant i \leqslant k$, where v_k is the first every vertex along P closer to v_m than to v_1. For $i = 1$, increment j, until $j = f(1)$. Calculate $S(1,j)$ and $U(1,j)$. For $i \geqslant 2$, by Observations 1b and 1c, we increment j from $f(i-1)$, $I(i,j)$ from $I(i-1, f(i-1))$ until $j = f(i)$. Compute $S(i,j)$ and $U(i,j)$ accordingly. This finishes the description of the algorithm. Since $I(i,j)$ and $f(i)$ are always non-decreasing during the algorithm, by Observation 1a and Lemma 3, it costs amortized $O(1)$ time per movement of j. The running time of the algorithm is $O(m)$ after preprocessing. The algorithm for scanning P from v_m is symmetric, thus $O(m)$ time after preprocessing.

Theorem 1. *The optimal solutions for the* EAT *problem with respect to every vertex on P can be computed in $O(n)$ time.*

3.2 Case 2: The Center Does Not Lie on P

It is possible that the center of the optimal solution for discrete ROAT is not on the longest path P. An example of this case is shown in Fig. 2(b).

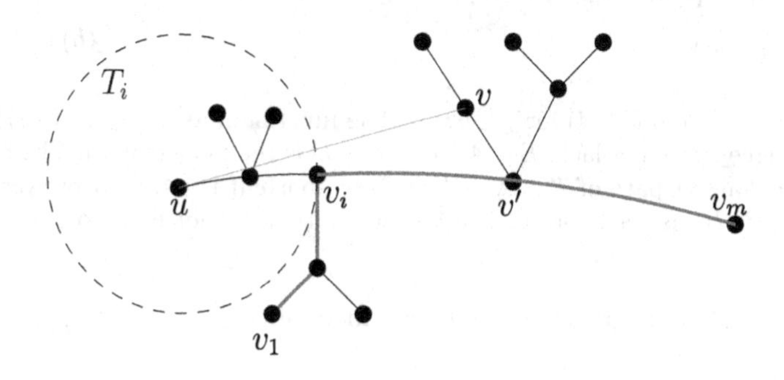

Fig. 3. Finding an optimal solution for a vertex not on P.

For a vertex u not on P, we can prove that there is an optimal edge for the EAT problem with respect to u with one endpoint at u (Lemma 1) and the other endpoint at a vertex on P.

Lemma 4. *Let u be a vertex in T_i, with $1 \leqslant i \leqslant m$. There is an optimal solution (u, v) for the* EAT *problem with respect to u such that v is on $P(v_i, v_l)$, where $l = \arg\max_{k \in \{1,m\}} \{d_P(v_i, v_k)\}$. This optimal shortcut for u can be found in $O(\log^2 n)$ time, after linear time preprocessing.*

Proof. Assume without loss of generality that $d_P(v_1, v_i) \leqslant d_P(v_i, v_m)$. The case when $d_P(v_1, v_i) > d_P(v_i, v_m)$ is symmetric.

From the assumption we know that v_m is a farthest vertex to u in T and $T(u, v_i) \cup P(v_i, v_m)$ is the shortest path from u to v_m in T. To simplify the description we arrange $T(u, v_i) \cup P(v_i, v_m)$ from left to right, as illustrated in Fig. 3. Let $T_u, T_{u_1}, \ldots, T_{u_\ell}, T_i, \ldots, T_m$ be the trees formed by removing edges in $T(u, v_i) \cup P(v_i, v_m)$ from T. Now assume that the optimal edge for the EAT problem with respect to u is the edge (u, v) where v is in a tree $T' \in \{T_{u_1}, \ldots, T_m\}$. Using exactly the same arguments as in the proof of Lemma 2, one can show that the edge (u, v') is at least as good as (u, v), where v' is the root of the tree T'.

Let $P_u = \langle u, \ldots, u_\ell, v_i, \ldots, v_m \rangle$ denote the path $T(u, v_i) \cup P(v_i, v_m)$ and set $w(v') = \max_{q \in T'} \{d_{T'}(v', q)\}$ as the weight of a vertex v' in P_u. Define $S(u, v')$ and $U(u, v')$ for P_u in the same way as we defined $S(i, j)$ and $U(i, j)$ for the longest path. Note that v_i has weight $d_P(v_1, v_i)$. Similar to $S(i, j)$ and $U(i, j)$,

$S(u, v')$ and $U(u, v')$ are monotonically non-decreasing and monotonically non-increasing, respectively, when v' moves to the right along the path. Note that when $v' = v_i$ then

$$S(u, v') \leqslant |uv_i| + d_P(v_i, v_1) \leqslant |uv_i| + d_P(v_i, v_m) = U(u, v').$$

From the monotonicity properties of $S(u, v')$ and $U(u, v')$, an optimal edge for u must have one endpoint on $P(v_i, v_m)$.

For the query time consider an optimal edge (u, v') for the EAT problem with respect to u. When v' is on P then the vertex-weighted distance from u to a vertex to the left of v_i is at most the vertex-weighted distance from u to v_i. Thus $S(u, v')$ is achieved at a vertex between v_i and v'. To find an optimal shortcut for u, we can treat $T(u, v_i)$ as an edge with length $d_T(u, v_i)$ and store the value of $d_T(u, v_i)$, for each v_i on P, during the preprocessing, which does not increase the total linear preprocessing time. With $d_T(u, v_i)$ precomputed, we can compute $S(u, v')$ in $O(\log n)$ time.

Thus for any vertex u not on P, an optimal shortcut can be computed in $O(\log^2 n)$ time. □

Combined with Theorem 1, we have

Theorem 2. *The discrete* ROAT *problem can be solved in $O(n \log^2 n)$ time.*

However, given a threshold value x, we can decide whether there is an optimal shortcut for the EAT problem with respect to u such that the eccentricity of u in the augmented tree is less than x in $O(\log n)$ time. By this observation, we can randomly permute all vertices not on the longest path and process them in sequence. For a vertex u_i in the sequence, we first test whether u_i's eccentricity after augmentation is smaller than the minimum eccentricity after augmentation for $u_1, \ldots, u_{i-1}$. If so, we compute this value for u_i. Otherwise we just proceed to the next vertex. The probability that u_i has the minimum eccentricity after augmentation among vertices $u_1, \ldots, u_i$ is $\frac{1}{i}$. The expected number of times that we compute an optimal shortcut for a vertex in the sequence is $\sum_{i=1}^{O(n)} \frac{1}{i} = \ln n$. Thus we can find the vertex not on P with minimum eccentricity for the EAT problem in expected $O(n \log n + \log n \cdot \log^2 n) = O(n \log n)$ time.

Theorem 3. *The discrete* ROAT *problem can be solved in expected $O(n \log n)$ time.*

4 A Linear Time Algorithm for Continuous ROAT

In this section we consider the continuous version of ROAT where the center of the augmented graph is allowed to be in the interior of an edge of T. Let (u, v) be an edge added to T and let c be the center of $T \cup \{(u, v)\}$. Let r be the radius of $T \cup \{(u, v)\}$. We have the following characterization of c.

Fact 1. *(Observation 1 in* [11]*)*
In $G(u, v) = T \cup (u, v)$, there are two vertices a and b such that:

1. $d_{G(u,v)}(c, a) = d_{G(u,v)}(c, b) = r$.
2. *There is a shortest path from c to a, denoted by* π_a*, and a shortest path from c to b, denoted by* π_b*, such that* π_a *and* π_b *are disjoint except at c.*

As in the last section, we first compute a longest path P for the input tree T. Let $P = \langle v_1, \ldots, v_m \rangle$ and let F the forest formed by removing the edges in P from T. The forest F consists of m trees $T_1, \ldots T_m$, with v_i in T_i, $1 \leqslant i \leqslant m$.

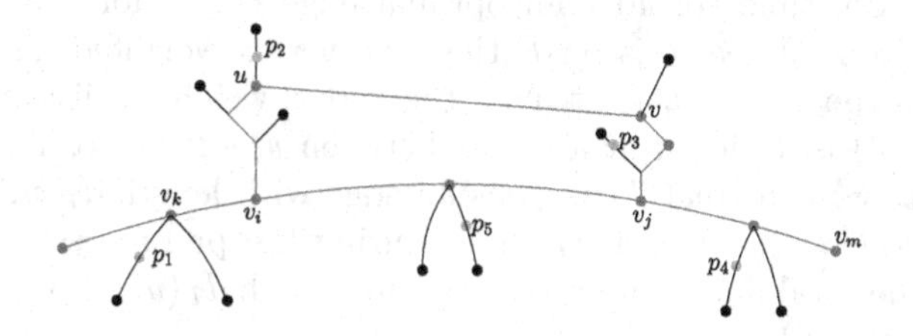

Fig. 4. Observation 2 states that for a shortcut (u, v) the center must lie on the parts highlighted in blue. (Color figure online)

Observation 2. Assume u is a vertex in T_i and v is a vertex in T_j. A center c of $T \cup \{(u, v)\}$ is a point on $P \cup T(v_i, u) \cup \{(u, v)\} \cup T(v_j, v)$.

Proof. We prove the observation by contradiction. Consider the subgraph $P \cup T(v_i, u) \cup \{(u, v)\} \cup T(v_j, v)$ of $T \cup \{(u, v)\}$, which is highlighted in blue in Fig. 4.

Assume $c = p_1$ and p_1 is a point on $T_k \setminus \{v_k\}$ where $1 \leqslant k < i$, see Fig. 4. Recall from Fact 1 that there exists two vertices a and b in T such that $d_{G(u,v)}(c, a) = d_{G(u,v)}(c, b) = r$. Consider the two paths π_a and π_b, as defined in Fact 1. Since both paths cannot all go through v_k, by Property 2. in Fact 1, a or b must lie in T_k and its path does not go through any point in $T(p_1, v_k)$. Assume without loss of generality that a lies in T_k. Then $d_{G(u,v)}(p_1, a) = d_T(p_1, a) < \min\{d_T(v_k, v_1), d_T(v_k, v_m)\} < r$, which is a contradiction. Symmetrically, c cannot be a point on $T_k \setminus \{v_k\}$ where $j < k \leqslant m$.

Next assume $c = p_2$ and p_2 is a point in $T_i \setminus T(v_i, u)$. As above, a must be a vertex in T_i and π_a doesn't go through any point in $T(p_2, v_i)$ while b must lie outside T_i, otherwise property 2. in Fact 1 is violated. The path going from p_2 to a has length less than $d_T(v_i, a) \leqslant r$, a contradiction. Similarly c cannot be a point in $T_j \setminus T(v_j, v)$ or a point on $T_k \setminus \{v_k\}$ where $i < k < j$. □

We are now ready to prove the main lemma of this section. It shows that there is an optimal shortcut which has both endpoints on the longest path P of T. Due to the space limitation, the proof is left to the full version of the paper.

Lemma 5. *Let* (u, v) *be a shortcut edge where* $u \in T_i$ *and* $v \in T_j$*. The radius of* $T \cup \{(v_i, v_j)\}$ *is less than or equal to the radius of* $T \cup \{(u, v)\}$*. That is,* (v_i, v_j) *is at least as good as* (u, v)*.*

Using Observation 2 and Lemma 5, we can reduce the continuous ROAT problem to the Radius-Optimally Augmenting vertex-Weighted Path in a metric space (ROAWP for short) problem, which we will explain below. Assume $P' = \langle v_1, \ldots, v_n \rangle$ is a path embedded in a metric space where each vertex v_k, $1 \leqslant k \leqslant n$, has a weight $w(k)$. Let $G'(i,j) = P' \cup (v_i, v_j)$ be the augmented graph after adding shortcut (v_i, v_j) to P'. The vertex-weighted eccentricity of a point p on $G'(i,j)$ is the maximum vertex-weighted distance from p to any vertex in $G'(i,j)$. The vertex-weighted center of $G'(i,j)$ is the point with minimum vertex-weighted eccentricity and its vertex-weighted eccentricity is the vertex-weighted radius of $G'(i,j)$. ROAWP problem is defined as:

Definition 1. *Given a vertex-weighted path* $P' = \langle v_1, \ldots, v_n \rangle$ *embedded in a metric space, where the weight of vertex* v_k, $1 \leqslant k \leqslant n$, *satisfies* $w(k) \leqslant \min\{d_{P'}(v_1, v_k), d_{P'}(v_n, v_k)\}$, *add one shortcut* (v_i, v_j), $1 \leqslant i < j \leqslant n$, *to* P' *so that the vertex-weighted radius of* $G'(i,j) = P' \cup (v_i, v_j)$ *is minimized.*

For the reduction we set $P' = P$ with vertex weights $w(k) = \max_{q \in T_k}\{d_T(v_k, q)\}$, $1 \leqslant k \leqslant m$. That is, $w(k)$ is the length of the longest shortest path from v_k to a point in T_k. Since P is a longest path in T, $w(k) \leqslant \min\{d_P(v_1, v_k), d_P(v_m, v_k)\}$. It follows that the continuous ROAT problem is reduced to the ROAWP problem.

4.1 Simplify the ROAWP Problem

To further simplify the ROAWP problem, we make a vertex-weight transformation as proposed by Bilò in [2]. Let

$$w^*(k) = \max\{w(j) - d_{P'}(v_j, v_k) | 1 \leqslant j \leqslant m\} \tag{1}$$

be the new transformed weight of v_k. Obviously $w(k) \leqslant w^*(k)$. From Eq. (1), we can see that $w^*(1) = w^*(m) = 0$.

Observation 3. For any $j, k \in [1, m]$,

$$w^*(k) \leqslant w^*(j) + d_{P'}(v_j, v_k).$$

Proof. Let $k' = \arg\max_j \{w(j) - d_{P'}(v_j, v_k)\}$,

$$\begin{aligned} w^*(k) &= w(k') - d_{P'}(v_k, v_{k'}) \\ &= w(k') - d_{P'}(v_k, v_{k'}) - d_{P'}(v_k, v_j) + d_{P'}(v_k, v_j) \\ &\leqslant w(k') - d_{P'}(v_{k'}, v_j) + d_{P'}(v_k, v_j) \\ &\leqslant w^*(j) + d_{P'}(v_j, v_k). \end{aligned}$$

□

The vertex-weight transformation can be done in $O(m)$ time by scanning P' from v_1 to v_m and then back to v_1. We call a vertex-weighted path $P' = \langle v_1, \ldots, v_m \rangle$ *monotone vertex-weighted* if Observation 3 is satisfied for any $j, k \in [1, m]$. *Monotone vertex-weighted* can be interpreted as follows. Assume p is any point on P' and lies on an edge (v_{j-1}, v_j). The vertex-weighted distances from p to $v_j, \ldots, v_m$ are non-decreasing. Similarly, the vertex-weighted distances from p to $v_{j-1}, \ldots, v_1$ are non-decreasing.

When the path in a ROAWP instance is monotone vertex-weighted, we call this instance a ROAMWP (MWP for Monotone vertex-Weighted Path) instance. We can reduce a ROAWP instance to a ROAMWP instance using the following lemma.

Lemma 6. *Let $P' = \langle v_1, \ldots, v_m \rangle$ be a vertex-weighted path in a metric space and let $w(k)$ be its vertex-weight function. Let P^* be P' with vertex-weight function $w^*(k) = \max\{w(j) - d_{P'}(v_j, v_k) | 1 \leqslant j \leqslant m\}$. The vertex-weighted radius of $G^*(i,j) = P^* \cup (v_i, v_j)$ equals the vertex-weighted radius of $G'(i,j) = P' \cup (v_i, v_j)$.*

Proof. Let c' be a vertex-weighted center of $G'(i,j)$ and let r' be its vertex-weighted radius. Let c^* be a vertex-weighted center of $G^*(i,j)$ and r^* be its vertex-weighted radius.

We first prove $r' \leqslant r^*$. When we regard c^* as a point on $G'(i,j)$, the vertex-weighted distance from c^* to a farthest vertex in $G'(i,j)$ is at least r'. So the vertex-weighted distance from c^* to this farthest vertex in $G^*(i,j)$ is at least r', since $w(k) \leqslant w^*(k)$. Thus $r' \leqslant r^*$.

Next we prove $r^* \leqslant r'$. When we regard c' as a point on $G^*(i,j)$, the vertex-weighted distance from c' to a farthest vertex in $G^*(i,j)$ is at least r^*. Let a' be this farthest vertex. Let a^* be the vertex such that $w^*(a') = w(a^*) - d_{P'}(a', a^*)$. The vertex-weighted distance from c' to a^* in $G'(i,j)$ is

$$
\begin{aligned}
d_{G'(i,j)}(c', a^*) + w(a^*) &= d_{G'(i,j)}(c', a^*) + w^*(a') + d_{P'}(a', a^*) \\
&\geqslant d_{G'(i,j)}(c', a') + w^*(a') \\
&\geqslant r^*.
\end{aligned}
$$

Thus $r' \geqslant d_{G'(i,j)}(c', a^*) + w(a^*) \geqslant r^*$. □

4.2 Solve ROAMWP in Linear Time

A monotone vertex-weighted path is very similar to a path with no vertex weights. In [11], Johnson and Wang give a linear time algorithm for the continuous ROAT problem where T is a path. Their algorithm is based on the observation that there are only a constant number of possible configurations (positions of c, π_a and π_b) for an optimal solution. For ROAMWP, there is the same set of possible configurations for an optimal solution as that for continuous ROAT.

For each possible configuration of an optimal solution, Johnson and Wang [11] give a linear time algorithm based on proving the monotonicity properties. For each possible configuration of an optimal solution for a monotone node-weighted path, we can prove the exact same monotonicity properties. The only difference is that we use vertex-weighted path length for ROAMWP instead of path length. Everything else is exactly the same. Thus we have

Lemma 7. *The* ROAMWP *problem can be solved in linear time.*

Reducing continuous ROAT problem to a ROAMWP problem takes linear time. The following theorem summarizes this section.

Theorem 4. *The continuous* ROAT *problem can be solved in linear time.*

References

1. Bender, M.A., Farach-Colton, M.: The LCA problem revisited. In: Gonnet, G.H., Viola, A. (eds.) LATIN 2000. LNCS, vol. 1776, pp. 88–94. Springer, Heidelberg (2000). https://doi.org/10.1007/10719839_9
2. Bilò, D.: Almost optimal algorithms for diameter-optimally augmenting trees. In: Proceedings of the 29th International Symposium on Algorithms and Computation (ISAAC), pp. 40:1–40:13 (2018)
3. Bilò, D., Gualà, L., Proietti, G.: Improved approximability and non-approximability results for graph diameter decreasing problems. Theoret. Comput. Sci. **417**, 12–22 (2012)
4. Carufel, J.D., Grimm, C., Maheshwari, A., Smid, M.H.M.: Minimizing the continuous diameter when augmenting paths and cycles with shortcuts. In: Proceedings of the 15th Scandinavian Symposium and Workshops on Algorithm Theory (SWAT), pp. 27:1–27:14 (2016)
5. De Carufel, J.-L., Grimm, C., Schirra, S., Smid, M.: Minimizing the continuous diameter when augmenting a tree with a shortcut. In: WADS 2017. LNCS, vol. 10389, pp. 301–312. Springer, Cham (2017). https://doi.org/10.1007/978-3-319-62127-2_26
6. Demaine, E.D., Zadimoghaddam, M.: Minimizing the diameter of a network using shortcut edges. In: Kaplan, H. (ed.) SWAT 2010. LNCS, vol. 6139, pp. 420–431. Springer, Heidelberg (2010). https://doi.org/10.1007/978-3-642-13731-0_39
7. Frati, F., Gaspers, S., Gudmundsson, J., Mathieson, L.: Augmenting graphs to minimize the diameter. Algorithmica **72**(4), 995–1010 (2015). https://doi.org/10.1007/s00453-014-9886-4
8. Gao, Y., Hare, D.R., Nastos, J.: The parametric complexity of graph diameter augmentation. Discret. Appl. Math. **161**(10–11), 1626–1631 (2013)
9. Große, U., Gudmundsson, J., Knauer, C., Smid, M., Stehn, F.: Fast algorithms for diameter-optimally augmenting paths. In: Halldórsson, M.M., Iwama, K., Kobayashi, N., Speckmann, B. (eds.) ICALP 2015. LNCS, vol. 9134, pp. 678–688. Springer, Heidelberg (2015). https://doi.org/10.1007/978-3-662-47672-7_55
10. Gudmundsson, J., Sha, Y.: Augmenting graphs to minimize the radius. Manuscript (2021)
11. Johnson, C., Wang, H.: A linear-time algorithm for radius-optimally augmenting paths in a metric space. In: Proceedings of the 15th International Symposium on Algorithms and Data Structures, pp. 466–480 (2019)
12. Oh, E., Ahn, H.: A near-optimal algorithm for finding an optimal shortcut of a tree. In: Proceedings of the 27th International Symposium on Algorithms and Computation (ISAAC), pp. 59:1–59:12 (2016)

13. Schoone, A.A., Bodlaender, H.L., van Leeuwen, J.: Diameter increase caused by edge deletion. J. Graph Theory **11**(3), 409–427 (1987)
14. Wang, H.: An improved algorithm for diameter-optimally augmenting paths in a metric space. Comput. Geom. **75**, 11–21 (2018)
15. Wang, H., Zhao, Y.: Algorithms for Diameters of Unicycle Graphs and Diameter-optimally Augmenting Trees (2020)
16. Wang, H., Zhao, Y.: A linear-time algorithm for discrete radius optimally augmenting paths in a metric space (2020)

Upper and Lower Bounds for Fully Retroactive Graph Problems

Monika Henzinger[1] and Xiaowei Wu[2(✉)]

[1] Faculty of Computer Science, University of Vienna, Vienna, Austria
monika.henzinger@univie.ac.at
[2] IOTSC, University of Macau, Macau, China
xiaoweiwu@um.edu.mo

Abstract. Classic dynamic data structure problems maintain a data structure subject to a sequence S of updates and they answer queries using the latest version of the data structure, i.e., the data structure after processing the whole sequence. To handle operations that change the sequence S of updates, Demaine et al. [7] introduced *retroactive data structures* (RDS). A retroactive operation modifies the update sequence S in a given position t, called *time*, and either creates or cancels an update in S at time t. A *fully retroactive* data structure supports queries at any time t: a query at time t is answered using only the updates of S up to time t. While efficient RDS have been proposed for classic data structures, e.g., stack, priority queue and binary search tree, the retroactive version of graph problems are rarely studied.

In this paper we study retroactive graph problems including connectivity, minimum spanning forest (MSF), maximum degree, etc. We show that under the OMv conjecture (proposed by Henzinger et al. [15]), there does not exist fully RDS maintaining connectivity or MSF, or incremental fully RDS maintaining the maximum degree with $O(n^{1-\epsilon})$ time per operation, for any constant $\epsilon > 0$. Furthermore, We provide RDS with *almost tight* time per operation. We give fully RDS for maintaining the maximum degree, connectivity and MSF in $\tilde{O}(n)$ time per operation. We also give an algorithm for the incremental (insertion-only) fully retroactive connectivity with $\tilde{O}(1)$ time per operation, showing that the lower bound cannot be extended to this setting.

We also study a restricted version of RDS, where the only change to S is the swap of neighboring updates and show that for this problem we can beat the above hardness result. This also implies the first non-trivial dynamic Reeb graph computation algorithm.

Keywords: Retroactive data structure · Dynamic connectivity

The full version of the paper can be found at https://arxiv.org/abs/1910.03332.
X. Wu—This work was done in part when the author was a postdoc at University of Vienna.

A. Lubiw et al. (Eds.): WADS 2021, LNCS 12808, pp. 471–484, 2021.
https://doi.org/10.1007/978-3-030-83508-8_34

1 Introduction

A dynamic data structure problem maintains a data structure on a set of elements subject to element insertions, deletions and modifications. An efficient dynamic algorithm updates the data structure after each element update, and supports queries on the latest version of the data structure. That is, an update can only append an operation to the end of the operation sequence, and a query can only be made on the data structure with all updates applied. However, in some applications, we are interested in modifying the update sequence in the middle. For example, if some past update on a database is mistaken and needs to be removed, we do not want to rollback the whole database by canceling all updates after the mistaken one. Besides, in some scenarios we are interested in querying the data structure when only part of the updates are applied, e.g., to answer questions like "which facebook user had the most friends in Jan 1st, 2015?". This motivates retroactive data structures (RDS) that were introduced by Demaine et al. [7]. They support (1) modifications to the historical sequence of updates performed on the data structure, and (2) queries on the data structure when only a prefix of the updates is applied.

Formally speaking, the data structure is defined by a sequence S of updates, each of which is associated with a time t. A RDS supports operations that create or cancel an update at any time t. There are $|S|+1$ versions of the data structure, on any of which a query can be made. Throughout this paper, we use *update* to denote a modification to the data structure, and *operation* to denote a retroactive action that creates or cancels an update. Depending on the queries supported, Demaine et al. [7] defined two classes of RDS: a *partially retroactive* data structure supports queries only at the present time, i.e., on the latest version of the data structure, while a *fully retroactive* data structure supports queries on any version of the data structure. For dynamic problems in which the ordering of updates is not important, e.g., maintaining a dictionary, standard dynamic algorithms are automatically partially retroactive. However, maintaining a fully RDS can be much more difficult, as a retroactive operation at time t can possibly change the outcome of *all* queries after time t. For example, an insertion of a very small key into a min-heap at time t can possibly change the output of every find-min query after time t. In general, there does not exist efficient transformation from partially RDS to fully retroactive ones. Demaine et al. [7] provided a general checkpointing method that converts a partially RDS into a fully retroactive one, with an $O(\sqrt{T})$ multiplicative overhead in the update and query time, where $T = |S|$. Indeed, the $O(\sqrt{T})$ multiplicative overhead is shown to be tight for some data structures [6], under some well-known computational hardness conjectures.

Prior Works. Demaine et al. [7] provided a partially retroactive priority queue with $O(\log T)$ update time and $O(1)$ query time, which implies a fully retroactive priority queue with $O(\sqrt{T}\log T)$ update and query time. The result was later improved by Demaine et al. [8], who proposed a fully retroactive priority queue with amortized polylogarithmic update and query time. They introduced a hierarchical checkpointing technique, which maintains a balanced binary tree with the set of updates as the leaves. Giora and Kaplan [13] considered the dynamic vertical ray shooting problem, and proposed a data structure that supports horizontal line segment insertions and deletions, and queries that report the first segment intersecting a vertical ray from a query point in worst case $O(\log T)$ time. Their data structure implies a fully retroactive binary search tree with $O(\log T)$ update and query time.

While dynamic graph problems flourished in the past decades, their retroactive versions are rarely studied. Dynamic algorithms maintaining connectivity [17,19], minimum spanning forest (MSF) [19,20] and maximal matching [2,4,26] with polylogarithmic update and query time are known, but their fully retroactive versions have not been studied yet. One exception is the empirical analysis of [1] on the fully retroactive minimum spanning tree (MST) problem. For the aforementioned problems, the dynamic data structures are equivalent to the partially retroactive ones. Thus by Demaine et al.'s reduction [7], there exist fully RDS for these problems, with $\tilde{O}(\sqrt{T})$ update and query time.[1] Note that, in general, the number of updates T can be much larger than the number of nodes and edges in the graph. Roditty and Zwick [23] proposed a fully RDS that supports queries of strong connectivity between two nodes at any version of the graph, subject to directed edge insertions and deletions. However, the retroactive operations are restricted to be *incremental*: each operation either creates an insertion of edge at the end of the update sequence, or cancels an existing update. Their algorithm answers each query in worst case $O(1)$ time and handles each update in amortized $O(m \cdot \alpha(m, n))$ time, where m is the number of edges in the graph and $\alpha()$ is the inverse Ackermann function [27]. Chen et al. [6] showed that there exist data structures for which a gap of $(\min\{n, \sqrt{T}\})^{1-o(1)}$ exists in the time per operation between partially and fully RDS, under some well-known conjectures. However, these data structure are not graph data structures, but rather unusual data structures.

Our Results. We study the fully RDS for graph problems, providing for a variety of fundamental graph problems efficient incremental fully RDS and almost matching upper and lower bounds for their fully dynamic fully retroactive counterparts. We start with some strong hardness results on the update and query time for fully RDS on several graph problems, assuming the online boolean matrix-vector multiplication (OMv) conjecture [15]. Our hardness results show that for many of the problems we study in this paper, it is difficult to get RDS with truly sublinear time per operation.

Theorem 1. *Assuming the OMv conjecture, there do not exist data structures for the following problems with $O(n^{1-\epsilon})$ update and query time subject to edge insertions/deletions:*

- *fully retroactive connectivity, maximal matching, MSF, maximum density;*
- *incremental fully retroactive maximum degree.*

Our hardness results hold even when the edges are unweighted. For maintaining a maximal matching and spanning forest, we assume that queries are

[1] Throughout this paper we use $\tilde{O}()$ to hide the polylogarithmic factors in T and n.

on the size of the matching and the forest, respectively. In the full version [16], we show that the same hardness result holds for fully RDS supporting queries on the existence of perfect matching. Moreover, some of our hardness results apply even to approximation algorithms. For the graph problems we study in this paper (in which the ordering of updates is not important, such as connectivity, maximal matching, and MSF), the partially retroactive setting is the same as the standard dynamic setting and can, thus, be solved in polylogarithmic time. Hence our hardness results imply a polynomial gap in the time per operation between the partially and fully RDS. Our hard instances consist of a sequence of $T = \Theta(n^2)$ operations and queries. Thus they also imply that under the OMv conjecture, getting an $O(T^{1/2-\epsilon})$ time per operation is impossible (for the aforementioned problems). Under the combinatorial boolean matrix multiplication conjecture, we show that our hardness results hold even when all operations are given before any query is made (which we refer to as the *offline* version of the problem), as long as the data structures are combinatorial.

We also provide RDS with almost tight time per operation. We first consider the *incremental* setting, in which a retroactive operation either creates an insertion, or cancels an existing insertion. In other words, the creation of a deletion is not supported. We provide incremental fully RDS for maintaining connectivity and spanning forest (SF) with polylogarithmic update and query time. Observe that the incremental partially retroactive setting is at least as hard as the (non-retroactive) fully dynamic setting, as the cancel operation in the retroactive setting serves the function of deletion in the dynamic setting. Our data structure for maintaining connectivity and spanning forest supports only unweighted edge insertions and deletions. However, we show that it can be extended to support weighted edge insertions and deletions, resulting in an $(1+\epsilon)$-approximation MSF with polylogarithmic update and query time.

Theorem 2. *There exist incremental fully RDS maintaining connectivity, spanning forest, and an $(1+\epsilon)$-approximation MSF with $\tilde{O}(1)$ amortized update time and $\tilde{O}(1)$ worst case query time.*

Note that while the incremental connectivity problem is equivalent to the union-find problem in the dynamic setting, their retroactive versions are different, at least as defined by Demaine et al. [7]. In the retroactive setting, an insertion of an edge at time t that connects two different connected components in the connectivity problem corresponds to a union operation between two equivalence classes in the union-find problem at time t. If we insert another edge connecting the same two components at time $t' > t$, then its corresponding operation in the union-find data structure of Demaine et al. is *illegal*, as two equivalence classes can not be united twice (at time t' and t). In other words, the set of retroactive operations allowed for the two problems are different. Consequently, the fully retroactive union-find data structure by Demaine et al. [7] with $O(\log T)$ time per operation can not be used to achieve the above result.

We also present data structures maintaining MSF and the maximum degree that supports (creation of) insertions and deletions of weighted edges. By Theorem 1, our data structures have almost tight time per operation.

Theorem 3. *There exist fully RDS maintaining connectivity, MSF and maximum degree of an undirected graph with amortized $\tilde{O}(1)$ update time and worst case $O(n \log T)$ query time.*

Our algorithmic results are obtained by maintaining a scapegoat tree [12] with $O(T)$ leaves, each of which is an interval defined by the times of two consecutive updates.[2] Each internal node stores a set of edges, and maintains a data structure (depending on the problem) to support the queries. The tree structure allows efficient retrieval of the edges that exist at time t by examining $O(\log T)$ internal nodes. Moreover, it can be shown that each edge is stored in $O(\log T)$ internal nodes. Consequently, for problems that admit linear time algorithms, e.g., maximal matching, each query can be answered in worst case $O(m \log T)$ time, where m is the maximum number of edges. For maintaining connectivity, MSF and the maximum degree, we show that the query time can be improved to $O(n \log T)$, by maintaining a sparse data structure in each internal node of the scapegoat tree. A similar (yet different) data structure was used by Demaine et al. [8] to maintain the set of retroactive operations sorted by time for their fully retroactive priority queue data structure. In their *checkpoint tree*, a scapegoat tree is maintained with the set of retroactive operations being the leaves. Each internal node u maintains a partially RDS induced by the operations (leaves) in the subtree rooted at u. Consequently, if an element is stored at some node u, it is also stored at the parent of u. In contrast, in our data structure, the set of elements stored at an internal node is disjoint from the set of the elements stored at its children. Moreover, since we do not maintain partially RDS in internal nodes, we do not need to maintain explicitly the set of invalid operations, e.g., a deletion of an edge that is inserted by an operation in another subtree. This property is crucial for efficient data structures on graph problems when edges are inserted and deleted multiple times. We summarize our results in Table 1 as follows (where Retro. stands for Retroactive).

Table 1. Summary of results. The complexity in each cell is for the amortized time per operation. The results in bold are almost tight.

	Incremental	Fully Retro.	Hardness
Maximum Degree	$\tilde{\mathbf{O}}(\mathbf{n})$	$\tilde{\mathbf{O}}(\mathbf{n})$	$\boldsymbol{\Omega}(\mathbf{n}^{1-o(1)})$ (Incremental)
Connectivity, SF	$\tilde{\mathbf{O}}(\mathbf{1})$	$\tilde{\mathbf{O}}(\mathbf{n})$	$\boldsymbol{\Omega}(\mathbf{n}^{1-o(1)})$ (Fully Retro.)
MSF	$\tilde{O}(1)$, $(1+\epsilon)$-approx.	$\tilde{\mathbf{O}}(\mathbf{n})$	$\boldsymbol{\Omega}(\mathbf{n}^{1-o(1)})$ (Fully Retro.)
Maximal Matching	$\tilde{O}(m)$	$\tilde{O}(m)$	$\Omega(n^{1-o(1)})$ (Fully Retro.)

As we will show in Sect. 5, in the (classic) dynamic setting, there exists a simple data structure that maintains the maximum degree of an unweighted graph in worst case $O(1)$ time. On the other hand, it is well-known that maintaining connectivity takes time $\Omega(\log n)$ [22]. In other words, maintaining maximum

[2] A similar data structure was mentioned in [7, Theorem 6]. However, they built a segment tree [3] on the leaves and some details on maintaining the tree were missing.

degree is "easier" than maintaining connectivity in the dynamic setting. However, Theorem 1 and Theorem 2 imply that in the incremental fully retroactive setting this relationship is reversed: maintaining the maximum degree cannot be done in truly sublinear time under the OMv conjecture, while the connectivity problem can be solved in polylogarithmic time. This interesting observation illustrates how different RDS can be, when compared to dynamic data structures.

Our study of RDS was motivated by an application in computational topology, specifically the problem of dynamically maintaining a Reeb graph [21]. However, for that problem a restricted version of the fully retroactive connectivity problem has to be solved. Specifically, no updates can be inserted or deleted in S, but *the order of two neighboring updates can be reversed.* We call such an operation a *swap operation.* Interestingly, under this restricted setting we can beat the lower bounds (Theorem 1) for the general retroactive setting. We give a $\tilde{O}(1)$ time data structure for this restricted version, leading to the first non-trivial dynamic Reeb graph algorithm. Indeed, our approach can be extended to a general class of problems, for which the answer only depends on the currently existing "elements" and *not* on the order of the updates.

Theorem 4. *Suppose for a dynamic version of a problem there exists a data structure with T_u update time, T_q query time, and space complexity $\mathcal{M}$. Then for any integer $1 \leq \tau \leq T$ and any fixed T updates S (each of which is associated with a time), there exists a fully RDS for the problem supporting swap operations with $O(T_u)$ update time and $O(T_q + (\tau - 1) \cdot T_u)$ query time. The data structure uses $O(T \cdot \mathcal{M}/\tau)$ space.*

Other Related Work. Persistence [10,11] is another concept of dynamic data structures that consider updates with times. The data structures maintain (and support queries on) several versions of the data structure simultaneously. Operations of a persistent data structure can be performed on any version of the data structure, which produces a new version. A key difference between persistent data structure and retroactive ones is that a retroactive operation at time t changes *all* later versions of a RDS, while in a persistent one each version is considered an unchangeable archive. Other efficient RDS, e.g., for dynamic point location and nearest neighbor search, can be found on [5,9,14,21].

2 Preliminaries

In a RDS, each update and query is associated with a time t, where t is a real number. We use $\mathsf{now} = +\infty$ to denote the present time. Each retroactive operation creates or cancels an update of the graph at time t, and each query at time t reveals some property of the graph at time t. Specifically, we use $\mathsf{Create}(\text{update}, t)$ to denote a retroactive operation that creates an update at time t and $\mathsf{Cancel}(t)$ to denote the retroactive operation that removes the update at time t. In this paper, updates are edge insertions $\mathsf{Insert}(e)$ and deletions $\mathsf{Delete}(e)$. Moreover, we assume that all operations are *legal.* For example, $\mathsf{Create}(\mathsf{Delete}(e), t)$ can only be issued when edge e exists at time t and is not deleted after time t; Cancel(t) can only be issued when there is an update at time t. We assume that the initial graph is empty, and all updates and queries take place at different times.

A fully RDS supports queries Query(parameters, t) at any time t, where the set of parameters can be empty. A query made at time t should be answered on the version of the graph at time t, on which only updates up to time t are applied. For example, for the connectivity problem, Query(u, v, t) answers whether u and v are connected by edges that exist at time t.

Throughout the whole paper, we use n to denote the number of nodes (which is fixed). We use T to denote the current number of updates (which is dynamic), excluding the updates that are cancelled. A RDS maintains a sequence of updates S sorted in ascending order of time. The size of S is T, which increases by one after each Create(update, t), and decreases by one after each Cancel(t). The set S defines $T + 1$ versions of the graph, and a query can be made on any of them. Note the difference between an operation and an update with the definition of S: S is a set of updates that define the versions of the graph, while operations modify S. Throughout this paper we assume that the word size of the RAM is $O(\log n)$, and T is polynomial[3] in n. Consequently, we have $O(\log T) = O(\log n)$ and we only need constant words to represent any time t. We also assume that the weights of edges are polynomial in n.

Incremental Fully Retroactive. In the incremental case, the retroactive operation Create(Delete$(e), t)$ does not exist, i.e., S contains only insertions of edges (at different times). Note that in the incremental case the Cancel(t) operation can still be issued, which removes one update (insertion) from S.

As we will show later, for maintaining connectivity, the incremental case is substantially easier than the general case; while for maintaining the maximum degree, even the incremental case can be very difficult. The following definition will be useful for our data structures.

Definition 1 (Lifespan). *For each edge e inserted at time t_a and whose earliest deletion after t_a is at time t_b (which is* now *if it is not deleted), let $L_e = (t_a, t_b]$ be the* lifespan *of e.*

While an edge can be inserted and deleted multiple times, to ease our notation we regard e as a new edge every time it is inserted. By definition, the set of edges existing at time t is given by $E_t = \{e : t \in L_e\}$,. A query made at time t should be answered based on the graph $G_t := (V, E_t)$.

3 Lower Bounds

We present the hardness result for maintaining fully retroactive connectivity based on the OMv conjecture in this section. That is, we prove Theorem 1 for

[3] Note that any data structure need to store the $|S| = T$ updates. Thus if T is too large then the space complexity would be already unacceptable. Alternatively we can assume that the word size is $O(\log T)$ as the parameters in the operations might have size $\Theta(\log T)$.

the fully retroactive connectivity problem. The proofs of other hardness results are included in the full version of the paper [16]. We first show that for almost all graph problems, "natural" fully retroactive algorithms can not have update and query time $o(\log T)$. Consider a simple fully RDS on a graph with $n = 2$ nodes. The data structure needs to support insertions and deletions of the edge between the two nodes, and queries of whether the edge exists at time t. We show that the problem is at least as hard as searching a key among T sorted elements. Thus no comparison-based[4] fully RDS has update and query time $o(\log T)$.

Let $k_1 < k_2 < \ldots < k_T$ be T points in time. For each $i = 1, 2, \ldots, T$, we insert an edge $e = (u, v)$ at time $t = k_i$ and delete the edge immediately. In other words, the edge e exists only at time $k_1, k_2, \ldots, k_T$. Assume that you are given a query operation with time parameter k, to check whether k is in $\{k_1, \ldots, k_T\}$, it suffices to query whether the edge exists at time k. Given that any comparison-based search requires $\Omega(\log T)$ time to find an element, we have an $\Omega(\log T)$ lower bound on the query time, for comparison-based fully retroactive algorithms of a large class of dynamic graph problems (including maximum degree, connectivity, maximal matching, etc.). The following lemma justifies the $O(\log T)$ factor that appears in the time per operation of our data structures.

Lemma 1. *Any comparison-based fully retroactive algorithm has $\Omega(\log T)$ time per operation.*

OMv Conjecture. In the Online Boolean Matrix-Vector Multiplication (OMv) problem, the algorithm is given an $n \times n$ boolean matrix M, while a sequence of n length-n boolean vectors $v_1, v_2, \ldots, v_n$ arrive online. The algorithm needs to output the vector Mv_i before seeing the next vector v_{i+1}. The OMv conjecture [15] states that there does not exit algorithm with $O(n^{3-\epsilon})$ running time for this problem, for any constant $\epsilon > 0$.

We give a reduction from the OMv problem to fully retroactive connectivity as follows. The reductions to other graph problems are similar. Given an instance of the OMv problem consisting of an $n \times n$ matrix M and an online sequence of n-dimensional vectors $\{v_i\}_{i \in [n]}$, let m_i be the i-th row of matrix M. Let $|x|$ denote the number of non-zero entries in a vector x. We construct a graph with $n+2$ nodes $a, b, u_1, \ldots, u_n$. We describe and construct a sequence of retroactive operations from the OMv instance as follows.

Recall that we assume all operations have different time. However, for convenience, we use the following description. By saying that we construct a set of retroactive operations S at time t, we fix an arbitrary order of the operations in S, and construct the operations one by one, at time $t, t+\epsilon, \ldots, t+(|S|-1)\epsilon$, where ϵ is arbitrarily small.

Fix any sequence $t_0 < t_1 < \ldots < t_n$ of $n+1$ points in time. We first describe the gadgets we construct for the rows of matrix M. At time t_1, we insert an edge between u_j and b for every $m_{1j} = 1$. That is, we construct a retroactive

[4] Given a query at time t, a comparison-based algorithm compare t with times of other updates to identify the one with time closest to t.

operation $\mathsf{Create}(\mathsf{Insert}(u_j, b), t_1)$ for every $j \in [n]$ with $m_{1j} = 1$, resulting in $|m_1|$ retroactive operations at time (very close to) t_1. Then for $i = 2, \ldots, n$, at time t_i, we create $|m_{i-1}| + |m_i|$ retroactive operations at time t_i as follows. We delete all edges incident to b (by operations $\mathsf{Create}(\mathsf{Delete}(u_j, b), t_i)$ for all $j \in [n]$ with $m_{i-1,j} = 1$), and create insertions of edges (u_j, b) for every $j \in [n]$ with $m_{ij} = 1$ (by operations $\mathsf{Create}(\mathsf{Insert}(u_j, b), t_i)$ for all $j \in [n]$ with $m_{ij} = 1$). Our construction of the graph and retroactive operations guarantee that at time $t \in (t_i, t_{i+1}]$, b is connected to u_j if and only if $m_{ij} = 1$. Next we describe the gadgets for the vectors $v_1, v_2, \ldots, v_n$.

At time t_0, we create an insertion of edge (a, u_j) for every $j \in [n]$ with $v_{1j} = 1$. Observe that $\mathsf{Query}(a, b, t) = 1$ for $t \in (t_i, t_{i+1}]$ if and only if there exist some u_j that is connected to both a and b at time t. By the above construction, that implies $m_i \cdot v_1 = 1$. Hence n connectivity queries, namely at $t_1, t_2, \ldots, t_n$, between a and b suffice to compute Mv_1. Given v_2, we modify the edges incident to a as follows. At time t_0, we delete all edges incident to a, and insert edge (a, u_j) for every $j \in [n]$ with $v_{2j} = 1$ (with $O(n)$ retroactive operations).

In other words, we change the edges between a and $\{u_j\}_{j\in[n]}$ at time t_0 based on v_2. Then we can compute Mv_2 by another n connectivity queries as discussed above. By repeating the above procedure for all vectors v_i, we can solve the OMv problem with $O(n^2)$ retroactive operations and queries, on a data structure with $O(n)$ nodes. Hence if there exists a fully RDS for the connectivity problem with $O(n^{1-\epsilon})$ update and query time, then the OMv problem can be solved in $O(n^{3-\epsilon})$ time, violating the OMv conjecture.

4 Incremental Fully Retroactive Connectivity and SF

In this section we propose an incremental fully RDS for connectivity and spanning forest with polylogarithmic update and query time. Recall that the edges are unweighted. We first present the data structure to support connectivity queries.

Formally, an incremental fully retroactive connectivity data structure supports the following retroactive operations:

- $\mathsf{Create}(\mathsf{Insert}(e), t)$: insert an edge e into the graph at time t;
- $\mathsf{Cancel}(t)$: cancel the insertion of edge at time t; and
- $\mathsf{Query}(u, v, t)$: return whether u and v are connected at time t.

Theorem 5. *There exists an incremental fully retroactive connectivity data structure with amortized $O(\frac{\log^4 n}{\log\log n})$ update time that answers each query with worst case $O(\log n)$ time.*

Proof. Recall that the set S (of updates) contains only insertions (each of them corresponds to a unique edge), while $\mathsf{Create}()$ and $\mathsf{Cancel}()$ modify S. Thus we can regard S as a dynamic set of edges, where each edge has weight equal to the time it is inserted. The set S defines an edge-weighted graph H, and the graph at time t is the subgraph induced by edges with weight at most t. It suffices to maintain a dynamic MSF on the graph H: each Create() inserts a weighted edge to H and each Cancel() deletes one from H.

We maintain a MSF on H using the algorithm by Holm et al. [20], and store the resulting MSF in a link-cut tree [25]. Given the MSF, we can answer Query(u, v, t) by looking at the edge with maximum weight t' on the path between u and v in the MSF, and answer "yes" iff $t' < t$, which can be done in $O(\log n)$ time. It is not difficult to show the correctness of the query. Suppose there exists a path connecting u and v using edges of weight at most t in H, then in the MSF, the maximum weight of an edge on the path between u and v must be at most t. Because otherwise we can remove that edge and include an edge with weight at most t, which violates the definition of MSF.

Obviously, every retroactive operation and query can be handled by a single update on the MSF, which can be done in amortized $O(\frac{\log^4 n}{\log\log n})$ time. ■

Next we describe the data structure and algorithm to maintain an incremental fully retroactive SF. To distinguish the SF from the MSF of H, we use MSF_H to denote the weighted spanning forest of H that we maintain. We use the same data structure (with minor changes) to support the following queries:

- Query(t): return a SF at time t;
- Query(size, t): return the size (number of edges) of a SF at time t.

Again, we maintain MSF_H on H: Query(t) can be trivially answered in $O(n)$ time by outputting all edges in the MSF_H with weight less than t. To support Query(size, t), we need to count the number of edges with weight less than t in MSF_H. We maintain an AVL tree that supports range query[5] on the weights of the edges of MSF_H. Since every retroactive operation changes MSF_H by at most one edge, the AVL tree can be maintained in $O(\log n)$ time per operation. We can answer Query(size, t) by querying the number of elements with value less than t in the AVL tree. In summary, we have the following.

Theorem 6. *There exists an incremental fully retroactive SF with amortized $O(\frac{\log^4 n}{\log\log n})$ update time that supports Query(t) in worst case $O(n)$ time and Query(size, t) in worst case $O(\log n)$ time.*

While our data structure supports only unweighted edge insertions and deletions, we show that it can be extended to the weighted case to maintain an $(1+\epsilon)$-approximate MSF. Using the techniques from Henzinger and King [18], we maintain an $(1+\epsilon)$-approximate MSF by partitioning the edges into weight classes. Basically, we round the edge weights up to powers of $1+\epsilon$, and maintain $O(\frac{1}{\epsilon}\log W)$ incremental fully RDS we described above, one for each weight class. Here we assume all edge weights are in $[1, W]$. Each insertion of a weighted edge translates into an insertion of an unweighted edge in the corresponding weight class. Queries for the approximation MSF made at time t can be answered by

[5] Please refer to https://www.geeksforgeeks.org/count-greater-nodes-in-avl-tree/ for an implementation.

collecting $O(\frac{1}{\epsilon}\log W)$ spanning forests (one from each data structure), and performing a static MSF algorithm, which takes time $O(\frac{n}{\epsilon}\log W)$.

In order to answer the total weight of the MSF more efficiently, we modify the data structure as follows. Each insertion of an edge of weight $(1+\epsilon)^i$ is translated to an insertion of an unweighted edge in each of the weight classes $j = i, i+1, \ldots, l$, where $l = \log_{1+\epsilon} W$. In other words, weight class j contains all edges of weight *at most* $(1+\epsilon)^j$. Then the query of the total weight at time t can be answered by $O(\frac{1}{\epsilon}\log W)$ queries $\mathsf{Query}(\text{size}, t)$ as follows. Let a_i be the size returned by $\mathsf{Query}(\text{size}, t)$ at weight class i, where $i = 0, 1, \ldots, l$. Then $a_0 + \sum_{i=1}^{l}(a_i - a_{i-1}) \cdot (1+\epsilon)^i$ is the total weight of an $(1+\epsilon)$-approximation MSF. Note that the query for the approximation MSF can still be answered by collecting $O(\frac{1}{\epsilon}\log W)$ spanning forests and performing a static MSF algorithm in $O(\frac{n}{\epsilon}\log W)$ time. In summary, the amortized update time is $O(\frac{\log^4 n}{\log\log n} \cdot \frac{1}{\epsilon}\log W)$, and the worst case query time is $O(\frac{n}{\epsilon}\log W)$ for the approximation MSF, $O(\log n \cdot \frac{1}{\epsilon}\log W)$ for its total weight.

5 Fully Retroactive Data Structures

In this section we present fully RDS for maintaining the maximum degree, connectivity and MSF. Recall that for maintaining the maximum degree and MSF, edges are weighted. Combined with the hardness results, the data structures we propose in this section achieve almost optimal (up to a polylogarithmic factor) time per operation. We first introduce a general framework for the fully RDS.

We present a dynamic balanced binary tree $\mathcal{T}$ that maintains the set of edges subject to insertions and deletions at different times. The balanced binary tree serves as the framework for several RDS we will introduce later. Depending on the problem, we maintain different (non-retroactive) dynamic data structures in the internal nodes. We implement the balanced binary tree using the scapegoat tree [12], which rarely rebuilds part of the tree to maintain balance.[6]

We show that the balanced binary tree $\mathcal{T}$ enables us to handle each retroactive operation by updating $O(\log T)$ internal nodes if no rebuild occurs. We rebuild the tree when it is not balanced and charge the cost of rebuild to the retroactive operations that are responsible for the imbalance, such that each operation is charged by $O(\log^2 T)$ updates of internal nodes.

Consider a sequence S of T updates and each update is associated with a time t. We order the updates in S in ascending order of their time, and we use $t_1 < t_2 < \ldots < t_T$ to denote these times. For completeness, let $t_0 = -\infty$ and $t_{T+1} = \mathsf{now}$. The scapegoat tree $\mathcal{T}$ we maintain has T leaf nodes $(t_i, t_{i+1}]$ for $i = 1, 2, \ldots, T$. For any node u, let $\mathcal{T}(u)$ denote the subtree rooted at u in $\mathcal{T}$. The scapegoat tree maintains the following invariant:

Invariant 51. *For each internal node u and its sibling v, $|\mathcal{T}(u)| \le 2 \cdot |\mathcal{T}(v)|$.*

[6] Other balanced search trees, e.g., AVL tree [24], maintain balance by rotating part of the tree, which will be expensive when we maintain a data structure in each internal node u depending on the set of leaves in $\mathcal{T}(u)$.

Whenever an internal node violates the invariant, the algorithm determines the internal node closest to the root that violates the invariant and rebuilds its subtree from scratch, fulfilling the invariant. The amortized cost of this rebuild is $O(\log T)$ per operation in $\mathcal{T}$.

A standard argument for balanced search tree implies that if the invariant is maintained, then the height of the tree is upper bounded by $O(\log T)$. We maintain the following data structures for each node u of the scapegoat tree:

- an interval I_u, which is the union of the intervals of the leaves of $\mathcal{T}(u)$.
- a data structure $\mathcal{D}(u)$ that stores the edges e such that (1) $I_u \subseteq L_e$; and (2) $I_w \nsubseteq L_e$, where w is the parent of u in $\mathcal{T}$. (Recall that L_e is the lifespan of edge e.) If u is the root of the tree then we only require that $I_u \subseteq L_e$. For convenience we also interpret $\mathcal{D}(u)$ as a set of edges. The exact choice of $\mathcal{D}(u)$ depends on the graph property that is maintained.

In other words, each internal node u maintains an interval I_u the subtree $\mathcal{T}(u)$ covers, and stores edge e if the interval of u is the maximal interval contained in L_e. The above data structure enables efficient retrieval of E_t, i.e., the set of edges existing at time t.

Lemma 2. *Fix any time $t \in (t_i, t_{i+1}]$. Let $(v_l, v_{l-1}, \ldots, v_0)$ be the path from the leaf node $v_l = (t_i, t_{i+1}]$ to the root v_0. We have $E_t = \bigcup_{i=0}^{l} \mathcal{D}(v_i)$ and $\mathcal{D}(v_i) \cap \mathcal{D}(v_j) = \emptyset$ for all $i \neq j$.*

Proof. First, for every $e \in E_t$ that exists at time t, we have $t \in L_e$, which implies that $v_l = (t_i, t_{i+1}] \subseteq L_e$. Thus e must be contained in some unique $\mathcal{D}(v_i)$. That is, $E_t \subseteq \bigcup_{i=0}^{l} \mathcal{D}(v_i)$. Specifically, e is contained in $\mathcal{D}(v_i)$ such that $I_{v_i} \subseteq L_e$ while $I_{v_{i-1}} \nsubseteq L_e$. Therefore the sets of edges $\mathcal{D}(v_0), \mathcal{D}(v_1), \ldots, \mathcal{D}(v_l)$ are disjoint. On the other hand, for any $e \in \mathcal{D}(v_i)$, we have $I_{v_i} \subseteq L_e$, which implies $t \in (t_i, t_{i+1}] \subseteq L_e$ and hence $e \in E_t$. ∎

Lemma 2 implies that with the tree $\mathcal{T}$, we can retrieve the edges E_t by looking at $O(\log T)$ internal nodes. In particular, $\mathsf{Query}(t)$ can be handled by data structures maintained by $O(\log T)$ nodes. For problems that admit linear time algorithms, e.g., connectivity and maximal matching, $\mathsf{Query}(t)$ can be handled in $O(\log T + |E_t|)$ time, by maintaining the set of edges $\mathcal{D}(u)$ in each internal node u. Next we show that the data structure maintains $O(\log T)$ copies of every edge. Consequently, the total size of the sets $\mathcal{D}(u)$ is bounded by $O(T \log T)$.

Lemma 3. *Each edge is contained in $O(\log T)$ internal nodes. Moreover, these internal nodes can be found in $O(\log T)$ time.*

Proof. Fix any edge e with $L_e = (t_a, t_b]$. By definition, if $\mathcal{D}(u)$ contains e for some internal node u, then $I_u \subseteq L_e$ and $I_w \nsubseteq L_e$. Thus w must be an ancestor of the leaf node $(t_{a-1}, t_a]$ or $(t_b, t_{b+1}]$, i.e., I_w intersects with L_e but is not contained in L_e. Therefore, every internal node u that contains e must be a child of some node on the path from $(t_{a-1}, t_a]$ to the root, or child of some node on the path from $(t_b, t_{b+1}]$ to the root. Since the height of tree of $O(\log T)$ and each internal node has two children, there are $O(\log T)$ internal nodes containing e and they can be found in $O(\log T)$ time. ■

Next we show how to handle retroactive operations by updating the tree $\mathcal{T}$. Intuitively, since each retroactive operation changes the lifespan of a single edge, by Lemma 3, the operation can be handled by updating $O(\log T)$ internal nodes. However, to maintain a balanced binary tree, sometimes we need to rebuild part of the tree, which increases the amortized update time.

Lemma 4. *Let t_{update} be the update time of the data structure maintained in an internal node. Each retroactive operation can be handled in amortized $O(\log^2 T \cdot t_{update})$ time.*

Due to space limit, we defer the proof of the above lemma to the full version of the paper [16], where we give data structures maintaining maximum degree, connectivity and MSF subject to retroactive operations. The data structures follow the above framework, while for different problems the data structures maintained by internal nodes are different.

Acknowledgment. Monika Henzinger acknowledges the Austrian Science Fund (FWF) and netIDEE SCIENCE project P 33775-N.

Xiaowei Wu is funded by the Science and Technology Development Fund, Macau SAR (File no. SKL-IOTSC-2021-2023), the Start-up Research Grant of University of Macau (File no. SRG2020-00020-IOTSC).

References

1. de Andrade Júnior, J.W., Duarte Seabra, R.: Fully retroactive minimum spanning tree problem. Comput. J. (2020)
2. Baswana, S., Gupta, M., Sen, S.: Fully dynamic maximal matching in o(log n) update time (corrected version). SIAM J. Comput. **47**(3), 617–650 (2018)
3. Bentley, J.L.: Algorithms for Klee's rectangle problems. Technical report, Computer (1977)
4. Bernstein, A., Forster, S., Henzinger, M.: A deamortization approach for dynamic spanner and dynamic maximal matching. In: SODA, pp. 1899–1918. SIAM (2019)
5. Blelloch, G.E.: Space-efficient dynamic orthogonal point location, segment intersection, and range reporting. In: SODA, pp. 894–903. SIAM (2008)
6. Chen, L., Demaine, E.D., Gu, Y., Williams, V.V., Xu, Y., Yu, Y.: Nearly optimal separation between partially and fully retroactive data structures. In: SWAT. LIPIcs, vol. 101, pp. 33:1–33:12. Schloss Dagstuhl - Leibniz-Zentrum fuer Informatik (2018)
7. Demaine, E.D., Iacono, J., Langerman, S.: Retroactive data structures. ACM Trans. Algorithms **3**(2), 13 (2007)
8. Demaine, E.D., Kaler, T., Liu, Q., Sidford, A., Yedidia, A.: Polylogarithmic fully retroactive priority queues via hierarchical checkpointing. In: Dehne, F., Sack, J.-R., Stege, U. (eds.) WADS 2015. LNCS, vol. 9214, pp. 263–275. Springer, Cham (2015). https://doi.org/10.1007/978-3-319-21840-3_22

9. Dickerson, M.T., Eppstein, D., Goodrich, M.T.: Cloning voronoi diagrams via retroactive data structures. In: de Berg, M., Meyer, U. (eds.) ESA 2010. LNCS, vol. 6346, pp. 362–373. Springer, Heidelberg (2010). https://doi.org/10.1007/978-3-642-15775-2_31
10. Driscoll, J.R., Sarnak, N., Sleator, D.D., Tarjan, R.E.: Making data structures persistent. J. Comput. Syst. Sci. **38**(1), 86–124 (1989)
11. Fiat, A., Kaplan, H.: Making data structures confluently persistent. J. Algorithms **48**(1), 16–58 (2003)
12. Galperin, I., Rivest, R.L.: Scapegoat trees. In: SODA, pp. 165–174. ACM/SIAM (1993)
13. Giyora, Y., Kaplan, H.: Optimal dynamic vertical ray shooting in rectilinear planar subdivisions. ACM Trans. Algorithms **5**(3), 28:1–28:51 (2009)
14. Goodrich, M.T., Simons, J.A.: Fully retroactive approximate range and nearest neighbor searching. In: Asano, T., Nakano, S., Okamoto, Y., Watanabe, O. (eds.) ISAAC 2011. LNCS, vol. 7074, pp. 292–301. Springer, Heidelberg (2011). https://doi.org/10.1007/978-3-642-25591-5_31
15. Henzinger, M., Krinninger, S., Nanongkai, D., Saranurak, T.: Unifying and strengthening hardness for dynamic problems via the online matrix-vector multiplication conjecture. In: STOC, pp. 21–30. ACM (2015)
16. Henzinger, M., Wu, X.: Upper and lower bounds for fully retroactive graph problems. arXiv preprint arXiv:1910.03332 (2019)
17. Henzinger, M.R., King, V.: Randomized fully dynamic graph algorithms with polylogarithmic time per operation. J. ACM **46**(4), 502–516 (1999)
18. Henzinger, M.R., King, V.: Maintaining minimum spanning forests in dynamic graphs. SIAM J. Comput. **31**(2), 364–374 (2001)
19. Holm, J., de Lichtenberg, K., Thorup, M.: Poly-logarithmic deterministic fully-dynamic algorithms for connectivity, minimum spanning tree, 2-edge, and biconnectivity. J. ACM **48**(4), 723–760 (2001)
20. Holm, J., Rotenberg, E., Wulff-Nilsen, C.: Faster fully-dynamic minimum spanning forest. In: Bansal, N., Finocchi, I. (eds.) ESA 2015. LNCS, vol. 9294, pp. 742–753. Springer, Heidelberg (2015). https://doi.org/10.1007/978-3-662-48350-3_62
21. Parsa, S.: Algorithms for the reeb graph and related concepts. Ph.D. thesis, Duke University (2014)
22. Patrascu, M., Demaine, E.D.: Logarithmic lower bounds in the cell-probe model. SIAM J. Comput. **35**(4), 932–963 (2006)
23. Roditty, L., Zwick, U.: A fully dynamic reachability algorithm for directed graphs with an almost linear update time. SIAM J. Comput. **45**(3), 712–733 (2016)
24. Sedgewick, R.: Algorithms. Addison-Wesley, Boston (1983)
25. Sleator, D.D., Tarjan, R.E.: A data structure for dynamic trees. J. Comput. Syst. Sci. **26**(3), 362–391 (1983)
26. Solomon, S.: Fully dynamic maximal matching in constant update time. In: FOCS, pp. 325–334. IEEE Computer Society (2016)
27. Tarjan, R.E.: Efficiency of a good but not linear set union algorithm. J. ACM **22**(2), 215–225 (1975)

Characterization of Super-Stable Matchings

Changyong Hu(✉) and Vijay K. Garg

University of Texas at Austin, Austin, TX 78705, USA
colinhu9@utexas.edu, garg@ece.utexas.edu

Abstract. An instance of the super-stable matching problem with incomplete lists and ties is an undirected bipartite graph $G = (A \cup B, E)$, with an adjacency list being a linearly ordered list of ties. Ties are subsets of vertices equally good for a given vertex. An edge $(x, y) \in E \backslash M$ is a blocking edge for a matching M if by getting matched to each other neither of the vertices x and y would become worse off. Thus, there is no disadvantage if the two vertices would like to match up. A matching M is super-stable if there is no blocking edge with respect to M. It has previously been shown that super-stable matchings form a distributive lattice [14,23] and the number of super-stable matchings can be exponential in the number of vertices. We give two compact representations of size $O(m)$ that can be used to construct all super-stable matchings, where m denotes the number of edges in the graph. The construction of the second representation takes $O(mn)$ time, where n denotes the number of vertices in the graph, and gives an explicit rotation poset similar to the rotation poset in the classical stable marriage problem. We also give a polyhedral characterization of the set of all super-stable matchings and prove that the super-stable matching polytope is integral, thus solving an open problem stated in the book by Gusfield and Irving [4].

Keywords: Super-stable matching · Distributive lattice · Matching polytope

1 Introduction

An instance of the super-stable matching problem with incomplete lists and ties is an undirected bipartite graph $G = (A \cup B, E)$, with an adjacency list being a linearly ordered list of ties. Ties are disjoint and may contain just one vertex. If vertices b_1 and b_2 are neighbors of vertex a in the graph G, then either (1) a strictly prefers b_1 to b_2, which we denote as $b_1 \succ_a b_2$; or (2) a is indifferent between b_1 and b_2, which means b_1 and b_2 are in a tie in a's adjacency list, and denote as $b_1 =_a b_2$; or (3) a strictly prefers b_2 to b_1. We say a weakly prefers b_1 to b_2 if either a strictly prefers b_1 to b_2 or a is indifferent between b_1 and b_2, which we denote as $b_1 \succeq_a b_2$. A matching M is a set of disjoint edges in the graph G. Let $e = (u, v)$ be an edge contained in the matching M. Then, we say

A. Lubiw et al. (Eds.): WADS 2021, LNCS 12808, pp. 485–498, 2021.
https://doi.org/10.1007/978-3-030-83508-8_35

that vertices u and v are matched in M and write $u = M(v)$ to denote that u is matched to v in M. An edge $(x, y) \in E \backslash M$ is a *blocking edge* for a matching M if by getting matched to each other neither of the vertices x and y would become worse off, i.e. x is either unmatched or x weakly prefers y to $M(x)$, and y is either unmatched or y weakly prefers x to $M(y)$. We abuse the notation $y \succeq_x M(x)$ for the case that x is unmatched in M. A matching is *super-stable* if there is no blocking edge with respect to it.

Super-stable matchings were first investigated by Irving [6], who gave three classes of stable matchings in the case of preference lists with ties, depending on the way of defining a *blocking edge* for a matching M. In the weakly stable matching problem an edge $e = (x, y)$ is blocking if by getting matched to each other, both x and y would become better off. In the strongly stable matching problem, an edge $e = (x, y)$ is blocking if one of x and y becomes better off and the other would not be worse off.

In this paper we study the problem of characterizing the set of all super-stable matchings. The problem was stated in the book by Gusfield and Irving [4] as one of the 12 open problems. The structure of the set of all stable matchings in the stable marriage problem without ties is well understood in Gusfield and Irving's book [4]. Recently, Kunysz et al. [11] gave compact representations for the set of all strongly stable matchings and showed that the construction can be done in $O(mn)$ time, where n and m denote the number of vertices and edges in the graph. Scott [22] investigated the structure of all super-stable matchings by defining an object that he called meta-rotation, which corresponds to one collection of rotations in some arbitrary tie-breaking instance of the original instance and the time complexity of the construction is $O(m^2)$.

We give two compact representations of the set of all super-stable matchings that can be constructed in, respectively, $O(nm^2)$ and $O(mn)$ time.

The first representation of the set of all super-stable matchings consists of $O(m)$ matchings, each of which is a man-optimal stable matching among all super-stable matchings that contains a given edge. We show that computing such matching for each edge can be reduced to computing a man-optimal super-stable matching in a reduced graph by deleting an appropriate subset of edges in graph G. The algorithm is described in Sect. 3.

Our second representation explicitly constructs rotations, which are differences between consecutive super-stable matchings in a maximal sequence of super-stable matchings starting with a man-optimal super-stable matching and ending with a woman-optimal super-stable matching. Unlike Scott's [22] meta-rotation, our rotation is the symmetric difference of two super-stable matchings, which could be a cycle or multiple cycles.

Our construction takes $O(mn)$ time, while Scott's [22] algorithm takes $O(m^2)$ time. We also show how to efficiently construct a partial order among rotations. This poset can be used to solve other problems connected to super-stable matchings such as the enumeration of all super-stable matchings and the maximum weight super-stable matching problem. Fleiner et al. [3] solve the weight super-stable matching by reducing it to the 2-SAT problem and the time complexity is $O(mn \log(W))$, where W is the maximum weight among all edges in G. By using the rotation poset constructed in this paper, the weighted problem can also be solved in $O(mn \log(W))$ time.

In this paper we also give a polyhedral characterization for the set of all super-stable matchings and prove that the super-stable matching polytope is integral. This result implies that the maximum weight super-stable matching problem can be solved in polynomial time. Though the complexity of solving LP is usually higher than combinatorial methods, like in [3], this gives an alternative direction to solve the weighted super-stable matching problem. Previously, it has been shown that the stable matching polytope and the strongly stable matching polytope are integral [11,21,25], we complete all three cases by proving that the super-stable matching polytope is integral as well.

We also proved a property called self-duality for the super-stable matching polytope, which also holds for the classical stable matching polytope [24] and the strongly stable matching polytope [11]. See details in our full version.

1.1 Related Works

Irving [6] gave an $O(m)$ algorithm to find a super-stable matching if it exists. Spieker [23] showed that super-stable matchings form a distributive lattice. Further properties of super-stable matchings were proved by Manlove in [14]. Scott [22] introduced the concept called *meta-rotation poset* for super-stable matchings and showed the one-to-one correspondence between super-stable matchings and closed subsets of the poset.

Irving [6] and Manlove [14] gave an $O(m^2)$ algorithm to find a strongly stable matching if it exists. Kavitha et al. [9] gave an $O(nm)$ algorithm for the strongly stable matching problem. Manlove [14] showed that strongly stable matchings form a distributive lattice. Kunysz et al. [12] gave a characterization of all strongly stable matchings and later Kunysz [11] gave a polyhedral description for the set of all strongly stable matchings and proved that the strongly stable matching polytope is integral.

For weakly stable matchings, it is not true that all weakly stable matchings of a given instance always have the same size. Weakly stable matching can be easily found by running the deferred-acceptance algorithm while breaking ties in an arbitrary manner. The problem of computing a maximum-size weakly stable matching is NP-hard, which has been proved by Iwama et al. [7]. Thus finding good approximations of the problem becomes very interesting. For the version when ties are allowed on both sides, the currently best approximation factor is 3/2 [10,16,17]. For the case when ties only occur on one side, there are a sequence of works pushing the approximation factor lower. Iwama et al. [8] gave an 25/17 approximation algorithm. Huang and Kavitha [5] improved it to 22/15. Later Radnai [20] improved the approximation factor to 41/28, then Dean et al. [2] pushed the approximation factor to 19/13. Most recent result by Lam and Plaxton [13] gave the currently best approximation factor of $1 + 1/e$.

2 Preliminaries

In this section we give some definitions and theorems that are useful in the following sections.

Theorem 1. [6,14] *There is an $O(m)$ algorithm to determine a man-optimal super-stable matching of the given instance or report that no such matching exists.*

Theorem 2. [14] *In a given instance of the super-stable matching problem, the same set of vertices are matched in all super-stable matchings.*

Lemma 1. [14] *Let M, N be two super-stable matchings in a given super-stable matching instance. Suppose that, for any agent p, $(p, q) \in M$ and $(p, q') \in N$, where p is indifferent between q and q', then $q = q'$.*

We recall some standard notations and definitions from the theory of matchings under preferences. For a given edge (m, w), any matching containing (m, w) is called an (m, w)-matching. Let us denote the set of all super-stable matchings of G by $\mathcal{M}_G$. Let $\mathcal{M}_G(m, w)$ be the set of all super-stable (m, w)-matchings in G.

For two super-stable matchings M and N, we say that M *dominates* N and write $M \succeq N$ if each man m weakly prefers $M(m)$ to $N(m)$. If M dominates N and there exists a man m who prefers $M(m)$ to $N(m)$, then we say M *strictly dominates* N, write $M \succ N$ and we call N a *successor* of M. Note that by Lemma 1, $M \succeq N$ implies $M \succ N$, assuming M is not equal to N.

3 Irreducible Super-Stable Matchings

In this section, we give our first representation via irreducible matchings. Birkhoff's representation theorem [1] for distributive lattices states that the elements of any finite distributive lattice can be represented as finite sets in such a way that the lattice operations correspond to unions and intersections of sets. The theorem gives a one-to-one correspondence between distributive lattices and partial orders. Our goal is to find the partial order that represents the set of all super-stable matchings.

Distributive lattice is closely related to rings of sets, which is a family of sets that is closed under set unions and set intersections. If the sets in a ring of sets are ordered by set inclusion, they form a distributive lattice. Theory regarding rings of sets and its application to representations of the set of stable matchings in the classical stable marriage problem is well studied by Irving and Gusfield [4]. Below we give a brief summary of this theory that serves as a preliminary for our algorithm.

Given a finite set B, the *base* set, a family $\mathcal{F} = \{F_0, F_1, \cdots, F_k\}$ of subsets of B is called a ring of sets over B if $\mathcal{F}$ is closed under set union and intersection. A ring of sets contains a unique minimal element and a unique maximal element.

For any element $a \in B$, we denote $\mathcal{F}(a)$ the set of all elements of $\mathcal{F}$ that contains a. It is obvious that $\mathcal{F}(a)$ is also a ring of sets over B. We define $F(a)$ to be the unique minimal element of $\mathcal{F}(a)$. An element $F \in \mathcal{F}$ that is $F(a)$ for some $a \in B$ is called *irreducible*. We denote $I(\mathcal{F})$ the set of all irreducible elements of $\mathcal{F}$. We view $(I(\mathcal{F}), \leq)$ as a partial order under the relation $\leq$ of set containment. We give the Birkhoff's representation theorem in the language of rings of sets below.

Theorem 3. [4] *i) There is a one-to-one correspondence between the closed subsets of $I(\mathcal{F})$ and the elements of $\mathcal{F}$.*
ii) If S and S' are closed subsets of $I(\mathcal{F})$ that generate $F = \bigcup S$ and $F' = \bigcup S'$ respectively, then $F \subseteq F'$ if and only if $S \subseteq S'$.

In the context of super-stable matchings, the base set B corresponds to the set of all acceptable pairs $(m, w) \in E$. We define the P-set of a super-stable matching M to be the set of all pairs (m, w), where w is either $M(m)$ or a woman whom m weakly prefers to $M(m)$, which corresponds to an element in $\mathcal{F}$. It is obvious that the unique minimal (man-optimal) super-stable matching in $\mathcal{M}_G(m, w)$, if nonempty, is *irreducible*.

Here we describe an $O(|E|)$ algorithm for computing a man-optimal super-stable (m, w)-matching in G. Algorithm 1 essentially constructs a reduced graph $G' \subseteq G$ by removing some edges from G (line 3 to line 13 in Algorithm 1). After that, the algorithm computes a man-optimal super-stable matching M' in the reduced graph G'. By adding back the edge (m, w), the new matching $M \cup (m, w)$ is super-stable in G.

Lemma 2. *Let M be a super-stable (m, w)-matching. Then $M' = M \backslash \{(m, w)\}$ is a super-stable matching in the reduced graph G'.*

Proof. We need to prove $M' \subseteq G'$ or equivalently none of edges removed from G is matched in M'. Suppose not, an edge (m', w') was removed from G and is matched in M'. Note that $m' \neq m$ and $w' \neq w$. Hence, it follows that there is an edge (m, w') or (m', w) which caused the removal of (m', w'). W.l.o.g, let's assume it is (m, w') which caused the removal of (m', w'). Then we have $w \preceq_m w'$ and $m \succeq_{w'} m'$. Obviously, (m, w') is a blocking pair, which leads to a contradiction of M being super-stable.

To prove super-stability of M' is easy. If there were an edge e blocking M', it would also block M.

Lemma 3. *Let M' be some super-stable matching in the reduced graph G' if exists. If $M' \cup (m, w)$ is a super-stable matching in G, then for each super-stable matching N' in G', $N' \cup (m, w)$ is a super-stable matching in G. If G' does not have any super-stable matching, then there is no super-stable (m, w)-matching.*

Proof. Let $M = M' \cup (m, w)$. Since M' is super-stable in G'. It follows that only the removed edges in $E \backslash E'$ can potentially block M. We have two cases. (i): any edge that is incident to m or w cannot block M. W.l.o.g, Suppose that for some w' that is incident to m, and (m, w') blocks M. Then we have $w' \succeq_m w$.

Algorithm 1: Computing man-optimal super-stable (m, w)-matching

```
1  Input: the graph G = (A ∪ B, E) and preference lists of G and an edge
   (m, w) ∈ E.
2  Output: man-optimal super-stable (m, w)-matching or deciding that no such
   matching exists.
3  G' ← G\{m, w} // remove m and w and all edges that are incident to them
4  for m' s.t. (m', w) ∈ E and m ⪯_w m' do
5      for w' s.t. (m', w') ∈ E and w ⪰_{m'} w' do
6          G' ← G'\(m', w')
7      end for
8  end for
9  for w' s.t. (m, w') ∈ E and w ⪯_m w' do
10     for m' s.t. (m', w') ∈ E and m ⪰_{w'} m' do
11         G' ← G'\(m', w')
12     end for
13 end for
14 compute man-optimal super-stable matching in G'.
15 if exists man-optimal super-stable matching M in G' and M ∪ (m, w) is
   super-stable in G
16     return M ∪ (m, w)
17 else
18     return no super-stable (m, w)-matching exists.
19 end if
```

By the construction of G', any edge (m', w') such that $m \succeq_{w'} m'$ was removed. Hence w' must be unmatched in M. From Theorem 2, w' is unmatched in any super-stable matching of G. Let us assume there exists some super-stable (m, w)-matching N. Then $N' = N \backslash (m, w)$ is super-stable in G'. Since w' is unmatched in N, (m, w') blocks N, contradiction. (ii): any edge (m', w') such that $m' \neq m$ and $w' \neq w$ cannot block M. By the construction of the reduced graph G', the removal of (m', w') was caused by some edge (m, w') or (m', w). W.l.o.g, some edge (m, w') caused the removal of (m', w'). Hence, if w' is matched in M, then $M(w') \succeq_{w'} m'$. (m', w') does not block M. In the case that w' is unmatched in M, w' is unmatched in any super-stable matching in G. Similar to Case 1, if there exists some super-stable (m, w)-matching N, then (m, w') blocks N, contradiction. By the same argument, if M is super-stable in G, for any other super-stable matching N' in G', M' and N' match the same set of vertices. No edges in $E \backslash E'$ can block $N' \cup (m, w)$.

Theorem 4. *Let (m, w) be an edge in G. There is an $O(m)$ algorithm for computing a man-optimal super-stable (m, w)-matching or deciding that no super-stable (m, w)-matching exists.*

Proof. Lemma 3 makes sure if Algorithm 1 outputs a matching M, then M is super-stable in G. Lemma 2 guarantees that if there exists any super-stable matching in G, then Algorithm 1 would never miss it.

Theorem 5. *$(I(\mathcal{M}_G), \leq)$ can be constructed in $O(nm^2)$ time.*

Proof. $I(\mathcal{M}_G)$ can be computed in $O(m^2)$ time by running Algorithm 1 for each edge $(m, w) \in E$. The set $I(\mathcal{M}_G)$ has at most m elements. By checking each pair of $I(\mathcal{M}_G)$, we can construct the partial order. Each check takes $O(n)$ time. Thus, the total time is $O(nm^2)$.

4 A Maximal Sequence of Super-Stable Matchings

Representation via irreducible matchings is intuitive, but the time complexity is high. In this section, we give another representation via rotation poset and the time complexity to construct this rotation poset is only $O(mn)$.

Rotation poset derives from the concept of *minimal differences* of a ring of sets. A chain $C = \{C_1, \cdots, C_q\}$ in $\mathcal{F}$ is an ordered set of elements of $\mathcal{F}$ such that C_i is an immediate predecessor of C_{i+1} for each $i \in [q]$. The maximal chain is a chain that begins at the minimal element of $\mathcal{F}$, F_0 and ends at the maximal element of $\mathcal{F}$, F_z. Let F_i and F_{i+1} be two elements of $\mathcal{F}$ such that F_i is an immediate predecessor of F_{i+1}. The difference $D = F_{i+1} \backslash F_i$ is called a *minimal difference* of $\mathcal{F}$. Note that for each two consecutive elements of a chain C, there is a minimal difference D, we say that C contains D. The following two theorems give another version of Birkhoff's representation theorem in the language of minimal differences. The reader can find more details in Irving and Gusfield's book [4].

Theorem 6. [4] *If F and F' are two elements in $\mathcal{F}$ such that $F \subset F'$, then every chain from F to F' in $\mathcal{F}$ contains exactly the same set of minimal differences (in a different order).*

Theorem 7. [4] *Let $D(\mathcal{F})$ denote the set of all minimal differences in $\mathcal{F}$. For two minimal differences D and D', $D \prec D'$ if and only if D appears before D' on every maximal chain in $\mathcal{F}$. There is a one-to-one correspondence between the elements of $\mathcal{F}$ and the closed subsets of $D(\mathcal{F})$.*

In the context of super-stable matchings, we want to compute a maximal sequence of super-stable matchings in $\mathcal{M}(G)$, i.e. a sequence $M_0 \succ M_1 \succ \cdots \succ M_z$ where M_0 is the man-optimal super-stable matching and M_z is the woman-optimal super-stable matching and for each $1 \leq i \leq z$, there is no super-stable matching M' such that $M_{i-1} \succ M' \succ M_i$. We call a matching M' a strict successor of a matching M if M' is a successor of M, i.e. $M \succ M'$ and there exists no super-stable matching M'' such that $M \succ M'' \succ M'$. We can solve this problem by computing a strict successor of any super-stable matching M.

Let M be a super-stable matching in G and m a vertex in A. Suppose that there exists a super-stable matching M' such that m gets a worse partner in M' than in M, i.e. $M(m) \succ_m M'(m)$. Let $w = M'(m)$, by Lemma 1, w must be matched in M and $m \succ_w M(w)$. Hence we are essentially searching for some vertex w such that $M(m) \succ_m w$ and $m \succ_w M(w)$. In Algorithm 2, the set E_c

contains for each man m highest ranked edges incident to him that satisfies the condition above. For each man m, the candidate edge (m, w) is not unique, there might be other edge (m, w') that forms a tie with (m, w). While in the case of strict preference list, the candidate edge is unique.

A strongly connected component S of a directed graph G is a subgraph S that is strongly connected, i.e. there is a path in S in each direction between each pair of vertices of S, and is maximal with this property: no additional edges or vertices from G can be included in the subgraph without breaking its property of being strongly connected. We say that $e = (m, w)$ is an outgoing edge of S if $m \in S$ and $w \notin S$. Let $S(m)$ denote the strongly connected component that contains m.

In Algorithm 2 given below we maintain a directed graph $G_d = (V, E_d)$, whose every edge $(m, w) \in E_d \cap M$ is directed from w to m and every other edge (m, w) is directed from m to w. G_d is a subgraph of G that contains the edges the algorithm traverses so far. The basic idea of this algorithm is that for each man m such that $M(m) \neq M_z(m)$, we traverse the preference list of m until we find some candidate edges defined above. We add the edges traversed into G_d and the candidate edges into G_c. For each strongly connected component S of G_d without outgoing edges, we try to find a perfect matching on S in $G_c = (V, E_c)$. If we are successful, we find a strict successor of M. Otherwise, we modify G_c and G_d by allowing edges of lower ranks.

4.1 Correctness of Algorithm 2

Due to the space limit, we defer the proof of Lemma 4, Lemma 5 and Lemma 6 in our full version. Lemma 4 proves that any edge removed from G_d (line 9 and line 30) never block any super-stable matching that the algorithm will output.

Lemma 4. *Let M be a super-stable matching in G. For any successor N of M such that N is also a super-stable matching in G and each $(m, w) \in M$, any edge (m, w') such that $w' \succeq_m w$ or (m', w) such that $m \succ_w m'$ cannot block N.*

Lemma 5. *No edge deleted in line 17 can belong to any super-stable matching N dominated by M.*

Lemma 6. *No edge deleted in line 23 can belong to any super-stable matching N dominated by M.*

Lemma 7. *The output matching M_i is super-stable and a strict successor of M_{i-1}.*

Proof. Note that the algorithm outputs M_i when the edge set E_c is a perfect matching in a strongly connected component S with no outgoing edges and $M_i = (M_{i-1} \backslash S) \cup (E_c \cap S)$. Suppose, for a contradiction, that M_i is blocked by some edge $(m, w) \in E_d$. There are four cases. (i): $m \notin S$ and $w \notin S$, it is obvious that (m, w) cannot block M_i, since it would block M_{i-1} as well. (ii): $m \in S$ and $w \notin S$, this is not possible, because this will imply S has an outgoing edge

Algorithm 2: Computing a maximal sequence of super-stable matchings

```
1  let M_0 be the (unique) man-optimal super-stable matching of G.
2  let M_z be the (unique) woman-optimal super-stable matching of G.
3  M ← M_0
4  let M' contain edge (m, M(m)) for each man m such that M(m) =_m M_z(m)
5  let E_d contain all edges of M
6  let G_d be the directed graph (V, E_d) such that each edge (m, w) ∈ E_d ∩ M is
   directed from w to m and every other edge (m, w) is directed from m to w
7  E' ← E\E_d
8  let E_c = M' and G_c = (V, E_c)
9  for each (m, w) ∈ M remove from E' each edge (m', w) such that m' ≺_w m and
   each edge (m, w') such that w' ⪰_m w
10 repeat
11     while (∃m ∈ A) deg_{G_c}(m) = 0 do
12         add the set E_m of top choices of m from E' to E_d
13         if outdeg(S(m)) = 0 then
14             add every edge (m, w) ∈ E_m such that m ≻_w M(w) and
15             M(m) ≻_m w to E_c
16             for each edge (m, w) of E_c that becomes strictly dominated by
17             some added edge (m', w) remove it from G_c
18             remove E_m from E'
19         end if
20     end while
21     for each m ∈ A such that outdeg(S(m)) = 0 do
22         delete all lowest ranked edge in E_c ∪ E' incident to any w ∈ S such
23         that w is multiple engaged
24     end for
25     while (∃S) outdeg(S) = 0 and E_c is a perfect matching on S do
26         M ← (E_c ∩ S) ∪ (M\S)
27         M_i ← M
28         output M_i
29         i ← i + 1
30         update G_c and G_d: E_c ∩ S contains only edges (m, M(m)) such that
31         M(m) =_m M_z(m); an edge (m, w) ∈ S stays in G_d only if w = M(m)
32         and rank_w(m) ≤ rank_M(w)
33     end while
34 until (∀m ∈ A) rank_M(m) = rank_{M_z}(m)
```

in E_d. (iii): $m \notin S$ and $w \in S$, then $M_i(m)(= M_{i-1}(m)) \succ_m w$, hence (m, w) would not block M_i. (iv): $m \in S$ and $w \in S$, if (m, w) never belong to E_c, then $M_i(w) \succ_w M_{i-1}(w) =_w m$, (m, w) can not block M_i; if (m, w) once belongs to E_c and got deleted later, then w always get a strictly better partner than m. We prove that no edge from E_d can block M_i. There might be some other edges $e \notin E_d$ that can potentially block M_i. These edges are deleted during the updating of E_d. Lemma 4 gives a proof that these set of edges cannot block any matching N that is dominated by M_{i-1}. Hence M_i is super-stable.

Next we prove that M_i is a strict successor of M_{i-1}. Suppose not and let m be any man in S and N a successor of M_{i-1} such that $M_{i-1}(m) \succ N(m) \succ M_i(m)$. If $(m, N(m)) \in E_c$ and is not deleted during the algorithm, then $(m, M_i(m))$ would not be in E_c, which is not true. Since N is a successor of M and is super-stable, by Lemma 5 and Lemma 6, the edge $(m, N(m))$ can never once belong to E_c. Let $w = N(m)$, by our updating rule of E_d, we have $N(w) \succeq_w M(w)$. While if $N(w) \succ_w M(w)$, then the edge (m, w) must once belong to E_c. Thus we have $N(w) =_w M(w)$, which violates Lemma 1.

Lemma 8. *If $M_i \neq M_z$, the algorithm always outputs a matching.*

Proof. The algorithm will end without outputting any matching if and only if in line 25 the while loop, it cannot find any strongly connected component with no outgoing edges. Note that every directed graph can be expressed as a directed acyclic graph of its strongly connected components. Hence, we can always find a strongly connected component without outgoing edges.

Theorem 8. *Algorithm 2 computes a maximal sequence of super-stable matchings.*

Proof. By Lemma 7 and Lemma 8, it is obvious that Algorithm 2 outputs a maximal sequence of super-stable matchings.

4.2 Running Time of Algorithm 2

Theorem 9. *The running time of Algorithm 2 is $O(mn)$.*

Proof. Each time we add new edges into E_d, we need to compute strongly connected components of G_d. Computing strongly connected component of any directed graph $G' = (V', E')$ can be done in $O(E)$ time. Each edge e of G is added to G_d at most once, and G_d is always a subgraph of G. Hence, a naive implementation takes $O(m^2)$ on computing strongly connected components of G_d. As mentioned in [12], Pearce [19] and Pearce and Kelly [18] sketch how to extend their algorithm and that of Marchetti-Spaccamela et al. [15] to compute strongly connected component dynamically. Their algorithm runs in $O(mn)$ if edges can only be added to the graph and not deleted. The edges in G_d can be deleted during the algorithm, but they are deleted only when E_c is perfect on a strongly connected component without outgoing edges. Thus, other strongly connected components are unchanged. Also as mentioned in [12], the edges remaining in the selected strongly connected component can be treated as they were added anew to the graph. Since the ranks of men increase as we output subsequent super-stable matchings, each edge can be added anew to G_d constant number of times. Thus, the amortized cost of edge insertion remains unchanged. The reader can easily check the other part of the algorithm takes at most $O(m)$ time. Hence, the total time is $O(mn)$.

4.3 Rotation Poset

We have shown all rotations $D(\mathcal{M}_G)$ can be found in time $O(mn)$ by Algorithm 2. It remains to show how to efficiently construct the precedence ration $\prec$ on $D(\mathcal{M}_G)$. Our construction is essentially the same as the construction given in [4] for the classical stable marriage problem. The only difference here is that one rotation for super-stable matchings can be one or multiple cycles, while one rotation for stable matchings in the classical stable marriage problem is always a cycle. The reader can find more details in [4]. Due to the space limit, we defer this section in our full version.

We summarize Sect. 4 with the following theorem.

Theorem 10. *The partial order $(D(\mathcal{M}_G), \prec)$ can be constructed in $O(mn)$.*

Proof. The construction of $D(\mathcal{M}_G)$ takes $O(mn)$ time by running Algorithm 2. The precedence relation can be constructed in $O(m)$ time. Hence, the time complexity is $O(mn)$.

5 The Super-Stable Matching Polytope

In this section, we give a polyhedral characterization of the set of all super-stable matchings and prove that the super-stable matching polytope is integral. The main result is the following theorem.

Theorem 11. *Let $G = (V, E)$ be a stable matching problem with ties where the graph G is bipartite, then the super-stable matching polytope $SUSM(G)$ is described by the following linear system:*

$$\sum_{u \in N(v)} x_{u,v} \leq 1, \qquad \forall v \in V, \tag{1a}$$

$$\sum_{i >_u v} x_{u,i} + \sum_{j >_v u} x_{j,v} + x_{u,v} \geq 1, \qquad \forall (u,v) \in E, \tag{1b}$$

$$x_{u,v} \geq 0, \qquad \forall (u,v) \in E \tag{1c}$$

where $N(v)$ denotes the set of neighbors of v in G, and $w >_u v$ means u prefers w to v.

Proof. Let x be a feasible solution. Define E^+ to be the set of edges (u, v) with $x_{u,v} > 0$, and V^+ the set of vertices covered by E^+. For each $u \in V^+$, let $N^*(u)$ be the maximal elements in $\{i : x_{u,i} > 0\}$. Note that there might be multiple maximal elements that form a tie.

We first show the following lemma.

Lemma 9. *For each vertex u and each vertex $v \in N^*(u)$, then u is the unique minimal element in $\{j : x_{j,v} > 0\}$ and that $\sum_{j \in N(v)} x_{j,v} = 1$.*

Proof. Indeed, (1b) implies

$$1 \leq \sum_{j >_v u} x_{j,v} + x_{u,v} = \sum_{j \in N(v)} x_{j,v} - \sum_{j <_v u} x_{j,v} - \sum_{\substack{j =_v u; \\ j \neq u}} x_{j,v} \leq 1 - \sum_{j <_v u} x_{j,v} - \sum_{\substack{j =_v u; \\ j \neq u}} x_{j,v} \leq 1 \quad (2)$$

Hence we have equality throughout in (2). This implies that $x_{j,v} = 0$ for each $\{j : j <_v u\}$ and each $\{j : j =_v u; j \neq u\}$ and that $\sum_{j \in N(v)} x_{j,v} = 1$. Since $x_{j,v} = 0$ for each $\{j : j =_v u; j \neq u\}$, v strictly prefers any other vertices in $\{j : x_{j,v} > 0\}$ over u, making u the unique minimal element in $\{j : x_{j,v} > 0\}$.

We then prove that for any v such that $v \in N^*(u)$ for some u, then u is unique. Suppose not, there is a vertex $u' \neq u$ and $v \in N^*(u')$. By Lemma 9, u is the unique minimal element in $\{j : x_{j,v} > 0\}$, and u' is the unique minimal element in $\{j : x_{j,v} > 0\}$, contradiction.

Now let U and W be the color classes of G. For any $u \in U \cap V^+$, there is at least one unique vertex $w \in N^*(u)$, such that $\sum_{j \in N(w)} x_{j,w} = 1$. Let $F_W(x)$ be the set of these vertices. Formally, $F_W(x) = \{w : w \in N^*(u), u \in U \cap V^+\}$. Then we have $|F_W(x)| \geq |U \cap V^+|$. We also have that

$$|F_W(x)| = \sum_{w \in F_W(x)} \sum_{j \in N(w)} x_{j,w} = \sum_{j \in U \cap V^+} \sum_{w \in F_W(x)} x_{j,w} \leq \sum_{j \in U \cap V^+} 1 = |U \cap V^+| \quad (3)$$

implying that $|F_W(x)| = |U \cap V^+|$. Hence, we conclude that for each $u \in U \cap V^+$, $|N^*(u)| = 1$, which implies that u has an unique maximal element in $\{i : x_{u,i} > 0\}$. Since $|N^*(u)| = 1$, we denote this unique vertex as $x^*(u)$. We then have the following corollary.

Corollary 1. *There is a bijection between $U \cap V^+$ and $F_W(x)$, and for each $u \in U \cap V^+$, $\sum_{i \in N(u)} x_{u,i} = 1$.*

Similarly, we may define $F_U(x) = \{u : u \in N^*(w), w \in W \cap V^+\}$ and we have

Corollary 2. *There is a bijection between $W \cap V^+$ and $F_U(x)$, and for each $w \in W \cap V^+$, $\sum_{j \in N(w)} x_{j,w} = 1$.*

Then we have $|U \cap V^+| = |F_W(x)| \leq |W \cap V^+|$ and $|W \cap V^+| = |F_U(x)| \leq |U \cap V^+|$, implying $|U \cap V^+| = |W \cap V^+| = |F_W(x)| = |F_U(x)|$. Then any $u \in U \cap V^+$ is also in $F_U(x)$, hence, u has an unique minimal element, denoted by $x_*(u)$.

The bijection between $U \cap V^+$ and $F_W(x)$ forms a perfect matching M in (V^+, E^+), i.e. the set of edges $\{(u, x^*(u)) : u \in U \cap V^+\}$. Similarly, the bijection between $W \cap V^+$ and $F_U(x)$ forms another perfect matching N, i.e. the set of edges $\{(x^*(w), w) : w \in W \cap V^+\}$.

Consider the vector $x' = x + \varepsilon\chi^M - \varepsilon\chi^N$, with ε close enough to 0 (positive or negative). we will show that x' is also feasible solution of (1a)–(1c). It is easy to see that x' satisfies (1a) and (1c). For each vertex $u \in U \cap V^+$, there is an unique maximal element $x^*(u)$ and $(u, x^*(u)) \in M$ and an unique minimal element $x_*(u)$ and $(u, x_*(u)) \in N$, implying $\sum_{i \in N(u)} x'_{u,i} = \sum_{i \in N(u)} x_{u,i} \leq 1$. To see that x' satisfies (1b), let (u, v) be an edge in E^+ attaining equality in (1b). The case that $(u, v) \in M$ or $(u, v) \in N$ is trivial. So assume that $(u, v) \notin M$ and $(u, v) \notin N$. The edge $(u, x^*(u)) \in M$ and $x^*(u) >_u v$. There is no other edge in $\{(u, i) : i \in N(u)\}$ belongs to M. We prove that there is no edge (j, v) in M and $j >_v u$ since if $(j, v) \in M$, j is the minimal element of v. Similarly, we can prove that there is exact one edge $(j, v) \in N$ and $j >_v u$. Concluding, $\sum_{i >_u v} x'_{u,i} + \sum_{j >_v u} x'_{j,v} + x'_{u,v} = \sum_{i >_u v} x_{u,i} + \sum_{j >_v u} x_{j,v} + x_{u,v} = 1$. Let x be an extreme point. The feasibility of x' implies that $\chi^M = \chi^N$, that is, $M = N$. So $E^+ = M$ since the maximal element is the same as the minimal element for each vertex, hence, $x = \chi^M$.

5.1 Partial Order Preference Lists

Partial order preference lists are generalisation of preference lists with ties in such a way that the preference list of each man or woman is an arbitrary partial order. It turns out that the linear system (1a)–(1c) can also describe the set of all super-stable matchings with partial order preference list. See more details in our full version.

5.2 The Strongly Stable Matching Polytope

Kunysz [11] gives a linear system that characterizes the set of all strongly stable matchings and proves this linear system is integral using the duality theory of linear programming. Here, we give an alternate and simpler proof that does not rely on the duality theory and uses only Hall's theorem. See the proof in our full version.

References

1. Birkhoff, G., et al.: Rings of sets. Duke Math. J. **3**(3), 443–454 (1937)
2. Dean, B., Jalasutram, R.: Factor revealing lps and stable matching with ties and incomplete lists. In: Proceedings of the 3rd International Workshop on Matching Under Preferences, pp. 42–53 (2015)
3. Fleiner, T., Irving, R.W., Manlove, D.F.: Efficient algorithms for generalized stable marriage and roommates problems. Theoret. Comput. Sci. **381**(1–3), 162–176 (2007)
4. Gusfield, D., Irving, R.W.: The Stable marriage problem - structure and algorithms. Foundations of computing series. MIT Press (1989)
5. Huang, C.-C., Kavitha, T.: An improved approximation algorithm for the stable marriage problem with one-sided ties. In: Lee, J., Vygen, J. (eds.) IPCO 2014. LNCS, vol. 8494, pp. 297–308. Springer, Cham (2014). https://doi.org/10.1007/978-3-319-07557-0_25

6. Irving, R.W.: Stable marriage and indifference. Discret. Appl. Math. **48**(3), 261–272 (1994)
7. Iwama, K., Miyazaki, S., Morita, Y., Manlove, D.: Stable marriage with incomplete lists and ties. In: Wiedermann, J., van Emde Boas, P., Nielsen, M. (eds.) ICALP 1999. LNCS, vol. 1644, pp. 443–452. Springer, Heidelberg (1999). https://doi.org/10.1007/3-540-48523-6_41
8. Iwama, K., Miyazaki, S., Yanagisawa, H.: A 25/17-approximation algorithm for the stable marriage problem with one-sided ties. Algorithmica **68**(3), 758–775 (2014). https://doi.org/10.1007/s00453-012-9699-2
9. Kavitha, T., Mehlhorn, K., Michail, D., Paluch, K.E.: Strongly stable matchings in time o (nm) and extension to the hospitals-residents problem. ACM Trans. Algorithms (TALG) **3**(2), 15-es (2007)
10. Király, Z.: Linear time local approximation algorithm for maximum stable marriage. Algorithms **6**(3), 471–484 (2013)
11. Kunysz, A.: An algorithm for the maximum weight strongly stable matching problem. In: 29th International Symposium on Algorithms and Computation (ISAAC 2018). Schloss Dagstuhl-Leibniz-Zentrum fuer Informatik (2018)
12. Kunysz, A., Paluch, K., Ghosal, P.: Characterisation of strongly stable matchings. In: Proceedings of the Twenty-Seventh Annual ACM-SIAM Symposium on Discrete Algorithms, pp. 107–119. SIAM (2016)
13. Lam, C.K., Plaxton, C.G.: A (1+ 1/e)-approximation algorithm for maximum stable matching with one-sided ties and incomplete lists. In: Proceedings of the Thirtieth Annual ACM-SIAM Symposium on Discrete Algorithms, pp. 2823–2840. SIAM (2019)
14. Manlove, D.F.: The structure of stable marriage with indifference. Discret. Appl. Math. **122**(1–3), 167–181 (2002)
15. Marchetti-Spaccamela, A., Nanni, U., Rohnert, H.: Maintaining a topological order under edge insertions. Inf. Process. Lett. **59**(1), 53–58 (1996)
16. McDermid, E.: A 3/2-approximation algorithm for general stable marriage. In: Albers, S., Marchetti-Spaccamela, A., Matias, Y., Nikoletseas, S., Thomas, W. (eds.) ICALP 2009. LNCS, vol. 5555, pp. 689–700. Springer, Heidelberg (2009). https://doi.org/10.1007/978-3-642-02927-1_57
17. Paluch, K.: Faster and simpler approximation of stable matchings. Algorithms **7**(2), 189–202 (2014)
18. Pearce, D.J., Kelly, P.H.: Online algorithms for topological order and strongly connected components. Technical report, Citeseer (2003)
19. Pearce, D.J.: Some directed graph algorithms and their application to pointer analysis. Ph.D. thesis, University of London (2005)
20. Radnai, A.: Approximation algorithms for the stable matching problem. Eötvös Lorand University (2014)
21. Rothblum, U.G.: Characterization of stable matchings as extreme points of a polytope. Math. Program. **54**(1–3), 57–67 (1992)
22. Scott, S.: A study of stable marriage problems with ties. Ph.D. thesis. University of Glasgow (2005)
23. Spieker, B.: The set of super-stable marriages forms a distributive lattice. Discret. Appl. Math. **58**(1), 79–84 (1995)
24. Teo, C.P., Sethuraman, J.: The geometry of fractional stable matchings and its applications. Math. Oper. Res. **23**(4), 874–891 (1998)
25. Vate, J.H.V.: Linear programming brings marital bliss. Oper. Res. Lett. **8**(3), 147–153 (1989)

Uniform Embeddings for Robinson Similarity Matrices

Jeannette Janssen(✉) and Zhiyuan Zhang

Dalhousie University, Halifax, NS, Canada
{Jeannette.Janssen,owen.zhang}@dal.ca

Abstract. A Robinson similarity matrix is a symmetric matrix where all entries in all rows and columns are increasing towards the diagonal. A Robinson matrix can be decomposed into the weighted sum of k adjacency matrices of a nested family of unit interval graphs. We study the problem of finding an embedding which gives a *simultaneous* unit interval embedding for all graphs in the family. This is called a *uniform embedding*. We give a necessary and sufficient condition for the existence of a uniform embedding, derived from paths in an associated graph. We also give an efficient combinatorial algorithm to find a uniform embedding or give proof that it does not exist, for the case where $k = 2$.

Keywords: Robinson similarity · Unit interval graph · Proper interval graph · Indifference graph

1 Introduction

In many different settings it occurs that a linearly ordered set of data items is given, together with a pair-wise similarity measure of these items, with the property that items are more similar if they are closer together in the ordering. A classic example of this setting is in archaeology, where sites are ordered according to their age, and the composition of the items found at the sites are more similar if the sites are closer in age. Other applications occur in evolutionary biology, sociology, text mining, and visualization. (See [12] for an overview.) The similarity between such an ordered set of items, when presented in the form of a matrix, will have the property that entries in each row and column increase towards the diagonal (when items are closer in the ordering), and decrease away from the diagonal. Such a matrix is called a Robinson matrix, or Robinson similarity matrix.

Formally, a *Robinson matrix* is a symmetric matrix where the entries $a_{i,j}$ satisfy the following condition:

$$\text{For all } u < v < w, a_{u,v} \geq a_{u,w} \text{ and } a_{v,w} \geq a_{u,w}. \tag{1.1}$$

Supported by an NSERC Discovery grant.

A. Lubiw et al. (Eds.): WADS 2021, LNCS 12808, pp. 499–512, 2021.
https://doi.org/10.1007/978-3-030-83508-8_36

In other words, a Robinson matrix is an asymmetric matrix where entries in each row and column are increasing towards the diagonal. See Example 1 (to follow) for examples. Robinson matrices are named after Robinson, who first mentioned such matrices in [14] in the context of archaeology.

Robinson wanted to solve the following question, that is also referred as the *seriation problem*: suppose a set of objects has an underlying linear order and given their pair-wise similarity, arrange the objects so that objects that are closer in the arrangement are more similar than pairs that are further apart. That is, find the ordering of the items for which the similarity values form a Robinson matrix. However, seriation only gives a *linear ordering* of the items. In this paper we focus on finding a linear representation of the items that takes into account the numerical value of the similarity. In the context of archaeology, this would mean that we are looking not only the order of the sites in terms of their age, but also of some indication of their age.

We assume similarities cannot be judged very precisely, and thus we focus on Robinson matrices where the entries are taken from a restricted set $\{0, 1, 2 \ldots, k\}$, where k indicates "very similar", and 0 indicates "not at all similar". We will assume throughout that all diagonal entries equal k. We are then looking for a linear embedding of the items so that the distance between pairs with the same similarity value are approximately similar. More precisely, we require that there exist threshold distances $d_1 > d_2 > \cdots > d_k > 0$ and an embedding of the items into $\mathbb{R}$, such that, if a pair of items has similarity level t, then the distance between their embedded values lies between threshold distances d_{t+1} and d_t. We will call this a *uniform embedding*; See Definition 1 for a formal definition.

A $\{0, 1\}$-valued symmetric matrix A is Robinson if and only if $A - I$ is the adjacency matrix of a *proper interval graph* [9]. The class of proper interval graphs equals the class of unit interval graphs [13], which equals the class of *indifference graphs*. A graph is an indifference graph if and only if there exists a linear embedding of the vertices with respect to a threshold distance $d > 0$ so that two vertices are adjacent if and only if their embedded values have distance at most d. An indifference graph embedding is therefore a uniform embedding for the associated binary Robinson matrix.

A Robinson matrix taking values in $\{0, 1, \ldots, k\}$ can be seen as the representation of a nested family of indifference graphs. Namely, any such matrix $A = (a_{u,v})$ can be written as $A = \sum_{t=1}^{k} A^{(t)}$, where for all $t \in [k]$, $A^{(t)} = (a_{u,v}^{(t)})$ is a binary matrix such that, $a_{u,v}^{(t)} = 1$ if $a_{u,v} \geq t$, and 0 otherwise. Clearly, each $A^{(t)}$ is Robinson and has all ones on the diagonal. Therefore $A^{(t)} - I$ is the adjacency matrix of an indifference graph $G^{(t)}$. These graphs are nested, i.e. for all $t < k$, $G^{(t+1)}$ is a subgraph of $G^{(t)}$. In this light, our problem can be restated as that of finding a *simultaneous* indifference graph embedding for all graphs $G^{(t)}$.

As shown in the following example, not every Robinson matrix has a uniform embedding.

Example 1. Consider the following matrices

$$A = \begin{bmatrix} 2 & 2 & 1 & 0 & 0 \\ 2 & 2 & 2 & 1 & 1 \\ 1 & 2 & 2 & 2 & 1 \\ 0 & 1 & 2 & 2 & 2 \\ 0 & 1 & 1 & 2 & 2 \end{bmatrix} \qquad B = \begin{bmatrix} 2 & 2 & 1 & 0 & 0 & 0 \\ 2 & 2 & 2 & 1 & 1 & 1 \\ 1 & 2 & 2 & 2 & 1 & 1 \\ 0 & 1 & 2 & 2 & 2 & 1 \\ 0 & 1 & 1 & 2 & 2 & 2 \\ 0 & 1 & 1 & 1 & 2 & 2 \end{bmatrix}$$

Matrix A has a linear embedding Π with threshold distances $d_1 = 8, d_2 = 6$ given by

$$\Pi = \langle 0, 5, 6.5, 11.75, 12.75 \rangle.$$

We can check that, for any pair (i, j), if $a_{i,j} = 2$ then $\Pi(i)$, $\Pi(j)$ have distance at most 6, if $a_{i,j} = 1$ then the distance between $\Pi(i)$ and $\Pi(j)$ lies between 6 and 8, and if $a_{i,j} = 0$ then this distance is greater than 8.

In contrast, matrix B does not have a uniform embedding. Suppose there exists such an embedding Π with threshold distances $d_1 > d_2 > 0$. Suppose also that Π is increasing; see Theorem 1 for the justification. Since $b_{1,4} = 0$ and $b_{4,6} = 1$, we have that $\Pi(4) - \Pi(1) > d_1$ and $\Pi(6) - \Pi(4) > d_2$. This implies that $\Pi(6) - \Pi(1) > d_1 + d_2$. On the other hand, we have that $b_{1,2} = 2$ and $b_{2,6} = 1$, so $\Pi(2) - \Pi(1) \leq d_2$ and $\Pi(6) - \Pi(2) \leq d_1$. This implies that $\Pi(6) - \Pi(1) \leq d_1 + d_2$. Combining the inequalities results in $d_1 + d_2 < \Pi(6) - \Pi(1) \leq d_1 + d_2$, a contradiction.

In this paper, we consider the problem of finding a uniform embedding of a Robinson matrix or giving proof that it does not exist. In Theorem 2 we give a condition for the existence of a uniform embedding in terms of the threshold distances $d_1, \ldots, d_k$. We then show that this condition is sufficient by giving an algorithm to find a uniform embedding, given threshold distances that meet the condition. We will also compare the complexity of verifying the given condition with the complexity of solving the inequality system defining a uniform embedding (Definition 1).

Finally, we consider the case where $k = 2$, so the problem is to find a uniform embedding for two nested indifference graphs. We give a combinatorial algorithm to either find a uniform embedding or give a substructure that shows it does not exist. The algorithm has complexity $O(N^{2.5})$, where $N = n(n-1)/2$ is the size of the input.

1.1 Related Works

In [3], the problem of finding a uniform embedding was studied for *diagonally increasing graphons*. A graphon is a symmetric function $w : [0, 1]^2 \rightarrow [0, 1]$. Graphons can be seen as generalizations of matrices. The results from [3] do not apply in the context of this paper. Namely, a matrix can be represented as a graphon, but the "boundaries" delineating the regions of $[0, 1]^2$ where the graphon takes a certain value $1 \leq t \leq k$ in this case are piecewise constant functions. The results from [3] apply only to boundary functions that are continuous and strictly increasing.

In [13], Roberts established the equivalence between the classes of unit interval graphs, proper interval class, indifference graphs, and *claw-free* interval graphs. Different proofs can be found in [2] and [7].

A lot of work has been done on the seriation problem. For binary matrices, this is equivalent to recognizing proper (or unit) interval graphs. Corneil [4] gives a linear time unit interval recognition algorithm, which improves on [5]. Atkins *et al.*[1] gave a spectral algorithm for the general seriation problem. Laurent and Seminaroti [10,11] give a combinatorial algorithm that generalizes the algorithm from [4]. A general overview of the seriation problem and its applications can be found in [12].

2 Uniform Embeddings

We start by formally defining a uniform embedding. First we introduce some notation. For an integer $t \in \mathbb{Z}_+$, let $[t] = \{1, 2, \ldots, t\}$. Let $\mathcal{S}^n[k]$ denote the set of all Robinson matrices with entries in $\{0, 1, 2, \ldots, k\}$, and let $\mathbb{D}^k$ be the set of threshold vectors $\mathbb{D}^k = \{\boldsymbol{d} \in \mathbb{R}^k : \boldsymbol{d} = (d_i)_{i \in [k]}, d_1 > \cdots > d_k > 0\}$.

Definition 1. *Given a matrix* $A \in \mathcal{S}^n[k]$ and a threshold vector $\boldsymbol{d} \in \mathbb{D}^k$, *a map* $\Pi : [n] \to \mathbb{R}$ *is a* uniform embedding of A with respect to $\boldsymbol{d}$ *if, for each pair* $u, v \in [n]$*:*

$$a_{u,v} = t \iff d_{t+1} < |\Pi(v) - \Pi(u)| \leq d_t \quad \textit{for } t \in \{0, \ldots, k\}, \tag{2.1}$$

where we define $d_{k+1} = -\infty$ *and* $d_0 = \infty$, *so that the lower bound for* $a_{u,v} = k$ *and the upper bound for* $a_{u,v} = 0$ *are trivially satisfied.*

The following theorem states that, if a uniform embedding exists, then we may assume it to have certain nice properties. The proof is straightforward but technical, and has been omitted here; it can be found in [8].

Theorem 1. *Let* $A \in \mathcal{S}^n[k]$. *If* A *has a uniform embedding, then there exist* $\boldsymbol{d} \in \mathbb{D}^k$ *and a uniform embedding* Π *with respect to* $\boldsymbol{d}$ *which is strictly monotone increasing, and which is such that the inequalities 2.1 are all strict. That is, for all pairs of* $u, v \in [n]$, $u < v$,

$$a_{u,v} = t \iff d_{t+1} < \Pi(v) - \Pi(u) < d_t, \tag{2.2}$$

where $d_{k+1} = 0$ *and* $d_0 = \infty$.

Note that the definition of d_{k+1} in Theorem 1 has changed from $-\infty$ to zero; this change enforces that Π is *strictly* increasing. In the later context, we will assume that any uniform embedding Π of matrix A with respect to any $\boldsymbol{d}$ is a map $\Pi : [n] \to \mathbb{R}$ which satisfies Eq. 2.2 (with the new definition of d_{k+1}). This will simplify the proofs and reduce the need to distinguish different cases. By Theorem 1, we can make this assumption without loss of generality.

3 Bounds, Walks, and Their Concatenation

The contradiction for matrix B in the Example 1 was derived from a cyclic sequence of vertices, namely $\langle 1,4,6,2,1\rangle$, and the bounds on the distance between successive pairs of this sequence. In this section, we show how walks in an associated graph generate a set of bounds which need to be satisfied by any linear embedding.

Definition 2. *Let $A \in \mathcal{S}^n[k]$ and fix $u, v \in [n]$. A vector $\boldsymbol{b} \in \mathbb{Z}^k$ is an* upper bound on (u,v) *if the inequality $\Pi(v) - \Pi(u) < \boldsymbol{b}^\top \boldsymbol{d}$ is implied by the inequality system (2.2) in the sense that, for any uniform embedding Π and threshold vector $\boldsymbol{d} \in \mathbb{D}^k$ satisfying inequality system (2.2), it holds that $\Pi(v) - \Pi(u) < \boldsymbol{b}^\top \boldsymbol{d}$. Similarly, the vector $\boldsymbol{b}$ is a* lower bound on (u,v) *if the inequality $\boldsymbol{b}^\top \boldsymbol{d} < \Pi(v) - \Pi(u)$ is implied by (2.2).*

It follows directly from inequality system (2.2) that, for any matrix $A \in \mathcal{S}^n[k]$, and any pair $u, v \in [n]$, $u < v$, the all-zero vector $\mathbf{0}$ is a lower bound on (u,v). Note that, if $\boldsymbol{b}$ is an upper bound on (u,v), then $-\boldsymbol{b}$ is a lower bound on (v,u), and vice versa.

We will see how new bounds can be obtained from walks in a corresponding graph. An original set of bounds, derived from edges, can be obtained directly from inequality system (2.2). This is made precise in the following definition. Let $\chi_i \in \mathbb{Z}^k$ denote the unit vector with 1 at the ith position and zero otherwise.

Definition 3. *Let $A = (a_{i,j}) \in \mathcal{S}^n[k]$. Let $u, v \in [n]$, $u < v$. Define*

$$\beta^+(u,v) = \begin{cases} \chi_t & \text{if } a_{u,v} = t \geq 1; \\ \text{undefined} & \text{if } a_{u,v} = 0. \end{cases}$$

and

$$\beta^-(u,v) = \begin{cases} \chi_{t+1} & \text{if } a_{u,v} = t \leq k-1; \\ \mathbf{0} & \text{if } a_{u,v} = k. \end{cases}$$

It follows immediately from Definition 3 and inequality system (2.2) that, for any $u, v \in [n]$, $u < v$, $\beta^+(u,v)$ is a lower bound on (u,v) (if $a_{u,v} \neq 0$), and $\beta^-(u,v)$ is a lower bound on (u,v).

We can consider $\langle u,v\rangle$ as a walk of length 1 in the complete graph with vertex set $[n]$. If $u < v$, then $\beta^+(u,v)$ is the upper bound defined by this walk, and $\beta^-(u,v)$ is the lower bound. We now extend this notion to unordered pairs, and, more generally, to longer walks. To accommodate the fact that there is no upper bound on $\Pi(v) - \Pi(u)$ if $a_{u,v} = 0$, we distinguish *edges* in this graph which are pairs $\{u,v\}$ such that $a_{u,v} > 0$, and *null-edges* which are pairs $\{u,v\}$ so that $a_{u,v} = 0$. In the following we will see that we can combine walks to obtain more bounds.

Definition 4. *A (u,v)-walk is a sequence $W = \langle w_0, w_1, \ldots, w_p\rangle$ where $w_i \in [n]$, $0 \leq i \leq p$, and $u = w_0$ and $v = w_p$. In other words, W is a walk in the* *complete graph with vertex set* $[n]$. *The walk* W *is an* upper-bound-walk *if for all* $1 \leq i \leq p$,

$$\{w_{i-1}, w_i\} \text{ is } \begin{cases} \text{an edge} & \text{if } w_{i-1} < w_i, \\ \text{an edge or a null-edge} & \text{if } w_{i-1} > w_i. \end{cases}$$

In other words, in an upper-bound-walk, null-edges are only traversed from larger to smaller vertices. Similarly, the walk W *is a* lower-bound-walk *if null-edges are only traversed to go from smaller to larger vertices.*

Now first define for all $u, v \in [n]$ *so that* $u < v$,

$$\beta^+(v, u) = -\beta^-(u, v) \text{ and } \beta^-(v, u) = -\beta^+(u, v). \tag{3.1}$$

Then for any walk $W = \langle w_0, w_1, \ldots, w_p \rangle$, *define*

$$\beta^+(W) = \textstyle\sum_{i=1}^p \beta^+(w_{i-1}, w_i) \text{ if } W \text{ is an upper-bound-walk, and} \tag{3.2}$$
$$\beta^-(W) = \textstyle\sum_{i=1}^p \beta^-(w_{i-1}, w_i) \text{ if } W \text{ is a lower-bound-walk.} \tag{3.3}$$

Given two walks $W_1 = \langle u_0, \ldots, u_s \rangle$ *and* $W_2 = \langle u_s, \ldots, u_p \rangle$, *denote* $W = W_1 + W_2 = \langle u_0, \ldots, u_p \rangle$ *as the* concatenation *of* W_1 *and* W_2.

For any walk $W = \langle w_0, w_1, \ldots, w_{p-1}, w_p \rangle$, define the *reverse* of W as $W^{\leftarrow} = \langle w_p, w_{p-1}, \ldots, w_1, w_0 \rangle$. Clearly, if W is an upper-bound-walk, then $W^{\leftarrow}$ is a lower-bound-walk, and vice versa. By (3.1), (3.2) and (3.3), we have that $\beta^+(W) = -\beta^-(W^{\leftarrow})$.

Lemma 1. *Let* $A \in \mathcal{S}^n[k]$. *For any* $u, v \in [n]$ *and any* (u, v)*-walk* W, *if* W *is an upper-bound-walk then* $\beta^+(W)$ *is an upper bound on* (u, v), *and if* W *is a* (u, v)*-lower-bound-walk then* $\beta^-(W)$ *is a lower bound on* (u, v).

The proof of this lemma follows by induction on the length of the walk, using the definitions. It can be found in [8].

4 A Sufficient and Necessary Condition

Section 3 introduced the necessary concepts to state the main theorem of this paper. We saw in the previous section that upper- and lower-bound-walks give bounds that must be satisfied by any uniform embedding. This hints at a condition for the existence of a uniform embedding: there must exist $\boldsymbol{d} \in \mathbb{D}^k$ so that each lower bound derived from a (u, v)-walk is smaller than each upper bound derived from a (u, v)-walk.

As it turns out, we only need to consider lower- and upper-bound-paths, that is, walks without repeated vertices. Given a matrix $A \in \mathcal{S}^n[k]$, let $\mathcal{L}_{u,v}$ be the set of all (u, v)-lower-bound-paths, and $\mathcal{U}_{u,v}$ be the set of all (u, v)-upper-bound-paths. Note that $\mathcal{U}_{u,v}$ and $\mathcal{L}_{u,v}$ are finite, whereas the set of all walks is infinite.

Define the inequality system:

For all $u, v \in [n]$, $u < v$, for any upper bound $\boldsymbol{b} = \beta^+(W_1)$ where $W_1 \in \mathcal{U}_{u,v}$ and any lower bound $\boldsymbol{a} = \beta^-(W_2)$ where $W_2 \in \mathcal{L}_{u,v}$,

$$\boldsymbol{a}^\top \boldsymbol{d} < \boldsymbol{b}^\top \boldsymbol{d}. \tag{4.1}$$

Theorem 2. *A Robinson matrix $A \in \mathcal{S}^n[k]$ has a uniform embedding if and only if there exists $\boldsymbol{d} \in \mathbb{D}^k$ satisfying inequality system (4.1).*

We can prove the necessity of Theorem 2 without other tools

Proof of the forward implication of Theorem 2: Suppose A has a uniform embedding. By Theorem 1, this implies that A has a uniform embedding Π with respect to a threshold vector $\boldsymbol{d} \in \mathbb{D}^k$ which satisfy inequality system (2.2). Let $u, v \in [n]$ with $u < v$. Let $W_1 \in \mathcal{L}_{u,v}$ and $W_2 \in \mathcal{U}_{u,v}$, and let $\boldsymbol{a} = \beta^-(W_1)$ and $\boldsymbol{b} = \beta^+(W_2)$. Then by Lemma 1 and Definition 2, $\boldsymbol{a}^\top \boldsymbol{d} < \Pi(v) - \Pi(u) < \boldsymbol{b}^\top \boldsymbol{d}$. □

For the converse, we will obtain an iterative procedure to construct a uniform embedding Π which satisfies inequality system Eq. 2.2. However, first we need to prove that condition (4.1) for paths implies that the same condition holds for all walks.

4.1 Cycles and Paths

In Sect. 3 we saw how walks can be used to generate new inequalities that are implied by the inequality system 2.2. In this section we show that, for the existence of a uniform embedding we need only to consider paths.

A (u, v)-upper-bound-walk $W = \langle u = w_0, w_1, \ldots, w_p = v\rangle$ is an *upper-bound-cycle* if $u = v$. and W contains no other repeated vertices. Note that the order in which the cycle is traversed determines whether or not it is an upper-bound-cycle.

Lemma 2. *Let $A \in \mathcal{S}^n[k]$ and $\boldsymbol{d} \in \mathbb{D}^k$. Let $C = \langle u_1, \ldots, u_p\rangle$, $u_1 = u_p = u$, be an upper-bound-cycle. If $\boldsymbol{d} \in \mathbb{D}^k$ satisfies Eq. 4.1, then $\beta^+(C)^\top \boldsymbol{d} > 0$.*

Proof: Suppose $v = u_i \in C$ for some $1 < i < p$, then $C = W_1 + W_2$ where $W_1 = \langle u_1, \ldots, u_i\rangle$ and $W_2 = \langle u_i, \ldots, u_p\rangle$. Then W_1 is a (u, v)-upper-bound-path, and W_2 is a (v, u)-upper-bound-path, so $W_2^{\leftarrow}$ is a (u, v)-lower-bound-path. Then by Definition 4,

$$\beta^+(C) = \beta^+(W_1) + \beta^+(W_2) = \beta^+(W_1) - \beta^-(W_2^{\leftarrow}).$$

By the choice of $\boldsymbol{d}$, $\beta^-(W_2^{\leftarrow})^\top \boldsymbol{d} < \beta^+(W_1)^\top \boldsymbol{d}$, and thus $\beta^+(C)^\top \boldsymbol{d} > 0$. □

Lemma 3. *Let $A \in \mathcal{S}^n[k]$ be a Robinson matrix and let $\boldsymbol{d} \in \mathbb{D}^k$, and suppose $\boldsymbol{d}$ satisfies (4.1). Suppose W is a (u, v)-upper-bound-walk W. Then there exists a (u, v)-upper-bound-path W' such that $\beta^+(W')^\top \boldsymbol{d} \leq \beta^+(W)^\top \boldsymbol{d}$. If W is a (u, v)-lower-bound-walk, then there exists a (u, v)-lower-bound-path W' such that $\beta^-(W')^\top \boldsymbol{d} \geq \beta^-(W)^\top \boldsymbol{d}$.*

The proof follows easily from the previous lemma and the well-known fact that each walk can be transformed into a path by successively removing cycles. Details of the proof can be found in [8]. We now have the following corollary.

Corollary 1. *Let $A \in \mathcal{S}^n[k]$ be a Robinson matrix and let $\boldsymbol{d} \in \mathbb{D}^k$. If $\boldsymbol{d}$ satisfies (4.1), then for every $u, v \in [n]$, for every (u, v)-upper-bound-walk W_1 and every (u, v)-lower-bound-walk W_2,*

$$\beta^-(W_1)^\top \boldsymbol{d} < \beta^+(W_2)^\top \boldsymbol{d}.$$

4.2 Finding a Uniform Embedding

In this section we prove the converse of Theorem 2. That is, given a matrix $A \in \mathcal{S}^n[k]$ we assume that there exists a $\boldsymbol{d} \in \mathbb{D}^k$ satisfying inequality system (4.1), and we show that there exists a uniform embedding Π with respect to this particular threshold vector $\boldsymbol{d}$. We present an iterative formula to calculate $\Pi : [n] \to \mathbb{R}$, given $\boldsymbol{d}$ and the sets $\mathcal{U}_{u,v}, \mathcal{L}_{u,v}$ for all $u, v \in [n]$, $u < v$. For brevity, let $\beta^+(\mathcal{U}_{u,v}) = \{\beta^+(W) : W \in \mathcal{U}_{u,v}\}$ and $\beta^-(\mathcal{L}_{u,v}) = \{\beta^-(W) : W \in \mathcal{L}_{u,v}\}$. Define Π as follows:

$$\begin{aligned} &\Pi(1) = 0 \\ &\Pi(v) = (ub_v + lb_v)/2, \quad \text{for } 2 \le v \le n, \end{aligned} \tag{4.2}$$

where ub_v, lb_v are defined iteratively using $\Pi(1), \ldots, \Pi(v-1)$ as:

$$\begin{aligned} ub_v &= \min_{i \in [v-1]} \left\{ \Pi(i) + \min\{\boldsymbol{b}^\top \boldsymbol{d} : \boldsymbol{b} \in \beta^+(\mathcal{U}_{i,v})\} \right\}, \\ lb_v &= \max_{i \in [v-1]} \left\{ \Pi(i) + \max\{\boldsymbol{a}^\top \boldsymbol{d} : \boldsymbol{a} \in \beta^-(\mathcal{L}_{i,v})\} \right\}. \end{aligned} \tag{4.3}$$

The following two lemmas show that Π defined as such is a uniform embedding of A with respect to $\boldsymbol{d}$.

Lemma 4. *The map Π as defined in (4.2) and (4.3) is strictly increasing.*

Proof: We prove by induction on v that Π is increasing on $[v]$. The base case, $v = 1$, is trivial. For the induction step, fix $v \ge 2$ and assume Π is increasing on $[v-1]$.

Note first that $lb_v \ge \Pi(v-1)$. Namely, $\langle v-1, v \rangle \in \mathcal{L}_{v-1,v}$, and thus either $\chi_t \in \beta^-(\mathcal{L}_{v-1,v})$ for some $t \in [k]$, or $\boldsymbol{0} \in \beta^-(\mathcal{L}_{v-1,v})$. Therefore, $\beta^-(\mathcal{L}_{v-1,v})$ contains at least one lower bound $\boldsymbol{a}$ so that $\boldsymbol{a}^\top \boldsymbol{d} \ge 0$.

We now show that $lb_v < ub_v$. This suffices to show that Π is strictly increasing: if $lb_v < ub_v$ then

$$\Pi(v) = (lb_v + ub_v)/2 > lb_v \ge \Pi(v-1).$$

Let u, w be the vertices attaining ub_v and lb_v respectively, and $\boldsymbol{b}_{\min} \in \beta^+(\mathcal{U}_{u,v})$ such that $\boldsymbol{b}_{\min}^\top \boldsymbol{d} = \min\{\boldsymbol{b}^\top \boldsymbol{d} : \boldsymbol{b} \in \beta^+(\mathcal{U}_{u,v})\}$, $\boldsymbol{a}_{\max} \in \beta^-(\mathcal{L}_{w,v})$ such that $\boldsymbol{a}_{\max}^\top \boldsymbol{d} = \max\{\boldsymbol{a}^\top \boldsymbol{d} : \boldsymbol{a} \in \beta^-(\mathcal{L}_{w,v})\}$, *i.e.*,

$$ub_v = \Pi(u) + \boldsymbol{b}_{\min}^\top \boldsymbol{d} = \min_{i \in [v-1]} \left\{ \Pi(i) + \min\{\boldsymbol{b}^\top \boldsymbol{d} : \boldsymbol{b} \in \beta^+(\mathcal{U}_{i,v})\} \right\},$$
$$lb_v = \Pi(w) + \boldsymbol{a}_{\max}^\top \boldsymbol{d} = \max_{j \in [v-1]} \left\{ \Pi(j) + \max\{\boldsymbol{a}^\top \boldsymbol{d} : \boldsymbol{a} \in \beta^-(\mathcal{L}_{j,v})\} \right\}.$$

Let W_B be a (u,v)-upper-bound-path such that $\boldsymbol{b}_{\min} = \beta^+(W_B)$ and W_A a (w,v)-lower-bound-path such that $\boldsymbol{a}_{\max} = \beta^-(W_A)$. Suppose first that $u = w$. Then, $\Pi(u) = \Pi(w)$, and $\boldsymbol{a}_{\max} \in \beta^-(\mathcal{L}_{u,v})$ and $\boldsymbol{b}_{\min} \in \beta^+(\mathcal{U}_{u,v})$. By the choice of $\boldsymbol{d}$, $\boldsymbol{a}_{\max}^\top \boldsymbol{d} < \boldsymbol{b}_{\min}^\top \boldsymbol{d}$, and thus $lb_v < ub_v$.

Suppose next that $u \neq w$. Then the concatenation $W_B + W_A^{\leftarrow}$ is a (u,w)-upper-bound-walk and $\beta^+(W_B + W_A^{\leftarrow}) = \boldsymbol{b}_{\min} - \boldsymbol{a}_{\max}$. By Lemma 1 this implies that $\Pi(w) - \Pi(u) < (\boldsymbol{b}_{\min} - \boldsymbol{a}_{\max})^\top \boldsymbol{d}$. This results in

$$lb_v = \Pi(w) + \boldsymbol{a}_{\max}^\top \boldsymbol{d} < \Pi(u) + \boldsymbol{b}_{\min}^\top \boldsymbol{d} = ub_v.$$

□

Lemma 5. *Given Robinson matrix $A \in \mathcal{S}^n[k]$, and let Π be defined as in (4.2). Then Π satisfies inequality system (2.2).*

Proof: Let $u, v \in [n]$ with $u < v$, and let $a_{u,v} = t$. We need to show that

$$d_{t+1} < \Pi(v) - \Pi(u) < d_t,$$

where $d_{k+1} = 0$ and $d_0 = \infty$.

We first prove the upper bound. If $t = 0$, then the inequality $\Pi(v) - \Pi(u) < d_0 = \infty$ is trivially satisfied. Suppose then that $t \neq 0$ (so uv is an edge). Then $\langle u, v \rangle$ is a (u,v)-upper-bound-path, so $\beta^+(u,v) \in \beta^+(\mathcal{U}_{u,v})$. By Definition 4, $\beta^+(u,v)^\top \boldsymbol{d} = d_t$. By Eq. (4.2),

$$\begin{aligned}
\Pi(v) < ub_v &= \min_{i \in [v-1]} \left\{ \Pi(i) + \min\{\boldsymbol{b}^\top \boldsymbol{d} : \boldsymbol{b} \in \beta^+(\mathcal{U}_{i,v})\} \right\}, \\
&\leq \Pi(u) + \min\{\boldsymbol{b}^\top \boldsymbol{d} : \boldsymbol{b} \in \beta^+(\mathcal{U}_{u,v})\} \\
&\leq \Pi(u) + \beta^+(u,v)^\top \boldsymbol{d} = \Pi(u) + d_t.
\end{aligned}$$

Next we prove the lower bound. If $t = k$, then the inequality $\Pi(v) - \Pi(u) > d_{k+1} = 0$ is satisfied since Π is strictly increasing. If $0 \leq t < k$, then $\langle u, v \rangle$ is a (u,v)-lower-bound-path. So $\beta^-(u,v) \in \beta^-(\mathcal{L}_{u,v})$, and

$$\begin{aligned}
\Pi(v) > lb_v &= \max_{i \in [v-1]} \left\{ \Pi(i) + \max\{\boldsymbol{b}^\top \boldsymbol{d} : \boldsymbol{b} \in \beta^-(\mathcal{L}_{i,v})\} \right\}, \\
&\geq \Pi(u) + \max\{\boldsymbol{b}^\top \boldsymbol{d} : \boldsymbol{b} \in \beta^-(\mathcal{L}_{u,v})\} \\
&\geq \Pi(u) + \beta^-(u,v)^\top \boldsymbol{d} = \Pi(u) + d_{t+1}.
\end{aligned}$$

□

Thus we have established that Π as defined in (4.2) is a uniform embedding.

5 Testing the Condition

According to Theorem 1, a uniform embedding exists if and only if the inequality system (2.2) has a solution. The existence of a uniform embedding can therefore be tested, and an embedding found, by using an linear program solver to determine feasibility of the system and, if feasible, find values for the variables d_i, $i \in [k]$ and $\Pi(u)$, $u \in [n]$. The condition for the existence of a uniform embedding as expressed in Theorem 2 involves solving another inequality system, namely (4.1). This system only contains the variables d_i, $i \in [k]$, but the number of inequalities equals the number of pairs of lower- and upper-bound-paths.

In this section, we will first give a bound on the number of inequalities, and compare the size of the two inequality systems. We then give an algorithm for generating all bounds that lead to inequalities for system (4.1), and discuss its complexity. Finally, we discuss the case where $k = 2$, and give a combinatorial algorithm to find a uniform embedding for a given matrix in $\mathcal{S}^n[k]$, or give proof that it does not exist.

5.1 A Partial Order on Bounds

Here we define a partial order on bounds and find out we only need the minimal/maximal elements of this partial order for inequality system (4.1). This will allow us to bound the size of this system.

To bound the number of equalities in (4.1), first note that any such inequality involves a (u, v)-upper-bound-path W_1 and a (u, v)-lower-bound-path W_2. Then $W_1 + W_2^{\leftarrow}$ is an upper-bound-cycle C, and the inequality can be rewritten as $\beta^+(C) > 0$. Thus we can rewrite (4.1). Let $\mathcal{C}$ be the set of upper-bound-cycles. Then $\boldsymbol{d}$ satisfies (4.1) if and only if,

$$\text{For all } C \in \mathcal{C}, \quad \beta^+(C)^\top \boldsymbol{d} > 0. \tag{5.1}$$

Any cycle in the complete graph with vertex set $[n]$ has length at most n. Any edge in the cycle can contribute at most one to the sum of the coefficients of the bound. Thus we have that, for any cycle C.

$$\beta^+(C) \in \mathbb{Z}_n^k := \{\boldsymbol{a} \in \mathbb{Z}^k : \sum_{i=1}^{k} |a_i| \leq n\}.$$

In particular, no coefficient of a path bound can have absolute value more than n. This implies that the number upper-bound-cycles, and thus the number of inequalities in (5.1) is at most $(2n)^k$. Thus, inequality system (5.1) has size $O(kn^k)$, while inequality system (2.2) has size $O(n^3)$.

However, using the partial order defined below we can give a tighter bound on the number of inequalities.

Definition 5. *Define the relation* $\preceq$ *on* $\mathbb{Z}^k$*, such that given any* $\boldsymbol{a} = (a_i), \boldsymbol{b} = (b_i) \in \mathbb{Z}^k$*,*

$$\boldsymbol{a} \preceq \boldsymbol{b} \quad \textit{if} \quad \sum_{i=1}^{t} a_i \leq \sum_{i=1}^{t} b_i \qquad \textit{for all } t \in [k].$$

Theorem 3. *Let* $\boldsymbol{a}, \boldsymbol{b} \in \mathbb{Z}^k$*, then* $\boldsymbol{a} \preceq \boldsymbol{b} \iff \boldsymbol{a}^\top \boldsymbol{d} \leq \boldsymbol{b}^\top \boldsymbol{d}$ *for all* $\boldsymbol{d} \in \mathbb{D}^k$.

The proof can be found in [8]. This theorem implies that $\preceq$ is indeed a partial order. More importantly, we have the following corollary.

Corollary 2. *Fix* $u, v \in [n]$*. If the inequalities of system (5.1) hold for all minimal elements of* $\{\beta^+(C) : C \in \mathcal{C}\}$ *under* $\preceq$*, then all inequalities of the system hold.*

This implies that the number of inequalities in (5.1) is bounded by the number of minimal elements (under $\preceq$), of $\mathbb{Z}_n^k$. The following lemma bounds this set for the special case where $k = 2$.

Lemma 6. *If* $k = 2$*, then inequality system (5.1), including only minimal bounds, has size at most* $2n$.

Proof: As argued above, the number of inequalities in this system is bounded by the number of minimal elements of $\mathbb{Z}_n^2$. We will bound this number by giving a decomposition of $\mathbb{Z}_n^2$ into $2n$ chains. The result then follows from Dilworth's theorem, and the fact that all minimal elements form an antichain.

Fix $t \in [n]$. Let S_t be the set of vectors $\boldsymbol{a} \in \mathbb{Z}^2$ so that $|a_1| + \cdots + |a_k| = t$. Consider the sets

$$S_t^1 = \{(-t+i, i)^\top : 0 \leq i \leq t\} \cup \{(i, t-i)^\top : 1 \leq i < t\}, \text{ and}$$
$$S_t^2 = \{(-t+i, -i)^\top : 1 \leq i \leq t\} \cup \{(i, -t+i)^\top : 1 \leq i \leq t\}.$$

Both S_t^1 and S_t^2 are chains under $\preceq$, and they form a partition of S_t. Since $\mathbb{Z}_n^2 = \cup_{t=1}^n S_t$, the result follows. □

5.2 Generating the Bounds

We employ a variation on the Floyd-Warshall algorithm [6] to enumerate all upper-bound-paths. See Algorithm 1 for the pseudocode. This also generates all lower-bound-paths, by reversal.

The complexity of this algorithm is dominated by the step where bounds are merged, and thus determined by the size of the set of bounds. If the minimality test in line 9 is implemented by looping through all elements of S, then the complexity of the bound-generation algorithm is $O(n^3 M^3)$, where M is the number of minimal elements in $\mathbb{Z}_n^k$.

The minimal upper-bound-paths give insight into the structure of the matrix that constrains the uniform embedding. If inequality system (5.1) does not have a solution, then any LP-solver will return a set of k inequalities which, taken together, show the impossibility of fulfilling all constraints. Each of these inequalities is derived from a cycle, and this collection of cycles can be interpreted as the bottleneck that prevents the existence of a uniform embedding.

Algorithm 1: Bound-Generation

```
input  : A Robinson matrix A ∈ S^k
output : Lookup table UBW, LBW defined on i, j ∈ [n]: where
         UBW(i, j) = all minimal elements of U_{i,j},
1  for i ∈ [n] do
2      for j = i, ..., n do
3          if a_{i,j} ≠ 0 then UBW(i, j) ← {⟨i, j⟩};
4          UBW(j, i) ← {⟨j, i⟩};
5  for s = 1, ..., n do
6      for i = 1, ..., n do
7          for j = i, ..., n do
8              foreach W_1 ∈ UBW(i, s) and W_2 ∈ UBW(s, j) do
9                  if W_1 + W_2 is minimal in UBW(i, j) ∪ {W_1 + W_2} then
10                     UBW(i, j) ← UBW(i, j) ∪ {W_1 + W_2};
11                     UBW(j, i) ← UBW(j, i) ∪ {(W_1 + W_2)^←};
12 return UBW;
```

5.3 A Combinatorial Algorithm for the Case $k = 2$

For $k = 2$, we can convert the bound generation, testing of condition (2.2), and construction of the uniform embedding into a combinatorial algorithm.

Consider inequality system (5.1) when $k = 2$. Each inequality is of the form $a_1 d_1 + a_2 d_2 > 0$, where $(a_1, a_2)^\top = \beta^+(C)$ for some upper-bound-cycle C. Depending on the sign of a_2, $-a_1/a_2$ will give either a lower bound (if $a_2 > 0$) or an upper bound (if $a_2 < 0$) on d_2/d_1. Thus we can find, in time linear in the number of minimal bounds, the largest lower bound and smallest upper bound on d_2/d_1. The inequality system has a solution if and only if the largest lower bound is smaller than the smallest upper bound.

If the bounds are incompatible and the system has no solution, then the two cycles giving the largest lower bound and smallest upper bound identify those entries of the matrix that cause the non-existence of a uniform embedding.

Combining the methods we have developed, we now give the steps of the algorithm solving the uniform embedding problem

Uniform Embedding Algorithm. Given a matrix $A \in \mathcal{S}^n[k]$, perform the following steps:

1. Generate all minimal upper bounds using the Bound-Generation algorithm (Algorithm 1). Use $\texttt{UBW}(v, v)$ to extract all minimal upper-bound-cycles.
2. For each minimal upper-bound cycle, find its associated bound $\beta^+(C)$, and convert the inequality $\beta^+(C)$ into a lower or upper bound on d_2/d_1. Only keep the cycles C_1 and C_2 generating the largest upper bound and the smallest upper bound encountered in each step.

3. If the largest lower bound is greater than or equal to the smallest upperbound on d_2/d_1, then print NO SOLUTION. **Exit** and return C_1 and C_2.
4. If the largest lower bound is smaller than the smallest upper bound on d_2/d_1, then choose d_1, d_2 so that d_2/d_1 lies between these bounds.
5. Compute a uniform embedding Π with respect to (d_1, d_2) using the formula given in (4.2) and (4.3). **Exit** and return Π.

The complexity of the algorithm is determined by the generation of bounds in the first step. We can use the partition of $\mathbb{Z}_n^2$ into chains S_t^1, S_t^2, $1 \leq t \leq n$, as given in Lemma 6. Given a bound $(a_1, a_2)^\top$, we can identify which set S_t^i the bound belongs to: $t = |a_1| + |a_2|$ and $i = 1$ if $a_2 > 0$ and $i = 2$ otherwise. Therefore, in line 9 of the Bound-Generation algorithm we only need to compare $W_1 + W_2$ with the unique minimal element of the set S_t^i it belongs to. This can be done in $O(1)$ time.

Since there are at most $2n$ minimal elements in $\texttt{UBW}(i, j)$, the loop starting in line 8 of the Bound-Generation algorithm takes $O(n^2)$ steps. Therefore, the Bound-Generation algorithm can be implemented to take $O(n^5)$ steps. Moreover, the algorithm can be easily modified to compute the upper-bound-paths as well as their associated bounds. The generation of the cycle inequalities in Step 2 is immediate from UBW, since every upper-bound-cycle will be included in $\texttt{UBW}(i, i)$, $i \in [n]$. Note that there will be duplications, since each cycle C will be included in $\texttt{UBW}(v, v)$ for any vertex $v \in C$. Therefore, generating the inequalities and the associated bounds on d_2/d_1 takes $O(n^2)$ steps.

If a threshold vector $(d_1, d_2)^\top$ can be found, then the uniform embedding can be found in $O(n^3)$ steps: there are n iterations, and each iteration involves computing lb_v and ub_v, which involves looping over all vertices $u < v$ and all bounds in $\texttt{UBW}(u, v)$.

The resulting complexity is somewhat higher than that of solving inequality system (2.2) with a state-of-the-art Linear Programming solver. Namely, using the method of Vaidya [15] , an LP with n variables and m inequalities can be solved in $O((n + m)^{1.5} n)$ time. In (2.2) there are $O(n^2)$ inequalities, so the LP solver has time complexity $O(n^4)$.

Finally, note that the input size of the problem is $N = n(n - 1)/2$, namely the number of upper diagonal entries of the matrix. Therefore, the complexity of the algorithm to find a uniform embedding for the case $k = 2$ has complexity $O(N^{2.5})$.

6 Conclusions

We gave a sufficient and necessary condition for the existence of a uniform embedding of a Robinson matrix, in the form of a system of inequalities constraining the threshold values $d_1, \ldots, d_k$. For Robinson matrices taking values in $\{0, 1, 2\}$, we gave a $O(N^{2.5})$ algorithm which returns a uniform embedding, or returns two cycles that identify the matrix entries that cause a contradiction in the inequalities defining a uniform embedding.

For Robinson matrices having more than three values, the condition for the existence of a uniform embedding involves solving an inequality system. An interesting question is whether the condition can be tested with a combinatorial algorithm, as in the case $k = 2$. In particular, we saw that the problem can also be formulated as that of finding simultaneous embeddings for a family of nested proper interval graphs. The $k = 2$ case shows that this is possible for a family of two. For $k = 3$, we can find conditions on d_1, d_2, d_3 so that any pair of graphs in the family has a simultaneous embedding. If these conditions are contradictory, then no uniform embedding of the family exists. But if they are not, can we then solve the uniform embedding question combinatorially?

References

1. Atkins, J.E., Boman, E.G., Hendrickson, B.: A spectral algorithm for seriation and the consecutive ones problem. SIAM J. Comput. **28**(1), 297–310 (1998)
2. Bogart, K.P., West, D.B.: A short proof that 'proper = unit'. Discret. Math. **201**(1), 21–23 (1999)
3. Chuangpishit, H., Ghandehari, M., Janssen, J.: Uniform linear embeddings of graphons. Eur. J. Comb. **61**, 47–68 (2017)
4. Corneil, D.G.: A simple 3-sweep LBF algorithm for the recognition of unit interval graphs. Discret. Appl. Math. **138**(3), 371–379 (2004)
5. Corneil, D.G., Kim, H., Natarajan, S., Olariu, S., Sprague, A.P.: Simple linear time recognition of unit interval graphs. Inf. Process. Lett. **55**(2), 99–104 (1995)
6. Floyd, R.W.: Algorithm 97: shortest path. Commun. ACM **5**(6), 345 (1962)
7. Gardi, F.: The Roberts characterization of proper and unit interval graphs. Discret. Math. **307**(22), 2906–2908 (2007)
8. Janssen, J., Zhang, Z.: Uniform embeddings for Robinson similarity matrices (2021). arXiv:2105.09197
9. Kendall, D.: Incidence matrices, interval graphs and seriation in archeology. Pac. J. Math. **28**(3), 565–570 (1969)
10. Laurent, M., Seminaroti, M.: A Lex-BF-based recognition algorithm for Robinsonian matrices. Discret. Appl. Math. **222**, 151–165 (2017)
11. Laurent, M., Seminaroti, M.: Similarity-first search: a new algorithm with application to Robinsonian matrix recognition. SIAM J. Discret. Math. **31**(3), 1765–1800 (2017)
12. Liiv, I.: Seriation and matrix reordering methods: an historical overview. Stat. Anal. Data Min ASA Data Sci. J. **3**(2), 70–91 (2010)
13. Roberts, F.S.: Indifference graphs, pp. 139–146. Academic Press (1969)
14. Robinson, W.S.: A method for chronologically ordering archaeological deposits. Am. Antiq. **16**(4), 293–301 (1951)
15. Vaidya, P.M.: Speeding-up linear programming using fast matrix multiplication. In: 30th Annual Symposium on Foundations of Computer Science, pp. 332–337 (1989)

Particle-Based Assembly Using Precise Global Control

Jakob Keller[1], Christian Rieck[1(✉)], Christian Scheffer[2], and Arne Schmidt[1]

[1] Department of Computer Science, TU Braunschweig, Braunschweig, Germany
{jkeller,rieck,aschmidt}@ibr.cs.tu-bs.de
[2] Department of Computer Science, University of Münster, Münster, Germany
christian.scheffer@uni-muenster.de

Abstract. In micro- and nano-scale systems, particles can be moved by using an external force such as gravity or a magnetic field. In the presence of adhesive particles that can attach to each other, the challenge is to decide whether a shape is constructible. Previous work provides a class of shapes for which constructibility can be decided efficiently, when particles move maximally into the same direction on actuation.

In this paper, we consider a stronger model. On actuation, each particle moves one unit step into the given direction. We prove that deciding constructibility is NP-hard for three-dimensional shapes, and that a maximum constructible shape can be approximated. The same approximation algorithm applies for 2D. We further present linear-time algorithms to decide whether a tree-shape in 2D or 3D is constructible. If scaling is allowed, we show that the c-scaled copy of every non-degenerate polyomino is constructible, for every $c \geq 2$.

Keywords: Programmable matter · Tile assembly · Tilt · Approximation · Hardness

1 Introduction

In recent years, the easier access to micro- and nano-scale systems has given rise to challenges that deal with programmable matter. In some of these applications, particles can be controlled by a global external force such as gravity or a magnetic field. On actuation, every particle moves into the same direction at unit speed. Assembly of particles into desired structures using maximal movements, i.e., every particle moves into a given direction until it hits an obstacle or another particle, has been investigated in [2,3,6,18]. However, it is also reasonable to expose the particles to these forces just for a limited amount of time, such that more precise movements become possible. Reconfiguration of a set of particles [2],

Due to space constraints, several technical details and proofs are omitted from this extended abstract. A full version of the paper can be found at [12].

A. Lubiw et al. (Eds.): WADS 2021, LNCS 12808, pp. 513–527, 2021.
https://doi.org/10.1007/978-3-030-83508-8_37

gathering all particles [5,13], or assembling patterned rectangles [8] are well studied problems.

In this paper, we consider the construction of tile-based structures (such as polyominoes in 2D or polycubes in 3D) through adhesive particles, that all move one step into the same direction on actuation. Whenever two particles come close, they stick together. In this model, we consider the problem of deciding whether a given shape is constructible. For definitions see Sect. 2.

SINGLE STEP TILT ASSEMBLY PROBLEM (STAP)
Given a shape P (a polyomino in 2D or a polycube in 3D), does there exist a sequence of tile moves to construct P?

We denote the problem in 2D and 3D by 2D-STAP and 3D-STAP, respectively. The optimization variant of STAP, called MAXSTAP, asks for a constructible subshape $P_{\max} \subset P$ of maximum size.

1.1 Our Contributions

We show the following results.

- 3D-STAP is NP-hard, see Theorem 2.
- In dimension $d = 2, 3$, there is an $\Omega(|P|^{-1/d})$-approximation algorithm for MAXSTAP, see Theorem 3 and Corollary 1.
- For tree-shapes in 2D and 3D, there is a linear-time algorithm for STAP, see Theorem 4 and Corollary 2.
- For every non-degenerate polyomino P, the 2-scaled copy of P is constructible, see Theorem 5, and for every non-degenerate polycube P, the 7-scaled copy of P is constructible, see Theorem 6.

1.2 Related Work

Self-assembly. Instead of relying on universal, external control of agents as discussed above, it is possible to use DNA as material. DNA-strands can either be used to fold into the desired shape [16], or to create building blocks based on Wang tiles [19] that can then again self-assemble into shapes in a non-deterministic way [17,20]. For more details on algorithmic self-assembly, we refer to the surveys by Doty [11], Patitz [15], and Woods [21].

Tilt Problems. The process of self-assembly works by diffusion and is non-deterministic. If a deterministic approach is desired, we can use global control to move particles into different directions. Using global control opens a wide field of problems, that we summarize briefly.

Mahadev et al. [13] show how to gather n particles within an obstacle environment in $O(n^3)$ actuation steps. Becker et al. [5] improve the runtime to depend only on the geometric complexity of the workspace, rather than on the number of particles. Closely related to this problem are occupancy, relocation and reconfiguration problems. For a given set of particles within an obstacle environment, these problems ask for a sequence of tilts such that (i) any particle reaches a designated position, (ii) a specific particle reaches a designated position, and (iii) every particle reaches its respective target position. If particles are allowed to move a unit step on actuation, Caballero et al. [9] show that (i) permits a linear-time algorithm, while the decision variants of (ii) and (iii) are NP-hard. More recently, Caballero et al. [10] show PSPACE-completeness for (ii). Becker et al. [4] show that (i) is NP-hard when particles have to move maximally. Balanza-Martinez et al. [2] give a tighter result for the maximal movement model, by proving PSPACE-completeness for all three problems.

All these papers have in common that particles do not stick to each other. Manzoor et al. [14] provide algorithms for assembling shapes, called *drop shapes*, from *sticky* particles under global control. In this assembly process, only one particle at a time is added by maximal movements to a seed assembly. They also show that the assembly process can be pipelined, i.e., the same shape is produced multiple times. Becker et al. [6] prove that every drop shape permits an amortized construction time of $O(1)$, provided that sufficiently many copies are constructed. However, every shape needs a custom-designed obstacle workspace. Balanza-Martinez et al. [3] designed a single obstacle workspace in which every drop shape can be constructed, if it fits in a $w \times h$ rectangle. By using not only single particles but whole subassemblies, the class of constructible shapes increases (see [2,18]). A crucial step is to decompose a given shape into two connected parts, that can then be pulled apart into a single direction, without causing collisions. Agarwal et al. [1] recently proved that this problem is NP-hard, even if a direction is given.

2 Preliminaries

Polyomino. Let $P \subset \mathbb{Z}^2$ be a finite set of N grid points in the plane. The grid points of $\mathbb{Z}^2$ are called *positions*. The embedded graph G_P is the grid graph induced by P, in which two vertices are adjacent if they are at unit distance. If G_P is connected, we obtain a *polyomino* by placing a unit square, called *tile*, centered on every vertex of G_P. Two tiles are *adjacent*, if their respective vertices in G_P are adjacent, i.e., they share an edge. A position $p \in \mathbb{Z}^2$ is *occupied*, if there is a tile that is placed on p, and *free* otherwise. The neighborhood $N[\cdot]$ of a tile or position is the respective set of adjacent positions. A polyomino P is *simple*, if the grid graph $\mathbb{Z}^2 \setminus G_P$ is connected. If there are at least two tiles $t_1, t_2 \in P$ that share a common point but for which there is no tile in $N[t_1] \cap N[t_2]$, P is called *degenerate*. If G_P is a tree, the respective polyomino is *tree-shaped.*

Workspace. The *workspace* is a rectangular region of $2N \times 2N$ positions, anchored at position $(0,0)$. The workspace contains a *seed tile* at position (N,N).

Construction Step. A tile can *move* within the workspace from one position p to an adjacent position q, as long as the neighborhood $N[p]$ is free. A *construction step* is a sequence σ of moves such that σ moves a tile to a position adjacent to an occupied position. Without loss of generality, we assume that every construction step starts at position $(0,0)$. Analogously, we define a *deconstruction step* as a sequence $\widetilde{\sigma}$ that moves a tile from its position to the position $(0,0)$. If there is a deconstruction step for a tile t, we call t *removable*.

Constructibility. Beginning with a seed tile, a polyomino P with N tiles is constructible, if and only if there is a *construction sequence* $\Sigma = (\sigma_1, \sigma_2, \ldots, \sigma_N)$ of N consecutive construction steps such that the resulting polyomino P', induced by successively adding tiles with Σ, is equal to P. Reversing Σ yields a *deconstruction sequence* $\widetilde{\Sigma}$, i.e., a sequence of tiles getting removed from P.

Note that the definitions are given for a 2D setting. It is straightforward to extend these to the 3D setting, by letting $P \subset \mathbb{Z}^3$, introducing two additional directions in that a tile can move, considering unit cubes instead of unit squares as tiles, and anchoring the workspace, that contains a seed tile at position (N, N, N), at position $(0, 0, 0)$.

Firstly, we restate a result by Becker et al. [6] for the full tilt assembly model in two dimensions, which can easily be adapted for the single step model considered in this paper, as well as it applies (in both models) to the 3D setting.

Theorem 1 (Theorem 2, [6]). *A polyomino P can be constructed if and only if it can be deconstructed using a sequence of tile removal steps that preserve connectivity. A construction sequence is a reversed deconstruction sequence.*

3 NP-Hardness of 3D-STAP

Becker et al. [6] showed that it is NP-hard to decide whether or not a polycube is constructible in the full tilt model. However, our model is more powerful and we cannot adapt their proof. In this section, we show that 3D-STAP is NP-hard. The proof is based on a reduction from the NP-hard problem PLANAR MONOTONE 3SAT [7]. This problem asks to decide whether a Boolean 3-CNF formula φ is satisfiable, for which in each clause the literals are either all positive or all negative, and for which the clause-variable incidence-graph is planar. Because of Theorem 1, we will argue that a polycube is deconstructible, if and only if a Boolean 3-CNF formula is satisfiable. Details of the omitted proofs can be found in the full version of the paper [12].

Outline of the NP-Hardness Reduction. For every instance φ of PLANAR MONOTONE 3SAT, we construct a polycube P_φ as an instance of 3D-STAP. We consider a rectilinear planar embedding of the variable-clause incidence-graph G_φ of φ where the variable vertices are placed on a line, and clauses containing positive and negative literals are placed on either side, respectively. For a symbolic overview of P_φ, consider Fig. 1. In P_φ, each variable and each clause of φ is represented by a variable gadget and a clause gadget, respectively. To realize the fact that a variable can be contained in several clauses, we introduce a conjunction gadget. An edge in G_φ is realized by a connector gadget. To guarantee connectivity during the deconstruction of the polycube P_φ, we need to

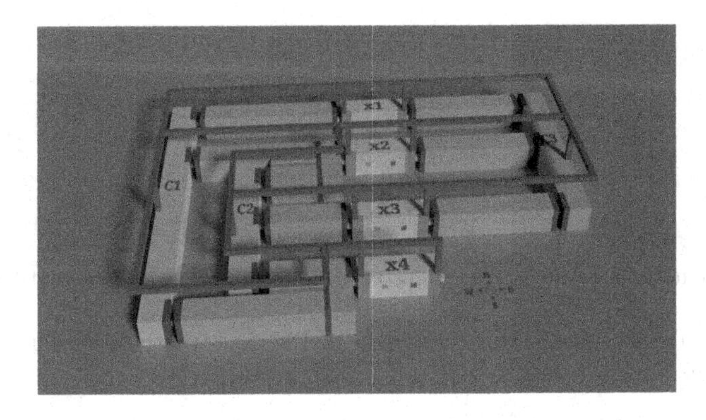

Fig. 1. Symbolic overview of the NP-hardness reduction. The instance is due to the PLANAR MONOTONE 3SAT formula $\varphi = (x_1 \vee x_2 \vee x_4) \wedge (x_2 \vee x_3 \vee x_4) \wedge (\overline{x_1} \vee \overline{x_2} \vee \overline{x_4})$. The variable gadgets are shown in white, while the positive and negative clause gadgets are shown in green and red, respectively. The gray cuboids represent connector gadgets. The 'colorful lines' are representing conjunction gadgets. All clauses and variables are connected by a blue frame above the construction. (Color figure online)

make sure that parts of the variable gadgets that are not participating in the satisfying assignment, are not disconnected in several parts. Therefore, we add a frame above the actual polycube, connecting all clauses with certain parts of the variable gadgets.

We can show that there is a deconstruction sequence for P_φ if and only if φ is satisfiable. By using a checkered tile arrangement within all gadgets, we can enforce a specific deconstruction sequence. On the one hand, we ensure that, due to the connectivity constraint, either the part of the variable that is connected to their positive or to their negative literal containing clauses can be deconstructed; this implies that, together with the conjunction gadgets, all clauses containing the respective literal can be deconstructed. On the other hand, the order of the deconstruction steps can be used to determine a valid variable assignment for φ.

Construction of the Gadgets. In the following, we describe polycubes serving as gadgets in the NP-hardness reduction. All gadgets are based on the following polycube that cannot be deconstructed if we restrict the deconstruction direction.

Indestructible Wall. A *wall* is the polycube depicted in Fig. 2(a). It consists of two layers, an odd dimensional *solid layer*, and a checkered *tooth layer*. This tooth layer consists of non-adjacent cubes (dark gray cubes) at even positions. This construction can easily be modified to construct a *k-wall*, see Fig. 2 for examples. Note that k can be at most 6, and that there exist several k-walls for $k \in \{3, 4\}$.

Recall that a deconstruction step is a sequence of moves that moves a tile of the polycube to position $(0, 0, 0)$. We want to show that a k-wall is *not deconstructible from a specific direction*, in particular from the solid layer. To do so, we

(a) Indestructible 1-wall.

(b) Indestructible 2-wall.

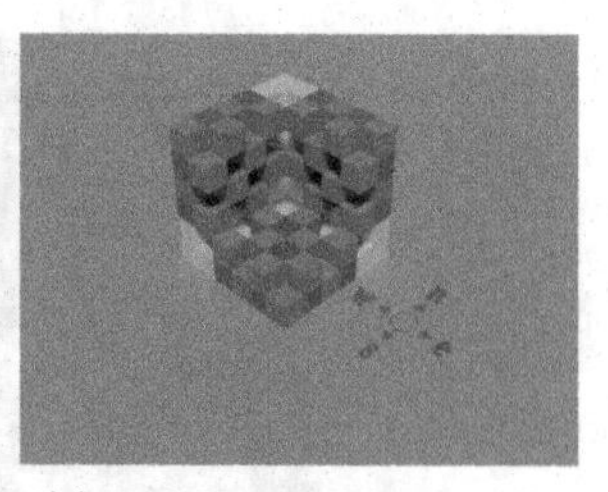

(c) Indestructible 3-wall.

Fig. 2. Indestructible walls. Red cubes indicate the respective positions of teeth. (Color figure online)

assume that the workspace is designed such that $(0, 0, 0)$ is below the wall and that there is no path that starts from above the wall, bypasses it and eventually reaches $(0, 0, 0)$.

Lemma 1. *A k-wall, $k \in \{1, \ldots, 6\}$, is not deconstructible from its solid layers.*

Proof. Suppose for the sake of a contradiction that a k-wall is deconstructible from its solid layers, and let $\widetilde{\Sigma} = (\sigma_1, \ldots, \sigma_N)$ be a deconstruction sequence. Because we only have access to the solid layers, σ_1 has to remove a cube from a solid layer. To maintain connectivity, only these cubes can be removed for that there is no cube at the respective position in the respective tooth layer. But then there cannot be a $\widetilde{\sigma}_i \in \widetilde{\Sigma}$ such that $\widetilde{\sigma}_{\underset{\sim}{i}}$ removes a cube from a tooth layer. This is a contradiction to the existence of $\widetilde{\underset{\sim}{\Sigma}}$. Thus, a wall is not deconstructible from the solid layer. □

We can show that the specific design of the tooth layers of a k-wall is necessary to guarantee the indestructability, i.e., if at least one tooth is missing, the resulting polycube is deconstructible from its solid layers.

Lemma 2. *There is at least one position p at a tooth layer of a k-wall, $k \in \{1, \ldots, 6\}$ such that the k-wall is deconstructible from its solid layers if p is free.*

Proof (Sketch). Remove the cube at the respective position of the missing tooth and all its adjacent cubes in the respective solid layer. This results in a hole, large enough that additional cubes from the tooth layers can be removed. □

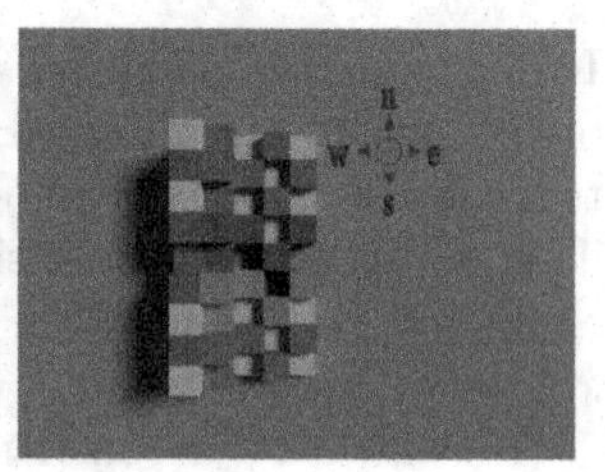

Fig. 3. Different views on a disconnected 2-wall. (Color figure online)

A simple observation is that these k-walls can arbitrarily be enlarged without losing the property of being indestructible. Of particular interest for the reduction are enlarged 6-walls, called *cuboids*, that will serve as *clause* and *connector gadgets*.

Another crucial observation is that because a k-wall is not deconstructible from its solid layers, we can leave out several cubes of the solid layers so that the remaining shape is disconnected into two parts, see Fig. 3. This insight will lead to a configuration that allows for a decision, i.e., that will serve as the variable gadget.

If two cuboids have to be connected, we place them at distance one to each other and add a single cube to connect them. Furthermore, we remove cubes at matching sides in a 3×3 area such that we can move cubes from the inside of one cuboid to the other through these holes, see Figs. 4(c) and 4(d) for illustration.

Variable Gadget. The variable gadget consists of two indestructible cuboids (Q_1 and Q_2) that share a solid layer, see Fig. 4(d) for an exploded illustration. As shown in Fig. 4, we remove cubes (similar to Fig. 3) to separate an L-shaped part of each cuboid (light blue cubes). These shapes are then reconnected by

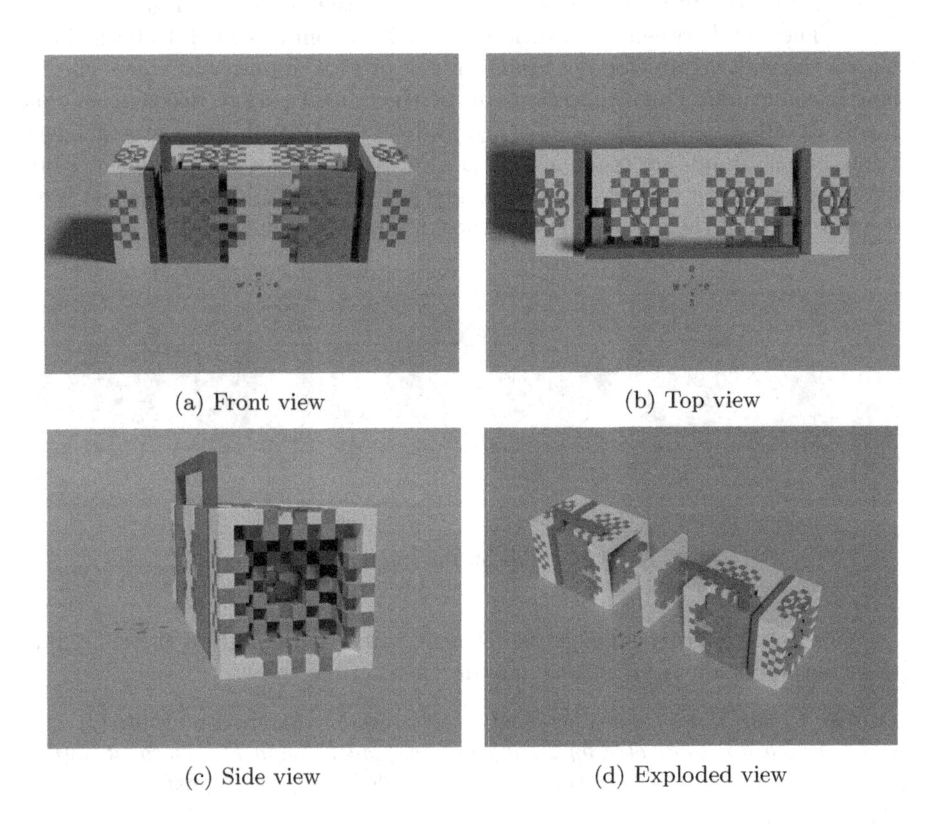

(a) Front view (b) Top view

(c) Side view (d) Exploded view

Fig. 4. Different views on the variable gadget. (Color figure online)

two bridges (green and orange cubes), see Fig. 4(a). Additionally, the L-shaped parts are connected by a thin frame above the cuboids (dark blue cubes).

Observation 1. *Solely removing the green and orange cubes of a variable gadget results in a disconnected shape.*

As a consequence of Observation 1, the forced choice of removing either the green or the orange cubes, can be used to determine an assignment for the respective Boolean variable. It remains to show how a variable gadget can be deconstructed, if additional cuboids are attached to each side.

Lemma 3. *Let P be a polycube that is put together by a variable gadget and one cuboid (Q_3 and Q_4) at each end, connected to the respective L-shaped parts. Then P is only deconstructible if at least Q_3 or Q_4 is deconstructible.*

As the last ingredient we need a gadget that realizes a conjunction. This gadget will be used to guarantee that a variable gadget can be completely deconstructed if and only if all clauses in which the respective variable participates, are satisfied.

Conjunction Gadget. As illustrated in Fig. 5, the conjunction gadget is T-shaped. The wall between the positions Q_1 and Q_2 contains teeth to both sides, whereas the wall at position Q_3 has teeth except for the positions where the T-shape is connected. This connection will be the crucial part to deconstruct this gadget. At all three positions (Q_1, Q_2, and Q_3) we attach connector gadgets leading either to another conjunction, to a variable, or to a clause gadget. Note that these connector gadgets have the same size as the conjunction gadget, i.e., the solid layers of the connectors and the conjunction gadget match.

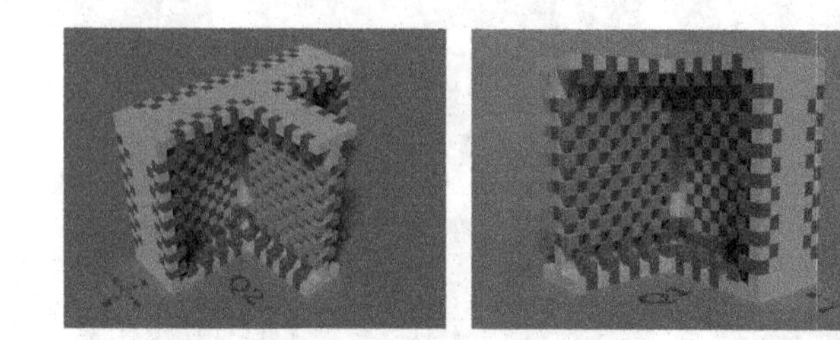
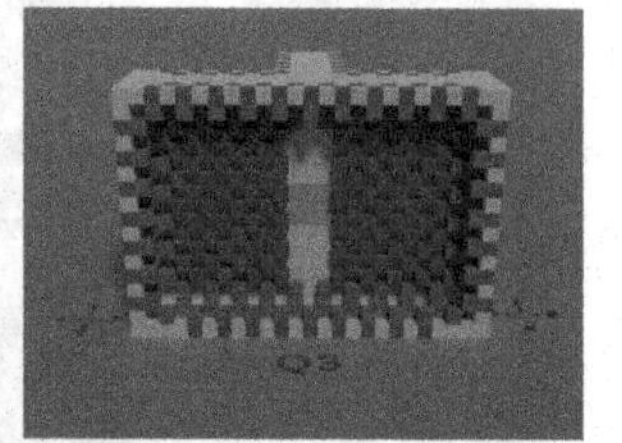

Fig. 5. Different views on the conjunction gadget.

We can show that this gadget is deconstructible if and only if the cuboid at Q_3, or both cuboids at Q_1 and Q_2 are deconstructible.

Lemma 4. *Let P be a polycube that is put together by three cuboids Q_1, Q_2, and Q_3 which are connected by a conjunction gadget. Then P is deconstructible if and only if Q_1 and Q_2 are both deconstructible, or Q_3 is deconstructible.*

Putting all these together yields the following.

Theorem 2. 3D-STAP *is NP-hard.*

4 Optimization Variant and Approximation

For polyominoes and polycubes that cannot be constructed, it is natural to consider the problem of maximizing a constructible subshape. We show that for each shape P, a portion of $\Omega(|P|^{(d-1)/d})$ can always be constructed, implying a $\Omega(|P|^{-1/d})$-approximation of MAXSTAP, where d denotes the dimension.

Definition 1. *The* maximum constructibility *of a polyomino P is the ratio $|P_{\max}|/|P|$ where $P_{\max} \subseteq P$ is a constructible polyomino of maximum size.*

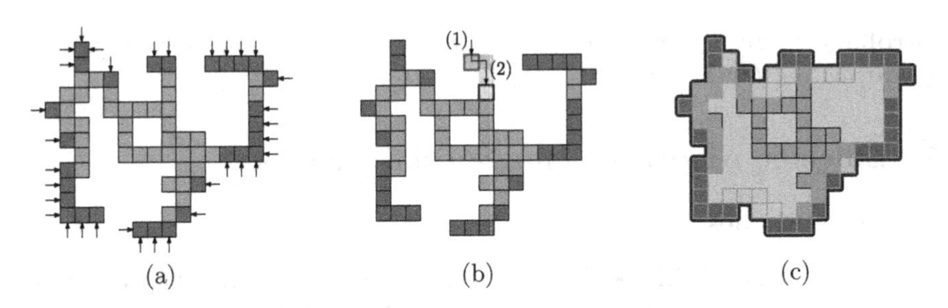

Fig. 6. (a) Boundary tiles (dark cyan) and non-boundary tiles (gray). (b) Every step of the algorithm adds a tile in two steps: (1) Move a new tile t to a boundary position p of P that is free. (2) Move t on P from p to a position that is adjacent to a tile of the current polyomino P'. (c) The curve B (blue), and the set T (orange). (Color figure online)

Theorem 3. *In dimension $d = 2, 3$, each d-dimensional polyomino P has a maximum constructibility of at least $\Omega(|P|^{-1/d})$.*

Proof. We prove the theorem by showing that greedily filling up accessible free positions leads to a polyomino $P' \subseteq P$ with $|P'|/|P| \in \Omega(|P|^{-1/d})$. A position p is *accessible* with respect to a polyomino P if we can move a tile t to this position such that t is never adjacent to a tile of P unless t lies on p. A tile $t \in P$ is a *boundary tile* of P if the respective position is accessible with respect to $P \setminus \{t\}$, see Fig. 6(a). If there is a boundary tile t of P that is not part of P', we add a new tile t' to P' in two steps, see Fig. 6(b): (1) We move t' to the position of t. (2) We move t' on P to a position adjacent to a tile of P'. This implies that the greedy algorithm ends up with a polyomino holding all boundary tiles of P.

Next, we show a lower bound on the boundary tiles of P of $\Omega(N^{d-1})$. We only consider $d = 2$, a similar surface-volume-argument holds for $d = 3$. Let B be the union of all edges lying between an accessible and a non-accessible position with respect to P, see blue curve in Fig. 6(c). B is a non-self-intersecting curve, by the definition of accessible positions. Thus, B partitions the plane into a bounded area A containing P and an unbounded area. Let T be the union of all positions from A sharing at least a corner with B, see the orange positions in Fig. 6(c).

Then $|T| \geq \sqrt{|A|}$. Note that not each position of T is occupied by P, see the light gray positions in Fig. 6(c). Let T' be the positions along B that share an edge with B. It is easy to see that $2|T'| \geq |T|$. Each position p from T' that is not a boundary tile from P is adjacent to a boundary tile p' from P. We call p' a *blocking tile* of p. Each boundary tile is a blocking tile for a constant number of positions $p \in T'$ that are not a boundary tile. Hence, there are $\Omega(|T'|)$ many positions from T' that are not a boundary tile implying that $\Omega(|T'|) = \Omega(|T|)$ tiles from T are boundary tiles. Because $P' \subset A$, we obtain $|P'| \leq |A|$ implying $|T| \in \Omega(\sqrt{|A|}) = \Omega(\sqrt{|P'|})$. Hence, P' has at least $\Omega(N^{1/2})$ boundary tiles. □

Theorem 3 implies the following.

Corollary 1. *In dimension $d = 2, 3$, the greedy algorithm is an $\Omega(|P|^{-1/d})$-approximation for* MaxSTAP.

5 Efficient Algorithms for Special Classes

5.1 Tree Shapes

We can show that STAP can be decided in linear time for tree-shaped polyominoes by a greedy algorithm. Because the removal of a tile with more than one neighbor results in splitting the polyomino in several parts, we are restricted to remove tiles with exactly one neighbor, i.e., leaves. If there are any tiles left, but no further tile can be removed, we conclude that the polyomino cannot be constructed.

We begin by stating two facts about removable tiles. Firstly, by removing a tile, other removable tiles do not lose their property to be removable. And secondly, if a tree-shaped polyomino is constructible, then after removing any removable tile, the resulting polyomino is also constructible.

Lemma 5. *Let P be a tree-shaped polyomino and $\mathcal{R}_P$ the set of removable tiles of P. For all $P' \subseteq P$ it holds that if $t \in \mathcal{R}_P \cap P'$, then $t \in \mathcal{R}_{P'}$.*

Lemma 6. *Let P be a constructible tree-shaped polyomino and t a removable tile. Then, $P \setminus \{t\}$ is also constructible.*

By using Lemma 6 iteratively, we obtain a simple strategy that decides whether a tree-shaped polyomino is constructible or not. By applying suitable subroutines and data structures, this yields a linear-time algorithm.

Theorem 4. *Let P be a tree-shaped polyomino consisting of N unit squares. We can decide in $O(N)$ time whether or not P is constructible.*

It is easy to see that the same holds true for tree-shaped polycubes.

Corollary 2. *Let P be a tree-shaped polycube consisting of N unit cubes. We can decide in $O(N)$ time whether or not P is constructible.*

Due to space constraints, the proofs can be found in the full version [12].

5.2 Scaled Shapes

Deciding constructibility of arbitrary polyominoes is more intricate than for tree-shaped polyominoes. On the one hand, it is not sufficient to restrict the search for removable tiles to *corner's* (tiles with exactly one horizontal and one vertical neighbor), because for successfully deconstructing a polyomino, it may be necessary to remove non-corner tiles first, see Fig. 7(a). On the other hand, removing non-corner tiles can result in an indestructible subshape, again see Fig. 7(a). Furthermore, even in simple polyominoes, the removal of a corner tile can result in an indestructible polyomino, see Fig. 7(b). Note that this is not the case if maximal movement is considered.

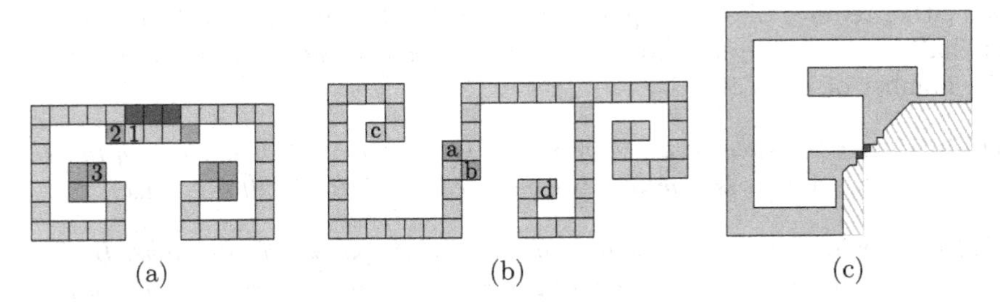

Fig. 7. (a) No corner tile (dark gray) can be removed, because they either do not have a deconstruction step or are essential for connectivity. Successively removing the tiles 1, 2, and 3 by suitable deconstruction steps, the obtained shape can easily be deconstructed. Removing the red tiles first results in an indestructible shape. (b) By removing a first, we can first remove the spiral starting with c, and the remaining shape afterwards. By removing b first, the spiral starting with d can be removed, but the remaining shape is indestructible. (c) For the deconstruction it is necessary to remove at least one of the red tiles. Independent from the scaling factor, both tiles block each other and we are only able to deconstruct a staircase part (hatched part) to the right and below the tiles, respectively. (Color figure online)

However, if scaling is allowed, we can show that the 2-scaled copy of a non-degenerate polyomino is constructible. Note that no scaling factor will suffice to guarantee constructibility of degenerate polyominoes, see Fig. 7(c). Recall that in a non-degenerate polyomino P, for every pair $t_1, t_2 \in P$ of tiles that share exactly one point, $N[t_1] \cap N[t_2] \neq \emptyset$, i.e., t_1, t_2 have a common neighbor.

Definition 2. *Let P be a polyomino and $c \in \mathbb{N}$. By P^c we denote the c-scaled copy of P, i.e., each tile in P is replaced by a $c \times c$ square of tiles.*

Definition 3. *Let P be a polyomino. We call P* c-empty *if for any pair (p_1, p_2) of free positions in the same connected component of $G_{\mathbb{Z}\setminus P}$, a square of size $c \times c$ can be moved from p_1 to p_2 without overlapping with P. We call P* weakly c-empty *if at any time at most one corner of the $c \times c$ square overlaps with a tile of P.*

For an illustration of Definition 3, see Fig. 8(b). For a 3-empty polyomino P, it is straightforward to see that as long as a tile $t \in P$ can be moved to a free position p that lies in the outer face, such that all surrounding positions of p are free as well, t can be removed from P. The intuition is the following: Consider a path of a 3×3-square, centered at that free position p, that connects p to the outside of the bounding box of P. The decomposition step for the tile t consists of all positions given by that path. Note that this still holds if we consider weakly 3-empty polyominoes.

We are able to show that we can always find such a removable tile, when P is the 2-scaled copy of a non-degenerate polyomino. One of the core ideas of our method is to make P weakly 3-empty. If P is weakly 3-empty, we consider a partition of P into horizontal *slabs*. Based on this partition, we can show that we can either remove tiles corresponding to a leaf of the dual graph of this partition, or that we can *cut open* a hole of P, i.e., we remove two adjacent tiles to reduce the number of holes of P by 1.

Definition 4. *A* slice *of a polyomino P is the set of all tiles sharing the same x-coordinate. A* slab *is a maximal connected set of tiles within a slice.*

Definition 5. *Let P be a polyomino, and $\mathcal{S}_P$ its partition into slabs. By $\mathcal{C}(\mathcal{S}_P)$ we refer to the edge-contact graph of $\mathcal{S}_P$, i.e., each slab is represented as a vertex, and two vertices are connected if and only if the union of both slabs is connected.*

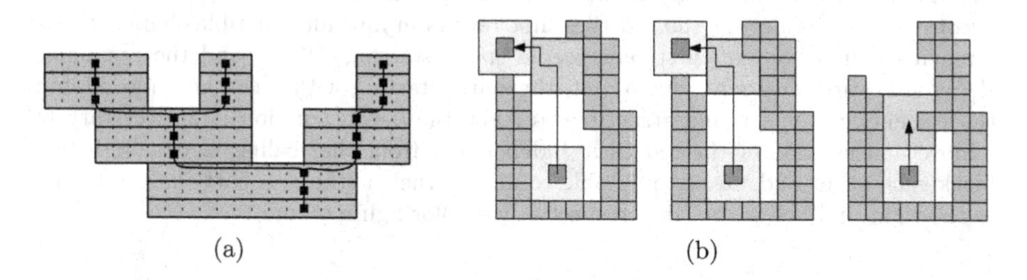

Fig. 8. (a) The 3-scaled copy P^3 of a polyomino P (gray tiles), its partition into horizontal slabs $\mathcal{S}_P$, and its dual graph $\mathcal{C}(\mathcal{S}_P)$ (black). (b) A polyomino that is 3-empty, weakly 3-empty, and not (weakly) 3-empty, respectively.

It is straightforward to see that the corresponding tiles of any leaf in $\mathcal{C}(\mathcal{S}_P)$ can be removed, if it lies in the outer face, and the polyomino is weakly 3-empty.

Lemma 7. *Let P be a non-degenerate polyomino that is weakly 3-empty. If $\mathcal{C}(\mathcal{S}_P)$ contains a degree-one vertex lying in the outer face, then the corresponding slab S can be removed from P.*

Theorem 5. *Let P be a non-degenerate polyomino. Then P^2 is constructible.*

Proof (Sketch). In a first phase we remove tiles from P^2 such that the remaining shape is weakly 3-empty. More precisely, we simply scratch off tiles within too narrow corridors.

In a second phase we deconstruct the remaining shape slab by slab. For this we make use of Lemma 7. If the shape contains any holes, we cut them open by removing two adjacent tiles. After this, we restart with the first phase. □

Because there are (even tree-shaped) non-constructible polyominoes, this result is tight. If the polyomino is already 3-empty, we can skip the first phase. Whenever we have to cut open a hole, we remove three (instead of two) tiles, such that the property of being 3-empty is preserved. This results in the following corollary.

Corollary 3. *Every non-degenerate, 3-empty polyomino is constructible.*

This idea can be adapted for 3D shapes scaled by a factor of 7, see the full version [12] for details.

Theorem 6. *Let P be a non-degenerate polycube. Then P^7 is constructible.*

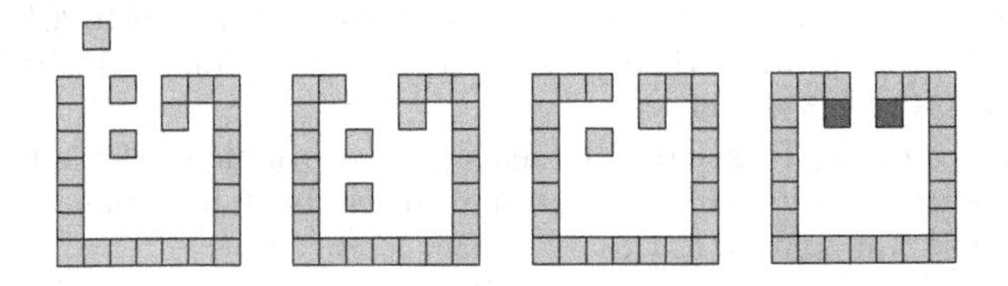

Fig. 9. The shape on the right is not constructible, if only one tile at a time is allowed to be controlled. If we can add multiple tiles at a time, or if we are allowed to position two (cyan colored) seed tiles, the polyomino is constructible. (Color figure online)

6 Conclusion and Future Work

We provided a number of algorithmic results for assembling shapes by connecting particles to a seed tile. For future research several interesting problems remain open. What is the computational complexity of 2D-STAP? This is also an open question in the full tilt model [6]. Does 3D-STAP remain NP-hard when restricted to the class of non-degenerate shapes? We conjecture this to be true.

In this paper, we added one tile after the other to the seed. If this assumption is relaxed, i.e., more than one tile at a time can be added to the workspace, it is easy to see that more shapes are constructible, see Fig. 9. Is there a classification of shapes that can be built in this model? This also leads to the question: "Which shapes are constructible by using pre-assembled shapes (e.g., trominoes, tetrominoes, etc.)?" By taking this a step further, we could also ask for a staged approach similar to [18], where whole subassemblies can attach to each other.

Another question arises by considering multiple seed tiles. What classes of shapes are constructible, if multiple seed tiles can be placed in advance, see Fig. 9.

Acknowledgements. We thank Linda Kleist for valuable discussions and suggestions that improved the presentation of this paper, Matthias Konitzny for the awesome 3D images, and the anonymous reviewers for providing helpful comments.

References

1. Agarwal, P.K., Aronov, B., Geft, T., Halperin, D.: On two-handed planar assembly partitioning. In: Proceedings of the Symposium on Discrete Algorithms (2021)
2. Balanza-Martinez, J., et al.: Hierarchical shape construction and complexity for slidable polyominoes under uniform external forces. In: Proceedings of the Symposium on Discrete Algorithms (2020)
3. Balanza-Martinez, J., et al.: Full tilt: universal constructors for general shapes with uniform external forces. In: Proceedings of the Symposium on Discrete Algorithms (2019)
4. Becker, A., Demaine, E.D., Fekete, S.P., Habibi, G., McLurkin, J.: Reconfiguring massive particle swarms with limited, global control. In: Flocchini, P., Gao, J., Kranakis, E., Meyer auf der Heide, F. (eds.) ALGOSENSORS 2013. LNCS, vol. 8243, pp. 51–66. Springer, Heidelberg (2014). https://doi.org/10.1007/978-3-642-45346-5_545346-5545346-55
5. Becker, A.T., et al.: Targeted drug delivery: algorithmic methods for collecting a swarm of particles with uniform, external forces. In: International Conference on Robotics and Automation (2020)
6. Becker, A.T., et al.: Tilt assembly: algorithms for micro-factories that build objects with uniform external forces. Algorithmica (2018)
7. de Berg, M., Khosravi, A.: Optimal binary space partitions for segments in the plane. Int. J. Comput. Geom. Appl. **22**, 187–205 (2012)
8. Caballero, D., Cantu, A.A., Gomez, T., Luchsinger, A., Schweller, R., Wylie, T.: Building patterned shapes in robot swarms with uniform control signals. In: Proceedings of the Canadian Conference on Computational Geometry (2020)
9. Caballero, D., Cantu, A.A., Gomez, T., Luchsinger, A., Schweller, R., Wylie, T.: Hardness of reconfiguring robot swarms with uniform external control in limited directions. J. Inf. Process. **28**, 782–790 (2020)
10. Caballero, D., Cantu, A.A., Gomez, T., Luchsinger, A., Schweller, R., Wylie, T.: Relocating units in robot swarms with uniform control signals is PSPACE-complete. In: Proceedings of the Canadian Conference on Computational Geometry (2020)
11. Doty, D.: Theory of algorithmic self-assembly. Commun. ACM **55**, 78–88 (2012)
12. Keller, J., Rieck, C., Scheffer, C., Schmidt, A.: Particle-based assembly using precise global control. arXiv:2105.05784 (2021)
13. Mahadev, A.V., Krupke, D., Reinhardt, J.M., Fekete, S.P., Becker, A.T.: Collecting a swarm in a grid environment using shared, global inputs. In: International Conference on Automation Science and Engineering (2016)
14. Manzoor, S., Sheckman, S., Lonsford, J., Kim, H., Kim, M.J., Becker, A.T.: Parallel self-assembly of polyominoes under uniform control inputs. Robot. Autom. Lett. **2**, 2040–2047 (2017)

15. Patitz, M.J.: An introduction to tile-based self-assembly and a survey of recent results. Natural Comput. **13**, 195–224 (2014)
16. Rothemund, P.W.: Design of DNA origami. In: International Conference on Computer-Aided Design (2005)
17. Rothemund, P.W., Winfree, E.: The Program-size complexity of self-assembled squares. In: Proceedings of the Symposium on Theory of Computing (2000)
18. Schmidt, A., Manzoor, S., Huang, L., Becker, A.T., Fekete, S.P.: Efficient parallel self-assembly under uniform control inputs. Robot. Autom. Lett. **3**, 3521–3528 (2018)
19. Wang, H.: Proving theorems by pattern recognition-II. Bell Syst. Tech. J. **40**, 1–41 (1961)
20. Winfree, E.: Algorithmic self-assembly of DNA. Ph.D. thesis, California Institute of Technology (1998)
21. Woods, D.: Intrinsic Universality and the Computational Power of Self-Assembly. Philosophical Transactions of the Royal Society A: Mathematical, Physical and Engineering Sciences (2015)

Independent Sets in Semi-random Hypergraphs

Yash Khanna, Anand Louis, and Rameesh Paul(✉)

Indian Institute of Science, Bangalore, India
{anandl,rameeshpaul}@iisc.ac.in

Abstract. A set of vertices in a hypergraph is called an independent set if no hyperedge is completely contained inside the set. Given a hypergraph, computing its largest size independent set is an NP-hard problem.

In this work, we study the independent set problem on hypergraphs in a natural semi-random family of instances. Our semi-random model is inspired by the Feige-Kilian model [9]. This popular model has also been studied in the works of [9,30,35] etc. McKenzie, Mehta, and Trevisan [30] gave algorithms for computing independent sets in such a semi-random family of graphs. The algorithms by McKenzie et al. [30] are based on rounding a "crude-SDP". We generalize their results and techniques to hypergraphs for an analogous family of hypergraph instances. Our algorithms are based on rounding the "crude-SDP" of McKenzie et al. [30], augmented with "Lasserre/SoS like" hierarchy of constraints. Analogous to the results of McKenzie et al. [30], we study the ranges of input parameters where we can recover the planted independent set or a large independent set.

Keywords: Planted independent sets · Semi-random models · Hypergraphs · Approximation algorithms · Semidefinite programming · Lasserre/SoS hierarchy · Beyond worst-case analysis.

1 Introduction

An independent set of a hypergraph $H = (V, E)$ is a subset of vertices such that no hyperedge is completely contained inside the subset. Computing a maximum independent set is a fundamental problem in the study of algorithms. The problem has applications in areas such as resource allocation in wireless networks [37], data clustering [36], computational biology [21], etc.

The problem of computing a maximum size independent set in graphs is well known to be NP-hard [16]. Håstad [13] showed that it is hard to approximate the maximum independent set in graphs to better than a factor of $n^{1-\varepsilon}$ for any $\varepsilon > 0$ unless $NP = ZPP$. Zuckerman [38] showed that there is no possible approximation ratio better than $n^{1-\varepsilon}$ unless $P = NP$. This hardness of approximation holds for the independent set problem on hypergraphs as well since it generalizes the independent set problem on graphs.

There has been a lot of work studying approximation algorithms of independent sets in graphs and hypergraphs, see Sect. 1.2 for a brief survey. Another direction of research related to intractable problems is to study families of "easier" instances of the problem. This includes studying various random and semi-random models of instances,

A. Lubiw et al. (Eds.): WADS 2021, LNCS 12808, pp. 528–542, 2021.
https://doi.org/10.1007/978-3-030-83508-8_38

instances satisfying certain properties, etc. We give a brief survey of the special class of graphs for which the independent set problem has been studied in Sect. 1.2.

The starting point in the study of random instances for the independent set problem in graphs was the $G(n, p)$ instances (Erdős–Rényi random graphs). Analysis of $G(n, p)$ [29] showed that a random graph doesn't have an independent set of size more than $(2 + o(1)) \log_{1/(1-p)} n$, w.h.p., for a large range of p. A simple algorithm can be used to compute an independent set of size $\log_{1/(1-p)} n$, w.h.p., but computing an independent set larger than this seems to be hard. Another popular model, the *planted solution model* considers the problem of recovering a hidden planted structure of size k in a graph with n vertices. For the planted clique (or independent set) model, Alon, Krivelevich, and Sudakov [1] showed that we can recover the planted clique as long as $k = \Omega(\sqrt{n})$ (for a constant p). Blum and Spencer [4] studied *semi-random models* of k-colorable graphs; such models allow an adversary to modify the instance without changing the planted structure. The model is defined by the set of actions allowed to the adversary. For the planted clique problem, in a rich adversarial semi-random model introduced by Feige and Kilian [9], the algorithm of [30] can recover the clique for $k = \Omega_p(n^{2/3})$ (see Sect. 1.2 for precise statement). We note that algorithms for independent set in graphs mentioned above hold more generally; the results here are stated assuming a constant value of p for the purpose of illustration.

The above is a broad classification of the models, and there are many other probabilistic generative models that fit at some intermediate hierarchy in this classification. A key advantage of studying random and semi-random instances is that it gives us insights into which aspects of the problem make it hard. Often algorithms for stronger models and stronger regimes of parameters may require using more advanced tools and techniques. For example, in the case of planted cliques/independent sets, for the regimes of $k \geqslant \Omega\left(\sqrt{n \log n}\right)$ we can recover the planted graph using combinatorial techniques, which essentially returns the vertices with top k degrees. However in regimes of $k = \Omega\left(\sqrt{n}\right)$ this approach no longer works, and the best-known algorithms [1] use spectral techniques. In the regime of semi-random instances of the problem, the best-known algorithms [30] are based on semidefinite programming.

1.1 Our Models and Results

Definition 1. *Given parameters n, k, r, and p, a hypergraph H is constructed as follows.*

1. *Let V be a set of n vertices. Fix an arbitrary subset $S \subset V$ of size k.*
2. *Add a hyperedge independently with probability p for each r–tuple of vertices $\{i_1, i_2, \ldots, i_r\}$, such that $\{i_1, i_2, \ldots, i_r\} \cap S \neq \emptyset$ and $\{i_1, i_2, \ldots, i_r\} \cap (V \setminus S) \neq \emptyset$. We denote the hypergraph induced by the collection of such r-tuples as $H[S, V \setminus S]$.*
3. *Allow a* monotone adversary *to add r-hyperedges arbitrarily to $H[S, V \setminus S]$ and hypergraph induced on $V \setminus S$ denoted by $H[V \setminus S]$.*

The model discussed above was introduced by Feige and Kilian [9] in the context of studying various graph partitioning problems. The work [30] studied an analogous model in the context of independent sets in graphs.

We study the ranges of parameters k, r, p (for a fixed n) in this model for which we can recover S efficiently. Our main results are informally stated below.

Theorem 1. (Informal version[1].). *There exists a deterministic algorithm which takes as input an instance of Definition 1 satisfying*

$$k = \Omega\left(\frac{n^{(r-1)/(r-0.5)}}{p^{3/(2r-1)}}\right),$$

has running time $n^{O(r)}$, and outputs a list of at most n independent sets, one of which is S, with high probability (over the randomness of the input).

Theorem 2. (Informal version of Theorem 3). *There exists a deterministic algorithm which takes as input $\varepsilon \in (0, 1)$ and an instance of Definition 1 satisfying*

$$k = \Omega\left(\frac{n^{(r-1)/(r-0.5)}}{\varepsilon^{1/(r-0.5)} p^{1/(2r-1)}}\right),$$

has running time $n^{O(r)}$, and outputs an independent set of size at least $(1-\varepsilon)k$, with high probability (over the randomness of the input).

Theorem 1 and Theorem 2 generalize to hypergraphs the analogous results for graphs by [30]. We state and prove the formal version of Theorem 2 in Theorem 3. We refer to the full version of this paper [19] for the formal version of Theorem 1. Our proofs of these theorems are based on rounding McKenzie et al. [30] "crude-SDP", augmented with "Lasserre/SoS like" hierarchy of constraints. The Lasserre/SoS hierarchy has been used in designing approximation algorithms for independent sets in hypergraphs in the works by Chlamtac [5] and Chlamtac and Singh [6], but the power of the Lasserre/SoS hierarchy for designing approximation algorithm for the independent set problem is yet to be fully understood.

1.2 Related Work

Independent Set Problem in Hypergraphs. The independent set problem in hypergraphs cannot be approximated to a factor better than $n^{1-\varepsilon}$ for any $\varepsilon > 0$ unless P = NP [38]. The work [14] gives a combinatorial algorithm to obtain an approximation ratio of $O\left(n/\left(\log^{(r-1)} n\right)^2\right)$ for a r-uniform hypergraph where $\log^{(r)} n$ denotes a r-fold repeated application of logarithm as $\log \ldots \log n$. This has been improved by Halldórsson in the work [11] where they study the problem on arbitrary weighted hypergraphs and give an $O(n/\log n)$ approximation algorithm that runs in $\mathsf{poly}(n, m)$ time where m denotes the number of hyperedges. From here onwards, a lot of work has been done in studying the problem in special classes of graphs. In this section, we do a brief survey of these results.

The problem has been extensively studied for 3-uniform hypergraphs which contain an independent set of size γn. Krivelevich, Nathaniel, and Sudakov [23] give an SDP-based algorithm that finds an independent set of size $\tilde{\Omega}\left(\min\left(n, n^{6\gamma-3}\right)\right)$ for $\gamma \geqslant 1/2$.

[1] The formal version of this theorem can be found in the full version of this paper [19].

The work Chlamtac [5] uses an SDP relaxation with the third level of the Lasserre/SoS hierarchy and returns an independent set of size $\Omega\left(n^{1/2-\gamma}\right)$. Chlamtac and Singh [6] gave an algorithm which computes an independent set of size $n^{\Omega(\gamma^2)}$ (where $\gamma \geqslant 0$ is a constant) using $\Theta(1/\gamma^2)$ levels of a mixed hierarchy which they called *the intermediate hierarchy*.

Halldórsson and Losievskaja [12] study the problem on bounded degree hypergraphs. For hypergraphs with degree bounded by Δ, the authors show that the classical greedy set cover algorithm can be analyzed to give $(\Delta + 1)/2$ approximation. The work [2] shows that the bounded degree case is Unique Games-hard to approximate within a factor of $O\left(\Delta/\log^2\Delta\right)$. In a recent work [3], the authors exhibit how to convert this inapproximability factor of $O\left(\Delta/\log^2\Delta\right)$ under UG-hardness to NP-hardness.

Random Models for Independent Set Problem. The model studied in this work is a generalization (to hypergraphs) of the planted independent set model on graphs studied in [30]. Their algorithm is based on rounding an SDP solution. However, instead of using a relaxation of the independent set problem, they used a crude-SDP (this idea was introduced in [22] and also used in many subsequent works [27]) which helps reveal the planted solution S. The main idea is to show that the expected ℓ_2^2 distance between vectors of S (the planted independent set) is "small". In other words, the SDP solution "clusters" the vectors of S. Their algorithm outputs an independent set of size $(1-\varepsilon)k$ for $k = \Omega\left(n^{2/3}/p^{1/3}\right)$ and for a larger value of k, i.e. when $k = \Omega\left(n^{2/3}/p\right)$, it outputs at most n independent sets, one of which is the planted one w.h.p. In this parameter range, they also consider a list decoding version, where when given a random vertex of S correctly picks S from this list. The proofs of Theorem 2 and Theorem 1 generalize the proofs of the corresponding results in [30].

The problem has also been studied in graphs in a weaker semi-random model [8] by Feige and Krauthgamer, which they call as sandwich model. They propose an algorithm based on Lovász theta function for the same which returns the planted clique for $k \geqslant \Omega(\sqrt{n})$ (for $p = 1/2$). Feige and Kilian [9] studied the problem in their semi-random model and they give an algorithm to recover an independent set of size αn for regimes of $p > (1+\varepsilon)\ln n/\alpha n$ and any $\varepsilon > 0$, where α is a constant. They also give efficient algorithms to recover a planted bisection and planted k-colorable graphs in semi-random models.

A closely related problem is about recovering planted clusters in random graphs known as the Stochastic Block Model (SBM) given by [15]. In [7] they study the hypergraph version of the problem where they partition a r-uniform random hypergraph $H(n, r, p, q)$ into k equally sized clusters with p as edge probability within a cluster and q as edge probability amongst clusters. They give a spectral algorithm that guarantees exact recovery when the number of clusters $k = \Theta(\sqrt{n})$. The work [10] studies this problem in more general models like the planted partition model for non-uniform hypergraphs. The work [20] gives an SDP based algorithm for the community detection problem in k-uniform hypergraphs.

Other Problems in Semi-random Models. In [27] they develop a general framework to study graph partition problems in a semi-random model similar (in strength) to the one by Feige and Kilian [9]. They give bi-criteria approximation algorithms for Sparsest cut, Uncut, Multi cut, Balanced Cut, and Small set expansion problems. In [28] they propose another semi-random model, which they call PIE (permutation invariant edges model) for the balanced cut problem. The works by Khanna, Louis, and Venkat [18,25,26] study the problems of graph expansion (vertex and edge) and the densest k-subgraph problem in semi-random models. The work by Khanna [17] studies the semi-random model with a planted clique while the rest of the graph is composed of small-sized bounded degree graphs, expanders, etc. stitched together by a random graph. These works also heavily rely on showing that the vectors corresponding to the planted structure are "clustered" together. Hence, using some basic geometric ideas, we can recover a large part of the planted portion.

1.3 Preliminaries and Notation

Our algorithms are based on the following "crude SDP".

SDP 1.

$$\max \sum_{\{i_1,i_2,\ldots,i_r\}\in\binom{V}{r}} \left\|x_{i_1,i_2,\ldots,i_r}\right\|^2$$

subject to

$$\|x_i\|^2 = 1 \qquad \forall i \in V \qquad (1)$$

$$\|x_e\|^2 = 0 \qquad \forall e \in E \qquad (2)$$

$$\langle x_I, x_J\rangle = \|x_{I\cup J}\|^2 \qquad \forall I, J(\neq \emptyset) \subseteq V,\ s.t\ |I \cup J| \leqslant r+1 \qquad (3)$$

$$\langle x_u, x_I\rangle \geqslant \langle x_u, x_J\rangle \qquad \forall u \in V, \forall I \subseteq J \subseteq V, |J| \leqslant r+1 \qquad (4)$$

$$1 - \left\|x_{u,v_1,\ldots,v_r}\right\|^2 \leqslant \sum_{i\in[r]}\left(1 - \left\|x_{u,v_i}\right\|^2\right) \qquad \forall \{u, v_1, \ldots, v_r\} \in \binom{V}{r+1}. \qquad (5)$$

The constraints in SDP 1 are inspired by the Lasserre/SoS hierarchy of constraints. The Lasserre/SoS hierarchy is a strengthened SDP relaxation for nonlinear $0-1$ programs attributed to the works of Shor [34], Nesterov [31], Jean B. Lasserre [24], and Parrilo [32]. We refer the reader to the survey by Thomas Rothvoß [33] for a detailed discussion.

We also introduce some basic notation that we will be using throughout this paper.

- Let $\partial(S)$ or the boundary of S denote $\binom{V}{r} \setminus \left(\binom{S}{r} \cup \binom{V\setminus S}{r}\right)$.
- Let the optimal solution of the above SDP be denoted by $\left\{x_I^*\right\}_{I\subset V, 1\leqslant|I|\leqslant r+1}$.
- Let $d(v)|_T$ be the degree of any vertex $v \in V$, when restricted to only count hyperedges in the set $\{v\} \cup T$.
- Throughout the paper, we will assume that $k \leqslant n/2$, and $r \geqslant 2$.

1.4 Proof Overview

In [30] they study a crude-SDP with the constraint $\langle x_i, x_j \rangle = 0, \forall \{i, j\} \in E$. Their crude SDP tries to cluster the vertices together, while the constraint $\langle x_i, x_j \rangle = 0, \{i, j\} \in E$ tries to ensure that no edges are contained in a cluster. Constraint 2 is a natural extension of this to hypergraphs. We add vectors for all subsets of vertices of size at most $r + 1$, and add consistency constraints 3 among them, as in the Lasserre/SoS hierarchy. However, we note that SDP 1 is different from a Lasserre/SoS relaxation since there is no natural interpretation of solution to this crude-SDP as a low-degree pseudo-distribution over independent sets in the hypergraph. However, we add the constraints in Eq. 3, 4 and 5 since our intended feasible solution x' constructed as,

$$x'_{i_1,i_2,\dots,i_l} = \begin{cases} \hat{e} & \text{if } \{i_1, i_2, \dots, i_l\} \in \binom{S}{l} \\ x^*_{i_1,i_2,\dots,i_l} & \text{if } \{i_1, i_2, \dots, i_l\} \in \binom{V \setminus S}{l} \\ 0 & \text{otherwise} \end{cases} \qquad \forall l \in [r+1] \tag{6}$$

where $\hat{e}$ denote a unit vector orthogonal to x^*_I, $\forall I \subseteq V \setminus S$, $|I| \leqslant r$. satisfies these constraints. The constraints in Eq. (4) and Eq. (5) are inspired from the locally consistent probability distributions viewpoint of a r-level Lasserre/SoS hierarchy [33]. A t-level vector in a Lasserre/SoS hierarchy can be interpreted as the probability of the joint event corresponding to indices of the vector. Constraint 4 corresponds to the fact that the probability of a sub event can only be larger than the probability of an event and constraint 5 corresponds to a union bound on the complement of the joint event (represented by $x_{u,v_1,\dots,v_r}$) given by sum of the complement of pairwise joint events $1 - x_{u,v_i}$ $\forall i \in [r]$.

In Sect. 2, we prove a lower bound on the contribution of the SDP mass in the optimal solution from the r-level vectors of S, i.e. $\left\{x^*_I\right\}_{I \subset S, |I|=r}$ (Corollary 1). The high-level idea of our proof is the same as that of [30]. However, we need some new ideas to extend them to hypergraphs. Using the approach of [30], we first lower bound the SDP mass from S and $S \times (V \setminus S)$ (Lemma 1). Therefore, upper bounding the contribution from $S \times (V \setminus S)$, will give us a lower bound on the contribution from S. In [30], $S \times (V \setminus S)$ is a random bipartite graph; they use Grothendieck's inequality and concentration bounds to upper bound the contribution from this part. In our setting, $S \times (V \setminus S)$ is a random hypergraph, and [30]'s techniques do not seem to be directly applicable here. Our main idea is to construct a random bipartite graph $G' = (U_1, U_2, E')$ based on this random bipartite hypergraph as follows (Construction 1). One side of the graph consists of vertices corresponding to subsets of S of cardinality at most $r - 1$, and other side consists of vertices corresponding to subsets of $V \setminus S$ of cardinality at most $r - 1$. We add an edge between two vertices if the union of the sets corresponding to them forms a hyperedge in our hypergraph. By our construction, $\sum_{\{a,b\} \in E'} \langle x_a, x_b \rangle$ is equal to the SDP mass from $S \times (V \setminus S)$ in our hypergraph. Moreover, since $S \times (V \setminus S)$ forms a random bipartite hypergraph, our construction gives us that G' is a random bipartite graph. Therefore, we can now proceed to bound the contribution from G' using [30]'s approach (Proposition 1).

Our proof of Theorem 2 (in Sect. 3) is a generalization of the proof of Theorem 1.1 of [30] to the case of hypergraphs and our higher-order SDP (SDP 1). Corollary 1 shows that the ℓ_2^2 lengths of the r-level vectors completely inside S are large. This in turn (by the SDP constraints) implies that there is a vertex $u \in S$ such that most of the $(r-1)$-level vectors in S have a large projection on x_u^* (Lemma 3). In [30] they use the SDP constraint $\langle x_u, x_v \rangle = 0, \forall \{u, v\} \in E$ to show that the set of vectors which have a large projection on x_u^* is an independent set. Therefore they proceed to bound the parameter regimes to obtain a small value of p and a $(1-\epsilon)k$ lower bound guarantee on the size of this set. However in our setting, for $r \geqslant 3$, we are unable to guarantee that this set of $(r-1)$ level vectors is an independent set. Therefore, we proceed by using Lemma 3 to show that there exists a vertex u such that a large fraction of the 1-level vectors in $\{x_v^* : v \in S\}$ have a large projection ($\geqslant \mathcal{R}'$) on x_u^*, along the lines of [30]. Let us consider the set of 1-level vectors that have a projection $\geqslant \mathcal{R}'$ on x_u^* (Definition 4). Showing that the r-level vectors consisting of vertices from this set have non-zero norms will suffice to guarantee that there are no hyperedges in this set. We use our "union-bound" SDP constraint 5 in our crude-SDP to establish this (Lemma 5). Choosing $\mathcal{R}'$ to be large enough ($\mathcal{R}' = 1 - 1/2r$) and using the SDP constraint 5, we establish a non-zero lower bound on $\left\langle x_u^*, x_{v_1,v_2,\ldots,v_r}^* \right\rangle$ for every r-tuple $(v_1, \ldots, v_r)$ consisting of vertices from the set. Now using SDP constraints 2 and 4, we can establish that the vertices inside the set do not form a hyperedge. For our choice of parameters in this theorem, the set of vertices corresponding to this set will contain at least $(1-\varepsilon)$ fraction of the vertices in S.

Our proof of Theorem 1 (in the full version of this paper [19]) is a generalization of the proof of Theorem 1.2 of [30] to the case of hypergraphs and our higher order SDP (SDP 1). In Lemma 3 we show that there exists a vertex $u \in S$ such that most of the $(r-1)$-level vectors in S have a large projection on x_u^*. Let us consider the set of $(r-1)$-level vectors which have a large projection ($\geqslant \mathcal{R}$) on x_u^* (Definition 4). The choice of p ensures that each vertex in $V \setminus S$ forms a hyperedge with at least one of the tuples corresponding to $(r-1)$ level vectors in the set w.h.p. Moreover, the choice of $\mathcal{R}$ ensures that the set can not contain two orthogonal vectors. Therefore, this ensures that the tuples in the set contain vertices only from S. Therefore, the union of the sets of vertices contained in the $(r-1)$-tuples corresponding to such $(r-1)$-level vectors would be a subset of S. A greedy algorithm can be used to recover the remaining vertices of S. Since we don't know this special vertex u, we perform this procedure on each vertex and return the set of independent sets obtained; one of these independent sets would be the planted one w.h.p. The whole procedure is presented in Algorithm 1. The range of p in this theorem is, however smaller than the range of p for which Theorem 2 is guaranteed to hold.

2 Bounding the Contribution from the Random Hypergraph

In this section, we bound the contribution of the SDP mass from the random portion of the hypergraph. As a result, we find a lower bound on the contribution of the vectors from our planted independent set S. The two key technical results (Proposition 1 and Corollary 1) which we prove in this section which generalize ([30], Lemma 2.1) to r-uniform hypergraphs are the following.

Proposition 1. *For* $k \geqslant \dfrac{r2^{2r+2}e^r}{3p}$,

$$\sum_{\{i_1,i_2,\ldots,i_r\}\in\partial(S)} \left\|x^*_{i_1,i_2,\ldots,i_r}\right\|^2 \leqslant \left(\frac{2^{3r-2}e^{3r/2-2}}{\sqrt{3}r^{r-5/2}}\right)\left(\sqrt{\frac{k}{p}}\right)n^{r-1}.$$

with high probability (over the randomness of the input).

Corollary 1. *For* $k \geqslant \dfrac{r2^{2r+2}e^r}{3p}$,

$$\sum_{\{i_1,i_2,\ldots,i_r\}\in\binom{S}{r}} \left\|x^*_{i_1,i_2,\ldots,i_r}\right\|^2 \geqslant \binom{k}{r} - \left(\frac{2^{3r-2}e^{3r/2-2}}{\sqrt{3}r^{r-5/2}}\right)\left(\sqrt{\frac{k}{p}}\right)n^{r-1}.$$

with high probability (over the randomness of the input).

The main lemma which connects the above two results is as follows.

Lemma 1.

$$\sum_{\{i_1i_2\ldots i_r\}\in\binom{S}{r}} \left\|x^*_{i_1,i_2,\ldots,i_r}\right\|^2 + \sum_{\{i_1,i_2,\ldots,i_r\}\in\partial(S)} \left\|x^*_{i_1,i_2,\ldots,i_r}\right\|^2 \geqslant \binom{k}{r}.$$

We refer to the full version of this paper [19] for the proof.

Note that the above lemma, which is similar to ([30], Lemma 2.2) helps us remove the dependence of the contribution of the vectors from $V \setminus S$, is the key lemma that allows us to work with an arbitrary subhypergraph $H[V \setminus S]$. Also, it makes our arguments invariant to any extra hyperedges added by an adversary.

Next, we proceed to prove Proposition 1. We begin by constructing a bipartite graph to simplify our calculations, as follows.

Construction 1. *We construct a bipartite graph* $G' \stackrel{\text{def}}{=} (U_1, U_2, E')$ *from the given input hypergraph* H *as follows.*

$$\text{Here } U_1 \stackrel{\text{def}}{=} (S) \cup \binom{S}{2} \cup \ldots \cup \binom{S}{r-1} \text{ and } U_2 \stackrel{\text{def}}{=} (V \setminus S) \cup \binom{V \setminus S}{2} \cup \ldots \cup \binom{V \setminus S}{r-1}.$$

Now for each hyperedge e *in our original hypergraph* H *(before the action of the monotone adversary on* $H[S, V \setminus S]$*) such that* $e \in E \cap \partial(S)$*, let* $I_e \stackrel{\text{def}}{=} e \cap S$ *and* $J_e \stackrel{\text{def}}{=} e \cap (V \setminus S)$*. We add an edge in the graph* G' *between the vertices* $I_e \in U_1$ *and* $J_e \in U_2$*. It is easy to see that there is a bijection between the random part of the hypergraph and* G'*.*

Let A denote the adjacency matrix of G' (of dimension $|U_1| + |U_2|$) and let m' denote the maximum number of edges in the random hypergraph.

In the next few lemmas, we setup up groundwork for using this construction in establishing our claims. We prove the following bounds on $|U_1|$, $|U_2|$ and m'. The proof uses some standard results on binomial coefficients, and we refer to the full version of paper [19] for these standard results.

Fact 1. *For all* $k \leqslant n/2, r \geqslant 2$ *we have,*

1. $1+|U_1| \leqslant r\left(\frac{2ek}{r}\right)^{r-1}$.
2. $1+|U_2| \leqslant r\left(\frac{2en}{r}\right)^{r-1}$.
3. $m' \leqslant \frac{(4e)^{r-2}kn^{r-1}}{r^{r-2}}$.
4. $m' \geqslant k\left(\frac{n}{2r}\right)^{r-1}$.

We refer to the full version of this paper [19] for the proof.

Definition 2. *We define a centered matrix* $B \in \mathbb{R}^{(|U_1|+|U_2|)\times(|U_1|+|U_2|)}$,

$$B_{I,J} \stackrel{\text{def}}{=} \begin{cases} p - A_{I,J} & \forall i \in [r-1], I \in \binom{S}{i}, J \in \binom{V\setminus S}{r-i}; \forall j \in [r-1], I \in \binom{V\setminus S}{j}, J \in \binom{S}{r-j} \\ 0 & \textit{otherwise}. \end{cases}$$

where A denotes the adjacency matrix of G' in Construction 1. Note that by construction, $\mathbb{E}[B] = 0$. We rewrite the contribution of the random hypergraph towards the SDP mass in terms of the matrix B using the next lemma.

Lemma 2.

$$\sum_{\{i_1,i_2,\ldots,i_r\}\in\partial(S)} \left\|x^*_{i_1,i_2,\ldots,i_r}\right\|^2 = \frac{1}{2p}\left(\sum_{u_1,u_2\in U_1\cup U_2} B_{u_1,u_2}\left\langle x^*_{u_1}, x^*_{u_2}\right\rangle\right).$$

We refer to the full version of this paper [19] for the proof.

It is important to note that the above lemma rewrites the mass of the SDP by vectors in the boundary of S (the random part) using the matrix B. The entries of B only depend on the initial set of random edges, thus any extra edges added by a monotone adversary can be ignored w.l.o.g.

We are now ready to prove Proposition 1. The proof uses some commonly used concentration inequalities. We refer to the full version of the paper [19] for the proof.

We define the following function for notational convenience.

Definition 3. *Let* $f(r) \stackrel{\text{def}}{=} \frac{r^{5/2}2^{3r-2}e^{3r/2-2}}{\sqrt{3}}$.

Proof (Proof of Corollary 1*).* The proof follows almost immediately from Proposition 1 and Lemma 1,

$$\sum_{\{i_1 i_2\ldots i_r\}\in\binom{S}{r}} \left\|x^*_{i_1 i_2\ldots i_r}\right\|^2 \geqslant \binom{k}{r} - \sum_{\{i_1,i_2,\ldots,i_r\}\in\partial(S)} \left\|x^*_{i_1,i_2,\ldots,i_r}\right\|^2 \geqslant \binom{k}{r} - \left(\frac{2^{3r-2}e^{3r/2-2}}{\sqrt{3}r^{r-5/2}}\right)\left(\sqrt{\frac{k}{p}}\right)n^{r-1}$$
$$= \binom{k}{r} - \frac{f(r)n^{r-1}\sqrt{k}}{r^r\sqrt{p}}.$$

3 Algorithm for Computing a Large Independent Set

In this section, we will prove a formal version of Theorem 2 which is a generalization of Theorem 1.1 of [30] to r-uniform hypergraphs (Lemma 3, Lemma 4 and proof of Theorem 2). We will crucially use the lower bound on the SDP mass from the vectors in S, i.e., Corollary 1. As a first step towards this, in Lemma 3, we show that there exists a vertex $u \in S$ for which the 1 level vectors x^*_v (corresponding to vertices in S) in the optimal solution have a large projection on x^*_u.

Lemma 3. *For* $k \geqslant \dfrac{r2^{2r+2}e^r}{3p}$, *there exists a vertex* $u \in S$ *such that, with high probability (over the randomness of the input).*

$$\mathbb{E}_{v \in S \setminus \{u\}} \langle x^*_u, x^*_v \rangle \geqslant \mathbb{E}_{\{i_1, i_2, \ldots, i_{r-1}\} \sim \binom{S \setminus \{u\}}{r-1}} \left\langle x^*_u, x^*_{i_1, i_2, \ldots, i_{r-1}} \right\rangle \geqslant 1 - \frac{f(r) n^{r-1}}{k^{r-0.5}\sqrt{p}}.$$

We refer to the full version of this paper [19] for the proof.

Lemma 3 shows that a large fraction of the 1-level vectors in S have a large projection on x^*_u. We start with the following definition,

Definition 4. *We denote the set of all l-tuples containing vertices from a set $T \subseteq V$ (where $l \leqslant |T|$) whose corresponding vectors have a projection at least $\mathcal{R}$ with the vector x^*_u by*

$$\mathcal{B}_u(l, \mathcal{R}, T) \stackrel{\text{def}}{=} \left\{ \{v_1, v_2, \ldots, v_l\} : \{v_1, v_2, \ldots, v_l\} \in \binom{T}{l} \text{ and } \left\langle x^*_u, x^*_{v_1, v_2, \ldots, v_l} \right\rangle \geqslant \mathcal{R} \right\}.$$

Note that the typical values of l of interest will be 1 in Theorem 2 and $r - 1$ in Theorem 1.

Lemma 4. *For* $k \geqslant \dfrac{r2^{2r+2}e^r}{3p}$, *there exists a vertex* $u \in S$ *such that*

$$\left| \mathcal{B}_u \left(1, 1 - \frac{1}{2r}, S \right) \right| \geqslant (k-1) \left(1 - \frac{2r f(r) n^{r-1}}{\sqrt{p} k^{r-0.5}} \right)$$

with high probability (over the randomness of the input).

We refer to the full version of this paper [19] for the proof.

In [30] they use the SDP constraint $\langle x_u, x_v \rangle = 0, \forall \{u, v\} \in E$ to show that the set of vectors which have a large projection on x^*_u is an independent set. Therefore they directly analyze the bound on the size of the set to obtain an independent set, in a range of p such that it covers at least $(1 - \varepsilon)$ fraction of vertices in S. However in our setting, we are unable to guarantee directly that this set of vectors is an independent set. We crucially use the Lasserre/SoS like SDP constraints 3 and 5 and an appropriately large value of $\mathcal{R}$ ($\mathcal{R} \geqslant 1 - \frac{1}{2r}$) to show that the set guaranteed in Lemma 4 is an independent set.

Lemma 5. *For $k \geqslant \frac{r2^{2r+2}e^r}{3p}$, there exists a vertex $u \in S$ such that $\mathcal{B}_u\left(1, 1-\frac{1}{2r}, V\right)$ is an independent set with high probability (over the randomness of the input).*

Proof. We consider the SDP constraint 5 and apply it to our optimal solution x^* . By using consistency constraints ($(\langle x_I, x_J\rangle = \langle x_{I'}.x_{J'}\rangle, \forall I \cup J = I' \cup J'$) (Eq. (3)) we can rewrite the constraint in Eq. (5) as,

$$1 - \left\|x^*_{u,i_1,\ldots,i_r}\right\|^2 \leqslant \sum_{l\in[r]} \left(1 - \left\langle x^*_u, x^*_{i_l}\right\rangle\right). \tag{7}$$

For $k \geqslant \frac{r2^{2r+3}e^r}{3p}$, if we pick any set of r vertices $\{i_1, \ldots, i_r\} \in \binom{V}{r}$ in $\mathcal{B}_u\left(1, 1-\frac{1}{2r}, V\right)$ (where u is the vertex guaranteed in Lemma 4) we know that $\left\langle x^*_u, x^*_{i_l}\right\rangle \geqslant 1 - \frac{1}{2r}, \forall l \in [r]$. By using Eq. (7) we have that,

$$\left\|x^*_{u,i_1,\ldots,i_r}\right\|^2 \geqslant 1 - \sum_{l\in[r]} \left(1 - \left\langle x^*_u, x^*_{i_l}\right\rangle\right) \geqslant 1 - \sum_{l\in[r]} \frac{1}{2r} \geqslant \frac{1}{2} > 0. \tag{8}$$

Now we examine the term $\left\|x^*_{i_1,i_2,\ldots,i_r}\right\|^2$ for these $\{i_1, \ldots, i_r\}$ and we have that,

$$\left\|x^*_{i_1,i_2,\ldots,i_r}\right\|^2 = \left\langle x^*_{i_1}, x^*_{i_2\ldots,i_r}\right\rangle \geqslant \left\langle x^*_{i_1}, x^*_{u,i_2\ldots,i_r}\right\rangle = \left\|x^*_{u,i_1,\ldots,i_r}\right\|^2 > 0$$

where the equality holds by consistency constraints, the first inequality above holds by constraint 4 and the last inequality holds by Eq. (8). Hence for any r-tuple $\{i_1, i_2, \ldots, i_r\} \subseteq \mathcal{B}_u\left(1, 1-\frac{1}{2r}, V\right)$, we have $\left\|x^*_{i_1,i_2,\ldots,i_r}\right\|^2 > 0$. Therefore by SDP constraint 2, it cannot form a hyperedge. Hence, the set of vertices in $\mathcal{B}_u\left(1, 1-\frac{1}{2r}, V\right)$ is an independent set.

Definition 5. *Let $\mathcal{S}_u$ denote the set of vertices formed by the union of all vertices by reading off the indices from the tuples of the set, $\mathcal{B}_u(l, r, V)$.*

Now, we have all the ingredients to prove our main result. We present the complete algorithm below and the proof of Theorem 2.

Algorithm 1:

Input: $H = (V, E)$, $l \in [r]$, and $\mathcal{R} \in (0, 1)$.
Output: A list of independent sets in H.
1: Solve SDP 1.
2: **for all** $u \in V$ **do**
3: Initialize $\mathcal{S}_u$ denote the union of set of vertices from the tuples in $\mathcal{B}_u(l, \mathcal{R}, V)$.
4: $\mathcal{S}'_u = \{u\} \cup \mathcal{S}_u$. If $\mathcal{S}'_u$ is not an independent set, Set $\mathcal{S}'_u = \emptyset$ and skip this iteration.
5: **for all** $v \in V \setminus \mathcal{S}_u$ **do**
6: Add vertex v to $\mathcal{S}'_u$ if $\mathcal{S}'_u \cup \{v\}$ is an independent set.
7: **end for**
8: **end for**
9: Return $\{\mathcal{S}'_u\}_{u \in V}$.

We set our parameters (n, p, k, ε) appropriately and show that the number of vertices in $\mathcal{B}_u$ along with the vertex u (denoted by $\mathcal{S}'_u$) cover $1 - \varepsilon$ fraction of vertices in S.

Theorem 3 (Formal version of Theorem 2**).** *There exists a deterministic algorithm which takes as input* $\varepsilon \in (0, 1)$ *and an instance of Definition 1 satisfying*

$$k \geqslant \max\left\{\frac{r2^{2r+2}e^r}{3p}, \frac{(2rf(r))^{1/(r-0.5)}n^{(r-1)/(r-0.5)}}{\varepsilon^{1/(r-0.5)}p^{1/(2r-1)}}\right\},$$

has running time $n^{O(r)}$*, and outputs an independent set of size at least* $(1-\varepsilon)k$*, with high probability (over the randomness of the input).*

Proof. We run the Algorithm 1 with the inputs, $H, l = 1$ and $\mathcal{R} = 1 - \dfrac{1}{2r}$ to get $\{\mathcal{S}'_u\}_{u \in V}$. In Lemma 4 we show that

$$\left|\mathcal{B}_u\left(1, 1 - \frac{1}{2r}, S\right)\right| \geqslant (k-1)\left(1 - \frac{2rf(r)n^{r-1}}{\sqrt{p}k^{r-0.5}}\right).$$

For a suitable choice of parameters we wish to have,

$$\left|\mathcal{B}_u\left(1, 1 - \frac{1}{2r}, S\right)\right| \geqslant (k-1)(1-\varepsilon). \tag{9}$$

We can then include in the vertex u to our independent set and we get

$$\begin{aligned}|\mathcal{S}'_u| \geqslant |\mathcal{S}_u| + 1 = \left|\mathcal{B}_u\left(1, 1 - \frac{1}{2r}, V\right)\right| + 1 \geqslant \left|\mathcal{B}_u\left(1, 1 - \frac{1}{2r}, S\right)\right| + 1 \\ \geqslant (k-1)(1-\varepsilon) + 1 \geqslant k(1-\varepsilon).\end{aligned}$$

We note that by setting $k \geqslant \dfrac{(2rf(r))^{1/(r-0.5)}n^{(r-1)/(r-0.5)}}{\varepsilon^{1/(r-0.5)}p^{1/(2r-1)}}$, Eq. 9 is satisfied and hence we can recover an independent set of size $(1-\varepsilon)k$ for all $\varepsilon \in (0, 1)$.

Acknowledgements. YK and RP thank Theo McKenzie for helpful discussions. AL was supported in part by SERB Award ECR/2017/003296 and a Pratiksha Trust Young Investigator Award.

References

1. Alon, N., Krivelevich, M., Sudakov, B.: Finding a large hidden clique in a random graph. In: Proceedings of the Eighth International Conference Random Structures and Algorithms (Poznan, 1997), vol. 13, pp. 457–466 (1998)
2. Austrin, P., Khot, S., Safra, M.: Inapproximability of vertex cover and independent set in bounded degree graphs. Theor. Comput. **7**, 27–43 (2011)
3. Bhangale, A., Khot, S.: UG-Hardness to NP-Hardness by losing half. In: Shpilka, A. (ed.) 34th Computational Complexity Conference (CCC 2019). Leibniz International Proceedings in Informatics (LIPIcs), pp. 3:1-3:20. Schloss Dagstuhl-Leibniz-Zentrum fuer Informatik, Dagstuhl, Germany (2019)
4. Blum, A., Spencer, J.: Coloring random and semi-random k-colorable graphs. J. Algorithms **19**(2), 204–234 (1995)
5. Chlamtac, E.: Approximation algorithms using hierarchies of semidefinite programming relaxations. In: 48th Annual IEEE Symposium on Foundations of Computer Science (FOCS 2007), pp. 691–701 (2007)
6. Chlamtac, E., Singh, G.: Improved approximation guarantees through higher levels of SDP hierarchies. In: Goel, A., Jansen, K., Rolim, J.D.P., Rubinfeld, R. (eds.) APPROX/RANDOM -2008. LNCS, vol. 5171, pp. 49–62. Springer, Heidelberg (2008). https://doi.org/10.1007/978-3-540-85363-3_5
7. Cole, S., Zhu, Y.: Exact recovery in the hypergraph stochastic block model: a spectral algorithm. Linear Algebra Appl. **593**, 45–73 (2020)
8. Feige, U., Krauthgamer, R.: Finding and certifying a large hidden clique in a semi-random graph. Tech. rep, ISR (1999)
9. Feige, U., Kilian, J.: Heuristics for semirandom graph problems. vol. 63, pp. 639–671 (2001), special issue on FOCS 98 (Palo Alto, CA)
10. Ghoshdastidar, D.: Consistency of spectral hypergraph partitioning under planted partition model. Ann. Statist. **45**(1), 289–315 (2017)
11. Halldórsson, M.M., Losievskaja, E.: Approximations of weighted independent set and hereditary subset problems. J. Graph Algorithms Appl. **4**(1), 16 (2000)
12. Halldórsson, M.M., Losievskaja, E.: Independent sets in bounded-degree hypergraphs. Discrete Appl. Math. **157**(8), 1773–1786 (2009)
13. Håstad, J.: Clique is hard to approximate within $n^{1-\epsilon}$. Electron. Colloquium Comput. Complex. 4(38) (1997). http://eccc.hpi-web.de/eccc-reports/1997/TR97-038/index.html
14. Hofmeister, T., Lefmann, H.: Approximating maximum independent sets in uniform hypergraphs. In: Brim, L., Gruska, J., Zlatuška, J. (eds.) MFCS 1998. LNCS, vol. 1450, pp. 562–570. Springer, Heidelberg (1998). https://doi.org/10.1007/BFb0055806
15. Holland, P.W., Laskey, K.B., Leinhardt, S.: Stochastic blockmodels: first steps. Soc. Networks **5**(2), 109–137 (1983)
16. Karp, R.M.: Reducibility among combinatorial problems, pp. 85–103. Springer, US, Boston, MA (1972). https://doi.org/10.1007/978-1-4684-2001-2_9
17. Khanna, Y.: Exact recovery of planted cliques in semi-random graphs. CoRR abs/2011.08447 (2020). https://arxiv.org/abs/2011.08447

18. Khanna, Y., Louis, A.: Planted models for the densest k-Subgraph problem. In: Saxena, N., Simon, S. (eds.) 40th IARCS Annual Conference on Foundations of Software Technology and Theoretical Computer Science (FSTTCS 2020). Leibniz International Proceedings in Informatics (LIPIcs), vol. 182, pp. 27:1–27:18. Schloss Dagstuhl-Leibniz-Zentrum für Informatik, Dagstuhl, Germany (2020)
19. Khanna, Y., Louis, A., Paul, R.: Independent sets in semi-random hypergraphs (2021)
20. Kim, C., Bandeira, A.S., Goemans, M.X.: Stochastic block model for hypergraphs: statistical limits and a semidefinite programming approach. CoRR abs/1807.02884 (2018). http://arxiv.org/abs/1807.02884
21. Klamt, S., Haus, U.U., Theis, F.: Hypergraphs and cellular networks. PLoS Comput. Biol. **5**(5), e1000385 6(2009)
22. Kolla, A., Makarychev, K., Makarychev, Y.: How to play unique games against a semi-random adversary: study of semi-random models of unique games. In: 2011 IEEE 52nd Annual Symposium on Foundations of Computer Science (2011)
23. Krivelevich, M., Nathaniel, R., Sudakov, B.: Approximating coloring and maximum independent sets in 3-uniform hypergraphs. J. Algorithms **41**(1), 99–113 (2001)
24. Lasserre, J.B.: Global optimization with polynomials and the problem of moments. SIAM J. Optim. **11**(3), 796–817 (2000)
25. Louis, A., Venkat, R.: Semi-random graphs with planted sparse vertex cuts: algorithms for exact and approximate recovery. In: Chatzigiannakis, I., Kaklamanis, C., Marx, D., Sannella, D. (eds.) 45th International Colloquium on Automata, Languages, and Programming, ICALP 2018, July 9–13 2018, Prague, Czech Republic. LIPIcs, vol. 107, pp. 101:1–101:15. Schloss Dagstuhl - Leibniz-Zentrum für Informatik (2018)
26. Louis, A., Venkat, R.: Planted models for k-way edge and vertex expansion. In: 39th IARCS Annual Conference on Foundations of Software Technology and Theoretical Computer Science, LIPIcs. Leibniz Int. Proc. Inform., vol. 150, pp. Art. No. 23, 15. Schloss Dagstuhl. Leibniz-Zent. Inform., Wadern (2019)
27. Makarychev, K., Makarychev, Y., Vijayaraghavan, A.: Approximation algorithms for semi-random partitioning problems. In: Proceedings of the Forty-Fourth Annual ACM Symposium on Theory of Computing, p. 367–384. STOC 2012, Association for Computing Machinery, New York, NY, USA (2012)
28. Makarychev, K., Makarychev, Y., Vijayaraghavan, A.: Constant factor approximation for balanced cut in the PIE model. In: STOC 2014–Proceedings of the 2014 ACM Symposium on Theory of Computing, pp. 41–49. ACM, New York (2014)
29. Matula, D.: The largest clique in a random graph. Tech. rep., Department of Computer Science, Southern Methodist University (1976). https://s2.smu.edu/~matula/Tech-Report76.pdf
30. McKenzie, T., Mehta, H., Trevisan, L.: A new algorithm for the robust semi-random independent set problem. In: Proceedings of the Fourteenth Annual ACM-SIAM Symposium on Discrete Algorithms, p. 738–746 (2020)
31. Nesterov, Y.: Squared functional systems and optimization problems. In: High performance optimization, Appl. Optim., vol. 33, pp. 405–440. Kluwer Acad. Publ., Dordrecht (2000)
32. Parrilo, P.A.: Semidefinite programming relaxations for semialgebraic problems. Math. Program. **96**(2), 293–320 (2003)
33. Rothvoß, T.: The lasserre hierarchy in approximation algorithms - Lecture Notes for the MAPSP Tutorial (2013). https://sites.math.washington.edu/~rothvoss/lecturenotes/lasserresurvey.pdf
34. Shor, N.: An approach to obtaining global extremums in polynomial mathematical programming problems. kibernetika, vol. 5, pp. 102–106 (1998). Nondifferentiable Optimization and Polynomial Problems (1987)
35. Steinhardt, J.: Does robustness imply tractability? a lower bound for planted clique in the semi-random model. Electron. Colloquium Comput. Complex. **24**, 69 (2017)

36. Yan, G.: Finding common ground among experts' opinions on data clustering: with applications in malware analysis. In: 2014 IEEE 30th International Conference on Data Engineering, pp. 15–27 (2014)
37. Zhang, H., Song, L., Han, Z., Zhang, Y.: Hypergraph theory in wireless communication networks. SpringerBriefs in Electrical and Computer Engineering, Springer, Cham (2018)
38. Zuckerman, D.: Linear degree extractors and the inapproximability of max clique and chromatic number. Theor. Comput. **3**(6), 103–128 (2007)

A Query-Efficient Quantum Algorithm for Maximum Matching on General Graphs

Shelby Kimmel[1] and R. Teal Witter[2(✉)]

[1] Department of Computer Science, Middlebury College, Middlebury, USA
skimmel@middlebury.edu
[2] Department of Computer Science and Engineering,
NYU Tandon School of Engineering, New York City, USA
rtealwitter@nyu.edu

Abstract. We design quantum algorithms for maximum matching. Working in the query model, in both adjacency matrix and adjacency list settings, we improve on the best known algorithms for general graphs, matching previously obtained results for bipartite graphs. In particular, for a graph with n vertices and m edges, our algorithm makes $O(n^{7/4})$ queries in the matrix model and $O(n^{3/4}(m+n)^{1/2})$ queries in the list model. Our approach combines Gabow's classical maximum matching algorithm [Gabow, *Fundamenta Informaticae*, '17] with the guessing tree method of Beigi and Taghavi [Beigi and Taghavi, *Quantum*, '20].

Keywords: Maximum matching · Quantum algorithm

1 Introduction

A matching is a set of non-adjacent edges in an undirected graph. In the maximum matching problem, one tries to find the matching with the largest number of edges. Finding the maximum matching in a graph is a problem that is both of fundamental and practical importance. Its practical applications range from kidney exchange to scheduling to characterizing chemical structures [7,14,17]. As a fundamental problem, it has stimulated a string of algorithmic developments, such as the use of blossoms and dual variables [6], which have been useful in the development of a broad range of algorithms. Additionally, maximum matching in general (bipartite and non-bipartite) graphs is notable for the difficulty researchers have had in finding a simple and correct algorithm for this seemingly straightforward problem [8,15].

We study maximum matching in the query setting: We are given a graph G as an adjacency matrix or adjacency list and the goal is to find a maximum matching with as few queries as possible. A query in the matrix model takes the form, "Do vertices x and y share an edge?" A query in the list model takes the form, "What is the ith vertex adjacent to vertex x?"

The best classical algorithms for maximum matching solve the problem in $O(m\sqrt{n})$ time for both bipartite and general graphs [8,9,15,18]. The query complexity of these classical algorithms is the trivial $O(n^2)$ in the matrix model and

A. Lubiw et al. (Eds.): WADS 2021, LNCS 12808, pp. 543–555, 2021.
https://doi.org/10.1007/978-3-030-83508-8_39

$O(m)$ in the list model. In fact, using an adversarial argument, it is easy to see that any classical algorithm must query all pairs of vertices or all edges to find a maximum matching in the worst case.

Using quantum computers, however, we can do better. Lin and Lin found a quantum algorithm that solves maximum matching on a bipartite graph in $O(n^{7/4})$ queries in the matrix model [13]. Beigi and Taghavi created an algorithm that uses $O(n^{3/4}\sqrt{m+n})$ queries in the list model for bipartite graphs [3], which in the worst case when $m = \Omega(n^2)$, matches the result of Lin and Lin. Both results use the guessing tree method: Lin and Lin introduced the method for functions with binary input and Beigi and Taghavi generalized it to functions with non-binary input.

Our contribution is a quantum maximum matching algorithm for general graphs that uses $O(n^{7/4})$ queries in the matrix model and $O(n^{3/4}\sqrt{m+n})$ in the list model, matching the prior results for bipartite graphs. We combine two powerful techniques to obtain our result: Beigi and Taghavi's guessing tree method and Gabow's relatively simple algorithm for maximum matching [3,8]. The key technical issues in combining these two approaches are a careful accounting of which steps of the classical algorithm actually require queries, slight modifications to the classical algorithm that help us bound the number of queries, and a well-chosen definition of the guessing scheme for the decision tree used in the guessing tree method.

The previous best known quantum algorithms for maximum matching on general graphs ran in trivial query complexity. Ambainis and Špalek designed algorithms for general maximum matching that run in $O(n^{5/2}\log n)$ time in the matrix model and $O(n^2(\sqrt{m/n}+\log n)\log n)$ time in the list model [2]. Dörn found an algorithm for general maximum matching that runs in $O(n^2\log^2 n)$ time in the matrix model and $O(n\sqrt{m}\log^2 n)$ time in the list model [5].

While our result unifies the cases of bipartite and general graphs, there remains a gap between our upper bound and the best known lower bound. Berzina et al. and Zhang found a lower bound for maximum matching of $O(n^{3/2})$ [4,19]. Interestingly, Zhang proved that Ambainis techniques (one of the most useful methods for finding quantum lower bounds) cannot improve the current lower bound [1,19].

1.1 Graph Theory

Given an undirected graph G, we denote by $V(G)$ the set of vertices and $E(G)$ be the set of edges of G. Call $n = |V(G)|$ the number of vertices in a graph and $m = |E(G)|$ the number of edges. We represent an edge between vertices x and y as xy.

We denote the *symmetric difference* of two graphs G_1 and G_2 as $G_1 \oplus G_2$. Then $V(G_1 \oplus G_2)$ is $V(G_1) \cup V(G_2)$ and $xy \in E(G_1 \oplus G_2)$ if and only if $xy \in E(G_1)$ but $xy \notin E(G_2)$ or $xy \in E(G_2)$ but $xy \notin E(G_1)$. We may think of the symmetric difference as the graph equivalent of addition modulo 2.

A *matching* M is a set of non-adjacent edges of G. That is, if xy is in M, then there is no other edge connected to x or y in M. The solid edges in Fig. 1

form a matching. A *maximum matching* on G is a matching with the most edges of any matching on G. We call a vertex a *free vertex* if it is not on any edge in matching M, while if a vertex is not free we called it *matched*. A *matched edge* is in a matching while an *unmatched edge* is not.

A *blossom* is a cycle of length $2k+1$ with k matched edges and $k+1$ unmatched edges. The edges alternate between matched and unmatched edges with the exception of the two edges connected to the root of the blossom. In Fig. 1, the blossom has $2(2)+1=5$ edges and the root is the vertex in the cycle closest to the left free vertex.

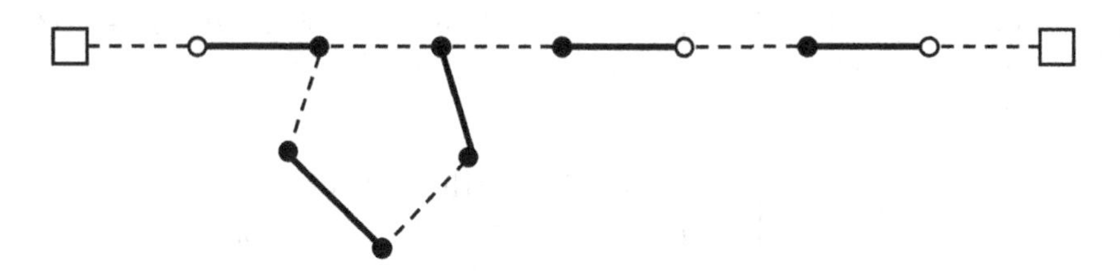

Fig. 1. Example graph with a matching where the solid lines are edges in the matching and the dotted lines are edges not in the matching but in the underlying graph. The free vertices are squares, the outer vertices (excluding the free vertices) are filled circles, and the inner vertices are hollow circles.

An *augmenting path* is a set of edges between two free vertices that alternates between matched and unmatched edges. In Fig. 1, the horizontal edges connecting the two free vertices (represented as squares) is not an augmenting path because there are two consecutive unmatched edges. A *sap* (shortest augmenting path) is an augmenting path with the fewest edges of any augmenting path. In Fig. 1, the augmenting path along the blossom between the free vertices forms a sap. We call a vertex *inner* with respect to an augmenting path if it is closer than its matched pair (the vertex with which it shares a matched edge) to the closest free vertex. Here 'closeness' is measured by the number of edges on the augmenting path between the vertex in question and the closest free vertex. Inner vertices are illustrated in Fig. 1 as hollow circles. All other vertices—including free vertices, all vertices on a blossom, and vertices adjacent to an edge equidistant between two free vertices—are *outer*. Whether a vertex is inner or outer may change as the augmenting paths grow: An inner vertex can become outer (e.g. if it becomes part of a blossom) but an outer vertex cannot become inner.

Notice that we can use the partial matching and sap in Fig. 1 to get a larger (in this case maximum) matching. We simply take the symmetric difference of the partial matching and augmenting path. That is, we include every unmatched edge (since it is in augmenting path but not the partial matching) and remove every matched edge (since it is in both the augmenting path and the partial matching). The result is a larger matching where each vertex with an edge in the partial matching has an edge in the larger matching and the previously free vertices also have matched edges.

1.2 Query Complexity

In both the list and matrix models, we learn the edges of G by querying (i.e. evaluating at various inputs) a function. We assume that G is a subgraph of the complete graph of n vertices, labeled by elements of $[n] = \{0, 1, \ldots, n-1\}$, where we do not know which edges of the complete graph are part of G and which are not.[1] Then in the case of the adjacency matrix, we have a function $E_M : [n] \times [n] \to \{0, 1\}$, where $E_M(x, y) = 1$ if and only if the edge $xy \in E(G)$.

In the case of the adjacency list, we have a function $E_L : [n] \times [n] \to [n] \cup \{\texttt{null}\}$ where

$$E_L(x,i) = \begin{cases} y & \text{if } y \text{ is the } i\text{th neighbor of } x \\ \texttt{null} & \text{if u has less than } i \text{ neighbors} \end{cases}.$$

Given access to one of these functions, the classical bounded error query complexity of maximum matching is the number of times we must evaluate the function in order to find a maximum matching with high probability.

In the quantum model, we are given access to unitaries called oracles that encode the information of the functions E_M and E_L. In the adjacency matrix model, we have access to an oracle O_M that acts on the Hilbert space $\mathbb{C}^n \times \mathbb{C}^n \times \mathbb{C}^2$ such that for an edge $e = xy$, and $b \in \{0, 1\}$, $O_M|e\rangle|b\rangle = |e\rangle|b \oplus E_M(e)\rangle$, where addition is modulo 2. In the adjacency list model, we have access to an oracle O_L that acts on the Hilbert space $\mathbb{C}^n \times \mathbb{C}^n \times \mathbb{C}^{n+1}$, where for a vertex x, index i, and $j \in [n+1]$, acts as $O_L|x, i\rangle|j\rangle = |x, i\rangle|j \oplus E_L(x, i)\rangle$, where addition is modulo $n + 1$.

Given access to one of these oracles, the quantum bounded error query complexity of maximum matching is the number of times we must apply the oracle (as part of a quantum algorithm) in order to find a maximum matching with high probability.

Given a classical query algorithm, one can create a decision tree that describes the sequence and outcomes of queries that are made throughout the algorithm. Each non-leaf vertex in the tree represents a query, and the outgoing edges from a vertex represent possible outcomes of the query. Sets of query outcomes may be grouped into a single edge (provided future decisions made by the algorithm are independent of which particular query outcome within the set occurred). Given such a decision tree, one can create a guessing scheme. A guessing scheme is a labeling of edges such that exactly one outgoing edge from each vertex is labelled as the guess. If the outcome of a query matches the guess, we say that the guessing scheme correctly guessed the outcome of that query. Otherwise, we say it was an incorrect guess.

[1] One can easily extend to the case that G is a subgraph of a multigraph; we consider complete graphs only for simplicity.

Given such a decision tree and guessing algorithm, it is possible to design a quantum algorithm:

Theorem 1. (Guessing Tree [3]). *For positive integers k, ℓ, and m, let $f : D_f \to [k]$ be a function with $D_f \subseteq [\ell]^m$. Let $\mathcal{T}$ be a decision tree for f with a guessing scheme and let T be the depth of $\mathcal{T}$. Define I as the maximum number of incorrect guesses in any path from the root to a leaf of $\mathcal{T}$. Then the bounded error quantum query complexity of evaluating f is upper bounded by $O(\sqrt{TI})$. The quantum space complexity is $O(m)$.*

See Beigi and Taghavi [3] for extensive applications of Theorem 1. Observe that the size of the image of the function f does not affect the query complexity or space complexity of the quantum algorithm that evaluates it. We use this fact to specify the maximum matching (all $O(n)$ edges) in the leaves of our decision tree.

2 Result

We use Gabow's algorithm to find a maximum matching in graph G. Gabow's algorithm runs in two phases. (The high level pseudocode is in Listing 21.) In the first phase, the algorithm finds all the edges in G that are on saps. In the second phase, the algorithm finds disjoint saps that are used to augment the partial matching. Since a maximal set of disjoint saps are found in each iteration, there are at most $O(\sqrt{n})$ iterations [9].

Listing 21. Gabow's Algorithm [8]

```
1   M ← ∅ /* M is the current partial matching */
2   loop
3     /* Phase 1 */
4     for every pair of vertices x,y do
5       if xy ∈ M then w(x,y) ← 2 else w(x,y) ← 0
6     Listing 2/3 (matrix/list model) to find pairs of vertices on saps
7     if no augmenting path is found then
8       break /* M has maximum cardinality */
9
10    /* Phase 2 */
11    Listing 4/5 (matrix/list) to create maximal set of disjoint saps P
12    augment M by the paths of P
```

The key idea behind the algorithm is the use of dual variables associated with each vertex, and which we denote using a function $d : V \to \mathbb{Z}$. Each dual variable is initialized to 1. A pair of vertices is tight if the sum of the dual variables $d(x)$ and $d(y)$ is $w(x, y)$. Recall from Listing 21 that $w(x, y)$ is 2 if xy is a matched edge and 0 otherwise. Intuitively, a pair of vertices is tight only if their shared edge could be part of a sap [8].

We use Gabow's maximum matching algorithm to construct a decision tree that finds a maximum matching. To apply Theorem 1 to the decision tree, we must design a guessing scheme. In the matrix model, we always guess that the edge we are querying is not present.

In the list model, when we are querying the ith vertex adjacent to x (call it y), our guess depends on the phase of the algorithm. In the first phase, we guess that x and y do *not* fit either of the following criteria:

- x and y are tight, x and y are not from the same blossom, and y has not yet been found (i.e. added to S, see Listing 23), or
- x and y are tight, x and y are not from the same blossom, and y is outer.

In the second phase, we guess that x and y do *not* fit either of the following criteria:

- x and y are tight, x and y do not share a matched edge, and y has not yet been found (i.e. added to S', see Listing 25), or
- x and y are tight, x and y do not share a matched edge, and x and y form a blossom.

If our query to the list returns `null`, that is, we have reached the end of a vertex's adjacency list, we say that our guess is incorrect.

In the list model, while there might be multiple outcomes of a single query that satisfy the correct guess conditions, we will see that the subsequent behavior of the algorithm is the same, so we group all such correct outcomes into a single edge in our decision tree, as described in Sect. 1.2.

Applying the above guessing scheme to Gabow's algorithm, we prove our main result:

Theorem 2. *Given a graph G with m edges and n vertices, there is a bounded error quantum algorithm that finds a maximum matching in $O(n^{7/4})$ queries in the matrix model and $O(n^{3/4}\sqrt{m+n})$ queries in the list model.*

In the remainder of this section, we explain enough of Gabow's algorithm to analyze the performance of the quantum algorithm and to prove Theorem 2. However, we do not address the correctness of Gabow's algorithm or provide sufficient details to understand why the algorithm is correct. Instead, we encourage interested readers to peruse Gabow's paper [8].

The choice to not make this paper self-contained is intentional: including the full details of Gabow's algorithm would double the length of this work without adding any novel contributions.

2.1 Breadth-First Search Subroutine

The first phase of Gabow's algorithm is a simplified search based on Edmonds' algorithm that explores G breadth-first [6]. The goal is to identify all the edges that are on saps. For this purpose, the algorithm maintains a subgraph S of G with the vertices and edges that have been explored. Initially, S consists of only free vertices. As the algorithm progresses, edges and vertices are added. We call the set of edges and vertices connected to a free vertex a *search tree.* The algorithm terminates once two search trees become connected i.e. there is an augmenting path from one free vertex to another.

The algorithm also maintains a record of the blossom that contains x, denoted by B_x. We initially set $B_x = x$ since every vertex is a trivial blossom and redefine B_x when merging blossoms. When all tight pairs of vertices have been checked and no sap has been found, the dual variables are adjusted to find new tight pairs of vertices. If the dual variables cannot be adjusted, there are no augmenting paths and the partial matching is maximum.

The execution of the simplified search based on Edmonds' algorithm depends on the data structure of the input graph. In the case of the matrix model described in Listing 22, we first identify vertices x and y that fit the criteria on Line 4. We then query the edge xy only if x and y satisfy either the if-statement on Line 5 or the if-statement on Line 8. If we reach neither Line 6 nor Line 9 then no query is made in that iteration. If we make a query on Line 6 or Line 9 and the edge is not present, our guess is correct. In order to bound the number of incorrect guesses, we bound the number of times we reach Line 7 and Line 10 which happens only if xy is present and is in the grow, blossom, or sap case.

Listing 22. Simplified Search based on Edmonds' Algorithm in the Matrix Model [8]

```
1   for every vertex x do d(x) ← 1
2   make every free vertex outer and add to V(S)
3   loop
4     if ∃ tight pair x,y with x outer , B_x ≠ B_y then
5       if y ∉ V(S) then /* grow step */
6         if xy ∈ E(G) /* query */ then
7           add xy, yy' to S where yy' ∈ M
8       else if y is outer then
9         if xy ∈ E(G) /* query */ then
10          if x and y in the same search tree then
11            /* blossom step */
12            merge all blossoms in fundamental cycle of xy
13          else /* xy forms a sap */
14            return /* continue in Listing 1 */
15    else
16      dual adjustment step
17      /* no queries are made, see Gabow Figure 2 for details */
```

In the case of the list model described in Listing 23, we query from an outer vertex x and find some adjacent vertex y. If x and y are not tight, x and y are not from the same blossom or neither of the criteria on Lines 9 and 11 apply, then our guess is correct. We bound the number of incorrect guesses by the number of times we reach Lines 7, 10, and 12, which happens only if we have reached the end of x's neighbors or x and y are in the grow, blossom, or sap case.

Observe that we can group the correct guesses in the list model into a single edge in the decision tree because the algorithm's behavior is the same in every case: continue to query neighbors of x.

Lemma 1. *The simplified search of Edmonds' algorithm makes at most $O(n)$ incorrect guesses in a single call.*

Proof. As discussed above, in both the matrix and list models, a guess is incorrect only if we are in the grow, blossom, or sap case (or in the list model at the end of a list). Therefore we bound the number of incorrect guesses by the number of times we can reach each case. In the grow case where $y \notin S$, we add both y and y' to S, where yy' is in the current partial matching M. Since this case only occurs when a vertex y is not in S, and there are at most n vertices in the graph, this case can trigger at most n incorrect guesses.

In the blossom case where x and y are in the same search tree, we have merged at least two blossoms. Each vertex is initially a blossom so we start with a total of n blossoms. Each time we merge two or more blossoms, we reduce the number of blossoms by at least one. Therefore we can merge blossoms at most n times, and so we can only make n incorrect guesses in this case.

In the case where xy completes a sap, we halt the algorithm and so this may happen at most once per call. In the list model, we can reach the end of a list at most n times so the number of incorrect guesses due to `null` outcomes is bounded by n.

Listing 23. Simplified Search based on Edmonds' Algorithm in the List Model

```
1  for every vertex x do d(x) ← 1
2  make every free vertex outer and add to V(S)
3  loop
4    for every outer vertex x do
5      for every vertex y adjacent to x do
6        if y is null then /* end of list */
7          break /* go to next x */
8        else if x and y are tight and B_x ≠ B_y then
9          if y ∉ V(S) then /* grow step */
10           add xy, yy' to S where yy' ∈ M
11         else if y is outer then
12           if x and y in the same search tree then
13             /* blossom step */
14             merge all blossoms in fundamental cycle of xy
15           else /* xy forms a sap */
16             return /* continue in Listing 1 */
17   dual adjustment step
18   /* no queries are made, see Gabow Fig. 2 for details */
```

2.2 Depth-First Search Subroutine

In the second phase of the algorithm—the path-preserving depth-first search—we identify disjoint saps. We define a subgraph H of the complete graph which we initialize with the edges between every pair of tight vertices in S. (While many edges in H were queried in the breadth-first subroutine, not all were; in particular, most edges between search trees have not yet been queried.) The algorithm explores H from each free vertex in order to find another free vertex.

Listing 24. Path-Preserving Depth-First Search in the Matrix Model [8]

```
1  initialize P to an empty set
2  for each free vertex f do
3    if f ∉ V(P) then
4      initialize S' to an empty graph
5      add f to S' as the root of a new search tree
6      find_ap(f)
7
8  procedure find_ap(x: /* x is an outer vertex */
9    for each edge xy ∈ E(H) \ M do
10     if y ∉ V(S') then
11       if xy ∈ E(G) /* query */ then
12         if y is free then /* y completes a sap */
13           add xy to S' and sap to P
14           terminate all current recursive calls to find_ap
15           remove all edges of sap from H
16           recursively remove all dangling edges from H
17         else /* grow step */
18           add xy, yy' to S' where yy' ∈ M
19           find_ap(y')
20           /* accessible only if y' is not on a sap */
21           remove y and y' from H
22       else
23         remove xy from H
24         recursively remove all dangling edges from H
25     else if blossom found then
26       if xy ∈ E(G) /* query */ then
27         blossom procedure /* see Gabow Fig. 4 for details */
28         /* calls find_ap(x) from each vertex x in blossom */
29       else
30         remove xy from H
31         recursively remove all dangling edges from H
```

While H contains edges on saps, one edge can be on more than one sap. This is a problem, as we need disjoint saps in order to augment the partial matching. To account for this, using recursive calls, the depth-first search explores H from a single free vertex and forms a new subgraph S' of visited vertices along the way. Once another free vertex is found from the starting free vertex, the algorithm processes the sap and terminates all current calls, disallowing edges of the present sap from being used in future saps and reinitializing S'. Then another call is made from a new free vertex. If the algorithm identifies a vertex on a blossom that has already been explored, new recursive calls are initiated from each vertex on the blossom.

We maintain the property that all edges in H are on as yet unidentified saps by deleting edges and vertices in several cases: When we find a sap, we remove all the edges and vertices along it. Thus no remaining sap in H can share an edge with one that was already found. When we query an edge that is not present, we remove it from H. When the recursive call does not find a sap containing vertex x, we remove x and its adjacent edges. After deletions, some *dangling* edges may remain in H. A dangling edge has an adjacent vertex with degree one (as a result of a deletion) that is not a free vertex. We remove dangling edges from H by recursively deleting the edge and adjacent vertex with degree one in addition to resulting dangling edges.

Listing 25. Path-Preserving Depth-First Search in the List Model

```
1  initialize P to an empty set
2  for each free vertex f do
3    if f ∉ V(P) then
4      initialize S' to an empty graph
5      add f to S' as the root of a new search tree
6      find_ap(f)
7
8  procedure find_ap(x) : /* x is an outer vertex */
9    for every vertex y adjacent to x do
10     if y is null then /* end of list */
11       break /* go to origin of current call to find_ap */
12     else if xy ∈ E(H) \ M then
13       if y ∉ V(S') then
14         if y is free then /* y completes a sap */
15           add xy to S' and sap to P
16           terminate all current recursive calls to find_ap
17           remove all edges of sap from H
18           recursively remove all dangling edges from H
19         else /* grow step */
20           add xy, yy' to S' where yy' ∈ M
21           find_ap(y')
22           /* accessible only if y' is not on a sap */
23           remove y and y' from H
24       else if blossom found then
25         blossom procedure /* see Gabow Figure 4 for details */
26         /* calls find_ap(x) from each vertex x in blossom */
```

Gabow's original version of the path-preserving depth-first search does not need to maintain the property that all edges in H are on as yet unidentified saps since other edges can be weeded out through the course of the algorithm. Since our goal is to bound costly "incorrect" queries, we cannot afford to wait to remove these edges and must preemptively do so. We need to ensure that this modification does not affect the correctness of the algorithm, but it is easy to see that the edges we remove from H (described in the previous paragraph) can not be part of any as yet undiscovered disjoint saps. Since the purpose of this subroutine is to discover a set of disjoint saps, this modification does not affect the correctness of this phase. This change might affect the runtime, but as we are concerned with query complexity rather than time complexity, we will not further analyze the runtime consequences.

The path-preserving depth-first search depends on the data structure of the input graph. In the case of the matrix model described in Listing 24, we identify vertices x and y that fit the criteria on Line 9 and either Line 10 or Line 25. We then query the edge xy on Line 11 or Line 26. If the edge is not present, our guess is correct. In order to bound the number of incorrect guesses, we bound the number of times we reach Line 12 and Line 27, which happens only if xy is present and completes a sap, triggers a grow step, or forms a blossom.

In the case of the list model described in Listing 25, we query from outer vertex x and find some adjacent vertex y. If x and y are not tight, x and y share a matched edge, or neither of the criteria on Lines 13 and 24 apply, then our guess is correct. While there might be multiple query outcomes that count as correct, the algorithm behaves the same in each case: continue to query the next neighbor of x. In order to bound the number of incorrect guesses, we bound the number of times we reach Lines 11, 13, and 25, which happens only if we have reached the end of x's neighbors or x and y complete a sap, trigger a grow step, or form a blossom.

Lemma 2. *The path-preserving depth-first search makes at most $O(n)$ incorrect guesses in a single call.*

Proof. In both the matrix and list models, a guess is incorrect only if we are in the sap, grow, or blossom case. Therefore we bound the number of incorrect guesses by the number of times we can reach each case. If y is a free vertex, we have found a sap and immediately remove x and y from H since they lie on a sap we have found. Thus we can bound the number of incorrect guesses in this case by the number of free vertices which is in turn bounded by n.

If y is not a free vertex, y may either be on a sap or not. Note that since xy is tight, it could be on a sap but if another edge further on the potential sap is not present or the potential sap overlaps with a sap already in P we say that y is not on a sap.

If y is not a free vertex and is on a sap, we remove x and y from H once the sap is found. Observe that there is a one-to-one correspondence between the edge xy and the vertex y. That is, since y is now in S', we will not process another edge zy for some vertex z. It follows that the number of incorrect guesses in this case is bounded by the number of vertices n.

If y is not a free vertex and is not on a sap, we will return from the call and remove y and y' from H (see Line 21 in Listing 24, Line 23 in Listing 25). We can safely remove these vertices because y' is not on a sap and for y to be on a sap, there would be two consecutive unmatched edges which is a contradiction. Then the number of incorrect guesses in this case is bounded by the number of vertices we can remove which is n.

If x and y form a blossom then we can bound the number of incorrect guesses by the number of times blossoms can be merged which is in turn bounded by n, the number of blossoms initially present. In the list model, we can reach the end of a list at most n times so the number of incorrect guesses due to `null` outcomes is bounded by n.

We now combine the two lemmas to prove our main result.

Proof (of Theorem 2). The guessing scheme is described above the statement of Theorem 2. We create a decision tree using Listing 21. The depth of the decision tree is the total number of queries we would need to make to learn the graph G. In the matrix model, this is n^2. In the list model, this is $m + n$ because we need to check each vertex and all the edges in its adjacency list. We can ensure this bound by keeping a classical record of our queries and query outcomes and, before querying the oracle, checking whether we have made this query before. By Lemma 1, Lemma 2, and the $O(\sqrt{n})$ bound on the number of iterations, the number of incorrect guesses is bounded by $O(n\sqrt{n})$. Then Theorem 2 follows from Theorem 1.

3 Conclusion

We used a classical maximum matching algorithm and the guessing tree method to give a $O(n^{7/4})$ query bound in the matrix model and $O(n^{3/4}\sqrt{m+n})$ query bound in the list model for maximum matching on quantum computers and general graphs. Our result narrows the gap between the previous trivial upper bounds of $O(n^2)$ and $O(m)$ and the quantum query complexity lower bound of $O(n^{3/2})$. An important open problem is to determine whether this algorithm is optimal. Progress on this question could be made by improving the lower bound, perhaps using the general adversary bound [10].

Another open problem is to bound the time complexity of the guessing tree method. Such a result would then allow us to compare the maximum matching algorithm described in this paper to existing quantum maximum matching algorithms that aim to minimize time complexity rather than query complexity. The time complexity of implementing the guessing tree method is currently unknown. The guessing tree algorithm is based on the dual adversary bound [3], and the quantum algorithm that results is an alternating sequence of input-dependent and input-independent unitaries, at least in the binary case [12,16]. While the input-dependent unitary is simply the oracle and may be applied in constant time, the time complexity of the input-independent unitary depends on finding an efficient implementation of a quantum walk on the decision tree. The guessing tree algorithm is similar to the *st*-connectivity span program algorithm, for which a relationship between query and time complexity is known [11]. The scaling between time and query complexity in that algorithm depends on the time complexity of implementing a quantum walk on the decision tree and on the spectral gap of the normalized Laplacian of the decision tree. It would be interesting if a similar relationship holds for the guessing tree algorithm, and if so, how it applies to the specific case of maximum matching.

References

1. Ambainis, A.: Quantum lower bounds by quntum arguments. J. Comput. Syst. Sci. **64**(4), 750–767 (2002). https://doi.org/10.1006/jcss.2002.1826, http://www.sciencedirect.com/science/article/pii/S002200000291826X
2. Ambainis, A., Špalek, R.: Quantum algorithms for matching and network flows. In: Durand, B., Thomas, W. (eds.) STACS 2006. LNCS, vol. 3884, pp. 172–183. Springer, Heidelberg (2006). https://doi.org/10.1007/11672142_13
3. Beigi, S., Taghavi, L.: Quantum speedup based on classical decision trees. Quantum 4, 241 (2020). https://doi.org/10.22331/q-2020-03-02-241, https://quantum-journal.org/papers/q-2020-03-02-241/, publisher: Verein zur Förderung des Open Access Publizierens in den Quantenwissenschaften
4. Berzina, A., Dubrovsky, A., Freivalds, R., Lace, L., Scegulnaja, O.: Quantum query complexity for some graph problems. In: Van Emde Boas, P., Pokorný, J., Bieliková, M., Štuller, J. (eds.) SOFSEM 2004. LNCS, vol. 2932, pp. 140–150. Springer, Heidelberg (2004). https://doi.org/10.1007/978-3-540-24618-3_11
5. Dörn, S.: Quantum algorithms for matching problems. Theor. Comput. Syst. **45**, 613–628 (2009). https://doi.org/10.1007/s00224-008-9118-x

6. Edmonds, J.: Paths, trees, and flowers. Can. J. Math. **17**, 449–467 (1965). https://doi.org/10.4153/CJM-1965-045-4, https://www.cambridge.org/core/journals/canadian-journal-of-mathematics/article/paths-trees-and-flowers/08B492B72322C4130AE800C0610E0E21
7. Fujii, M., Kasami, T., Ninomiya, K.: Optimal sequencing of two equivalent processors. SIAM J. Appl. Math. **17**(4), 784–789 (1969). https://www.jstor.org/stable/2099319
8. Gabow, H.N.: The weighted matching approach to maximum cardinality matching. Fundamenta Informaticae **154**(1–4), 109–130 (2017). https://doi.org/10.3233/FI-2017-1555, https://content.iospress.com/articles/fundamenta-informaticae/fi1555, publisher: IOS Press
9. Hopcroft, J.E., Karp, R.M.: An n^(5/2) algorithm for maximum matchings in bipartite graphs. SIAM J. Comput. Philadelphia **2**(4), 7 (1973). http://dx.doi.org.ezproxy.middlebury.edu/10.1137/0202019, http://search.proquest.com/docview/919736551/abstract/79AD5CB7D4BA4C4EPQ/1, num Pages: 7 Place: Philadelphia, United States, Philadelphia Publisher: Society for Industrial and Applied Mathematics
10. Hoyer, P., Lee, T., Spalek, R.: Negative weights make adversaries stronger. In: Proceedings of the Thirty-Ninth Annual ACM Symposium on Theory of Computing, pp. 526–535. STOC 2007, Association for Computing Machinery, San Diego, California, USA (2007). https://doi.org/10.1145/1250790.1250867, https://doi.org/10.1145/1250790.1250867
11. Jeffery, S., Kimmel, S.: Quantum algorithms for graph connectivity and formula evaluation. Quantum 1, 26 (2017). https://doi.org/10.22331/q-2017-08-17-26, https://quantum-journal.org/papers/q-2017-08-17-26/
12. Lee, T., Mittal, R., Reichardt, B.W., Spalek, R., Szegedy, M.: Quantum query complexity of state conversion. In: 2011 IEEE 52nd Annual Symposium on Foundations of Computer Science, pp. 344–353 (2011). https://doi.org/10.1109/FOCS.2011.75, http://arxiv.org/abs/1011.3020, arXiv: 1011.3020
13. Lin, C., Lin, H.H.: Upper bounds on quantum query complexity inspired by the Elitzur-Vaidman bomb tester. Theor. Comput. **12**(18), 1–35 (2016). https://doi.org/10.4086/toc.2016.v012a018
14. May, J.W.: Cheminformatics for genome-scale metabolic reconstructions. Ph.D. Thesis, Cambridge University (2015). https://doi.org/10.17863/CAM.15987
15. Micali, S., Vazirani, V.V.: An O(sqrt(—v—)—E—) algorithm for finding maximum matching in general graphs. In: 21st Annual Symposium on Foundations of Computer Science (sfcs 1980), pp. 17–27 (Oct 1980). https://doi.org/10.1109/SFCS.1980.12, ISSN: 0272-5428
16. Reichardt, B.W.: Span programs and quantum query complexity: the general adversary bound is nearly tight for every boolean function. In: 2009 50th Annual IEEE Symposium on Foundations of Computer Science, pp. 544–551 (2009). https://doi.org/10.1109/FOCS.2009.55, http://arxiv.org/abs/0904.2759, arXiv: 0904.2759
17. Roth, A.E., Sonmez, T., Unver, M.U.: Pairwise Kidney Exchange. J. Econ. Theor. 125(2), 151–188 (2005). https://www.hbs.edu/faculty/Pages/item.aspx?num=19520
18. Vazirani, V.V.: A simplification of the MV matching algorithm and its proof. arXiv:1210.4594 [cs] (2013). http://arxiv.org/abs/1210.4594, arXiv: 1210.4594
19. Zhang, S.: On the power of ambainis lower bounds. Theor. Comput. Sci. **339**(2), 241–256 (2005). https://doi.org/10.1016/j.tcs.2005.01.019, http://www.sciencedirect.com/science/article/pii/S0304397505001234

Support Optimality and Adaptive Cuckoo Filters

Tsvi Kopelowitz[1], Samuel McCauley[2(✉)], and Ely Porat[1]

[1] Bar-Ilan University, Ramat Gan, Israel
poratelyQcs.biu.ac.il
[2] Williams College, Williamstown, MA 01267, USA
srm2@williams.edu

Abstract. Filters (such as Bloom Filters) are a fundamental data structure that speed up network routing and measurement operations by storing a compressed representation of a set. Filters are very space efficient, but can make bounded one-sided errors: with tunable probability ϵ, they may report that a query element is stored in the filter when it is not. This is called a ***false positive***. Recent research has focused on designing methods for ***dynamically adapting*** filters to false positives, thereby reducing the number of false positives when some elements are queried repeatedly.

Ideally, an adaptive filter would incur a false positive with bounded probability ϵ for each new query element, and would incur $o(\epsilon)$ total false positives over all repeated queries to that element. We call such a filter ***support optimal***.

In this paper we design a new Adaptive Cuckoo Filter, and show that it is support optimal (up to additive logarithmic terms) over any n queries when storing a set of size n.

We complement these bounds with experiments that show that our data structure is effective at fixing false positives on network trace datasets, outperforming previous Adaptive Cuckoo Filters.

Finally, we investigate adversarial adaptivity, a stronger notion of adaptivity in which an adaptive adversary repeatedly queries the filter, using the result of previous queries to drive the false positive rate as high as possible. We prove a lower bound showing that a broad family of filters, including all known Adaptive Cuckoo Filters, can be forced by such an adversary to incur a large number of false positives.

1 Introduction

A ***filter*** is a data structure that supports membership queries for a set of elements $S = x_1, \dots x_n$ from a universe U. The answer to each filter query is `present` or `absent`. Typically, a filter has a ***correctness*** guarantee: if an element $q \in S$, the filter must return `present` to the query with probability 1.

This work was supported in part by ISF grants no. 1278/16 and 1926/19, by a BSF grant no. 2018364, and by an ERC grant MPM under the EU's Horizon 2020 Research and Innovation Programme (grant no. 683064).

A. Lubiw et al. (Eds.): WADS 2021, LNCS 12808, pp. 556–570, 2021.
https://doi.org/10.1007/978-3-030-83508-8_40

There is also a ***performance*** guarantee: if an element $q \notin S$, the filter must return `present` with tunable probability at most ϵ. If a query on an element $q \notin S$ returns `present` then q is called a ***false positive***. Typically, filters use a small amount of space.

A filter's small size means that the filter can be stored in an efficiently accessible location. Meanwhile, the no-false-negative guarantee implies that if the filter returns $q \notin S$ for a query q, then there is no need for accessing the actual data, which is typically stored in a medium with expensive access cost. This ability to filter out queries to items not in S in a small-size structure has found a wide variety of applications both in theory [8,11] and in practice, e.g. [5,10].

There are several different kinds of filters. The Bloom filter [4] was the first filter data structure to be designed; it is still very popular due to its simplicity and efficiency. Later filters were designed to provide better worst-case lookup times and space guarantees [2,14,17], improved practical performance [9,18], and improved cache performance [3].

In this paper, we focus on filters that achieve space very close to the optimal $n \log 1/\epsilon$ bits [6,12], and that store elements from a large universe $|U| \gg n$.

Fixing False Positives. A well-known issue with many existing filters is that they cannot ***adapt*** to queries: if a query $q \notin S$ is a false positive, all subsequent queries $q' = q$ will be false positives. The focus of this paper is designing filters that do adapt to false positive queries, so that if a query q is a false positive, the filter undergoes structural changes so that a later query to q is unlikely to be a false positive. An element q is said to be ***fixed*** if q was previously a false positive, but is no longer a false positive. Similarly, q is ***broken*** if q was previously fixed, but is now a false positive.

Related Work. Bender et al. [2] analyzed how to fix false positives against an adversary. They give a data structure such that if queries are generated by an adversary trying to maximize the false positive rate, each query to a filter is a false positive with probability at most ϵ, even if the query element was queried before. This requirement essentially provides concentration bounds: over n queries, their filter incurs ϵn false positives, even if the queries are maliciously chosen based on previous false positives.

However, the benefit of adaptivity goes beyond resisting an adversary. As shown experimentally by Mitzenmacher et al. [13], adapting to queries can significantly decrease the number of false positives—in fact, if queries are repeated sufficiently frequently, the performance can be much better than $O(\epsilon n)$. In particular, network trace data consists of a structured sequence of queries—can we give a data structure that performs particularly well on this kind of data?

Most recently, Bender et al. [1] compared adaptivity to cache-based strategies, finding that adaptivity leads to significantly better practical performance.

Support Optimality. Ideally, an adaptive filter would incur a false positive with probability ϵ for each new query, and incur no further queries asymptotically. Thus, every new false positive is fixed, and this fixing is unlikely to break previously-fixed false positives. In particular, let $q_1, \ldots q_n$ be a predetermined sequence of queries[1] to a filter $\mathcal{F}$, and let $Q = \bigcup_{i=1}^{n}\{q_i\}$ be the set of unique queries in the sequence. We say that $\mathcal{F}$ is ***support optimal*** if the expected number of false positives when querying $q_1, \ldots q_n$ is $\epsilon|Q|(1 + o(1))$. In this paper we give a support-optimal filter up to additive polylogarithmic terms, and show that it significantly improves practical performance.

1.1 Results

We discuss three data structures in this paper: two versions of the Adaptive Cuckoo Filter originally presented in [13] (which we call the ***Cyclic ACF*** and ***Swapping ACF***), and a Cuckoo Filter augmented with a new method of achieving adaptivity, which we call the ***Cuckooing ACF***.

The first contribution of this paper is the Cuckooing ACF, a support-optimal filter which can be implemented using almost-trivial changes to current Cuckoo Filter implementations.

In Sect. 3, we analyze the Cuckooing ACF and prove that it is support optimal over any n queries, up to additive polylogarithmic terms. This gives a significant performance improvement over previous filters even for large Q, and the difference becomes more dramatic for small Q. For example, for the case of a repeated single query ($|Q| = 1$), static filters incur ϵn expected false positives, whereas the Cuckooing ACF incurs $O(\log^4 n)$ expected false positives.

We show that despite their strong practical performance, the Cyclic ACF and Swapping ACF are not support optimal–even if there are a constant number of queries ($|Q| = O(1)$), they may incur $\Omega(n)$ false positives, whereas the Cuckooing ACF incurs at most $O(\log^4 n)$. Thus, from the standpoint of support optimality, cuckooing is a better method for achieving adaptivity.

In Sect. 4, we provide experimental results that show that the theory bears out in practice: the Cuckooing ACF attains a low false positive rate on network trace datasets, which contain many repeated queries. The performance is not only stronger than a vanilla Cuckoo Filter, but also improves upon the performance of a Cyclic ACF and a Swapping ACF of the same size. This shows that the Cuckooing ACF is effective at fixing false positives in a practical sense. These results also emphasize the benefit of a simple adaptive filter: not only is the resulting data structure easier to implement, the simplicity entails less space usage compared to previous Adaptive Cuckoo Filters, leading to a significant performance improvement.

Finally, in Sect. 5, we prove lower bounds that demonstrate that a broad family of filters cannot be adaptive in the adversarial sense of Bender et al. [2]; this includes the Cyclic ACF, the Swapping ACF, and the Cuckooing ACF. This lower bound motivates the concept of support optimality: a support optimal filter achieves strong performance on real datasets without achieving adversarial

[1] Note that the filter does not have access to this sequence ahead of time; it must process the queries online.

adaptivity. Our proof also gives insight into the structure of adaptive filters—specifically, it shows that a space-efficient filter must have variable-sized fingerprints in order to be adversarially adaptive.

2 Three Adaptive Cuckoo Filters

In this section describe a new kind of filter, the Cuckooing ACF. We then discuss the Cyclic ACF and the Swapping ACF, both originally introduced in [13].

2.1 ACF Parameters and Internal State

We begin by defining a more general data structure which we call the ***adaptive cuckoo filter*** (ACF). As the name suggests, the Cyclic ACF, the Swapping ACF, and the Cuckooing ACF are adaptive cuckoo filters.

An ACF $\mathcal{F}$ has integer parameters $f, k, b > 0$, an additional parameter $\gamma > 1$, and supports storing n elements from a universe U with a false positive rate ϵ. The internal representation of a filter $\mathcal{F}$ consists of k hash tables, each of $N = \gamma n/bk$ bins,[2] where each bin consists of b slots of f bits; thus, the space usage of $\mathcal{F}$ is $N \cdot b \cdot f \cdot k$ bits. The parameter γ determines how densely elements are packed, trading off between insert time and space; often $\gamma \approx 1.05$ is used.

The hash tables are accessed using $k + 1$ hash functions: k ***location hash*** functions $h_1^\ell, \ldots, h_k^\ell : U \to \{0, \ldots, N-1\}$ that hash from U to a bit string of length $\log N$,[3] and a single ***fingerprint hash*** h^f mapping each $x \in U$ to an f-bit ***fingerprint***. The range and domain of h^f depend on which ACF is used and may depend on the internal state of $\mathcal{F}$; we provide details below. Following previous results on filters [2–4,7,9,13], this paper assumes free access to uniform random hash functions.[4]

When a set S is stored in $\mathcal{F}$, for each element $x_i \in S$, the fingerprint of x_i is stored in one of the slots of bin $B(x_i)$ in the β_ith hash table; this bin is defined using a location hash: $B(x_i) = h_{\beta_i}^\ell(x)$ for some integer $0 \le \beta_i < k$. We say a slot σ is ***occupied*** if the fingerprint of some $x_i \in S$ is stored in σ; otherwise σ is ***empty***. We call β_i the ***hash index*** of x_i.

Since an ACF stores each element using a hash index, we can keep track of the internal state of a filter using the hash index of each element. Thus, we use $C = (C[1], C[2], \ldots, C[n]) = (\beta_1, \ldots, \beta_n)$ to define the ***configuration*** of $\mathcal{F}$. This fully defines the internal representation of a Cuckooing ACF. The internal representation of a Cyclic ACF also depends on s metadata bits stored for each element, and the internal representation of a Swapping ACF also depends on which slot within the bin is used to store each element.

---

[2] We assume γn is an integer multiple of bk for simplicity.

[3] When treating the hash value as a bit string we assume that N is a power of two for simplicity; this assumption is not necessary for the implementation.

[4] While such strong hashes are not usable in practice, this analysis is generally predictive of experimental results (see i.e. [9,13,16]).

Suppose S is stored using hash indices $\beta_1, \dots \beta_n$ under some configuration C. Then query $q \notin S$ ***collides*** with an element $x_i \in S$ under C when $h^\ell_{\beta_i}(x_i) = h^\ell_{\beta_i}(q)$ and q and x_i have the same fingerprint.

2.2 Cuckoo Filter Operations

We begin by describing how inserts and queries work for an ACF. The Cuckooing ACF and Swapping ACF insert and query using these methods; the Cyclic ACF uses a generalization of these methods.

Insert. Suppose an element x_i is inserted into a set S of size $i-1$ currently stored with filter $\mathcal{F}$ in configuration C, where elements $S = x_1, \dots x_{i-1}$ have hash indices $\beta_1, \dots \beta_{i-1}$. Assume that $\mathcal{F}$ can store up to $n \geq i$ elements. The insertion algorithm finds a valid configuration C' of $\mathcal{F}$ on S such that there exists a hash index $\beta'_i \in \{0, \dots k-1\}$ for which bin $h^\ell_{\beta'_i}$ has an empty slot. This may involve updating the hash indices of other elements; for $1 \leq j < i$ let β'_j be the hash index of x'_j under C'. We describe how to determine C' below.

If there is already an available empty slot, the filter stores the element immediately in that slot. Specifically, if there exists a $\beta \in \{0, \dots, k-1\}$ where bin $h^\ell_\beta(x_i)$ in hash table β contains an empty slot, the filter sets $\beta'_i = \beta$, and stores the fingerprint of x_i in the empty slot. All other slots remain unchanged: $\beta'_j = \beta_j$ for all $1 \leq j < i$.

Now, consider the case where there is no available empty slot. Then the ACF makes room by shifting elements as one would in cuckoo hashing [15]. The filter selects a hash index β_i arbitrarily from $\{0, \dots, k-1\}$. Since all slots in bin $h^\ell_{\beta'_i}(x_i)$ are occupied in C, the filter ***moves*** the fingerprint of some element x_j stored in a slot in $h^\ell_{\beta'_i}(x_j) = h^\ell_{\beta'_i}(x_i)$, leaving an empty slot in which x_i can be stored. If $h^\ell_\beta(x_j)$ contains an empty slot for some $\beta \in \{0, \dots, k-1\}$ (i.e. if x_j can be stored in an empty slot), one such empty slot is arbitrarily selected to store x_j. Otherwise, the filter increments $\beta'_j = \beta_j + 1 \pmod{k}$ and recurses, moving an element stored in $h^\ell_{\beta'_j}(x_j)$ as necessary.

The move the elements as described above, the ACF must be able to access the set S during an insert in order to rehash each x_j. We follow all past work on adaptive filters [1,2,13] in assuming that an external dictionary can be accessed, enabling an element to be rehashed while inserting or fixing.

If this recursive process takes too many steps (more than $\Theta(\log n)$ elements are moved), the filter chooses new hashes and rebuilds from scratch. If $\mathcal{F}$ uses $N = \Omega(n)$ hash slots, then over n inserts, the probability of a rebuild is $O(1/n)$ [15].

Query. On a query q, a filter $\mathcal{F}$ in configuration C returns `present` if there exists a β and a slot index $\sigma \in \{1, \dots b\}$ such that slot σ in bin $h^\ell_\beta(q)$ of table β is occupied and stores the fingerprint of q. This immediately guarantees correctness of the filter (queries to $x_i \in S$ always return `present`) and, via a union bound over the elements of S, a false positive rate of at most $n/(N2^f)$. The filter achieves a desired false positive rate ϵ by setting $f = \log(n/(N\epsilon)) = \log(bk/\epsilon\gamma)$.

Fixing False Positives. If an ACF returns `present` on a false positive query q (the filter knows that $q \notin S$ from the external dictionary storing S), the ACF modifies its configuration to attempt to fix q, so that subsequent queries to q return `absent`. Each type of ACF has its own method for fixing false positives, which we describe below. Notice that the process of modifying the configuration may cause some query $q' \notin S$ to become a false positive, even if q' was fixed some time in the past.

2.3 Cuckooing ACF

The primary data structure contribution of this paper is the ***Cuckooing ACF***. This data structure is a standard Cuckoo Filter [9] with an added operation to fix false positives; inserts and queries work exactly as described in Sect. 2.2.

Let q be a false positive under configuration C; we define how the Cuckooing ACF finds a new configuration C' with hash indices $\beta'_1, \ldots \beta'_n$ to attempt to fix q. For each $x_i \in S$ that collides with q under C, the filter moves x_i recursively as it would during an insert. Specifically, the filter sets the new hash index $\beta'_i = \beta_i + 1 \pmod k$; if bin $h^\ell_{\beta'_i}(x_i)$ in table β'_i does not contain an empty slot, an element x_j stored in $h^\ell_{\beta'_i}(x_i)$ under C is moved recursively. If $\Omega(\log n)$ steps are taken, the filter is rebuilt. Standard cuckoo hashing analysis shows that for any false positive on a Cuckooing ACF with $\gamma = 1 + \Omega(1)$ the probability of a rebuild is $O(1/n^2)$ [15].

2.4 Cyclic ACF

The ***cyclic ACF*** of Mitzenmacher et al. [13] is an ACF where each slot contains s additional ***hash selector*** bits. The cyclic ACF generally has $b = 1$; thus, the total space used by a Cyclic ACF is $kN(f + s)$. Usually, s is a small constant.

In the Cyclic ACF, the fingerprint hash maps $U \times \{0, \ldots, 2^s - 1\} \rightarrow \{0, \ldots, 2^f - 1\}$. In particular, the hash selector bits are used to determine the fingerprint of an element stored in a given slot.

When an element x_i is initially inserted, the insertion process continues as in Sect. 2.2, with fingerprint $h^f(x_i, 0)$. When an empty slot σ is found that can store x_i, the hash selector bits of σ are set to 0, and $h^f(x_i, 0)$ is stored in σ.

To query an element q, for each location hash h^ℓ_β, with $\beta \in \{0, \ldots, k-1\}$, the filter looks at the slot h^ℓ_β of table β. The s hash selector bits stored in the slot contain a value $0 \le \alpha \le 2^s - 1$. The filter compares $h^f(q, \alpha)$ with the fingerprint stored in the slot; the filter returns `present` if they are equal. Otherwise the filter increments β and repeats. If no collisions are found for all $0 \le \beta \le k - 1$, the filter returns `absent`.

If a query q is a false positive, the Cyclic ACF fixes the query as follows. Let x_i be the element that collides with q, let σ be the slot storing x_i, and let α be the value of the s hash selector bits stored in σ. Then the filter sets the hash selector bits of σ to store value $\alpha + 1$, and stores $h^f(x_i, \alpha + 1)$ in σ. If multiple $x \in S$ collide with q, this procedure is repeated for each such x.

2.5 Swapping ACF

The idea of the Swapping ACF [13] is to have elements hash to bins with $b > 1$ slots, and to have the fingerprint of an item depend on its slot. In this way, false positives can be (potentially) fixed by moving elements to a different slot.

Inserts proceed as described in Sect. 2.2. However, in the Swapping ACF, the fingerprint hash maps $U \times \{0, \ldots, b-1\} \rightarrow \{0, \ldots, 2^f - 1\}$. During an insert, an element's slot must be determined before its fingerprint can be calculated.

If a query q is a false positive under configuration C, the filter can fix the query as follows. Let x_i be the element that collides with q and let $b(x_i) = h^{\ell}_{\beta_i}(x_i)$ be the bin currently storing x_i. Let $\sigma_i \in \{0, \ldots, b-1\}$ be the index of the slot in $b(x_i)$ currently storing x_i; thus x_i is stored in slot $h^{\ell}_{\beta_i}(x_i) \cdot b + \sigma_x$.

The filter picks a slot index $\sigma' \in \{0, \ldots, \sigma_i - 1, \sigma_i + 1, \ldots b-1\}$, selected at random from the slots in $b(x_i)$, excluding the slot currently storing x_i. Let x_j be the element currently stored in that slot if it exists. The filter then swaps the elements: it stores fingerprint $h^f(x_i, \sigma')$ in slot $h^{\ell}_{\beta_i}(x_i) \cdot b + \sigma'$, and fingerprint $h^f(x_j, \sigma_i)$ in slot $h^{\ell}_{\beta_i}(x_i) \cdot b + \sigma_i$ (if x_j does not exist, σ_i becomes unoccupied).

3 Bounding the False Positive Rate by the Number of Distinct Queries

In this section we show that the Cuckooing ACF is support optimal: it achieves strong performance against skewed datasets, where the queries are taken from a relatively small set of elements.

Our analysis focuses on a Cuckooing ACF with $k = 2$ hash tables, $b = 1$ slots per bin, and $N = n$ slots per hash table[5] (corresponding to the classic Cuckoo Hashing analysis). The experiments in Sect. 4 indicate that our analysis likely extends to broader parameter ranges. However, formally completing the analysis for all parameters would require significant new structural insights in our proofs (e.g. Lemma 1); we leave this to future work.

Theorem 1. *Consider a sequence of at most n queries $q_1, \ldots q_n$ to a Cuckooing ACF $\mathcal{F}$ with $k = 2$ hash tables, $N = n$ slots per table, and fingerprints of length $f = \log 1/\epsilon$ bits. Let $Q = \bigcup_{i=1}^{n} \{q_i\}$. Then the expected number of false positives incurred by $\mathcal{F}$ while querying $q_1, \ldots, q_n$ is $\epsilon|Q| + O(\epsilon^2|Q| + \log^4 n)$.*

Thus, for any sequence of n queries with a support of size $|Q| = \omega(\log^4 n/\epsilon)$, the Cuckooing ACF is support optimal.

In contrast, for a worst-case input sequence, the Cyclic ACF and the Swapping ACF do not perform much better than a Cuckoo Filter. Taking the Cyclic ACF as an example, consider a sequence of n queries, each chosen uniformly at random from a randomly-selected set of size $|Q| = 1/\epsilon^{2^s}$. Each of these queries collides with some $x \in S$ under *every* choice of hash selector bits with probability $\Omega(\epsilon^{2^s})$. Thus, over n queries, the Cyclic ACF incurs $\Omega(n)$ false positives for constant ϵ and s, compared to $O(\log^4 n)$ false positives for the Cuckooing ACF via Theorem 1. See the proof of Theorem 2 for a more detailed explanation.

[5] That is to say, $\gamma = 2$.

3.1 Proof Sketch of Theorem 1

We sketch the proof of Theorem 1, but do not include the details of the proof due to space. A full proof can be seen in the full version of the paper.

Without loss of generality we assume that each false positive query only collides in one of the hash tables. Since $k = 2$, fixing a query that collides in both hash tables can be simulated by fixing each hash table separately.

To simplify notation, we define $B(i, C) = h^{\ell}_{C[i]}(x_i)$ to be the slot storing x_i under configuration C, and $B'(i, C) = h^{\ell}_{1-C[i]}(x_i)$ to be the alternate slot for x_i.

Let C_0 be the configuration of $\mathcal{F}$ before the first query q_1, and for $1 \leq i \leq n$ let C_i be the configuration after query q_i. For each $1 \leq i \leq n$, if q_i is a false positive under C_{i-1}, let k_i be the number of elements moved when fixing query q_i; otherwise let $k_i = 0$. We denote the sequence of elements moved when fixing q_i as $x_{i_1}, x_{i_2}, \ldots, x_{i_{k_i}}$. Thus, q_i collides with x_{i_1} under C_{i-1}. We call the sequence of slots affected by these movements $B(i_1, C_{i-1}), B(i_2, C_{i-1}), \ldots B(i_{k_i}, C_{i-1}), B'(i_{k_i}, C_{i-1})$ the ***path*** on C_{i-1} of q_i.

We say that q_i ***loops*** if one of the moved elements repeats; i.e. there exist $1 \leq \ell_1 < \ell_2 \leq k_i$ such that $i_{\ell_1} = i_{\ell_2}$. Interestingly, classic cuckoo hashing analysis generally only needs to bound the number of queries that loop twice, as only twice-looping queries force a rebuild. However, even a query that loops once cannot be fixed in a Cuckooing ACF, so we must bound how frequently this happens in our analysis.

Let the ***initial false positives*** be the queries in Q that are false positives for $\mathcal{F}$ in configuration C_0.

We start with a structural lemma: the elements moved when fixing any query consist of a (possibly empty) sequence of elements stored in the slot they occupied in C_0, followed by a (possibly empty) sequence of elements not stored in the slot they occupied in C_0.

Lemma 1. *If a query q_i on a configuration C_{i-1} moves an element x_{i_ℓ} satisfying $C_{i-1}[i_\ell] \neq C_0[i_\ell]$, and q_i does not loop, then all j with $\ell \leq j \leq k_i$ satisfy $C_{i-1}[i_j] \neq C_0[i_j]$.*

Proof. This proof is by induction on j; the base case $j = \ell$ is immediate.

Assume by induction that $C_{i-1}[i_{j-1}] \neq C_0[i_{j-1}]$ for some $j > \ell$. Since q_i does not loop, when $x_{i_{j-1}}$ is moved, it cannot have been moved previously while fixing q_i, and thus must be stored in slot $B(i_{j-1}, C_{i-1})$. Then after $x_{i_{j-1}}$ is moved it must be stored in slot $B(i_{j-1}, C_0)$; this must be equal to the slot storing x_{i_j}. Because q_i does not loop, x_{i_j} must be stored where it was when the fixing began; i.e. in $B(i_j, C_{i-1})$. Thus $B(i_j, C_{i-1}) = B(i_{j-1}, C_0)$, so $C_0[i_j] \neq C_{i-1}[i_j]$, as otherwise x_{i_j} and $x_{i_{j-1}}$ would be stored in the same slot in C_0.

Lemma 1 immediately gives structure to the problem in two key ways. First, it limits how queries can break one another: if q is a false positive, but is not an initial false positive, then there must be some initial false positive q_i that caused q to become a false positive. We do not need to worry about non-initial false positives causing other, new false positives. Second, it ties the behavior of all elements to how they behave on the initial configuration C_0. This means that we can make statements about how queries interact using C_0; we do not need to reset our analysis every time the filter configuration changes.

Let us summarize how to obtain the $\epsilon|Q| + O(\epsilon^2|Q| + \log^4 n)$ bound in Theorem 1. The initial false positives immediately give us a cost of $\epsilon|Q|$ false positives; we must bound the cost of all other queries (including repeated queries to the initial false positives) by $O(\epsilon^2|Q| + \log^4 n)$.

We give a set of four criteria that constitute ***costly queries***. For example, a query q_i is costly if it hashes to the path of some initial false positive q_j on C_0—this means that q_i can break q_j. Another example is if q_i loops—in that case, the Cuckooing ACF cannot fix it.

We begin by showing that all false positives are either costly queries, or are initial false positives. (Lemma 1 is the basic building block of this proof.)

The remainder of the proof uses a potential function analysis, where the potential of a configuration of the filter is the number of pairs (q, x_i), where q is a query, and x_i is an element of S stored in its original position (i.e. its position in C_0). One important property of this potential function is that if a query is not costly, it has no amortized cost (if it is a false positive, the potential function decreases by at least 1, offsetting the false positive cost incurred by the query)—again, Lemma 1 is crucial in showing this step. This is where we bound the cost of repeated queries to initial false positives—each such false positive can be charged to the (costly) query that broke it.

Then, we begin analyzing the impact of the costly queries on the potential function. We show that the expected amortized cost of a costly query q_i is $O(1 + \epsilon|Q|k_i/n)$—that is to say, it depends on the length of the path of q_i.

We complete our analysis by first bounding k_i for each q_i (conditioned on q_i being costly), and finally bounding the total cost of all costly queries. The number of elements moved by each query are not independent—for example, it is possible (though extremely unlikely) that all $x \in S$ are stored in $n + 1$ slots, where for each j the second hash of x_j is equal to the first hash of x_{j+1}. In this case, we would have $\mathbb{E}[k_i] = \Omega(n)$ for all false positive queries. To avoid these cases, we must show that all of the paths of costly queries are fairly small and do not intersect with high probability, allowing us to treat them independently. These bounds are the source of the $O(\log^4 n)$ term in the final bound.

4 Experiments

In this section, we examine how the Cuckooing ACF performs on network trace datasets. There are two main takeaways from this section. First, the design of the Cuckooing ACF results in better practical performance than previous adaptive filters on network trace datasets. Second, the analysis of Sect. 3 extends to practice: an adaptive cuckoo filter with practical parameter settings (including very high load factor) still achieves strong performance.

Our experiments use three network traces from the CAIDA 2014 dataset, as in the experiments of Mitzenmacher et al. [13]: equinix-chicago.dirA.20140619

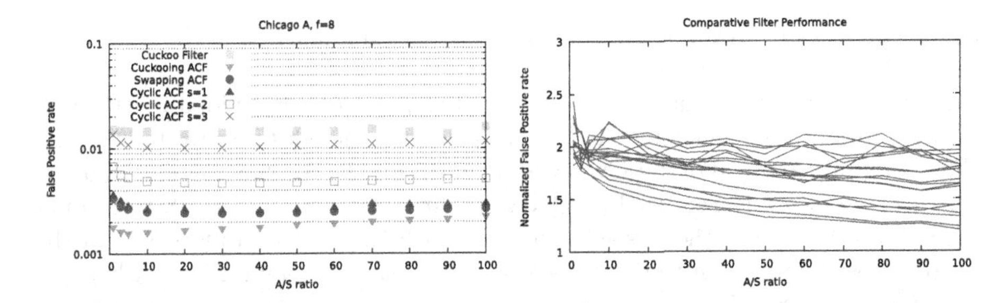

Fig. 1. We examine the false positive rate of each adaptive filter, varying the ratio of the number of queries to the number of stored elements. The right hand figure normalizes the false positive rate by the number of false positives incurred by the Cuckooing ACF. It summarizes the results for the Swapping ACF and the Cyclic ACF with $s = 1$, for all three datasets, for $f = 8, 12, 16$.

(which we call "Chicago A") equinix-chicago.dirB.20140619-432600 ("Chicago B"), and equinix-sanjose.dirA.20140320-130400 ("San Jose"). Let A be the set of query elements. We perform tests for different $|A|/|S|$ ratios; specifically $|A|/|S| = \{1, 3, 5, 10, 20, 30, 40, 50, 60, 70, 80, 90, 100\}$.

We begin by setting $n = |S|$ using the prescribed $|A|/|S|$ ratio and the total number of unique flows in the trace. The first n unique flows seen in the trace are inserted into each filter. The remaining flows in the trace (those not in S) are used as queries. We consider six data structures in our experiments:

- a classic Cuckoo Filter, with $k = 4$ hash tables and $b = 1$ slots per bucket;
- the Cuckooing ACF, with $k = 4$ hash tables and $b = 1$ slots per bucket;
- three implementations of the Cyclic ACF described in [13], with $s = 1$, $s = 2$, and $s = 3$ hash selector bits. To ensure a fair comparison in space usage, each hash selector bit used is accounted for with a corresponding decrease in the number of fingerprint bits; and
- a Swapping ACF with $b = 4$ slots per bin and $k = 2$ hash tables.

All filters are at 95% occupancy in all of our experiments (i.e. $\gamma = 1/.95$). We give results for fingerprints of length $f = 8$ bits on Chicago A, and summarize key results for Chicago B and San Jose, as well as results with $f = 12$ and $f = 16$ bits on all datasets. All results given are average performance over 10 trials.

4.1 Experimental Results

The left hand plot in Fig. 1 shows that the Cuckooing ACF has the strongest performance of all adaptive filters on the Chicago A dataset with $f = 8$. Its performance is particularly strong for low values of $|A|/|S|$—that is to say, its performance is strong when the number of unique queries is small relative to n (as one may expect given Theorem 1).

We ran further experiments, using fingerprints of size $f = 8$, $f = 12$, and $f = 16$ on Chicago A, Chicago B, and San Jose datasets, achieving similar (in fact slightly better) results. These experiments are summarized in the right hand plot of Fig. 1, and provided in full in the full version of the paper. The y-axis in this figure indicates the false positive rate of the given filter divided by the false positive rate of the Cuckooing ACF. This plot only includes the two best filters: the Swapping ACF, and the Cyclic ACF with $s = 1$. We run the experiments for three fingerprint sizes $\{8, 12, 16\}$ on all three datasets, giving 18 total lines in the plot. Note that there is some overlap with the left hand plot—one of the bottommost two lines in the plot corresponds to the Swapping ACF with $f = 8$ on Chicago A. Specifically, the Cuckooing ACF does even better with larger fingerprints like $f = 12$ and $f = 16$ compared to $f = 8$.

Overall, the Cuckooing ACF always performs at least as well as every other cuckoo filter on these datasets, often outperforming them by nearly a factor of 2.

5 Adversarial Adaptivity

Previous work leaves a dichotomy: the Adaptive Cuckoo Filters of Mitzenmacher et al. [13] work well in practice, whereas the "Broom Filter" of Bender et al. [2] is effective even against an adversary that tries to "learn" a filter's internal state. In Sects. 3 and 4 we showed that the Cuckooing ACF is practical while retaining theoretical bounds. But our theoretical bounds are not adversarial; they are based on the number of unique queries made to the filter. Can our results be taken further—is there an ACF that adapts effectively even against an adversary?

In this section we give a general lower bound showing that an adversary can obtain a false positive rate of $\Omega(1)$ against any space-efficient ACF. This result is closely tied to a key structural distinction: the Broom Filter is difficult to implement because the length of the stored fingerprint may be different for each element. Our lower bound shows that this flexibility is, in fact, necessary in order to achieve adaptivity.

5.1 Definition

Bender et al. [2] defined a notion of adaptivity that captures a worst-case adversary attempting to maximize the filter's false positive rate. We summarize this model in this subsection, and refer readers to [2] for a more thorough discussion.

In the ***adaptivity game***, an adversary generates a sequence of queries. After each query q, the adversary and filter both learn if q was a false positive. The filter may change its internal representation in response. The adversary will use whether or not q was a false positive to determine the further queries.

At any time, the adversary may name a special element $\hat{q}$—the adversary is asserting that this query is likely to be a false positive. The adversary "wins" if $\hat{q}$ is a false positive, and the filter "wins" if $\hat{q}$ is not a false positive.

The ***sustained false positive*** rate of a filter is the maximum probability ϵ with which the adversary can win the adaptivity game. We call a filter $\mathcal{F}$ ***adaptive*** if $\mathcal{F}$ can achieve a sustained false positive rate of ϵ for *any* constant ϵ.

5.2 Lower Bounds

To begin, we note that the Cyclic ACF is not adaptive. A nearly-identical proof shows that the Swapping ACF is not adaptive.

Theorem 2. *Let $\mathcal{F}$ be a Cyclic ACF with $k = O(1)$ hash tables, each with $N = \Theta(n)$ slots. Then there exists an adversarial strategy, making $\Theta(2^s/\epsilon^{2^s})$ queries, which wins the adaptivity game against $\mathcal{F}$ with probability $\Omega(1)$. Thus the sustained false positive rate of $\mathcal{F}$ is $\Omega(1)$.*

Proof. The adversary begins by selecting a query element q_1 at random. The adversary queries q_1 2^s times. If q_1 is a false positive every time it is queried, the adversary sets $\hat{q} \leftarrow q_1$; otherwise the adversary chooses a new query element q_2 and repeats. This process is repeated until $O(1/\epsilon^{2^s})$ query elements have been chosen, requiring $O(2^s/\epsilon^{2^s})$ queries overall.

We show that the adversary finds a $\hat{q}$ with probability $\Omega(1)$, and that $\hat{q}$ will be a false positive with probability 1.

Each time q collides with an element $x_i \in S$, the hash selector bits associated with x_i are incremented; thus, if q does not collide with x_i on the jth query, it will not collide on the j'th query for $j' > j$. Then if q is a false positive on all 2^s collisions, there is an $x_j \in S$ such that $h^f(q, \alpha) = h^f(x_j, \alpha)$ for all $\alpha \in \{0, \ldots, 2^s - 1\}$. We immediately obtain that any $\hat{q}$ found by the adversary is a false positive with probability 1.

For a given query q_i and a given $x_j \in S$, the probability that $h^{\ell}_{\beta_j}(q_i) = h^{\ell}_{\beta_j}(x_j)$ is $1/n$. The probability that, for all α, $h^f(q_i, \alpha) = h^f(x_j, \alpha)$ is $1/\epsilon^{2^s}$.

After the algorithm has made $1/\epsilon^{2^s}$ queries, the probability that there exists a query q' and an $x^* \in S$ such that $h^f(q_i, \alpha) = h^f(x_j, \alpha)$ for all α is $1 - (1 - \epsilon^{2^s}/n)^{n/\epsilon^{2^s}} = \Omega(1)$. Thus, the adversary finds a $\hat{q}$ with constant probability.

One might think that hashing elements to another bucket (as in the Cuckooing ACF) is sufficient to make a filter adaptive. The reason Theorem 2 gives such a strong lower bound for the Cyclic ACF is that when we move an element to the next fingerprint, it is still a false positive with probability ϵ. A constant number of these movements still leaves a significant probability that a false positive is not yet fixed. In contrast, when a colliding element is moved in the Cuckooing ACF, it still collides with the query with probability only ϵ/n—this seems low enough that almost all queries are successfully fixed after only a single movement.

Nonetheless, the adversary can use a birthday attack to find a small set of elements that cannot all be simultaneously fixed.

We obtain lower bounds for a fairly broad class of filters, where the total total information stored (i.e. hash index plus location plus fingerprint) for each element is at most $\log(n/\epsilon) + O(1)$ bits. This stands in contrast to the Broom Filter of Bender et al. [2], which is adaptive and which stores an *average* of $\log(n/\epsilon) + O(1)$ bits—in short, this proof shows that the nonuniformity of hash lengths in [2] is crucial to achieving adaptivity.

Definition 1. *A deterministic* k*-adaptive filter* $\mathcal{F}$ *on* n *elements with false positive rate* ϵ *is a filter satisfying the following:*

- $\mathcal{F}$ *has access to* k *uniform random hash functions* $h_0, \ldots, h_{k-1}$*. Each hash has length at most* $\log(N/\epsilon)$*, for some* $N = O(n/k)$ *with* $N \geq n/k$*.*
- *For every configuration* C *of* $\mathcal{F}$*, each* $x_i \in S$ *is stored using at least one hash* $h_{C[i]}(x)$*,* $0 \leq C[i] \leq k-1$*.*
- *The filter answers* `present` *to a query* q *on configuration* C *if there exists an* x_i *such that* $h_{C[i]}(q) = h_{C[i]}(x_i)$*. Otherwise, it answers* `absent`*.*
- *On a false positive* q*,* $\mathcal{F}$ *updates* C *to a new configuration* C' *in round-robin order. In particular, if a query* q *collides with an element* $x_i \in S$ *stored using* h_β*, then* x_i *is stored in* C' *using* $h_{\beta'}$ *satisfying* $\beta' = \beta + 1 \pmod k$*.*

By setting each hash h_i in Definition 1 so that for any $i \in \{0, \ldots, k-1\}$ and $x \in U$, $h_i(x)$ is the concatenation of $h_i^\ell(x)$ and $h^f(x, i)$, the Cuckooing ACF is a deterministic k-adaptive filter. By setting $h_{(i,\alpha)}(x)$ to be the concatenation of $h_i^\ell(x)$ and $h^f(x, \alpha)$, the Cyclic ACF is a deterministic $k2^s$-adaptive filter.[6]

The round-robin ordering requirement stands out as being a bit artificial, but our proof can fairly easily be generalized to handle other deterministic methods to update the configuration.

Theorem 3. *There exists an adversarial strategy making* $O(n)$ *queries such that, for any deterministic* k*-adaptive filter* $\mathcal{F}$ *with* $k < \log n/(6 \log\log n)$ *and* $\epsilon > 1/(n^{1/k})$*, the adversary wins the adaptivity game with probability* $\Omega(1)$*.*

Querying to Find a Mutually Unfixable Set. The proof of Theorem 3 begins with the adversary searching for a structure that "blocks" the filter, preventing it from fixing a false positive.

Consider a stored element $x_i \in S$, and fix a filter $\mathcal{F}$ with k hash functions $h_0, \ldots h_{k-1}$. A set of queries K is called ***mutually unfixable for*** x_i if,

- for all $\beta \in \{0, \ldots, k-1\}$, there exists a $q' \in K$ with $h_\beta(x_i) = h_\beta(q')$, and
- for all $q' \in K$ there exists a $\beta, \in \{0, \ldots, k-1\}$ such that $h_\beta(x_i) = h_\beta(q')$.

The goal of our adversary is to find a mutually unfixable set of the queries, since for any configuration, at least one element in such a set is a false positive.

We briefly sketch the remaining details of our adversary to prove Theorem 3. The full proof can be found in the full version of the paper.

The adversary begins by choosing a set Q of size $(1+1/k)N/(\epsilon n^{1/k})$ uniformly at random from U. We show that if Q is this size, then with constant probability Q will contain $\Theta(1)$ mutually unfixable sets, each of size $O(k)$.

[6] The Cyclic ACF does not quite satisfy Definition 1 since its hashes are not independent. However, this only makes it easier for an adversary to find false positives.

The adversary then queries members of Q for $2k$ rounds; any query that is a false positive during the second set of k rounds is stored in a set Q_d. We show that Q_d will be the union of some mutually unfixable subsets of Q.

Finally, the adversary repeatedly selects k elements from Q_d and randomly selects a permutation P on these elements. The adversary queries these elements in order, twice. We show that, over $O(n)$ total queries, the adversary will (with constant probability) find k elements corresponding to a mutually unfixable set, and query them in an order such that each is a false positive every time it is queried. Then, the adversary can find a false positive $\hat{q}$ with constant probability.

References

1. Bender, M.A., Das, R., Farach-Colton, M., Mo, T., Tench, D., Ping Wang, Y.: Mitigating false positives in filters: to adapt or to cache? In: Symposium on Algorithmic Principles of Computer Systems (APOCS), pp. 16–24. ACM-SIAM (2021)
2. Bender, M.A., Farach-Colton, M., Goswami, M., Johnson, R., McCauley, S., Singh, S.: Bloom filters, adaptivity, and the dictionary problem. In: Foundations of Computer Science (FOCS), pp. 182–193. IEEE (2018)
3. Bender, M.A., et al.: Don't thrash: how to cache your hash on flash. Proc. VLDB Endow. **5**(11), 1627–1637 (2012)
4. Bloom, B.H.: Space/time trade-offs in hash coding with allowable errors. Commun. ACM **13**(7), 422–426 (1970)
5. Broder, A., Mitzenmacher, M.: Network applications of bloom filters: a survey. Internet Math. **1**(4), 485–509 (2004)
6. Carter, L., Floyd, R., Gill, J., Markowsky, G., Wegman, M.: Exact and approximate membership testers. In: Symposium on Theory of Computing (STOC), pp. 59–65 (1978)
7. Eppstein, D.: Cuckoo filter: simplification and analysis. In: Scandinavian Symposium and Workshops on Algorithm Theory (SWAT), vol. 53, pp. 8:1–8:12 (2016)
8. Eppstein, D., Goodrich, M.T., Mitzenmacher, M., Torres, M.R.: 2–3 cuckoo filters for faster triangle listing and set intersection. In: Principles of Database Systems (PODS), pp. 247–260. ACM (2017)
9. Fan, B., Andersen, D.G., Kaminsky, M., Mitzenmacher, M.D.: Cuckoo filter: practically better than Bloom. In: International Conference on Emerging Networking Experiments and Technologies (CoNEXT), pp. 75–88. ACM (2014)
10. Geravand, S., Ahmadi, M.: Bloom filter applications in network security: a state-of-the-art survey. Comput. Netw. **57**(18), 4047–4064 (2013)
11. Jiang, S., Larsen, K.G.: A faster external memory priority queue with decrease keys. In: Symposium on Discrete Algorithms (SODA), pp. 1331–1343. ACM-SIAM (2019)
12. Lovett, S., Porat, E.: A lower bound for dynamic approximate membership data structures. In: Foundations of Computer Science (FOCS), pp. 797–804. IEEE (2010)
13. Mitzenmacher, M., Pontarelli, S., Reviriego, P.: Adaptive cuckoo filters. In: Workshop on Algorithm Engineering and Experiments (ALENEX), pp. 36–47 (2018)
14. Pagh, A., Pagh, R., Rao, S.S.: An optimal bloom filter replacement. In: Symposium on Discrete Algorithms (SODA), pp. 823–829. ACM-SIAM (2005)
15. Pagh, R., Rodler, F.F.: Cuckoo hashing. J. Algorithms **51**(2), 122–144 (2004)

16. Pandey, P., Bender, M.A., Johnson, R., Patro, R.: A general-purpose counting filter: making every bit count. In: International Conference on Management of Data (SIGMOD), pp. 775–787. ACM (2017)
17. Porat, E.: An optimal bloom filter replacement based on matrix solving. In: Frid, A., Morozov, A., Rybalchenko, A., Wagner, K.W. (eds.) CSR 2009. LNCS, vol. 5675, pp. 263–273. Springer, Heidelberg (2009). https://doi.org/10.1007/978-3-642-03351-3_25
18. Wang, M., Zhou, M., Shi, S., Qian, C.: Vacuum filters: more space-efficient and faster replacement for bloom and cuckoo filters. Proc. VLDB Endow. **13**(2), 197–210 (2019)

Computing the Union Join and Subset Graph of Acyclic Hypergraphs in Subquadratic Time

Arne Leitert[(✉)]

Department of Computer Science, Central Washington University, Ellensburg, WA, USA
arne.leitert@cwu.edu

Abstract. We investigate the two problems of computing the union join graph as well as computing the subset graph for acyclic hypergraphs and their subclasses. In the *union join graph* G of an acyclic hypergraph H, each vertex of G represents a hyperedge of H and two vertices of G are adjacent if there exits a join tree T for H such that the corresponding hyperedges are adjacent in T. The *subset graph* of a hypergraph H is a directed graph where each vertex represents a hyperedge of H and there is a directed edge from a vertex u to a vertex v if the hyperedge corresponding to u is a subset of the hyperedge corresponding to v.

For a given hypergraph $H = (V, \mathcal{E})$, let $n = |V|$, $m = |\mathcal{E}|$, and $N = \sum_{E \in \mathcal{E}} |E|$. We show that, if the Strong Exponential Time Hypothesis is true, both problems cannot be solved in $\mathcal{O}(N^{2-\varepsilon})$ time for α-acyclic hypergraphs and any constant $\varepsilon > 0$, even if the created graph is sparse. Additionally, we present algorithms that solve both problems in $\mathcal{O}(N^2/\log N + |G|)$ time for α-acyclic hypergraphs, in $\mathcal{O}(N \log(n+m) + |G|)$ time for β-acyclic hypergraphs, and in $\mathcal{O}(N + |G|)$ time for γ-acyclic hypergraphs as well as for interval hypergraphs, where $|G|$ is the size of the computed graph.

1 Introduction

A *hypergraph* $H = (V, \mathcal{E})$ is a generalisation of a graph in which each edge $E \in \mathcal{E}$, called *hyperedge*, can contain an arbitrary positive number of vertices from V. One may also see a hypergraph H as a family $\mathcal{E}$ of subsets of some set V. Indeed, we say that the family $\mathcal{F}$ of sets *forms* the hypergraph $H = (V, \mathcal{E})$ if $V = \bigcup_{S \in \mathcal{F}} S$ and $\mathcal{E} = \mathcal{F}$. We use $n = |V|$, $m = |\mathcal{E}|$, and $N = \sum_{E \in \mathcal{E}} |E|$ to respectively denote the cardinality of the vertex set, the cardinality of the hyperedge set, and the total size of all hyperedges of H.

1.1 Acyclic Hypergraphs

A tree T is called a *join tree* for H if the hyperedges of H are the nodes of T and, for each vertex $v \in V$, the hyperedges containing v induce a subtree of T. That

A. Lubiw et al. (Eds.): WADS 2021, LNCS 12808, pp. 571–584, 2021.
https://doi.org/10.1007/978-3-030-83508-8_41

is, if $v \in E_i \cap E_j$, then v is contained in each hyperedge (i.e., node of T) on the path from E_i to E_j in T. A hypergraph is *acyclic* if it admits a join tree. There is a linear-time algorithm which determines if a given hypergraph is acyclic and, in that case, constructs a corresponding join tree for it [21].

Acyclic hypergraphs have various applications. They are, for example, a desired structure when designing relational databases [1]. There is also a close relation between acyclic hypergraphs and chordal as well as dually chordal graphs. Namely, a graph is chordal if and only if its maximal cliques form an acyclic hypergraph [12], and a graph is dually chordal if and only if its closed neighbourhoods form an acyclic hypergraph [5].

Tree-decompositions are another application. The idea is to decompose a graph $G = (V, E)$ into multiple induced subgraphs, usually called *bags*, where each vertex can be in multiple bags. The set of bags $\mathcal{B}$ forms a tree T in such a way that the following requirements are fulfilled: Each vertex is in at least one bag, each edge is in at least one bag, and T is a join tree for the hypergraph $(V, \mathcal{B})$. Usually tree-decompositions are considered with additional restrictions. The most known is called *tree-width*; it limits the maximum cardinality of each bag. For a graph class with bounded tree-width, many NP-complete problems can be solved in polynomial or even linear time. Alternatively, one may limit the distances between vertices inside a bag. Such a tree-decomposition can be used, for example, for constructing tree-spanners [7,8] and efficient routing schemes [6].

An inclusion-maximal subset of vertices of a graph G is called an *atom* if it induces a connected subgraph of G without a clique separator. It is known that the atoms of a graph form an acyclic hypergraph [16]. The corresponding join tree is then called *atom tree*.

The most general acyclic hypergraphs are called *α-acyclic* (i.e.., each acyclic hypergraph is α-acyclic). They are closely related to chordal graphs and to dually chordal graphs. Subclasses of α-acyclic hypergraphs are *β-acyclic* hypergraphs which are closely related to strongly chordal graphs and *γ-acyclic* hypergraphs which are closely related to ptolemaic graphs (graphs that are chordal and distance-hereditary). We also consider *interval* hypergraphs. These are acyclic hypergraphs for which one of their join trees forms a path. As the name suggests, they are closely related to interval graphs. We give formal definitions and more information about each subclass later in their respective sections. See Brandstädt and Dragan [4] for a summary of known properties of acyclic hypergraphs as well as their relations to various graph classes.

1.2 Union Join Graph

Note that the join tree of an acyclic hypergraph is not always unique. For example, each tree with n nodes is a valid join tree for the hypergraph formed by $\{\{0, 1\}, \{0, 2\}, \ldots, \{0, n\}\}$. The *union join graph* G of a given acyclic hypergraph H is the union of all its join trees. That is, each vertex of G represents a hyperedge of H and two vertices of G are adjacent if there exits a join tree T for H such that the corresponding hyperedges are adjacent in T. The union join graph of a hypergraph H may also be called *clique graph* if H represents the maximal cliques of a chordal graph [11,14], or *atom graph* if H represents the atoms of some graph [15]. In [2], Berry and Simonet present algorithms which compute the union join graph of an acyclic hypergraph in $\mathcal{O}(Nm)$ time.

1.3 Subset Graph

The *subset graph* of a hypergraph H is a directed graph G where each vertex represents a hyperedge of H and there is a directed edge from a vertex u to a vertex v if the hyperedge corresponding to u is a subset of the hyperedge corresponding to v. Pritchard presents an algorithm in [20] that computes the subset graph for a given hypergraph in $\mathcal{O}(N^2/\log N)$ time. They also show that any subset graph has at most $\mathcal{O}(N^2/\log^2 N)$ many edges. There are various publications that present algorithms for special cases and different computational models; see for example [9,19] and the work cited therein.

The *Strong Exponential Time Hypothesis*, *SETH* for short, states that there is no algorithm that solves the Boolean satisfiability problem (without limitation on clause size) for some constant $\varepsilon > 0$ in $\mathcal{O}((2-\varepsilon)^n)$ time where n is the number of variables in the given instance. A function $f(n)$ is called *truly subquadratic* if $f(n) \in \mathcal{O}(n^{2-\varepsilon})$ for some constant $\varepsilon > 0$. Borassi et al. [3] show that, if SETH holds, then there is no algorithm to compute the subset graph of an arbitrary hypergraph in truly subquadratic time, even if the output is sparse. Note that the results in [3] and [20] are not conflicting, since $N^{2-\varepsilon} \in o(N^2/\log N)$.

1.4 Our Contribution

In this paper, we investigate the two problems of computing the union join graph as well as computing the subset graph for acyclic hypergraphs and their subclasses. We show in Sect. 3 that there is a close relation between both problems by presenting reductions in both directions. It then follows that the result by Borassi et al. still holds when restricted to α-acyclic hypergraphs and also applies to computing a union join graph. We then develop efficient algorithms to solve both problems for acyclic hypergraphs and their subclasses. In particular, we show that, if $|G|$ denotes the size of the computed graph G, then both problems can be solved in $\mathcal{O}(N^2/\log N + |G|)$ time for α-acyclic hypergraphs (Sect. 3.2), in $\mathcal{O}(N \log(n+m) + |G|)$ time for β-acyclic hypergaphs (Sect. 4.1), and in $\mathcal{O}(N+|G|)$ time for γ-acyclic hypergraphs (Sect. 4.2) as well as for interval hypergraphs (Sect. 4.3).

2 Preliminaries

Let $H = (V, \mathcal{E})$ be a hypergraph. The *incidence graph* $\mathcal{I}(H) = (U_V \cup U_{\mathcal{E}}, E_{\mathcal{I}})$ of H is a bipartite graph were U_V represents the vertices of H, $U_{\mathcal{E}}$ represents the hyperedges of H, and there is an edge between two vertices $u_v \in U_V$ and $u_E \in U_{\mathcal{E}}$ if the corresponding vertex v (of H) is in the corresponding hyperedge E. That is, $U_V = \{ u_v \mid v \in V \}$, $U_{\mathcal{E}} = \{ u_E \mid E \in \mathcal{E} \}$, and $E_{\mathcal{I}} = \{ u_v u_E \mid v \in E \}$. Note that $|E_{\mathcal{I}}| = N$. If not stated or constructed otherwise, the incidence graphs of all hypergraphs occurring in this paper are connected, finite, undirected, and without multiple edges. Additionally, whenever a hypergraph is given, it is given as its incidence graph; hence, the input size is in $\Theta(N)$. We say two hyperedges of H are *distinct* if they are represented by two different vertices in $\mathcal{I}(H)$, even if both hyperedges contain the same vertices.

A sequence $\langle v_1, v_2, \ldots, v_k \rangle$ of vertices of H *forms a path* in H if, for each i with $1 \leq i < k$, H contains a hyperedge E with $v_i, v_{i+1} \in E$. Let X, Y, and Z be sets of vertices of H. X *separates* Y form Z if $X \neq \emptyset$ and each sequence of vertices that forms a path from Y to Z in H contains a vertex from X.

Let T be the join tree of some acyclic hypergraph H and let E_i and E_j be two hyperedges of H which are adjacent in T. We then call the set $S = E_i \cap E_j$ a *separator* of H with respect to T. If T is rooted and E_i is the parent of E_j, we call $S^{\uparrow}(E_j) := E_i \cap E_j$ the *up-separator* of E_j. Note that each separator corresponds to an edge of T and vice versa. We call the hypergraph formed by the set of all separators of H the *separator hypergraph* $\mathcal{S}(H)$ for H with respect to T. It follows from properties (ii) and (iii) of Lemma 5 (see Sect. 3) that $\mathcal{S}(H)$ is always the same for a given H, independent of the used join tree.

3 α-Acyclic Hypergraphs

In this section, we investigate the problems of computing a union join graph and computing a subset graph for the most general case of acyclic hypergraphs. We first show that computing these graphs cannot be done in truly subquadratic time if the SETH is true. For that, we use a problem called *Sperner Family* problem. It asks whether a family of sets contains two sets S and S' such that $S \subseteq S'$. If the SETH is true, then there is no algorithm that solves it truly subquadratic time [3]. Afterwards, we give an algorithm that allows to quickly compute the union join graph if a fast algorithm for the subset graph problem is given.

3.1 Hardness Results

Let $\mathcal{F} = \{S_1, S_2, \ldots, S_m\}$ be a family of sets. We create an acyclic hypergraph H from $\mathcal{F}$ as follows. Create a new vertex u (i.e., u is not contained in any set S_i) and, for each set S_i, create a hyperedge $E_i = S_i \cup \{u\}$. Additionally, create a hyperedge $\mathcal{S}$ which is the union of all hyperedges E_i. Formally, we have that $H = (V, \mathcal{E})$ with $V = \mathcal{S}$ and $\mathcal{E} = \{ E_i \mid S_i \in \mathcal{F} \} \cup \{\mathcal{S}\}$. One can create a join tree T for H by starting with $\mathcal{S}$ and then making each hyperedge E_i adjacent to it. Thus, H is acyclic. Note that one can create H and T from $\mathcal{F}$ in linear time.

For the remainder of this subsection, assume that we are given a family $\mathcal{F}$, a hypergraph H, and a corresponding join tree T for H as defined above. Our results in this subsection are based on the following observation.

Lemma 1. *$\mathcal{F}$ contains two distinct sets S_i and S_j with $S_i \subseteq S_j$ if and only if there is a join tree for H that contains the edge E_iE_j.*

Proof. First, assume that $\mathcal{F}$ contains two distinct sets S_i and S_j with $S_i \subseteq S_j$. In that case, we can create a new join tree T' as follows. Remove the edge $E_i\mathcal{S}$ from T and make E_i adjacent to E_j instead. Since $S_i \subseteq S_j$, each element $x \in E_i \cap \mathcal{S}$ is also contained in E_j. Thus, T' is a join tree for H and contains the edge E_iE_j.

Next, assume that there is a join tree T' for H with the edge E_iE_j. Without loss of generality, let E_j be closer to $\mathcal{S}$ in T' than E_i. Recall that $E_i \subseteq \mathcal{S}$. Therefore, by properties of join trees, each vertex in E_i is also in E_j. It then directly follows from the construction of H that $S_i \subseteq S_j$. □

We use the Sperner Family problem to show that there is no truly subquadratic-time algorithm to compute the union join graph of a given acyclic hypergraph. To do so, we first show the following.

Lemma 2. *If the SETH is true, then there is no algorithm which decides in $\mathcal{O}(N^{2-\varepsilon})$ time whether or not a given acyclic hypergraph has a unique join tree.*

Proof. Recall that we can create a join tree T for H by making each hyperedge E_i adjacent to the hyperedge $\mathcal{S}$. To prove Lemma 2, we show that $\mathcal{F}$ contains two distinct sets S_i and S_j with $S_i \subseteq S_j$ if and only if T is not a unique join tree for H.

First, assume that $\mathcal{F}$ contains two such sets S_i and S_j. In that case, Lemma 1 implies that there is a join tree T' for H with the edge E_iE_j. Since E_iE_j is not an edge in T, T is not unique. Next, assume that T is not unique. Then, there is a join tree T' and a hyperedge E_i such that E_i is not adjacent to $\mathcal{S}$ in T'. Hence, E_i is adjacent to some hyperedge E_j that is closer to $\mathcal{S}$ in T' than E_i. Since $E_i \subseteq \mathcal{S}$, properties of join trees imply that $E_i \subseteq E_j$. Subsequently, due to Lemma 1, $S_i \subseteq S_j$.

It follows that a truly subquadratic-time algorithm which determines if an acyclic hypergraph has a unique join tree would imply an equally fast algorithm to solve the Sperner Family problem for any family of sets. □

Note that, by definition of a union join graph, H has a unique join tree if and only if the union join graph of H is a tree. Therefore, we get the following.

Theorem 3. *If the SETH is true, then there is no algorithm which constructs the union join graph of a given acyclic hypergraph in $\mathcal{O}(N^{2-\varepsilon})$ time, even if that graph is sparse.*

We now show that computing the subset graph of an acyclic hypergraph is as hard as computing the subset graph for a general family of sets.

Theorem 4. *If the SETH is true, then there is no algorithm which constructs the subset graph of a given acyclic hypergraph in truly subquadratic time.*

Proof. Let G be the subset graph for H and $G_{\mathcal{F}}$ be the subset graph for $\mathcal{F}$. Since, by construction of H, $E_i \subseteq E_j$ if and only if $S_i \subseteq S_j$, G contains the edge (E_i, E_j) if and only if $G_{\mathcal{F}}$ contains the edge (S_i, S_j). We can therefore construct $G_{\mathcal{F}}$ from G by simply removing the vertex representing $\mathcal{S}$ from G (and its incident edges).

Recall that we can construct H from $\mathcal{F}$ in linear time. Therefore, a truly subquadratic-time algorithm to construct the subset graph of a given acyclic hypergraph would imply an equally fast algorithm to construct a subset graph of a given family of sets. □

3.2 Union Join Graph via Subset Graph

In the previous subsection, we show how to compute the subset graph using the union join graph of an acyclic hypergraph. We now present an algorithm that computes the union join graph of a given acyclic hypergraph with the help of a subset graph. The runtime of our algorithm then depends on the runtime required to compute that subset graph.

For the remainder of this subsection, assume that we are given an acyclic hypergraph $H = (V, \mathcal{E})$ and let G be the union join graph of H (with for us unknown edges). Lemma 5 below gives various characterisations for G.

Lemma 5. *For any distinct $E_i, E_j \in \mathcal{E}$, the following are equivalent.*

(i) *E_iE_j is an edge of G.*
(ii) *H has a join tree with the edge E_iE_j.*
(iii) *Each join tree T of H has an edge $E_i'E_j'$ on the path from E_i to E_j in T such that $E_i \cap E_j = E_i' \cap E_j'$.*
(iv) *Each join tree T of H has a separator S on the path P_{ij} from E_i to E_j in T with $S \subseteq S_i$ and $S \subseteq S_j$ where S_i and S_i are the separators in P_{ij} which are respectively closest to E_i and E_j.*
(v) *$E_i \cap E_j$ separates $E_i \setminus E_j$ from $E_j \setminus E_i$.*

Most of the properties in Lemma 5 repeat, generalise, or paraphrase existing results (see [2,11,14]). Property (iv) is, to the best of our knowledge, a new observation. For completeness, however, we prove all of them.

Proof. By definition of G, properties (i) and (ii) are equivalent. It follows from properties of join trees that (ii) implies (v).

We next show that (v) implies (iii). Assume that E_i and E_j are not adjacent in a join tree T. Then there is a path $\langle E_i = X_1, X_2, \ldots, X_k = E_j \rangle$ of hyperedges from E_i to E_j in T. For each p with $1 \leq p < k$, let $S_p = X_p \cap X_{p+1}$ be the separator corresponding to the edge X_pX_{p+1} of T. By properties of join trees, $E_i \cap E_j \subseteq S_p$ for each S_p. Now assume that each S_p contains a vertex $v_p \notin E_i \cap E_j$. Then, $\langle v_1, v_2, \ldots, v_{k-1} \rangle$ would form a path in H from $v_1 \in E_i \setminus E_j$ to $v_{k-1} \in E_j \setminus E_i$. That contradicts with property (v). Therefore, there is at least one separator S_p with $S_p \subseteq E_i \cap E_j$, i.e., there is an edge X_pX_{p+1} in T with $E_i \cap E_j = X_p \cap X_{p+1}$.

To show that (iii) implies (ii), consider a join tree T where E_i and E_j are not adjacent. We can create a join tree T' by removing the edge $E_i'E_j'$ and adding the edge E_iE_j instead. Since E_i and E_j are on different sides of $E_i'E_j'$ in T, T' is also a tree. Additionally, because $E_i \cap E_j = E_i' \cap E_j'$, T' is a valid join tree for H.

It remains to show that (iv) is equivalent to (iii). We first assume property (iii). Let $S = E_i \cap E_j$ be a separator on the path from E_i to E_j in some join tree T. Since, by properties of join trees, each vertex in $S = E_i \cap E_j$ is also in S_i and S_j, it follows that $S \subseteq S_i$ and $S \subseteq S_j$. Now assume property (iv). Because $S \subseteq S_i \subseteq E_i$ and $S \subseteq S_j \subseteq E_j$, it is also the case that $S \subseteq E_i \cap E_j$. Since S is on the path from E_i to E_j in T, each vertex that is in both E_i and E_j also has to be in S, i.e., $S \supseteq E_i \cap E_j$. Therefore, $S = E_i \cap E_j$. □

Based on Lemma 5, we can construct G as follows. Compute a join tree T for H, the separator hypergraph $\mathcal{S}(H)$ (with respect to T), and its subset graph $G_{\mathcal{S}}$. Next, use $G_{\mathcal{S}}$ to find all triples S_i, S_j, S of separators which satisfy property (iv) of Lemma 5. Since their corresponding hyperedges are then adjacent in some join tree of H, make the corresponding vertices adjacent in G.

Before analysing our approach further, we address some needed preprocessing. Assume that H contains two hyperedges E_i and E_j which are not adjacent in T, but are adjacent in some other join tree. There might then be multiple separators S on the path from E_i to E_j in T which satisfy property (iv) of Lemma 5. Our algorithm would, therefore, add the edge E_iE_j to G multiple times, once for each such S. While it is easy to remove redundant edges from G afterwards, we still want to ensure that the time needed to create and remove these edges does not become too much. To achieve that, Algorithm 1 modifies T such that each hyperedge becomes adjacent to its highest possible ancestor in T. As by-product, Algorithm 1 also computes the up-separator of each hyperedge (and, thus, the separator hypergraph $\mathcal{S}(H)$).

Algorithm 1. Modifies the join tree of a given acyclic hypergraph such that each hyperedge becomes adjacent to its highest possible ancestor.

Input: An acyclic hypergraph $H = (V, \mathcal{E})$ and a join tree T for H.
Output: A modified join tree T' for H and the separator hypergraph $\mathcal{S}(H)$.

1 Root T in an arbitrary hyperedge R and then run a pre-order on T. Let $\sigma = \langle R = E_1, E_2, \ldots, E_m \rangle$ be the resulting order.
2 For each vertex v, set $\lambda(v) := \min\{\, i \mid v \in E_i \,\}$.
3 **for** $i := 2$ **to** m **do**
4 | Set $S^{\uparrow}(E_i) := \{\, v \in E_i \mid \lambda(v) < i \,\}$.
5 | Let $j = \max\{\, \lambda(v) \mid v \in S^{\uparrow}(E_i) \,\}$ and make E_j the parent of E_i.
6 Let $\mathcal{S}(H)$ be the hypergraph formed by the family $\{\, S^{\uparrow}(E_i) \mid E_i \in \mathcal{E}, E_i \neq R \,\}$.

Lemma 6. *Algorithm 1 runs in linear time.*

Proof. Line 1 runs in $\mathcal{O}(m)$ time, since the nodes of T are the hyperedges of H. Recall that H is given as an incidence graph $\mathcal{I}(H)$. Hence, the following are equivalent (with respect to runtime): (i) for each vertex, iterating over all hyperedges containing it; (ii) for each hyperedge, iterating over all vertices it contains; and (iii) iterating over all edges of $\mathcal{I}(H)$. Therefore, line 2, line 4, and line 5 (and subsequently Algorithm 1) run in $\mathcal{O}(N)$ total time. □

Lemma 7. *The tree T' created by Algorithm 1 is a valid join tree for H.*

Proof. Let T_i be the tree after processing E_i, i.e., $T = T_1$ and $T_m = T'$. Thus, T_1 is a valid join tree for H. Assume, by induction, that T_{i-1} (with $i \geq 2$) is a valid join tree for H too. Recall that, by definition of join trees, the set of hyperedges containing a vertex v form a subtree T_v of T. The roots of all such T_v where $v \in S^{\uparrow}(E_i)$ are ancestors of E_i in T and, thus, form a path. By definition of j (line 5), E_j is the lowest of such roots in T. It therefore follows that $S^{\uparrow}(E_i) \subseteq E_j$. Subsequently, for each $v \in S^{\uparrow}(E_i)$, the hyperedges containing v still form a subtree of T_i after changing the parent of E_i if they did so in T_{i-1}. Note that each subtree T_u of a vertex $u \notin S^{\uparrow}(E_i)$ remains unchanged, since it does not contain the edge $E_i E_k$. Therefore, for each vertex, the hyperedges containing it form a subtree of T_i and, thus, T_i is a join tree for H. □

Lemma 8. *Let E_i and E_j be two hyperedges of H, T' be the tree computed by Algorithm 1, and P_{ij} be the path from E_i to E_j in T'. Additionally, let S_i and S_j be the separators on P_{ij} which are closest to E_i and E_j, respectively. There are at most two separators S on P_{ij} such that $S \subseteq S_i$ and $S \subseteq S_j$.*

Proof. Let E_k be the lowest common ancestor of E_i and E_j in T'. Although T' has a potentially different structure than T, it is still the case that the parent of a hyperedge in T' was an ancestor of it in T. Thus, $k \leq i, j$. Note that P_{ij} goes through E_k and let P_{ik} and P_{kj} be the respective subpaths of P_{ij}. If P_{ij} contains more than two separators S as defined in Lemma 8, at least two of them are either part of P_{ik} or P_{kj}. Without loss of generality, let them be on P_{kj} and let S be the lowest such separator. Additionally, let X be the hyperedge directly below S, i.e., $S^{\uparrow}(X) = S$. It follows that X is not adjacent to E_k in T'.

Since $S \subseteq S_i$, each vertex in S is in all hyperedges on the path from X to E_i in T', including E_k. Therefore, $S \subseteq E_k$ and $\max\{\, \lambda(v) \mid v \in S \,\} \leq k$. That is a contradiction, since Algorithm 1 would have made X adjacent to E_k or one of its ancestors. □

Algorithm 2 now implements the approach described above. It also uses Algorithm 1 as preprocessing. Therefore, due to Lemma 8, the algorithm adds each edge $E_i E_j$ at most two times into G.

Theorem 9. *Algorithm 2 computes the union join graph G of a given acyclic hypergraph H in $\mathcal{O}(T_{\mathcal{A}}(H) + N + |G|)$ time where $T_{\mathcal{A}}(H)$ is the runtime of a given algorithm $\mathcal{A}$ with the separator hypergraph of H as input.*

Algorithm 2. Computes the union join graph of an acyclic hypergraph.

Input: An acyclic hypergraph $H = (V, \mathcal{E})$ and an algorithm $\mathcal{A}$ that computes the subset graph for a given family of sets.
Output: The union join graph G of H.

1 Find a join tree for H (see [21]) and call Algorithm 1. Let T be the resulting join tree and $\mathcal{S}$ the resulting family of separators (i.e., the hyperedges of $\mathcal{S}(H)$).
2 Use algorithm $\mathcal{A}$ to compute the subset graph $G_{\mathcal{S}}$ of $\mathcal{S}$.
3 Create a new graph $G = (\mathcal{E}, E_G)$ with $E_G = \emptyset$.
4 **foreach** $S \in \mathcal{S}$ **do**
5 | Use $G_{\mathcal{S}}$ to determine all separators S' with $S \subseteq S'$ (including S itself).
6 | For each such S', let EE' be the edge of T which S' represents and let E be the hyperedge farther away from S in T. Add E to a set $\mathbb{E}$ of hyperedges. If S and S' represent the same edge of T, also add E'.
7 | Partition $\mathbb{E}$ into two sets $\mathbb{E}_1$ and $\mathbb{E}_2$ based on which side of S they are in T.
8 | For each pair E_1, E_2 with $E_1 \in \mathbb{E}_1$ and $E_2 \in \mathbb{E}_2$, add E_1E_2 into E_G.

Proof (Correctness). Let E_i and E_j be two hyperedges of H. Additionally, let S_i and S_j be the separators on the path from E_i to E_j in T (computed in line 1) which are closest to E_i and E_j, respectively. We show the correctness of Algorithm 2 by showing that E_iE_j is an edge of G if and only if there is a join tree for H with the edge E_iE_j.

First, assume that there is a join tree for H with the edge E_iE_j. Lemma 5 then implies that there is a separator $S \in \mathcal{S}$ such that $S \subseteq S_i$, $S \subseteq S_j$, and E_i and E_j are on different sides of S in T. Therefore, when processing S, the algorithm finds S_i and S_j (line 5) and consequently adds E_i and E_j into $\mathbb{E}$ (line 6). Since both hyperedges are on different sides of S, Algorithm 2 then also adds the edge E_iE_j to G (line 8).

We now assume that E_iE_j is an edge of G. Note that Algorithm 2 only adds edges to G in line 8. Thus, there is a separator $S \in \mathcal{S}$ for which the algorithm adds E_iE_j to G. For that S, one of E_i and E_j is in $\mathbb{E}_1$ and the other is in $\mathbb{E}_2$ (line 8) and, hence, E_i and E_j are on different sides of S in T (line 7). This implies that $S \subseteq S_i$ and $S \subseteq S_j$ (line 5 and line 7). Therefore, by Lemma 5, there is a join tree for H with the edge E_iE_j. □

Proof (Complexity). Creating a join tree for a given acyclic hypergraph H can be implemented in $\mathcal{O}(N)$ time [21]. Modifying that join tree (thereby computing T) and computing $\mathcal{S}(H)$ using Algorithm 1 can also be done in $\mathcal{O}(N)$ time (Lemma 6). Thus, line 1 runs in total $\mathcal{O}(N)$ time. Computing the subset graph $G_{\mathcal{S}}$ in line 2 requires $\mathcal{O}(T_{\mathcal{A}}(H))$ time. Since the hyperedges of H form the vertices of G and since G is created without edges, line 3 runs in $\mathcal{O}(m)$ time.

We show next that a single iteration of the loop starting in line 4 runs in $\mathcal{O}(|\mathbb{E}_1| \cdot |\mathbb{E}_2|)$ time. That is, the runtime for a single iteration is (asymptotically) equivalent to the number of edges of G created. Note that each iteration creates at least one such edge, namely the edge in T that S represents. Additionally,

Lemma 5 and Lemma 8 imply that each edge E_iE_j is added at most twice to G. Therefore, line 4 to line 8 run in $\mathcal{O}(|G|)$ total time.

For a separator $S \in \mathcal{S}$, let $\mathbb{S}$ denote the set of separators S' with $S \subseteq S'$. Since the subset graph $G_{\mathcal{S}}$ is given, one can compute $\mathbb{S}$ (line 5) in $\mathcal{O}(|\mathbb{S}|)$ time by determining all incoming edges of S in $G_{\mathcal{S}}$. For each $S' \in \mathbb{S}$, the algorithm adds, in line 6, exactly one hyperedge into $\mathbb{E}$ plus one additional hyperedge for S. Thus, $|\mathbb{E}| = |\mathbb{S}| + 1$.

One can determine the hyperedges E and E' that form a separator S', which one is farther from S, and on which side of S they are in T as follows. When creating S', add a reference to both hyperedges and include which is the parent and which is the child in T. Now assume that each S' is also a node of T adjacent to E and E'. Root T in an arbitrary hyperedge, run a pre-order and post-order on T, and let $\text{pre}(x)$ and $\text{post}(x)$ be the indices of a node x in that respective order. For two distinct nodes x and y of T (representing either separators or hyperedges), x is then a descendant of y if and only if $\text{pre}(x) > \text{pre}(y)$ and $\text{post}(x) < \text{post}(y)$. There are four cases when determining which of E and E' to add into $\mathbb{E}$: if S and S' represent the same edge of T, add both hyperedges; if S' is a descendant of S, add the child-hyperedge; if S' is an ancestor of S, add the parent-hyperedge; and if S' is neither an ancestor nor a descendant of S, add the child-hyperedge. Clearly, one side of S contains all its descendants and the other side all remaining hyperedges and separators. That allows us, after a $\mathcal{O}(m)$-time preprocessing, to determine in constant time on which side of S a give a hyperedge is. Therefore, line 6 and line 7 run in $\mathcal{O}(|\mathbb{E}|)$ time.

Line 8 clearly runs in $\mathcal{O}(|\mathbb{E}_1| \cdot |\mathbb{E}_2|)$ time. Recall that $|\mathbb{S}| + 1 = |\mathbb{E}| = |\mathbb{E}_1| + |\mathbb{E}_2|$. Therefore, a single iteration of the loop starting in line 4 also runs in $\mathcal{O}(|\mathbb{E}_1| \cdot |\mathbb{E}_2|)$ time. □

Recall that there is an algorithm which computes the subset graph for any given hypergraph in $\mathcal{O}(N^2/\log N)$ time [20]. Thus, we have the following.

Theorem 10. *There is an algorithm that computes the union join graph G of an acyclic hypergraph in $\mathcal{O}(N^2/\log N + |G|)$ time.*

The upper bound of at most $\Theta(N^2/\log^2 N)$ many edges for any subset graph [20] does not apply to union join graphs. Consider a hypergraph $H = (V, \mathcal{E})$ with $V = \{u, v_1, \ldots, v_n\}$ and $\mathcal{E} = \{ E_i \mid 1 \leq i \leq n \}$ where $E_i = \{u, v_i\}$. Note that $N = 2n$ and that each tree with $\mathcal{E}$ as nodes is a valid join tree for H. Hence, the union join graph of H is a complete graph with $\Theta(N^2)$ edges.

4 Subclasses of Acyclic Hypergraphs

In this section, we summarise our results for subclasses of acyclic hypergraphs. Detailed discussions of these results are omitted due to space limitations. A pre-print of the full paper is available online [17].

4.1 β-Acyclic Hypergraphs

A hypergraph $H = (V, \mathcal{E})$ is *β-acyclic* if each subset of $\mathcal{E}$ forms an acyclic hypergraph. See [10] for more definitions.

A matrix is *binary* if its entries are either 0 or 1. One can use a binary $n \times m$ matrix M to represent a given hypergraph $H = (V, \mathcal{E})$ as follows. Let each row i represent a vertex $v_i \in V$ and each column j represent a hyperedge $E_j \in \mathcal{E}$. An entry $M_{i,j}$ is then 1 if and only if $v_i \in E_j$. That matrix is called the *incidence matrix* of H. A matrix is *doubly lexically ordered* if rows and columns are permuted in such a way that rows vectors and columns are both in non-decreasing lexicographic order (rows from left to right and columns from top to bottom). Within a row, priorities of entries are decreasing from right to left, and, within a column, priorities of entries are decreasing from bottom to top. One can compute such an ordering in $\mathcal{O}(N \log(n + m))$ time [18].

Assume now that we are given a β-acyclic hypergraph H and a doubly lexical ordering σ for some incidence matrix M for H, even though we are not given M itself. For two hyperedges E_i and E_j of H, we say $E_i \preceq E_j$ if the column of E_i is lexicographically smaller than or equal to the column of E_j with respect to σ. Then, we can observe the following.

Lemma 11. *Let E_i and E_j be two hyperedges of H and let v be the vertex in E_i which is earliest in the doubly lexical ordering (i.e., highest in M). Then, $E_i \subseteq E_j$ if and only if $E_i \preceq E_j$ and $v \in E_j$.*

Lemma 11 allows to compute the subset graph G of a β-acyclic hypergraph as follows. First, find doubly lexicographical ordering of vertices and hyperedges. For each hyperedge E, determine all hyperedges E' with $E \preceq E'$ which contain v as defined in Lemma 11, and add the edge (E, E') into G.

Theorem 12. *There is an algorithm that computes the subset graph G of a given β-acyclic hypergraph in $\mathcal{O}(N \log(n + m) + |G|)$ time.*

Using Theorem 12 together with Algorithm 2 allows us to conclude this subsection as follows.

Theorem 13. *If a hypergraph is β-acyclic, then its separator hypergraph is β-acyclic, too. Therefore, there is an algorithm that computes the union join graph G of a given β-acyclic hypergraph in $\mathcal{O}(N \log(n + m) + |G|)$ time.*

4.2 γ-Acyclic Hypergraphs

A hypergraph is *γ-acyclic* if, for all distinct hyperedges E_i and E_j, $E_i \cap E_j \neq \emptyset$ implies $E_i \cap E_j$ separates $E_i \setminus E_j$ from $E_j \setminus E_i$. Fagin [10] gives various other definitions. The *line graph* $L(H)$ of a hypergraph H is the intersection graph of its hyperedges. That is, $L(H) = (\mathcal{E}, \mathcal{E}_L)$ with $\mathcal{E}_L = \{\, E_iE_j \mid E_i, E_j \in \mathcal{E}; E_i \cap E_j \neq \emptyset \,\}$. Based on these definitions and Lemma 5, we can observe the following.

Theorem 14. *An acyclic hypergraph is γ-acyclic if and only if its line graph is isomorphic to its union join graph. Therefore, there is an algorithm that computes the union join graph G of a given γ-acyclic hypergraph in $\mathcal{O}(N + |G|)$ time.*

Consider a hypergraph $H = (V, \mathcal{E})$, let $\mathcal{E}'$ be a subset of $\mathcal{E}$, and let $\mathfrak{X}$ be the intersection of all hyperedges in $\mathcal{E}'$. We then define $\mathcal{X}$ as the set of all such $\mathfrak{X}$ which are non-empty, i.e., $\mathcal{X} = \bigcup_{\mathcal{E}' \subseteq \mathcal{E}} \left\{ \mathfrak{X} \;\middle|\; \mathfrak{X} = \bigcap_{E \in \mathcal{E}'} E, \mathfrak{X} \neq \emptyset \right\}$. The *Bachman diagram* $\mathcal{B}(H)$ of H is a directed graph with the node set $\mathcal{X}$ such that there is an edge from $\mathfrak{X}$ to $\mathfrak{Y}$ if $\mathfrak{X} \supset \mathfrak{Y}$ and there is no $\mathfrak{Z}$ with $\mathfrak{X} \supset \mathfrak{Z} \supset \mathfrak{Y}$. Note that, if H contains two distinct hyperedges E_i and E_j with the same vertices, they are represented by the same node in $\mathcal{B}(H)$. It is known [10] that a hypergraph is γ-acyclic if and only if its Bachman diagram forms a tree.

Let $\phi(E)$ be the node of $\mathcal{B}(H)$ which represents the hyperedge E. We can then make the following observation: For two hyperedges E_i and E_j of H, $E_i \subseteq E_j$ if and only if there is a path from $\phi(E_j)$ to $\phi(E_i)$ in $\mathcal{B}(H)$. Using a technique from [22] and a strong connection between γ-acyclic hypergraphs and distance-hereditary graphs allow us to compute a simplified version of a Bachman diagram in $\mathcal{O}(N)$ time. Together with the observation above, we are then able to achieve the following result.

Theorem 15. *There is an algorithm that computes the subset graph G of a given γ-acyclic hypergraph in $\mathcal{O}(N + |G|)$ time.*

4.3 Interval Hypergraphs

An acyclic hypergraph $H = (V, \mathcal{E})$ is an *interval hypergraph* if it admits a join tree that forms a path. That is, there is an order $\sigma = \langle E_1, E_2, \ldots, E_m \rangle$ for the hyperedges of H such that, for each vertex $v \in V$, $v \in E_i \cap E_j$ implies that $v \in E_k$ for all k with $i \leq k \leq j$. One can recognise interval hypergraphs and compute a corresponding order σ in linear time [13].

To compute the subset graph and union join graph, we first determine for each vertex v the index $\phi(v)$ of the left-most hyperedge containing it (with respect to σ). Next, we compute the separators between consecutive hyperedges (see Algorithm 1). Let S_i denote the separator between E_{i-1} and E_i and let $\phi(S_i) = \max_{v \in S_i} \phi(v)$. Then, for each E_j with $j < i$, it holds that (i) $E_j \supseteq E_i$ if and only if $|E_i| = |S_i|$ and $j \geq \phi(S_i)$, and (ii) $E_i E_j$ is an edge of the union join graph of H if and only if $j \geq \phi(S_i)$. Running the same approach again using the reverse of σ therefore allows to compute the subset graph and union join graph in $\mathcal{O}(N + |G|)$ time.

Theorem 16. *There are algorithms that compute the union join graph and subset graph, respectively, of a given interval hypergraph in $\mathcal{O}(N + |G|)$ time where $|G|$ is the size of the computed graph.*

Acknowledgements. We would like to thank Feodor F. Dragan and Rachel Walker for stimulating discussions.

References

1. Beeri, C., Fagin, R., Maier, D., Yannakakis, M.: On the desirability of acyclic database schemes. J. ACM **30**(3), 479–513 (1983). https://doi.org/10.1145/2402.322389
2. Berry, A., Simonet, G.: Computing the atom graph of a graph and the union join graph of a hypergraph. CoRR abs/1607.02911 (2016)
3. Borassi, M., Crescenzi, P., Habib, M.: Into the square: on the complexity of some quadratic-time solvable problems. Electron. Notes Theor. Comput. Sci. **322**, 51–67 (2016). https://doi.org/10.1016/j.entcs.2016.03.005
4. Brandstädt, A., Dragan, F.F.: Tree-structured graphs. In: Handbook of Graph Theory, Combinatorial Optimization, and Algorithms, pp. 751–826. CRC Press (2015)
5. Brandstädt, A., Dragan, F.F., Chepoi, V., Voloshin, V.I.: Dually chordal graphs. SIAM J. Discret. Math. **11**(3), 437–455 (1998). https://doi.org/10.1137/S0895480193253415
6. Dourisboure, Y.: Compact routing schemes for generalised chordal graphs. J. Graph Algorithms Appl. **9**(2), 277–297 (2005). https://doi.org/10.7155/jgaa.00109
7. Dourisboure, Y., Dragan, F.F., Gavoille, C., Yan, C.: Spanners for bounded tree-length graphs. Theoret. Comput. Sci. **383**(1), 34–44 (2007). https://doi.org/10.1016/j.tcs.2007.03.058
8. Dragan, F.F., Köhler, E.: An approximation algorithm for the tree t-spanner problem on unweighted graphs via generalized chordal graphs. Algorithmica **69**(4), 884–905 (2013). https://doi.org/10.1007/s00453-013-9765-4
9. Elmasry, A.: Computing the subset partial order for dense families of sets. Inf. Process. Lett. **109**(18), 1082–1086 (2009). https://doi.org/10.1016/j.ipl.2009.07.001
10. Fagin, R.: Degrees of acyclicity for hypergraphs and relational database schemes. J. ACM **30**(3), 514–550 (1983). https://doi.org/10.1145/2402.322390
11. Galinier, P., Habib, M., Paul, C.: Chordal graphs and their clique graphs. In: Nagl, M. (ed.) WG 1995. LNCS, vol. 1017, pp. 358–371. Springer, Heidelberg (1995). https://doi.org/10.1007/3-540-60618-1_88
12. Gavril, F.: The intersection graphs of subtrees in trees are exactly the chordal graphs. J. Comb. Theory Ser. B **16**(1), 47–56 (1974). https://doi.org/10.1016/0095-8956(74)90094-X
13. Habib, M., McConnell, R.M., Paul, C., Viennot, L.: LEX-BFS and partition refinement, with applications to transitive orientation, interval graph recognition and consecutive ones testing. Theor. Comput. Sci. **234**(1–2), 59–84 (2000). https://doi.org/10.1016/S0304-3975(97)00241-7
14. Habib, M., Stacho, J.: Reduced clique graphs of chordal graphs. Eur. J. Comb. **33**(5), 712–735 (2012). https://doi.org/10.1016/j.ejc.2011.09.031
15. Kaba, B., Pinet, N., Lelandais, G., Sigayret, A., Berry, A.: Clustering gene expression data using graph separators. Silico Biol. **7**(4–5), 433–452 (2007)
16. Leimer, H.: Optimal decomposition by clique separators. Discret. Math. **113**(1–3), 99–123 (1993). https://doi.org/10.1016/0012-365X(93)90510-Z
17. Leitert, A.: Computing the union join and subset graph of acyclic hypergraphs in subquadratic time. CoRR abs/2104.06636 (2021)
18. Paige, R., Tarjan, R.E.: Three partition refinement algorithms. SIAM J. Comput. **16**(6), 973–989 (1987). https://doi.org/10.1137/0216062

19. Pritchard, P.: Opportunistic algorithms for eliminating supersets. Acta Inform. **28**(8), 733–754 (1991). https://doi.org/10.1007/BF01261654
20. Pritchard, P.: On computing the subset graph of a collection of sets. J. Algorithms **33**(2), 187–203 (1999). https://doi.org/10.1006/jagm.1999.1032
21. Tarjan, R.E., Yannakakis, M.: Simple linear-time algorithms to test chordality of graphs, test acyclicity of hypergraphs, and selectively reduce acyclic hypergraphs. SIAM J. Comput. **13**(3), 566–579 (1984). https://doi.org/10.1137/0213035
22. Uehara, R., Uno, Y.: Laminar structure of ptolemaic graphs with applications. Discret. Appl. Math. **157**(7), 1533–1543 (2009). https://doi.org/10.1016/j.dam.2008.09.006

Algorithms for the Line-Constrained Disk Coverage and Related Problems

Logan Pedersen and Haitao Wang(✉)

Department of Computer Science, Utah State University, Logan, UT 84322, USA
logan.pedersen@aggiemail.usu.edu, haitao.wang@usu.edu

Abstract. Given a set P of n points and a set S of m weighted disks in the plane, the disk coverage problem asks for a subset of disks of minimum total weight that cover all points of P. The problem is NP-hard. In this paper, we consider a line-constrained version in which all disks are centered on a line L (while points of P can be anywhere in the plane). We present an $O((m+n)\log(m+n)+\kappa\log m)$ time algorithm for the problem, where κ is the number of pairs of disks that intersect. For the unit-disk case where all disks have the same radius, the running time can be reduced to $O((n+m)\log(m+n))$. In addition, we solve in $O((m+n)\log(m+n))$ time the L_∞ and L_1 cases of the problem, in which the disks are squares and diamonds, respectively. Using our techniques, we further solve two other geometric coverage problems. Given in the plane a set P of n points and a set S of n weighted half-planes, we solve in $O(n^4\log n)$ time the problem of finding a subset of half-planes to cover P so that their total weight is minimized. This improves the previous best algorithm of $O(n^5)$ time by almost a linear factor. If all half-planes are lower ones, our algorithm runs in $O(n^2\log n)$ time, which improves the previous best algorithm of $O(n^4)$ time by almost a quadratic factor.

Keywords: Disk coverage · Line-constrained · Half-plane coverage · Geometric coverage · Facility location

1 Introduction

Given a set P of n points and a set S of m disks in the plane such that each disk has a weight, the *disk coverage* problem asks for a subset of disks of minimum total weight that cover all points of P. We assume that the union of all disks covers all points of P. The problem is known to be NP-hard [11] and approximation algorithms have been proposed, e.g., [17,19].

In this paper, we consider a line-constrained version of the problem in which all disks (possibly with different radii) have their centers on a line L, say, the x-axis. To the best of our knowledge, this line-constrained problem was not particularly studied before. We present an $O((m+n)\log(m+n)+\kappa\log m)$ time

This research was supported in part by NSF under Grant CCF-2005323. A full version of this paper is available at https://arxiv.org/abs/2104.14680.

A. Lubiw et al. (Eds.): WADS 2021, LNCS 12808, pp. 585–598, 2021.
https://doi.org/10.1007/978-3-030-83508-8_42

algorithm, where κ is the number of pairs of disks that intersect (and thus $\kappa \leq m(m-1)/2$; e.g., if the disks are disjoint, then $\kappa = 0$ and the algorithm runs in $O((m+n)\log(m+n))$ time). For the *unit-disk case* where all disks have the same radius, the running time can be reduced to $O((n+m)\log(m+n))$. We also solve in $O((m+n)\log(m+n))$ time the L_∞ and L_1 cases of the problem, in which the disks are squares and diamonds, respectively. As a by-product, we obtain an $O((m+n)\log(m+n))$ time algorithm for the 1D version of the problem where all points of P are on L and the disks are line segments of L. In addition, we show that the problem has an $\Omega((m+n)\log(m+n))$ time lower bound in the algebraic decision tree model even for the 1D case. This implies that our algorithms for the 1D, L_∞, L_1, and unit-disk cases are all optimal.

Our algorithms potentially have applications, e.g., in facility locations. For example, suppose we want to build some facilities along a railway which is represented by L (although an entire railway may not be a straight line, it may be considered straight in a local region) to provide service for some customers that are represented by the points of P. The center of a disk represents a candidate location for building a facility that can serve the customers covered by the disk and the cost for building the facility is the weight of the disk. The problem is to determine the best locations to build facilities so that all customers can be served and the total cost is minimized. This is exactly an instance of our problem.

Although the problems are line-constrained, our techniques can actually be used to solve other geometric coverage problems. If all disks of S have the same radius and the set of disk centers are separated from P by a line ℓ, the problem is called *line-separable unit-disk coverage.* The unweighted case of the problem where the weights of all disks are 1 has been studied in the literature [2,9,10]. In particular, the fastest algorithm was given by Claude et al. [9] and the runtime is $O(n\log n + nm)$. The algorithm, however, does not work for the weighted case. Our algorithm for the line-constrained L_2 case can be used to solve the weighted case in $O(nm\log(m+n))$ time or in $O((m+n)\log(m+n) + \kappa\log m)$ time, where κ is the number of pairs of disks that intersect on the side of ℓ that contains P. More interestingly, we can use the algorithm to solve the following *half-plane coverage problem.* Given in the plane a set P of n points and a set S of m weighted half-planes, find a subset of the half-planes to cover all points of P so that their total weight is minimized. For the *lower-only case* where all half-planes are lower ones, Chan and Grant [8] gave an $O(mn^2(m+n))$ time algorithm. In light of the observation that a half-plane is a special disk of infinite radius, our line-separable unit-disk coverage algorithm can be applied to solve the problem in $O(nm\log(m+n))$ time or in $O(n\log n + m^2\log m)$ time. This improves the result of [8] by almost a quadratic factor (note that the techniques of [8] are applicable to more general problem settings such as downward shadows of x-monotone curves). For the general case where both upper and lower half-planes are present, Har-Peled and Lee [13] proposed an algorithm of $O(n^5)$ time when $m = n$. By using our lower-only case algorithm, we solve the problem in $O(n^3m\log(m+n))$ time or in $O(n^3\log n + n^2m^2\log m)$ time. Hence, our result improves the one in [13] by almost a linear factor.

1.1 Related Work

Our problem is a new type of set cover. The general set cover problem, which is fundamental and has been studied extensively, is hard to solve, even approximately [12,14,18]. Many set cover problems in geometric settings, often called geometric coverage problems, are also NP-hard, e.g., [8,13]. As mentioned above, if the line-constrained condition is dropped, then the disk coverage problem becomes NP-hard, even if all disks are unit disks with the same weight [11]. Polynomial time approximation schemes (PTAS) exist for the unweighted problem [19] as well as the weighted unit-disk case [17].

Alt et al. [1] studied a problem closely related to ours, with the same input, consisting of P, S, and L, and the objective is also to find a subset of disks of minimum total weight that cover all points of P. But the difference is that S is comprised of all possible disks centered at L and the weight of each disk is defined as r^α with r being the radius of the disk and α being a given constant at least 1. Alt et al. [1] gave an $O(n^4 \log n)$ time algorithm for any L_p metric and any $\alpha \geq 1$, an $O(n^2 \log n)$ time algorithm for any L_p metric and $\alpha = 1$, and an $O(n^3 \log n)$ time algorithm for the L_∞ metric and any $\alpha \geq 1$. Recently, Pedersen and Wang [20] improved all these results by providing an $O(n^2)$ time algorithm for any L_p metric and any $\alpha \geq 1$. A 1D variation of the problem was studied in the literature where points of P are all on L and another set Q of m points is given on L as the only candidate centers for disks. Bilò et al. [5] first showed that the problem is solvable in polynomial time. Lev-Tov and Peleg [16] gave an algorithm of $O((n+m)^3)$ time for any $\alpha \geq 1$. Biniaz et al. [6] recently proposed an $O((n+m)^2)$ time algorithm for the case $\alpha = 1$. Pedersen and Wang [20] solved the problem in $O(n(n+m) + m \log m)$ time for any $\alpha \geq 1$.

Other line-constrained problems have also been studied in the literature, e.g., [15,21].

1.2 Our Approach

We first solve the 1D problem by a simple dynamic programming algorithm. Then, for the "1.5D" problem (i.e., points of P are in the plane), an observation is that if the points of P are sorted by their x-coordinates, then the sorted list can be partitioned into sublists such that there exists an optimal solution in which each disk covers a sublist. Based on the observation, we reduce the 1.5D problem to an instance of the 1D problem with a set P' of n points and a set S' of segments. But two challenges arise.

The first challenge is to give a small bound on $|S'|$. A naive method shows that $|S'| \leq n \cdot m$. In the unit-disk case and the L_1 case, we prove that $|S'|$ can be reduced to m by similar methods. In the L_∞ case, we show that $|S'|$ can be bounded by $2(n+m)$. The most challenging case is the L_2 case. By a number of observations, we prove that $|S'| \leq 2(n+m) + \kappa$.

The second challenge is to compute the set S' (P', which actually consists of all projections of the points of P onto L, can be easily obtained in $O(n)$ time). Our algorithms for computing S' for all cases use the sweeping technique. The algorithms for the unit-disk case and the L_1 case are relatively easy, while those for the L_∞ and L_2 cases require much more effort. Although the two algorithms for L_∞ and L_2 are similar in spirit, the intersections of the disks in the L_2 case bring more difficulties and make the algorithm more involved and less efficient. In summary, computing S' can be done in $O((n+m)\log(n+m))$ time for all cases except the L_2 case which takes $O((n+m)\log(n+m)+\kappa\log m)$ time.

Outline. The rest of the paper is organized as follows. We define notation in Sect. 2. The algorithms for the L_∞ and L_2 cases are given in Sect. 3. Due to the space limit, lemma proofs, algorithms for the unit-disk, and L_1 cases, the lower bound proof (which is based on a reduction from the element uniqueness problem), algorithms for the line-separable disk coverage and half-plane coverage problems are all omitted but can be found in the full paper.

2 Preliminaries

We assume that L is the x-axis. We also assume that all points of P are above or on L because if a point p_i is below L, then we could obtain the same optimal solution by replacing p_i with its symmetric point with respect to L. For ease of exposition, we make a general position assumption that no two points of P have the same x-coordinate and no point of P lies on the boundary of a disk of S.

For any point p in the plane, we use $x(p)$ and $y(p)$ to refer to its x-coordinate and y-coordinate, respectively. We sort all points of P by their x-coordinates, and let $p_1, p_2, \ldots, p_n$ be the sorted list from left to right on L. For any $1 \leq i \leq j \leq n$, let $P[i,j]$ denote the subset $\{p_i, p_{i+1}, \ldots, p_j\}$. Sometimes we use indices to refer to points of P, e.g., point i refers to p_i.

We sort all disks of S by the x-coordinates of their centers from left to right, and let $s_1, s_2, \ldots, s_m$ be the sorted list. For each s_i, let c_i denote its center and w_i denote its weight. We assume that each w_i is positive (otherwise one could always include s_i in the solution). For each disk s_i, let l_i and r_i refer to its leftmost and rightmost points, respectively.

We often talk about the relative positions of two geometric objects O_1 and O_2 (e.g., two points, or a point and a line). We say that O_1 is to the *left* of O_2 if $x(p) \leq x(p')$ holds for any point $p \in O_1$ and any point $p' \in O_2$, and *strictly left* means $x(p) < x(p')$. Similarly, we can define *right, above, below,* etc.

For convenience, we use p_0 (resp., p_{n+1}) to denote a point on L strictly to the left (resp. right) of all points of P and all disks of S. We use the term *optimal solution subset* to refer to a subset of S used in an optimal solution.

In the 1D problem, each disk $s_i \in S$ is a line segment on L. The problem can be solved by a straightforward dynamic programming algorithm of $O((n+m)\log(n+m))$ time. The details are omitted but can be found in the full paper.

3 The L_∞ and L_2 Cases

In this section, we give our algorithms for the L_∞ and L_2 cases. The algorithms are similar in the high level. However, the L_2 case is more involved in the low level computations. In Sect. 3.1, we present a high-level algorithmic scheme that works for both metrics. Then, we complete the algorithms for the L_∞ and L_2 cases in Sects. 3.2 and 3.3, respectively.

3.1 An Algorithmic Scheme for L_∞ and L_2 Metrics

In this subsection, unless otherwise stated, all statements are applicable to both metrics. Note that a disk in the L_∞ metric is a square.

For a disk $s_k \in S$, we say that a subsequence $P[i,j]$ of P with $1 \leq i \leq j \leq n$ is a *maximal subsequence covered* by s_k if all points of $P[i,j]$ are covered by s_k but neither p_{i-1} nor p_{j+1} is covered by s_k (it is well defined due to p_0 and p_{n+1}). Let $F(s_k)$ be the set of all maximal subsequences covered by s_k. Note that the subsequences of $F(s_k)$ are pairwise disjoint.

Lemma 1. *Suppose S_{opt} is an optimal solution subset and s_k is a disk of S_{opt}. Then, there is a subsequence $P[i,j]$ in $F(s_k)$ such that the following hold.*

1. *$P[i,j]$ has a point that is not covered by any disk in $S_{opt} \setminus \{s_k\}$.*
2. *For any point $p \in P$ that is covered by s_k but is not in $P[i,j]$, p is covered by a disk in $S_{opt} \setminus \{s_k\}$.*

In light of Lemma 1, we reduce the problem to an instance of the 1D problem with a point set P' and a line segment set S', as follows.

For each point of P, we vertically project it on L, and the set P' is comprised of all such projected points. Thus P' has exactly n points. For any $1 \leq i \leq j \leq n$, we use $P'[i,j]$ to denote the projections of the points of $P[i,j]$. For each point $p_i \in P$, we use p'_i to denote its projection point in P'.

The set S' is defined as follows. For each disk $s_k \in S$ and each subsequence $P[i,j] \in F(s_k)$, we create a segment for S', denoted by $s[i,j]$, with left endpoint at p'_i and right endpoint at p'_j. Thus, $s[i,j]$ covers exactly the points of $P'[i,j]$. We set the weight of $s[i,j]$ to w_k. Note that if $s[i,j]$ is already in S', which is defined by another disk s_h, then we only need to update its weight to w_k in case $w_k < w_h$ (so each segment appears only once in S'). We say that $s[i,j]$ is defined by s_k (resp., s_h) if its weight is equal to w_k (resp., w_h).

By Lemma 1, we intend to say that an optimal solution OPT' to the 1D problem on P' and S' corresponds to an optimal solution OPT to the original problem on P and S as follows: if a segment $s[i,j] \in S'$ is included in OPT', then we include the disk that defines $s[i,j]$ in OPT. However, since a disk of S may define multiple segments of S', to guarantee the correctness of the correspondence, we need to show that OPT' is a *valid solution*: no two segments in OPT' are defined by the same disk of S. For this, we have the following lemma.

Lemma 2. *Any optimal solution on P' and S' is a valid solution.*

With our algorithm for the 1D problem, we have the following result.

Lemma 3. *If the set S' is computed, then an optimal solution can be found in $O((n + |S'|) \log(n + |S'|))$ time.*

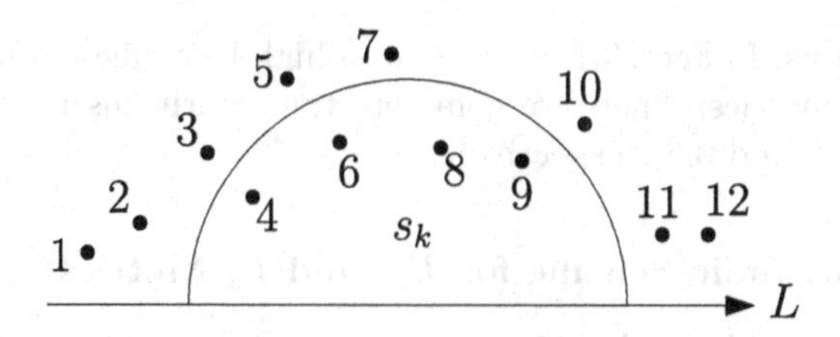

Fig. 1. Illustrating the definition of bounding couples: the numbers are the indices of the points of P. In this example, $p_l(s_k)$ is point 2 and $p_r(s_k)$ is point 11, and the bounding couples are: $(2, 3)$, $(3, 5)$, $(5, 7)$, $(7, 10)$, $(10, 11)$.

It remains to determine the size of S' and compute S'. An obvious answer is that $|S'|$ is bounded by $m \cdot \lceil n/2 \rceil$ because each disk can have at most $\lceil n/2 \rceil$ maximal sequences of P, and a trivial algorithm can compute S' in $O(nm \log(m+n))$ time by scanning the sorted list P for each disk. Therefore, by Lemma 3, we can solve the problem in both L_∞ and L_2 metrics in $O(nm \log(m+n))$ time.

With more geometric observations, we will prove the following two lemmas.

Lemma 4. *In the L_∞ metric, $|S'| \leq 2(n+m)$ and S' can be computed in $O((n+m)\log(n+m))$ time.*

Lemma 5. *In the L_2 metric, $|S'| \leq 2(n+m)+\kappa$ and S' can be computed in $O((n+m)\log(n+m)+\kappa \log m)$ time, where κ is the number of pairs of disks of S that intersect each other.*

With Lemma 3, we can solve the L_∞ case in $O((n+m)\log(n+m))$ time and the L_2 case in $O((n+m)\log(n+m)+\kappa \log m)$ time.

Bounding Couples. Before moving on, we introduce a new concept *bounding couples*, which will be used to prove Lemmas 4 and 5 later.

Consider a disk $s_k \in S$. Let $p_l(s_k)$ denote the rightmost point of $P \cup \{p_0, p_{n+1}\}$ strictly to the left of l_k; similarly, let $p_r(s_k)$ denote the leftmost point of $P \cup \{p_0, p_{n+1}\}$ strictly to the right of r_k. Let $P(s_k)$ denote the subset of points of P between $p_l(s_k)$ and $p_r(s_k)$ inclusively that are outside s_k. We sort the points of $P(s_k)$ by their x-coordinates, and we call each adjacent pair of points (or their indices) in the sorted list a *bounding couple* (e.g., see Fig. 1). Let $C(s_k)$ denote the set of all bounding couples of s_k, and for each bounding couple of $C(s_k)$, we assign w_k to it as the weight. Let $\mathcal{C} = \bigcup_{1 \leq k \leq m} C(s_k)$, and if the same bounding couple is defined by multiple disks, we only keep the copy in $\mathcal{C}$ with the minimum weight. Also, we consider a bounding couple (i, j) as an ordered pair with $i < j$, and i is considered as the left end of the couple while j is the right end.

The reason why we define bounding couples is that if $P[i, j]$ is a maximal subsequence of P covered by s_k then $(i-1, j+1)$ is a bounding couple. On the other hand, if (i, j) is a bounding couple of $C(s_k)$, then $P[i+1, j-1]$ is a maximal subsequence of P covered by s_k unless $j = i+1$. Hence, each bounding couple (i, j) of $\mathcal{C}$ with $j \neq i+1$ corresponds to a segment in the set S', and $|S'| \leq |\mathcal{C}|$.

Observe that $\mathcal{C}$ has at most $n-1$ couples (i, j) with $j = i + 1$, and given $\mathcal{C}$, we can obtain S' in additional $O(|\mathcal{C}|)$ time. According to our above discussion, to prove Lemmas 4 and 5, it suffices to prove the following two lemmas.

Lemma 6. *In the L_∞ metric, $|\mathcal{C}| \leq 2(n+m)$ and $\mathcal{C}$ can be computed in $O((n+m)\log(n+m))$ time.*

Lemma 7. *In the L_2 metric, $|\mathcal{C}| \leq 2(n+m)+\kappa$ and $\mathcal{C}$ can be computed in $O((n+m)\log(n+m)+\kappa\log m)$ time.*

Consider a bounding couple (i, j) of $\mathcal{C}$, defined by a disk s_k. We call it a *left bounding couple* if $p_i = p_l(s_k)$, a *right bounding couple* if $p_j = p_r(s_k)$, and a *middle bounding couple* otherwise (e.g., in Fig. 1, $(2, 3)$ is the left bounding couple, $(10, 11)$ is the right bounding couple, and the rest are middle bounding couples). Note that a disk can define at most one left bounding couple and at most one right bounding couple. Therefore, the number of left and right bounding couples in $\mathcal{C}$ is at most $2m$. It remains to bound the number of middle bounding couples of $\mathcal{C}$. We will prove Lemma 6 and 7 in Sects. 3.2 and 3.3, respectively.

3.2 The L_∞ Metric

In this section, our goal is to prove Lemma 6. In the L_∞ metric, every disk is a square that has four axis-parallel edges. We use l_k and r_k to particularly refer to the left and right endpoints of the upper edge of s_k, respectively.

For a point p_i and a square s_k, we say that p_i is *vertically above (resp., below)* the upper edge of s_k if p_i is above (resp., below) the upper edge of s_k and $x(l_k) \leq x(p_i) \leq x(r_k)$. Due to our general position assumption, p_i is not on the boundary of s_k, and thus p_i above/below the upper edge of s_k implies that p_i is strictly above/below the edge. Also, since no point of P is below L, a point $p_i \in P$ is in s_k if and only if p_i is vertically below the upper edge of s_k. If p_i is vertically above the upper edge of s_k, we also say that p_i is vertically above s_k or s_k is vertically below p_i. The following lemma proves an upper bound for $|\mathcal{C}|$.

Lemma 8. $|\mathcal{C}| \leq 2(n+m)$.

We proceed to compute the set $\mathcal{C}$. The following lemma gives an algorithm to compute all left and right bounding couples of $\mathcal{C}$.

Lemma 9. *All left and right bounding couples of $\mathcal{C}$ can be computed in $O((n+m)\log(n+m))$ time.*

Computing the Middle Bounding Couples We now compute all middle bounding couples of $\mathcal{C}$. We sweep a vertical line l from left to right, and an event happens if l encounters a point in $P \cup \{l_k, r_k | \ 1 \leq k \leq m\}$. Let H be the set of disks that intersect l. During the sweeping, we maintain the following information and invariants (e.g., see Fig. 2).

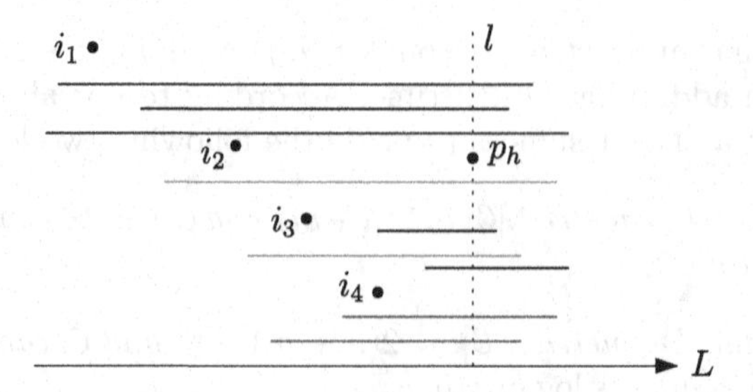

Fig. 2. In this example, $P(l) = \{p_{i_1}, p_{i_2}, p_{i_3}, p_{i_4}\}$. Each horizontal segment represents the upper edge of a disk. $H(i_1)$ consists of two blue disks and $H(i_4)$ consists of two red disks. H_0 consists of three black disks. After processing the event at p_h, i_2, i_3, and i_4 will be removed from $P(l)$ and p_h will be inserted, so after the event $P(l) = \{p_{i_1}, p_h\}$. (i_2, h), (i_3, h), (i_4, h) will be reported as middle bounding couples. (Color figure online)

1. A sequence $P(l) = \{p_{i_1}, p_{i_2}, \ldots, p_{i_t}\}$ of t points of P, which are to the left of l and ordered from northwest to southeast. $P(l)$ is stored in a balanced binary search tree $T(P(l))$.
2. A collection $\mathcal{H}$ of $t+1$ subsets of H: $H(i_j)$ for $j = 0, 1, \ldots, t$, which form a partition of H, defined as follows.
   $H(i_t)$ is the subset of disks of H that are vertically below p_{i_t}. For each $j = t-1, t-2, \ldots, 1$, $H(i_j)$ is the subset of disks of $H \setminus \bigcup_{k=j+1}^{t} H(i_k)$ that are vertically below p_{i_j}. $H(i_0) = H \setminus \bigcup_{j=1}^{t} H(i_j)$. While $H(i_0)$ may be empty, none of $H(i_j)$ for $1 \leq j \leq t$ is empty.
   Each $H(i_j)$ is maintained by a balanced binary search tree $T(H(i_j))$ ordered by the y-coordinates of the upper edges of the disks. We have all disks stored in leaves of $T(H(i_j))$, and each internal node v of the tree also stores a weight equal to the minimum weight of all disks in the leaves of the subtree at v.
3. For each point $p_{i_j} \in P(l)$, among all points of P strictly between p_{i_j} and l, no point is vertically above any disk of $H(i_j)$.
4. Among all points of P strictly to the left of l, no point is vertically above any disk of $H(i_0)$.

In summary, our algorithm maintains the following trees: $T(P(l))$, $T(H(i_j))$ for all $j \in [0, t]$. Initially when l is to the left of all disks and points of P, we have $H = \emptyset$ and $P(l) = \emptyset$. We next describe how to process events.

If l encounters the left endpoint l_k of a disk s_k, we insert s_k to $H(i_0)$. The time for processing this event is $O(\log m)$ since $|H(i_0)| \leq m$.

If l encounters the right endpoint r_k of a disk s_k, we need to determine which set $H(i_j)$ of $\mathcal{H}$ contains s_k. For this, we associate each right endpoint with its disk in the preprocessing so that it can keep track of which set of $\mathcal{H}$ contains the disk. Using this mechanism, we can determine the set $H(i_j)$ that contains s_k in constant time. We then remove s_k from $T(H(i_j))$. If $H(i_j)$ becomes empty and $j \neq 0$, then we remove p_{i_j} from $P(l)$. One can verify that all algorithm invariants still hold. The time for processing this event is $O(\log(m+n))$.

If l encounters a point p_h of P, which is a major event we need to handle, we process it as follows. We search $T(P(l))$ to find the first point p_{i_j} of $P(l)$ below p_h (e.g., $j = 3$ in Fig. 2). We remove the points p_{i_k} for all $k \in [j, t]$ from $P(l)$.

Lemma 10. *For each point p_{i_k} with $k \in [j, t]$, (i_k, h) is a middle bounding couple defined by and only by the disks of $H(i_k)$ (i.e., $H(i_k)$ consists of all disks of S that define (i_k, h) as a middle bounding couple).*

By Lemma 10, for each $k \in [j, t]$, we report (i_k, h) as a middle bounding couple with weight equal to the minimum weight of all disks of $H(i_k)$, which is stored at the root of $T(H(i_k))$.

Next, we process the point $p_{i_{j-1}}$, for which we have the following lemma. The proof technique is similar to that for Lemma 10, so we omit it.

Lemma 11. *If p_h is vertically below the lowest disk of $H(i_{j-1})$, then (i_{j-1}, h) is not a middle bounding couple; otherwise, (i_{j-1}, h) is a middle bounding couple defined by and only by disks of H_{j-1} that are vertically below p_h.*

By Lemma 11, we first check whether p_h is vertically below the lowest disk of $H(i_{j-1})$. If yes, we do nothing. Otherwise, we report (i_{j-1}, h) as a middle bounding couple with weight equal to the minimum weight of all disks of $H(i_{j-1})$ vertically below p_h, which can be computed in $O(\log m)$ time by using weights at the internal nodes of $T(H(i_{j-1}))$. We further have the following lemma.

Lemma 12. *If all disks of $H(i_{j-1})$ are vertically below p_h, then there does not exist a middle bounding couple (i_{j-1}, b) with $b > h$.*

We check whether p_h is above the highest disk of $H(i_{j-1})$ using the tree $T(H(i_{j-1}))$. If yes, then the above lemma tells that there will be no more middle bounding couples involving i_{j-1} any more, and thus we remove $p_{i_{j-1}}$ from $P(l)$.

The following lemma implies that all middle bounding couples with p_h as the right end have been computed.

Lemma 13. *For any middle bounding couple (b, h), b must be in $\{i_{j-1}, i_j, \ldots, i_t\}$.*

Next, we add p_h to the end of the current sequence $P(l)$ (note that the points p_{i_k} for all $k \in [j, t]$ and possibly $p_{i_{j-1}}$ have been removed from $P(l)$; e.g., see Fig. 2). Finally, we need to compute the tree $T(H(h))$ for the set $H(h)$, which is comprised of all disks of H vertically below p_h since p_h is the lowest point of $P(l)$. We compute $T(H(h))$ as follows.

First, starting from an empty tree, for each $k = t, t-1, \ldots, j$ in this order, we merge $T(H(h))$ with the tree $T(H(i_k))$. Notice that the upper edge of each disk in $T(H(i_k))$ is higher than the upper edges of all disks of $T(H(h))$. Therefore, each such merge operation can be done in $O(\log m)$ time. Second, for the tree $T(H(i_{j-1}))$, we perform a split operation to split the disks into those with upper edges above p_h and those below p_h, and then merge those below p_h with $T(H(h))$ while keeping those above p_h in $T(H(i_{j-1}))$. The above split and merge operations can be done in $O(\log m)$ time. Third, we remove those disks below p_h from $H(i_0)$ and insert them to $T(H(h))$. This is done by repeatedly removing the lowest disk s from $H(i_0)$ and inserting it to $T(H(h))$ until the upper edge of s is higher than p_h. This completes our construction of the tree $T(H(h))$.

The above describes our algorithm for processing the event at p_h. One can verify that all algorithm invariants still hold. The runtime of this step is $O((1+k_1+k_2)\log m)$, where k_1 is the number of points removed from $P(l)$ (the number of merge operations is at most k_1) and k_2 is the number of disks of $H(i_0)$ got removed for constructing $T(H(h))$. As we sweep the line l from left to right, once a point is removed from $P(l)$, it will not be inserted again, and thus the total sum of k_1 in the entire algorithm is at most n. Also, once a disk is removed from $H(i_0)$, it will never be inserted again, and thus the total sum of k_2 in the entire algorithm is at most m. Hence, the overall time of the algorithm is $O((n+m)\log(n+m))$. This proves Lemma 6.

3.3 The L_2 Metric

In this section we prove Lemma 7. Recall our general position assumption that no point of P is on the boundary of a disk of S. Also recall that all points of P are above L. In the L_2 metric, the two extreme points l_k and r_k of a disk s_k are unique. For a point $p_i \in P$ and a disk $s_k \in S$, we say that p_i is *vertically above* s_k if p_i is outside s_k and $x(l_k) \leq x(p_i) \leq x(r_k)$, and p_i is *vertically below* s_k if p_i is inside s_k. We also say that s_k is vertically below p_i if p_i is vertically above s_k. Lemma 14 gives an upper bound for $|\mathcal{C}|$.

Lemma 14. $|\mathcal{C}| \leq 2(n+m)+\kappa$.

We next describe our algorithm for computing the set $\mathcal{C}$. For each disk s_k, we refer to the half-circle of the boundary of s_k above L as the *arc* of s_k. Note that every two arcs of S intersect at most once. Below, depending on the context, s_k may also refer to its arc. Lemma 15 computes the left and right bounding couples.

Lemma 15. *All left and right bounding couples of* $\mathcal{C}$ *can be computed in* $O((n+m)\log(n+m)+\kappa\log m)$ *time.*

To compute the middle bounding pairs of $\mathcal{C}$, the algorithm is similar in spirit to that for the L_∞ case. However, it is more involved and requires new techniques due to the nature of the L_2 metric as well as the intersections of the disks of S. We sweep a vertical line l from left to right; an event happens if l encounters a point in $P \cup \{l_k, r_k \mid 1 \leq k \leq m\}$ or an intersection of two disk arcs. Let H be the set of arcs that intersect l. During the sweeping, we maintain the following information and invariants (e.g., see Fig. 3).

1. A sequence $P(l) = \{p_{i_1}, p_{i_2}, \ldots, p_{i_t}\}$ of t points to the left of l that are sorted from left to right. $P(l)$ is maintained by a balanced binary search tree $T(P(l))$.

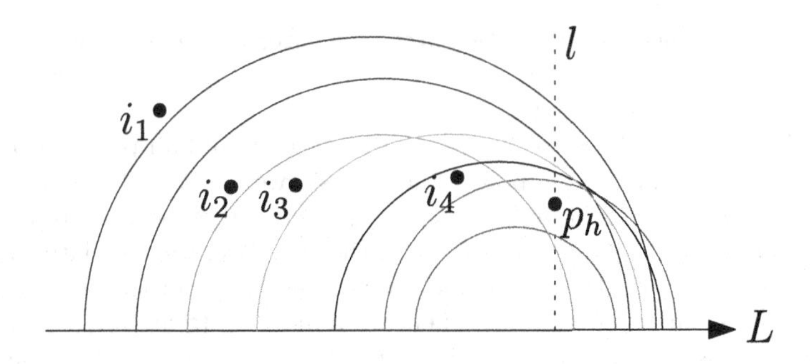

Fig. 3. In this example, $P(l) = \{p_{i_1}, p_{i_2}, p_{i_3}, p_{i_4}\}$. $H(i_1)$ consists of the two blue arcs and $H(i_4)$ consists of the two red arcs. $H(i_0)$ consists of the only black arc. After processing the event at $p_h \in P$, (i_2, h) and (i_4, h) will be reported as middle bounding couples, point i_2 will be removed from $P(l)$, and p_h will be inserted to $P(l)$. (Color figure online)

2. A collection $\mathcal{H}$ of $t+1$ subsets of H: $H(i_j)$ for $j = 0, 1, \dots, t$, which form a partition of H, defined as follows. $H(i_t)$ is the set of disks of H vertically below p_{i_t}. For each $j = t-1, t-2, \dots, 1$, $H(i_j)$ is the set of disks of $H \backslash \bigcup_{k=j+1}^{t} H(i_k)$ vertically below p_{i_j}. $H(i_0) = H \setminus \bigcup_{j=1}^{t} H(i_j)$. While $H(i_0)$ may be empty, none of $H(i_j)$ for $j \geq 1$ is empty.
   Each $H(i_j)$ for $0 \leq j \leq t$ is maintained by a balanced binary search tree $T(H(i_j))$ ordered by the y-coordinates of the intersections of l with the arcs of the disks. We have all disks stored in the leaves of the tree, and each internal node v of the tree stores a weight that is equal to the minimum weight of all disks in the leaves of the subtree rooted at v.
   For each subset $H' \subseteq H$, the arc of H' whose intersection with l is the lowest is called *the lowest arc* of H'. We maintain a set H^* consisting of the lowest arcs of all sets $H(i_k)$ for $1 \leq k \leq t$. So $|H^*| = t$. We use a binary search tree $T(H^*)$ to store disks of H^*, ordered by the y-coordinates of their intersections with l.
3. For each point $p_{i_j} \in P(l)$, among all points of P strictly between p_{i_j} and l, no point is vertically above any disk of $H(i_j)$.
4. Among all points of P strictly to the left of l, no point is vertically above any disk of $H(i_0)$.

Remark. Our algorithm invariants are essentially the same as those in the L_∞ case. One difference is that the points of $P(l)$ are not sorted simultaneously by y-coordinates, which is due to that the arcs of S may cross each other (in contrast, in the L_∞ case the upper edges of the squares are parallel). For the same reason, for two sets $H(i_k)$ and $H(i_j)$ with $1 \leq k < j \leq t$, it may not be the case that all arcs of $H(i_k)$ are above all arcs of $H(i_j)$ at l. Therefore, we need an additional set H^* to guide our algorithm, as will be clear later.

In our sweeping algorithm, we use similar techniques as the line segment intersection algorithm [3,4,7] to determine and handle arc intersections of S (we are able to do so because every two arcs of S intersect at most once), and the time on handling them is $O((m+\kappa)\log m)$. Below we will not explicitly explain how to handle arc intersections.

Initially $H = \emptyset$ and l is to the left of all arcs of S and all points of P.

If l encounters the left endpoint of an arc s_k, we insert s_k to $H(i_0)$.

If l encounters the right endpoint r_k of an arc s_k, then we need to determine which set of $\mathcal{H}$ contains s_k. For this, as in the L_∞ case, we associate each right endpoint with the arc. Using this mechanism, we can find the set $H(i_j)$ of $\mathcal{H}$ that contains s_k in constant time. Then, we remove s_k from $H(i_j)$. If $j = 0$, we are done for this event. Otherwise, if s_k was the lowest arc of $H(i_j)$ before the above remove operation, then s_k is also in H^* and we remove it from H^*. If the new set $H(i_j)$ becomes empty, then we remove p_{i_j} from $P(l)$. Otherwise, we find the new lowest arc from $H(i_j)$ and insert it to H^*. Processing this event takes $O(\log(n+m))$ time using the trees $T(H^*)$, $T(P(l))$, and $T(H(i_j))$.

If l encounters an intersection q of two arcs s_a and s_b, in addition to the processing work for computing the arc intersections, we do the following. Using the right endpoints, we find the two sets of $\mathcal{H}$ that contain s_a and s_b, respectively. If s_a and s_b are from the same set $H(i_j) \in \mathcal{H}$, then we switch their order in the tree $T(H(i_j))$. Otherwise, if s_a is the lowest arc in its set and s_b is also the lowest arc in its set, then both s_a and s_b are in H^*, so we switch their order in $T(H^*)$. The time for processing this event is $O(\log m)$.

If l encounters a point p_h of P, which is a major event to handle, we process it as follows. As in the L_∞ case, our goal is to determine the middle bounding couples (i,h) with $p_i \in P(l)$.

Using $T(H^*)$, we find the lowest arc s_k of H^*. Let $H(i_j)$ for some $j \in [1,t]$ be the set that contains s_k, i.e., s_k is the lowest arc of $H(i_j)$. If p_h is above s_k, then we can show that (i_j,h) is a middle bounding couple defined by and only by the arcs of $H(i_j)$ below p_h (e.g., see Fig. 3). The proof is similar to Lemma 10, so we omit the details. Hence, we report (i_j,h) as a middle bounding couple with weight equal to the minimum weight of all arcs of $H(i_j)$ below p_h, which can be found in $O(\log m)$ time using $T(H(i_j))$. Then, we split $T(H(i_j))$ into two trees by p_h such that the arcs above p_h are still in $T(H(i_j))$ and those below p_h are stored in another tree (we will discuss later how to use this tree). Next we remove s_k from H^*. If the new set $H(i_j)$ after the split operation is not empty, then we find its lowest arc and insert it into H^*; otherwise, we remove p_{i_j} from $P(l)$. We then continue the same algorithm on the next lowest arc of H^*.

The above discusses the case where p_h is above s_k. If p_h is not above s_k, we are done with processing the arcs of H^*. We can show that all middle bounding couples (b,h) with h as the right end have been computed. The proof is similar to Lemma 13, and we omit it.

Finally, we add p_h to the rear of $P(l)$. As in the L_∞ case, we need to compute the tree $T(H(h))$ for the set $H(h)$, which is comprised of all arcs of H below p_h, as follows.

Initially we have an empty tree $T(H(h))$. Let H' be the subset of the arcs of H^* vertically below p_h; here H^* refers to the original set at the beginning of the event for p_h. The set H' has already been computed above. Let $\mathcal{H}'$ be the subcollection of $\mathcal{H}$ whose lowest arcs are in H'. We process the subsets $H(i_j)$ of $\mathcal{H}'$ in the inverse order of their indices (for this, after identifying $\mathcal{H}'$, we can sort the subsets $H(i_j)$ of $\mathcal{H}'$ by their indices in $O(|H'| \log m)$ time; note that $|H'| = |\mathcal{H}'|$), i.e., the subset of $\mathcal{H}'$ with the largest index is processed first.

Suppose we are processing a subset $H(i_j)$ of $\mathcal{H}'$. Let s be the lowest arc of $H(i_j)$. Recall that we have performed a split operation on the tree $T(H(i_j))$ to obtain another tree consisting of all arcs of $H(i_j)$ below p_h, and we use $H'(i_j)$ to denote the set of those arcs and use $T(H'(i_j))$ to denote the tree. If $T(H(h))$ is empty, then we simply set $T(H(h)) = T(H'(i_j))$. Otherwise, we find the highest arc s' of $T(H(h))$ at l. If s is above s' at l, then every arc of $T(H'(i_j))$ is above all arcs of $T(H(h))$ at l and thus we simply perform a merge operation to merge $T(H'(i_j))$ with $T(H(h))$ (and we use $T(H(h))$ to refer to the new merged tree). Otherwise, we call (s, s') an *order-violation pair*. In this case, we do the following. We remove s from $T(H'(i_j))$ and insert it to $T(H(h))$. If $T(H'(i_j))$ becomes empty, then we finish processing $H(i_j)$. Otherwise, we find the new lowest arc of $T(H'(i_j))$, still denoted by s, and then process s in the same way as above.

The above describes our algorithm for processing a subset $H(i_j)$ of $\mathcal{H}'$. Once all subsets of $\mathcal{H}'$ are processed, the tree $T(H(h))$ for the set $H(h)$ is obtained.

After processing the arcs of H^* as above, we also need to consider the arcs of $H(i_0)$. For this, we scan the arcs from low to high using $T(H(i_0))$, and for each arc s, if s is above p_h, then we stop the procedure; otherwise, we remove s from $T(H(i_0))$ and insert it to $T(H(h))$.

This finishes our algorithm for processing the event at p_h. One can verify that the time complexity of this step is $O((1 + k_1 + k_2 + k_3) \cdot \log m)$ time, where k_1 is the number of middle bounding couples reported (the number of merge and split operations is at most k_1; also, $|H'| = k_1$), k_2 is the number of arcs of $H(i_0)$ got removed for constructing $T(H(h))$, and k_3 is the number of order-violation pairs. By Lemma 14, the total sum of k_1 is at most $2(n + m) + \kappa$ in the entire algorithm. As in the L_∞ case, the total sum of k_2 is at most m in the entire algorithm. The following lemma proves that the total sum of k_3 is at most κ. Therefore, the overall time of the algorithm is $O((n + m) \log(n + m) + \kappa \log m)$.

Lemma 16. *The total number of order-violation pairs in the entire algorithm is at most κ.*

References

1. Alt, H., et al.: Minimum-cost coverage of point sets by disks. In: Proceedings of the 22nd Annual Symposium on Computational Geometry (SoCG), pp. 449–458 (2006)
2. Ambühl, C., Erlebach, T., Mihalák, M., Nunkesser, M.: Constant-factor approximation for minimum-weight (connected) dominating sets in unit disk graphs. In: Proceedings of the 9th International Conference on Approximation Algorithms for Combinatorial Optimization Problems (APPROX), and the 10th International Conference on Randomization and Computation (RANDOM), pp. 3–14 (2006)

3. Bentley, J., Ottmann, T.: Algorithms for reporting and counting geometric intersections. IEEE Trans. Comput. **28**(9), 643–647 (1979)
4. de Berg, M., Cheong, O., van Kreveld, M., Overmars, M.: Computational Geometry – Algorithms and Applications, 3rd edn. Springer-Verlag, Berlin (2008)
5. Bilò, V., Caragiannis, I., Kaklamanis, C., Kanellopoulos, P.: Geometric clustering to minimize the sum of cluster sizes. In: Proceedings of the 13th European Symposium on Algorithms, pp. 460–471 (2005)
6. Biniaz, A., Bose, P., Carmi, P., Maheshwari, A., Munro, I., Smid, M.: Faster algorithms for some optimization problems on collinear points. In: Proceedings of the 34th International Symposium on Computational Geometry (SoCG), pp. 1–14 (2018)
7. Brown, K.: Comments on Algorithms for reporting and counting geometric intersections. IEEE Trans. Comput. **30**, 147–148 (1981)
8. Chan, T., Grant, E.: Exact algorithms and APX-hardness results for geometric packing and covering problems. Comput. Geom. Theory Appl. **47**, 112–124 (2014)
9. Claude, F., et al.: An improved line-separable algorithm for discrete unit disk cover. Discrete Math. Algorithms Appl. **2**, 77–88 (2010)
10. Claude, F., Dorrigiv, R., Durocher, S., Fraser, R., López-Ortiz, A., Salinger, A.: Practical discrete unit disk cover using an exact line-separable algorithm. In: Proceedings of the 20th International Symposium on Algorithm and Computation (ISAAC), pp. 45–54 (2009)
11. Feder, T., Greene, D.: Optimal algorithms for approximate clustering. In: Proceedings of the 20th Annual ACM Symposium on Theory of Computing (STOC), pp. 434–444 (1988)
12. Feige, U.: A threshold of ln n for approximating set cover. J. ACM **45**, 634–652 (1998)
13. Har-Peled, S., Lee, M.: Weighted geometric set cover problems revisited. J. Comput. Geom. **3**, 65–85 (2012)
14. Hochbaum, D., Maass, W.: Fast approximation algorithms for a nonconvex covering problem. J. Algorithms **3**, 305–323 (1987)
15. Karmakar, A., Das, S., Nandy, S., Bhattacharya, B.: Some variations on constrained minimum enclosing circle problem. J. Comb. Optim. **25**(2), 176–190 (2013)
16. Lev-Tov, N., Peleg, D.: Polynomial time approximation schemes for base station coverage with minimum total radii. Comput. Netw. **47**, 489–501 (2005)
17. Li, J., Jin, Y.: A PTAS for the weighted unit disk cover problem. In: Proceedings of the 42nd International Colloquium on Automata, Languages and Programming (ICALP), pp. 898–909 (2015)
18. Lund, C., Yannakakis, M.: On the hardness of approximating minimization problems. J. ACM **41**, 960–981 (1994)
19. Mustafa, N., Ray, S.: PTAS for geometric hitting set problems via local search. In: Proceedings of the 25th Annual Symposium on Computational Geometry (SoCG), pp. 17–22 (2009)
20. Pedersen, L., Wang, H.: On the coverage of points in the plane by disks centered at a line. In: Proceedings of the 30th Canadian Conference on Computational Geometry (CCCG), pp. 158–164 (2018)
21. Wang, H., Zhang, J.: Line-constrained *k*-median, *k*-means, and *k*-center problems in the plane. Int. J. Comput. Geom. Appl. **26**, 185–210 (2016)

A Universal Cycle for Strings with Fixed-Content (Which Are Also Known as Multiset Permutations)

J. Sawada[1(✉)] and A. Williams[2]

[1] School of Computer Science, University of Guelph, Guelph, Canada
jsawada@uoguelph.ca
[2] Computer Science, Williams College, Williamstown, USA
aaron.williams@williams.edu

Abstract. We develop the first universal cycle construction for strings with fixed-content (also known as multiset permutations) using shorthand representation. The construction runs a necklace concatenation algorithm on cool-lex order for fixed-content strings, and is implemented to generate the universal cycle in amortized $O(1)$-time per symbol. This generalizes two previous results: a universal cycle for shorthand permutations by Ruskey, Holroyd, and Williams [*Algorithmica* 64 (2012)] and a universal cycle for shorthand fixed-weight binary strings by Ruskey, Sawada, and Williams [*SIAM J. on Disc. Math.* 26 (2012)]. A consequence of our construction is the first shift Gray code for fixed-content necklaces.

Keywords: De bruijn cycle · Universal cycle · Fixed-content · Multiset permutation · Parikh vector · Necklace · Gray code · Cool-lex

1 Introduction

A *universal cycle* for a set $\mathbf{S}$ of length n strings, is a circular string of length $|\mathbf{S}|$ where every string in $\mathbf{S}$ appears exactly once as a substring. Universal cycles generalize *de Bruijn sequendces*, in which $\mathbf{S}$ is the set of all k-ary strings of length n. See [7,8] and debruijnsequence.org for recent surveys on the rich history of these objects.

Universal cycles for many interesting sets are known to exist [3,4,11,12,14,19–23]. But they do not exist for other common sets, such as permutations, or binary strings with fixed-weight (the number of 1s is fixed), as explained in the first paragraph of [10]. Fortunately, in these two cases, a *shorthand* notation can instead be used, since both permutations and fixed-weight strings can be represented by their length $n-1$ prefixes, as the final symbol is redundant. When $\mathbf{S}$ is the set of these *shorthand* representations, universal cycle constructions are known (for permutations [10,17], for fixed-weight binary strings [16]) and the cycles can be generated efficiently.

J. Sawada—Research supported by NSERC.

A. Lubiw et al. (Eds.): WADS 2021, LNCS 12808, pp. 599–612, 2021.
https://doi.org/10.1007/978-3-030-83508-8_43

Example 1 Consider the set $\mathbf{S}_1 = \{12, 13, 21, 23, 31, 32\}$ of shorthand permutations of order $n = 3$. Observe that 231321 is a universal cycle for $\mathbf{S}_1$.

Example 2 Consider the set $\mathbf{S}_2 = \{0001, 0010, 0100, 1000, 0011, 0110, 0101, 1001, 1010, 1100\}$ of shorthand fixed-weight strings for $n = 5$ weight 2. Observe that 1010011000 is a universal cycle for $\mathbf{S}_2$.

A set of strings with *fixed-content* consists of every arrangement of a multiset of symbols, which is the *content* of the set. These sets generalize permutations of order n, whose content is $\{1, 2, \ldots, n\}$, and binary strings of length n with fixed-weight d, whose content is d 1s and $n - d$ 0s. These sets are also known as *multiset permutations*, and consist of strings with the same *Parikh vector*. We consider the construction of universal cycles for fixed-content strings, using their shorthand representation.

Main Result. The first known construction of universal cycles for strings with fixed-content. The construction is based on a known concatenation construction applied to the cool-lex order of necklaces to generate the universal cycles in $O(1)$-amortized time per symbol.

Along the way, we develop an algorithm to list necklaces with fixed-content in a shift Gray code order in $O(n)$-amortized time per necklace.

The rest of the paper is presented as follows. Section 2 discusses preliminary concepts, including fixed-content necklaces and cool-lex order. Section 3 provides a new recursive algorithm for generating fixed-content necklaces in cool-lex order. Section 4 presents our universal cycle construction for strings with fixed-content.

Our universal cycle construction for fixed weight strings is implemented in C in the Appendix, and its output can be viewed at debruijnsequence.org. We also mention that the shorthand representation used in this article is not the only representation that could be considered; see [13] and [2] for other representations used for permutations.

2 Preliminaries

In this section, we introduce the basic concepts and notation used in the construction of our universal cycle.

Let S be a multiset over the alphabet $\{1, 2, \ldots, k\}$, denoting the fixed-content of our strings with $n = |S|$, and let $\boldsymbol{S}(S)$ denote the set of all strings with fixed-content S. Let $\alpha = a_1 a_2 \cdots a_n$ be a string. Let α^t denote the string composed of t copies of α. The *period* of α is the smallest value j such that $\alpha = (a_1 \cdots a_j)^t$ for some integer t; we say $a_1 \cdots a_j$ is the *aperiodic prefix* of α. If α has period n (it is the same as its aperiodic prefix), we say it is *aperiodic*; otherwise we say it is *periodic*.

2.1 Necklaces with Fixed-Content

A *necklace* is defined to be the lexicographically smallest string in an equivalence class of strings under rotation. Let $\boldsymbol{N}(S)$ denote the set of all necklaces with fixed-content S. The number of fixed-content necklaces can be deduced using Pólya theory as discussed in [9]. In the following formula, it is assumed that the content S is composed of $n_i \geq 1$ occurrences of each symbol i, $|S| = n$, and $k \geq 1$:

$$\boldsymbol{N}(S) = \frac{1}{n} \sum_{j | gcd(n_1, n_2, \ldots, n_k)} \phi(j) \frac{(n/j)!}{(n_1/j)! \cdots (n_k/j)!} \tag{1}$$

where Euler's totient function $\phi(j)$ denotes the number of positive integers less than or equal to j that are relatively prime to j.

There exists a $O(1)$-amortized time algorithm to list $\boldsymbol{N}(S)$ [18].

2.2 Cool-Lex Order

Cool-lex order for fixed-content strings was introduced in [24]. The order is a *Gray code*, meaning that successive strings differ by a simple operation. More specifically, it is a *prefix-shift Gray code*, meaning that successive strings differ by a single prefix-shift. A *prefix-shift* removes a single symbol and reinserts it as the first symbol; in a linked list representation, this corresponds to a removing a node and reinserting it as the head. The order is also *cyclic* meaning that a prefix-shift also transforms the last string in the order into the first. The set $\boldsymbol{S}(\{1, 1, 2, 2, 3, 3\})$ is listed in cool-lex order on the left side of Fig. 1, with examples of the prefix-shift operation given in the caption.

One of the most notable features of cool-lex order is that it has a simple *successor rule*. In other words, the prefix-shift that creates the next fixed-content string in the order is relatively easy to specify. To describe the rule, let *the non-decreasing prefix* of a string be its longest prefix with no decreases. In other words, if $a_1 a_2 \cdots a_n$ is the string, then its non-decreasing prefix is $a_1 a_2 \ldots a_p$ with $a_i \leq a_{i+1}$ for $1 \leq i < p$ and either $p = n$ or $a_p > a_{p+1}$. Now we can describe the cool-lex successor rule.

Cool-lex Successor Rule
If the non-decreasing prefix of $a_1 a_2 \cdots a_n$ has length $p < n$, then the next string in cool-lex order is obtained by a prefix-shift of length $p + 1$ if $p = n - 1$ or $a_p > a_{p+2}$, and otherwise by a prefix-shift of length $p + 2$.

This rule specifies every transition in the cyclic order except one: If the string itself is non-decreasing, then the next string is obtained by a prefix-shift of length n.

Another benefit of cool-lex order is that its relative order provides shift Gray codes for other sets of strings. This phenomenon was discussed for fixed-weight sets in [15], and for fixed-content sets in [25]. In particular, this occurs for necklaces, as illustrated

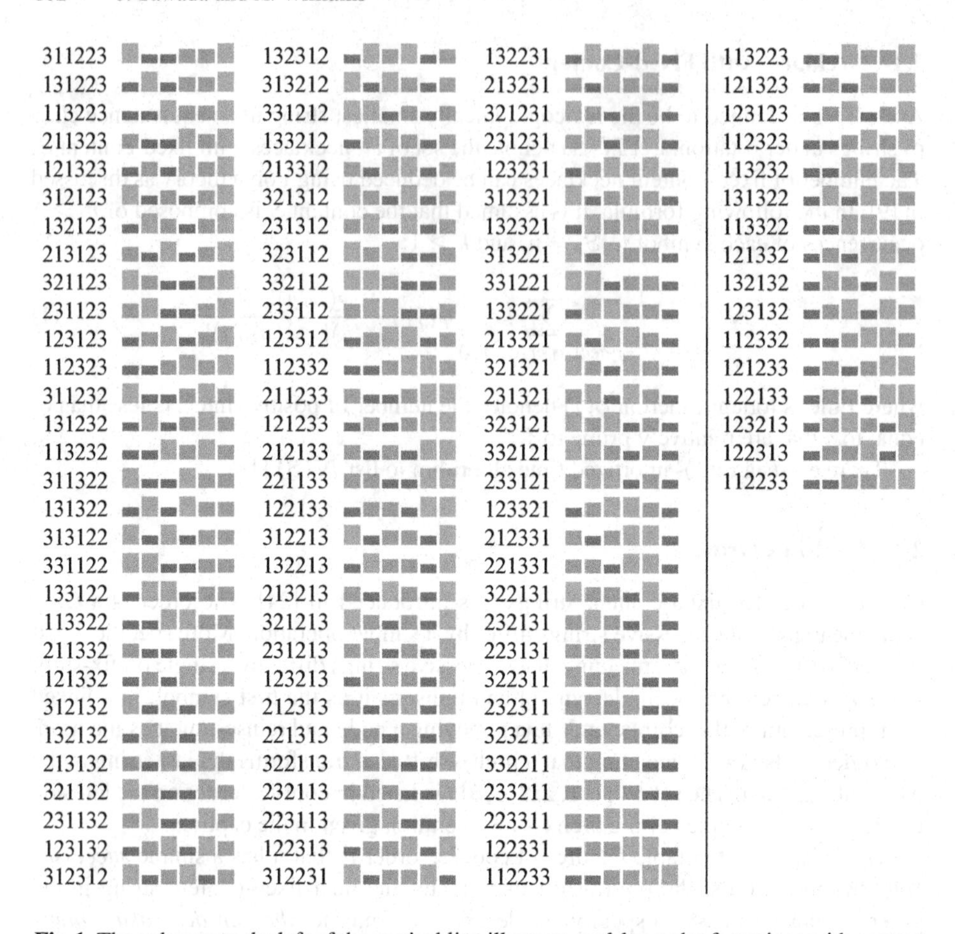

Fig. 1. The columns to the left of the vertical line illustrate cool-lex order for strings with content $S = \{1, 1, 2, 2, 3, 3\}$. Observe that each string is obtained from the previous by a prefix-shift. For example, the third string 113$\underline{2}$23 is transformed into the fourth string $\underline{2}$11323 by moving the underlined symbol to the left into the first position (or equivalently by rotating the prefix 113$\underline{2}$ one position to the right to obtain $\underline{2}$113). The order is also cyclic in this regard, since the last string is transformed into the first by a prefix-shift. The column to the right of the vertical line illustrates the necklaces with content S as they appear in cool-lex order. Observe that each necklace is obtained from the previous one by a shift. For example, the first necklace 113$\underline{2}$23 is transformed into the second necklace 1$\underline{2}$1323 by moving the underlined symbol two positions to the left (or equivalently by rotating the substring 13$\underline{2}$ one position to the right to obtain $\underline{2}$13). The order is again cyclic in this regard, since the last string is transformed into the first by a shift.

in Fig. 1 for $S = \{1, 1, 2, 2, 3, 3\}$. By adapting the techniques from [15,25], we obtain the successor rule given below for necklaces with fixed-content. Our shift notation is discussed after the rule is presented, with an example given in the caption of Fig. 1.

Cool-lex Successor Rule for Fixed-Content Necklaces

Let $\alpha = a_1a_2 \cdots a_n = \lambda\gamma \in \mathbf{N}(S)$, where $|\alpha| = |S| = n$, λ is α's non-decreasing prefix, and $m = |\lambda|$. The necklace following α in cool-lex order is denoted $\mathsf{next}(\alpha)$ and is obtained from α by the shift in the following cumulative cases

$$\mathsf{next}(\alpha) = \begin{cases} \mathsf{lshift}_\alpha(m) & \text{if } m = n & (2a) \\ \mathsf{lshift}_\alpha(m+1) & \text{if } m = n-1 \text{ or } a_m > a_{m+2} \text{ or } \beta \notin \mathbf{N}(S) & (2b) \\ \mathsf{lshift}_\alpha(m+2) & \text{otherwise} & (2c) \end{cases}$$

where β is obtained from α by swapping the two symbols after λ (if they exist).

Now we define the lshift operation used in (2). If $\alpha = a_1a_2 \cdots a_n$ is a necklace, then $\mathsf{lshift}_\alpha(i)$ bubbles a_i as far to the left as possible while always maintaining that the result is still a necklace. In particular, $\mathsf{lshift}_\alpha(i) = \alpha$ if the first swap results in a non-necklace. For example, consider the necklace $\alpha = a_1a_2a_3a_4a_5a_6 = 122313$. We can determine the result of $\mathsf{lshift}_\alpha(4)$ by bubbling the symbol $a_4 = 3$ to the left, starting from α, as follows:

$$\begin{aligned} a_1a_2a_3a_4a_5a_6 &= 122313 \text{ is a necklace;} \\ a_1a_2a_4a_3a_5a_6 &= 123213 \text{ is a necklace;} \\ a_1a_4a_2a_3a_5a_6 &= 132213 \text{ is not a necklace.} \end{aligned}$$

The result is the last necklace in this list. Hence, $\mathsf{lshift}_\alpha(4) = a_1a_2a_4a_3a_5a_6 = 123213$. To motivate the next section, it is important to note that this calculation involved testing if multiple strings were necklaces. This means that without further optimization, the necklace successor rule runs in $O(n^2)$ time. Note that $O(n)$ time and $O(n)$ space is sufficient for testing whether or not a string is a necklace [1].

2.3 Necklace-Prefix Algorithm

Perhaps the most well-known de Bruijn sequence is the so-called *granddaddy de Bruijn sequence*; it is the lexicographically smallest k-ary de Bruijn sequence of order n. It can be generated very elegantly using an approach that is often referred to as the *FKM construction* or *FKM algorithm*, due to its discoverers [5,6]. As discussed in [16], the authors of this article prefer to describe the construction using a slightly different approach called the *necklace-prefix algorithm*. The latter approach constructs the granddaddy de Bruijn sequence in a nearly identical manner, but it is often more well-suited for creating other sequences.

The necklace-prefix algorithm takes an order of strings, filters out the non-necklaces, reduces the remaining necklaces to their aperiodic prefix, and concatenates the prefixes. Amazingly, the granddaddy de Bruijn sequence is created by applying the necklace-prefix algorithm to the k-ary strings of length n in lexicographic order. This is illustrated in Fig. 2 for $n = 2$ and $k = 4$. The approach has been generalized to other sets in [22].

	necklaces	aperiodic prefixes
lexicographic order ↓	00	0
	01	01
	02	02
	03	03
	11	1
	12	12
	13	13
	22	2
	23	23
	33	3
	(a)	(b)

granddaddy de Bruijn sequence

3 · 0 · 01 · 02 · 03 · 1 · 12 · 13 · 2 · 23

(c)

Fig. 2. The necklace-prefix algorithm applied to the 4-ary strings of length 2 constructs the granddaddy de Bruijn sequence for $n = 2$ and $k = 4$. The algorithm starts with the lexicographic order of 4-ary strings of length 2 (which are not shown), then reduces the order to the necklaces in column (a), and their aperiodic prefixes in column (b), and concatenates these prefixes to get the granddaddy de Bruijn sequence 0010203112132233 in (c).

Unfortunately, the magic runs out when we consider fixed-content strings, even in their shorthand representatives. As an illustration, note that the lexicographic order of necklaces with content $S = \{1, 1, 2, 2, 3, 3\}$ places the following necklaces consecutively,

$$\ldots 113322, 121233, \ldots,$$

and so, the necklace-prefix algorithm genereates $\cdots 113\mathbf{22121}233 \cdots$. The bold substring of length $n-1 = 5$ is not shorthand for a string with the content S because it has too many 2's. The cause of the issue is also clear: The leftmost 2 moves several positions to the left from $1133\underline{2}2$ to $1\underline{2}1233$. This issue leads us to instead use a reversed version of cool-lex order, since this will ensure that individual symbols move at most one position to the left between successive necklaces.

3 Recursive Algorithm

In [24], a recursive description is given to list all strings with fixed-content S in cool-lex order. In that description, the focus is on strings in reverse lexicographic order, whereas, we will focus on lexicographic order. In this section, we restate this recurrence using the original terminology and then apply it to generate necklaces with fixed-content S.

The *tail* of S, denoted $tail(S)$, is the unique non-decreasing string composed of *all* the elements of S. A *scut* [1] of S is any non-decreasing string α composed of *some* of the elements of S such that α is not a suffix of $tail(S)$, but every proper suffix of α is a suffix of $tail(S)$. Let $\alpha_i(S)$ (or simply, α_i) denote the i-th scut of S when the scuts are

[1] In nature, a scut is a short tail. Here, it is a suffix of $tail(S)$ with a small symbol prepended.

listed in decreasing order of the first symbol, then by decreasing length. Let R_i denote the multi-set S with the content of $\alpha_i(S)$ removed.

Example 3 Consider $S = \{1,1,2,2,3,3\}$. Then $tail(S) = 112233$ and the scuts of S in decreasing order of the first symbol, then decreasing length, are:

$$23,\ 2,\ 1233,\ 133,\ 13,\ 1.$$

Note $\alpha_4(S) = 133$ and $R_4 = \{1,1,2,2,3,3\} \setminus \{1,3,3\} = \{1,2,2\}$.

If S is a multiset with j scuts, then the following recurrence $\mathcal{C}(S,\gamma)$ (simplified from Definition 2.4 in [24]) produces a listing for all strings of the form $\beta\gamma$ where β has content S as they appear in cool-lex order:

$$\mathcal{C}(S,\gamma) = \mathcal{C}(R_1,\alpha_1\gamma), \mathcal{C}(R_2,\alpha_2\gamma), \ldots, \mathcal{C}(R_j,\alpha_j\gamma), tail(S)\gamma.$$

Note that $\mathcal{C}(S,\epsilon)$ will produce a listing of all strings with fixed-content S. Recall Fig. 1 illustrating the cool-lex order for $\mathbf{S}(\{1,1,2,2,3,3\})$. This is the same listing generated by $\mathcal{C}(\{1,1,2,2,3,3\},\epsilon)$. In particular observe that the strings are ordered by suffixes corresponding to the scuts: 23, 2, 1233, 133, 13, 1.

We now focus on how to modify this recurrence to list the necklaces with fixed-content S as they appear in cool-lex order.

Lemma 1. *If $a_1a_2\cdots a_n$ is a necklace that contains a smallest index t such that $a_t > a_{t+1}$, then $a_1\cdots a_{t-1}a_{t+1}a_ta_{t+2}\cdots a_n$ is a necklace.*

Proof. Let $\beta = b_1b_2\cdots b_n = a_1\cdots a_{t-1}a_{t+1}a_ta_{t+2}\cdots a_n$. Let β_j denote the rotation of β starting at b_j and let α_j denote the rotation of $\alpha = a_1a_2\cdots a_n$ starting at a_j. If $\beta \le \beta_j$ for each $2 \le j \le n$, then β is a necklace. Since α is a necklace, each $a_i \ge a_1$ and thus each $b_i \ge b_1$. Since $b_1\cdots b_{t-1}$ is non-decreasing it is straightforward to observe that $\beta_j > \beta$ for $2 \le j \le t+1$. Now consider the prefix of length t for β_j where $t+2 \le j \le n$. This prefix is the same as the length t prefix of α_j. If this prefix if less than or equal to $b_1\cdots b_t$, then it must be strictly less than $a_1\cdots a_t$ since $a_t > b_t$. But this contradicts the fact that α is a necklace. Thus this prefix must be strictly greater than $b_1\cdots b_t$. Thus $\beta_j \ge \beta$ for each $2 \le j \le n$ and hence β is a necklace.

Lemma 2. *$\mathcal{C}(S,\gamma)$ contains a necklace if and only if $tail(S)\gamma$ is a necklace.*

Proof. ($\Leftarrow$) $tail(S)\gamma$ is in $\mathcal{C}(S,\gamma)$ by definition. Thus if $tail(S)\gamma$ is a necklace then $\mathcal{C}(S,\gamma)$ contains a necklace. ($\Rightarrow$) If $\mathcal{C}(S,\gamma)$ contains necklace then it must be of the form $\lambda\gamma$ where λ has content S. If $\lambda = tail(S)$, then we are done. Otherwise, repeated application of Lemma 1 implies that $tail(S)\gamma$ is a necklace.

Based on Lemmas 1–2, the recurrence $\mathcal{C}(S,\gamma)$ can be updated to list only the necklaces as follows (where $\langle\,\rangle$ denotes an empty list).

$$\mathcal{N}(S,\gamma) = \begin{cases} \langle\,\rangle & \text{if } tail(S)\gamma \text{ is not a necklace;} \\ \mathcal{N}(R_1,\alpha_1\gamma), \ldots, \mathcal{N}(R_j,\alpha_j\gamma), tail(S)\gamma & \text{otherwise,} \end{cases}$$

Algorithm 1. Recursive algorithm to list all necklaces with content S as they appear in cool-lex order. The string $a_1a_2\cdots a_n$ is intialized to $tail(S)$, and the initial call is COOL(n).

```
1: procedure COOL( t)
2:     i ← t
3:     while a_i ≠ a_1 do
4:         while a_i = a_{i-1} do  i ← i-1
5:         for j from i to t do
6:             SWAP(j-1, j)
7:             if a_1 a_2 ··· a_n is a necklace then COOL(j-1)
8:         for j from t down to i do  SWAP(j-1, j)
9:         i ← i-1
10:    VISIT( )
```

The function COOL(t) in Algorithm 1 implements the above recurrence. Given content S, by initializing the global string $a_1a_2\cdots a_n$ to $tail(S)$, the initial call COOL(n) generates all necklaces with fixed-content S. The parameter t passed in the function COOL(t) indicates how the string $a_1a_2\cdots a_n$ is partitioned into the two pieces based on $\mathcal{N}(S',\gamma)$: $a_1a_2\cdots a_t = tail(S')$ and $a_{t+1}\cdots a_n = \gamma$. Each call COOL($t$) corresponding to $\mathcal{N}(S',\gamma)$ iterates through the scuts of S' in the proper order. This is done by scanning $tail(S') = a_1\cdots a_t$ from right to left until we reach an index i where $a_i \neq a_{i-1}$ (Line 4). To produce all scuts starting with a_{i-1}, and their corresponding recursive calls if a necklace can be produced, we iteratively shift this symbol through positions $i, i+1, \ldots, t$ obtaining a new scut for each swap (Lines 5–7). Once all scuts starting with a_{i-1} have been processed we restore $a_1\cdots a_t$ to $tail(S')$ (Line 8). We repeat this approach by continuing to traverse $tail(S)$ from right to left until we reach a symbol that is the same as a_1 (Line 3). The function VISIT() outputs the string $a_1a_2\cdots a_n$, and the function SWAP(i, j) swaps the symbols at index i and j in $a_1a_2\cdots a_n$.

When analyzing this algorithm, if every string tested in Line 7 was a necklace, then the work done by each necklace test could be assigned to the following recursive call. Since each recursive call generates at least one necklace, and since the necklace testing can be done in $O(n)$-time [1], the overall algorithm would run in $O(n)$-amortized time algorithm. However, within each recursive call, there could be a number of negative necklace tests. For instance, consider the string $\alpha = 112233112233$ and the call to COOL(6). This results in necklace tests for the following 6 strings, none of which are necklaces since the rotation starting with the suffix 112233 is smaller than string in question:

$$112323112233, 112332112233, 121233112233,$$
$$122133112233, 122313112233, 122331112233.$$

Fortunately there exists a simple optimization: once a string tested on Line 7 is **not** a necklace, then by applying Lemma 1 (and further observing the definition of a necklace) none of the following strings tested will be either. This optimization can be applied to COOL(t) by replacing Line 7 with the following fragment:

```
if a1a2 ··· an is a necklace then COOL(j−1)
else
    for s from j down to i do  SWAP(s−1, s)
    VISIT()
    return
```

This optimization ensures that at most one necklace test is negative per recursive call.

Theorem 1. *Let S denote a multi-set from the elements $1, 2, \ldots, k$ If $a_1 a_2 \cdots a_n$ is initialized to $tail(S)$, then a call to the optimized* COOL*(n) lists all necklaces with fixed-content S in cool-lex order in $O(n)$-amortized time per string.*

4 Constructing a Shorthand Universal Cycle for Fixed-Content

In this section, we provide the first explicit construction of a shorthand universal cycle for fixed-content strings. If the content of the strings is the multiset of symbols S, then the shorthand universal cycle is obtained by the applying the necklace-prefix algorithm to cool-lex order for S. More precisely, we use reverse cool-lex order. This order starts with the unique non-decreasing string. Then successive strings differ by our notion of a right-shift, which removes and reinserts a single symbol further to the right, while the intermediate symbols move one position to the left. Also recall that the relative order of the symbols has been inverted in our presentation, with respect to the original presentation of fixed-content cool-lex [24], so that we can use the traditional notion of a necklace being a string in its lexicographically least rotation.

Let $\mathcal{U}(S)$ denote the string resulting from the necklace-prefix algorithm applied to $\boldsymbol{S}(S)$ when listed in reverse cool-lex order. That is, $\mathcal{U}(S)$ is the concatenation of the aperiodic prefixes of necklaces with content S in reverse cool-lex order. An example of $\mathcal{U}(S)$ is provided in Fig. 3 for $S = \{1, 1, 2, 2, 3, 3\}$. Let $\boldsymbol{S}^{-1}(S)$ denote the shorthand representations of the strings in $\boldsymbol{S}(S)$.

Now we prove a preliminary result in Theorem 2, followed immediately by our main result in Theorem 3. Let $\mathcal{U}^{+}(S)$ be the result of concatenating the necklaces with content S in reverse cool-lex order. In other words, $\mathcal{U}^{+}(S)$ is the same as $\mathcal{U}(S)$, however, the periodic necklaces are not reduced to their aperiodic prefix.

Theorem 2. *The circular string $\mathcal{U}^{+}(S)$ contains every string in $\boldsymbol{S}^{-1}(S)$ at least once as a substring.*

Proof. Consider a string in $\boldsymbol{S}(S)$ whose last symbol is $x \in \{1, 2, \ldots, k\}$. At least one rotation of this string is a necklace. Thus, it can be written as $\mathbf{p}\mathbf{q}x$, such that $\mathbf{q}x\mathbf{p} \in \boldsymbol{N}(S)$. We need to find the string's shorthand reprsentation, $\mathbf{p}\mathbf{q}$, as a substring in the universal cycle $\mathcal{U}^{+}(S)$. In all of our cases, we will use the fact that $\mathbf{q}x\mathbf{p}$ is a necklace.

We first consider two special cases.

- If $\mathbf{p}$ is empty, then the desired substring $\mathbf{p}\mathbf{q} = \mathbf{q}$ is contained in the necklace $\mathbf{q}x\mathbf{p} = \mathbf{q}x$, and hence is in $\mathcal{U}^{+}(S)$.
- If $\mathbf{q}$ is empty, then the desired substring $\mathbf{p}\mathbf{q} = \mathbf{p}$ is contained in the necklace $\mathbf{q}x\mathbf{p} = x\mathbf{p}$, and hence is in $\mathcal{U}^{+}(S)$.

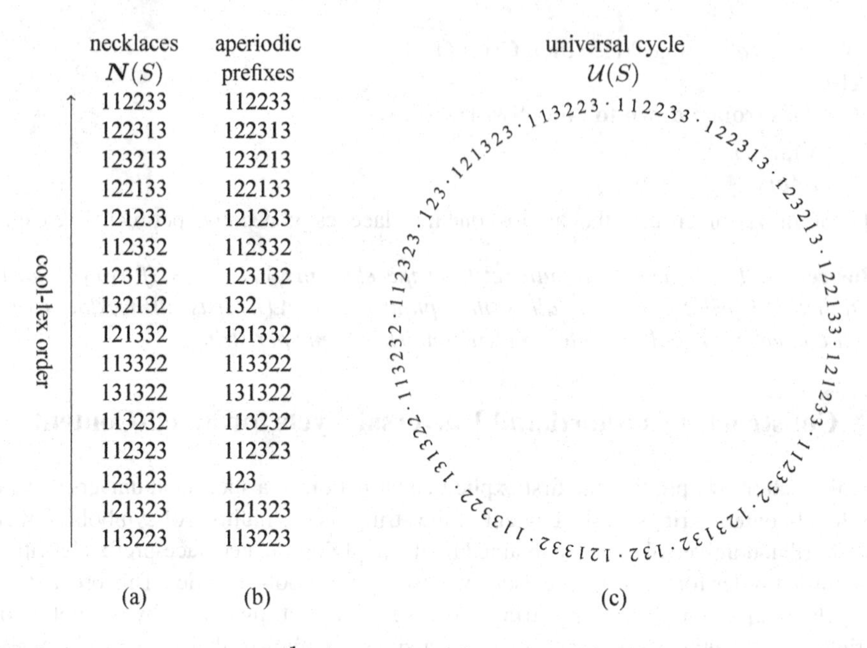

necklaces
$N(S)$
aperiodic
prefixes
universal cycle
$\mathcal{U}(S)$
112233 112233
122313 122313
123213 123213
122133 122133
121233 121233
112332 112332
123132 123132
132132 132
121332 121332
113322 113322
131322 131322
113232 113232
112323 112323
123123 123
121323 121323
113223 113223
cool-lex order
112233 · 122313 · 123213 · 122133 · 121233 · 112332 · 123132 · 132 · 121332 · 113322 · 131322 · 113232 · 112323 · 123 · 121323 · 113223
(a)
(b)
(c)

Fig. 3. A universal cycle for $\boldsymbol{S}^{-1}(S)$, where $S = \{1, 1, 2, 2, 3, 3\}$. The cycle uses the shorthand representation, and is constructed using the necklace-prefix algorithm on reverse cool-lex order. The fixed-content necklaces over S are given in reverse cool-lex order in column (a), they are reduced to their aperiodic prefix in column (b), and their concatenation gives the universal cycle in column (c).

In the remaining cases, we need to search across two necklaces to find the desired substring $\mathbf{pq}$. In other words, we must find a necklace $\alpha \in \boldsymbol{N}(S)$, such that $\mathbf{pq}$ is a substring of $\mathsf{next}(\alpha) \cdot \alpha$, and thereby a substring of $\mathcal{U}^+(S)$. Specifically, we will find $\alpha \in \boldsymbol{N}(S)$ with prefix $\mathbf{q}$, such that $\mathsf{next}(\alpha)$ has suffix $\mathbf{p}$. The next point handles the easiest remaining case.

- If $\mathbf{q}$ is not non-decreasing (i.e. its non-decreasing prefix is a strict prefix), then the necklace $\alpha = \mathbf{q}x\mathbf{p}$ again suffixes. To see why this is true, observe that α has prefix $\mathbf{q}$, and we claim that $\mathsf{next}(\alpha)$ has suffix $\mathbf{p}$. This is because $\mathsf{next}(\alpha)$ is obtained from α by shifting either a symbol in $\mathbf{q}$, or x, to the left by 2. Hence, the suffix $\mathbf{p}$ carries over from α to $\mathsf{next}(\alpha)$.

In the remaining cases, $\mathbf{p}$ and $\mathbf{q}$ are both non-empty, and $\mathbf{q}$ is non-decreasing by itself. We proceed with two representative cases, based on the non-decreasing prefixes of $\mathbf{q}x\mathbf{p}$ and $\mathbf{qp}$. In both cases, we create α by starting from the necklace $\mathbf{q}x\mathbf{p}$, and shifting x far enough to the right, so that it becomes the symbol that is shifted to the left in $\mathsf{next}(\alpha)$.

- If the non-decreasing prefix in $\mathbf{q}x\mathbf{p}$ is $\mathbf{q}$, then we start with $\alpha = \mathbf{q}x\mathbf{p}$, and repeatedly update the necklace until we have the desired properties. Let m be the length of the non-decreasing prefix, with regard to the successor rule in (2), and note that x is in the $(m+1)$st position in α. if $m = n - 1$, or $a_m > a_{m+2}$, or β is not a necklace, then we stop and use the current value of α. Otherwise, bubble x one position to the right to create a new value of α. Notice that α is a necklace due, to the fact that β is a necklace. Furthermore, x is again one symbol to the right of the non-decreasing prefix of α due to the fact that we had $a_m \leq a_{m+2}$ with respect to the previous value of α. Therefore, we can repeat the above steps until we find a suitable α. Observe that $\mathsf{next}(\alpha)$ will shift x to the left by (2b).

- If the non-decreasing prefix in $\mathbf{q}x\mathbf{p}$ is $\mathbf{q}x$, and the non-decreasing prefix in $\mathbf{qp}$ is $\mathbf{q}$, then the last symbol of $\mathbf{q}$ is larger than the first symbol in $\mathbf{p}$. Let α be the result of bubbling x one position to the right in $\mathbf{q}x\mathbf{p}$. Thus, the non-decreasing prefix of α is precisely $\mathbf{q}$. Let $m = |\mathbf{q}|$, with regard to the successor rule in (2), and note that x is in the $(m+1)$st position in α. Furthermore, $a_m \leq a_{m+2}$ based on the assumptions of this case, and $\beta = \mathbf{q}x\mathbf{p}$ is the necklace we started from. Therefore, α suffices since $\mathsf{next}(\alpha)$ will shift x to the left by (2c).

Theorem 3. *The string $\mathcal{U}(S)$ is universal cycle for fixed-content strings using their shorthand representation. In other words, every string in $\boldsymbol{S}^{-1}(S)$ appears in $\mathcal{U}(S)$ exactly once as a substring.*

Proof. Observe that $\mathcal{U}(S)$ has the correct length. That is, $|\mathcal{U}(S)| = |\boldsymbol{S}^{-1}(S)|$. This is due to the fact that every necklace contributes equally to both quantities. More precisely, a necklace whose period is p will contribute p symbols to $\mathcal{U}(S)$, and its p unique rotations to $\boldsymbol{S}^{-1}(S)$. Due to this equality, we only need to prove that each string in $\boldsymbol{S}^{-1}(S)$ appears in $\mathcal{U}(S)$ at least once as a substring. Because of Theorem 2, we can accomplish this by showing that $\mathcal{U}(S)$ has the same set of substrings of length $n-1$ as $\mathcal{U}^{+}(S)$. In other words, we can prove the result by showing that no substrings are lost when we reduce each necklace to its aperiodic prefix in the concatenation.

Consider an arbitrary periodic necklace whose aperiodic prefix is γ. Since it is periodic, we can write it as γ^r for some $r > 1$. First we prove that $\mathsf{next}(\gamma^r)$ is aperiodic. To see why this is true, observe that $\mathsf{next}(\gamma^r)$ will contain a prefix that is lexicographically smaller than γ. Thus, $\mathsf{next}(\gamma^r)$ is aperiodic. Similarly, $\mathsf{prev}(\gamma^r)$ is aperiodic.

Now we compare the local area around γ^r in $\mathcal{U}^{+}(S)$, and the local area around its reduction tp γ in $\mathcal{U}(S)$.

$$\underbrace{\cdots \mathsf{next}(\gamma^r) \cdot \gamma^r \cdot \mathsf{prev}(\gamma^r) \cdots}_{\mathcal{U}^{+}(S)} \qquad \underbrace{\cdots \mathsf{next}(\gamma^r) \cdot \gamma \cdot \mathsf{prev}(\gamma^r) \cdots}_{\mathcal{U}(S)}$$

We claim that the two concatenations have the same set of substrings of length $n-1$. To establish this fact, let the length of the aperiodic prefix of the necklace γ^r be $t = |\gamma| = n/r$. Now observe two points. First, $\mathsf{next}(\gamma^r)$ and γ^r share a suffix of length at least $n-t-2$. This is due to the cool-lex successor rule in (2) and because the length of the non-decreasing prefix in γ^r is at most t. Second, γ^r and $\mathsf{prev}(\gamma^r)$ must share a prefix of length at least 1, since they are both necklaces. Therefore, the concatenation in $\mathcal{U}^{+}(S)$ has at least $(n-t-2)+t+1 = n-1$ consecutive symbols from γ^r, which means that it contains γ^r in shorthand. Furthermore, the substrings of length $n-1$ in $\mathcal{U}^{+}(S)$ before and after this shorthand copy of γ^r are identical to those in $\mathcal{U}(S)$.

4.1 Efficiency

To construct the reverse of the universal cycle $\mathcal{U}(S)$, we can directly apply Algorithm 1 to list $\mathbf{N}(S)$ in cool-lex order with a simple modification. Instead of outputting the current necklace $\alpha = a_1 a_2 \cdots a_n$, the function VISIT( )

▷ determines the length p of the aperiodic prefix of α and then
▷ outputs $a_p a_{p-1} \cdots a_1$.

Since the aperiodic prefix of α can be determined in $O(n)$ (see [1]), the modified algorithm still runs in $O(n)$-amortized time per symbol. Since the total length of $\mathcal{U}(S)$ is proportional to $n|\mathbf{N}(S)|$ (see Sect. 5 in [18] which implies $|\mathcal{U}(S)| \geq n|\mathbf{N}(S)|/2$) we obtain the following theorem.

Theorem 4. *The universal cycle $\mathcal{U}(S)$ for fixed-content strings using their shorthand representation can be generated in $O(1)$-amortized time per symbol using $O(n)$ space.*

Appendix - Universal Cycles for Fixed-Content Strings Using Shorthand Representation in $O(1)$ Time Per Symbol

```
#include <stdio.h>
int N, K, a[100];
//------------------------------------------------------------------
// If a[1..n] is a necklace return its period p; otherwise return 0
//------------------------------------------------------------------
int Necklace() {
    int i, p=1;

    for (i=2; i<=N; i++) {
        if (a[i-p] > a[i]) return 0;
        if (a[i-p] < a[i]) p = i;
    }
    if (N % p != 0) return 0;
    return p;
}
//--------------------------------
void Visit() {
    int i;
    for (i=Necklace(); i>=1; i--) printf("%d ", a[i]);
}
//--------------------------------
void Swap(int i, int j) {
    int temp;
    temp = a[i];   a[i] = a[j];   a[j] = temp;
}
//---------------------------------
void Gen(int t) {
    int i,j,s;

    i = t;
    while (a[i] != a[1]) {
        while (a[i] == a[i-1]) i--;
        for (j=i; j<=t; j++) {
            Swap(j-1,j);
            if (Necklace()) Gen(j-1);
            else {
                for (s=j; s>=i; s--) Swap(s-1,s);
                Visit();
                return;
            }
        }
```

```
            for (j=t; j>=i; j--) Swap(j-1,j);
            i--;
        }
        Visit();
}
//--------------------------------
int main() {
        int i,j,tmp;

        printf("Enter K: "); scanf("%d", &K);
        N = 0;
        for (i=1; i<=K; i++) {
            printf("N_%d: ", i);  scanf("%d", &tmp);
            for (j=1; j<=tmp; j++) a[N+j] = i;
            N += tmp;
        }
        Gen(N);
}
```

References

1. Booth, K.S.: Lexicographically least circular substrings. Inf. Process. Lett. **10**(4/5), 240–242 (1980)
2. Cantwell, A., Geraci, J., Godbole, A., Padilla, C.: Graph universal cycles of combinatorial objects. Adv. Appl. Math. **127**, (2021)
3. Chung, F., Diaconis, P., Graham, R.: Universal cycles for combinatorial structures. Discret. Math. **110**(1), 43–59 (1992)
4. Diaconis, P., Graham, R.L.: Products of universal cycles. In: Demaine, E., Demaine, M., Rodgers, T. (eds.) A Lifetime of Puzzles, pp. 35–55. A K Peters/CRC Press (2008)
5. Fredricksen, H., Kessler, I.: An algorithm for generating necklaces of beads in two colors. Discret. Math. **61**(2), 181–188 (1986)
6. Fredricksen, H., Maiorana, J.: Necklaces of beads in k colors and k-ary de Bruijn sequences. Discret. Math. **23**, 207–210 (1978)
7. Gabric, D., Sawada, J., Williams, A., Wong, D.: A framework for constructing de Bruijn sequences via simple successor rules. Discret. Math. **341**(11), 2977–2987 (2018)
8. Gabric, D., Sawada, J., Williams, A., Wong, D.: A successor rule framework for constructing k -ary de Bruijn sequences and universal cycles. IEEE Trans. Inf. Theory **66**(1), 679–687 (2020)
9. Gilbert, E.N., Riordan, J.: Symmetry types of periodic sequences. Illinois J. Math. **5**(4), 657–665 (1961)
10. Holroyd, A.E., Ruskey, F., Williams, A.: Shorthand universal cycles for permutations. Algorithmica **64**(2), 215–245 (2012)
11. Horan, V., Hurlbert, G.: Universal cycles for weak orders. SIAM J. Discret. Math. **27**(3), 1360–1371 (2013)
12. Jackson, B., Stevens, B., Hurlbert, G.: Research problems on Gray codes and universal cycles. Discret. Math. **309**(17), 5341–5348 (2009)
13. Johnson, J.R.: Universal cycles for permutations. Discret. Math. **309**(17), 5264–5270 (2009)
14. Leitner, A., Godbole, A.: Universal cycles of classes of restricted words. Discret. Math. **310**(23), 3303–3309 (2010)
15. Ruskey, F., Sawada, J., Williams, A.: Binary bubble languages and cool-lex Gray codes. J. Comb. Theory Ser. A **119**(1), 155–169 (2012)
16. Ruskey, F., Sawada, J., Williams, A.: De Bruijn sequences for fixed-weight binary strings. SIAM J. Discret. Math. **26**(2), 605–617 (2012)

17. Ruskey, F., Williams, A.: An explicit universal cycle for the (n-1)-permutations of an n-set. ACM Trans. Algorithms (TALG) **6**(3), 45 (2010)
18. Sawada, J.: A fast algorithm to generate necklaces with fixed content. Theor. Comput. Sci. **301**(1), 477–489 (2003)
19. Sawada, J., Stevens, B., Williams, A.: De Bruijn Sequences for the Binary Strings with Maximum Density. In: Katoh, N., Kumar, A. (eds.) WALCOM 2011. LNCS, vol. 6552, pp. 182–190. Springer, Heidelberg (2011). https://doi.org/10.1007/978-3-642-19094-0_19
20. Sawada, J., Williams, A., Wong, D.: Universal Cycles for Weight-Range Binary Strings. In: Lecroq, T., Mouchard, L. (eds.) IWOCA 2013. LNCS, vol. 8288, pp. 388–401. Springer, Heidelberg (2013). https://doi.org/10.1007/978-3-642-45278-9_33
21. Sawada, J., Williams, A., Wong, D.: The lexicographically smallest universal cycle for binary strings with minimum specified weight. J. Discret. Algorithms 28, 31–40 (2014). StringMasters 2012, 2013 Special Issue (Volume 1)
22. Sawada, J., Williams, A., Wong, D.: Generalizing the classic greedy and necklace constructions of de Bruijn sequences and universal cycles. Electron. J. Comb. 23(1), P1.24 (2016)
23. Sawada, J., Wong, D.: Efficient universal cycle constructions for weak orders. Discret. Math. **343**(10), (2020)
24. Williams, A.: Loopless generation of multiset permutations using a constant number of variables by prefix shifts. In: Proceedings of the Twentieth Annual ACM-SIAM Symposium on Discrete Algorithms, SODA 2009, pp. 987–996. SIAM (2009) SODA 2009, pp. 987–996. SIAM (2009)
25. Williams, A.: Shift Gray codes. PhD thesis, University of Victoria (2009)

Routing on Heavy-Path WSPD-Spanners

Prosenjit Bose and Tyler Tuttle(✉)

School of Computer Science, Carleton University, Ottawa, Canada
jit@scs.carleton.ca, TylerTuttle@cmail.carleton.ca

Abstract. Using the Well-Separated Pair Decomposition (WSPD) of Callahan and Kosaraju [JACM, 42(1):67–90, 1995], we present a construction of a $1 + 2/s + 2/(s-1)$-spanner of size $O(s^d n)$ on a set of n points in $\mathbf{R}^d$ that we call a heavy-path WSPD-spanner, where $s > 2$ is the separation ratio. We also show that this graph has a hop spanning ratio of at most $2 \lg n + 1$. The heavy-path WSPD-spanner is amenable to local routing. We present a memoryless local routing algorithm for heavy-path WSPD-spanners. The routing ratio is at most $1 + 4/s + 1/(s-1)$ and at least $1 + 4/s$ and the number of edges on a path found by the algorithm is bounded by $2 \lg n + 1$. A total of $O(s^d n \log n)$ bits of information is distributed among the vertices of the spanner in the form of routing tables to aid the routing algorithm.

Keywords: Well-Separated Pair Decomposition · Spanner · Routing

1 Introduction

The Well-Separated Pair Decomposition (WSPD) of a set P of n points in $\mathbf{R}^d$ is a versatile structure that has found many applications [12]. Among these applications is the ability to construct a $(1 + \varepsilon)$-spanner for any $\varepsilon > 0$. A t-spanner in this context is a weighted graph whose vertex set is P, whose edges are weighted by the Euclidean distance between their endpoints and between every pair of points $x, y \in P$, there exists a path in the graph whose weight is at most t times the length of the segment xy. This path is often referred to as a *spanner path*. Typically, a t-spanner has a linear number of edges and serves as an approximation of the complete graph.

A t-spanner guarantees the existence of a short path between any pair of vertices, but in most applications, the mere existence of a short path is not sufficient. One requires that the path actually be computed. There exist many algorithms to compute shortest paths in weighted graphs, such as Dijkstra's algorithm [14]. However, Dijkstra's algorithm requires full knowledge of the graph, in the sense that the routing algorithm needs to know the vertex set and edge set of the graph.

The problem offers different challenges in the *online* setting, where the routing algorithm has no knowledge of the graph at the outset. As such, the algorithm starts with the knowledge of the source and destination vertices and explores the

Research supported in part by NSERC.

A. Lubiw et al. (Eds.): WADS 2021, LNCS 12808, pp. 613–626, 2021.
https://doi.org/10.1007/978-3-030-83508-8_44

graph while trying to reach the destination. A route or path is constructed incrementally. After each step, the algorithm acquires the information stored at each vertex, called the routing table. The routing table can store the neighbourhood of the vertex, and some additional bits of information.

If the *only* information that the algorithm uses to make its forwarding decision is the information stored at the current vertex and knowledge of the destination vertex, then the algorithm is called *local*. In addition, if the algorithm has no memory and each step is a local decision, then the algorithm is called *memoryless*. If no routing table is stored at a vertex, then an online local memoryless routing algorithm cannot identify a short path in general or any path for that matter [7]. An adversary can fool such an algorithm to cycle and never reach its destination. As such, the goal is to store as little information as possible in the routing tables to be able to compute a short path.

The ratio of the length of the path found by a routing algorithm and the Euclidean distance between the source and destination vertex is called the *routing ratio*. Note that the routing ratio is an upper bound on the spanning ratio. In the literature, online local memoryless (or $O(1)$-memory) routing algorithms have been designed for various families of geometric spanners such as Θ-graphs [13,19], Yao-graphs [22,25], or Delaunay Graphs [5,6,8,10].

In this article, we focus on the design of an online local memoryless routing algorithm on a spanner obtained from a WSPD of a point set P of n points in $\mathbf{R}^d$. Given a WSPD of P, one can construct many different $(1+\varepsilon)$-spanners since there is some flexibility in the structure. Therefore, we present a specific construction of a $(1+\varepsilon)$-spanner from a WSPD, which we call a *Heavy-Path WSPD-spanner*. We show that a Heavy-Path WSPD-spanner has spanning ratio $1+2/s+2/(s-1)$ and $O(s^d n)$ size, where $s > 2$ is the separation ratio. The larger the value of s, the closer ε is to zero. Our spanner also has the property that spanner paths have at most $2\lg n + 1$ edges.

Our main contribution is a memoryless local routing algorithm for heavy-path WSPD-spanners. The routing ratio is at most $1+4/s+1/(s-1)$ and at least $1+4/s$ which is close to the spanning ratio. Moreover, the number of edges on the path is also bounded by $2\lg n + 1$. Each vertex v of the graph stores $O(\deg(v)\log n)$ bits of information to aid the routing algorithm, where $\deg(v)$ is the degree of v. This implies a total of $O(s^d n \log n)$ bits in total.

Our result is a four-fold improvement on previous work in this area [4,9,11]. Our routing algorithm works on d-dimensional point sets whereas all previous results are focused on 2 dimensions. Our algorithm uses fewer bits in total for the routing tables. Our routing algorithm has the lowest routing ratio. Finally, our routing algorithm returns a spanner path that uses at most $2\lg n + 1$ edges. We also feel that our algorithm is simpler than the previous algorithms, however, it is difficult to quantify simplicity. See Table 1 for a comparison. A full and detailed description of these results appears in Tuttle's masters thesis [24].

Now, we will describe the data structures that are needed to construct the heavy path WSPD spanner. In particular we will define compressed quadtrees, which are used to construct a WSPD, and WSPDs themselves.

Table 1. Routing algorithms for WSPD spanners. B is the number of bits needed to store a bounding box, Δ is the ratio of the largest distance between two points to the smallest, and s is the separation ratio. An algorithm is k-local if it uses information about vertices at most k hops away to make routing decisions.

Algorithm	Routing tables	Routing Ratio	Hop	Memory
[11] (2-local)	$O(s^2n^2B)$	$1+6/(s-2)+4/s$	$O(n)$	memoryless
[11] (1-local)	$O(s^2nB)$	$1+8/(s-2)+4/s+8/s^2$	$O(n)$	memoryless
[4]	$O(s^2n\log\Delta)$	$1+O(1/s)$	$O(n)$	$O(\log\Delta)$
This article	$O(s^dn\log n)$	$1+4/s+1/(s-1)$	$2\lg n+1$	memoryless

1.1 Compressed Quadtrees

A quadtree is a tree data structure for storing spatial data. Let S be a set of n points in $\mathbf{R}^d$. If $n=1$, then the quadtree for S is a single node that stores the lone point of S. If $n>1$, to construct a quadtree for S we need a hypercube that contains S. Let C be a hypercube that contains S. We can assume that this is given to us, but if not it is simple to construct such a hypercube in time $O(dn)$.

Subdivide C into 2^d smaller hypercubes $C_i,\dots C_{2^d}$ by bisecting it along each dimension. For each nonempty C_i, recursively construct a quadtree on the points in C_i. The root of the quadtree stores the hypercube C. Each of the recursively constructed quadtrees is a child of the root.

In a (uncompressed) quadtree, each node a is associated with a hypercube $C(a)$ that contains all the points in the subtree of a. If a is at level i in the tree, then the side length of $C(a)$ is $2^{-i}L$, where L is the side length of hypercube associated to the root.

A quadtree is a tree with n leaves. Each internal node has at least one child, and at most 2^d children. The fact that a node can have only a single child means the height of the tree can be unbounded. We can fix this in the following way. If a quadtree has a long chain of internal nodes with only one child, then compress them all into a single edge. The resulting structure is called a compressed quadtree, and the height is now linear with respect to the number of points in the worst case.

In a compressed quadtree, a node no longer corresponds to just one hypercube. Instead, each node a corresponds to two hypercubes. A node in a compressed quadtree might correspond to an entire path in the uncompressed quadtree. We store the hypercube $C_L(a)$ that corresponds to the shallowest node on that path, and the hypercube $C_S(a)$ that corresponds to the deepest node on that path. If $p(a)$ denotes the parent of a, then $C_L(a)$ is obtained by splitting $C_S(p(a))$ along each dimension. Let $S(a)$ denote the set of points stored in the leaves of the subtree rooted at a. For any compressed quadtree node, we have $S(a)\subset C_S(a)\subseteq C_L(a)$. The two hypercubes $C_S(a)$ and $C_L(a)$ can be equal if the node a does not correspond to a compressed chain of nodes in the quadtree.

Theorem 1 ([1, Section 19.2.5]). *Let S be a set of n points in $\mathbf{R}^d$. A compressed quadtree for S can be constructed in $O(dn\log n)$ time.*

See [1] for a tof different quadtree variants. For our application, we will need the following property of compressed quadtrees.

Lemma 1. *Let T be a compressed quadtree, and let a be a non-root node of T. The node a corresponds to two hypercubes, $C_L(a)$ and $C_S(a)$. Let $\ell(a)$ be the diagonal length of $C_S(a)$. Note that this is an upper bound on the diameter of the points stored in the subtree of a. We have $\ell(a) \leq (1/2)\ell(p(a))$, where $p(a)$ is the parent of a in T.*

1.2 The Well-Separated Pair Decomposition

Let S and S' be two point sets in $\mathbf{R}^d$. We say that S and S' are well-separated with respect to $s > 2$ if $d(S, S') \geq s \cdot \max\{\text{diam}S, \text{diam}S'\}$, where $d(S, S') = \min\{|pq| : p \in S, q \in S'\}$ and diam$S$ is the diameter of S, the maximum distance between two points in S. The number s is called the separation ratio.

The following lemma [12] about well-separated pairs will make precise the idea that distances in one set are small compared to distances between sets.

Lemma 2. *Let S and S' be well-separated point sets with respect to $s > 2$. Then for any points $p, p' \in S$ and $q, q' \in S'$: $|pp'| \leq (1/s)|pq|$, and $|p'q'| \leq (1+2/s)|pq|$.*

A well-separated pair decomposition (WSPD) of S is a sequence $\{A_1, B_1\}, \ldots, \{A_m, B_m\}$ of pairs of subsets of S such that (a) $A_i \cap B_i = \varnothing$ for all i (b) for each pair p, q of points in S there is exactly one i such that $p \in A_i$ and $q \in B_i$ (or $p \in B_i$ and $q \in A_i$) (c) A_i and B_i are s-well separated for all i.

Given a compressed quadtree T, we can construct a WSPD with a recursive algorithm [16, Section 3.1.1]. The following theorem summarizes the construction.

Theorem 2 ([16, Theorem 3.10]). *Given a set S of n points in $\mathbf{R}^d$, a WSPD with separation ratio $s > 2$ with $O(s^d n)$ pairs can be constructed using a compressed quadtree in time $O(d(n \log n + s^d n))$.*

The WSPD that results from this algorithm has an important property that we will need in Sect. 2.1, so we will state it now.

Lemma 3. *Let T be a compressed quadtree for some point set P, and let W be a WSPD computed using T. Every pair in W has the form $\{S(a), S(b)\}$ for some nodes a, b of T. Let p, q be any two points of P and let $\{S(a), S(b)\}$ be the pair that separates them. If c is a node that stores both p and q in its subtree, then a and b are both descendants of c.*

In the construction of a WSPD, we used compressed quadtrees. There are alternative tree structures that have been used instead. For example, the fair split tree of Callahan and Kosaraju [12]. The reason compressed quadtrees are used is so that we can use Lemma 1. The diameter of the hypercube representing a compressed quadtree node is a constant fraction of the diameter of its parent. In a fair split tree this fraction depends on the dimension of the point set since a split is only done along one dimension, instead of along all dimensions simultaneously. This property of compressed quadtrees will be used in the analysis of the routing algorithm.

2 The Heavy Path WSPD Spanner

We now describe the heavy path decomposition of a tree. Let T be a rooted tree. If a is a node of T, then the size of a is the number of leaves in the subtree rooted at a. It is worth noting here that a node a is considered to be an ancestor and a descendant of itself. For each internal node a, choose one child of maximal size (breaking ties arbitrarily), and mark the edge from a to that child as heavy. The other edges are marked light. If b is a child of a and the edge from a to b is heavy, then we call b the heavy child of a. Otherwise b is a light child of a. What results is a decomposition of the tree into heavy paths, one for each leaf node. The heavy path decomposition of a tree with n leaves can be computed in $O(n)$ time [23].

For an internal node a, let $r(a)$ be the leaf node defined by following the unique heavy path down the tree starting from a. This leaf is called the representative of a. Let $h(a)$ be the node defined by following the heavy path up the tree, again starting from a, until the edge to the parent is no longer heavy.

Lemma 4 ([23, Lemma 1]). *The number of light edges on any root-to-leaf path in a heavy path decomposition of a compressed quadtree is at most* $\lg n$, *where* n *is the number of leaves in the compressed quadtree.*

Lemma 5. *Let* T *be a tree, and let* a *be an internal node of* T. *Compute a heavy path decomposition of* T. *Let* $r(a)$ *be the representative of* a. *Then for every node* b *on the path from* $h(a)$ *to* $r(a)$, *we have* $r(b) = r(a)$.

2.1 Constructing a Heavy Path WSPD Spanner

Constructing a spanner graph given a WSPD is simple. For each pair in the WSPD, choose an arbitrary point from each set and add an edge between those two points. The result is a t-spanner for $t = (s + 4)/(s - 4)$, where $s > 4$ is the separation ratio of the WSPD [21]. Since we can choose these points in any manner, we are free to decide on a scheme that benefits our application. In this article, we will choose the points using a method based on the heavy path decomposition. The spanner that we construct will be called a heavy path WSPD spanner. This construction originally appeared in Arya et al. [2,3] for the fair split tree. Here we present it for WSPDs built from a compressed quadtree.

Let T be the compressed quadtree used to compute the WSPD. Compute a heavy path decomposition of T. For each pair $\{A, B\}$ in the WSPD, there is a corresponding pair $\{a, b\}$ of nodes in T. The edge that we add to the graph will be between the points $r(a)$ and $r(b)$.

Now we will prove that this graph is a $(1 + 2/s + 2/(s - 1))$-spanner. To construct a path between two points p and q, consider Algorithm 1.1. Let $\{S(a), S(b)\}$ be the WSPD pair that separates p from q. The algorithm adds an edge between $r(a)$ and $r(b)$, and recursively constructs a path from p to $r(a)$ and from $r(b)$ to q.

To analyze the spanning ratio, we will first consider a special case, where q is the representative of a node storing p. In other words, $h(q)$ is an ancestor of

Algorithm 1.1 Constructing a short path in a heavy path WSPD spanner

```
Input: Two points p and q in a heavy path WSPD spanner
Output: A path between p and q
procedure BuildPath(p, q)
    if p = q then                                                  ▷ Base case
        return ∅
    else
        let {S(a), S(b)} be the WSPD pair that separates p from q
        return BuildPath(p, r(a)) ∪ r(a)r(b) ∪ BuildPath(r(b), q)
    end if
end procedure
```

$h(p)$. In this case, in every call made to BuildPath that does not immediately return, at least one of the two recursive calls will be to the base case.

Lemma 6. *Let S be a set of points in $\mathbf{R}^d$, and let T be a compressed quadtree for the points of S. Construct a heavy path WSPD spanner for S. Let p and q be points stored in the leaves of T such that q is the representative of some node containing p. In a call to* BuildPath*(p, q), at most one edge is added at each level of recursion.*

Consider an initial call to BuildPath(p, q), where q is not necessarily the representative of an ancestor of p. Two recursive calls are made, to BuildPath$(p, r(a))$ and BuildPath$(r(b), q)$. Both of these calls satisfy the conditions for Lemma 6.

We can also bound the length of the edges being added to the path, as a function of the recursion depth.

Lemma 7. *Let S be a set of points in $\mathbf{R}^d$, and let T be a compressed quadtree for the points of S. Construct a heavy path WSPD spanner for S. Let p and q be points of S. Consider the series of recursive calls made during a call to* BuildPath*(p, q). If the level of recursion of some call is k, the length of the edge added during that call is at most $(1/s)^k|pq|$.*

Using these two lemmas, we can bound the spanning ratio of the path between p and q. Note that this implies the graph is connected.

Theorem 3. *Let S be a set of n points in $\mathbf{R}^d$. The heavy path WSPD spanner G for S has a spanning ratio of at most $1 + 2/s + 2/(s-1)$.*

Proof. Let p and q be points of S. Algorithm 1.1 constructs a path from p to q. The edge added from $r(a)$ to $r(b)$ has length at most $(1 + 2/s)|pq|$ by Lemma 2. The length of the path from p to $r(a)$ can be bounded using Lemma 6 and Lemma 7. There is at most one edge being added at each level of recursion, and the length of the edge being added at level k is at most $(1/s)^k|pq|$. Let M be the maximum recursion depth. Therefore, the length of the path from p to $r(a)$ is at most

$$\sum_{k=1}^{M}\left(\frac{1}{s}\right)^k |pq| \leq \frac{1}{s-1}|pq|.$$

The length of the path from $r(b)$ to q can be bounded in the same way. Therefore the total length of the path is at most

$$\frac{1}{s-1}|pq| + \left(1+\frac{2}{s}\right)|pq| + \frac{1}{s-1}|pq| = \left(1+\frac{2}{s}+\frac{2}{s-1}\right)|pq|.$$

In addition to bounding the length of the path, we can bound the number of edges on the path from p to q, using Lemma 4. This is because every edge on the spanner path "traverses" at least one light edge. The diameter of a spanner is the maximum number of edges over all the shortest paths between any pair of points in the spanner.

Lemma 8. *For two points p and q in a heavy path WSPD spanner, the number of edges on the path from p to q found by* BUILDPATH*(p, q) is at most $2\lg n + 1$. In other words, the heavy path WSPD spanner is a $(2\lg n + 1)$-hop spanner.*

Proof. Let $\{S(a), S(b)\}$ be the WSPD pair that separates p from q. Consider the subpath $p = p_1, p_2, \ldots, p_k = r(a)$. For each edge $p_i p_{i+1}$, we know that p_i is contained in the subtree of $h(p_{i+1})$, where $h(p_{i+1})$ is the shallowest node in the compressed quadtree that p_{i+1} is the representative of.

The sequence of nodes $h(p_1), h(p_2), \ldots, h(p_k)$ must then all lie on the same root-to-leaf path (that is, the path from p to the root). Since all of these nodes have different representatives, by Lemma 4 there can be at most $\lg n$ of them.

The same is true for the subpath between $r(b)$ and q, and then adding the edge between $r(a)$ and $r(b)$ gives an upper bound of $2\lg n + 1$ edges on the spanner path.

We end this section with a theorem that summarizes the entire construction of a heavy path WSPD spanner and all its properties. Note that the BUILDPATH algorithm that finds a path between two points in a heavy path WSPD spanner is not a local algorithm since it requires knowing the pair $\{S(a), S(b)\}$ from p.

Theorem 4. *Let S be a set of n points in $\mathbf{R}^d$, and let $s > 2$. In $O(d(n \log n + s^d n))$ time, we can construct a graph G called a heavy path WSPD spanner with the following properties:*

- *The number of edges in G is $O(s^d n)$.*
- *G is a $(1 + 2/s + 2/(s-1))$-spanner.*
- *G is a $(2\lg n + 1)$-hop spanner.*

Additionally, between any two points there is a single path (found by algorithm BUILDPATH*) that achieves both the spanning and hop-spanning ratio.*

3 Local Routing in Euclidean Space

In this section, we present a local routing algorithm for heavy path WSPD spanners. Let S be a set of points, T be a compressed quadtree for S, W be a WSPD computed using T, and G be a heavy path WSPD spanner constructed as described in the previous section. We now have all the tools to describe the routing algorithm. First, we will explain what we need to store at each vertex. Then, we can present the routing algorithm and analyze it.

3.1 Routing Tables

First, we describe a labelling scheme for the nodes of T. The vertices of G will store these labels. The message will only use the label of the destination to route. In other words, the algorithm is memoryless.

Each leaf will get a unique label in the range $1, 2, \ldots, n$. Perform a depth-first traversal of T, and label the leaves in the order that they are visited. The label of an internal node will be the set of all the labels in the leaves of that node's subtree. We call this the DFS labelling scheme. The labelling scheme ensures that this set will be an interval.

Lemma 9. *Let T be a tree, and label its leaves using a depth first search. Let a be a node of T. Let I be the set of labels of the leaves in the subtree rooted at a. The labels form a contiguous subset of $\{1, 2, \ldots, n\}$. That is, if i is the minimum label and j is the maximum label in I, then $I = \{i, i+1, \ldots, j-1, j\}$.*

Since we only need to store the minimum and maximum labels of each interval, we only need $2 \lg n$ bits to store the label of an internal node.

In a depth-first search, the children of a node can be visited in any order. If we always visit the child with the largest subtree first (i.e., always follow the heavy edge), and then visit the other children in an arbitrary order, then we can save some memory as shown in the following lemma. We call this type of DFS labelling scheme a heavy path DFS labelling scheme.

Lemma 10. *Let T be a compressed quadtree and let a be an internal node of T. In a heavy path DFS labelling of T, the label of $r(a)$ is the minimum label of all the points stored in the subtree of a. That is, if the label of a is $[x, y]$, then the label of $r(a)$ is x.*

We now describe the information that needs to be stored in the routing table of a vertex u. First, store the label of u. Second, for each neighbour v of u, let $\{S(a_v), S(b_v)\}$ be the WSPD pair that generated the edge between u and v, where $v \in S(b_v)$, i.e., v is the leaf in the subtree of b_v with minimum label. Store the labels (defined by the heavy path DFS labelling of T) of v, b_v, and $h(v)$. Recall that $h(v)$ is the shallowest node in T for which v is a representative. Notice that the label of a_v is not stored, as it is never used by the routing algorithm.

Lemma 11. *The total size of the routing tables is $O(s^d n \log n)$.*

Proof. The label of a point is a single integer in the range $\{1,\ldots,n\}$, and the label of an internal node is two integers in the same range, so in total we need to store $5\lg n$ bits for each neighbour of u. This can be improved by applying Lemma 10. We know that v is the representative of b_v, since the edge uv was generated by the pair $\{a_v, b_v\}$. We also know that v is the representative of $h(v)$, by the definition of $h(v)$. So if x is the label of v, then the labels of b_v and $h(v)$ are of the form $[x,y]$ and $[x,z]$, respectively. Since three of the integers being stored are equal, we actually only need $3\lg n$ bits.

The total size of the routing table at a vertex u of G is $(3\deg(u)+1)\lg n$, where $\deg(u)$ is the number of neighbours of u in G. The total size of the routing tables stored in the entire graph is

$$\sum_{u\in P}(3\deg(u)+1)\lg n = (6m+n)\lg n$$

where m is the number of edges in the spanner. Since we know $m = O(s^d n)$, the total size of the routing table is therefore $O(s^d n\log n)$.

3.2 Routing in a Heavy Path WSPD Spanner

We now present the routing algorithm. Let p be the starting vertex, and let q be the destination vertex. We can assume that the label of the destination q is stored with the message. No other information needs to be stored with the message (that is, the algorithm is memoryless). The algorithm proceeds in two stages: the ascending stage and the descending stage. We first check if we are in the descending stage of the algorithm. If so, perform a descending step. If not, perform an ascending step. We will refer to this algorithm as the heavy path routing algorithm.

1. [*Descending step*] If u has a neighbour v (with WSPD pair $\{a_v, b_v\}$) such that $q \in b_v$, then forward the message to v.
2. [*Ascending step*] Otherwise, find the representative of the parent of $h(u)$, and forward the message to that vertex.

The proof that this routing algorithm guarantees delivery is split into two stages. Let $\{S(a), S(b)\}$ be the WSPD pair separating p from q. First we will prove that the ascending step will be applied until the message reaches $r(a)$. Then, we will prove that the descending step will be applied until the message reaches its destination. That is, the routing algorithm can be split into two "stages": a series of ascending steps followed by a series of descending steps.

First we need to show that it is possible to implement an ascending step using only the information stored in the vertices of G.

Lemma 12. *The representative of the parent of $h(u)$ is a neighbour of u in G, and can be found using only the information in the routing table at u.*

Another way of viewing Lemma 12 is that one application of the ascending step will move the message one light edge "up" in the quadtree.

The next two lemmas prove that, from p, the routing algorithm will repeatedly apply an ascending step until $r(a)$ is reached. This part of the algorithm is called the ascending stage.

Lemma 13. *Starting from p, repeated application of the ascending step will forward the message to $r(a)$.*

Lemma 14. *The ascending step is always applied if u is in a, but not equal to $r(a)$.*

Once the message reaches $r(a)$, the descending step will be applied until the destination is reached.

Lemma 15. *If u is on the path constructed in Algorithm 1.1 and not in $S(a)$, then the descending step is applied and will forward the message to the next point on the spanner path.*

Putting the ascending and descending steps together will therefore successfully route a message from p to q.

Theorem 5. *The heavy path routing algorithm will successfully route a message in a heavy path WSPD spanner, with information stored in each vertex as outlined in Sect. 3.1.*

3.3 Analysis of the Local Routing Algorithm

In this section we will bound the routing ratio of the heavy path routing algorithm. First we will bound the length of the path found in the descending stage, as it is much easier to do.

Lemma 16. *Let p and q be points in a heavy path WSPD spanner. The length of the path constructed during the descending stage of the heavy path routing algorithm is at most $(1 + 2/s + 1/(s-1))|pq|$.*

Proof. Lemma 15 implies that the descending stage finds a path from $r(a)$ to q, where $r(a)$ is the representative of the set containing p in the WSPD pair $\{S(a), S(b)\}$ that separates p from q.

From Theorem 3, we know that the edge $r(a)r(b)$ has length at most $(1 + 2/s)|pq|$, and that the subpath constructed in Algorithm 1.1 from $r(b)$ to q has length at most $1/(s-1)|pq|$. Lemma 15 implies that the path from $r(a)$ to q found by the heavy path routing algorithm is the same as the spanner path, so its length is at most $1 + 2/s + 1/(s-1)$ times $|pq|$, because that is the length of the spanner path from $r(a)$ to q.

The only thing that remains to be bounded is the length of the path constructed during the ascending stage.

Lemma 17. *Let $p = p_1, p_2, \ldots, p_k = r(a)$ be the points visited during the ascending stage. For any point p_i, the points p_1 through p_{i-1} are all stored in the subtree rooted at $h(p_i)$.*

We can bound the length of the path constructed during the ascending stage using the previous lemma.

Lemma 18. *The length of the path constructed during the ascending stage of the algorithm is no more than $(2/s)|pq|$.*

Proof. First, note that if a is the parent of b in the quadtree, then $\ell(a) \geq (1/2)\ell(b)$, by Lemma 1. The path from p to $r(a)$ is contained in the subtree rooted at a, so the length of every edge on the path is at most $\ell(a)$. That path, minus the last edge, is contained in the subtree rooted at one of the children of a by Lemma 17, so the length of all but the last edge is at most $(1/2)\ell(a)$.

Repeating this argument will show that the length of the entire path is not more than $\ell(a) + (1/2)\ell(a) + (1/2)^2\ell(a) + \cdots = 2\ell(a)$. By the condition for checking well-separatedness in the WSPD construction algorithm, $\ell(a) \leq (1/s)d(a,b) \leq (1/s)|pq|$. Therefore, the length of the path from p to $r(a)$ is at most $(2/s)|pq|$.

Theorem 6. *The routing ratio of the heavy path routing algorithm is at most $1 + 4/s + 1/(s-1)$.*

Proof. By Lemma 18, the total length of the path constructed during the ascending stage is no more than $(2/s)|pq|$. The path constructed during the descending step is equal to the spanner path from $r(a)$ to q, and so we can use the bound on the spanning ratio to bound this part of the path. By Lemma 16 this is at most $(1 + 2/s + 1/(s-1))|pq|$. Therefore the length of the path from p to q is

$$\frac{2}{s}|pq| + \left(1 + \frac{2}{s} + \frac{1}{s-1}\right)|pq| = \left(1 + \frac{4}{s} + \frac{1}{s-1}\right)|pq|.$$

Theorem 7. *The routing ratio of the heavy path routing algorithm is at least $1 + 4/s$ in the worst case.*

Similar to Lemma 8, we can bound the number of edges on the spanner path. In fact, the proof works almost identically to the proof of Lemma 8.

Lemma 19. *Starting at a point p, a message can be forwarded to any other point q after forwarding only $2\lg n + 1$ times.*

Proof. Let $\{S(a), S(b)\}$ be the WSPD pair that separates p from q. Consider the subpath $p = p_0, p_1, \ldots, p_k = r(a)$ found during the ascending stage. There will be one edge added to this path for each light edge on the path from p to a in the compressed quadtree. By Lemma 4, this is at most $\lg n$.

Since the path constructed during the descending stage follows the spanner path, Lemma 8 implies that the number of forwards during the descending stage is at most $\lg n + 1$. Therefore the number of forwards for the entire routing algorithm is at most $2\lg n + 1$.

The results of this section are summarized in the following theorem.

Theorem 8. *Let G be a heavy path WSPD spanner for a set S of points in $\mathbf{R}^d$, and let p and q be points of S. There exists a local, memoryless routing algorithm that can find a path from p to q, such that:*

- *The number of bits stored at each vertex u is $(3\deg(u) + 1)\lg n$*
- *The length of the path found from p to q is at most $(1 + 4/s + 1/(s-1))|pq|$*
- *The number of edges on the path is at most $2\lg n + 1$*

4 Conclusions

We presented a spanner construction scheme based on the WSPD of a point set P of n points in $\mathbf{R}^d$ that has spanning ratio $1 + 2/s + 2/(s-1)$, has spanning paths with at most $2\lg n + 1$ edges and $O(s^d n)$ size, where $s > 2$ is the separation ratio. We call these spanners heavy-path WSPD-spanners. We then presented a memoryless local routing algorithm for heavy-path WSPD-spanners with routing ratio at most $1 + 4/s + 1/(s-1)$ and at least $1 + 4/s$ which is close to the spanning ratio. The number of edges on the spanner path is bounded by $2\lg n + 1$ and a total of $O(s^d n \log n)$ bits are used to store all the routing tables. These are currently the best known bounds for routing algorithms on WSPD-spanners.

The spanner construction and routing algorithm presented in this paper can be modified to work in a more general setting than d-dimensional Euclidean space. Specifically it is possible to show that a variant of the heavy-path WSPD-spanner can be constructed for metric spaces of bounded doubling dimension. Although the compressed quadtree cannot be constructed in this setting, the net tree of Har-Peled and Mendel [17] can be used instead. This construction is also amenable to a memoryless competitive online local routing algorithm, however, the algorithm is more complicated and the doubling dimension factors into all of the bounds for both the construction and routing algorithms. For details of this generalisation, refer to Tuttle [24].

Another avenue to explore is to consider other spanner constructions based on WSPDs to try to improve some of our bounds. The spanner construction that we presented is designed to allow for a fairly simple routing algorithm. Recall that given a WSPD of point set, one can construct many different $(1+\varepsilon)$-spanners since there is some flexibility in the structure. There are various spanner construction schemes that can produce WSPD-spanners with other desirable properties such as bounded degree or $o(\log n)$ hop distance. For details on the various constructions of spanners from WSPDs, see Narasimhan and Smid [21]. These different constructions may improve some of our bounds at the cost of a slightly more elaborate and complicated routing scheme.

Finally, WSPDs have been defined for the unit disk graph [15]. Using these WSPDs, local routing in the unit disk graph is possible as was shown by Kaplan et al. [18] and Mulzer & Willert [20]. Our algorithm routes on spanners constructed directly from a WSPD, as such, our result does not immediately transfer to this setting. We leave as an open problem to determine whether our routing scheme can be adapted to this setting and provide a better trade-off.

References

1. Aluru, S.: Quadtrees and octrees. In: Mehta, D.P., Sahni, S. (eds.) Handbook of Data Structures and Applications, chapter 19, 1 edn, pp. 1–26. Chapman and Hall/CRC (2004). https://doi.org/10.1201/9781420035179. ISBN 1-58488-435-5
2. Arya, S., Mount, D.M., Smid, M.: Randomized and deterministic algorithms for geometric spanners of small diameter. In: Proceedings of the 35th Annual Symposium on Foundations of Computer Science, pp. 703–712 (1994). https://doi.org/10.1109/SFCS.1994.365722. ISBN 0-8186-6580-7
3. Arya, S., Das, G., Mount, D.M., Salowe, J.S., Smid, M.: Euclidean spanners: short, thin and lanky. In: Proceedings of the Twenty-Seventh Annual ACM Symposium on Theory of Computing, pp. 489–498 (1995). https://doi.org/10.1145/225058.225191. ISBN 978-0-89791-718-6
4. Baharifard, F., Farhadi, M., Zarrabi-Zadeh, H.: Routing in well-separated pair decomposition spanners. In: Proceedings of the 1st Iranian Conference on Computational Geometry, pp. 25–28 (2018)
5. Bonichon, N., Bose, P., De Carufel, J.L., Perković, L., Van Renssen, A.: Upper and lower bounds for online routing on delaunay triangulations. Discrete Comput. Geom. **58**(2), 482–504 (2017). https://doi.org/10.1137/110832458
6. Bonichon, N., Bose, P., De Carufel, J.L., Despré, V., Hill, D., Smid, M.: Improved routing on the Delaunay triangulation. In: 26th Annual European Symposium on Algorithms, pp. 1–13 (2018). https://doi.org/10.4230/LIPIcs.ESA.2018.22. ISBN 978-3-95977-081-1
7. Bose, P., Morin, P.: Online routing in triangulations. SIAM J. Comput. **33**(4), 937–951 (2004). https://doi.org/10.1137/S0097539700369387
8. Bose, P., Fagerberg, R., Van Renssen, A., Verdonschot, S.: Optimal local routing on Delaunay triangulations defined by empty equilateral triangles. SIAM J. Comput. **44**(6), 1626–1649 (2015)
9. Bose, P., De Carufel, J.-L., Dujmović, V., Paradis, F.: Local routing in spanners based on WSPDs. In: WADS 2017. LNCS, vol. 10389, pp. 205–216. Springer, Cham (2017). https://doi.org/10.1007/978-3-319-62127-2_18
10. Bose, P., De Carufel, J.L., Devillers, O.: Olivier: expected complexity of routing in θ_6 and half-θ_6 graphs. J. Comput. Geom. **11**(1), 212–234 (2020)
11. Bose, P., De Carufel, J.L., Dujmović, V., Paradis, F.: Local routing in spanners based on WSPDs. J. Comput. Geom. **12**(1), 1–34 (2021)
12. Callahan, P.B., Kosaraju, S.R.: A decomposition of multidimensional point sets with applications to k-nearest-neighbors and n-body potential fields. J. ACM **42**(1), 67–90 (1995). https://doi.org/10.1145/200836.200853
13. Clarkson, K.L.: Approximation algorithms for shortest path motion planning. In: Proceedings of the 19th Annual ACM Symposium on Theory of Computing, pp. 56–65 (1987). https://doi.org/10.1145/28395.28402. ISBN 978-0-89791-221-1
14. Dijkstra, E.W.: A note on two problems in connexion with graphs. Numer. Math. **1**, 269–271 (1959). https://doi.org/10.1007/BF01386390
15. Gao, J., Zhang, L.: Well-separated pair decomposition for the unit-disk graph metric and its applications. SIAM J. Comput. **35**(1), 151–169 (2006). https://doi.org/10.1137/S0097539703436357
16. Har-Peled, S.: Geometric Approximation Algorithms. Number 173 in Mathematical Surveys and Monographs. American Mathematical Society, USA (2011). ISBN 978-0-8218-4911-8

17. Har-Peled, S., Mendel, M.: Fast construction of nets in low-dimensional metrics and their applications. SIAM J. Comput. **35**(5), 1148–1184 (2006). https://doi.org/10.1137/S0097539704446281
18. Kaplan, H., Mulzer, W., Roditty, L., Seiferth, P.: Routing in unit disk graphs. Algorithmica **80**(3), 830–848 (2017). https://doi.org/10.1007/s00453-017-0308-2
19. Keil, J.M., Gutwin, C.A.: Classes of graphs which approximate the complete Euclidean graph. Discret. Comput. Geom. **7**(1), 13–28 (1992). https://doi.org/10.1007/BF02187821
20. Mulzer, W., Willert, M.: Compact routing in unit disk graphs. In: ISAAC, vol. 181 of LIPIcs, pp. 1–14. Schloss Dagstuhl - Leibniz-Zentrum für Informatik (2020)
21. Narasimhan, G., Smid, M.: Geometric Spanner Networks. Cambridge University Press, Cambridge (2007). ISBN 978-0-521-81513-0
22. Ruppert, J., Seidel, R.: Approximating the d-dimensional complete Euclidean graph. In: Proceedings of the 3rd Canadian Conference on Computational Geometry (CCCG 1991), pp. 207–210 (1991)
23. Sleator, D.D., Tarjan, R.E.: A data structure for dynamic trees. J. Comput. Syst. Sci. **26**(3), 362–391 (1983). https://doi.org/10.1016/0022-0000(83)90006-5
24. Tuttle, T.G.: Routing on heavy path WSPD spanners. Master's thesis, Carleton University, Ottawa, Canada (2020)
25. Yao, A.C.: On constructing minimum spanning trees in k-dimensional space and related problems. SIAM J. Comput. **11**, 721–736 (1981)

Mapping Multiple Regions to the Grid with Bounded Hausdorff Distance

Ivor van der Hoog, Mees van de Kerkhof, Marc van Kreveld, Maarten Löffler, Frank Staals, Jérôme Urhausen(✉), and Jordi L. Vermeulen

Department of Information and Computing Sciences, Utrecht University, Utrecht, The Netherlands
j.e.urhausen@uu.nl

Abstract. We study a problem motivated by digital geometry: given a set of disjoint geometric regions, assign each region R_i a set of grid cells P_i, so that P_i is connected, similar to R_i, and does not touch any grid cell assigned to another region. Similarity is measured using the Hausdorff distance. We analyze the achievable Hausdorff distance in terms of the number of input regions, and prove asymptotically tight bounds for several classes of input regions.

Keywords: Computational geometry · Digital geometry · Hausdorff distance · Simple polygons

1 Introduction

Digital geometry is concerned with the proper representation of geometric objects and their relationships using a grid of pixels. This greatly simplifies both representation and many operations, but the downside is that common properties of geometric objects no longer hold. For example, it may be that two digitized lines intersect in multiple connected components. One objective of digital geometry is how to *consistently* digitize a set of geometric objects. Another objective is the presentation of vector objects with bounded error, using subsets of pixels.

Early results in digital geometry were mostly concerned with consistency and arose in computer vision. For a survey, see Klette and Rosenfeld [11,12]. More recently, also error bounds under the Hausdorff distance have been studied. Chun et al. [5] investigate the problem of digitizing rays originating in the origin to digital rays such that certain properties are satisfied. They show that rays can be represented on the $n \times n$ grid in a consistent manner with Hausdorff distance $O(\log n)$. This bound is tight in the worst case. By ignoring one of the consistency conditions, the distance bound improves to $O(1)$. Their research is extended by Christ et al. [3] to line segments (not necessarily starting in the origin), who obtain the logarithmic distance bound in this case as well. A possible extension to curved rays was developed by Chun et al. [4]. Other results with a digital geometry flavor within the algorithms community are those on snap rounding [6,7,10], integer hulls [1,9], and discrete schematization [13].

A. Lubiw et al. (Eds.): WADS 2021, LNCS 12808, pp. 627–640, 2021.
https://doi.org/10.1007/978-3-030-83508-8_45

In a recent paper, Bouts et al. [2] showed that any simple polygon, no matter how detailed, can be represented by a simply connected set of unit pixels such that the Hausdorff distance to and from the input is bounded by $3\sqrt{2}/2$.

Contribution. We extend the result from [2] to multiple regions, see Fig. 1. We investigate several restrictions on the class of regions and we show that stricter restrictions allow for pixel representations with a smaller symmetric Hausdorff distance. All our bounds are tight. We express our bounds in the number of input regions. Our results are shown in Table 1; they are fundamental results on the error that may be incurred when converting vector to grid representations, a common operation in computer graphics and GIS.

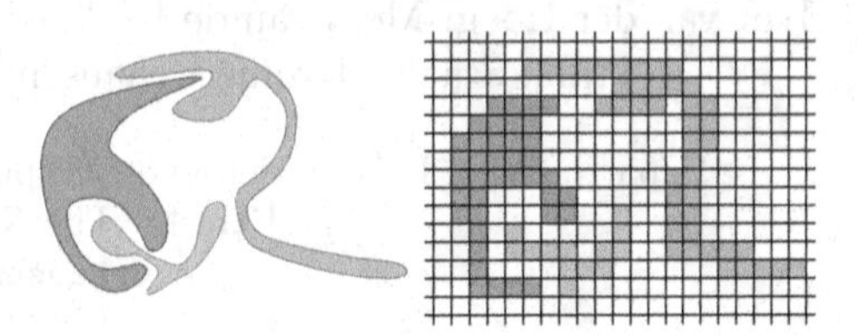

Fig. 1. Three disjoint simply connected regions and a representation by simply connected sets of disjoint pixels.

We do not make any assumptions on the resolution of the input. If the minimum distance between any pair of polygons is at least some constant (e.g., $4\sqrt{2}$ is enough), then we can realize a constant Hausdorff bound in all cases by applying the results from Bouts et al. [2] separately on each polygon. We consider the case where no such assumptions are made.

Table 1. Worst-case bounds on Hausdorff distances for m regions; β is constant.

Region class	Points	Convex β-fat	Convex	Two regions	Three regions
Hausdorff distance	$\Theta(\sqrt{m})$	$\Theta(\sqrt{m})$	$\Theta(m)$	$\Theta(1)$	unbounded

Notation and Definitions. We denote by Γ the (infinite) unit grid, whose unit squares are referred to as *pixels*. The *(symmetric) Hausdorff distance* between two sets $A, B \subset \mathbb{R}^2$ is defined as $H(A, B) = \max\{\max_{a \in A}(\min_{b \in B}(|ab|)), \max_{b \in B}(\min_{a \in A}(|ab|))\}$, where $|ab|$ is the distance between the points a and b. Further we denote by $H'(A, B) = \max\{H(A, B), H(\partial A, \partial B)\}$ the (symmetric) Hausdorff distance between the sets themselves and between their boundaries. See Fig. 2 for an example where the distinction between $H(\cdot, \cdot)$ and $H'(\cdot, \cdot)$ is important.

Let $\mathcal{R} = \{R_1, R_2, \ldots R_m\}$ be a set of m disjoint simply connected regions in the plane. In this paper, we show how to assign a subset of the pixels Two such *grid polygons* are disjoint if they do not meet in any edge or vertex of the grid. A grid polygon is connected $P_i \subset \Gamma$ to each region $R_i \in \mathcal{R}$, such that the result is a set of m disjoint simply connected regions. if its

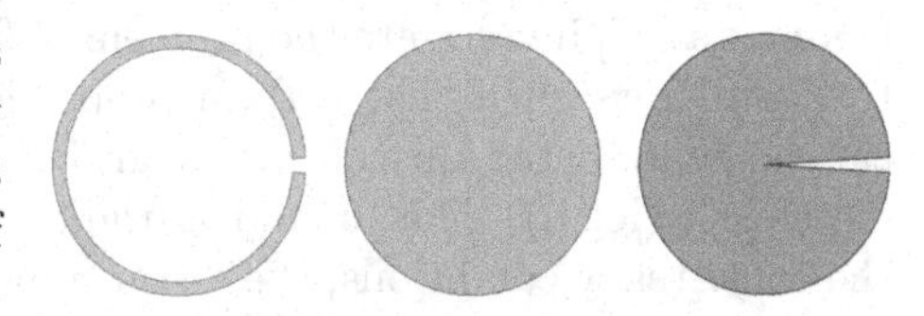

Fig. 2. The Hausdorff distance between the green and red regions is large while the Hausdorff distance between their boundaries is small. The inverse is true for the red and purple regions. (Color figure online)

pixels are connected by edge adjacency, and simply connected if it is connected and its complement is also connected by edge adjacency. We call the set $\{P_1, P_2, \ldots, P_m\}$ of such grid polygons a *valid assignment* for $\mathcal{R}$.

Overview. We are interested in finding for any set of regions $\mathcal{R}$ a valid assignment such that for all i the (symmetric) Hausdorff distance between R_i and P_i is at most h, and the (symmetric) Hausdorff distance between their boundaries is also at most h. In general, a worst-case bound on h will be a function of m. We study this problem under several restrictions on $\mathcal{R}$; refer to Table 1. For each class of restrictions, first we show that there is a set of regions in that class for which any valid assignment contains at least one region R_i with a grid polygon P_i where $H'(R_i, P_i) = \Omega(h)$. Second we show that for any set of regions in that class, we can find a valid assignment such that for all regions $R_i \in \mathcal{R}$ with corresponding grid polygon P_i, we have $H'(R_i, P_i) = O(h)$. Hence, our bounds are asymptotically tight.

We may interpret a solution to our problem as a *coloring* of Γ: each pixel $q \in \Gamma$ is assigned one color in $C = \{c_1, \ldots c_m\} \cup \{b\}$, where c_i is the color of the input region R_i and b is the background color.

Our upper bound constructions all follow a similar scheme. Let Γ_k be a coarsening of the grid Γ whose cells have $k \times k$ pixels. We call these cells *superpixels*. We will determine for each region from $\mathcal{R}$ which superpixels it contains and which ones it properly intersects. If a region R_i contains a superpixel, then all pixels of Γ in that superpixel will be part of P_i. If R_i properly intersects a superpixel, we ensure that at least one, but not all pixels in that superpixel will be part of P_i. A superpixel not intersecting R_i will have no pixels in P_i. The main challenge is then finding a scheme by which each grid polygon becomes simply connected yet all remain disjoint. It is then relatively straightforward to see that $H'(R_i, P_i) \leq k\sqrt{2}$.

2 Input Regions are Points

In this section we first consider the simplest possible case, namely, $\mathcal{R}$ is a set of points. We will construct a map that assigns points to pixels such that the symmetric Hausdorff distance between each point and its corresponding pixel is bounded. For a lower bound, consider a set of m points $\mathcal{R}$ that all lie within a single pixel. If we want to assign each point to a unique pixel, we clearly need to use m different pixels. Any set of m pixels has diameter $\Omega(\sqrt{m})$, so at least one of the point regions will be mapped to a pixel at distance $\Omega(\sqrt{m})$.

We now present a scheme that maps any set of m points $\mathcal{R}$ to a set of pixels, such that the symmetric Hausdorff distance between any point and its pixel is at most $O(\sqrt{m})$. Let Γ_k be a coarsening of Γ with $k = 2\lceil\sqrt{m}\rceil$. Associate each region in $\mathcal{R}$ with the superpixel that contains it. Each superpixel has the space to accommodate m disjoint pixels without using the bottom row and right column by using exactly the odd numbered rows and columns. Any assignment of the points to these pixels is easily seen to have Hausdorff distance $O(\sqrt{m})$.

Theorem 1. *If $\mathcal{R}$ is a set of m points, a valid assignment exists such that for each region $R_i \in \mathcal{R}$ with a corresponding region P_i, we have $H'(R_i, P_i) = O(\sqrt{m})$. Furthermore, there exists a set $\mathcal{R}$ of m points such that for every valid assignment we have $H(R_i, P_i) = \Omega(\sqrt{m})$.*

3 Input Regions are Convex β-fat Regions

A connected region R is β-fat if for some point t in R, the ratio of the radius of the smallest t-centered circle containing R and the radius of the largest t-centered circle contained in R, is β (or larger) [14]. Observe that the only regions that are 1-fat are points and disks, as points are β-fat regions for *any* $\beta \geq 1$ by convention. In this section we consider the class $\mathcal{R}$ of convex β-fat regions for a constant β. From Sect. 2 it follows that for any m, there exists a set of m regions for which the (symmetric) Hausdorff distance between $\mathcal{R}$ and any valid assignment is $\Omega(\sqrt{m})$.

Let $\mathcal{R}$ be a set of convex β-fat regions and let Γ_k be a coarsening of Γ with $k = 2\lceil\sqrt{m}\rceil + 3$. We present an algorithm that maps $\mathcal{R}$ to a set of grid polygons $\mathcal{P}$, such that the symmetric Hausdorff distance between any region R_i and its assigned region P_i is at most $O(\beta\sqrt{m})$.

Lemma 1. *Let R be a convex β-fat region, and let p be a point in R. Either R has diameter less than $16\beta k$, or R contains a superpixel within distance $16\beta k$ from p.*

This leads to the following algorithm with two cases for each region R_i, depending on the set of superpixels $\mathcal{S}_i$ contained in R_i.

Case 1: $\mathcal{S}_i$ is empty. We select any superpixel S intersected by R_i and we assign R_i to a unique pixel in S while using neither the topmost, bottommost, leftmost, or rightmost rows and columns, similar to the procedure in Sect. 2. This pixel has a distance of at most $16\beta k + \sqrt{2}k$ to any point on R_i since R_i has diameter smaller than $16\beta k$ by Lemma 1. This also means that for each such region R_i, we have $H'(R_i, P_i) \leq 32\beta k$.

Case 2: $\mathcal{S}_i$ is not empty. We need two steps. First we assign all pixels in each superpixel of $\mathcal{S}_i$ to R_i. Note that $\mathcal{S}_i$ is not necessarily connected, as can be seen in Fig. 3 (left). Nonetheless we can connect the superpixels in the second step using Lemma 2 below.

Lemma 2. *Let S_1 and S_2 be two superpixels in different connected components of $\mathcal{S}_i$. Let v_1 be the center of S_1 and v_2 the center of S_2. The path consisting of pixels that either intersect or border the line segment $\overline{v_1 v_2}$ must be entirely contained in R_i, and at least at twice the unit distance from the border of R_i.*

Proof. The line segment between v_1 and v_2 is contained within R_i by convexity. Similarly, the line segment from any vertex of S_1 to a vertex of S_2 is contained in R_i and necessarily also in the bounded slab that bounds these sixteen edges. Such a slab is at least as wide as S_1 and S_2 (hence it is at least $16\beta k$ pixels wide).

The line segment between v_1 and v_2 forms the spine of this slab, any pixel that intersects or borders this spine has at most two unit distance to this spine and hence is contained within the slab and via transitivity in R_i. Moreover, since the slab is at least $16\beta k$ wide, and since each pixel has distance at most two from the spine, each pixel in the path is at much more than distance two from the border of the slab and via transitivity the border of R_i. □

Let S_1 and S_2 be two superpixels in different connected components of the superpixels contained in R_i. We connect S_1 and S_2 with a path of pixels according to Lemma 2. Since this path is entirely contained in R_i and since there are at least two pixels between a pixel in this path and the border of R_i, no other region will attempt to color the pixels in this path. We repeat this process until for each region the assigned pixels form a connected grid polygon and whenever we enclose an area between superpixels with these paths, we make sure to assign all the pixels in this area to R_i; by the convexity of R_i all these pixels are contained in R_i. This provides our pixel assignment P_i.

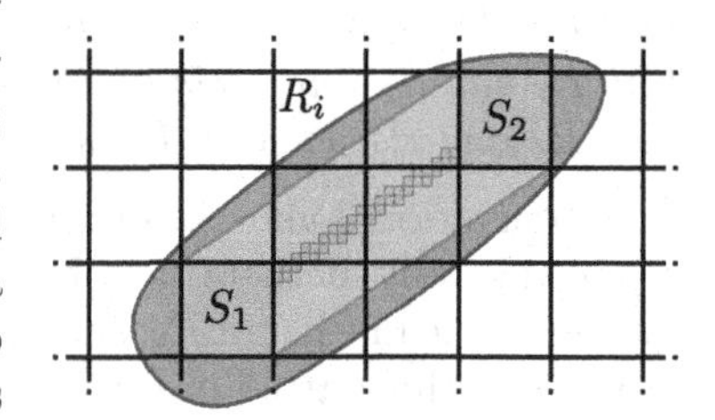

Fig. 3. A convex β-fat region R_i (purple), and the region formed by sweeping a superpixel from S_1 to S_2 (green). P_i (red) consists of S_1, S_2, and all pixels on the segment between the centers of S_1 and S_2. (Color figure online)

What remains to be proven, is that for each region R_i with non-empty $\mathcal{S}_i$, $H'(R_i, P_i) \leq 32\beta k$ holds. First, we prove that for each (boundary) point p of P_i, there is a (boundary) point q of R_i within distance $32\beta k$. By construction, we know $P_i \subseteq R_i$, so the claim holds for interior points. Now, let $p \in \partial P_i$. We assume for the sake of contradiction that there is no point of ∂R_i within distance $\sqrt{2}k$. As p is contained within R_i, we have that R_i contains the superpixels containing p, a contradiction. Second, we prove the inverse. For a point q of R_i, Lemma 1 guarantees that R_i contains a superpixel S within distance $16\beta k$ of q. Then $S \subseteq P_i$ holds, proving the claim. As $P_i \subseteq R_i$, this also proves that for each boundary point q of R_i, there is a boundary point p of P_i within distance $16\beta k$.

Theorem 2. *If $\mathcal{R}$ is a set of m β-fat convex regions for a constant β, a valid assignment exists such that for each region $R_i \in \mathcal{R}$ with a corresponding region P_i, we have $H'(R_i, P_i) = O(\sqrt{m})$. Furthermore, for any $\beta \geq 1$, there exists a set $\mathcal{R}$ of m β-fat regions such that for every valid assignment $H(R_i, P_i) = \Omega(\sqrt{m})$.*

4 Input Regions are Convex Regions

When $\mathcal{R}$ is a set of convex regions, we can easily show that the coloring has a lower-bound Hausdorff distance of $\Omega(m)$: we can place m horizontal line segments of length $\Omega(m)$ that all pass through the same pixels. Then $\mathcal{P}$ must have its elements on disjoint lines of pixels, giving Hausdorff distance at least $\Omega(m)$ for the outer regions. Each P_i must extend sufficiently far left and right. Since all P_i are connected, they will intersect a common vertical line. The topmost or bottommost intersection with this line belongs to a grid polygon with Hausdorff distance $\Omega(m)$. (Note that if the P_i need not be connected, $O(\sqrt{m})$ Hausdorff distance can always be realized.)

We will describe an algorithm that, given a set of convex regions $\mathcal{R}$, gives a set of disjoint orthoconvex grid polygons $\mathcal{P}$ such that, for all i, $H'(R_i, P_i) = O(m)$.

Observation 3. *Let $R_1, R_2 \in \mathcal{R}$ be two disjoint convex regions, and let ℓ be a horizontal line that intersects R_1 left of R_2. Then any horizontal line intersecting both R_1 and R_2 intersects R_1 left of R_2. Similarly, all vertical lines that intersect both R_1 and R_2 do so in the same order.*

Observation 3 allows us to define two partial orders $\preceq_x$ and $\preceq_y$ on $\mathcal{R}$: $R_i \preceq_x R_j$ if and only if there is a horizontal line intersecting both regions and R_i intersects the line left of R_j; since the regions are convex we get a partial order [8]. We extend this partial order as follows: first we add transitive arrows, where we recursively add the inequality $R_i \preceq_x R_j$ if there exists a region R_k with $R_i \preceq_x R_k$ and $R_k \preceq_x R_j$ and we denote this partial order by $\Pi_x(\mathcal{R})$. We then transform $\Pi_x(\mathcal{R})$ into a linear order $X_\mathcal{R} : \mathcal{R} \rightarrow [1, m]$ in any manner. A linear order $Y_\mathcal{R} : \mathcal{R} \rightarrow [1, m]$ is defined symmetrically.

Given $X_\mathcal{R}$ and $Y_\mathcal{R}$, we assign a coloring. Let Γ_k be a coarsening of Γ with $k = 2m$. For any superpixel $S \in \Gamma_k$, we denote by $S[x, y]$ the pixel that is the $(2x)^{\text{th}}$ from the left and $(2y)^{\text{th}}$ from the bottom within S. Additionally the horizontal and vertical lines induced by Γ_k are called *major lines*. Each region R_i that intersects at most one major horizontal line and at most one major vertical line is a *small region*. Each region R_i that intersects at least two major horizontal lines or at least two major vertical major lines is a *large region*. Our assignment of regions to pixels, illustrated in Fig. 4, is:

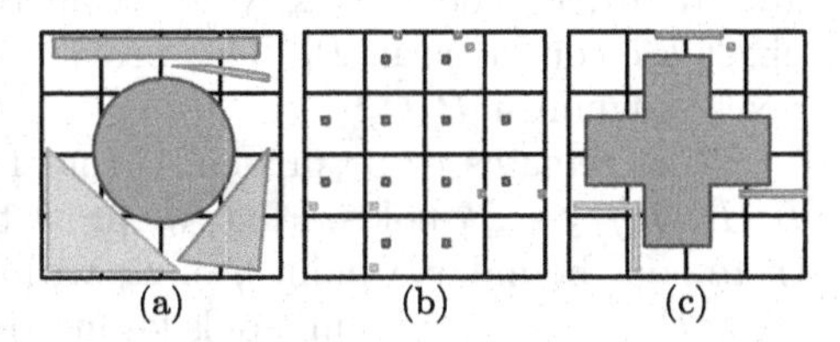

Fig. 4. The coloring algorithm for convex regions. (a) The input of four convex regions, overlaid onto a superpixel grid with $k = 10$. (b) The pixels colored in Step 1 and 2 of the algorithm. (c) The final coloring obtained after Steps 3 and 4.

1. For each small region R_i we choose one superpixel S containing a point of R_i and color the pixel $p(S, R_i) = S[X_\mathcal{R}(R_i), Y_\mathcal{R}(R_i)]$ with c_i (this single pixel will be P_i).
2. For each superpixel S and each large region R_i intersecting S that also intersects the two major horizontal *lines* incident to S, or the two major vertical *lines* incident to S, we color $p(S, R_i) = S[X_\mathcal{R}(R_i), Y_\mathcal{R}(R_i)]$ with c_i. Note that region R_i need not intersect two opposite *edges* of S.
3. For any two pixels that are colored with c_i in edge-adjacent superpixels (R_i must be large), we color all pixels in the row or column between them with c_i.
4. For any four superpixels that share a common vertex, if they each contain a pixel colored with c_i in Step 1, we color all pixels in the square between these pixels with c_i.

Let $\mathcal{P}$ be the set of polygons induced by this grid coloring.

Lemma 3. *Each polygon $P_i \in \mathcal{P}$ is simply connected.*

Proof. If R_i is small, P_i is a single pixel and thus simply connected. If R_i is large, it intersects a connected set of superpixels, and our algorithm connects all of these together, so P_i is connected. The resulting grid polygon P_i cannot contain holes: the presence of a hole would imply that the set of superpixels intersected by R_i contains a hole, which is not possible due to R_i being simply connected and convex.

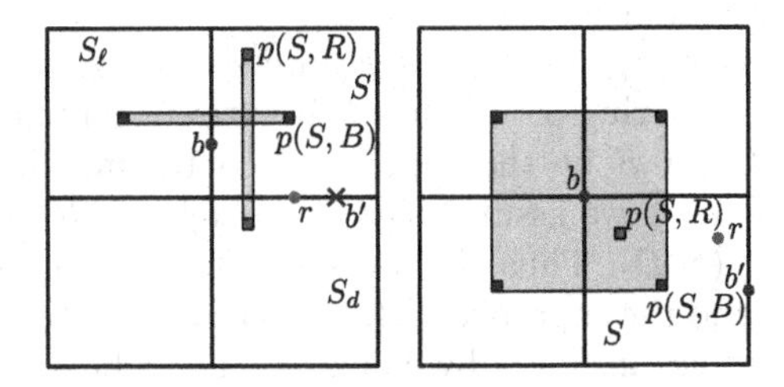

Fig. 5. The cases for the proof of Lemma 4.

Our algorithm actually produces orthoconvex polygons (refer to the full version for details).

Lemma 4. *The polygons in $\mathcal{P}$ are pairwise disjoint.*

Proof. Assume by contradiction that the colorings of two regions R and B intersect. Then the intersection was created during one of the four coloring steps. In steps 1 and 2, we assign each color to single pixels per superpixel in unique rows and columns, hence they cannot create two colorings that intersect.

Let the colorings of R and B intersect after step 3. This implies that R and B are both large regions. The intersection occurs between a vertical and horizontal pixel sequence in a super pixel S. Assume without loss of generality that the vertical sequence belongs to R and the horizontal sequence belongs to B. Consider the case that the pixel $p(S, R)$ assigned to R in S in step 2 is to the top-left of $p(S, B)$ (See Fig. 5); the other three cases are symmetric. Then the intersection occurs between the column sequence connecting $p(S, R)$ to $p(S_d, R)$ and the row sequence connecting $p(S, B)$ to $p(S_\ell, B)$, where S_d is the superpixel directly below S and S_ℓ is the superpixel directly to the left of S.

Since B is large and assigned a pixel in S it intersects both horizontal major lines incident to S or both vertical major lines incident to S. The same applies for R. We first consider the case where B does not intersect the major line through the bottom edge of S, and hence it must intersect both vertical lines. That is, B spans the vertical slab defined by S and does so in or above S. Since R intersects the cell S_d below S it then follows that $R \preceq_y B$. However, since

$p(S,R)$ lies above $p(S,R)$ we also have $B \preceq_y R$. Since $B \neq R$ we thus obtain a contradiction.

Thus, B intersects the horizontal major line ℓ through the bottom edge of S. Since R is convex, and intersects both S and S_d it intersects the bottom edge of S (and thus ℓ) in a point r. Symmetrically, B intersects the left edge of S in a point b. If B also intersects the horizontal line ℓ in some point b' this point cannot be left of r, as this would immediately imply that $B \preceq_x R$, contradicting the assignment of $p(S,R)$ and $p(S,B)$. So b' lies right of r. However, then the vertical ray starting at r pointing upwards intersects the segment connecting b and b'. Since B is convex, this segment is contained in B. This implies $R \preceq_y B$, which again contradicts the assignment of $p(S,R)$ and $p(S,B)$. It follows that step 3 does not create intersecting colorings.

Finally, let (the colorings of) R and B intersect only after step 4. Without loss of generality, the coloring of a region R is entirely contained in the coloring of a large region B. Let S be the superpixel containing the lone pixel of R. Without loss of generality we assume that the pixel $p(S,R)$ assigned to R in S is to the top-left of $p(S,B)$. Thus, B intersects S, the superpixel above S, the superpixel left of S, and the superpixel left and above S. The point b where these four superpixels meet lies inside B by convexity. Let r be any point in $R \cap S$.

As B is a large region it needs to intersect two opposite major lines incident to S. Assume that B intersects the vertical major lines, in particular the one incident to the right edge of S in a point b'. The vertical line through r intersects the segment between b and b'. The point r is above that segment, because the opposite would imply $R \preceq_y B$. As a consequence r is also right of the segment between b and b', which implies that the horizontal line through r intersects this segment left of R, a contradiction. The case where B intersects the major horizontal line through the bottom edge of S is symmetric. □

If a region R_i intersects a superpixel S, then P_i has a pixel in S or in at least one of the eight adjacent superpixels. Conversely, if P_i contains a pixel in S, we know that R_i intersects S. This gives a bound on the Hausdorff distance between the regions and the grid polygons. For the boundaries, note that if R_i contains a superpixel S and all four edge-adjacent superpixels, then P_i contains S. Furthermore, if P_i contains a superpixel S, then R_i also contains S. Together this gives a bound on the Hausdorff distance between the boundaries. Since superpixels have size $\Theta(m)$, the Hausdorff distance between R_i and P_i and between their boundaries is at most $O(m)$. We thus obtain the following result.

Theorem 4. *If $\mathcal{R}$ is a set of m convex regions, a valid assignment exists such that for each region $R_i \in \mathcal{R}$ with a corresponding region P_i, we have $H'(R_i, P_i) = O(m)$. Furthermore, there exists a set $\mathcal{R}$ of m convex regions such that for every valid assignment, there exists some $1 \leq i \leq m$ with $H(R_i, P_i) = \Omega(m)$.*

5 Input Regions are General Regions

When the input regions are arbitrary, we see a sharp contrast between the case $m \leq 2$, where constant Hausdorff distance can be realized, and the case $m \geq 3$, where the Hausdorff distance may be unbounded. The fact that a single region can be represented as a grid polygon with constant Hausdorff distance was shown before by Bouts et al. [2]. In Sect. 5.1 we show that the same result holds for two regions. In Sect. 5.2 we show that for three regions, no bounded Hausdorff distance bound exists that applies to all inputs.

5.1 Two Regions

Our result for two arbitrary regions is based on a combination of two previous results: mapping a polygon to the grid with constant Hausdorff distance by Bouts et al. [2], and a result on the Painter's Problem in [15]. We briefly explain the former result in our framework using superpixels first (see Fig. 6), and then extend it to our case with two regions using the latter result.

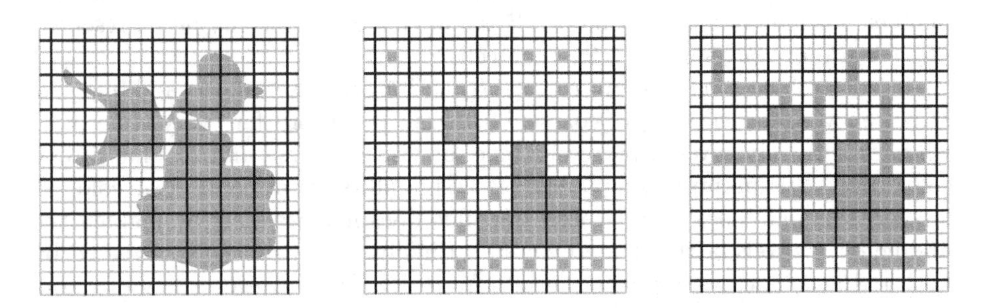

Fig. 6. Left, a region with Γ and Γ_3. Middle, the set P' of pixels chosen in the first selection. Right, the set P of pixels chosen after the spanning tree pixels are added.

Assume we have a region R that we want to represent by a grid polygon P. Consider the grid coarsening Γ_3, which has superpixels of 3×3 pixels. For every superpixel fully covered by R, choose all nine pixels in P. For every superpixel visited but not covered by R, take the middle pixel. Take nothing from superpixels not visited by R. Let the chosen pixels be P'.

Observe that P' forms a set of grid polygons that has no interior boundary cycles. Also observe that all superpixels for which at least one pixel is in P' is a connected (but not necessarily simply connected) part of Γ_3.

We make P' into one simply connected grid polygon P by using a (minimum) spanning tree on the components of P'. We will add pixels from visited superpixels only, and only ones adjacent to the already chosen center pixel. Two separate components will always be connected using one or two pixels.

Since the boundary of P does not intersect the interior of fully covered superpixels and visited superpixels always have a piece of boundary of P, it is easy to see that $H(R_i, P_i) = \Theta(1)$ and $H(\partial R_i, \partial P_i) = \Theta(1)$. This result is an alternative to the one by Bouts et al., albeit with worse constants.

A Painter's Problem instance takes a grid, and for each cell, the color white, blue, red, or purple. White indicates the absence of red and blue while purple indicates the presence of both red and blue. The question is whether two disjoint simply connected regions for red and blue exist that are consistent with all specifications of the cells, or, in the terminology of [15], "admits a painting". Since red cells can simply be colored red and blue cells blue, the problem boils down to recoloring the purple cells with red and blue pieces. The red and blue pieces in a cell provide a panel, and all panels together make up a painting. They prove:

Lemma 5 (Theorem 2 in [15]**).** *If a partially 2-colored grid admits a painting, then it admits a 5-painting.*

In a 5-painting each cell contains at most 5 components. The components make sure that the overall red and blue parts are connected across the whole painting. Additionally [15] show that each cell has at most 3 intervals of alternating red and blue along each side. This implies that there are only a constant number of configurations within a cell, so all configurations can be represented using a grid of constant size c for each cell.

In our problem, we have two regions R_1 and R_2 that we call red and blue, for consistency. We create a grid coarsening Γ_{c+2}. We record for every superpixel whether it is fully covered by red or blue, or visited by red and/or blue. If one color covers a superpixel completely, we assign all of its pixels to that color. If a color, say, red, visits a superpixel but blue does not, we start by making the middle $c \times c$ pixels of that superpixel red. Finally, for all superpixels visited by both red and blue, we apply the results from [15]. Since the recording of colors with panels comes from disjoint simply connected regions, namely, our input, we know that the 2-colored grid of superpixels admits a painting with connected regions/colors, so it admits one as specified in Lemma 5.

Once we choose a coloring of pixels in each 2-colored superpixel according to the panels, it remains to make the red set and blue set of pixels simply connected. The method from [15] did not produce any cycles in the 2-colored superpixels, the visited 1-colored superpixels are separate connected components of $c \times c$ pixels in the middle, and the covered 1-colored superpixels cannot create cycles either. We create a single red component by making a spanning tree of the red components. To achieve this, we only need to use pixels in the outer ring of the visited 1-colored superpixels. Then we do the same with blue. Since we add pixels of the same color to 1-colored superpixels, we will never try to color a pixel in both colors or create crossings. We then obtain the following result:

Theorem 5. *If $\mathcal{R}$ consists of two disjoint, simply connected regions, a valid assignment exists such that for each region $R_i \in \mathcal{R}$ with corresponding P_i, we have $H'(R_i, P_i) = \Theta(1)$.*

5.2 Three or More Regions

In the following we argue that the Hausdorff distance between an input of at least three general regions and any corresponding grid polygons is unbounded.

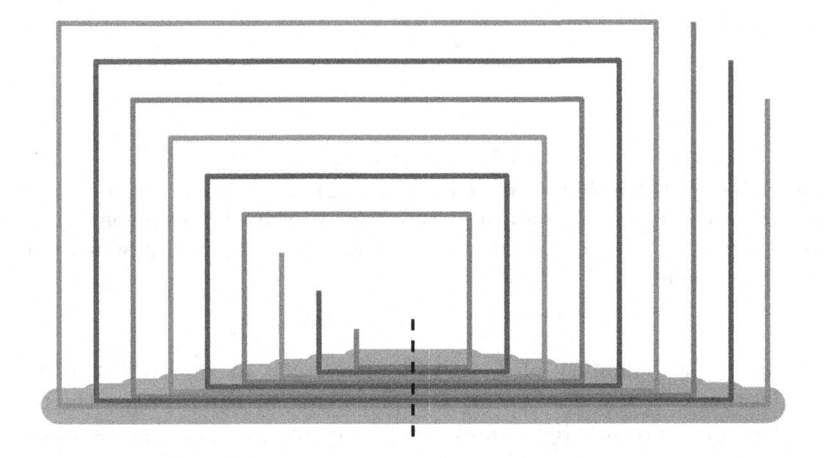

Fig. 7. The regions for $h = 3$; $\mathcal{I}$ is highlighted. The dashed line subdivides the boundary of $\mathcal{I}$ into its left and right part.

Formally, for a given integer $h > 0$, we show a construction of regions $\mathcal{R} = \{R, B, G\}$ for which there are no corresponding grid polygons with Hausdorff distance smaller than h. We only sketch the main idea here, see the full version for details.

We construct regions $\mathcal{R} = \{R, B, G\}$ that form nested spirals with a long bottleneck of height 1. The bottleneck is traversed from left to right h times by each of R, B, and G. If we remove the parts of R, B, and G inside the bottleneck, we get $3h + 3$ connected components in total. This is illustrated in Fig. 7 for $h = 3$. Outside the horizontal strip of height 1 containing the bottleneck, the three regions are more than $2h$ apart. We define the part of the plane within distance h of at least one of the bottom horizontal segments of the regions $\mathcal{R}$ as $\mathcal{I}$. All region components must be connected inside $\mathcal{I}$. Inside $\mathcal{I}$, it is possible that the grid polygons make different connections than those in $\mathcal{R}$. However, we argue that no matter how these connections are made, the grid polygons P_R, P_B, and P_G, together have to pass through $\mathcal{I}$ from left to right at least $h + 2$ times, thus requiring $\mathcal{I}$ to have height at least $2h + 3$. However, the available vertical space is only $2h + 1$ if the Hausdorff distance must stay below h, allowing $h + 1$ connections of pixel polygons. Hence, we obtain a contradiction.

The most involved part is to argue that P_R, P_B, and P_G, together have to pass through $\mathcal{I}$ at least $h + 2$ times. This argument critically depends on the following Lemma (see Fig. 8 for an illustration).

Lemma 6. *Given an alternating sequence $V = r_1, b_1, g_1, ..., r_k, b_k, g_k$ of $3k$ 3-colored points on a line, any planar drawing below the line connecting points of the same color induces a partition of the points into at least $2k + 1$ components.*

The idea is that $\mathcal{I}$ splits the regions in $\mathcal{R}$ (and thus their corresponding grid polygons) into $3h + 3$ connected components. However, the regions intersect the

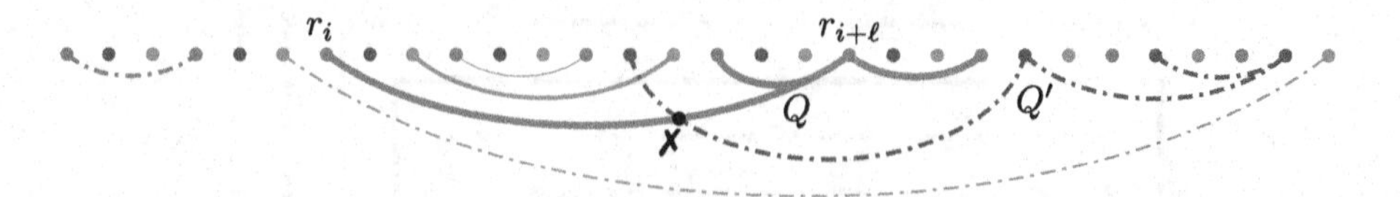

Fig. 8. A set Q that includes two red points r_i and $r_{i+\ell}$ splits V into two disjoint subsequences V_1 and V_2, that have at most one set, namely Q, in common. If there was a second such a set Q', the grid polygons corresponding to Q and Q' would intersect. (Color figure online)

right half of the boundary of $\mathcal{I}$ only $3h$ times, and in an order in which the colors alternate, we can use Lemma 6 to show that we can decrease the number of connected components by at most $h-1$ by connecting the regions incident to "the right side" of $\mathcal{I}$ to other regions on the right side of $\mathcal{I}$. The same holds for the regions on the left side of $\mathcal{I}$. It thus follows that the remaining $3h-2(h-1) = h+2$ of the reduction in the number of connected components (after all, in the end there are only three regions left) must be achieved by connecting regions incident to "the left side of" $\mathcal{I}$ to "the right side" of $\mathcal{I}$. Therefore, P_R, P_B, and P_G pass through $\mathcal{I}$ at least $h+2$ times as claimed. Therefore, this allows us to obtain the following result:

Theorem 6. *For all integer $h > 0$ there exist three regions $\mathcal{R} = \{R_1, R_2, R_3\}$, for which there is no valid assignment to grid polygons P_1, P_2, P_3 so that all regions $R_i \in \mathcal{R}$ have $H(R_i, P_i) < h$.*

6 Conclusion

In this paper we have shown what Hausdorff distance bounds can be attained when mapping disjoint simply connected regions to the unit grid. We expressed our bounds in terms of the number of regions and obtained different results depending on the shape and size characteristics of the regions, and showed that they are worst-case optimal. The result in Sect. 5.1 generalizes a result of Bouts et al. [2] and the result in Sect. 5.2 shows that a result by Van Goethem et al. [15] cannot be generalized from two to three colors. Our results are slightly more general than we expressed them: for example, the bound for point regions in fact holds for any set of regions that each have constant diameter.

We assumed that our regions all had the same shape and size characteristics. In some cases it is interesting to see what happens in combinations. In particular, suppose we have one general region R_0 and m point regions $R_1, \ldots, R_m$; what Hausdorff bounds can be attained? It turns out that we get a trade-off: we can realize a Hausdorff distance of $O(\sqrt{m})$ for the point regions and for R_0, but we can also realize a Hausdorff distance of $O(1)$ for R_0 but then some point region will have a Hausdorff distance of $\Theta(m)$. Figure 9 illustrates this. We may map the points to the grid first using the $O(\sqrt{m})$ bound, and then map R_0, or we can map the points to the grid in a constant width strip close to the boundary

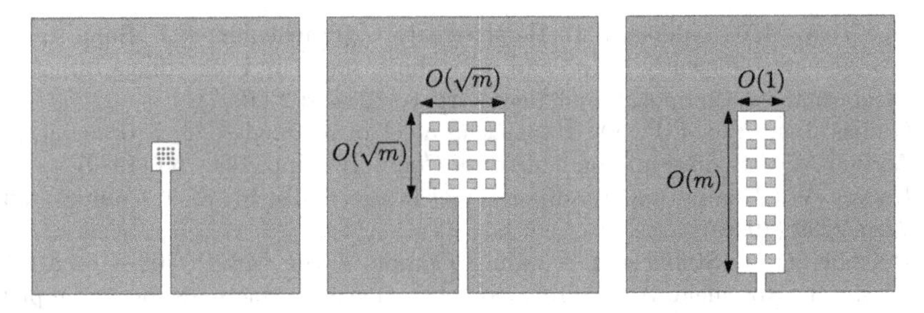

Fig. 9. Left, an instance with one general region (purple) and m point regions. Middle and right, two possible realizations for different Hausdorff bounds.

of R_0. Note that in the former case, we could have left a spacing of three pixels between the mappings of the point regions. Then the point regions still attain the $O(\sqrt{m})$ bound, while $H(R_0, P_0) = O(1)$ by using the extra space to allow P_0 to reach every necessary place. However, $H(\partial R_0, \partial P_0)$ will still be $\Theta(\sqrt{m})$, so we do not improve $H'(R_0, P_0)$.

While we concentrated on worst-case optimal bounds, our constructive proofs of the upper bounds will often give visually unfortunate output. Also, for a given instance we may not achieve $O(1)$ Hausdorff distance for m point, β-fat convex, or convex regions even when constant would be possible for that instance. This leads to the following two open problems. Firstly, can we realize visually reasonable output when this is possible for an instance (and how do we define this)? Secondly, can we realize a Hausdorff distance that is at most a constant factor worse than the best possible for each instance, in polynomial time?

Acknowledgments. This research was partially supported by the Dutch Research Council under the projects 612.001.651, 614.001.504 and 628.011.005.

References

1. Althaus, E., Eisenbrand, F., Funke, S., Mehlhorn, K.: Point containment in the integer hull of a polyhedron. In: Proceedings 15th Annual ACM-SIAM Symposium on Discrete Algorithms, pp. 929–933 (2004)
2. Bouts, Q.W., Kostitsyna, I., van Kreveld, M., Meulemans, W., Sonke, W., Verbeek, K.: Mapping polygons to the grid with small Hausdorff and Fréchet distance. In: Proceedings 24th Annual European Symposium on Algorithms, pp. 1–16 (2016)
3. Christ, T., Pálvölgyi, D., Stojaković, M.: Consistent digital line segments. Discret. Comput. Geom. **47**(4), 691–710 (2012)
4. Chun, J., Kikuchi, K., Tokuyama, T.: Consistent digital curved rays. In: Abstracts 34th European Workshop on Computational Geometry (2019)
5. Chun, J., Korman, M., Nöllenburg, M., Tokuyama, T.: Consistent digital rays. Discret. Comput. Geom. **42**(3), 359–378 (2009)
6. de Berg, M., Halperin, D., Overmars, M.: An intersection-sensitive algorithm for snap rounding. Comput. Geom. **36**(3), 159–165 (2007)

7. Goodrich, M.T., Guibas, L.J., Hershberger, J., Tanenbaum, P.J.: Snap rounding line segments efficiently in two and three dimensions. In: Proceedings 13th Annual Symposium on Computational Geometry, pp. 284–293 (1997)
8. Guibas, L.J., Yao, F.F.: On translating a set of rectangles. In: Proceedings 12th Annual ACM Symposium on Theory of Computing, pp. 154–160 (1980)
9. Harvey, W.: Computing two-dimensional integer hulls. SIAM J. Comput. **28**(6), 2285–2299 (1999)
10. Hershberger, J.: Stable snap rounding. Comput. Geom. **46**(4), 403–416 (2013)
11. Klette, R., Rosenfeld, A.: Digital Geometry: Geometric methods for digital picture analysis. Elsevier (2004)
12. Klette, R., Rosenfeld, A.: Digital straightness - a review. Discret. Appl. Math. **139**(1–3), 197–230 (2004)
13. Löffler, M., Meulemans, W.: Discretized approaches to schematization. In: Proceedings 29th Canadian Conference on Computational Geometry (2017)
14. Löffler, M., Simons, J.A., Strash, D.: Dynamic planar point location with sub-logarithmic local updates. In: Proceedings 13th International Symposium on Algorithms and Data Structures, pp. 499–511 (2013)
15. van Goethem, A., Kostitsyna, I., van Kreveld, M., Meulemans, W., Sondag, M., Wulms, J.: The painter's problem: covering a grid with colored connected polygons. In: Frati, F., Ma, K.-L. (eds.) GD 2017. LNCS, vol. 10692, pp. 492–505. Springer, Cham (2018). https://doi.org/10.1007/978-3-319-73915-1_38

Diverse Partitions of Colored Points

Marc van Kreveld[1], Bettina Speckmann[2], and Jérôme Urhausen[1](✉)

[1] Department of Information and Computing Sciences, Utrecht University, Utrecht, The Netherlands
{m.j.vankreveld,j.e.urhausen}@uu.nl
[2] Department of Mathematics and Computer Science, TU Eindhoven, Eindhoven, The Netherlands
b.speckmann@tue.nl

Abstract. Imagine that a set of objects is represented by points in space and that different types or classes of objects are represented by colors. We study the algorithmic problem of creating convex or Voronoi partitions of space with maximally diverse cells, using two classic diversity measures: the *richness* (number of different colors) and the *Shannon index.* The diversity of a partition is the sum of the diversity scores of its cells. Hence, we wish to compute either a *diverse convex partition* (DCP) or a *diverse Voronoi partition* (DVP), which maximizes the diversity score of the partition. Surprisingly, computing a DVP is NP-hard already in 1D and for only four colors, while DCP can easily be computed with dynamic programming. We show that DVP can be solved in polynomial time in 1D if a discrete set of candidate positions for the Voronoi sites is part of the input. These results apply to both the richness and the Shannon index. For richness, we also present a polynomial-time algorithm to compute a Voronoi partition whose diversity is at least $1-\varepsilon$ times the optimal diversity. In 2D, we show that both DCP and DVP are NP-hard, for richness as diversity measure. The reductions use constantly many colors for DVP and polynomially many colors for DCP.

Keywords: Computational geometry · Voronoi diagrams · Diversity · Colored points · Convex subdivision · Np-completeness · Species richness · Shannon index

1 Introduction

Imagine that a data set consists of objects that have different types, or classes, like genre of a book or species of a tree. As an abstraction, we represent such different types by different colors, and we are interested in *diverse* subsets. There

---

Research on the topic of this paper was initiated at the 3rd Workshop on Applied Geometric Algorithms (AGA 2018) in Langbroek, The Netherlands. Marc van Kreveld and Jérôme Urhausen were partially supported by the Dutch Research Council (NWO) under project no. 612.001.651. Bettina Speckmann was partially supported by the Dutch Research Council (NWO) under project no. 639.023.208.

A. Lubiw et al. (Eds.): WADS 2021, LNCS 12808, pp. 641–654, 2021.
https://doi.org/10.1007/978-3-030-83508-8_46

are many different ways to partition objects, and also many different ways to define the diversity of a partition. In this paper we study a fundamental *geometric* variant, namely the diversity of groups of colored points that are induced by general convex partitions of space, or those induced by a Voronoi diagram.

Two common ways to define the diversity of a set are the *(species) richness* and the *Shannon index*. The richness is the number of different colors in the set, while the Shannon index is defined as $-\sum_{i=1}^{h} \rho_i \log \rho_i$, where h is the number of colors and ρ_i is the proportion of objects of color i in the set. For example, the Shannon index of the set {red, green, blue} is $-(\frac{1}{3}\log\frac{1}{3} + \frac{1}{3}\log\frac{1}{3} + \frac{1}{3}\log\frac{1}{3}) = \log 3 \approx 1.585$, whereas the Shannon index of the set {red, green, blue, blue} is $-(\frac{1}{4}\log\frac{1}{4} + \frac{1}{4}\log\frac{1}{4} + \frac{1}{2}\log\frac{1}{2}) = 1.5$. Hence the first set is more diverse. When we have a partition of space and the objects are points, we can view the points in each cell as a set, and hence we can define the *diversity score* of a cell. The diversity score of the partition is the sum of the diversity scores of its cells.

Besides general convex partitions, a meaningful subclass of convex partitions in this context are *Voronoi partitions*, that is, partitions of space which are induced by the Voronoi diagram of a set of *sites*. The Voronoi site can serve as the representative of the points contained in its cell. Conversely, each point is represented by the site closest to it. Using richness, a diverse site represents many colors; using the Shannon index, a diverse site also represents many colors, which are additionally present in roughly equal proportions. Intuitively the Shannon index of a region increases if we add a point with a new color or if we equalize the proportions of the existing colors.

Formal Problem Statement. Our input is a set P of n points in h different colors and a number $k \in \mathbb{N}$. For any partition into k cells, the diversity d_i in a cell is the *score* of that cell, and $\sum d_i$ is the *score* of the partition. Our goal is to compute either a *diverse convex partition* (**DCP**) or a *diverse Voronoi partition* (**DVP**) with k cells which maximizes the overall diversity score according to the richness measure or the Shannon index. Since k is given, maximizing the total diversity and average diversity is equivalent. In the case of diverse Voronoi partitions, the problem is to find a set $S = \{s_1, \dots, s_k\}$ of k sites such that the sum of the diversity scores over all Voronoi cells is maximized. See Fig. 1 for an example of a convex partition (left) and a Voronoi partition (right, white disks represent Voronoi sites).

Results and Organization. We study diverse convex partitions (DCP) and diverse Voronoi partitions (DVP) both on the line (1D) and in the plane (2D). We begin in Sect. 2 by surveying related research on diversity and on partitioning problems for colored points. In Sect. 3 we illustrate how convex and Voronoi partitions differ in 1D and also show how to test if a given convex partition can be realized by a Voronoi partition. It is straightforward to compute a DCP in 1D using dynamic programming. Surprisingly, Sect. 4 shows that computing a DVP is NP-hard already in 1D and for only 4 colors. This result holds for both the richness and the Shannon index. For richness, the NP-hardness can be extended to 2D using 12 colors. We also show that a DVP can be computed in polynomial time in 1D if a discrete set of m candidate positions for the Voronoi sites is part

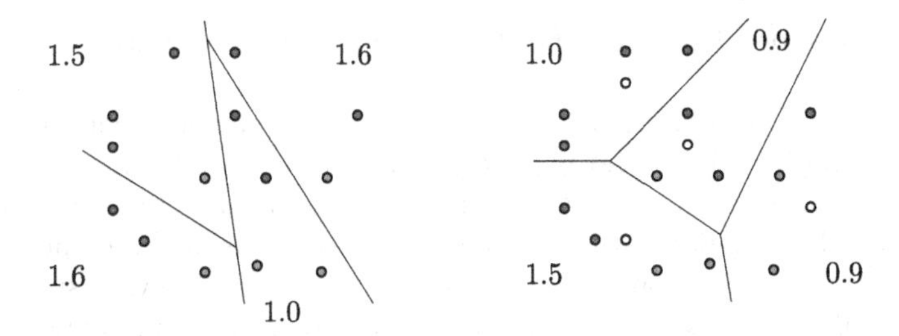

Fig. 1. A colored point set with a convex partition (left, diversity score 11 for richness, 5.7 for Shannon index) and a Voronoi partition (right, diversity score 9 for richness, 4.3 for Shannon index; white disks denote sites). Each cell is annotated with its (rounded) Shannon index. (Color figure online)

of the input. In Sect. 5 we show how to compute—in polynomial time for any constant $\varepsilon > 0$—a 1D Voronoi partition whose richness diversity is at least $1 - \varepsilon$ times the richness of the DVP. Finally, in Sect. 6 we show that computing a DCP is NP-hard in 2D for richness. We reduce from Maximum Independent Set in Orthogonal Line Segments using a colored grid structure which allows us to limit the possible shapes of convex sets.

2 Related Work

Diversity as a scientific concept is used to characterize sets; it is related to entropy, variety and representation [8]. Diversity plays an important role, for example, in ecology ([15,19]; diversity of species), information retrieval ([6,20]; diversification of query results), evolutionary algorithms ([14]; diversity in the population), and machine learning ([17]; diversity of classifiers). There are many measures of diversity, including species richness, the Shannon index, the Simpson index, and Rényi entropy. All these measures relate to sets of objects in classes, and have no spatial aspect. In a seminal paper, Whittaker [19] argued the need for differentiation in local (small-scale) diversity and regional (larger-scale) diversity. This requires a partition of a larger region into several smaller regions, and subsequently the analysis of diversity in the larger region and all of the smaller regions. The combined diversity measure is referred to as β-diversity; there are multiple different definitions in use [18]. The partitions we compute maximize the average or sum of local diversities over arguably reasonable geometric partitions.

The measure richness leads to algorithmic problems with a classic combinatorial and geometric structure. The Shannon index, however, gives rise to new challenges, which can be observed in the NP-hardness proof for DVP in 1D, and the fact that the approximation algorithm does not easily generalize. For richness, a DCP has at most n different summed diversity scores; for the Shannon index this is exponential. Furthermore, the richness of a cell cannot decrease if we grow the cell and collect more points; the Shannon index can decrease when extra points make the distribution of colors less balanced.

Partitioning problems for sets of points are extensively studied in computational geometry and related areas. There is a variety of specific problem formulations and, correspondingly, a multitude of related work. Here we highlight some results on bi-colored and multi-colored point sets and convex partitions.

Kaneko and Kano [13] present a survey of results for red and blue points in the plane. This includes computing convex partitions of the plane, such that each cell contains a red and b blue points, or, alternatively, a specific ratio. Bespamyatnikh *et al.* [4] study equitable partitions of red-blue point sets using convex sets. This line of inquiry has been further extended by Bereg *et al.* [1,3], Chierichetti *et al.* [7], and Holmsen *et al.* [12]. Bereg *et al.* [2] define coarseness of bicolored point sets as a measure of how mixed the two colors are. They give efficient algorithms for 2-partitions in the plane, and for point sets in convex position. Dumitrescu and Pach [9] partition multi-colored point sets into uni-colored subsets, whose convex hulls are disjoint; this minimizes diversity. Majumder *et al.* [16] consider the same problem, however, they partition the multi-colored input with axis-parallel lines. In d dimensions, Blagojević *et al.* [5] show that dn d-colored points can always be partitioned into n sets with disjoint convex hulls and evenly distributed colors, thus maximizing diversity.

Diverse Voronoi Partition is in some sense a clustering problem reminiscent of k-means clustering, which aims to represent multiple points by a single point, following a nearest neighbor rule. While k-means clustering minimizes the sum of squared distances to nearest representatives, DVP aims to maximize the diversity of colors for each representative. DVP also bears resemblance to multi-criteria facility location, where sites need to be placed to optimize two or more often conflicting criteria (see the extensive survey by Farahani *et al.* [10]). In the case of DVP these criteria are distance and diversity; we are not aware of diversity being used as a criterion in facility location, nor in partitioning in general.

3 Convex Versus Voronoi Partitions in 1D

In this section we explore the difference between convex and Voronoi partitions in 1D. Consider the example in Fig. 2 of 15 colored points, 5 in each of $h = 3$ colors. It is easy to see that there is a convex partition with a perfect richness score of 15 using $k = 5$ cells (intervals). To achieve the same score with a Voronoi partition, we need to place 5 sites such that the induced *boundaries* between Voronoi cells lie between the same input points as the corresponding boundaries of the convex partition. We capture this restriction on the Voronoi partition using so-called *b-intervals.* A b-interval is the open interval between two consecutive input points. A Voronoi partition that realizes a richness score of 15 must place 5 sites in such a way that each boundary between Voronoi cells lies in the corresponding b-interval. A careful inspection shows that this cannot be done. The middle site s_3 must be sufficiently far to the left to ensure that the second boundary is correctly placed (between the second green and the third red point), and at the same time sufficiently far to the right to ensure that the third boundary is correctly placed (between the third green and the fourth red point). It is impossible to move the other sites s_1, s_2, s_4, and s_5 to realize this.

Testing Realizability for Voronoi Partitions. Using b-intervals it is straightforward to test—using linear programming—if a given convex partition can be realized as a Voronoi partition (see also [11]). The convex partition directly induces the b-intervals. Let b_i be the midpoint of the interval between s_i and s_{i+1}, for $1 \leq i < k$. Then it must hold that $s_1 \leq s_2 \leq \cdots \leq s_k$, and $b_i = (s_i + s_{i+1})/2$. To ensure that the Voronoi cell boundaries b_i lie inside their respective b-intervals, we use another $2k - 2$ linear inequalities. Altogether, we have a system of $5k - 5$ linear inequalities whose solution—if it exists—gives a Voronoi partition.

Perfect Partitions. If the input S consists of exactly n/h points per color, we can ask if there is a *perfect partition* using $k = n/h$ sites that together have a richness score of n. The unique perfect convex partition, if it exists, can be found in $O(n)$ time if the points are given in sorted order. Constructing the corresponding system of linear inequalities also takes $O(n)$ time. Solving this linear program, and hence testing for a perfect Voronoi partition, takes polynomial time in $k = n/h$.

4 Diverse Voronoi Partition in 1D

4.1 NP-Hardness When Richness is the Diversity Measure

We prove that the decision version of DVP (D-DVP) is NP-complete, even in 1D. D-DVP has an extra parameter z and asks if a diversity score of at least z can be realized with a Voronoi partition using k points. We first argue containment in NP. For a given instance of D-DVP, there are only exponentially many partitions into subsets, each defined by $k - 1$ b-intervals. We can test for each of these partitions if they can be realised as a Voronoi partition with k sites in polynomial time using linear programming (see Sect. 3).

For hardness we reduce from Subset Sum: for a set $A = \{a_1, \ldots, a_r\}$ of integers and an integer b, is there a subset $A' \subseteq A$ such that $\sum_{a_j \in A'} a_j = b$? We first define a few terms. A point $p \in P$ is *represented* by a site $s \in S$ if s is the site closest to p. For each color $i \in \{1, \ldots, h\}$, P_i is the subset of points of color i, and a point $p \in P_i$ is *scored* if each other point $p' \in P_i$ that is represented by the same site is to the right of p. That is, for each site only the leftmost point of each color that it represents is scored. A point is *unscored* if it is not scored. Hence, an optimal set S of sites maximizes the number of scored points. Our reduction uses only four colors, so we define point sets P_1, P_2, P_3, and P_4 from an instance of Subset Sum with total size $n = 8r + 14$.

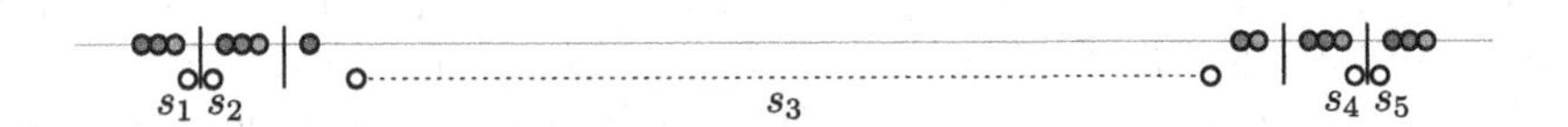

Fig. 2. Points that admit a perfect convex partition but not a perfect Voronoi partition. (Color figure online)

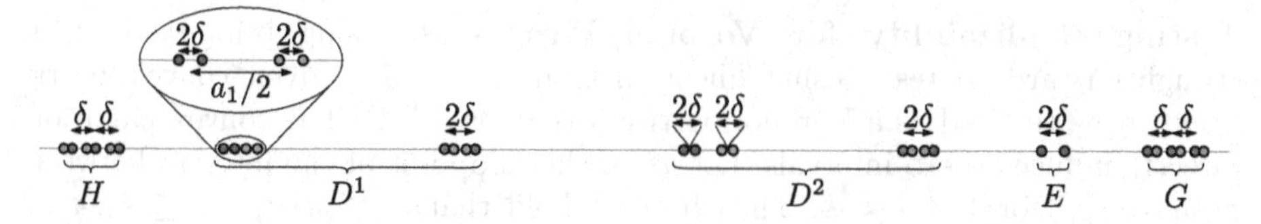

Fig. 3. Reducing SUBSET SUM to D-DVP for $a_1 = 1$, $a_2 = 2$ and $b = 2$ using P_1 (blue), P_2 (green), P_3 (red), and P_4 (yellow). Touching points are at ε distance, δ is not drawn to scale. (Color figure online)

The Construction. Let $0 < \delta \ll 1$ be a small real and let $0 < \varepsilon \ll \delta$ be an even smaller real. We can take $\delta = 1/r^2$ and $\varepsilon = 1/r^4$. Later we can multiply the coordinates of the constructed points by r^4 and thus obtain a set of integer positions with polynomial size. Let $a = \sum_{i=1}^{r} a_i$ be the total sum of the integers of the SUBSET SUM instance.

We construct the set P using the values a_i and b. The goal is that the new D-DVP instance has a solution if and only if the SUBSET SUM instance has a solution. We describe P from left to right. First, there is a starting gadget H of six points. Then we have a gadget D^j for each a_j, consisting of eight points (these gadgets can be in any order). Next, we have a subset sum gadget E of two points to represent b, and finally we have an ending gadget G of six points. $P = H \cup D^1 \cup \cdots \cup D^r \cup E \cup G$. Figure 3 shows an example for $A = \{1, 2\}$ and $b = 2$, so $P = H \cup D^1 \cup D^2 \cup E \cup G$. Intuitively, each cell in a solution of the D-DVP instance contains exactly one (blue) point from P_1 and one (green) point from P_2. In each gadget D^j there is a choice to either separate two (red) points from P_3 or two (yellow) points from P_4. This choice corresponds to not choosing or choosing a_j in the subset sum so far.

To start the construction we define a set H of six points in two colors. We set $H_1 = \{-2\delta, -\delta, 0\} \subset P_1$ and $H_2 = \{-2\delta - \varepsilon, -\delta - \varepsilon, -\varepsilon\} \subset P_2$. The set H thus forms three groups of two points of different colors. We can only score all points in H with three sites s_{-2}, s_{-1}, s_0 if we have $-\delta < s_0 < 2\delta - 2\varepsilon$. So, in order to score all six points with three sites, the rightmost of those sites, s_0, needs to be close to zero.

For each $a_j \in A$ we create a set D^j of eight points that will encode whether a_j is chosen in the subset A' or not. Let $D^j = D^j_1 \cup D^j_2 \cup D^j_3 \cup D^j_4$, with $D^j_i \subset P_i$ (the points in D^j_i have color i). Let $D^j_1 = \{(4j-1)a - \delta, (4j-1)a + \delta\}$, $D^j_2 = \{(4j-1)a - \delta + \varepsilon, (4j-1)a + \delta - \varepsilon\}$, $D^j_3 = \{(4j-3)a - \delta, (4j-3)a + \delta\}$ and $D^j_4 = \{(4j-3)a + a_j/2 - \delta, (4j-3)a + a_j/2 + \delta\}$. The distances between D^j_3 and D^j_4 are roughly $a_j/2$.

We define a set $E \subset P$ that encodes that we want the subset sum to be b. We define $E \subset P_3$ with $E = \{4ra + (a-b)/2 - \delta, 4ra + (a-b)/2 + \delta\}$.

Finally we define a set G of six points, similar to H. It can only be scored fully by three sites if the leftmost of the sites is close to $(4r+1)a$. We set $G_1 = \{(4r+1)a, (4r+1)a + \delta, (4r+1)a + 2\delta\} \subset P_1$ and $G_2 = \{(4r+1)a + \varepsilon, (4r+1)a + \delta + \varepsilon, (4r+1)a + 2\delta + \varepsilon\} \subset P_2$.

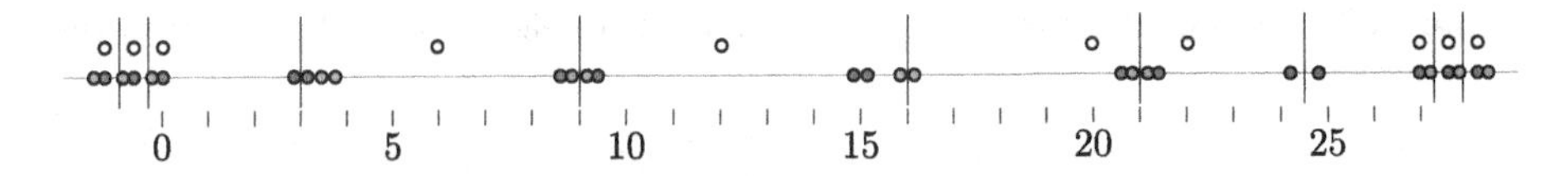

Fig. 4. D-DVP instance: $a_1 = 1$, $a_2 = 2$ and $b = 2$. Sites and boundaries for a score of $7r + 14$. (Color figure online)

Equivalence. The instance of D-DVP has $n = 8r + 14$ points and asks to place $k = 2r + 6$ sites to realize a score of $z = 7r + 14$, which can be achieved if and only if the corresponding SUBSET SUM instance has a solution. Intuitively, we want to use the sites to create boundaries that separate the first three pairs of points, the last three pairs of points, either D_3^j or D_4^j, and also E. Separating D_3^j corresponds to not choosing a_j in a subset and separating D_4^j corresponds to choosing a_j. If we choose the correct a_j, the boundary between the last site chosen for D^r and the first site chosen for G will "magically" separate the points in E. Then, only one point of each D^j is not scored. See Fig. 4 for an example. The proof of correctness argues that essentially there are no other options: the SUBSET SUM instance has a solution if and only if the D-DVP instance can score $7r + 14$.

Theorem 1. *Deciding if a diverse Voronoi partition of n colored points in four colors in 1D of richness diversity at least z exists, using k cells, is NP-complete.*

The extension to 2D is not immediate; the proof uses 12 colors and points on three parallel lines. The description is omitted from this version of the paper.

4.2 NP-Hardness When Shannon Index is the Diversity Measure

We prove that deciding DVP is NP-complete in 1D, also when using the Shannon index as diversity measure. For ease of argument, we allow points of different colors to share locations. For containment in NP, we note that we need logarithms of $1..n$ only, and we can still guess the partition and approximate its total Shannon index sufficiently precisely. For NP-hardness, we reduce from SUBSET SUM as before: for a set $A = \{a_1, \ldots, a_r\}$ of integers and an integer b, is there a subset $A' \subseteq A$ such that $\sum_{a_j \in A'} a_j = b$? The set of points we construct is similar to the one using richness, but the proof arguments are more complex. We place all yellow, red, and blue points at exactly the same positions as before, and set $\varepsilon = 0$ so that each green point coincides with the nearest blue point. We use two more red points in the start gadget and two more in the end gadget, to coincide with the first two blue points and the last two blue points, as shown in Fig. 5.

Equivalence. The decision question corresponding to SUBSET SUM is: using $2r + 6$ sites, can we get a score of at least $(\ell_3 + \ell_5)r + 6\ell_3$ in their Voronoi cells, where $\ell_3 = \log(3) \approx 1,58$ is the score of a cell with three points of different colors and $\ell_5 = \log(5) - 2/5 \approx 1,92$ is the score of a cell with two points of one color and three other points of different colors?

Table 1. The Shannon index for distributions of up to eight points that occur in the construction.

Cell	Score	Cell	Score	Cell	Score
$[\emptyset]$	0	$[1,1,1]$	1.585	$[2,1,1,1]$	1.922
$[1]$	0	$[2,1,1]$	1.5	$[2,2,1,1]$	1.918
$[2]$	0	$[2,2,1]$	1.522	$[2,2,2,1]$	1.950
$[1,1]$	1	$[2,2,2]$	1.585	$[2,2,2,2]$	2
$[2,1]$	0.918	$[3,3,2]$	1.561	$[3,2,2,1]$	1.906
$[2,2]$	1	$[3,3,3]$	1.585		

Table 1 gives the Shannon index for all possible cells in this instance with up to eight points, where $[x_1, \ldots, x_m]$ denotes a cell with m different colors and x_i points per color. Note that any cell, even with more points, has a score of at most 2, the maximum with four colors.

We prove that using $2r+6$ *convex* cells, the maximum possible score is $r(\ell_3+\ell_5)+6\ell_3$, and this can only be achieved with $r+6$ cells of the form $[1,1,1]$ and r cells of the form $[2,1,1,1]$. A diverse *Voronoi* partition can realize this too if and only if the SUBSET SUM instance has a solution.

Let A be the subdivision of the DVP instance into $2r+6$ convex cells that yields the maximum total Shannon index score. Let us count the number of cells of this optimal assignment using classes. C_0 contains the cells that contain at most 1 color, and C_1 (C_{ℓ_3}) contain the cells with points of precisely two (resp., three) colors. The cells with precisely four colors appear in two classes: C_{ℓ_5} and C_2 contain the cells with four colors that contain exactly one, respectively two or more blue point(s). Let $c_i = |C_i|$. Each cell with three or more colors contains at least one blue point, since blue and green points coincide. As there are $2r+6$ cells and $2r+6$ blue points, the following inequalities hold:

$$2r+6 \geq c_1 + c_{\ell_3} + c_{\ell_5} + c_2 \quad (1)$$

$$2r+6 \geq c_{\ell_3} + c_{\ell_5} + 2c_2 \quad (2)$$

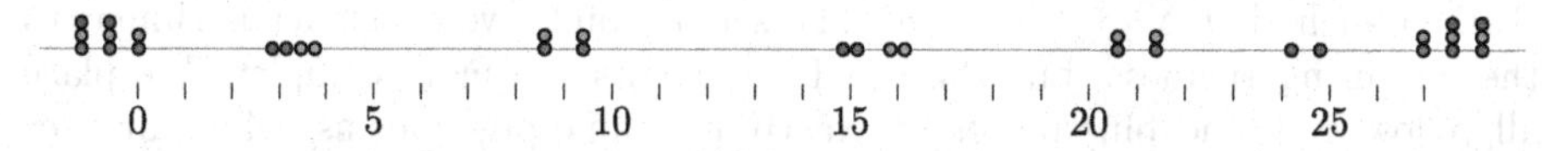

Fig. 5. The set of points constructed for $a_1 = 1$, $a_2 = 2$ and $b = 2$ to prove NP-completeness when using the Shannon index for diversity, corresponding to Fig. 4. (Color figure online)

$$
\begin{aligned}
\text{Max Score} &\leq c_1 + \ell_3 c_{\ell_3} + \ell_5 c_{\ell_5} + 2c_2 \\
&\overset{(1)}{\leq} (2r + 6 - c_{\ell_3} - c_{\ell_5} - c_2) + \ell_3 c_{\ell_3} + \ell_5 c_{\ell_5} + 2c_2 \\
&= 2r + 6 + (\ell_3 - 1)c_{\ell_3} + (\ell_5 - 1)c_{\ell_5} + c_2 \\
&\overset{(2)}{\leq} 2r + 6 + (\ell_3 - 1)c_{\ell_3} + (\ell_5 - 1)c_{\ell_5} + (2r + 6 - c_{\ell_3} - c_{\ell_5})/2 \\
&= 3r + 9 + (\ell_3 - 3/2)c_{\ell_3} + (\ell_5 - 3/2)c_{\ell_5} \qquad (3)
\end{aligned}
$$

As we have $\ell_5 > \ell_3 > 3/2$ and $c_{\ell_3} + c_{\ell_5} \leq 2r + 6$, in order to upper-bound expression (3) we maximize c_{ℓ_5} first and c_{ℓ_3} second. Each cell in class C_{ℓ_5} has points of four colors and exactly one blue point. No cell in C_{ℓ_5} can contain a yellow point p and a red point q with $p < q$, otherwise the cell would contain two blue points. The yellow points appear in adjacent pairs and there are r such pairs. Thus $c_{\ell_5} \leq r$ holds, and equality is attainable. Given $c_{\ell_5} = r$, we get $c_{\ell_3} \leq r + 6$, and also here equality is attainable in the construction. So

$$
\begin{aligned}
\text{Max Score} &\leq 3r + 9 + (\ell_3 - 3/2)c_{\ell_3} + (\ell_5 - 3/2)c_{\ell_5} \\
&\leq 3r + 9 + (\ell_3 - 3/2)(r + 6) + (\ell_5 - 3/2)r \\
&= (\ell_3 + \ell_5)r + 6\ell_3
\end{aligned}
$$

This concludes the proof and shows:

Theorem 2. *Deciding if a diverse Voronoi partition of n colored points in four colors in 1D of Shannon index at least z exists, using k cells, is NP-complete.*

4.3 Polynomial-Time Solution for Discrete Candidate Sites

If we assume that there is a fixed set of candidate positions for the sites, then optimal diverse Voronoi partitions can be computed in 1D by dynamic programming. The description is omitted from this version of the paper.

Theorem 3. *A diverse Voronoi partition of n points in h colors in sorted order in 1D into k cells using m discrete candidate positions can be computed in $O(km^3 + hm^3 + n)$ time, for richness or Shannon index, or in $O(km^3 + m^2 h \log h + n)$ time for richness.*

5 Approximation for Diverse Voronoi Partition in 1D

In this section we show that for any constant $\varepsilon > 0$, we can compute a Voronoi partition whose diversity (richness) is at least $1 - \varepsilon$ times the optimal diversity in polynomial time. We will first use more sites than allowed in order to separate subproblems, which we solve optimally using linear programming. We combine the subsolutions using dynamic programming, and then remove sites to the desired number without deteriorating the solution too much.

Let $P = \{p_1, \ldots, p_n\}$, k, and $0 < \varepsilon < 1$ be given. Let $e = \lceil 2/\varepsilon \rceil$, and let δ be a small number, for example $\min_{i=1,..,n-1}(p_{i+1} - p_i)/4$. For $i = 1, \ldots, n-1$ we define $m_i = (p_i + p_{i+1})/2$ as the middle between p_i and its right neighbor p_{i+1}.

Our goal is to subdivide P into $g = \lceil k/e \rceil$ subsets and then place e sites optimally within each subset. For each of the $\binom{n+1}{2}$ non-empty convex subsets of points, we calculate a specific score $s(i,j)$ that is specified for a convex subset $p_i, \ldots, p_j$, where $2 \leq i < j \leq n-1$, as follows. Place two sites, one at $m_{i-1}+\delta$ and one at $m_j - \delta$; these are fixed. Then we place another e sites in between these two sites in an optimal way, maximizing the score. We do this by placing the $e+1$ boundaries between the $e+2$ sites, and then checking whether these boundaries are realizable by a Voronoi partition, using linear programming. There are $O((j-i)^{e+1}) = O(n^{e+1})$ choices to be checked. We store the maximum in a table for $s(i,j)$. For the convex subsets $p_1, \ldots, p_j$ or $p_i, \ldots, p_n$ we compute the score slightly differently, because they do not need the leftmost or rightmost extra site, and because the last convex subset may have fewer than e sites remaining. For the last convex subset, we have $k \mod e = k - e(g-1)$ sites, to be precise, so we compute $s(i,n)$ for all i and $k \mod e$ sites, plus one extra at $m_{i-1} + \delta$.

We then find the optimal subdivision of P into g convex subsets, such that the sum of the scores of all the subsets is maximal. We do this using dynamic programming to compute a function $f(h,j)$, representing the optimal score for the points $p_1, \ldots, p_j$ by using h convex subsets that partition $p_1, \ldots, p_j$, and each convex subset is scored with the $s(.,.)$ function. Since the application of dynamic programming is standard, we simply state:

$$f(h,j) = \max_{\ell < j} \left(f(h-1,\ell) + s(\ell+1,j) \right) .$$

The value $f(g,n)$ then gives the maximum sum of scores when subdividing P into g subsets; a set S' of $k + 2(g-1)$ sites attains this.

Lemma 1. *The score of the calculated sites S' is at least the score* Opt *of an optimal solution S^* with k sites.*

The remainder of the algorithm is simple: we determine the score of each site, choose the site with the lowest score, and remove it. After $2(g-1)$ iterations, we have a set of k sites, which we show to be a $(1-\varepsilon)$-approximation of the best possible with k sites.

Lemma 2. *Reducing the set S' of sites to a set with k sites costs no more than ε times the score of S'*

Proof. When removing a site the overall score drops by at most the score of the cell of the site that was removed. So one by one, we choose the site whose associated cell has the lowest score and remove it. Let Y_0 be the score of the set of sites S'. For $w = |S'| = k + 2g - 2$, the score drops by at most Y_0/w for removing the first cell, thus the score after removing that cell is larger than $Y_1 = Y_0 \frac{w-1}{w}$. Iteratively, we define $Y_i = Y_{i-1} \frac{w-i}{w-(i-1)}$. Using a telescope sum we get $Y_i = Y_0 \frac{w-i}{w}$. Thus the score after removing $2g-2$ sites is at least

$Y_{2g-2} = Y_0 \frac{w-2g+2}{w}$. So the score Y' of the remaining k sites is at least $Y_{2g-2} \geq \text{OPT}\frac{(k+2g-2)-2g+2}{k+2g-2} = \text{OPT}\frac{k}{k+2\lceil k/e \rceil -2} \geq \text{OPT}\frac{k}{k+2(k/e+1)-2} = \text{OPT}\frac{1}{1+2/e}$. Thus $Y'(1+2/e) \geq \text{OPT} \iff Y'(1+\varepsilon) \geq \text{OPT} \implies Y' \geq \text{OPT}(1-\varepsilon)$. □

Theorem 4 directly follows.

Theorem 4. *Let P be a set of n points in 1D, let k be a positive integer, and let $\varepsilon > 0$ be a constant. There is a polynomial-time $(1-\varepsilon)$-approximation algorithm for computing a diverse Voronoi partition based on richness diversity.*

6 Diverse Convex Partition is NP-Hard in 2D

We show that the following diverse convex partition problem is NP-hard: given a set of colored points in the plane, partition the plane into the minimum number of convex regions so that the total diversity score according to richness is the same as the number of points (*full score*). We only sketch the structure and main ideas of the proof here.

First of all, we use a scaffolding structure that is a regular pattern of points in four colors, so that unit squares would provide a full score partition with the minimum number of convex sets. We let the *4-fold colored grid point set* be the point set with all points $(i \pm \delta, j \pm \delta)$ with colors: green if i and j are even; yellow if i is odd and j is even; blue if i is even and j is odd; red if i and j are odd. The four equal-colored points close to an integer grid point are called a 4-group. We choose a rectangular portion $[0 : w] \times [0 : h]$ of the 4-fold colored grid point set and call it G; see Fig. 6. At the edges of this rectangular region there are 2-groups and 1-groups. In total G has $4wh$ colored points. We let $n = \max(w, h)$ and $\delta = 1/n^2$. In an $n \times n$ unit grid, it is known that the shortest strictly positive distance from a line segment between grid points to another grid point is $\Omega(1/n)$. However, the 4-group points are just $\sqrt{2}/n^2$ away from their closest grid point. As a consequence, a convex partition with full score is severely limited: no connection between two differently colored points can separate a 4-group of yet a different color, implying that the four points of a 4-group must be at corners of different cells in a partition with full score.

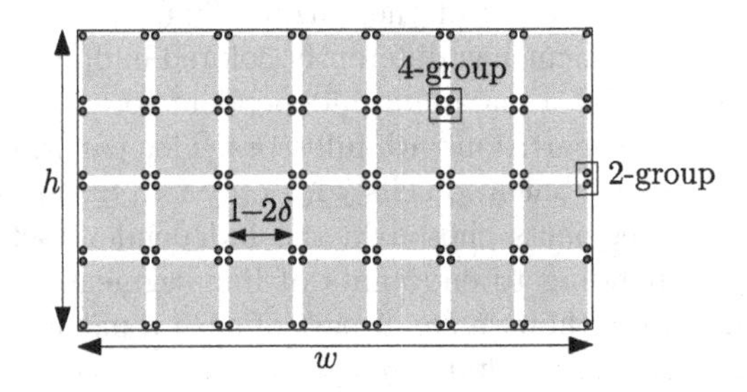

Fig. 6. The set G of $4wh$ colored points, a 4-group, and a 2-group. (Color figure online)

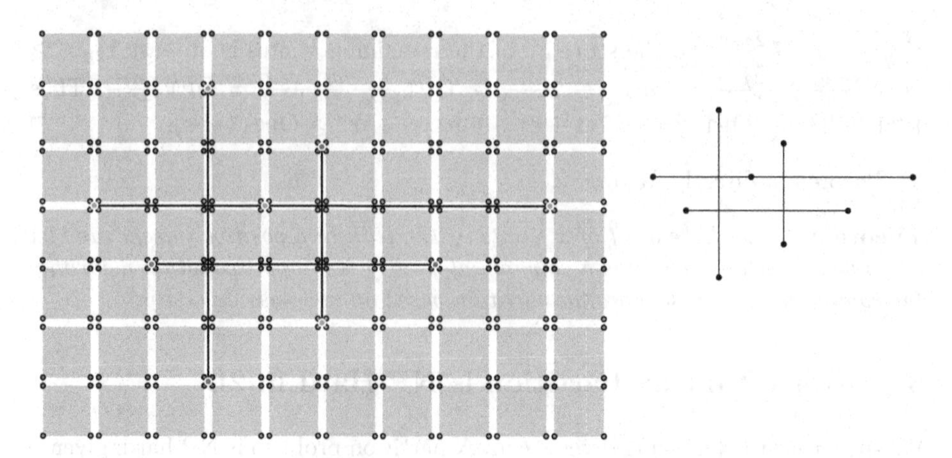

Fig. 7. Reduction from an instance of MAXIMUM INDEPENDENT SET IN ORTHOGONAL LINE SEGMENTS to an instance of 2D DIVERSE CONVEX PARTITION. (Color figure online)

A *4-set* is a set of four different-colored points of G whose convex hull does not contain any other points of G. We observe that G has a diverse convex partition into wh convex sets with diversity score $4wh$, the best possible, by choosing wh 4-sets that are squares of side length $1 - 2\delta$. These squares are called *g-squares*; they are shown in grey in the figures. Other 4-sets are possible, but we can show that they do not lead to a convex partition with diversity score $4wh$ using wh cells.

We reduce from MAXIMUM INDEPENDENT SET (MIS) IN ORTHOGONAL LINE SEGMENTS. This problem is NP-hard if we allow two horizontal or two vertical line segments to intersect in a joint endpoint. We can assume that all endpoints of the line segments have different x- and y-coordinates, with the exception of: (i) the two endpoints of one line segment, and (ii) pairs of parallel line segments that share an endpoint.

The line segments of an instance are locally scaled in x- and y-direction to an equivalent instance, so that the endpoints have suitable integer coordinates and line segments have odd length, see Fig. 7. These endpoints lie in the 4-fold integer grid and adopt the color of the surrounding 4-group. The odd length ensures that each line segment has differently colored endpoints.

It remains to argue that a maximum independent set corresponds to certain groups in a diverse convex partition with full score. This partition will have many g-squares with four colors, but also cells with only two colors corresponding to the endpoints of line segments chosen in the independent set, and cells with only one color corresponding to endpoints of line segments not chosen in the independent set. Figure 8 shows a small part of such a partition.

To ensure that there are no alternative solutions we need polynomially many colors that will rule out cells that would extend beyond a single unit square or a chosen line segment. Note that in Fig. 8, there are other convex partitions

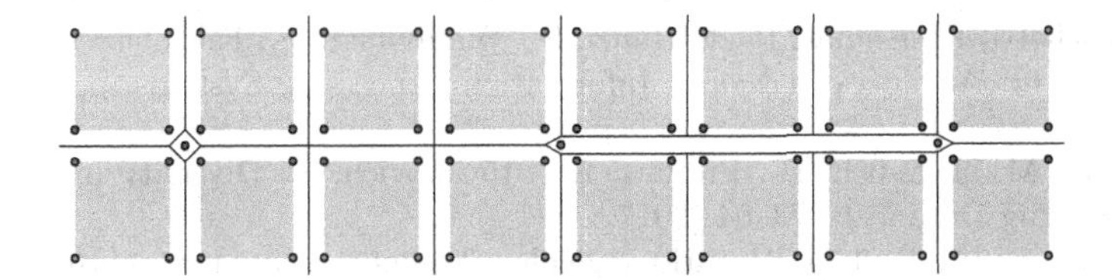

Fig. 8. A convex partition that includes convex cells corresponding to a line segment chosen in the independent set, and an isolated endpoint. (Color figure online)

with the same score and the same number of cells (the isolated red point can be combined with one of eight points roughly one unit away into a cell with two colored points), but these alternative solutions will still include the maximum independent set.

Theorem 5. *Computing a minimum-size convex partition of full richness diversity score is NP-hard.*

7 Conclusions

We have cast the non-spatial concept of diversity into a geometric setting and introduced the computational problem of computing geometric partitions with high diversity. We used two versions of diversity: richness and the Shannon index. While richness leads to computational problems common in computational geometry, the Shannon index needs new proof ingredients. The main open problems are finding a $(1-\varepsilon)$-approximation scheme for 1D DVP for the Shannon index and an NP-hardness proof for 2D DCP for richness using constantly many colors. NP-hardness of 2D DCP for the Shannon index is also unresolved.

As an alternative to high within-cell diversity, a partition could aim for a high diversity *between* different cells. Then statistics like the Jaccard index could be used to measure differences between cells. We leave this extension to future work.

References

1. Bereg, S., Bose, P., Kirkpatrick, D.: Equitable subdivisions within polygonal regions. Comput. Geom. **34**(1), 20–27 (2006)
2. Bereg, S., Díaz-Báñez, J.M., Lara, D., Pérez-Lantero, P., Seara, C., Urrutia, J.: On the coarseness of bicolored point sets. Comput. Geom. **46**(1), 65–77 (2013)
3. Bereg, S., et al.: Balanced partitions of 3-colored geometric sets in the plane. Discret. Appl. Math. **181**, 21–32 (2015)
4. Bespamyatnikh, S., Kirkpatrick, D., Snoeyink, J.: Generalizing ham sandwich cuts to equitable subdivisions. Discret. Comput. Geom. **24**(4), 605–622 (2000)
5. Blagojević, P.V., Rote, G., Steinmeyer, J.K., Ziegler, G.M.: Convex equipartitions of colored point sets. Discret. Comput. Geom. **61**(2), 355–363 (2019)
6. Borodin, A., Jain, A., Lee, H.C., Ye, Y.: Max-sum diversification, monotone submodular functions, and dynamic updates. ACM Trans. Algorithms (TALG) **13**(3), 1–25 (2017)

7. Chierichetti, F., Kumar, R., Lattanzi, S., Vassilvitskii, S.: Fair clustering through fairlets. In: Advances in Neural Information Processing Systems, pp. 5029–5037 (2017)
8. Drosou, M., Jagadish, H., Pitoura, E., Stoyanovich, J.: Diversity in big data: a review. Big Data **5**(2), 73–84 (2017)
9. Dumitrescu, A., Pach, J.: Partitioning colored point sets into monochromatic parts. Int. J. Comput. Geom. Appl. **12**(05), 401–412 (2002)
10. Farahani, R.Z., SteadieSeifi, M., Asgari, N.: Multiple criteria facility location problems: a survey. Appl. Math. Model. **34**(7), 1689–1709 (2010)
11. Hartvigsen, D.: Recognizing Voronoi diagrams with linear programming. ORSA J. Comput. **4**(4), 369–374 (1992)
12. Holmsen, A.F., Kynčl, J., Valculescu, C.: Near equipartitions of colored point sets. Comput. Geom. **65**, 35–42 (2017)
13. Kaneko, A., Kano, M.: Discrete geometry on red and blue points in the plane, a survey. In: Aronov, B., Basu, S., Pach, J., Sharir, M. (eds.) Discrete and Computational Geometry, The Goodman-Pollack Festschrift, pp. 551–570. Springer, Heidelberg (2003). https://doi.org/10.1007/978-3-642-55566-4_25
14. Lacevic, B., Amaldi, E.: On population diversity measures in Euclidean space. In: IEEE Congress on Evolutionary Computation, pp. 1–8 (2010)
15. Magurran, A.E.: Ecological Diversity and its Measurement. Princeton University Press, Princeton (1988)
16. Majumder, S., Nandy, S.C., Bhattacharya, B.B.: Separating multi-color points on a plane with fewest axis-parallel lines. Fundamenta Informaticae **99**(3), 315–324 (2010)
17. Tang, E.K., Suganthan, P.N., Yao, X.: An analysis of diversity measures. Mach. Learn. **65**(1), 247–271 (2006)
18. Tuomisto, H.: A diversity of beta diversities: straightening up a concept gone awry. Part 1. Defining beta diversity as a function of alpha and gamma diversity. Ecography **33**(1), 2–22 (2010)
19. Whittaker, R.H.: Vegetation of the Siskiyou mountains, Oregon and California. Ecol. Monogr. **30**(3), 279–338 (1960)
20. Zheng, K., Wang, H., Qi, Z., Li, J., Gao, H.: A survey of query result diversification. Knowl. Inf. Syst. **51**(1), 1–36 (2016). https://doi.org/10.1007/s10115-016-0990-4

Reverse Shortest Path Problem for Unit-Disk Graphs

Haitao Wang and Yiming Zhao(✉)

Department of Computer Science, Utah State University, Logan, UT 84322, USA
{haitao.wang,yiming.zhao}@usu.edu

Abstract. Given a set P of n points in the plane, a unit-disk graph $G_r(P)$ with respect to a radius r is an undirected graph whose vertex set is P such that an edge connects two points $p, q \in P$ if the Euclidean distance between p and q is at most r. The length of any path in $G_r(P)$ is the number of edges of the path. Given a value $\lambda > 0$ and two points s and t of P, we consider the following *reverse shortest path problem*: finding the smallest r such that the shortest path length between s and t in $G_r(P)$ is at most λ. It was known previously that the problem can be solved in $O(n^{4/3} \log^3 n)$ time. In this paper, we present an algorithm of $O(\lfloor \lambda \rfloor \cdot n \log n)$ time and another algorithm of $O(n^{5/4} \log^2 n)$ time.

1 Introduction

Given a set P of n points in the plane and a radius r, a unit-disk graph $G_r(P)$ is an undirected graph whose vertex set is P such that an edge connects two points $p, q \in P$ if the Euclidean distance between p and q is at most r. Alternatively, $G_r(P)$ may be viewed as the intersection graph of the set of congruous disks of radius $r/2$ centered at the points of P (i.e., disks are vertices and two disks have an edge if they intersect). The *length* of a path in $G_r(P)$ is defined to be the number of edges of the path. For any two points p and q of P, their *distance* in $G_r(P)$ is defined as the length of a shortest path from p to q in $G_r(P)$.

Finding shortest paths in unit-disk graphs (e.g., single-source shortest paths and all-pairs shortest paths) has been extensively studied, e.g., [5,6,14,18,20]. In this paper, we consider the following *reverse shortest path problem*: Given a value $\lambda > 0$ and two points s and t of P, find the smallest value r such that the distance between s and t in $G_r(P)$ is at most λ. As the length of any path in $G_r(P)$ is an integer, the length of a path of $G_r(P)$ is at most λ if and only if the length of the path is at most $\lfloor \lambda \rfloor$; therefore, we can replace λ in the problem by $\lfloor \lambda \rfloor$. In the following, we simply assume that λ is an integer.

Using the distance selection algorithm of Katz and Sharir [15], Cabello and Jejčič [5] pointed out a straightforward algorithm of $O(n^{4/3} \log^3 n)$ time for the problem and asked whether better algorithms exist. In this paper, we present an algorithm of $O(\lambda \cdot n \log n)$ time and another algorithm of $O(n^{5/4} \log^2 n)$ time. Thus, the first algorithm is preferable when λ is relatively small.

This research was supported in part by NSF under Grant CCF-2005323. A full version of this paper is available at https://arxiv.org/abs/2104.14476.

A. Lubiw et al. (Eds.): WADS 2021, LNCS 12808, pp. 655–668, 2021.
https://doi.org/10.1007/978-3-030-83508-8_47

1.1 Related Work

Unit-disk graphs is an important class of geometric intersection graphs, and a vast amount of problems have been studied in unit-disk graphs due to many of their applications, e.g., in wireless sensor networks.

Finding a shortest path in the unit-disk graph $G_r(P)$ has been well-studied. Although $G_r(P)$ may have $\Omega(n^2)$ edges, it is possible to find a shortest path in $G_r(P)$ between two given points of P in sub-quadratic time using certain geometric properties of P. Roditty and Segal [18] first gave an algorithm of $O(n^{\frac{4}{3}+\epsilon})$ time, where and throughout the paper ϵ is an arbitrarily small positive constant. The algorithm also works for the *weighted case* where the weight of each edge of $G_r(P)$ is defined to be the Euclidean distance of the two vertices connected by the edge; in contrast, in the *unweighted case* the length of each edge of $G_r(P)$ is one. Cabello and Jejčič [5] proposed an algorithm of $O(n \log n)$ time for the unweighted case. They also gave an $O(n^{1+\epsilon})$ time algorithm for the weighted case by using a dynamic data structure for bichromatic closest pairs [1]. Using the improved (and randomized) result of Kaplan et al. [14] for the dynamic bichromatic closest pairs, the weighted case can be solved in $O(n \log^{12+o(1)} n)$ expected time. Recently, Wang and Xue [20] derived a new (deterministic) algorithm for the weighted case and the runtime is $O(n \log^2 n)$. In addition, Chan and Skrepetos [6] gave an $O(n)$-time algorithm for the unweighted case, provided that the points of P are presorted by both the x- and y-coordinates.

In addition to the shortest path problem, many other problems of unit-disk graphs have also been studied, i.e. clique [8], independent set [16], distance oracle [7,13], diameter [6,7,13], etc. Comparing to general graphs, many problems can be solved efficiently in unit-disk graphs by exploiting their underlying geometric structures, although there are still problems that are NP-hard for unit-disk graphs and other geometric intersection graphs, e.g., [3,8].

Note that reverse/inverse shortest path problems have been studied in the literature in various problem settings. Roughly speaking, the problems are to modify the graph (e.g., modify some edge weights) so that certain desired constraints related to shortest paths in the graph can be satisfied, e.g., [4,21]. Our reverse shortest path problem in unit-disk graphs may find applications in scenarios like the following. Consider $G_r(P)$ as a unit-disk intersection graph representing a wireless sensor network in which each disk represents a sensor and two sensors can communicate with each other (e.g., directly transmit a message) if there is an edge connecting them in the graph. The radius of a disk is proportional to the energy of the sensor. For two specific sensors s and t, suppose we want to know the minimum energy for all sensors so that s and t can transmit messages to each other within λ steps for a given value λ. It is easy to see that this is equivalent to our reverse shortest path problem.

1.2 Our Approach

Let r^* denote the optimal radius for the reverse shortest path problem, i.e., the smallest r such that the distance of s and t in $G_r(P)$ is at most λ. Our goal is to compute r^*. Given a value r, *the decision problem* is to decide whether $r \geq r^*$. It is not difficult to see that $r \geq r^*$ if and only if the distance of s and t in $G_r(P)$ is at most λ. Therefore, the decision problem can be solved in $O(n \log n)$ time by using the shortest path algorithm for the unweighted unit-disk graphs [5,6]. More efficiently, with $O(n \log n)$-time preprocessing (to sort the points of P), given any r, whether $r \geq r^*$ can be decided in $O(n)$ time by the algorithm of Chan and Skrepetos [6].

Observe that r^* must be equal to the distance of two points of P. Therefore, we can find r^* by doing binary search on the set of pairwise distances of all points of P. Given any $k \in [1, n(n-1)/2]$, the distance selection algorithm of Katz and Sharir [15] can compute the k-th smallest distance among all pairs of points of P in $O(n^{4/3} \log^2 n)$ time. Using this algorithm, the binary search can find r^* in $O(n^{4/3} \log^3 n)$ time. This is the algorithm mentioned in [5].

Our new algorithm is based on parametric search [9,17], by parameterizing the decision algorithm of Chan and Skrepetos [6] (which we refer to as the CS algorithm). More specifically, the CS algorithm first builds a grid in the plane and then runs the breadth-first-search (BFS) algorithm with the help of the grid; in the i-th step of the BFS, the algorithm finds the set of points of P whose distances from s in $G_r(P)$ are equal to i. Although we do not know r^*, we run the CS algorithm on a parameter r in an interval $(r_1, r_2]$ such that each step of the algorithm behaves the same as the algorithm running on r^*. The algorithm terminates after t is reached, which will happen within λ steps. In each step, we use the decision algorithm to compare r^* with certain *critical values*, and the results of these comparisons will shrink the interval $(r_1, r_2]$. Once the algorithm terminates, r^* is equal to r_2 of the current interval $(r_1, r_2]$. With the linear-time decision algorithm (i.e., the CS algorithm [6]), each step runs in $O(n \log n)$ time. The total time of the algorithm is $O(\lambda \cdot n \log n)$.

The above algorithm is only interesting when λ is relatively small. In the worst case, however, λ can be $\Theta(n)$, which would make the running time become $O(n^2 \log n)$. Next, by combining the strategies of the above two algorithms, we derive a better algorithm. The main idea is to partition the cells of the grid in the CS algorithm into two types: *large cells*, which contain at least $n^{3/4}$ points of P each, and *small cells* otherwise. For small cells, we process them using the above binary search algorithm with the distance selection algorithm [15]; for large cells, we process them using the above parametric search techniques. This works out due to the following observation. On the one hand, the number of large cells is relatively small (at most $O(n^{1/4})$) and thus the number of steps using the parametric search is also small. On the other hand, each small cell contains relatively few points of P (at most $O(n^{3/4})$) and thus the total time we spend on the distance selection algorithm is not big. The threshold value $n^{3/4}$ is carefully chosen so that the total time for processing the two types of cells is minimized. In addition, instead of applying the distance selection algorithm [15] directly, we find that it suffices to use only a subroutine of that algorithm, which not only simplifies the algorithm but also reduces the total time by a logarithmic factor. All these efforts lead to an $O(n^{5/4} \log^2 n)$ time algorithm to compute r^*.

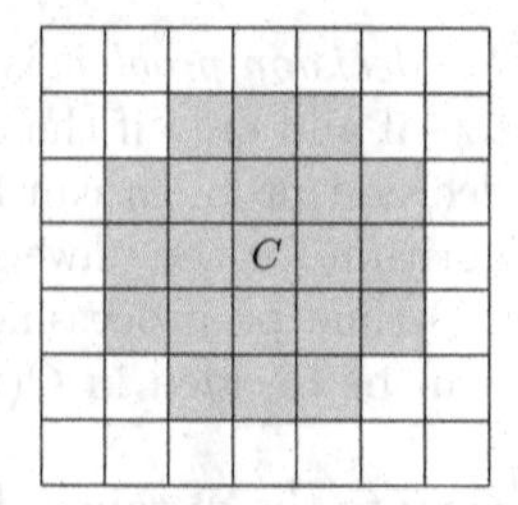

Fig. 1. The grey cells are all neighbor cells of C.

Outline. In the following, Sect. 2 defines notation and reviews the CS algorithm. Our first algorithm is presented in Sect. 3 while the second algorithm is described in Sect. 4. Section 5 concludes with remarks on a more general problem and the weighted case of the problem. Due to the space limit, many proofs are omitted but can be found in our full paper.

2 Preliminaries

For any radius r, we use $d_r(p,q)$ to denote the distance of two vertices p and q in $G_r(P)$. Note that $d_r(p,q) \leq d_{r'}(p,q)$ if $r \geq r'$.

For any two points p and q in the plane, let $|p-q|$ denote their (Euclidean) distance. For any subset P' of P and any region R in the plane, we use $P'(R)$ or $P' \cap R$ to refer to the subset of points P' contained in R. For any point p, let $x(p)$ and $y(p)$ denote its x- and y-coordinates, respectively.

We next review the CS algorithm [6]. Suppose we have a sorted list of P by x-coordinate and another sorted list of P by y-coordinate. Given a radius r, the algorithm can compute in $O(n)$ time the distances from s to all other points of P in $G_r(P)$.

The first step is to compute a grid $\Psi_r(P)$ of square cells whose side lengths are $r/\sqrt{2}$. A cell C' of $\Psi_r(P)$ is a *neighbor* of another cell C if the minimum distance between a point of C and a point of C' is at most r. Note that the number of neighbors of each cell of $\Psi_r(P)$ is $O(1)$ (e.g., see Fig. 1) and the distance between any two points in each cell is at most r.

Next, starting from the point s, the algorithm runs BFS in $G_r(P)$ with the help of the grid $\Psi_r(P)$. Define S_i as the subset of points of P whose distances in $G_r(P)$ from s are equal to i. Initially, $S_0 = \{s\}$. Given S_{i-1}, the i-th step of the BFS is to compute S_i by using S_{i-1} and the grid $\Psi_r(P)$, as follows. If a point p is not in $\bigcup_{j=0}^{i-1} S_j$, we say that p has not been *discovered* yet. For each cell C that contains at least one point of S_{i-1}, we need to find points that are not discovered yet and at distances at most r from the points of $S_{i-1} \cap C$ (i.e., the points of S_{i-1} in C); clearly, these points are either in C or in the neighbor cells of C. For points of $P(C)$, since every two points of C are within distance r from each other, we add all points of $P(C)$ that have not been discovered to S_i. For each neighbor cell C' of C, we need to solve the following subproblem: find the points of $P(C')$ that are not discovered yet and within distance at most r from the points of $S_{i-1} \cap C$. Since C' and C are separated by either a vertical line or a horizontal line, we essentially have the following subproblem.

Subproblem 1. *Given a set of n_r red points below a horizontal line ℓ and a set of n_b blue points above ℓ, both sorted by x-coordinate, determine for each blue point whether there is a red point at distance at most r from it.*

The subproblem can be solved in $O(n_r + n_b)$ time as follows. For each red point p, the circle of radius r centered at p has at most one arc above ℓ (we say that this arc is defined by p). Let Γ be the set of these arcs defined by all red points. Since all arcs of Γ have the same radius and all red points are below ℓ, every two arcs intersect at most once and the arcs above ℓ are x-monotone. Further, as all red points are sorted already by x-coordinate, the upper envelope of Γ, denoted by $\mathcal{U}$, can be computed in $O(n_r)$ time by an algorithm similar in spirit to Graham's scan. Then, it suffices to determine whether each blue point is below $\mathcal{U}$, which can be done in $O(n_r + n_b)$ time by a linear scan. More specifically, we can first sort the vertices of $\mathcal{U}$ and all blue points. After that, for each blue point p, we know the arc of $\mathcal{U}$ that spans p (i.e., $x(p)$ is between the x-coordinates of the two endpoints of the arc), and thus we only need to check whether p is below the arc. In summary, solving the subproblem involves three subroutines: (1) compute $\mathcal{U}$; (2) sort all vertices of $\mathcal{U}$ with all blue points; (3) for each blue point p, determine whether it is below the arc of $\mathcal{U}$ that spans p.

The above computes the set S_i. Note that if $S_i = \emptyset$, then we can stop the algorithm because all points of P that can be reached from s in $G_r(P)$ have been computed. For the running time, notice that points of P in each cell of the grid $\Psi_r(P)$ can be involved in at most two steps of the BFS. Further, since each grid cell has $O(1)$ neighbors, the total time of the BFS algorithm is $O(n)$.

In order to achieve $O(n)$ time for the overall algorithm, the grid $\Psi_r(P)$ must be implicitly constructed. The CS algorithm [6] does not provide any details about that. There are various ways to do so. Below we present our method, which will facilitate our algorithm in the next section.

The grid $\Psi_r(P)$ we are going to build is a rectangle that is partitioned into square cells of side lengths $r/\sqrt{2}$ by $O(n)$ horizontal and vertical lines. These partition lines will be explicitly computed. Let P' be the subset of points of P located in $\Psi_r(P)$. P' has the following property: for each $p \in P \backslash P'$, p cannot be reached from s in $G_r(P)$, i.e., the distances from s to the points of $P \backslash P'$ in $G_r(P)$ are infinite. Let $\mathcal{C}$ denote the set of cells of $\Psi_r(P)$ that contain at least one point of P. For each cell $C \in \mathcal{C}$, let $N(C)$ denote the set of neighbors of C in $\mathcal{C}$. The information computed in Lemma 1 suffices for implementing the above BFS algorithm in linear time.

Lemma 1. *Both P' and $\mathcal{C}$, along with all vertical and horizontal partition lines of $\Psi_r(P)$, can be computed in $O(n)$ time. Further, with $O(n)$ time preprocessing, the following can be achieved:*

1. *Given any point $p \in P'$, the cell of $\mathcal{C}$ that contains p can be obtained in $O(1)$ time.*
2. *Given any cell $C \in \mathcal{C}$, the neighbor set $N(C)$ can be obtained in $O(|N(C)|)$ time.*
3. *Given any cell $C \in \mathcal{C}$, the subset $P(C)$ of P can be obtained in $O(|P(C)|)$ time.*

To make the description concise, in the following, whenever we say "compute the grid $\Psi_r(P)$" we mean "compute the grid information of Lemma 1"; similarly, by "using the grid $\Psi_r(P)$", we mean "using the grid information of Lemma 1".

3 The First Algorithm

In this section, we present our $O(\lambda \cdot n \log n)$ time algorithm for the reverse shortest path problem. Our goal is to compute r^*, the optimal radius of the disks.

Our algorithm uses parametric search [9,17]. But different than the traditional parametric search where parallel algorithms are used, our decision algorithm (i.e., the CS algorithm for the shortest path problem [6]) is inherently serial. We run the CS algorithm with a parameter r in an interval $(r_1, r_2]$ by simulating the algorithm on the unknown r^*; at each step of the algorithm, the decision algorithm will be invoked on certain *critical values* r to compare r and r^*, and the algorithm will proceed accordingly based on the results of the comparisons. The interval $(r_1, r_2]$ always contains r^* and will keep shrinking during the algorithm (note that "shrinking" includes the case that the interval does not change). Initially, we set $r_1 = 0$ and $r_2 = \infty$. Hence, $(r_1, r_2]$ contains r^*.

Recall that the CS algorithm has two major steps: build the grid and then run BFS with the help of the grid. Correspondingly, our algorithm also first builds a grid and then runs BFS accordingly using the grid.

3.1 Building the Grid

The first step is to build a grid $\Psi(P)$. Our goal is to shrink $(r_1, r_2]$ so that it contains r^* and if $r^* \neq r_2$ (and thus $r^* \in (r_1, r_2)$), then for any $r \in (r_1, r_2)$, $\Psi_r(P)$ has the same *combinatorial structure* as $\Psi_{r^*}(P)$, i.e., both grids have the same number of columns and the same number of rows, and a point of P is in the cell of the i-th row and j-th column of $\Psi_{r^*}(P)$ if and only if it is also in the cell of the i-th row and j-th column of $\Psi_r(P)$. To this end, we have the following lemma.

Lemma 2. *An interval $(r_1, r_2]$ containing r^* can be computed in $O(n \log n)$ time so that if $r^* \neq r_2$, then for any $r \in (r_1, r_2)$, the grid $\Psi_r(P)$ has the same combinatorial structure as $\Psi_{r^*}(P)$.*

Let $(r_1, r_2]$ be the interval computed by Lemma 2. We pick any value r in (r_1, r_2) and compute the grid information of $\Psi_r(P)$ by Lemma 1. By Lemma 2, these information is the same as that of $\Psi_{r^*}(P)$ if $r^* \neq r_2$. Below we will use $\Psi(P)$ to refer to the grid information computed above.

3.2 Running BFS

For a fixed radius r, we use $S_i(r)$ to denote the set of points of P whose distances from s is equal to i in $G_r(P)$, which is computed in the i-th step of the BFS algorithm if we run the CS algorithm with respect to r. Initially, we have $S_0(r) = \{s\}$. Below, using the interval $(r_1, r_2]$ obtained in Lemma 2, we run the BFS as in the CS algorithm with a parameter $r \in (r_1, r_2)$, by simulating the algorithm for r^*. The algorithm maintains an invariant that the i-th step computes a subset $S_i \subseteq P$ and shrinks $(r_1, r_2]$ so that it contains r^* and if $r^* \neq r_2$ (and thus $r^* \in (r_1, r_2)$), then $S_i = S_i(r) = S_i(r^*)$ for any $r \in (r_1, r_2)$. Initially, we set $S_0 = \{s\}$ and thus the invariant holds as $S_0(r) = \{s\}$ for any r. As will be seen later, the algorithm stops within λ steps and each step takes $O(n \log n)$ time.

Consider the i-th step. Assume that we have S_{i-1} and $(r_1, r_2]$, and the invariant holds, i.e., $(r_1, r_2]$ contains r^* and if $r^* \neq r_2$, then $S_{i-1} = S_{i-1}(r) = S_{i-1}(r^*)$ for any $r \in (r_1, r_2)$. Using the grid $\Psi(P)$, we obtain the grid cells containing the points of S_{i-1}. For each such cell C, for points of P in C, we have Lemma 3.

Lemma 3. *Suppose $r^* \neq r_2$. Then, for each point $p \in P(C)$ that has not been discovered by the algorithm, i.e., $p \notin \bigcup_{j=1}^{i-1} S_j$, p is in $S_i(r)$ for any $r \in (r_1, r_2)$.*

Proof. Let q be a point of S_{i-1} in C. By our algorithm invariant, $(r_1, r_2]$ contains r^*. Since $r^* \neq r_2$, $r^* \in (r_1, r_2)$. Let r be any value of (r_1, r_2). In light of Lemma 2, both p and q are in the same cell of $\Psi_r(P)$, and thus $|p - q| \leq r$. By our algorithm invariant, $S_j = S_j(r)$ for all $0 \leq j \leq i - 1$. Since $p \notin \bigcup_{j=1}^{i-1} S_j$, we have $p \notin \bigcup_{j=1}^{i-1} S_j(r)$. As $q \in S_{i-1}(r)$ and $|p - q| \leq r$, we obtain that $p \in S_i(r)$. □

Due to Lemma 3, we add to S_i the points of $P(C)$ that have not been discovered yet. Next, for each neighbor C' of C, we need to solve Subproblem 1; we use $\mathcal{I}$ to denote the set of all instances of this subproblem in the i-th step of the BFS. Consider one such instance. Recall that solving it for a fixed r involves three subroutines. First, compute the upper envelope $\mathcal{U}$ of the arcs of Γ above ℓ of all red points. Second, sort all vertices of $\mathcal{U}$ with all blue points. Third, for each blue point p, determine whether it is below the arc of $\mathcal{U}$ that spans p. To solve our problem, we parameterize each subroutine with a parameter r so that the behavior of the algorithm is consistent with that for $r = r^*$ if $r^* \neq r_2$.

Computing the Upper Envelope. We use $\Gamma(r)$ to denote the set of arcs above ℓ defined by the red points with respect to the radius r; similarly, define $\mathcal{U}(r)$ as the upper envelope of $\Gamma(r)$. The goal of the first subroutine is to shrink the interval $(r_1, r_2]$ such that it contains r^* and if $r^* \neq r_2$, then $\mathcal{U}(r^*)$ has the same combinatorial structure as $\mathcal{U}(r)$ for any $r \in (r_1, r_2)$, i.e., the set of red points that define the arcs on $\mathcal{U}(r)$ is exactly the set of red points that define the arcs on $\mathcal{U}(r^*)$ with the same order. Note that the order of the arcs on $\mathcal{U}(r)$ is consistent with the x-coordinate order of the red points defining these arcs [6].

To this end, we have the following observation. Consider $\mathcal{U}(r)$ for an arbitrary r. If r changes, the combinatorial structure of $\mathcal{U}(r)$ does not change until one

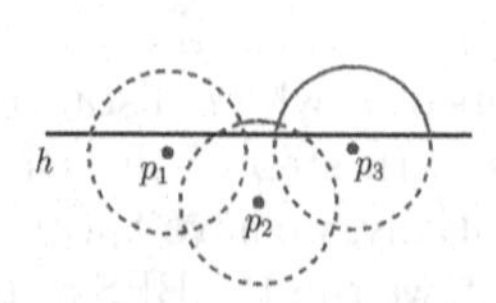

(a) The upper envelope is comprised of three arcs centered at p_1, p_2 and p_3.

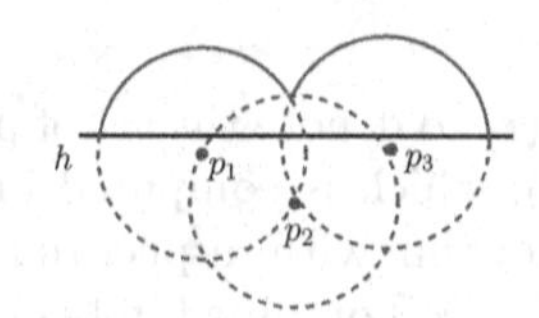

(b) The moment when the three arcs have a common intersection, which is a vertex of the upper envelope.

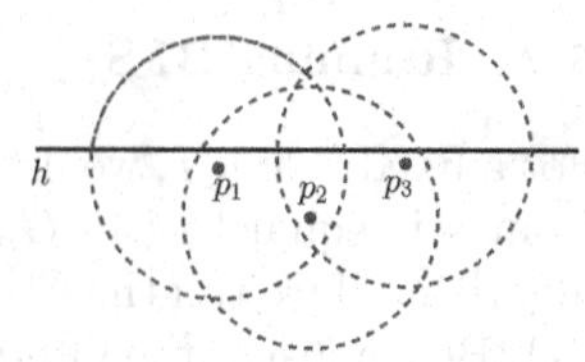

(c) The middle arc centered at p_2 disappears from the upper envelope.

Fig. 2. The change of the combinatorial structure of the upper envelope $\mathcal{U}(r)$ (the red solid arcs) as r increases. (Color figure online)

arc (e.g., defined by a red point p_2) disappears from $\mathcal{U}(r)$ (e.g., see Fig. 2). Let p_1 and p_3 be the red points defining neighboring left and right arcs of the arc defined by p_2 on $\mathcal{U}(r)$, respectively. Then, at the moment when p_2 disappears from $\mathcal{U}(r)$, the three arcs defined by p_1, p_2, and p_3 intersect at a common point q, which is equidistant to the three points. Further, since q is currently on $\mathcal{U}(r)$, there is no red point that is closer to q than p_i for $i = 1, 2, 3$, and the distance from q to each p_i, $i = 1, 2, 3$, is equal to the current value of r. Hence, q is a vertex of the Voronoi diagram of the red points. This implies that as r changes, the combinatorial structure of $\mathcal{U}(r)$ does not change until possibly when r is equal to the distance $|q - p|$, where q is a vertex of the Voronoi diagram of all red points and p is a nearest red point of q.

Based on the above observation, our algorithm works as follows. We build the Voronoi diagram for all red points, which takes $O(n_r \log n_r)$ time [12,19]. For each vertex v of the diagram, we add $|v - p|$ to the set Q (initially $Q = \emptyset$), where p is a nearest red point of v (p is available from the diagram). Note that $|Q| = O(n_r)$, and we refer to each value of Q as a *critical value*. Next, we sort Q, and then do binary search on Q using the decision algorithm to find the smallest value r'_2 of Q with $r'_2 \geq r^*$ as well as the largest value r'_1 of Q smaller than r^*, which can be done in $O(n \log n_r)$ time (note that $n_r \leq n$). By definition, $(r'_1, r'_2]$ contains r^* and (r'_1, r'_2) does not contain any value of Q. According to the above observation, if $r^* \neq r'_2$, then the combinatorial structure of $\mathcal{U}(r^*)$ is the same as that of $\mathcal{U}(r)$ for any $r \in (r'_1, r'_2)$.

We analyze the running time of this subroutine for all instances of $\mathcal{I}$. Clearly, the total time for all instances is bounded by $O(|\mathcal{I}| \cdot n \log n)$, which is $O(n^2 \log n)$ as $|\mathcal{I}| = O(n)$. We can reduce the time to $O(n \log n)$ by considering the critical values of all instances of $\mathcal{I}$ altogether. Specifically, let Q now be the set of critical values of all instances of $\mathcal{I}$. Then, $|Q| = O(n)$. We sort Q and do binary search on Q to find r'_1 and r'_2 as defined above with respect to the new Q. Now, for each instance of $\mathcal{I}$, if $r^* \neq r'_2$, then the combinatorial structure of $\mathcal{U}(r^*)$ is the same as that of $\mathcal{U}(r)$ for any $r \in (r'_1, r'_2)$. The total time for all instances of $\mathcal{I}$ is now bounded by $O(n \log n)$. Finally, we update $r_1 = \max\{r_1, r'_1\}$ and $r_2 = \min\{r_2, r'_2\}$. As $r^* \in (r'_1, r'_2]$, the new interval $(r_1, r_2]$ still contains

r^*. Further, as $(r_1, r_2) \subseteq (r_1', r_2')$, for each instance of $\mathcal{I}$, if $r^* \neq r_2$, then the combinatorial structure of $\mathcal{U}(r^*)$ is the same as that of $\mathcal{U}(r)$ for any $r \in (r_1, r_2)$.

Sorting the Upper Envelope Vertices and Blue Points. The goal of the second subroutine is to shrink the interval $(r_1, r_2]$ such that it contains r^* and if $r^* \neq r_2$, then the sorted list of all vertices of $\mathcal{U}(r^*)$ and all blue points by their x-coordinates is the same as the sorted list of all vertices of $\mathcal{U}(r)$ and all blue points for any $r \in (r_1, r_2)$. Recall that after the first subroutine, the interval $(r_1, r_2]$ contains r^*, and if $r^* \neq r_2$, then the combinatorial structure of $\mathcal{U}(r^*)$ is the same as that of $\mathcal{U}(r)$ for any $r \in (r_1, r_2)$.

To sort all vertices of $\mathcal{U}(r^*)$ and all blue points, we apply Cole's parametric search [9] with AKS sorting network [2], using the CS algorithm as the decision algorithm; the running time is bounded by $O(n \log n)$ as the number of vertices of $\mathcal{U}(r^*)$ is $O(n_r)$ and the number of blue points is $O(n_b)$ (and $n_r + n_b = O(n)$). To see why this works, it suffices to argue that the "root" of each comparison involved in the sorting can be obtained in $O(1)$ time (more specifically, the root refers to the value of $r \in (r_1, r_2)$ at which the two operands involved in the comparison are equal). Indeed, the comparisons can be divided into three types based on their operands: (1) a comparison between the x-coordinates of two blue points; (2) a comparison between the x-coordinates of two vertices of $\mathcal{U}(r^*)$; (3) a comparison between the x-coordinates of a blue point and a vertex of $\mathcal{U}(r^*)$. For the first type, as blue points are fixed, independent of the parameter r, it is trivial to handle. For the second type, as the combinatorial structure of $\mathcal{U}(r)$ does not change for all $r \in (r_1, r_2)$, each such comparison can be resolved by taking any value of $r \in (r_1, r_2)$ and then comparing the two vertices under r. The third type is a little more involved. Consider the comparison of the x-coordinates of a blue point q and a vertex v of $\mathcal{U}(r^*)$. Note that v is the intersection of arcs of two circles of radius r and centered at two red points, say p_1 and p_2, respectively. Observe that v is on the bisector of p_1 and p_2 (e.g., see Fig. 3). Furthermore, when r changes, v moves on the bisector of p_1 and p_2, while the position of the blue point q does not change. Hence, the root of the comparison, i.e., the value r (if exists) in (r_1, r_2) such that $x(q) = x(v)$ can be obtained in constant time by elementary geometry (e.g., see Fig. 4). Note that if such r does not exist in (r_1, r_2), then either $x(q) < x(v)$ holds for all $r \in (r_1, r_2)$ or $x(q) > x(v)$ holds for all $r \in (r_1, r_2)$, which can be easily determined. As such, with Cole's parametric search [9] and the linear time decision algorithm (i.e., the CS algorithm), we can obtain a sorted list of the upper envelope vertices and the blue points by x-coordinate; the algorithm shrinks $(r_1, r_2]$ so that the new interval $(r_1, r_2]$ contains r^* and if $r^* \neq r_2$, then the above sorted list is fixed for all $r \in (r_1, r_2)$.

Since the running time of the above sorting algorithm is $O(n \log n)$, as before for the first subroutine, the sorting for all problem instances of $\mathcal{I}$ takes $O(n^2 \log n)$ time. To reduce the time, as before, we sort all elements in all instances of $\mathcal{I}$ altogether, which takes $O(n \log n)$ time in total. Specifically, in each problem instance, we need to sort a set of blue points and vertices of upper envelopes of a set of red points. We put all blue points and the upper envelopes

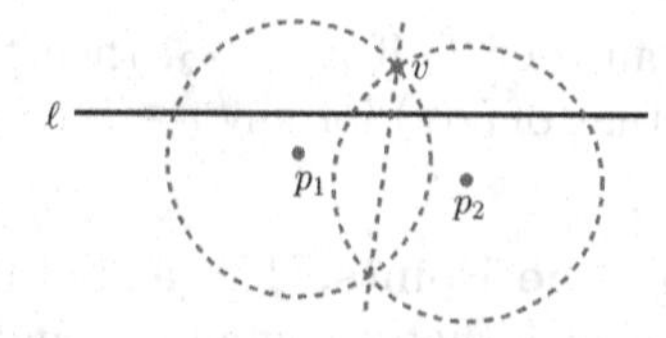

Fig. 3. Illustrating a vertex v of the upper envelope, which is defined by two red points p_1 and p_2. The red solid segment is the bisector of p_1 and p_2. (Color figure online)

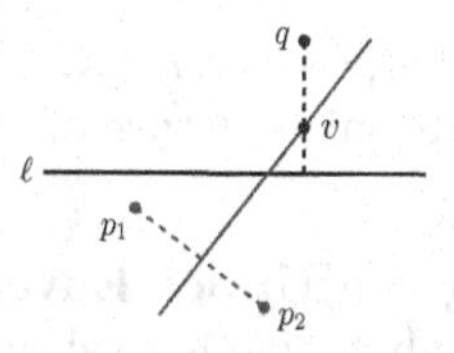

Fig. 4. Illustrating the scenario where $x(q) = x(v)$, where v is on the bisector (the red solid segment) of p_1 and p_2. (Color figure online)

of all red points of all problem instances of $\mathcal{I}$ in one coordinate system and apply the sorting algorithm as above. One difference is that we now have a new type of comparisons: compare the x-coordinate of a vertex v_1 of the upper envelope from one problem instance with the x-coordinate of a vertex v_2 of the upper envelope from another problem instance. In this case, when r changes, both v_1 and v_2 moves on the bisectors of their defining red points. But we can still find in constant time a root r (if exists) in (r_1, r_2) for the comparison by elementary geometry. As such, we can complete the sorting for all problem instances of $\mathcal{I}$ in $O(n \log n)$ time in total, for the total number of all blue points and red points in all problem instances of $\mathcal{I}$ is $O(n)$. Again, the interval $(r_1, r_2]$ will be shrunk. This finishes the second subroutine.

Deciding Whether Each Blue Point is Below the Upper Envelope. We now have an interval $(r_1, r_2]$ containing r^* such that if $r^* \neq r_2$, then each blue point q is spanned by an arc $\alpha_q(r)$ of $\mathcal{U}(r)$ defined by the same red point for all $r \in (r_1, r_2)$ (note that $\alpha_q(r)$ moves as r changes, for r is the radius of the arc). Each blue point q is below the upper envelope $\mathcal{U}(r)$ if and only if q is below the arc $\alpha_q(r)$. The goal of the third subroutine is to shrink the interval $(r_1, r_2]$ so that the new interval $(r_1, r_2]$ still contains r^* and if $r^* \neq r_2$, then for each blue point q, the relative position of q with respect to $\alpha_q(r)$ (i.e., whether q is above or below $\alpha_q(r)$) is fixed for all $r \in (r_1, r_2)$. To this end, we proceed as follows.

As r changes in (r_1, r_2), $\alpha_q(r)$ changes while q does not. For each blue point q, we compute in constant time a critical value r (if exists) in (r_1, r_2) such that q is on α_q, and we add r to the set Q ($Q = \emptyset$ initially). Note that if such value r does not exist in (r_1, r_2), then either q is above $\alpha_q(r)$ for all $r \in (r_1, r_2)$ or q is below $\alpha_q(r)$ for all $r \in (r_1, r_2)$, which can be easily determined. The size of Q is at most n_b. Then, we sort Q, and do binary search on Q with our decision algorithm to find the smallest value r_2' of Q with $r_2' \geq r^*$ and the largest value r_1' of Q with $r_1' < r^*$. We then update $r_1 = \max\{r_1, r_1'\}$ and $r_2 = \min\{r_2, r_2'\}$. The new interval $(r_1, r_2]$ still contains r^* and (r_1, r_2) does not contain any value of Q. Hence, if $r^* \neq r_2$, then for each blue point q, the relative position of q with respect to $\alpha_q(r)$ is fixed for all $r \in (r_1, r_2)$. As such, the new interval $(r_1, r_2]$ satisfies the goal of the third subroutine as mentioned above.

Finally, we pick an arbitrary $r \in (r_1, r_2)$, and for each blue point q, if q is below the arc $\alpha_q(r)$, then we add q to the set S_i.

The running time of the above algorithm is $O(n \log n_b)$. Thus the total time of the third subroutine is $O(n^2 \log n)$ for all problem instances of $\mathcal{I}$. To reduce the time, we again consider the subroutine of all instances of $\mathcal{I}$ altogether. More specifically, we put all critical values r in all problem instances of $\mathcal{I}$ in Q. Thus, the size of Q is $O(n)$. We then run the same algorithm as above using the new set Q. The total time is bounded by $O(n \log n)$.

Terminating the Algorithm. This finishes the i-th step of the BFS, which computes a set S_i along with an interval $(r_1, r_2]$. According to the above discussion, $(r_1, r_2]$ contains r^* and if $r^* \neq r_2$ (and thus $r^* \in (r_1, r_2)$), then $S_i = S_i(r^*) = S_i(r)$ for all $r \in (r_1, r_2)$.

If the point t is in S_i and $i \leq \lambda$, then we stop the algorithm. In this case, we have the following lemma.

Lemma 4. *If $t \in S_i$ and $i \leq \lambda$, then $r^* = r_2$.*

Proof. Assume to the contrary that $r^* \neq r_2$. Then, since $r^* \in (r_1, r_2]$, we have $r^* \in (r_1, r_2)$. Let $r' = (r_1 + r^*)/2$. Clearly, $r' \in (r_1, r_2)$ and $r' < r^*$. As $r' \in (r_1, r_2)$, $S_i = S_i(r')$ by our algorithm invariant. Since $t \in S_i(r')$, we obtain that $d_{r'}(s, t) = i \leq \lambda$. This incurs contradiction as $r' < r^*$ and r^* is the minimum value r with $d_r(s, t) \leq \lambda$. □

If $t \notin S_i$ and $i = \lambda$, then we also stop the algorithm. In this case, we have the following lemma.

Lemma 5. *If $t \notin S_i$ and $i = \lambda$, then $r^* = r_2$.*

Since initially $i = 0$ and $S_0 = \{s\}$, the above implies that the BFS algorithm will stop in at most λ steps. As each step takes $O(n \log n)$ time, r^* can be computed in $O(\lambda \cdot n \log n)$ time.

Theorem 1. *The reverse shortest path problem for unit-disk graphs can be solved in $O(\lambda \cdot n \log n)$ time.*

4 The Second Algorithm

In this section, we present our second algorithm for the reverse shortest path problem. As discussed in Sect. 1, the main idea is to combine the strategies of the first algorithm in Sect. 3 and the naive binary search algorithm using the distance selection algorithm [15].

First of all, we still build in $O(n \log n)$ time the grid $\Psi(P)$ as in Sect. 3.1, and thus the information of Lemma 2 is available for the grid. More specifically, we obtain an interval $(r_1, r_2]$ such that if $r^* \neq r_2$, then the combinatorial data structure of $\Psi_r(P)$ is fixed for all $r \in (r_1, r_2)$, implying that $\mathcal{C}$, P', $N(C)$ and $P(C)$ for each $C \in \mathcal{C}$ are fixed for all $r \in (r_1, r_2)$. Next, we will run the BFS algorithm, but in a different way than before.

A cell C of $\mathcal{C}$ is called a *large cell* if $|P(C)| \geq n^{3/4}$ and a *small cell* otherwise. Clearly, the number of large cells is at most $n^{1/4}$. For all pairs of cells (C, C') with $C \in \mathcal{C}$ and $C' \in N(C)$, we call (C, C') a *small-cell pair* if both C and C' are small cells and a *large-cell pair* otherwise (i.e., at least one cell is a large cell). As $|N(C)| = O(1)$ for each cell C and the number of large cells is at most $n^{1/4}$, the total number of large-cell pairs is $O(n^{1/4})$.

Recall that each step of the BFS algorithm of our first algorithm in Sect. 3.2 boils down to solving instances of Subproblem 1, and each such instance involves a cell pair (C, C') with $C \in \mathcal{C}$ and $C' \in N(C)$. If (C, C') is a large-cell pair, we will run the same algorithm as in Sect. 3.2. Otherwise, we will use the original CS algorithm to solve it, which takes only linear time. For this, using the distance selection algorithm [15], we preprocess all these small-cell pairs before starting the BFS algorithm by the following lemma.

Lemma 6. *An interval $(r'_1, r'_2]$ containing r^* can be computed in $O(n^{5/4} \log^2 n)$ time with the following property: for any small cell pair (C, C') with $C \in \mathcal{C}$ and $C' \in N(C)$, for any two points p and p' with $p \in P(C)$ and $p' \in P(C')$, either $|p - p'| < r$ holds for all $r \in (r'_1, r'_2)$ or $|p - p'| > r$ holds for all $r \in (r'_1, r'_2)$.*

With the interval $(r'_1, r'_2]$ computed by the above lemma, we update $r_1 = \max\{r_1, r'_1\}$ and $r_2 = \min\{r_2, r'_2\}$. By definition, $(r_1, r_2] \subseteq (r'_1, r'_2]$. Hence, the interval $(r_1, r_2]$ also has the same property as $(r'_1, r'_2]$ in Lemma 6.

Next, we run the BFS algorithm as in Sect. 3.2. To solve each instance of Subproblem 1, if one of the two involved cells is a large cell (we refer to this case as the *large-cell instance*), then we use the same algorithm as before, i.e., parametric search; otherwise (i.e., both involved cells are small cells; we refer to this case as *small-cell instance*), due to the preprocessing of Lemma 6, we can solve the subproblem directly using the original CS algorithm by picking an arbitrary value $r \in (r_1, r_2)$. In this way, the time for solving all small-cell instances in the entire BFS algorithm is $O(n)$. For each large-cell instance, it can be solved in $O(n \log n)$ time as discussed in Sect. 3.2. As the number of large cells of $\mathcal{C}$ is at most $n^{1/4}$ and $|N(C)| = O(1)$ for each cell $C \in \mathcal{C}$, the total number of large-cell instances of Subproblem 1 is at most $O(n^{1/4})$. Hence, the total time for solving the large-cell instances in the entire BFS algorithm is $O(n^{5/4} \log n)$. The proof of the following lemma is in our full paper, which presents the details of the new BFS algorithm sketched above.

Lemma 7. *The BFS algorithm, which computes r^*, can be implemented in $O(n^{5/4} \log n)$ time.*

Combining with the algorithm of Lemma 6, the overall time of the algorithm for computing r^* is $O(n^{5/4} \log^2 n)$. We thus obtain the following theorem.

Theorem 2. *The reverse shortest path problem for unit-disk graphs can be solved in $O(n^{5/4} \log^2 n)$ time.*

5 Concluding Remarks

In this paper, we propose two algorithms for the reverse shortest path problem for (unweighted) unit-disk graphs with time complexities of $O(\lambda \cdot n \log n)$ and $O(n^{5/4} \log^2 n)$, respectively. Interestingly, our second algorithm breaks the $O(n^{4/3})$ time barrier for certain geometric problems [10,11].

Our techniques can be extended to solve a more general problem: Given a source point $s \in P$ and a value λ, compute the smallest value r^* such that the lengths of shortest paths from s to all vertices of $G_r(P)$ are at most λ, i.e., $\max_{t \in P} d_{r^*}(s,t) \leq \lambda$. The decision problem (i.e., deciding whether $r \geq r^*$ for any r) now becomes deciding whether $\max_{t \in P} d_r(s,t) \leq \lambda$. The algorithm of Chan and Skrepetos [6] is actually for finding shortest paths from s to all vertices of $G_r(P)$, and thus we can solve the decision problem by using the algorithm of Chan and Skrepetos [6] in the same way as before but with an additional last step to compute the value $\max_{t \in P} d_r(s,t)$ (the total running time is still $O(n)$ after the $O(n \log n)$-time preprocessing). As such, to compute r^*, we can follow the same algorithm scheme as before but instead use the above new decision algorithm. In addition, we make the following changes to the first algorithm (the second algorithm is changed accordingly). After the i-th step of the BFS, which computes a set S_i along with an interval $(r_1, r_2]$. If all points of P have been discovered after this step and $i \leq \lfloor \lambda \rfloor$, then we have $r^* = r_2$ and stop the algorithm; the proof is similar to Lemma 4. We also stop the algorithm with $r^* = r_2$ if $i = \lfloor \lambda \rfloor$ and not all points of P have been discovered; the proof is similar to Lemma 5. As before, the algorithm will stop in at most $\lfloor \lambda \rfloor$ steps. In this way, the first algorithm can compute λ^* in $O(\lfloor \lambda \rfloor \cdot n \log n)$ time. Analogously, the second algorithm can compute λ^* in $O(n^{5/4} \log^2 n)$ time.

Other than further improving our result, future work also includes studying the weighted case of the problem. A straightforward solution is again doing binary search on all pairwise distances of all points of P using the $O(n^{4/3} \log^2 n)$ time distance selection algorithm [15] with the shortest path algorithms for the weight unit-disk graphs [5,20] as decision algorithms. The total time of the algorithm is $O(n^{4/3} \log^3 n)$. A logarithmic factor can be reduced using our techniques in Lemma 6 (i.e., instead of applying the distance selection algorithm [15] directly, use a subroutine of it), resulting in an $O(n^{4/3} \log^2 n)$ time algorithm. It would be interesting to see whether better solutions exist, e.g., using parametric search, and in particular, whether the $O(n^{4/3})$ time barrier can be broken.

References

1. Agarwal, P., Efrat, A., Sharir, M.: Vertical decomposition of shallow levels in 3-dimensional arrangements and its applications. SIAM J. Comput. **29**, 912–953 (1999)
2. Ajtai, M., Komlós, J., Szemerédi, E.: An $O(n \log n)$ sorting network. In: Proceedings of the 15th Annual ACM Symposium on Theory of Computing (STOC), pp. 1–9 (1983)

3. de Berg, M., Bodlaender, H., Kisfaludi-Bak, S., Marx, D., van der Zanden, T.: A framework for ETH-tight algorithms and lower bounds in geometric intersection graphs. In: Proceedings of the 50th Annual ACM Symposium on Theory of Computing (STOC), pp. 574–586 (2018)
4. Burton, D., Toint, P.: On an instance of the inverse shortest paths problem. Math. Program. **53**, 45–61 (1992)
5. Cabello, S., Jejčič, M.: Shortest paths in intersection graphs of unit disks. Comput. Geom.: Theory Appl. **48**, 360–367 (2015)
6. Chan, T., Skrepetos, D.: All-pairs shortest paths in unit-disk graphs in slightly subquadratic time. In: Proceedings of the 27th International Symposium on Algorithms and Computation (ISAAC), pp. 24:1–24:13 (2016)
7. Chan, T., Skrepetos, D.: Approximate shortest paths and distance oracles in weighted unit-disk graphs. In: Proceedings of the 34th International Symposium on Computational Geometry (SoCG), pp. 24:1–24:13 (2018)
8. Clark, B., Colbourn, C., Johnson, D.: Unit disk graphs. Discret. Math. **86**, 165–177 (1990)
9. Cole, R.: Slowing down sorting networks to obtain faster sorting algorithms. J. ACM **34**, 200–208 (1987)
10. Erickson, J.: On the relative complexities of some geometric problems. In: Proceedings of the 7th Canadian Conference on Computational Geometry (CCCG), pp. 85–90 (1995)
11. Erickson, J.: New lower bounds for Hopcroft's problem. Discret. Comput. Geom. **16**, 389–418 (1996)
12. Fortune, S.: A sweepline algorithm for Voronoi diagrams. Algorithmica **2**, 153–174 (1987)
13. Gao, J., Zhang, L.: Well-separated pair decomposition for the unit-disk graph metric and its applications. SIAM J. Comput. **35**, 151–169 (2005)
14. Kaplan, H., Mulzer, W., Roditty, L., Seiferth, P., Sharir, M.: Dynamic planar Voronoi diagrams for general distance functions and their algorithmic applications. In: Proceedings of the 28th Annual ACM-SIAM Symposium on Discrete Algorithms (SODA), pp. 2495–2504 (2017)
15. Katz, M., Sharir, M.: An expander-based approach to geometric optimization. SIAM J. Comput. **26**, 1384–1408 (1997)
16. Matsui, T.: Approximation algorithms for maximum independent set problems and fractional coloring problems on unit disk graphs. In: Akiyama, J., Kano, M., Urabe, M. (eds.) JCDCG 1998. LNCS, vol. 1763, pp. 194–200. Springer, Heidelberg (2000). https://doi.org/10.1007/978-3-540-46515-7_16
17. Megiddo, N.: Applying parallel computation algorithms in the design of serial algorithms. J. ACM **30**, 852–865 (1983)
18. Roditty, L., Segal, M.: On bounded leg shortest paths problems. Algorithmica **59**, 583–600 (2011)
19. Shamos, M., Hoey, D.: Closest-point problems. In: Proceedings of the 16th Annual Symposium on Foundations of Computer Science (FOCS), pp. 151–162 (1975)
20. Wang, H., Xue, J.: Near-optimal algorithms for shortest paths in weighted unit-disk graphs. Discret. Comput. Geom. **64**, 1141–1166 (2020)
21. Zhang, J., Lin, Y.: Computation of the reverse shortest-path problem. J. Glob. Optim. **25**, 243–261 (2003)

Correction to: Algorithms and Data Structures

Anna Lubiw, Mohammad Salavatipour, and Meng He

Correction to:
A. Lubiw et al. (Eds.): ***Algorithms and Data Structures*****,**
LNCS 12808, https://doi.org/10.1007/978-3-030-83508-8

The original version of this book was revised. The originally published omitted one volume editor of the book. This has now been corrected.

The updated version of the book can be found at
https://doi.org/10.1007/978-3-030-83508-8

A. Lubiw et al. (Eds.): WADS 2021, LNCS 12808, p. C1, 2022.
https://doi.org/10.1007/978-3-030-83508-8_48

Author Index

Printed in the United States
by Baker & Taylor Publisher Services